MYSTERIOUS UNIVERSE:

A HANDBOOK OF ASTRONOMICAL ANOMALIES

Compiled by

WILLIAM R. CORLISS

Illustrated by John C. Holden

Published and Distributed by

The Sourcebook Project G en Arm, MD 21057

ISBN 0-915554-05-4

LC 78-65616

NOTICE

This handbook is one of a series published by the Sourcebook Project, P. O. Box 107, Glen Arm, MD 21057. The handbooks and closely related loose-leaf sourcebooks may be purchased directly from the Project. Write for details and prices.

First printing: March 1979

Printed in the United States of America

TABLE OF CONTENTS

PREFACE

This book's title is an intentional echo of Sir James Jeans' classic THE MYSTERIOUS UNIVERSE. Jeans' book was exciting astronomy: vast island universes suspended in the immensity of space; stars beyond counting with ample space for other inhabited worlds. But the astronomy of my youth is no more. Astronomy is still exciting; but the universe no longer seems so orderly. Rather it is a maelstrom of chemical and physical surprises: forces, energies, and radiations that Jeans never drempt of. There may be biological surprises, too. Astronomical anomalies are accumulating rapidly. Questions are coming faster than answers. More than ever, it is a mysterious universe.

The primary objective of this handbook is to provide libraries and individuals with a wide selection of reliable descriptions of unusual astronomical phenomena. To meet this goal, I have analyzed hundreds of volumes of astronomical journals as well as the complete files of Nature and Science. The result of this research is an incomparable collection of information on the frontiers of astronomy. From this assemblage, I have selected the most interesting and controversial for this book.

My criteria for selecting "anomalous" data were: (1) the information contradicted current astronomical theories, or (2) the article raised personal questions not answered adequately to my knowledge. Usually both criteria were satisfied simultaneously. Much of the information printed herein will prove controversial, particularly that selected from amateur astronomical publications. It will soon become obvious to the reader that secondary objectives are the posing of challenges to establishment science and the stimulation of useful controversy.

I make no claim to completeness because new and relevant material is being discovered constantly as my search of the literature continues. Indeed, the near-infinite mine of observatory reports, university theses, and foreign journals has scarcely been touched. Even so, I have collected much more intriguing astronomical information than I can conveniently publish here. However, the complete master file is being published in looseleaf "sourcebooks." The Sourcebook Project welcomes inquiries concerning these cumulative sourcebooks.

The looseleaf sourcebooks were, in fact, the first publications of the Sourcebook Project. Although thousands of these notebooks have been sold to libraries, feedback from librarians indicated that casebound books would be more acceptable. Such suggestions were a major factor in the decision to publish selections from our collection in a series of casebound handbooks.

My hope is that this handbook will become a standard reference work on the frontiers of astronomy. To this end, I have utilized reports taken primarily from scientific journals. The screening provided by editors and referees of these publications helps to minimize hoaxes and errors. In expectation that establishment astronomy is too conservative and too confined by dogmas, I have introduced a handful of articles from fringe periodicals and books that are doubtless considered offbeat and "wild" by most professional astronomers.

Most of the 120 illustrations consist of line drawings by John C. Holden and are based on sketches and photos in the original articles. Since many of the articles are decades old, it was impossible to ferret out all original photos. Actually, in the older literature, line drawings predominate in any case.

Since the bulk of this handbook consists of direct quotations from original sources, I hasten to acknowledge the many writers of papers, letters-to-the-editor, and sundry publications who have contributed these descriptions of phenomena. When lengthy quotations are taken from publications still protected by copyright, permissions have been obtained.

William R. Corliss

Glen Arm, Maryland
January 2, 1979

Chapter 1

THE SUN

INTRODUCTION

Early astronomers considered the sun to be "perfect"; that is, unblemished and constant. The telescopic observations of sunspots by Galileo in 1610 proved that the sun departed from ideality. Over the following centuries, many other imperfections accumulated. Some are associated with the corona and prominences and seem trivial in nature, marking the normal boundary between a highly turbulent body and the near-vacuum of space. Other solar phenomena, such as the recently discovered surface oscillations are apparently connected in some unknown way with the sun's vital processes. Radiation detectors also tell us that the sun emits anomalous radiation beyond the visible spectrum. Solar neutrinos, in particular, pose a major puzzle because our model of how the sun works requires the emission of many more neutrinos than we observe.

The sun as our nearest star dominates the entire solar system. The eleven-year cycle of solar activity, in fact, is blamed for many cyclic terrestrial phenomena, from stock market prices to the weather. Phenomena of the other planets, such as luminescence, also seem keyed to solar radiation levels. Much more controversial is the apparent link between the positions of the planets and the cycle of solar activity. Known physical forces originating with the planets, such as gravity, do not seem nearly powerful enough to affect the huge solar machine. The mutual interactions between the sun and its family of planets are apparently far from being completely understood.

• Unusual Events and Features Observed on the Sun

AN ACCOUNT OF A VERY SINGULAR PHENOMENON SEEN IN THE DISK OF THE SUN .....

Anonymous; *Annual Register,* 9:120-121, 1766.

The 9th of August, 1762, M. de Rostan, of the economic society at Berne, and of the medicophysical society at Basle, whilst he was taking the sun's altitudes with a quadrant, at Lausanne, to verify a meridian, observed that the sun gave but a faint pale light, which he attributed to the vapours of the Leman lake: however, happening to direct a fourteen foot telescope, armed with a micrometer, to the sun, he was surprised to see the eastern side of the sun, as it were eclipsed about three digits, taking in a kind of nebulosity, which environed the opaque body, by which the sun was eclipsed. In the space of about two hours and a half, the south side of the said body, whatever it was, appeared detached from the limb of the sun; but the limb, or more properly, the northern extremity of this body, which had the shape of a spindle, in breadth about three of the sun's digits, and nine in length, did not quit the sun's northern limb. This spindle kept continually advancing on the sun's body, from east towards west, with no more than about half the velocity with which the ordinary solar spots move; for it did not disappear till the 7th of September, after having reached the sun's western limb. M. Rostan, during that time, observed it almost every day; that is to say, for near a month; and, by means of a camera obscura, he delineated the figure of it, which he sent to the royal academy of sciences at Paris.

The same phaenomenon was observed at Sole, in the bishopric of Basle, situated about five-and-forty German leagues northward of Lausanne. M. Coste, a friend of M. de Rostan, observed it there, with a telescope of eleven feet, and found it of the same spindle-like form, as M. de Rostan, only it was not quite so broad; which probably might be owing to this, that growing near the end of its apparition, the body began to turn about, and present its edge. A more remarkable circumstance is, that at Sole it did not answer to the same point of the sun as it did at Lausanne: it therefore had a considerable parallax: but what so very extraordinary a body, placed between the fun and us, should be, is not easy to divine. It was no spot, since its motion was greatly too slow; nor was it a planet or comet, its figure seemingly proving the contrary. In a word, we know of nothing to have recourse to in the heavens, whereby to explain this phaenomenon; and, what adds to the oddness of it, M. Meffier, who constantly observed the sun at Paris during the same time, saw nothing of such an appearance.

INDENTATION OBSERVED IN THE SOLAR DISK

Codde, Marius, and Payan, A.: *L'Astronomie*, 3:426-428, 1887.

Compiler's Translation: Just before the eclipse I had agreed to meet with M. Leotard at the hill of Notre Dame de la Garde to ascertain whether we would be able to witness the end of the eclipse. Our eastern horizon is very mountainous, and we were almost certain to see nothing.

M. Leotard being late, I had only time enough to go alone to the observatory of the society. I waited for the sun, which made its appearance at 5^h 23^m, but the eclipse had already ended at 5^h 17^m.

Thinking no further of the eclipse, I trained the telescope on the sun and, examining the limb, noticed that delicate waviness which is always visible at sunrise and sunset.

My attention was suddenly drawn by a phenomenon to which at first I attached no great importance: I saw on the southwest limb a small notch, which persisted about 20 seconds. This indentation was circular in form, and, to speak in clock-terms, it had a size equal to the space between the 36th and the 37th minute. I believe this is very close to the same place from which the moon's edge had just previously vanished.

Notch observed on the solar disc, August 19, 1887

M. Leotard, who had come up alone, had seen the sun rise at 5^h 18^m without noticing anything unusual.

But our colleague in the Society, M. Payan, who had gone up to Bell-de-Mai, 2 km north of the city, had seen the same phenomenon as

I, at the same position and at the same time, with the aid of a smoked glass.

We cannot explain the phenomenon and submit it for your opinion. The Sun had not risen at the same point on the horizon for the two observers, and the phenomenon having persisted for 20 or 30 seconds, certainly was caused by nothing terrestrial. (Marius Codde)

On the day of the eclipse, I was at Belle-de-Mai. The sunrise had begun at $5^h\ 23^m$; two minutes later, the sun was completely above the horizon. I saw no part of the sun eclipsed, unless you would accept a small notch, quite unusual, on the lower limb of the disc and at the left. On each side of this notch I saw a dim shadow stretching along the limb of the sun. The size of the indentation was about 1 minute of arc. Duration of the phenomenon 30 seconds. (A. Payan)

THE CORONA

Young, C. A.; *The Sun,* D. Appleton and Co., New York, 1896.

Unlike the chromosphere, which seems first to have been observed, as was mentioned in the previous chapter, only a little more than a century ago, the corona has been known from antiquity, and is described by Philostratus and Plutarch in almost the same terms we should ourselves employ. And yet our knowledge of it remains very limited. The chromosphere and prominences we can now reach and study, comparatively at our leisure, by the help of the spectroscope; but the corona is still inaccessible, except during the short and precious moments of a total eclipse---in all, not more than a few days in a century---so that our knowledge of its cause and nature can grow but slowly at the best.

The character of the phenomenon is such also as to make its accurate observation exceedingly difficult; slight differences in the transparency of the atmosphere, in the sensitiveness of the observer's eye, a preoccupation of the mind by some feature which first happens to strike the attention, or a peculiarity in the manner of representing what one sees, will often make the descriptions and drawings of two observers, side by side, so discrepant that one would hardly imagine they could refer to the same object. For instance, in 1870, two naval officers on the deck of the same vessel made drawings of the corona, one of which represented it as a six-rayed star, while the other showed it as composed of two ovals crossing at right angles. In 1878 the writer, on comparing notes immediately after the eclipse with other members of his party, found that about half of them saw the corona principally extended to the east and west, while the other half, himself among them, were just as positive that it brushed out mainly to the north and south. The photographs, and other data since collected, show that the principal extension was undoubtedly along the east-and-west line, but that there were much better outlined streamers, though shorter and less brilliant, directed from the solar poles. Some eyes were more impressed by definiteness of form, others by size and luminosity.

A "north-south" corona

Obviously, conclusions must be drawn from ocular impressions only with the greatest caution. Photographs are, of course, more to be trusted, as far as they go; but, even with them, a slight difference in the sensitiveness of the plate, in the exposure, or in the development, will make a great difference in the resulting picture. Neither can any photograph ever bring out everything which is visible to the eye. An exposure, sufficient to exhibit well the fainter details, will spoil the brighter features, and vice versa. Moreover, it may, and not seldom does, bring out features that the eye can not see because their light is mainly ultra-violet.

We can do no better than to refer one, who is curious to see how various are the representations of this wonderful object, to Mr. Ranyard's magnificent work upon the observations made during total solar eclipses, published as Volume XLI of the "Memoirs of the Royal Astronomical Society of Great Britain." In it he has reproduced nearly a hundred different drawings and photographs of the corona, as seen during the eclipses since 1850. The steel engravings of the eclipses of 1870 and 1871, based upon the photographs then made, are among the most accurate and beautiful representations of the corona anywhere to be found. We have copied a few of his woodcuts, which give an idea of the more remarkable features of the phenomenon, and exhibit the differences between its character and appearance on different occasions; we have added also a picture of the corona as seen in 1878, in which we have combined the sketches of several observers with our own impressions. Woodcuts, however, are not competent to bring out the peculiar filmy, nebulous character of many of the details, which can be fairly represented only by steel engraving.

The drawing of Liais shows the "petal"-like forms which have been noticed in the corona at other times, but seem to have been especially prominent in the eclipse of 1857. The figures of the corona of 1860, by Secchi and Tempel, show how widely observers only a few miles apart will differ in their impressions. (pp. 238-242)

Corona of 1857

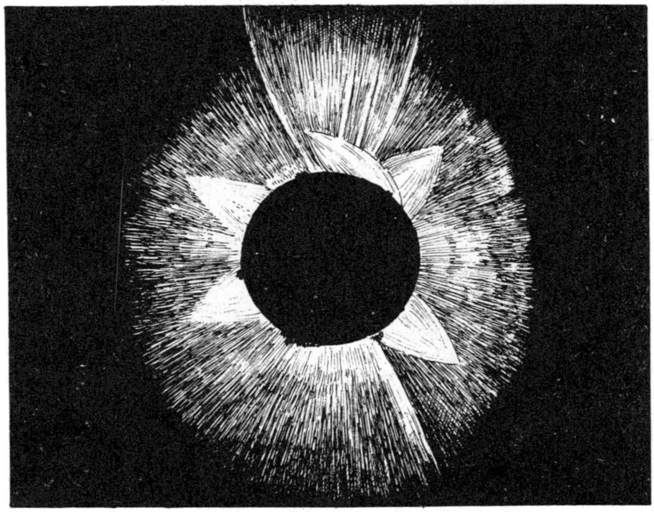

Corona of 1860

A COMETARY STRUCTURE IN THE CORONA OF APRIL 16, 1893

Scharberle, J. M.; *Astronomy and Astrophysics*, 13:307, 1894.

In the October number of this journal I called attention to a comet-like structure near the Sun during the total eclipse of last April, as shown on all the Lick Observatory photographs of the outer corona.

The form and position of this object is shown in accompanying sketch. The straight, slender, nearly radial streamer, from the Moon's outline to the structure in question, is conspicuously visible and distinctly isolated from the more inclined neighboring streamers not shown in the sketch. The drawing is made from an original negative taken with the 40 ft. telescope. On the Dallmeyer negatives the tail-end of the structure, which for the 40-ft. telescope falls outside of the limits of the 18 x 22 inch plate, can be traced for more than a degree from the nucleus or head of the object. Until I have seen copies of the results obtained at the other eclipse stations, I do not wish to express an opinion as to the true nature of this object. Copies of our own photographs were distributed more than eight months ago, but repeated requests for copies of the results obtained at the other stations have thus far been in vain.

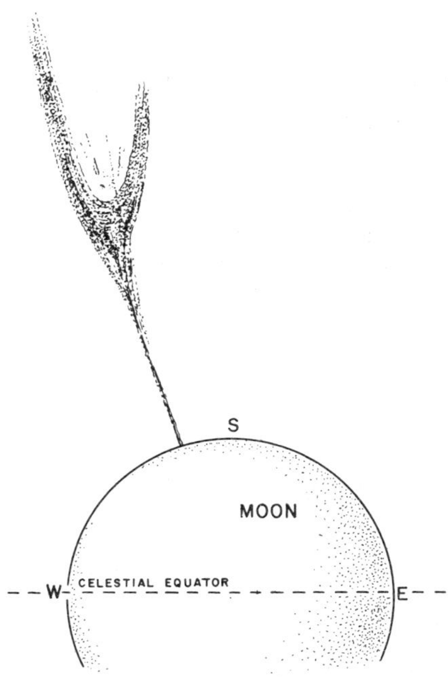

The 1893 cometary structure

RAPID CHANGE ON THE SURFACE OF THE SUN

Anonymous; *Popular Astronomy*, 8:109, 1900.

Caroline E. Furness in charge of the Observatory at Vassar College, Poughkeepsie, New York, was examining the surface of the Sun, with the telescope, Jan. 22, at about 3 o'clock in the afternoon, and noticed an interesting change going on at one point in the surface. When first looked at the solar surface was everywhere clear of spots. Fifteen minutes later, a very small spot was seen near the central region. The image of the Sun was thrown upon a screen and made about fifteen inches in diameter. The spot then appeared like a pin-head. While watching it closely a second fainter one appeared near it, and presently two or three attendant darkenings which looked very much like the ordinary nodules, except that they were dark enough to stand out quite conspicuously against the mottled disc of the Sun. In ten minutes these dark bodies had faded so as to be scarcely visible, and only the first spot remained. As the Sun was low the observations could not be continued longer, and the next morning there was no trace of any disturbance.

It is true that important changes in the solar surfaces are constantly going on, yet it is somewhat rare to notice those that are so peculiar and so rapid.

DARK MARKINGS IN SOLAR PROMINENCES

Daunt, R.; *British Astronomical Association, Journal*, 16:280-281, 1906.

I was much interested in Mr. Newbegin's remarks on dark markings in prominences in his paper in the last number of the "Journal," and Mr. Maunder's comments thereon. I have myself seen the same phenomenon on three occasions lately, but, being a beginner in spectroscopic work, I hardly liked attaching too much importance to what I saw, owing to my inexperience.

However, on April 6 last, so striking an example occurred that I am fully convinced of its reality.

At the base of a prominence situated about -25° W., there was a dome-shaped marking perfectly black, and which looked like something welling up from the limb of the sun, though cut off by the chromosphere.

This marking had every appearance of there being something tangible there. It was well defined, and the edges perfectly clear cut.

These markings to which I refer look perfectly black in contrast to the prominences, and are very much darker than, and have quite a different appearance to, any gaps in the form of a prominence or to the space enclosed between, let us say, two curling hydrogen jets.

My instrument is one of Thorp's prism grating spectroscopes which I obtained last November; its definition is excellent.

SUN'S DISK SHOWS UNUSUAL FEATURE

Williamson, Adrian; *Popular Astronomy,* 42:286-287, 1934.

On April 6, about 5:00 p.m., using a 5-3/8-inch Alvan Clark refractor, with dark red glass filter on a solar eyepiece, I was observing the sun for sunspots. I found no sunspot at that time but was surprised and interested to discover an irregular white line, appearing against the red disc of the sun, near but definitely within the outer circumference, at a position corresponding to about 2:30 or 3:00 o'clock on the dial of a clock. I watched the line some time, or rather a number of times, over a period of about a half an hour. The lower part of the line seemed to change somewhat in shape and position during that period.

I will be glad to know whether you have heard of any other observers noticing this unusual marking on the sun at the time referred to. I am sure that the appearance of the object was approximately as above reported. I called my wife on account of the unusual character of the marking, and she made one or two observations of it with me, verifying my own observation.

My surmise is that what I saw was a hydrogen prominence of such unusual intensity as to be visible within the circumference of the sun, against the darker color of the surrounding area as seen through the red glass filter of the solar eyepiece.

BULGES ON THE SUN'S LIMB

Hoffleit, Dorrit; *Sky and Telescope*, 13:77, 1954.

A recent number of the AAVSO Solar Division Bulletin has two articles on distortions of the solar limb. Although gifted observers like Secchi reported limb distortions in the 19th century, modern observations have often been received skeptically, mainly because the instruments used were too small and the seeing conditions unfavorable for revealing real effects as small as one second of arc. The first of the AAVSO papers is a preliminary note by Harry Bondy on distortions observed on five dates in 1951. One of them was seen by two, another by four observers widely separated geographically. Although the significance of these distortions is not understood, they did occur at times and in areas of other solar activity. Their dimensions are not given, but the author states that the over-all heights were too small to be recorded photographically.

The other paper, by Dr. M. Waldmeier, of Zurich, describes a remarkable distortion on March 6, 1953. It appears to have been of a different type, for at this time no other solar activity such as flares or sunspots was evident, and it was larger and of much shorter duration than those previously recorded. The object on the sun's limb, unique in Dr. Waldmeier's 20 years' observing experience, was discovered while he was adjusting his coelostat half an hour after sunrise on a day when observing conditions were exceptionally fine. Even while he was tracing the outline of the sun, the size of the "nose" diminished. He estimated that his observations lasted some 40 seconds; they began at or after maximum, and hence the entire phenomenon could have lasted a few minutes. The height of the nose or wart was about 5,000 kilometers. It could not have been caused by any terrestrial atmospheric effects. In brightness and color it was indistinguishable from the adjoining portions of the limb.

No purely solar process seems to account for the occurrence, writes Dr. Waldmeier. The region of the distortion was checked with the coronagraph at Arosa, where the observations were made, but there were no unusual corona phenomena and no prominence activity. He suggests cautiously that it might have been caused by the fall of meteoritic material into the sun, or perhaps the close passage of a cosmic body.

PECULIAR SOLAR OBSERVATIONS EXPLAINED

Anonymous; *Sky and Telescope*, 19:472, 1960.

On January 16, 1959, Swedish astronomer Yngve Ohman was observing the sun at Stockholm Observatory through a telescope equipped with a hydrogen-alpha filter. His attention was caught by a rapidly moving dark object crossing the sun, which on passing the solar limb appeared as a bright surge, visible for about two seconds and extending about two minutes of arc outward from the sun. A month later, a similar observation in white light was made by a colleague of Dr. Ohman's.

It was suspected that these apparent prominences might be caused by very distant jet planes passing in front of a fairly low sun. The Swedish Air Force co-operated in testing this hypothesis, and two transits of a jet plane, distant about 10 kilometers, were observed. In each case apparent prominences were seen, lasting somewhat over one second, and having a brightness similar to that of the solar surface. Dr. Ohman's account appeared in the Observatory for December.

SUDDEN EXTINCTION OF THE LIGHT OF A SOLAR PROMINENCE

Trouvelot, L.; *American Journal of Science*, 3:15:85-88, 1878.

On the 26th of June, 1874, while making my daily observations of the sun with the spectroscope at the Harvard College Observatory, I saw an unusual phenomenon, which may be worth recording. The narrow slit of the instrument was directed on the preceding side, about 270°, just above a group of spots which was then very near the limb, when I saw a brilliant protuberance partly projected on the spectrum, on the side of the rays of less refrangibility. In shape, this hydrogen flame resembled an elongated comma, having its acute extremity directed toward the sun, where it terminated just a little above the chromosphere. The chromosphere under this protuberance formed several slender and acute aigrette-shaped flames, none of which, however, reached it. The large prominence, which was slightly inclined to the limb, had a height of 3' 37", and about 3° in its greatest width.

. . . .

I had been observing this phenomenon for eight or ten minutes, when, while looking at it with the slit wide open, the flame suddenly vanished, at 10h 30m, no traces of it remaining. As no motion of any kind, no extension, no contraction, could be perceived before or at the moment this phenomenon took place, and as the light did not go out of it gradually, but as suddenly as a flash of lightning, it does not seem that a change of position was the cause of its disappearance, but rather because the light which rendered it visible abandoned it in an instant.

According to theory, this protuberance was moving rapidly away from the earth at the moment of the observation, as it was projected upon the less refrangible side of the spectrum; yet this would fail to explain its sudden disappearance, since for this it should have moved out of sight with an unconceivable velocity.

For over half an hour I watched attentively the same spot in expectation of seeing the flame reappear; employing for this a narrow and wide slit in succession, but with entirely negative results. I saw no more traces of it, although the small aigrette-shaped flames of the chromosphere, which were still visible, indicated the exact place where it had vanished, and where very probably it still existed, but now as a dark protuberance.

On several occasions I have seen the light abandon a protuberance gradually, but never so suddenly and on such a grand scale; and sometimes I have seen also the light gradually illuminating protuberances which were invisible before, something after the manner of clouds in our atmosphere, lighting up and fading into darkness by the appearance or disappearance of the sun. Of course, the illumination of dark solar protuberances cannot be conceived as being due to the reflection of light, as in the case of the clouds in our atmosphere: it is the protuberance itself which is rendered luminous by some change taking place in it. These observations would seem to indicate that on the sun there are sometimes dark and non-luminous protuberances, which may cause the spots of absorption often observed in the vicinity of spots.

. . . .

SOLAR OSCILLATION PROBLEM GOES, NEUTRINO PROBLEM STAYS

Anonymous; *New Scientist*, 76:221, 1977.

The Birmingham group which had previously detected an oscillation of the Sun with a period of 2hr 40min, thus causing considerable controversy, and also found shorter period oscillations (40 and 57 mins) presented their latest results at a meeting of the Solar Physics Study Panel at the Royal Society recently. And their evidence for the controversial long period oscillation is considerably weaker.

This 2hr 40min oscillation had aroused considerable interest for two reasons: first because it was not explained as easily as the short 5 to 50-minute solar oscillations and secondly because it was detected not only by the Birmingham group but also by an independent group from Russia. Also the most direct interpretation of the oscillation---as a coherent "bouncing" of the solar surface in and out---implied that the Sun had the same density throughout, an impossible condition to implement theoretically.

Although there were reports at the latest meeting that the Russians are still detecting this oscillation for the third year running, the evidence presented by the Birmingham group was very inconclusive and left many astronomers in the audience doubtful.

Solar buffs had earlier suggested that the oscillation was an observational artifact: the result of super-granules on the Sun's surface passing through the field of view as the Sun rotates. But the Birmingham group argue that such an effect requires a fixed field of view; and the Birmingham group do not have such a field. However, the Russian group do, and so may be seeing only this supergranulation effect.

With the absence of fresh evidence from Birmingham the long period oscillation may have ceased to be a problem for solar physicists. A recent article by P. Goldreich and D. A. Keeley from Caltech suggested such phenomena should be ignored until stronger evidence for their existence is presented. That advice is now likely to be heeded.

SOLAR OSCILLATIONS: WEAK BUT ALIVE AND WELL

Anonymous; *New Scientist*, 78:290, 1978.

Recently published data on long-period oscillations of the Sun suggest that these controversial solar vibrations are still very much in evidence. The Russian group at the Crimean Astrophysical Observatory, V. A. Kotov, A. B. Severny and T. T. Tsap, has just reported the details of four years observations of the 160-minute-period oscillation of the Sun from 1974 to 1977 (Monthly Notices of the Royal Astronomical Society, vol 183, p 61).

Their initial results, published in 1976, confirmed the detection of this oscillation by a University of Birmingham group. Recent results from Birmingham suggested that the oscillation was no longer detectable, but the Russians claim to have followed the oscillation over four years.

They have also shown that the amplitude or strength of the oscillation has been gradually reducing from 1.54 m/s in 1974 to about half that value in 1977, which may account for the recent negative results from Birmingham. In addition the Russians have followed the phase of the oscillations, which, although it has drifted by 15 minutes each year, has been consistent throughout this time. They have detected the oscillation in the solar magnetic field as well as in the visible light from the Sun.

Several other periodic oscillations have been detected by the Crimea group in the range 134 to 175 minutes, all of which can be interpreted as being higher harmonics of a gravity wave (like ocean waves) rather than different modes of a radial wave (like seismic waves) (New Scientist, vol 76, p 618). Such an interpretation ties in with the work of Professor Henry Hill of Arizona, who measures shorter-period solar oscillations.

However, to produce the waves with these frequencies requires the standard solar model to have a slightly higher abundance of heavy elements, which in turn predicts a much higher output of solar neutrinos; in fact four times more than currently predicted and 20 times more than currently measured. If solar oscillations were seismic waves the neutrino problem would go away as the seismic wave explanation changes the solar model considerably. Either way there is a problem and the Russian observations help tip the evidence in favour of gravity waves, a standard solar model but an increased neutrino problem.

ANOMALOUS SOLAR ROTATION IN THE EARLY 17TH CENTURY

Eddy, John A., et al; *Science*, 198:824-829, 1977.

Abstract. The character of solar rotation has been examined for two periods in the early 17th century for which detailed sunspot drawings are available: A. D. 1625 through 1626 and 1642 through 1644. The first period occurred 20 years before the start of the Maunder sunspot minimum, 1645 through 1715; the second occurred just at its commencement. Solar rotation in the earlier period was much like that of today. In the later period, the equatorial velocity of the sun was faster by 3 to 5 percent and the differential rotation was enhanced by a factor of 3. The equatorial acceleration with declining solar activity is in the same sense as that found in recent Doppler data. It seems likely that the change in rotation of the solar surface between 1625 and 1645 was associated with the onset of the Maunder Minimum.

• Anomalous Solar Radiation

AN EXCEPTIONAL INCREASE OF COSMIC RAYS

Duperier, A.; *Nature*, 151:308-309, 1943.

In a previous communication, an account was given of the hourly records of the intensity of cosmic rays which are being obtained by making use of a battery of Geiger counters registering threefold coincidences. No absorbing screens are used.

The records generally show that the changes in cosmic rays from day to day are more or less what can be expected from well-known geophysical influences. Nevertheless, last August a transient change was recorded which apparently cannot be ascribed to any of these influences. The points on the diagram of Fig. 1 represent the hourly counting rate for the period August 12-15, 1942. The scale is in percentages from the normal value. The rapid and quite unusual increase which began at

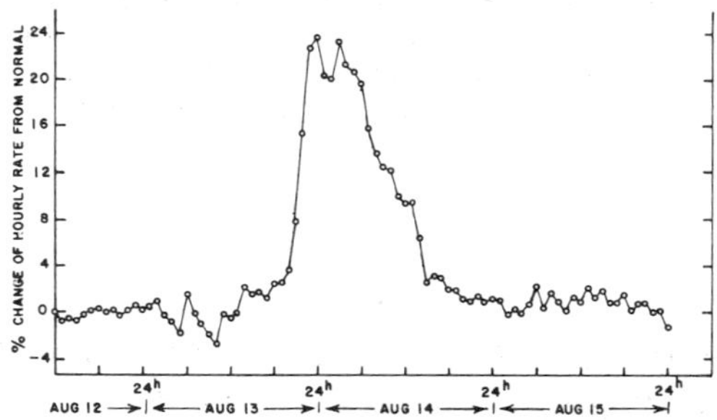

about 18h. (G. M. T.) August 13 was preceded by fluctuations in the counting rate which are not common in the absence of magnetic activity. When these fluctuations were being observed it was considered advisable, in order to test the apparatus, to substitute a new pair of thyratrons for the first pair in the counting circuit. The point corresponding to 12h. on the diagram had already been observed after putting in the new valves. These were permanently retained, and the other parts of the apparatus had remained untouched.

As the diagram shows, the results of the first few observations made with the new thyratrons and those which follow the great increase are entirely comparable with the results of previous observations. To what extent this is true may be inferred from the diagrams in Fig. 2. [Not reproduced] The larger circles represent the averages of the hourly counting rate per day, and the smaller circles the corresponding average atmospheric pressures. Furthermore, the old pair of thyratrons was replaced a month later and again the results did not show any marked difference. In any event, it would be difficult to explain how so sudden, great and regular a change of pulse-rate could be the result of changing the valves. Thus everything seems to point to the conclusion that the relation in time between the substitution of valves and the sudden increase in counting rate was accidental. It thus becomes probable that the observed increase of pulse-rate actually represents an increase of cosmic rays.

However, it seems that this phenomenon cannot be explained by any unusual geomagnetic activity. From information kindly given by the Royal Observatory, Greenwich, it appears that the magnetic field remained perfectly calm during the period August 13-14, and that the only disturbances recorded, on August 10, 15-17, and 22-24, were slight, with a range in the horizontal magnetic force H of 110γ, equivalent to about -0.6 per cent of the H value.

. . . .

So far as is known to me, there has never before been recorded so great an increase of cosmic rays which has extended for such a length of time. On August 17, 1941, our apparatus registered a quite abrupt increase of more than 30 per cent, which lasted only from 1h. until 3h. Also last March two sudden increases of about 7 per cent were recorded in Cheltenham and London, which lasted not more than four hours each. But these were probably associated with magnetic storms.

SOLAR FLARE ACTIVITY:EVIDENCE FOR LARGE-SCALE CHANGES IN THE PAST

Zook, Herbert A., et al; *Icarus*, 32:106-126, 1977.

Abstract. An analysis of radar and photographic meteor data and of spacecraft meteoroid penetration data indicates that there probably has not been a large increase in meteoroid impact rates in the last 10^4 year. The solar flare tracks observed in the glass linings of meteoroid impact pits on lunar rock 15205 are therefore reanalyzed assuming a meteoroid flux that is constant in time. Based on this assumption, the data suggest that the production rate of Fe-group solar flare tracks may have varied by as much as a factor of 50 on a time scale of about 10^4 yr. No independently obtained data are known to require conflict with this interpretation. Confidence in this conclusion is somewhat qualified by the experimental and analytical uncertainties involved, but the conclusion nevertheless remains the present "best" explanation for the observed data trends.

AN UNEXPECTED EFFECT IN SOLAR COSMIC RAY DATA RELATED TO 29.5 DAYS

Dodson, Helen W., and Hedeman, E. Ruth; *Journal of Geophysical Research,* 69:3965-3971, 1964.

Abstract. Data relating to the detection of solar protons in the neighborhood of the earth (1952-1963) and to neutron counts (1958-1963) have been distributed on the basis of the mean synodic solar rotation period, 27.3 days, and the approximate synodic month, 29.5 days. In the latter, apparent departures from random distribution are especially marked. At the present time it is not clear whether the 29.5-day 'effect' is related to the sun or the moon or is only a statistical accident.

SOLAR NEUTRINOS: WHERE ARE THEY?

Hammond, Allen L.; *Science,* 175:505, 1972 (Copyright 1972 by the American Association for the Advancement of Science)

An experiment that monitors the flux of neutrinos received from the sun is yielding information that may upset current theories of nuclear processes within the sun and similar stars. Earlier results from this solar experiment being conducted by Raymond Davis, Jr., and his colleagues at Brookhaven National Laboratory pointed to a rate of neutrino production that was lower than the rates predicted from models of the sun (see Science, 10 September 1971, pp. 1011). The discrepancy between theory and experiment might have been explained, in the opinion of some observers, by uncertainties in the experiment and drastic but not inconceivable modifications in the solar models. The first results from recent trials with improved detectors and additional shielding---improvements that eliminate some of the background radiation in the experiment---now indicate a still lower neutrino flux, which, if substantiated, will be extremely unsettling to theorists. Already the solar models and the understanding of nuclear processes in astrophysical contexts are undergoing reexamination.

To detect neutrinos, the Brookhaven scientists have placed a large tank of tetrachloroethylene about 1.5 kilometers underground in a mine shaft. The capture of neutrinos by chlorine-37 atoms (an extremely rare event since neutrinos interact with matter very weakly) converts the chlorine to argon-37, which is then extracted from the tank. A counting system is used to monitor the decay of the radioactive argon, which has a half-life of about 35 days. The flux of neutrinos is then calculated from the measured decay rate. In the early runs, the Brookhaven team detected about five decay events per 35-day counting period, a rate corresponding to the capture of about $(1.5 \pm 1) \times 10^{-36}$ neutrinos per second per chlorine-37 atom. By comparison, several theoretical models predict about 9×10^{-36} captures per second per target atom, a factor of 6 higher.

Several uncertainties affect the experimental result. Neutrons from spontaneous fission of uranium-238 or from alpha particle reactions (uranium and thorium emit alpha particles when they decay) in the surrounding rocks could also produce argon-37 from the chlorine target atoms. The atmosphere contains small amounts of argon-37 from nuclear weapon tests, and thus argon from the air in the mine might leak into the tank or contaminate the absorbers used in extracting the argon from the tank. Cosmic ray muons that penetrate the 1.5 km to the mine shaft can convert the chlorine in the tank to argon-37 by various nuclear processes. All of these background sources, however, would lead to a spuriously high rate for the neutrino capture. The Brookhaven team now believe that they have eliminated some of these possible sources of error.

In the most recent two runs, for example, the experimental chamber was flooded with 400,000 gallons of water to moderate or slow down any neutrons to energies below 1×10^6 electron volts (1 Mev), the threshold for the production of argon-37. Even more significant, Davis believes, are improvements in the proportional counters used to record argon-37 decays, improvements which now allow the experimenters to distinguish between argon-37 decays and other charged particle events that produce a pulse from the detectors. The Brookhaven counters are extremely small and respond very rapidly, so that the pulse shape as well as the energy of the pulse can be determined. Argon-37 decays produce a characteristic rapidly rising pulse, whereas cosmic rays, beta rays, or Compton electrons, because they tend to ionize molecules strung out across the counter, show a more slowly rising pulse.

The remaining uncertainties include residual background of argon-37 produced by cosmic rays and possible contamination from argon in the air. Davis monitors the cosmic ray-flux higher up in the mine, but must extrapolate to calculate what the flux would be at the depths of the mine. He also monitors the amount of argon-37 present in the air in the mine, which, in the latest run, was unusually high. In fact, counts of a sample of the air would have yielded more argon-37 decay than that experimentally recorded. It therefore appears that some of the apparent neutrino flux might be due to background sources; even if there were no contribution from these sources, however, the measured neutrino capture rate is strikingly low.

In their latest experiment, the Brookhaven scientists find a rate below 1.0×10^{-36} captures per second per target atom. In theory most of the neutrinos expected would come from the decay of boron-8 in a minor branch of the proton-proton nuclear synthesis chain that is believed to give the sun its power. But even the neutrinos expected from the proton-electron-proton step of the nuclear synthesis reaction and from the decay of the beryllium-7 isotope, according to conventional theories, would result in a neutrino capture rate of 1.1×10^{-36} per second per atom, a rate comparable to the present uncertainty of the experiment. The experimental result is still tentative, because with such low capture rates (approaching one argon-37 decay per month), the possibility of statistical variations from 1 month to the next cannot be ruled out. Nonetheless, the results are discouraging for solar physicists.

The implication of Davis' result, if sustained by further experiment, is that either existing solar models or some fundamental nuclear theory as applied to the sun are seriously in error. Among the questions arising are whether the measured opacity of the sun is more than twice too large, or whether heavy elements such as carbon, nitrogen, and oxygen are perhaps 20 times less abundant in the interior of the sun than at the surface. Neither of these possibilities is considered to be very likely by solar physicists such as Icko Iben, of the Massachusetts Institute of Technology. More probable, in some assessments, is the chance that the extrapolation of laboratory data on nuclear physics to the reactions taking place in the sun is incorrect. John Bahcall of the California Institute of Technology has suggested (see Physical Review Letters, 31 January 1972, p. 316) that the discrepancy might be due to still undiscovered properties of the neutrino. He proposes that if these particles have a finite mass (they are generally considered to have no mass), they might be unstable and decay in the 500 seconds required for their travel from the sun to the earth. Bahcall suggests additional neutrino capture experiments to test this possibility. Other suggestions are likely to appear in coming months as physicists try to explain the absence of the missing solar neutrinos.

SOLAR OSCILLATION PROBLEM GOES, NEUTRINO PROBLEM STAYS

Anonymous; *New Scientist*, 76:221, 1977.

However the other perennial thorn in the solar physicist's side, the neutrino problem, remains. Professor Ian Roxburgh of Queen Mary College, London, giving an invited review on the solar interior, told the Royal Society meeting that the latest results reported from Davis' neutrino experiment in Arizona gave the neutrino flux as 1.4 ± 0.4 snu (solar neutrino units). Classic solar theory suggests there should be 5 to 6 snu; so in Roxburgh's view the Davis results put a new strong constraint on solar modelling.

The constraint can be accommodated by assuming the Sun has a slightly cooler central temperature---because the number of high energy neutrinos expected by Davis is strongly dependent on the temperature. There would also be a correspondingly bigger core. In turn that implies that the "convection zone"---the sub-surface shell of the Sun where heat gets out mostly by convection---is more important than had been believed. That makes life difficult for solar theorists, for the analytical problem of convection in a rotating spherical body remains unsolved. [Excerpt]

• THe Effect of the Sun on Solar-System Phenomena

THE SUN SINCE THE BRONZE AGE

Eddy, John A.; *American Geophysical Union Paper,* International Symposium on Solar-Terrestrial Physics, 1976.

Introduction. I was originally given the subject "Outstanding and Unusual Solar-Terrestrial Events in History", which was meant, I suppose, as license to bring before you a kind of solar-terrestrial sideshow of historical marvels: the Astounding Aurora of 1716, the Mysterious Dark Day of 1790, the Colossal White Light Flare of 1859, and so on. I for one would enjoy that, and it would probably be appropriate to end a symposium as long and large as this one on such a carnival note.

The difficulty is that we do not know, in the summer of 1976, what is outstanding in solar behavior and what is commonplace. We cannot distinguish the unusual from the usual. In particular, if we look beyond the immediate present, earlier than the last several hundred years, we discover that what we have always thought were common in modern solar history appear in longer view to be the freaks. We find that what we thought to be anomalies---like the Maunder Minimum---are really rather ordinary. We must allow that in the perspective of but a thousand years the aurora borealis is unusual, as may be the solar corona, and perhaps the solar chromosphere, prominences, and flares. It now seems quite possible that the common 11-year sunspot cycle is but a temporary feature of the most recent solar history, or that it gets switched off and on in a program that seems almost random.

This quandary, as you may know, is new in solar physics. A few years ago most of us were confident that the sun was a regular and repeatable star of near perfect constancy. We believed in a kind of solar "uniformitarianism", by which concept the modern behavior of the sun was taken to represent its normal course in a much longer span---certainly of hundreds or thousands of years. Many of us made use of Schove's reconstruction of an early sunspot cycle, which was built on the assumptions of solar regularity and uniformitarianism.

Some of us are now concerned that these superannuated assumptions of constancy and regularity may have long misled us, both in solar physics and in related disciplines. We urge caution in making the 11-year cycle the basis of physical understanding of solar activity or of terrestrial effects, since, when you look hard at the historical record there is little or no evidence that the 11-year cycle existed at all before very modern times---perhaps about A. D. 1700. What now seem more important are gross changes of behavior which the sun experiences on more ponderous scale---of 100's to 1000's of years. During these excursions, of which there have been about 12 in the last 5000 years, the sun has been both a good deal less active, and probably more active than anything we have seen in the modern era. At first look, moreover, the pattern of occurrence of these major solar changes is probably not periodic, but possibly bimodal or stochastic.

Needless to say, solar-terrestrial physics is directly affected by these new realizations of solar behavior. Probably most affected are solar-weather and solar-climate studies, which for an embarrassingly long time have been hung up on searches for relationships on the shortest time scales---with daily and annual sunspot number and with the 11-year solar cycle. Have so many in science ever worked so long in a mine of such low yield? I think it time we admitted that no convincing and enduring correlations of an important nature have ever been found. Surely a hundred years of frustration are enough to suggest that we try a different approach.

The New Solar Physics tells us that the 11-year cycle is but a ripple on an ocean of great and sweeping tides. It suggests we step back and look instead at the longer-term changes, when the sun drifts in and out of eras like the Maunder Minimum. It says that these changes may be the more fundamental on the sun, the more indicative of changes in the sun's energetic, radiative output, and the more important in terrestrial effect. And when we look at the record of climate we find indeed their clear and unmistakable signature.

Evidence which has prompted this solar reappraisal has come from two sources: a re-evaluation of the historical record which has confirmed an unappreciated period of solar "anomaly" in the 17th and early 18th centuries (the Maunder Minimum), and a subsequent re-evaluation of the ^{14}C record, which extends solar history to about 5000 B. C., almost halfway to the end of the last glaciation, and well beyond the reach of the written word.

Radiocarbon. The ^{14}C isotope is produced in the upper atmosphere of the earth as a result of bombardment by galactic cosmic rays. The cosmic ray flux is not constant, and thus the production rate of the isotope varies with time. Among the important modulators is solar activity, which affects terrestrial ^{14}C production in an inverse way; when solar activity is high, the earth is more shielded from galactic cosmic rays and ^{14}C production goes down; when solar activity is low we receive an increased flux of galactic cosmic rays and the ^{14}C production rate increases. The solar modulation of galactic nucleonic flux is well established, as is its effect on ^{14}C production. Other effects are also important---among them the variable shielding introduced by the changing strength of the earth's magnetic field, which varies by about a factor two in a period of roughly 10,000 years.

If we had a record of how much ^{14}C was present in the atmosphere in the past we could in principle deduce the history of solar activity. Such a record exists in carbonaceous fossil material, and most usefully in trees, where ^{14}C is assimilated as CO_2 in the process of photosynthesis. Individual tree rings preserve a record of the prevailing ^{14}C:^{12}C abundance ratio in the lower atmosphere at the time they were formed. The record can be read in living trees, such as the bristlecone pine, to about 3000 B.C., and extended in well-preserved dead wood to beyond 5000 B.C.

In interpreting the tree ring record for evidence of changes in solar activity we must allow for several important effects. Of fundamental importance is an appreciable delay in the atmospheric reservoir between instantaneous changes in ^{14}C production in the upper atmosphere and resultant ^{14}C abundance variations in the biosphere. This lag is on the order of 10 to 50 years. It tends to smear and wash out short-term changes such as the 11-year solar cycle, and to displace all effects in time. In tree rings formed this year, for example, is the smeared record of nucleonic flux variations of 10 to 50 years ago. Thus we find the Maunder Minimum (A.D. 1645-1715) in tree rings formed somewhat later than the historically established time of the real drop in solar activity and aurorae.

Figure 1 is a compilation of ^{14}C data by Lin, Fan, Damon, and Wallick, who have assembled tree-ring derived ^{14}C results from a number of laboratories. Plotted is the deviation of relative ^{14}C abundance from the 1890 normal, expressed in parts per thousand with positive deviation (increased ^{14}C) downward, to agree in sense with solar activity. The 1890 norm ($\Delta^{14}C = 0$) is shown as a dashed, horizontal line. The observations have been fitted with a sinusoidal curve derived by Lin, Fan, Damon and Wallick. They point out that it matches very well the smoothed curve of changing magnetic moment of the earth which is obtained from paleomagnetic data. The strength of the earth's dipole moment reached a maximum in about A.D. 200, at which time we should expect ^{14}C production to minimize, as indeed the data show. Half a cycle earlier, about 5000 B.C., the earth's magnetic moment was at a minimum; at that time we should expect maximum galactic cosmic ray flux and a maximum in ^{14}C production, as is shown. Thus, to a first approximation, the overall envelope of the observed ^{14}C curve is explained as the result of slow and apparently cyclic changes in the strength of the terrestrial magnetic field.

Some of the remaining structure on the compiled ^{14}C curve is probably observational error, but we can expect the significant observed deviations from the smoothed sinusoidal curve to be of likely solar ori-

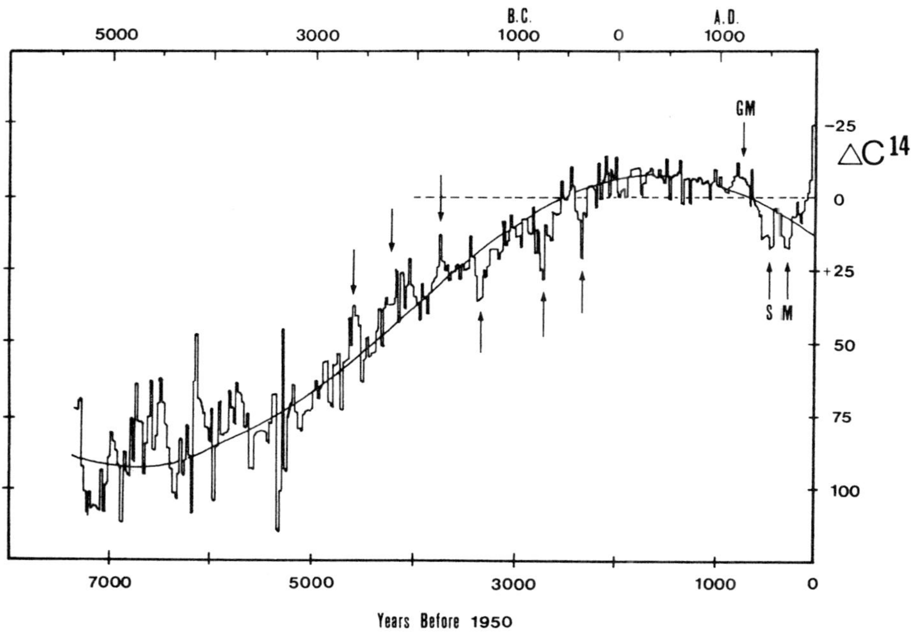

Carbon-14 relative abundance using 1890 norm. (Fig. 1)

gin, as has been pointed out by many authors. Thus the two dips (increased ^{14}C) at the recent end of the curve, labelled "S" and "M" are the probable signature of marked decreases in solar activity, and the opposite excursion about A. D. 1200, labelled "GM", the result of a marked and prolonged increase. Other major excursions can be readily identified. In a recent review Damon has shown that the increased amplitude of excursions in the earliest part of the record (about 5000 to 7000 years B. P.) is not observational noise but an effect of the weaker geomagnetic shielding at the time, which tends to increase the relative effect of solar modulation. Thus the excursions in this era, including marked maxima at about 6000 and 6500 B. P. and a remarkable minimum at about 7200 B. P., are probably real solar effects.

A more expanded plot of ^{14}C data covering only the Christian era, also from Damon, is shown in Figure 2. (Figures 2 and 3 omitted) Again the sinusoidal archaeomagnetic curve is shown as a solid curve, which can be taken as an approximate baseline in identifying other meaningful excursions. We see again the same features noted in Figure 1. Also apparent are less certain features of a more minor nature: an apparent minimum in solar activity about A. D. 650-750 which seems confirmed in catalogs of aurorae and naked-eye sunspots and a possible maximum about A. D. 100, in the Roman era.

In both Figures 1 and 2 the abrupt drop in ^{14}C concentration (upward spike) at the most modern end of the curve is the Suess effect: the result of the introduction of significant CO_2 in the atmosphere through the combustion of fossil fuels. The Suess effect overwhelms and presumably destroys the solar information in the modern radiocarbon record. Thus the radiocarbon data after the middle or late 19th century cannot be directly related to levels of solar activity, or used to provide a present standard of solar behavior in assessing the past. This is no reason why we cannot judge with certainty whether the modern era represents normal or abnormal solar behavior, although the change in radiocarbon in the pre-Suess-effect 19th century suggests that we are in or moving toward another Grand Maximum, as does the overall envelope of the sunspot curve. This would say that we are now in the middle or rising part of abnormally high solar activity, of the sort which has pertained only perhaps 10% of the time in the past five millenia.

The Maunder Minimum. A yardstick which is useful in scaling the solar significance of the radiocarbon record is the Maunder Minimum, A. D. 1645-1715, marked "M" in Figures 1 and 2. Unlike the other excursions pointed out earlier in the curve, the Maunder Minimum comes late enough---after the development of the telescope---that we have adequate historical records to describe with some certainty the behavior of the sun at the time. In this sense the Maunder Minimum is the Rosetta stone which has allowed us to translate the quantitative solar information in the radiocarbon record.

The 1645-1715 period was a time of unique solar behavior in recent historical time, and it probably qualifies as an "Outstanding and Unusual Solar-Terrestrial Event." Eddy has verified that during the long span sunspots were very rare, as shown in Figure 3. For 70 years solar activity hovered at a level somewhat lower than that characteristic of the minima of the present 11-year cycle, and for periods of up to 10 years no sunspots were seen at all. None was found in the whole northern hemisphere of the sun for 32 years. The possibility that the sunspot dearth was an artificiality of inadequate observers or poor technique seems untenable when one considers the advances made in other areas of astronomy and the exquisite and detailed drawings of the sun and sunspots made before and during the period. Reports of aurorae throughout Europe fell sharply during the Maunder Minimum and rose abruptly after it. The solar electron corona was either severely weakened or absent altogether; observers of the sun at total eclipses during the Maunder Minimum described a narrow ring of light around the moon, reddish in color and of uniform breadth---which fits the description of Fraunhofer corona (or zodiacal light) were the continuum corona stripped away. Spots were reported on the sun from time to time, but usually as isolated features and always at low latitudes. This pattern of appearance suggests, literally, a "prolonged sunspot minimum", as Maunder first described the period, but it seems impossible to determine whether or not the 11-year cycle continued to operate at a suppressed or nearly invisible level.

Nor is it certain whether the 11-year cycle operated in the 1610-1645 period, after the introduction of the telescope and before the onset of the Maunder Minimum. In truth, 1700, or perhaps 1750, are the earliest dates for which we have unambiguous evidence of an 11-year cycle. When Galileo first turned his telescope on the sun, in about 1611, the surface was probably more spotted than at any time in the ensuing century, and, we may assume, the sun was probably near a moderate maximum of activity. The numbers soon fell, however, as best we can determine from a far from continuous record. Rudolf Wolf assigned probable dates of maxima and minima of a continued 11-year cycle for the 1610-1700 period, but these were largely extrapolations in which he felt little confidence. He was also unsure, by the way, of the reconstructed 1700-1750 sunspot numbers which we use today. Eddy, Gilman, and Trotter have shown that solar rotation was truly anomalous in the period just before spots disappeared in the Maunder Minimum: equatorial regions rotated about 3% faster than at present and the differential rotation was enhanced by about a factor 3.

A History of the Sun in the Last 5000 Years. We may presume that the 15th century period labelled "S" in Figures 1 and 2 was another era of solar behavior much like the Maunder Minimum, since the ^{14}C record at the time seems almost identical to that of the 1645-1715 period. Historical records are poorer for this earlier, Sporer Minimum, but its reality seems confirmed in a paucity of auroral counts, an absence of naked-eye sunspot reports, and corona-less descriptions of the eclipsed sun. There were again probably almost no sunspots, and, we may presume, a similar dearth of flares, and prominences. By the same reasoning the marked change in $\Delta^{14}C$ between about 1100 and 1300 (an upward feature labelled "GM" in Figures 1 and 2) suggests a time of prolonged high solar activity, probably higher than what we have seen in modern times, although a definitive comparison is made difficult by the Suess effect in the modern ^{14}C record. During this Medieval Maximum, auroral reports were higher than in preceding or succeeding centuries, and there was a marked increase in the frequency of reports of naked-eye sunspots.

....

The names presumptiously assigned the solar features in Table 1 are meant for easy, preliminary identification; for features occurring earlier than the Maunder and Sporer minima they describe the general historical period in which the apparent anomaly falls: for example, the "Egyptian" solar minimum (feature 9) between about 1420 and 1260 B.C. occurred during the "Golden Age" of the New Kingdom of Ancient Egypt and the minima immediately following (feature 8) fell shortly after the time of Homer.

I see no striking periodic behavior in these representations of long-term solar changes, although we must allow that more subtle cyclic features could be masked by noise in the ^{14}C record or destroyed in the process of selection. We find more minima than maxima in most of the period covered, and they do not alternate. The Sporer and Maunder Minima may be parts of a single minimum in a long cycle of about 2500 years, of which the Egyptian, Homeric and Grecian minima (features 7, 8 and 9) are one full cycle away. J. R. Bray, who has pioneered the study of long-term solar change, has noted a cycle of this length in earlier, more preliminary radiocarbon data. More recently Damon has subjected the data in Figure 1 to power spectrum analysis to search for obvious cyclic effects. He divided the data into 2000 year periods and found, interestingly, that statistically significant periods appeared, but of different length in different epochs, as though solar activity were subject to some kind of frequency modulation. In the first 2000 years B.P., Damon found significant power at periods of 56, 69, 182, and 400 years; between 2000 and 4000 B.P. the significant periods were 286 and 500 years, and from 4000 to 6000 B. P. they were 100, 286, and 1000 years. These are very preliminary findings but they suggest that the pattern of long-term solar behavior is not what purists would call well-behaved. At this point I prefer to describe long-term solar variability as meandering.

Table 1. Major Solar Excursions since 5000 Yrs. B. P.

Feature (Fig. 4)	Beginning & End in Radiocarbon Record		Probable Extent in Real Time		Amplitude: Δ ^{14}C	Corrected
1. Modern Maximum	AD 1800?	---	AD 1780?	---	?	
2. Maunder Minimum	AD 1660	AD 1770	AD 1640	AD 1710	-1.0	-1.0
3. Sporer Minimum	AD 1420	AD 1570	AD 1400	AD 1510	-1.0	-1.1
4. Medieval Maximum	AD 1140	AD 1340	AD 1120	AD 1280	0.7	0.8
5. Medieval Minimum	AD 660	AD 770	AD 640	AD 710	-0.6	-0.7
6. Roman Maximum	AD 1	AD 140	20 BC	AD 80	0.6	0.7
7. Grecian Minimum	420 BC	300 BC	440 BC	360 BC	-2.0	-2.1
8. Homeric Minimum	800 BC	580 BC	820 BC	640 BC	-2.1	-2.0
9. Egyptian Minimum	1400 BC	1200 BC	1420 BC	1260 BC	-1.5	-1.4
10. Stonehenge Maximum	1850 BC	1700 BC	1870 BC	1760 BC	1.6	1.3
11. Pyramid Maximum	2350 BC	2000 BC	2370 BC	2060 BC	1.4	1.1
12. Sumerian Maximum	2700 BC	2550 BC	2720 BC	2610 BC	1.7	1.3

An Interpretation of the Major ^{14}C Excursions. Figure 4b interprets the schematic ^{14}C data of Figure 4a as a direct representation of solar activity. The interpretation rests on the established correspondence between post-1650 (A. D.) radiocarbon data and historical observations of the sun from the Maunder Minimum through the onset of the Suess effect (ca. 1850). For these historically accessible periods, the ^{14}C residuum (the difference between observed radiocarbon deviations and the sinusoidal terrestrial magnetic curve) followed very closely the observed envelope of the annual sunspot number. We have therefore assumed that the general, long-term level of solar activity (or the envelope of the curve of annual sunspot numbers) can be read almost directly in the radiocarbon residuum: bottoming out in departures like the Maunder Minimum and maximizing when the radiocarbon residuum reaches the large negative levels of the A. D. 1100-1300 Medieval Maximum. For the

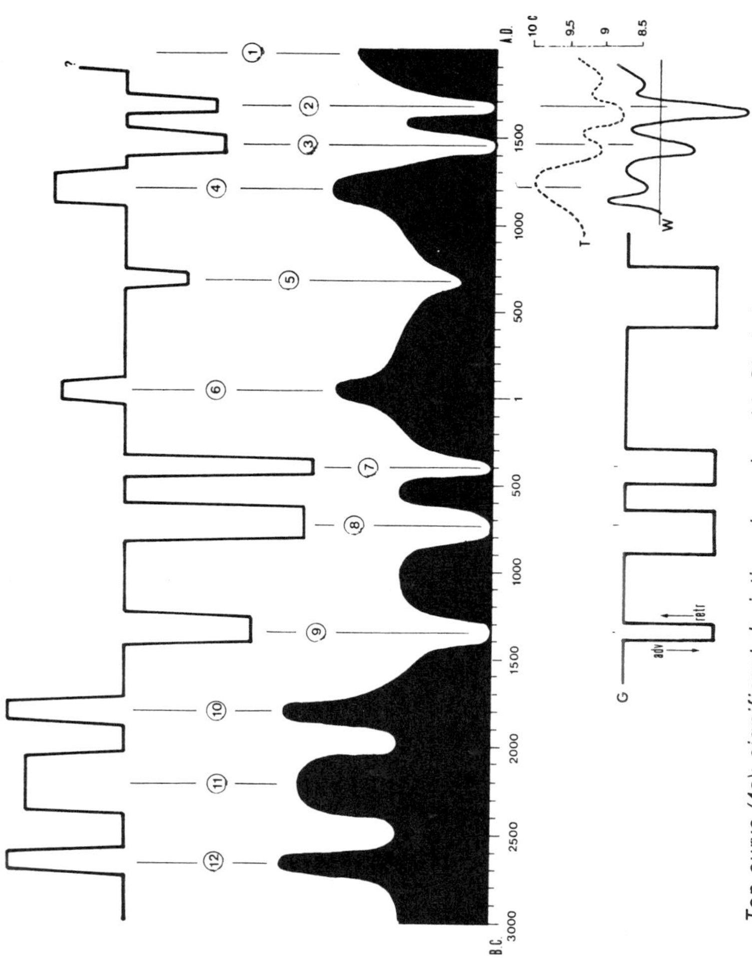

Top curve (4a): significant deviations in carbon-14. Circled numbers refer to Table 1. Middle curve (4b): interpretation of top curve as long-term envelope of possible sunspot cycles. Curve G: advances and retreats of Alpine glaciers. Curve T: estimate of mean annual temperature in England. Curve W: winter severity index for Paris-London area (downward is colder) (Fig. 4)

modern end of the curve which is affected by the Suess effect, we have simply used the observed envelope of sunspot number, which indicates a continued rise in the level of solar activity from A. D. 1715 at least through the 1959 maximum. The rounded, connecting curve in most of Figure 4b in an arbitrary and wholly artistic connection between the maxima and minima of Table 1 and Figure 4a.

The "floor" imposed on the interpreted curve in Figure 4c acknowledges that solar activity has a zero level below which it cannot go and which was nearly reached during the Maunder Minimum. That there are deeper minima in the first two millenia B. C. (Figure 4a) is interpreted as the result of the longer persistence of these three, earlier events from their clumping in time, since the radiocarbon data necessarily reflect a temporal integration in the atmospheric reservoir.

In the reconstructed solar activity curve is a possible explanation for two historical enigmas of solar and solar-terrestrial history: the "auroral turn-on" in the early 18th century and the apparent absence of reports of the structured corona before the same general date. If we accept that activity-related aurorae and the solar corona are both threshold phenomena which correspond to a certain minimum long-term level of solar activity, then their absence or suppression in much of early history seems a logical result of the apparent pattern of excursions of solar activity. By this interpretation we would expect frequent aurorae and a prominent and extended electron corona only during periods like the present, which are times of maxima in Table 1 and Figure 4. In the past three millenia these conditions have applied very infrequently---perhaps no more than 10% of the time: for several centuries during the Medieval Maximum and an even shorter interval during the Roman Maximum. When these limited opportunities are combined with sociological trends in the rise of civilization, and the difficulties of securing evidence from ever more ancient times, the enigmas of the missing aurorae and coronae largely vanish. Ironically, these spectacular displays of nature would seem to have been withheld or suppressed during some of the more vigorous times of learning on the earth, including the era of early Greek interest in science and natural philosophy.

The Sun and Climate History. We must allow that these massive solar changes---not ripples, but tides and tidal waves---could have had pronounced effect on terrestrial climate, and through regional and global climate change, on the course of civilization itself. By the same reasoning it does not seem unlikely that day to day and year to year changes, and the 11-year cycle, could be such minor perturbations in the life of the sun that their imprint, if any, on earth and climate could be lost in more energetic and self-generated changes in the atmosphere itself. The close correspondence of the Maunder Minimum, the Sporer Minimum, and the Medieval Maximum of solar behavior with the long-term record of climate has been pointed out before. It is particularly striking when one allows for the 40-year delay between the tree ring record of ^{14}C and the initiating changes in the upper atmosphere. Times of depressed solar activity correspond to times of global cold: the Maunder and Sporer minima match the two coldest extremes of the Little Ice Age, when global temperatures were depressed 0.5 to 1°C. High levels of solar activity seem to relate to periods of high global temperatures: the Medieval Maximum to the Middle Ages Warm Epoch, or Climatic Optimum.

The correspondence is no less striking when the earlier solar record is compared with even earlier climate history, as best as it is known. In part C of Figure 4 we show this comparison, on the same time scale as the rest of the figure. The step function G depicts the advance (downward) and retreat (upward) of Alpine glaciers, taken from the climate summary of Le Roy Ladurie. Curves T and W are temperatures (scale at right) and estimates of winter severity (colder downward) for England and Paris-London, respectively, from the historically reconstructed data of H. H. Lamb. The correspondence, feature for feature, is I think, almost the fit of a key in a lock. Wherever a dip in solar activity occurs (as in features 2, 3, 5, 7, 8, and 9) the climate swings coldward, and glaciers advance. When we have a prolonged maximum of solar activity (as in features 4 and 6) glaciers retreat and the earth warms. We should recognize that we deal here with very coarse data, particularly in the record of reconstructed climate, and we should also be warned that these "climate" curves may represent only regional (European) trends. Bray, however, has demonstrated the global applicability of many of these same climate epochs, and indeed has pointed out the same long-term sun-climate correspondence shown here.

The physical connection with solar changes could be through the recognized increase in ultraviolet solar flux with solar activity, and the effect of that increase on chemical processes in the upper atmosphere. Were that the case, however, I would expect more obvious correlation of shorter-term solar activity and weather. It could also come about through known changes in the particle flux from the sun and some triggered reaction necessary to amplify the wholly inadequate energies in these fluxes. I am not ready to entertain either of these more complex mechanisms until we examine the simplest and most straightforward process: namely, that the total radiative output of the sun, or solar constant, is slowly and ponderously changing, and that these possible meandering changes are reflected in sign and magnitude in the overall envelope of solar activity. By this notion the curves of Figure 4 are proposed as records of the solar constant, with peak-to-peak amplitudes of perhaps 1%, the amount that seems adequate in global climate models to change the terrestrial temperature by 1° or 2°C. Long-term changes of this amount in the solar constant, by the way, would be very difficult to detect directly, and would be unnoticeable in observations of other G stars.

This proposed association is based on a hunch, on an admitted distaste for trigger mechanisms, and on a preliminary finding that the average value of the measured solar constant increased steadily in the first half of the 20th century---by about 0.25%, which is the right amount to explain the established increase in world temperature during the same span. During the same half century the envelope of sunspot number was also monotonically increasing. It may be significant that while the solar constant was presumably rising, between about 1908 and 1955, its measured fluctuations did not seem to follow the 11-year cycle, although we can question whether the measurements were adequate to sense these more noise-limited changes. More recent and precise measurements of the solar constant, from spacecraft in six-month periods in 1969 and 1975, also failed to detect significant short-term changes.

If the solar constant does not follow the wiggles in daily or annual sunspot number, how can it follow the envelope? A simple answer is that the solar constant may not follow the sunspot number at all; rather, <u>the sunspot number may follow changes in the solar constant, through a kind of amplitude modulation of an otherwise more uniform cycle</u>. A mechanism for this modulation exists in the solar dynamo, which we now think responsible for the maintenance of the 11-year sunspot cycle. By this hypothesis, were the flow of radiation through the outer solar atmosphere perfectly constant, we might expect a sunspot cycle whose peaks were almost uniform in amplitude. If the flow of radiation were slowly increased, we would expect an overall enhancement of sunspot production, which would be most visible in <u>retrospect</u>, in the run of heights of the 11-year peaks. If the flow of radiation were slightly reduced, the peaks of the cycle would be depressed. And if the radiation fell below some critical level, perhaps only a drop of 1% or less, the amplitude of the cycle might be damped so much that the cycle would shut down, or appear to shut down, as during the Maunder Minimum, and presumably the Sporer

Minimum and the earlier cases we have pointed out.

An intriguing consequence of this hypothesis is that individual ups and downs of the 11-year cycle, or of shorter-term solar variability, are almost wholly unrelated to the problem: they would tell little of changes in the solar output and predict almost nothing of consequence in terrestrial meteorology. If one sought a solar-weather connection of periods shorter than climatic time scales he would be always frustrated in what he found, and driven to ever more elaborate restrictions and ever more intricate mechanisms, much as pre-Copernican astronomers were driven into epicycles. And that, I would submit, may be just exactly what has happened in the past century of solar-weather research.

. . . .

AN UNUSUAL SOLAR ACTIVITY CORRELATION

M., T. L.; *British Astronomical Association, Journal,* 58:288, 1948.

In the Comptes Rendus of the French Academy (1947 Dec. 22), F. Link and V. Vanysek produce some evidence of a connection between the number of comets discovered and the sunspot cycle. They show a flat minimum of comet discoveries near sunspot maximum and connect it with a rainfall maximum. But on repeating the investigation for minor planet discoveries they find no clear correlation and are led to suggest that the sun's radiation may have an effect on the brightness of comets. One might feel however that the difference lies much more in the method of discovery of comets and minor planets respectively; anyone who has tried the two would probably agree that the weather has much more effect on whether or not a comet is noticed, whereas the seasonal effect will predominate in the case of asteroids.

SOLAR ROTATION AND THE PERIHELION ADVANCE OF THE PLANETS

Roxburgh, Ian W.; *Icarus*, 3:92-97, 1964.

Abstract. The hypothesis that the inside of the Sun is rotating much more rapidly than the surface layers is examined and found to be consistent with present ideas on the formation of the solar system and the internal constitution of the Sun. The angular velocity of the inner region is estimated and it is shown that the rotational distortion of the Sun produces a perihelion advance of the planets. If the angular velocity inside the Sun has the same magnitude as in a typical rapidly rotating star, then the anomalous advance of the perihelion of Mercury, usually counted as one of the crucial tests of general relativity, can be explained by the gravitational effect of the rotating Sun.

WHY THE SOLAR CYCLE ALTERS THE PLANETS' BRIGHTNESS

Anonymous; *New Scientist,* 52:146, 1971.

Do the planets vary in brightness as solar radiation changes in intensity and distribution in the course of the solar cycle? Two astronomers working at NASA's Goddard Space Flight Center, Maryland, are convinced that they do, but hesitate over the mechanism responsible for causing the phenomenon (Solar Physics, vol 19, p 257).

V. K. Balasubrahmanyan and D. Venkatesan have analysed luminosity data for the giant planets and found an increase in brightness corresponding to the most active period in the solar cycle, and a decrease toward solar minimum. Jupiter, Saturn, Uranus, and Neptune change in brightness as much as 20 per cent, according to a study by W. Becker 20 years ago. More recent observations of Uranus and Neptune over a short period (six years) indicate a variation of 2 per cent in the blue region of the spectrum.

If the magnitude of the variations had been smaller, it would have been reasonable to suppose that the solar "constant" representing the Sun's energy output--- 1.36 kW/sq. m, according to a study hot off the press (ibid, p 3)---is not quite as constant as previously thought. The German astronomers D. Labs and H. Neckel are willing to entertain the possibility of a 1 per cent variation (ibid, p 11). But for the relatively large-magnitude fluctuations actually observed, one must seek another explanation.

The blue magnitude variations of Uranus and Neptune may be ascribed in part to changes in light-reflecting properties as the planet rotates. But the bulk of the variation remains something of a mystery. If one argues that the solar constant would be expected to decrease when the solar disc is most densely covered with sunspots, which are cooler than the normal photosphere, then it follows that there should be a slight drop toward solar maximum---in contradiction to the observed magnitude brightening.

A more promising explanation may lie in terms of the highly variable solar radiation in the extreme ultraviolet (as opposed to the relatively stable visible-region output). Suppose that, like the terrestrial ionosphere, the atmospheres of the outer planets expand under ultraviolet bombardment, at its peak under solar maximum. Then the larger scattering region of the expanded atmosphere might very well account for the observed planetary brightening. Such conjecture cannot be confirmed without a great many more observations, short- and long-term, of the puzzling variations in planetary luminosities.

PERTURBATIONS OF PIONEER 6 TELEMETRY SIGNAL DURING SOLAR OCCULATIONS

Chastel, Arnaud A., and Heyvaerts, Jean F.; *Nature*, 249:21-22, 1974.

Attention has been drawn recently to unexplained perturbations in the telemetry signal of Pioneer 6 (2,300 MHz) during solar occultation. The results shown in Fig. 1 present the following odd features: [Fig. omitted]

(1) An anomalous redshift is added to a normal linear redshift due to the spacecraft oscillator. This residual redshift which is symmetrical with respect to the centre of the Sun is of the order of $z = 5.18^{-8}$ at four solar radii.

(2) The bandwidth increases sharply when the telemetry signal grazes the Sun.

(3) There are some extremely sharp pulses in the bandwidth. In Fig. 2 [not reproduced] we show that these pulses are clearly associated with a sharp increase of the redshift and correspond to solar flares.

The existence of the redshift is particularly puzzling because it cannot be attributed to gravitational effects nor to the usual Doppler effect.

. . . .

• Purported Effects of the Planets upon the Sun

THE REACTION OF THE PLANETS UPON THE SUN

Anonymous; *Popular Astronomy*, 23:324-325, 1915.

In the Smithsonian Report for 1913 is given a translation of a lecture by P. Puiseux, astronomer at the Paris Observatory, delivered at the Conservatoire des Arts et Metiers, February 23, 1913. The subject of the lecture was "The reaction of the planets upon the sun," and the speaker after an introductory discussion of the general phenomena of the sun, took up the question of the possible relation between the sunspot cycle and the positions of the planets.

Considering the periods and masses of the planets, the most favorable planet to study first would appear to be Jupiter, but the sun-spot cycle is 11.13 years while the period of Jupiter is 11.86 years, a difference of eight months which in the course of nine cycles would change a complete coincidence into absolute discordance. The result is no more favorable in the case of the other planets, none of the periods fitting into the 11 year cycle.

On the other hand, when the question is considered whether the planets may not produce tidal waves upon the extremely mobile surface of the sun, moving over it with the corresponding synodical rotation period and capable of producing visible disturbances, apparently reliable favorable evidence is found in the case of the earth, Mercury, Venus and Jupiter. The deforming factor or tide-producing factor varies as the product of the mass by the inverse cube of the distance. The following are the deforming factors for the different planets expressed in terms of the earth's factor as unity:

Mercury	1.04	Jupiter	2.20
Venus	2.09	Saturn	0.106
Earth	1.00	Uranus	0.019
Mars	0.03	Neptune	0.001

It at once appears that Jupiter and Venus have the greatest disturbing power. Mercury and Earth coming next in order, the effect of the other planets being much less.

As the earth is the most favorably situated that is considered first. We can count all the spots which appear on the hemisphere of the sun nearest us. If the earth has no influence upon the sun, the spots on the eastern and western halves of the disk should, in the long run, be equal in number. If the presence of the earth above the horizon of a region of the sun's surface favors the development of a spot there, such action will require time to make its results manifest and so more spots should appear on the western half than on the eastern half of the disk. If the presence of the earth above the horizon is unfavorable to the existence of spots, the number of spots will be less on the western half of the disk than on the eastern.

From Mrs. Maunder's researches on the photographs of the sun taken in the English observatories in the interval 1889 to 1901, it is found that during that cycle both the total areas and the number of spots were greater on the eastern half than on the western half of the disk, i.e., the earth's influence was to cause spots to disappear. The same result was obtained from both the northern and southern hemispheres.

The astronomers at the Kew observatory in 1872 obtained exactly the opposite result, and also found that spotted areas tend to increase opposite to Mercury and Venus. Jupiter, upon which the greatest hope was placed, gave no definite results. The data at the disposal of the Kew observers was comparatively meager and Mrs. Maunder's results should be given greater weight.

In 1911 Professor A. Schuster studied the results from the Greenwich photographs for the years 1874 to 1909. He considered only the births of spots which lasted over the interval between the plates of two successive days, excluding as more subject to error those whose births occurred within 30° in longitude of the eastern edge of the sun's disk. There remained 4,271 spots to consider. For each planet the sun was divided into 12 equivalent vertical zones, the solar meridian passing through the planet forming the boundary between zones 6 and 7 on the hemisphere toward the planet and between 12 and 1 on the farther side. The number of spots seen for the first time in each zone was counted and used to form a plot having as abscissae the zone numbers. The results are somewhat irregular. When the spots counted when the earth is east and west of the meridian are considered separately each of the planets Mercury, Jupiter and Venus seems to produce a minimum of spots just where the other may produce a maximum. When the above distinction is not made the results seem more accordant. For all there is a minimum upon zone three, where the planet is just rising, and a maximum on zone eight, which has already passed the meridian. Schuster considers that the similarity of march of the three curves for divisions 3 and 8 is sufficiently characteristic for rendering very probable the reality of a planetary influence. When two or more planets were in conjunction no marked difference was evident. Schuster thinks that a planet may have merely an exciting action, effective only in putting into play a force already existing in the sun, so that a second planet on conjunction might not have any additional effect.

Mr. F. J. M. Stratton also in 1911 studied the disappearances of spots as well as the appearances, using the same series of the Greenwich photographs as Schuster used. He considered only Jupiter and Venus and divided the sun's surface into 24 equal zones instead of 12. The plots were very irregular. Generally there was no similarity in their contour, even for the same planet, between the two hemispheres; neither was there between the same hemispheres for different planets. There is one single coincidence, perhaps, which seems not due to chance. There is a maximum of ephemeral spots noted in the zones the meridians of which either Jupiter or Venus had passed three hours previously.

On the whole the impression made by Professor Puiseux's address, although he speaks quite optimistically, is that the evidence of planetary influence upon the sun is extremely uncertain.

A CRITICAL TEST OF THE PLANETARY HYPOTHESIS OF SUN SPOTS

Alter, Dinsmore; *Monthly Weather Review*, 57:143–146, 1929.

With Schwabe's discovery that there is a cycle in sunspot numbers, nearly equal to the period of the planet Jupiter, it is obvious that a study of the possibility of the tides produced by the planet Jupiter, on the sun, as a cause of the phenomenon would follow. The first paper on such a hypothesis seems to be that by Fritz in 1866. This was followed by a series of papers on the same and related subjects by him and by Wolf, throughout the remainder of the century. It is a great pity that this most important of all series of sun-spot studies is not more generally available. Fortunately synopses and resumes are common. The Monthly Weather Review has recently published a translation by Mr. Reed of one of Fritz's more important papers, a proceeding in which the Editor is to be commended and urged to continue.

However, the most important single analytical study of the possibilities of the hypothesis was that by E. W. Brown. In this paper Professor Brown compares the observed epochs of maxima and minima of all available data with the curve formed by a rough combination of the tides of Jupiter and Saturn. The agreement is so excellent that usually it would be accepted as sufficient evidence of the truth of an hypothesis. In this case, however, the fact that the tidal forces are negligible in comparison with the sun's own gravitational field made even such strong evidence seem insufficient and the hypothesis found very few followers.

Many other hypotheses have been advanced but uniformly have met with unsurmountable objections. For a number of years the feeling has been growing that the variations can not be accounted for by any fairly constant periodicities and that investigations were a waste of time.

Recently the writer made another analysis of the sun-spot numbers, applying the new correlation periodogram. This periodogram, which compares the actual shape of two sections of the curve, and not merely one Fourier term at a time, is more sensitive than other methods in determining, and weighing the probability of actual physical existence of irregularly shaped repetitions of data. In that paper the conclusion was reached that if a series of periodicities do exist which account for the main variations in sun-spot numbers they must be harmonics of approximately 252 years.

About this time a conversation with Professor Brown led to the computation of an ephemeris based on Brown's early paper. The ephemeris fitted the data during the 30 years since the paper had appeared as accurately as during the years used in the original paper. It is given here as Figure 1. Such a fulfillment of predictions was unique in the history of the subject.

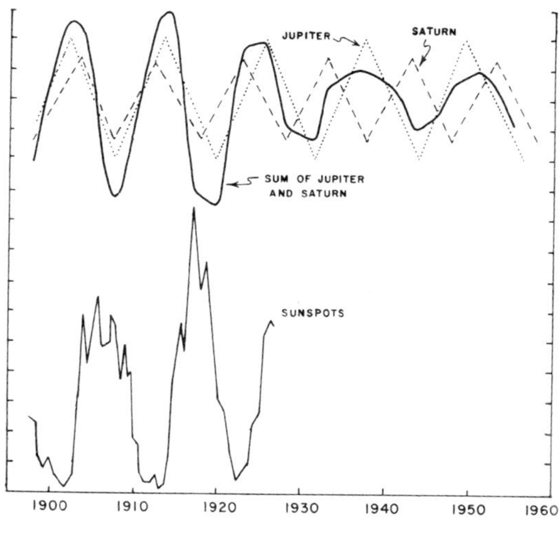

Tidal forces due to Jupiter and Saturn compared to sunspot numbers.

Following this, the writer examined the 252-year period to find whether it bore any relationship to the planetary tidal periods. It was found that 250 years, well within the limits of error of his work, was an almost perfect least common multiple of the tidal periods of all the planets from Jupiter outward.

During the past few years modern physical theory and the spectroscopic study of the solar atmosphere have removed the great objection which has existed with respect to the tidal hypothesis, by showing that the reversing layer is in almost perfect equilibrium between the inward force of gravity and the outward force of light pressure. Under such a condition, long thought to exist only when the rate of rotation was rapid enough, very small tidal forces might conceivably produce very marked disturbances. If so, they would be a common feature for all stars surrounded by planets.

In all this work the tides of the inner planets, equal in force to those of the outer, were neglected. Sun spots develop slowly and often last for many months. A period of two or three months could not be expected to show a variation comparable with that which would be produced by an equal force acting through a longer period. However, there should be some such variation present, its magnitude being a function of the force and of the length of period through which it acted. An examination of the data for such periods might form a critical test of the hypothesis.

The calculation of the periods and magnitudes of such tidal forces is standard and need not be repeated here. The periods are the synodic periods of the three planets and in addition the period of revolution of Mercury, due to its large eccentricity. The eccentricity of Mercury will also cause a variable amplitude to its combinations with Venus and the earth. The period due to the combination of Venus and the earth may be considered as a constant one. We can calculate in advance exactly where such periodicities should show peaks in the periodogram. If they are found there regularly, the truth of the hypothesis is established. If they are not found, or if they are often missing, the hypothesis is not entirely voiced, but the evidence against it becomes serious.

In the calculations it is assumed that the density of the solar atmosphere is small enough that each molecule may be assumed as independent of the others. In this case the forces due to each of the planets may be projected to get a resultant force in considering the tides raised. Forces do not appreciably affect the heights of tides along lines perpendicular to them, in such a case.

. . . .

In the table below where the terms, or periods, are segregated the magnitudes stated are the contributions of the terms to the square of <u>R</u>.

Cause of term	Designation	Length of period	Magnitude
Mercury-Venus	A	$72^d.28$	1.512 to 0.430.
Venus-earth	B	291.96	0.902.
Mercury-earth	C	57.94	0.451 times mag. of A
Eccentricity of Mercury	D	87.97	0.541.

The magnitudes of the contributions to R^2 show that only the first two terms, designated as A and B need be considered. The forces from these terms will average about the same, but the term B is of a period almost exactly four times as long as term A. For this reason we will expect its effect on the number of spots to be considerably greater than that of any of the other terms.

Obviously it is impossible to look for a small amplitude term of period 72.28 days with data which are monthly averages. Fortunately five-day averages, or more accurately, averages over one seventy-third of a year are available for the years 1876-1911. Doctor Elsa Frenkel (now Dagobert) in 1913 investigated, by means of Schuster's periodogram, the possibilities of short periods. She arrived at the conclusion that two highly variable periods exist, one of length averaging 68.5 days, the other in the neighborhood of 200 days. The method used, apparently showed the shorter to vary in length from 50 to 100 days. Ten-day averages are about the proper length for our search, accordingly means were made of each consecutive pair of her data. Since her paper is available, the data will not be reprinted here.

We are looking for terms of small amplitude, therefore we must choose the lags for the periodogram, such that there will be many products for each point computed. It was decided to compute a periodogram from lags, varying by two seventy-thirds of a year between 609 and 688 of such intervals. In this stretch we can compute the position of 11 crests of the A term and 2 crests of the B term. The number of products of data formed for each point varies from 705 for the 609 interval lag to 626 for the 688 interval lag. Even such a large number of products will leave accidental errors of somewhat troublesome proportions.

An interesting feature of any sharp peaks to be found in the periodogram, is that they must be followed at an interval of one rotation of the sun by a satellite. If to-day we were to observe an unusual number of sun spots, since many of them would persist, we would have another maximum in about 26 days, due to the adopted method of counting. Such a rotational pseudo period could easily be differentiated from a true one by the fact that after one or more rotations of the sun it would die out, and when such a period would show again in another part of the periodogram, it would bear no phase relationship to the former set, unless its primary were due to a real periodicity. On the other hand, peaks due to a real periodicity would follow regularly at intervals equal to the period length throughout the whole stretch of the periodogram. This last feature is one of the strongest advantages of the correlation periodogram in a search for periodicities.

The 292-day period is long enough that the variation due to the 11-year cycle will mask it, unless the latter is eliminated. The 72.28 day period, however, should show sharp peaks, displaced in the direction of ascent of the 11-year cycle. Each of these should be followed by a satellite. Since our smallest interval for the periodogram is 10 days, the satellites must follow sometimes at 20-day intervals and at others at 30-day intervals. If we find in approximately the computed places, the crests of the A term and find them followed at the proper intervals by their satellites, and if we find few or no peaks other than those computed in advance, the point will be established that the planetary tides are a major factor in production of sun spots. The divergence from such ideal condition will mark the uncertainty of the conclusion. Afterward it will be necessary to eliminate the 11-year cycle from the results and see whether the 292-day period appears.

It would be natural to expect that if tides do affect the number of sun spots, the part of the sun undergoing the maximum tidal agitation would be that which would show the most spots. There are two such nearly equal areas opposite each other on the sun. If we had used 5-day means we would expect the peaks of the shorter terms to be fish-tailed with a separation of one-half a solar rotation. The 10-day means are nearly enough the half rotation period that we will get merely a broadening and sometimes a displacement either way. This half rotation period can not extend to the interval at which the satellites follow, for they merely, in a lesser degree, retrace what has already occurred in the primary peak one rotation period earlier.

In this paper we are not interested in the actual correlation coefficients. It is sufficient to take terms which are proportional to them. The means of the products formed for each periodogram point are practically so proportional and can be computed in much less time than the coefficients themselves. The amount of work to be done in forming and adding more than 50,000 products is tremendous and any saving of time becomes worth while, so long as it can in no way invalidate the results.

The periodogram as computed, and without elimination of the 11-year term is given as Figure 2. The abscissae are in units of two seventy-thirds of a year and the ordinates in means of the products used in forming the points. Since the lags vary from a little less than one and a half of the 11-year cycles to a little more, the means of the products are all very large negatives. An arrow, with the letter A over it, points to each place where a crest of the 11-year term has been computed and one with an S over it follows each A by 26 days, to indicate the computed position of its satellite. [Figs. 2 and 3 omitted]

Three misses are seen at a glance. The ones at 612 and 648 are extra peaks, the only ones in the whole periodogram. The one at 634 is, of course, a badly displaced A peak. The other 10 A peaks all fall sufficiently close to the computed positions to be called fits. Eight of these peaks come within one periodogram point of the computed positions and are, therefore, exact. Each of the 11 A peaks is followed by its satellite, in only one case at an interval different from the computed 20 to 30 days. The 612 and 648 peaks seem rather large to be accidental but no explanation of them has been found.

All that remains now is to eliminate the 11-year variation from the periodogram to see whether the 292-day period shows and whether its amplitude is greater than the 72-day, as it should be; and whether its crests fall at the proper places. Four times the A period is 289.12 days, almost exactly equal to the long 291.96-day period. If, therefore, we form averages of 29 consecutive points, both these periods are eliminated. Curvature of the 11-year cycle during this interval will cause the means not to fall exactly on the 11-year curve, but the errors caused are not large and can bear no possible relationship to the periods for which we are searching. These means can be subtracted from the periodogram points and will leave us any short variations. Changes in the curvature of the 11-year variation, during the stretch of the periodogram, will somewhat warp our resultant modified periodogram, but can neither introduce nor hide any significant features. Figure 3 is this modified periodogram. In it the A period and its satellites are marked as in the first figure. The computed crests for the B term are shown by arrows under that letter. Both fall in the proper places and the amplitude is found to be around two and a half times that of the 72-day term. The 648 peak now resembles a displaced A peak, rather than an extra one, as we considered it from the original periodogram.

The agreements between hypothesis and observation in this paper and in Brown's early paper are so excellent that there seems to be no question that planetary action, presumably tidal, is the principal controlling factor in the number of sunspots observed. This leads to several interesting speculations regarding the effects of tides of stars on the sun and of the light changes in certain variable stars. The periods considered in this paper have forces equal to those of the outer planets, yet they produce much smaller disturbances. Evidently the variation must continue through a few years if it is to show its maximum effect. Apparently, from the approximate agreement of magnitude ratios in Brown's calculations, periods of a length such as those of Jupiter and Saturn have time to get in their full effect. A star of mass equal to that of the sun would have to approach to within 53 astronomical units to have an equal tidal effect. This is about four thousand times as close as any star known at present. If the sun were distended to one thousand times its present radius, far larger than any known giant, the distance would still have to be four hundred times as close as Alpha Cantauri. Perhaps the writer is in error, for he is no authority on the subject, but he can not see that there would be any tidal difference in the cases of equilibrium due on the one hand to light pressure and gravitation and on the other hand to rotational speed and gravitation. It would seem to him that a star to have had an effect in the evolution of planets would have had to approach within an unreasonably small distance. Material left behind by a pulsating giant during its return from its maximum diameter would seem a more likely source of planetary material. Once the material for one such planet had gathered, it would begin to produce serious tidal disturbances on the giant and probably, for a time at least, cause more material to be lost and larger planets to be formed. Later, as the star contracted, especially if its oscillations in diameter decreased, planets formed would be smaller. The subject matter of this paragraph is offered very tentatively, and the writer has seriously considered whether its possibilities are great enough to include it in a paper whose body is so much more definite.

A by-product of this investigation is to show that monthly or semimonthly means, used to compute a correlation periodogram, will in this manner locate the dates of sun-spot maxima and minima more accurately than do the methods now used. There is no single point to stand out and catch the eye, and the result depends upon a consideration of every variation in shape during the cycle.

INFLUENCE OF THE PLANET MERCURY ON SUNSPOTS

Bigg, E. K.; *Astronomical Journal,* 72:463–466, 1967.

Introduction. It is the current belief that the formation and disappearance of sunspots are due wholly to internal influences in the sun, the earlier theories of external influences---planetary positions, in particular---having been largely discredited.

Recently there have been two papers, by Jose (1965) and by Wood and Wood (1965), which have revived the planetary theory. The first paper shows that there is a basic periodicity in sunspot numbers corresponding very closely to the motion of the center of the solar system;

and the second shows that the well-known 11-yr mean sunspot periodicity coincides with a strict periodicity in the rate of change of solar acceleration caused by planetary motions.

A defect of the first paper is the difficulty of establishing with the small amount of accurate data---little more than one cycle---that is available, the constancy of the 178-yr periodicity and therefore its relation to planetary influences. In the second paper, the coincidence is shown to be striking, but while the planetary periodicity is strict, the sunspot periodicity is obviously variable, and this weakens the implication that planets are involved.

The purpose of this paper is to apply simple tests to see if the planet Mercury influences sunspot numbers. It is a natural choice in a search for individual planetary influences, for its eccentric orbit leads to a 3-to-1 variation in tide-raising force on the sun, thus providing a physical basis for an effect.

Principle of the Method. The problem is very similar to the experimental one in which a weak radio signal of exactly known frequency is sought in the presence of a wide spectrum of "noise" or other signals.

The chance of detecting it is enhanced by using a narrow-band receiver, so that the noise is suppressed as far as possible. The receiver may initially be tuned to a series of equally spaced frequencies and the output power or peak amplitude (with or without filtering) measured for each. A probable identification of the expected signal would be claimed if the output was higher at its predicted frequency than at a large number of "control" frequencies.

Fine tuning would then be used to verify the identification. The maximum output should occur at the predicted frequency within the limits of error set by the receiver's ability to discriminate between two frequencies close together.

If the nature of the signal was such that a characteristic waveform might be expected, a further test would be to compare the mean waveform from one time sequence with that during another. If only noise were present, correlation between the waveforms would be slight.

If all these tests were successful it would be reasonable to claim a positive identification even though the signal was small and its waveform noisy.

The statistical analysis described below follows exactly these steps.

. . . .

Discussion and Conclusions. The tests which have been applied to see whether Mercury influences sunspot numbers do not individually show any effect large enough to be entirely convincing. Considered together, it is difficult to believe that the effects could be due to chance.

Positive identification of a weak radio signal of known frequency would normally be claimed if the analogous tests yielded comparable results. Although logically statistical tests should be regarded in exactly the same way, a curious bias exists against accepting their findings when confidence levels are not overwhelmingly high. It is for this reason that the relation between the statistics and familiar experimental concepts have been so much emphasized here.

Some remarks by Brier (1965) on the large effects produced by relatively small periodic forces are relevant to a consideration of the nature of Mercury's influence. He has shown that when some quantity is slowly approaching a critical point---for example, a "no sunspot-

sunspot" transition---the timing of the transition event may be considerably changed by the application of a periodically impressed influence which is small compared with that actually producing the transition. Thus it may be more appropriate to suppose that the planet's position alters the timing of spot formation or disappearance than to regard it as a direct cause of these events.

The original hypothesis used as a basis for the investigation---that Mercury's tide-raising force may influence sunspot formation or disappearance---is not immediately supported by Fig. 4, which suggests some more complex phenomenon. However, the complexity may arise from other causes, such as interactions with other planets, or of continuing changes within modified spot regions.

The way in which the other planets appear to enhance or suppress the Mercury effect suggests that much larger planetary influences might exist than those discussed here, but so far they have not been detected.

The important point of this work, however, is to show that, even though they may be small, planetary effects do exist. Current theories of sunspot formation and development entirely ignore the possibilities of such external influences. A more detailed knowledge of them could be very useful in providing constraints on such theories, and a better understanding of solar properties.

. . . .

ON THE PLANETARY THEORY OF SUNSPOTS

Okal, Emile, and Anderson, Don L.; *Nature*, 253:511-513, 1975.

It has been proposed that sunspot activity is affected by positions of the planets, and calculations have been presented, which purport to show that planetary tides on the Sun vary in the same way as the sunspot variations. We believe that the apparent agreement of the sunspot cycle with planetary tidal effects is an artefact of the calculation.

The calculation in question was used to compute the absolute value of the difference in tidal potential between Earth-Venus conjunctions and oppositions at the sub-Jupiter solar point. The effect of Mercury, one of the strongest tide-raising planets was ignored on the basis that its period is short compared with that of sunspot activity. The absolute value of the tide and the effect of partial line-ups of Venus (or the Earth) with Jupiter were not computed. Furthermore, it is not clear that the absolute difference between opposition and conjunction tidal potentials has any physical meaning.

Here we compute the full tidal problem for Mercury, Venus, Earth and Jupiter, the tide-raising planets, taking into account the complete orbital elements, including eccentricity, inclination and their variation with time.

At any given time, the tidal potential at a given point M on the surface of the Sun caused by the planets is proportional to where m_i is the

$$T = \sum \frac{m_i}{d_i^3} \ (\cos^2 z_i - 1/3)$$

mass of the planet, d_i its distance from the centre of the Sun and z_i the angle from the planet to point $\underline{M}$ as seen from the centre of the Sun. We restrict ourselves to Mercury, Venus, Earth and Jupiter. Mars, Saturn and the other outer planets can easily be shown to have trivial contributions compared to the above planets.

Table 1. Comparison of tidal and sunspot dates

Tidal peak	Sunspot peak	Dts*
1809	1816	-7
1822	1830	-8
1833	1837	-4
1845	1848	-3
1857	1860	-3
1869	1871	-2
1881	1884	-3
1892	1894	-2
1905	1906	-1
1916	1918	-2
1927	1928	-1
1939	1938	1
1951	1948	3
1963	1958	5
1974	1969	5

*Dts is the difference between tidal and sunspot peaks in years.

In the plane of the ecliptic, the potential depends on the longitude Φ through a first order polynomial of sin 2Φ and cos 2Φ . Over a period of 10 d (our sampling interval) the Sun rotates about halfway around its axis and therefore any given point attached to its surface is subject to the whole variation of the tidal potential. So we characterised our problem by the maximum value (over Φ) of this polynomial. We note that Mercury is the slowest planet around the Sun relative to a point on the Sun's surface and its contribution, contrary to previous arguments should therefore not be neglected.

The positions of the planets were computed from the best available planetary elements for 65, 536 epochs at intervals of 10 d, or roughly 1, 800 yr, starting in the year 1800. A fast fourier transform (FFT) technique was then used to extract the power spectrum for comparison with the solar activity spectrum.

Figure 1 [not reproduced] shows the tidal potential as a function of time for the years 1800–1808, both including Mercury (upper trace) and excluding it (lower trace). The high frequency effect is due to the eccentricity of Mercury's orbit and has a period of 0. 24 yr.

Figure 2 [not reproduced], an extension of the lower trace in Fig. 1, shows the tidal potential for the 25-yr period 1800–1825, excluding Mercury. The long period, beat-type phenomenon ($\sim$11. 9 yr) arises because of the eccentricity of Jupiter's orbit. By contrast with the sunspot cycle, the tidal pattern repeats almost exactly every 11. 9 yr and shows no evidence of a beat of $\sim$ 100 yr; successive peaks in the tidal envelope are of almost exactly the same amplitude. Wood's samples (• in Fig. 2) have a spacing too large to provide a valid description of the tidal potential even excluding Mercury; such a sampling leads to aliasing of the lower frequencies. Figure 3 [not reproduced] shows the

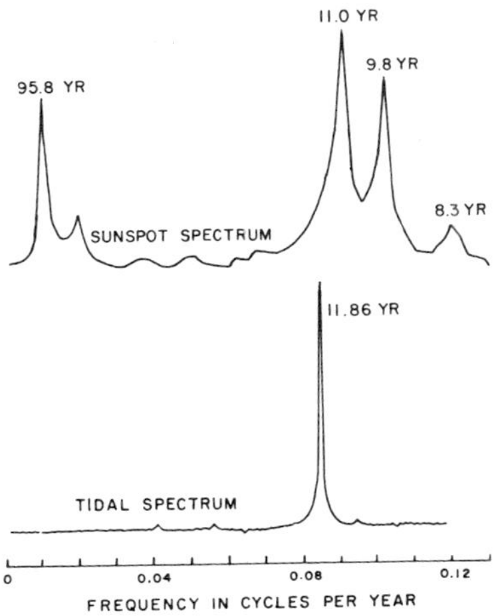

Comparison of sunspot and tidal spectra. (Fig. 4)

power spectrum obtained from an FFT analysis of the whole 1,800-yr four-planet potential. The fundamental periods are those of the alignments of pairs of planets. Alignments of three or four planets come as beats of these primary values and consequently do not appear on the spectrum. The periods of Jupiter and Mercury and some of their harmonics show up because of the eccentricity of their orbits. The lower frequency part of the spectrum is very flat, Jupiter's being the longest period involved.

Figure 4 (frequency scaled) shows an enlargement of the lower frequency part of the spectrum, superimposed with Cohen and Lintz's sunspot spectrum. They showed a strong peak occurring at 11 yr, the familiar sunspot cycle, and smaller peaks at about 9.8, 95.8 and 8.3 yr. In addition they demonstrated that the longer period, ~ 180-yr cycle proposed for the solar sunspot spectrum arises from the beat of the 11 and 9.8-yr cycles and is not an intrinsic periodicity. This removes the basis for one of the other planetary theories of sunspots. Figure 4 shows no planetary peaks at 8.3, 9.8 or 95.8 yr, peaks which are prominent in the sunspot spectrum.

The origin of the 11.08-yr tidal period claimed by Wood is not clear. Such a peak does not appear in our spectrum. In Wood's simplified three-planet system, it could only be the consequence of a Jupiter-Earth-Venus alignment. But if 11.08 yr is indeed a multiple of the Venus-Earth 0.799-yr alignment period, it is not a multiple of the Jupiter-Earth synoptic period, and therefore is not a fundamental period of the problem. This disrepancy between the tidal and the sunspot activity periods is further demonstrated in Table 1, which gives sunspot peak dates and approximate dates of envelope maxima of the tidal potential, excluding Mercury. The average period between envelope peaks is 11.8 yr, the orbital period of Jupiter. From the beginning of the nineteenth century to the present the discrepancy between tidal peak dates and sunspot peak dates has slipped from approximately -7 yr to + 5 yr. This is the difference between the 11.86-yr Jupiter period and the average sunspot period, 11.05 yr taken over 165 yr (14 cycles). Over a limited period of time, such as 1892 to 1939, the peak years agree to within a year or two but this is to be expected when comparing two periodic functions of nearly the same period. The next tidal envelope maxima occur in 1987 and 1998. In the incomplete tidal theory maxima are predicted for 1982, 1993 and 2003.

A further look at our tidal potential values shows no drastic effect expected in 1982 when planets are supposed to align on the same side of the Sun. Indeed, better alignment will be achieved in 1990. Even then, no special tidal effect occurs because alignment of the outer planets has no pronounced effects on the tides. Alignment of the tide-raising planets within 10 degrees is a common phenomenon, occurring approximately every 10.4 yr and is not associated with drastic tidal effects.

PLANETARY ALIGNMENTS, SOLAR ACTIVITY AND CLIMATIC CHANGE

Gribbin, John; *Nature,* 246:453-454, 1973.

There may be a causal chain linking changes in the alignments of the planets, through their tidal effect on the activity of the Sun, with climatic change on Earth. Although the separate components of this chain have been discussed before, it seems that they have not previously been linked in this way to form a coherent whole. If these ideas are correct, they provide the best basis yet found for predicting climatic changes in the immediate future.

It is convenient to assess the evidence for such a chain in the reverse direction---that is, starting from the Earth. There is good evidence that the rate at which the terrestrial length of day changes is influenced by the mean level of solar activity and although the question of a direct link between specific events on the Sun and specific changes in the length of day seems more contentious it is clearly relevant to consider by what mechanism the longer term variations in solar activity affect such a gross property as the spin of the Earth. Since large scale movements of air masses associated with seasonal changes in the atmosphere are already known to produce a component in the variation of the length of day with a period of $\sim$ 1 yr, it seems reasonable to look for the atmospheric link between changes in the activity of the Sun and changes in the spin of the Earth.

Solar Storms and Weather. In spite of the difficulty of establishing that individual large solar storms produce detectable discontinuous changes in the Earth's spin rate, there is little doubt that an increase in the level of solar cosmic rays arriving at the Earth can cause a short term change in the weather at high latitudes. Roberts and colleagues have shown that troughs at high latitudes are amplified when the level of cosmic rays reaching the ionosphere (revealed, for example, by auroral activity) increases abruptly.

This discovery has obvious implications for improving short term local weather forecasting in some parts of the world. But as far as climatic changes are concerned it is necessary to look at data averaged over at least one cycle of solar activity. King has pointed out that "it is now well established that the density and temperature of the atmosphere above about 100 km vary markedly with solar activity during.... the 11-yr solar cycles", and he has shown that there is a positive correlation between solar radiation changes which occur during the solar cycle and meteorological variations. For example, as the level of solar activity, revealed by the number of sunspots, decreases, there is a decrease in mean temperatures and an increase in winter rainfall at Beirut. These changes seem to be related to changes in the altitude of the 500 mbar surface. But even observing variations over an 11-yr period---or over several solar cycles---is not sufficient to determine patterns of climatic change.

Historical Data. Fortunately, there is a great deal of meteorological data available, and records of "good" and "bad" seasons go back for almost a thousand years, in England at least. Lamb has used some of these data to determine variations of climate in the northern hemisphere and he explains them in terms of successive movements of climatic zones to south and north over periods of hundreds of years. According to Winstanley, 1930 marked the peak of a northward excursion of climatic zones, and we are now experiencing a southward shift which has, among other effects, caused the serious drought in Africa just south of the Sahara in recent years. He foresees continued southward movement of both the Sahara and the Rajasthan Deserts until at least 2030 assuming that the present variation is part of a 200-yr cycle. The evidence for a cycle of approximately that period comes both from climatic records and from estimates of variations in the radiocarbon content of the atmosphere; the cycle last peaked, before 1930, in the early part of the eighteenth century. It is also possible that 1930 marked the "peak" of a 700-yr cycle, although the data are not sufficient to show conclusively that such a cycle really exists.

It seems obvious to ask if there is any link between these longer term effects and the mean level of solar activity. The changes in atmospheric radiocarbon levels are particularly tantalising, since they hint at changes in the mean cosmic ray flux with period of approximately 200 yr; unfortunately, however, the only record of solar activity available which goes back for more than a few years is sunspot number, and as King points out that is a relatively crude index. Nevertheless, when sunspot numbers are plotted for this century it is immediately striking, in the present context, that the maximum number of spots, at the time of maximum solar activity, was itself a minimum in the cycle which peaked just before 1930, at the "peak" of the 200-yr climatic cycle mentioned by King. But good data for the numbers of sunspots at the peaks of sunspot cycles go back only to 1820---fourteen complete cycles---and there are no numerical data for cycles before 1750. Similarly, climatic data become more qualitative as one searches further into the past, and detailed quantitative data are only available for years since the mid-nineteenth century.

Possible Implications for Prediction. Attempting to establish a correlation between the two sets of data might thus seem to be foolhardy, and indeed although such attempts have been made they have met with some opposition. But the urgency of climatic research has been highlighted by recent events in Africa and India, and although it would still perhaps be foolhardy to make any conclusions drawn from comparison of these data sets the basis for immediate action, it does seem to be worthwhile to present these conclusions for discussion at the present time, when there is no completely satisfactory theory to explain climatic changes. Lamb has discussed previous attempts to relate climatic changes to long solar cycles. One of the most important of these is the discovery of a correlation between climatic series, solar variation and radiocarbon levels over more than 2,000 yr; but in addition to the difficulty of relating past climatic changes to past changes in solar activity there is the pressing problem of predicting future climatic trends. Assuming that the correlations outlined by Lamb and discussed in detail in the references he cites do have a secure basis, is there any hope of predicting future solar activity and thus estimating future climatic effects?

Wood has noted that the maximum smoothed monthly number of sunspots for each solar cycle seems to repeat after about 170 to 180 yr. He made this discovery when analysing the available sunspot data for evidence of an influence which might be attributed to the tidal effects of the planets on the Sun; I have mentioned elsewhere that if this is a periodic effect then it might be linked to the 179-yr period between the alignments of the outer planets. Certainly, Wood has produced convincing evidence that there is a close link between the height of the tide raised on the Sun by various planetary alignments and the dates of sunspot maxima and minima; there is also a weaker link between the variation of tidal height and sunspot number. That makes it possible to predict future solar activity levels, and the dates of maxima and minima, rather more accurately than simply extrapolating the average 11.1-yr cycle. And if the level of solar activity affects climate, as is strongly suggested by King's work, that in turn provides a means for predicting climatic change. Perhaps the similarity between the approximately 200-yr variation of climate and atmospheric radiocarbon levels mentioned by Winstanley and the 170 to 180 (or 179?) year variation in solar activity is more than a coincidence.

Further speculation along these lines would be inappropriate here, even granted the need for investigating all possibilities when concerned with predicting climatic change. But it does seem that climatologists and others concerned by large scale effects of changes in the atmospheric circulation such as those reported by Winstanley could do worse than to examine the predictions of solar activity made by Wood on the basis of planetary alignments.

Chapter 2

VULCAN: THE LOST INTRAMERCURIAL PLANET

INTRODUCTION

It is astounding that astronomers attach so much importance to the tiny specks of matter that circle the sun. This preoccupation with the planets must be the heritage from many generations of sky-watching astrologers. Of what terrestrial import is a miniscule bit of matter in orbit between Mercury and the sun? Being so near the sun, it would be almost impossible to see and (supposedly) of no physical consequence to earthly events; yet anyone who discovered an intramercurial planet would be famous forever.

A French physician named Lescarbault claimed to have discovered an intramercurial planet in 1859. The new planet filled an astronomical need. It helped explain the advance of Mercury's perihelion and did not throw Bode's Law out of joint. Lescarbault's sighting was therefore almost expected and even welcome. Many famous astronomers subsequently saw this intramercurial planet; and it was given an honored place in the astronomical tables under the name Vulcan.

Still, Vulcan was elusive. Organized searches failed to find it during the late 1800s. It had vanished---if indeed it ever existed. The earlier sightings were blamed on misidentifications and instrumental errors. Vulcan became an embarrassment. When Einstein's Special Relativity Theory accounted for Mercury's anomalous motion, reports of intramercurial planets dropped to a near-zero level.

A few reports of objects that could be Vulcan still appear, even in recent years. And why not? The solar system contains much debris; and some small chunks may circulate close to the sun.

• The Early Observations of Vulcan

A SUPPOSED NEW INTERIOR PLANET

Anonymous; *Royal Astronomical Society, Monthly Notices,* 20:98-101, 1859.

The conclusions arrived at by the distinguished Director of the Imperial Observatory at Paris in his recent investigations of the corrections required in the theory of Mercury, and which he communicated in so striking a manner to the world by his letter addressed to M. Faye, appear to have already received signal confirmation. An abstract of this remarkable letter was printed in the first number of the present volume, and it may now be acceptable to the readers of the Monthly Notices to be made acquainted with the circumstances of the observation recently announced.

From various accounts received from Paris it appears that not long after the publication of M. Le Verrier's assertion of the probable existence of a planet or planets interior to Mercury it began to be rumoured in Paris that more than one person had already seen the planet, but that, as is usual, the rumour assumed different forms and was coupled with different names. Towards the close of the year, however, a communication was made of so definite a kind to M. Le Verrier that he considered his official position required that he should at once probe it to the bottom, and take such steps as should prevent the success of any attempt to palm off a fraud on the public. Noting the name and address of the asserted observer he proceeded on the last Saturday of the year to the nearest railway station to Orgeres, in the department Eure et Loire, and accompanied by a friend whom he took as a desirable witness of his proceedings, made his way direct and unannounced to the house of M. Lescarbault, residing at Orgeres and practising as a country physician. A very graphic account of what here passed is given by the Abbe Moigno in a recent number of his useful publication, Cosmos, and is stated to be given nearly as recounted by M. LeVerrier to an assembly of friends on his return. M. Lescarbault was subjected to a severe cross-examination by his unknown visitor, who pressed him hard from step to step till he had obtained such material and verbal evidence as no longer permitted him to doubt the reality of the observation or the good faith of the observer. Not content to leave the question of good faith, where so much rested upon it, dependent on the replies of the individual concerned, he was careful to obtain collateral evidence of the high character and worth of Dr. Lescarbault, from such other persons of station in the neighbourhood as should at once satisfy himself and others; and we believe that the precautions thus taken have been accepted as sufficient by all concerned.

At the first sitting of the Academy of Sciences after his return M. Le Verrier announced that on the afternoon of the 26th of March, 1859, a small dark body had been observed to transit a portion of the sun's disk by M. Lescarbault, and which bore all the appearance of being a new interior planet. And at the same time he stated that the observer had made such observations as led to the conclusion that the supposed new planet's distance from the sun was about 0.1427, its period less than 20 days, its ascending node situated at about 13° of longitude, and its inclination between 12° and 13°.

M. Lescarbault has since addressed a written account of his observation to M. Le Verrier which has been published, in which he states that having witnessed the transit of Mercury in 1845, he had himself at once inferred that the body was an interior planet, and that, living in retirement, he had kept his discovery to himself, in the hope of being able to come to some conclusion respecting its distance from the sun, by calculation, or by being so fortunate as again to observe its transit; but that being only a moderate geometer, and much pressed by professional engagements, the problem had baffled him, and he had at last been induced to break silence on reading an account of M. Le Verrier's theoretical conclusions, printed in the publication of Abbe Noigno, to which he was accustomed to subscribe.

M. Lescarbault first detected the body when a little way advanced on to the disk, and inferred the time of first contact by noting the interval which elapsed while it advanced over what he estimated to be an equal space. He then repeatedly measured the angle from the zenith or nadir of the point of the limb to which the body was nearest and its distance from the limb, and watched its passage off the sun. Correcting his measured angles for the angle between the pole and zenith, it would appear that the first contact took place at 19° 4' and the last contact at 52° 42' from the sun's north point measured towards the west, the greatest distance from the limb being 41".3, and the times $4^h\ 0^m$ and $5^h\ 47^m.2$ Orgeres local time, little more or less. The estimated diameter was about one-fourth that of Mercury when last on the sun; and as this was 11".6, the inference from the statement would be about 3" for the new body, if so small a quantity can be admitted as probable.

At the meeting of the Society on January 13, Mr. Carrington exhibited two diagrams in illustration of the original observation and of the inferences to be made from it, and pointed out in detail the very simple process required in such a case for arriving at the approximate elements of the orbit supposed circular. The concluded elements were not sensibly different from those given by M. Le Verrier, except in the case of the inclination, which was concluded to be about 11° 51'. The inclination to the orbit of Mercury was inferred to be about 7°.

The remark of principal immediate importance was that, inasmuch as the observation of March 26, 1859 was made about 7-1/2 days before the earth passed through the calculated line of nodes, and an equal space thereafter was similarly available, it followed that for the space of 15 days at each conjunction the orbit was projected on the sun's disk; and that as the revolution was performed in less than 20 days, it was more probable than not that, unless the original observation were more in fault than supposed, the body might be re-observed in the spring of this year. The attention of observers was particularly called to this circumstance. The days referred to are

March 25 to April 10 at descending node,
and Sept. 27 to Oct. 14 at ascending node.

The singular merit of M. Lescarbault's observation will be recognised by all who examine the attendant circumstances; and astronomers of all countries will unite in applauding this second triumphant conclusion to the theoretical inquiries of M. Le Verrier. The redetection of this new member of the solar system must next engage the cooperation of observers; and it is hoped that the astronomer at Madras will unite his efforts with those of European and American observers for the purpose.

In the 10th number of Professor Wolf's Mittheilungen uber die Sonnenflecken several cases are quoted of the observation of planetary bodies in transit over the sun, some of which are evidently of another charact er, but the following deserving of attention:---

On January 18, 1798, Dangos saw a round sharply defined spot traverse the sun in about 2 hours.

On February 12, 1820, Stark saw about noon a singular and well-defined circular spot with indications of an atmosphere, which was not visible in the evening of the same day.

At the end of February, 1762, Staudacher saw a black round spot, which he missed the next day, and of which he made a note, "Might this have been a new planet?"

At the beginning of May, 1764, Hoffmann remarked a large round black spot which appeared to have independent motion.

On June 6, 1761, Scheuter at Crefeld believed that he saw traverse the sun a round and well-defined spot, one-fourth the size of Venus and as black (as that planet when in transit?).

On October 9, 1819, Stark saw a well-defined and truly circular spot about the size of Mercury, which he could not find again in the evening.

On October 10, 1802, Fritsch saw a small round spot pass over the sun.

On November 19, 1762, Lichtenberg soon after sunrise saw a large round black spot, which manifestly moved before the sun, and made its exit after traversing a chord of 70°.

The observations of Stark on Oct. 9, 1819, and of Fritsch on Oct. 10, 1802, may prove to be transits at the ascending node of Lescarbault's planet as the recorded dates fall within the days of visibility. Professor Wolf quotes also an observation of Julius Schmidt on Oct. 11, 1847, but Schmidt himself remarks that the rapidly moving point seen by him was probably an insect or distant bird, and there are no sufficient grounds for retaining his note as of further value.

THE NEW PLANET

Anonymous; *Scientific American,* 10:323, 1864.

Since the brief paragraph was in type announcing the discovery of a new planet, we have received Mr. Beswick's computation of its elements:

Epoch, February 12, 1864.

Distance	0.4053
Longitude of node	321° 48'
Inclination	10.77°
Period in days	94.1326
Daily motion	3.824390°
Apparent diameter	8"
Mass (Earth -1.00)	0.11

Mr. Beswick saw this planet cross the sun's disk on February 12, 1864, at 8h. 20m. in the morning. It was then 10' 20" from the eastern limb, and 14' 20" from the southern limb of the sun. Its motion was exactly 711".66 in 100 minutes.

The whole time of transit was exactly 4 hours 33.5 minutes. The segment of the sun's disk which the apparent path of the body cut off was a complete diameter.

Its progress across the sun's face exceeded that of Venus and was less than that of Mercury. Its figure was that of a dark, round, and well-defined spot. Its size was 8", or a little less than that of Mercury at its greatest diameter. Indeed the whole appearance---figure, density, velocity and regularity of motion---was indicative of its being the transit of a planetary body whose path is included between the orbits of Mercury and Venus.

Mr. Beswick cites six previous observations by astronomers of the transit of a dark body across the sun's disk, which correspond exactly with the periods of this planet as computed by him:

"The first recorded transit of an unknown body is that of Lescarbault, who saw a round body transit the sun's disk on March 26, 1859, at about four o'clock in the afternoon. He and Leverrier computed its orbit, and announced it as an inter-Mercurial planet, whose period is less than twenty days, and its distance about 0.1427. And Leverrier has given to this planet the name of Vulcan. But the planet's place is unknown, as it has never been seen since.

"Now, in computing back the times of inferior transit of the body seen by me February 12, 1864, I find that it would transit the same on the very day and hour when Lescarbault saw a body cross the sun's disk. For between February 12, 1864, and March 26, 1859, there are exactly 14 inferior transits, or

$$\frac{1784.9 \text{ days.}}{14 \text{ transits.}} = 126.817 \text{ days.}$$

A circle whose radius is 15°, with the sun as a center, will include the position of this planet up to May 10th. And during its Eastern elongation, which will continue to May 17, it is possible that the planet may show a phase distinct enough to be visible just before sunrise, to a good observer with a good instrument. Its next inferior transit across the sun's disk will be early in the morning of June 18, 1864.

So that the planet seen by Lescarbault in France, and named Vulcan by Leverrier, is doubtless the same planet as the one whose orbit I have computed, and which was seen by me in 1864. But its distance from the sun is 0.4053 instead of 0.1427, or nearly three times as distant as Leverrier supposed it to be; and its path, which lies between Mercury and Venus, instead of Mercury and the Sun, is 94.1326 days instead of less than 20 days, or nearly five times as great as Leverrier supposed it to be."

The next observation in going back was that of Schmidt, October 11, 1847, the interval being exactly equal to 33 inferior transits. The next was by both Stark and Steinhubel, June 20, 1820, the interval being just 80 transits. It was seen by Stark at the next earlier transit than that of June 20, 1820, on the 9th of October 1819. Fritsch saw an unknown body transit the sun on the 10th of October 1802, and this interval proves to be equal to exactly 49 transits. Finally the earliest known mention of this body was by Schentan and Crefield on the 6th of June, 1764; 111 transits earlier than that observed by Fritsch.

....

PLANETARY SPOTS PASSING OVER THE SUN

Denning, William F.; *Astronomical Register,* 9:287, 1871.

"I have received from two different observers accounts of observations of planetary spots passing over the sun. I do not think these observations have ever been published, and so send them to you for publication, thinking that they may prove interesting, although I cannot vouch for their authenticity. On August 1, 1858, 4h., Mr. Robert Wilson, of Manchester, observed the partial transit of a circular opaque body over the sun. He watched it from 4h. to 5h. 30m., when the observation was interrupted. Its motion was from east to west, across the solar disc, and it presented an appearance very analogous to a planet in transit. The other observation is by Mr. William Waite, of London, who, in a letter to me, states:---'It may interest you to know that, some years ago, I saw, what I suppose must have been, a planetary body in transit across the sun. A dark speck in the lower limb of the sun caught my eye just about sunset; thinking it to be a sun spot, I got a glass to look at it, but found it to be a globular body of the apparent bigness of an ordinary sized marble, and intensely black. I had not time to notice in which direction it was moving, as the sun dipped almost immediately. Unfortunately, I am unable to recollect what year it was in, but it must have been between June, 1860 and June, 1863, and I imagine the season was either spring or autumn, as the house fronted nearly due west, and the sun was setting just opposite. If such things have been seen, they may be seen again; and every additional observation tending to increase such probability leads me to trouble you with this rather lame tale.' It is hardly necessary for me to comment on the foregoing, so I leave your readers to form their own opinion as to what amount of reliance it is entitled to. I would remark, however, that the expression (in regard to the object seen by Mr. Waite) as to its size, is a rather vague one. With reference to the object seen by Mr. Wilson, it would seem that it could not have been the planet that was seen by Lescarbault, inasmuch as it was seen in August 1, whereas it would appear that the latter body can only be observed in transit during the intervals from March 20, April 10, September 27, and October 14, that is, if the rough date supplied by Lescarbault is reliable."

• The Great Debate and More Sightings

AN INTRA-MERCURIAL PLANET (?)
Anonymous; *Nature*, 14:418, 1876.

At the sitting of the Paris Academy of Sciences on the 28th ultimo, M. Leverrier announced that he had received a letter from Prof. Rudolf Wolf, of Zurich, in which it was stated that three observers situated in three different places had witnessed, on April 4, the passage of a round spot over the sun's disc. The three localities were---in Germany (near Munster), Greece (Athens), and Switzerland (Zurich). The date is subsequent to the observation of Dr. Lescarbault by 6,219 days, which figure is the product of 148 into 42.02 (printed 40.02 in L'Institut, whence this notice is taken), and it may be conjectured that, if the object were a planet, it had made this number of revolutions of 42.02 days.

Such a body would have a mean distance from the sun equal to 0.2365 of the earth's mean distance, with a maximum elongation in a nearly circular orbit of about 13-1/2 degrees, the period of revolution being almost precisely half that of Mercury.

We await details of the observations before examining how far the date 1876, April 4, can be made to agree with similar ones already upon record, supposing all to refer to a single body revolving under the conditions named.

THE INTRA-MERCURIAL PLANET OR PLANETS
Hind, J. R.; *Nature*, 14:469-470, 1876.

The question of the existence of one or more planetary bodies revolving within the orbit of Mercury is again revived by Weber's observation of a round black spot just within the sun's eastern limb, on the afternoon of April 4 in the present year which had not been visible on the same morning, and early on the following day had disappeared. The position at 2-3/4' only from this limb is one, where an ordinary spot would not be expected to exhibit a circular outline; and a round black disk, in such a position more especially, must instantly attract the attention of a practised observer. On April 4 clouds unfortunately prevented lengthened observation, and in Weber's notice there is no reference to any perceptible motion during the short time the spot could be watched.

This observation resembles others already upon record, made by persons equally worthy of credit, which it is hardly possible to explain except on the hypothesis that one or more planetary bodies exist with mean distance less than Mercury, the rate of motion where motion has been detected by the most reliable observers, not being consistent with greater distance from the sun. While it is certain that comets with perihelia within the earth's orbit have transited the solar disk, and notwithstanding such transits may have been more frequent than is generally supposed, the appearance of the spots now in question seems, at least in several of the best authenticated cases, to negative any idea of their being due to the passage of comets across the sun, near their nodes. At the same time there are several instances where the form of the spots would perhaps accord better with the assumption of a cometary transit, unless we can admit that the deviation from circular contour is attributable to an optical cause.

It may be remembered that the attention of astronomers was first seriously directed to the possible existence of a planet or planets interior to the orbit of Mercury, by M. Leverrier's announcement that the motion of the perihelion of this planet was not explained by known causes of perturbation, but that an excess of 38 seconds in the century must be admitted beyond the value derived from theory, to produce an agreement between calculation and observation in the discussion of the long series of observed transits across the sun's disk. The unexplained motion of the line of apsides might, as M. Leverrier remarked, be due to the existence of a single interior planet of a mass which would depend upon its mean distance. With a distance of 0.17 (period of revolution, 25.6 days) the mass would be precisely equal to that of Mercury, and it would vary inversely with the distance. Or it might be due to a group of small planets circulating within the orbit of Mercury.

Having before us the whole of the recorded observations of the presence of suspicious spots upon the sun's disk, we shall soon discover that they hardly admit of explanation on the hypothesis of a single planet, even if we assume a small inclination of the orbit of this planet to the ecliptic, a condition which, while it would greatly extend the transit-limits, must at the same time render the transits so frequent that it is in a high degree improbable the planet could have so long escaped certain detection. Some few of the observations, as just remarked, we may perhaps refer to comets in transit; it remains to endeavour to ascertain from observations not thus explained what period or periods will best represent them, with the view to being warned of the probable times of future transits.

This subject has engaged the attention of M. Leverrier during the last few weeks, or since he became cognisant of Weber's observation last April, the notification of which was long delayed. It appears that the observations of Stark and Steinheibel, 1820, February 12, Lescarbault, 1859, March 26, and that of Weber, may refer to the same planetary body if the revolution be supposed 28.0077 days; this being the sidereal revolution with respect to the node, the synodical period would be 30.33 days; the corresponding mean distance from the sun is 0.18, and the maximum elongation 10-1/2 degrees. Such a planet would again be in conjunction with the sun on October 2nd or 3rd of the present year; and if Lescarbault's observation affords any approximation to the position of the line of nodes would pass across the sun's disk, and for this reason M. Leverrier has directed attention to the importance of a close watch upon the same, during these days, such watch, if possible, to extend to distant meridians, so as to insure pretty continuous observation through the forty-eight hours, Paris time. He has already advised American observatories through Prof. Henry, Secretary of the Smithsonian Institution, and it is to be hoped the chance of making an important discovery at this time, may be made known to observers in the East. It will be seen that the aid of the telegraph is indispensable, in order to secure complete evidence of the existence or non-existence of the hypothetical planet this autumn.

Other observations may be reconciled with a period of similar length, but the planet to which they may be supposed to refer cannot be identical with the above. Thus if Mr. Lummis's sketch of the path of the small round black spot, which he remarked upon the sun at Manchester on the morning of March 20, 1862, is reliable in the hurried and otherwise disadvantageous circumstances under which it was made, the ascending node was almost diametrically opposite to that of Lescarbault's planet, elements which have been attributed to MM. Valz and Radau, and exhibiting a near agreement in the position of the line of nodes, being certainly erroneous. Again, one of the most interesting observations bearing upon the existence of an intra-mercurial planet is that made about the end of June or beginning of July 1847 in this country, which can hardly be supposed to refer to either of the objects seen by Lescarbault and Lummis respectively. The exact date of this observation is unfortunately lost beyond recovery.

Mr. B. Scott, the City Chamberlain, observing the sun's disk near London, a short time before sunset late in June or on one of the first days in July, remarked upon it a perfectly circular black disk, and was so confident of the unusual character of the spot that he was on the point of making known his observation through one of the London daily journals on the evening of the same day, when unfortunately an astronomical friend, under the impression that an ordinary spot had been observed, dissuaded Mr. Scott from so doing. It thus happened that the matter dropped until the announcement in 1860 of Lescarbault's observation on March 26 in the preceding year, when Mr. Scott, in a communication addressed to the Times, drew attention to his experience in the summer of 1847. It was then discovered that he had not been the only observer of the strange object. Mr. Wray, the well-known optician, then resident at Whitby, had remarked a small circular black spot upon the sun late one afternoon at the end of June or early in July, though he also had, in 1860, lost the exact date. Both these gentlemen have furnished the writer with every other particular of their observations. That they refer to the same object can hardly be doubted. Mr. Wray had it under observation for forty minutes, when the sun sank into a bank of cloud and was not again visible that day. In this interval the spot appeared to have moved about five minutes of arc, and when last perceived was so near the western limb of the sun that Mr. Wray believes if the cloud had not interfered, in about ten minutes he would have witnessed the egress. This circular spot, the diameter of which he judged to be about six seconds of arc, was not visible early on the following morning, though other spots of ordinary form which were present on the disk remained nearly unchanged. Mr. Scott was observing with a refractor of about 4-1/2 inches aperture, Mr. Wray with a fine 6-feet Newtonian reflector of equal aperture, which he was employing at the time in a study of the varying aspect of the solar spots.

Notwithstanding the unfortunate loss of the date of these observations, such particulars as are available are still of value as certifying the existence of such objects in transit; there is no observation of the kind resting upon more excellent authority.

A letter from Prof. Heis, of Munster, the author of the "Atlas Caelestis Nova," received while closing these remarks, gives full details respecting Weber's observation. The spot was intensely black, perfectly round, and smaller than the planet Mercury in transit. Prof. Heis expresses the utmost confidence in this observation by his friend, who has long been accustomed to examine the solar disk.

AN INTRA-MERCURIAL PLANET

Russell, F. A. R.; *Nature*, 14:505, 1876.

The discussion as to the existence of a planet within the orbit of Mercury leads me to communicate an observation made many years ago, which I believe nothing but the existence of an unknown planet between us and the sun can explain. On Sunday, January 29, 1860, the sun rose in a fog in London, so that he could be steadily looked at as if through a dark glass. Soon after eight o'clock a perfectly round black object was seen by four persons, including myself, clearly defined upon the lower half, according to my recollection, of the sun's disc. It passed slowly across his face and made its egress at about half-past nine A. M. In apparent size it was equal to the representations I have seen of Mercury in transit.

THE INTRA-MERCURIAL PLANET QUESTION

Anonymous; *Nature*, 14:533, 1876.

M. Leverrier made a further communication to the Paris Academy, on the 2nd inst., with reference to this subject. Having collected in his previous communications, chiefly from the original authorities, such observations as could be supposed to bear upon it in any way, he finally selects for discussion those only which, in addition to the roundness and blackness of the spots, have distinct mention of sensible change of position upon the sun's disk on the day of observation. There are ten cases under this head in the months of January, February, March, May, and June, or possibly beginning of July, and October. M. Leverrier remarks it is inadmissible that a body projected upon the sun on February 12, which is the date of the observation by Steinheibel mentioned in the correspondence between Olbers and Bessel, could repass at the end of March or beginning of October, i.e., when arriving in the line of nodes of the objects seen by Lescarbault and Lummis. This could only happen if the first body moved in an orbit very little inclined to the ecliptic, but in this case the necessary frequency of the transits must have led to its being more often observed. For the present, therefore, he confines himself to treating five observations in October and March, where motion like that of a planet in transit are recorded. His data stand thus:---

Decuppis,	1839,	Oct.	2.0	Helioc. long.,	8°.60
Fritsch,	1802,	Oct.	10.0	"	16.46
Sidebotham,	1849,	March	12.18	"	172.01
Lummis,	1862,	March	19.87	"	179.86
Lescarbault,	1859,	March	26.22	"	186.60

And it is found that these five longitudes are represented with all the precision permitted by the nature of the observations by the formula (v = helioc. longitude)---

$$v = 121^{\circ}.49 + 10^{\circ}.9017834\underline{j} - 0^{\circ}.52 \cos. v,$$

$\underline{j}$ being reckoned in days from 1750.0.

The differences between calculation and observation are:---

1839	3°.6	1849	3°.5
1802	-3.6	1862	0.8
		1859	-4.6

None of the residuals exceeding a half-day's motion, M. Leverrier thinks it permissible to infer that the five observations appertain to the transits of the same body.

With the above motion the period of revolution is 33.0225 days, and the semi-axis major 0.201.

The existence of an intra-Mercurial body announced by theory, being, according to M. Leverrier, beyond doubt; to use his own words, "nous voila desormais en possession de donnees permettant des a present de constituer une premiere theorie qui conduira a retrouver la planete avec facilite et a la faire rentrer dans le systeme regulier des corps celestes." In conclusion he states that he is now occupied in determining the epochs of the next following transits over the sun's disk.

SUPPOSED DISCOVERY OF VULCAN

Swift, Lewis; *Observatory*, 2:161-162, 1878.

During the total phase of the late solar eclipse, as observed at Denver, Colorado, I had a view of a celestial object not down in Argelander's charts, which to my mind, without any doubt, is the long-sought Vulcan.

It was in field with a star of the same magnitude, probably θ Cancri.

As soon as totality was over, I recorded in my notebook the following:---"Saw two stars, about 3° S.W. of Sun, apparently of the 5th mag., some 12' apart, pointing towards Sun; both red." I immediately acquainted Prof. Hough, formerly director of the Dudley Observatory, and my two assistants of the discovery, and, as soon as possible on the same evening, Mr. Burnham, the celebrated double-star observer, and many others. None of them knew of any star in that position answering the description.

On my way home the thought occurred to me that the distance between them was about equal, perhaps a little greater than half that between Mizar and Alcor. On arriving at home, I found Webb gives their distance as 11-1/2', which would make the distance between θ Cancri and the new object some 6' or 7' instead of 12, as estimated (hastily and roughly, of course) at the time.

The next morning I learned by telegraph that Prof. Watson, at Separation, Wyoming Territory, saw a star 2-1/2° S. W. of the Sun and of 4-1/2 mag., not down in any chart.

The two observations are therefore confirmatory each of the other.

As totality commenced some four minutes earlier with him than with me, of course he antedates me by that amount.

Telescope 4-1/2-in. aperture; comet eyepiece; power 25; field 1-1/2°; periscopic and very superior.

STELLAR OBJECTS SEEN DURING THE ECLIPSE OF 1869

Hind, J. R.; *Nature*, 18:663-664, 1878.

It will be remembered by the reader who has interested himself in the published reports of observations of recent total eclipses of the sun, that during the totality on August 7, 1869, at a point in Iowa, called St. Paul's Junction, several observers attached to a party organised by Mr. W. S. Gilman, of New York, remarked below the sun what they termed "a little brilliant," and that one of the number using a small telescope, reported having seen just before the sun disappeared and as he came out again a minute crescent, in a similar direction from the moon. Commodore Sands, then Superintendent of the Naval Observatory, Washington, expressed his regret that these objects should not have been seen by Mr. Gilman himself, who had experience of the use of the telescope and was using a larger instrument than the others who had optical aid---but his "plan of operations" did not permit of it. The facts are thus stated:---A few moments after the corona formed, a small but exceedingly bright point, like a star, was noted independently by four of the party, two of whom it is mentioned were observing without telescopes; it appeared near the limits of the corona, below the moon's disc, and with one exception the observers located it a little to the right of an "anvil-shaped" prominence, or "at about 230° from the north point, reckoning by the east," and it is added that each of the observers felt quite positive that what he saw was truly a star. With respect to the small crescent Mr. Gilman reports that about half a minute preceding totality another member of the party, Mr. Vincent, came to him exclaiming that he saw a miniature-crescent-shaped star under the moon, and asking him to verify the observation, but, interested in his own work, he did not at the moment do so; on Mr. Vincent returning more urgent than ever, Mr. Gilman says he did look in a hurried manner but saw nothing in the few seconds he gave to the search; he afterwards states, however, that he does not think he looked so far away from the moon as the crescent was located in a drawing made immediately after the eclipse by Mr. Vincent; in this drawing it was placed "at one and a half times the moon's diameter from its limb, and to the left of a perpendicular down to the horizon." Mr. Gilman adds he could not connect this crescent with the small star of the other observers, indeed Mr. Vincent estimated the object seen by him at three times as far removed from the moon's limb as the small star, which would assign for the latter a distance of about half a degree, corresponding very well to the expression used by the four observers who noted it, that it was near the limits of the corona. Dr. B. A. Gould, now Director of the Observatory at Cordoba, who observed this eclipse at a different station, gave some attention to a search for any object near the sun which might be an intra-Mercurial planet, and he states he saw the star π Cancri, but did not meet with any other stellar body. This star being at the time in a similar direction from the moon's centre, to "the little brilliant" of the Iowa observers, there has been a pretty general opinion that it was the object remarked by them, and, in conversation with Dr. Gould several years since, I found him tolerably well satisfied that he had thus sufficiently explained their observations. But the discovery, or rather discoveries, of Prof. Watson, lend a new interest to them, and a more strict examination of the circumstances may not be out of place here. The position of St. Paul's Junction is stated to be in latitude 42° 47' 30" N., and longitude 19° 5' 45" W. of Washington. The totality was observed to commence at 5h. 48m. 4.6s., ending at 5h. 51m. 34s. M. T. at Washington, so that the middle occurred at 10h. 58m. 22s. M. T. at Greenwich, which agrees exactly with a calculation made with the <u>Nautical Almanac</u> elements. We will assume 10h. 59m. G. M. T. as the time to which the observations of the brilliant point apply. Correcting the moon's place for the effect of parallax, we find her apparent position at this time to be in right ascension, 9h. 11m. 26.7s., and north declination 16° 13' 58"; her augmented semi-diameter was 16' 37". We must assume that both star-like object and crescent were on an angle of 230°, the latter one-and-a-half times the moon's diameter from her limb, and the former at one-third of this distance, whence, referring to the moon's centre, we have, for the crescent, $\Delta \underline{a} =$ -3m. 32s., $\Delta\delta =$ -42'.7, and for the bright little star, $\Delta \underline{a} =$ -1m. 46s., $\Delta\delta =$ -21'.4; and thus,

	R. A.	Decl.
	h. m. s.	° '
Stellar object	9 9 41	15 52.6
Crescent	9 7 55	15 31.2

The former was therefore 28' and the latter 56' south of the ecliptic.

Now with regard to the star, the presence of which has been supposed to explain the observation of the four observers who noted "the little brilliant," there has been some slight confusion. In a note inserted in the last "Annual Report" of the Royal Astronomical Society it is stated that the object seen "has been satisfactorily identified as the star π^1 Cancri," which is assuredly a mistake. π^1, according to Argelander, is only a seventh magnitude, which is hardly to be glimpsed with the most acute sight in the darkest winter sky. For π^1 no doubt we must read π^2, or 82 Cancri. But this star, also, is of a degree of brightness wholly insufficient to allow of it being possible to discern it at all so near the sun's place without some optical aid in the still illuminated sky-ground, much less to be caught up as a brilliant point of light, with the naked eye; the "Durchmusterung" extimate is 5.8m., which is confirmed by the careful estimations of the second Radcliffe Catalogue, where we find it rated 5.9m., or, in round magnitudes, a sixth. It should be mentioned that the apparent place of 82 Cancri was in right ascension 9h. 7m. 59.5s., declination 15° 28' 56", agreeing nearly with that we have found for the minute crescent, but 33' from the small star. It appears probable, in view of Prof. Watson's discovery, that Dr. Gould may have mistaken an intra-Mercurial planet for π^2 Cancri, and if the statements of the four observers at St. Paul's Junction (one of them, by the way, a lady) are accepted, it can hardly be doubted that they also were attracted by an unknown object, since in that part of the sky there is no star which could be visible as they describe it. It is significant that Regulus, upwards of 12° from the sun's place, was only noticed as a "glimpse star" at St. Paul's Junction.

THE INTRA-MERCURIAL PLANETS

Flammarion, Camille; *Popular Science Monthly*, 14:714-721, 1879.

A good deal of noise was made a few months ago about a discovery that an American astronomer believed he had made during the recent solar eclipse of July 29, 1878. At the moment of totality, while the bright disk of the sun was completely hidden by the black disk of the moon, and after the eye had become habituated to this sudden darkness, the American astronomer made a search to find whether there might not be, in the vicinity of the sun, a planet answering to the theoretical planet Vulcan, whose existence was announced by Leverrier after he had mathematically analyzed the motion of Mercury. As every one knows, during total eclipses of the sun, our atmosphere being no longer illumined, night comes on as though at the bidding of an enchanter, and the brighter stars make their appearance in the heavens. It is this sudden metamorphosis of nature that most forcibly impresses primitive races; it was this which, on the occasion of the earliest eclipse recorded in history---the eclipse of Thales---put a stop to the famous war of the Medes and the Persians, by sending the chill of mortal terror into the hearts of the warriors as they were on the point of engaging in battle; this it was which, at the last eclipse, led a negro, suddenly frenzied and convinced that a deluge was about to be sent down by an angered Deity, to strangle his own wife and children; finally, it is this same feature of the solar eclipse which makes the deepest impression on the mind of the astronomer who has made all his preparations for observation, but is so full of emotion in view of the grandeur of the spectacle, that he can only with difficulty analyze with his wonted coolness, and during the few minutes of total darkness, the details of the phenomenon.

The American astronomer Professor Watson, the discoverer of a great number of small planets, has declared that his one thought during the recent eclipse was to look for the intra-Mercurial planet. In announcing this observation to the Academy of Sciences, the Director of the Paris Observatory, M. Mouchez, expressed himself thus:

> The accomplished astronomer of Ann Arbor has seen a heavenly body of the fourth magnitude, situated two degrees distant from the sun, and whose position was: right ascension, $8^h\ 26^m$; declination north, $18^o\ 0'$.
>
> The star nearest to this position is Theta Cancri ($8^h\ 24^m$ and $18^o\ 30'$), and it is of the fifth magnitude. This difference of magnitude and of position justifies us in supposing that in all probability it was the planet Vulcan which was again seen by Mr. Watson. The Academy can not but receive with great pleasure this observation, which is only a new tribute to the scientific glory of Leverrier. M. Gaillot has calculated an orbit and an ephemeris. The time of revolution is only twenty-four days. The planet is at its greatest distance from the sun today (August 5th); to-morrow it will be at thirty-eight minutes of time. This distance is great enough to allow of our observing it, if not at Paris, at least at other more favored localities.

Watson's own account of the observation is as follows:

> At the recent total eclipse of the sun I was occupied exclusively in a search for any intra-Mercurial planet which might be visible. For this purpose I employed an excellent four-inch refractor, by Alvan Clark & Sons, mounted equatorially with a magnifying power of forty-five. There were no circles originally attached to the instrument, and accordingly I placed on it circles of hard wood, the declination circle being five inches and the hour circle four and three quarter inches in diameter. On these I pasted circles of cardboard, and pointers were provided so that I could mark with a sharp pencil the position corresponding to any particular pointing of the instrument. This method does not compare in accuracy with graduated circles and verniers, but it has the advantage, and a very important one in the present case, of avoiding the uncertainty which might be attributed to erroneous readings of the circles. To read the divided circles would require considerable time, while the pointings can be marked on the paper disks in a few moments. And, besides, while a doubt might be raised as to the correctness of the recorded circle readings, no such doubt can exist in reference to the positions marked on these paper circles. The chronometer times corresponding to each pointing were recorded, and the designation of the object observed was also marked on the paper disks, so that there is no difficulty in identifying the several marks.

He then goes on to give the particulars of his sweeps over the regions east and west of the sun, which were without result till, at last---

> Between the sun and θ Cancri....I came across a star estimated at the time to be of four and a half magnitude, which shone with a ruddy light and certainly had a larger disk than the spurious disk of a star. The focus of the eye-piece had been carefully adjusted beforehand and securely clamped, and the definition was excellent. I proceeded, therefore, to mark its position on the paper circles, and to record the time of observation. It was designated by a. The place of the sun had been recorded a few minutes previously, and marked S_1. Placing my eye again at the telescope, I assured myself that it had not been disturbed, and proceeded with the search. I noticed particularly that the object in question did not present any elongation such as would be probable were it a comet in that position.

This body he holds to be Leverrier's intra-Mercurial planet. Its place is given as follows: $8^h\ 26^m\ 24^s$; declination 18° 16'. It will be seen that this position differs from that given above, the declination being here 18° 16', instead of 18° 0'.

On August 23d the observer added a new correction: "In consequence of having employed an inexact value for the correction of the chronometer, an error has crept into the results. The true position is this: right ascension, $8^h\ 27^m\ 35^s$; declination, 18° 16'."

Here we have a fresh difference in the first figure. The result is, first, that the orbit calculated immediately upon receipt of the telegram was made too hastily and on an insufficient basis.

According to the American observer, the definitive differences between the planet and the sun were: in right ascension, $8^m\ 21^s$; in declination, 0° 22'. But in this same letter of August 23d he announces that he observed another star, also of the fourth magnitude, which presented the following differences from the sun: in right ascension, $27^m\ 18^s$; in declination, 0° 35'. Whence results for this second star the position: right ascension, $8^h\ 8^m\ 38^s$; declination, 18° 3'.

A fourth datum sent to the London Royal Astronomical Society again corrects these positions as follow:

$8^h\ 27^m\ 24^s$	18° 16'
$8^h\ 9^m\ 24^s$	18° 3'

We will remind our readers that the right ascension of a star is its distance east or west from the first point of Aries, measured along the celestial equator, and its declination is its distance above or below that equator. They are the longitudes and latitudes of the heavens, corresponding to those of earth, and they serve to determine the positions of stars as earth longitudes and latitudes serve to fix in geography the exact positions of cities.

And now comes another American astronomer, Mr. Swift, who also announces that during the same total eclipse he observed simultaneously in the field of his telescope two stars, of which the one was Theta Cancri, and the other a planet shining with the luster of a star of the fifth magnitude, and whose position he estimates approximately to be, right ascension $8^h\ 26^m\ 40^s$, declination 18° 30' 35", a position very near to that of the star seen and determined by Mr. Watson.

The eclipse of July 29th last, one of the most remarkable of the present century for the duration of totality, was observed all along the zone of centrality (which passed across North America), by a great number of able astronomers, both American and English. Nearly all of these searched for a new star in the neighborhood of the sun, but, with the exception of the two named above, all declare that they saw nothing beyond the stars rendered momentarily visible by the obscuration of the sun's light.

Are we thence to conclude that the testimony of these two observers must be rejected? By no means. But between this and the conclusion that the two stars signalized by Watson are in fact two planets traveling between Mercury and the sun, is a long way. Of these two stars, the second, it is supposed, can not be the star Zeta Cancri, whose position is $8^h\ 5^m\ 12^s$ and $18^o\ 2'$.

The difference of three minutes is no doubt very great, but when we take account of the haste of observation, and the doubt expressed by the observer himself with regard to its exactness and the possible derangement of his telescope by the action of the wind, this star must not be dismissed without seeing whether or not it will explain the observations. Now, supposing an error of three minutes more or less, the position of the first star becomes $8^h\ 24^m$ and $18^o\ 16'$, and this is very nearly the position of the star Theta Cancri.

All that is needed to show how probable is this explanation is to take up a celestial atlas and to locate the sun at the point where it was at the moment of the eclipse, i.e., in front of Delta Cancri (which was visible through the aureolar corona of the eclipsed sun, at its eastern margin).

We have reproduced in the figure the aspect of the heavens during totality. [Figure not reproduced]

1 is Mercury,	5 is Procyon,
2 is Regulus,	6 is Pollux,
3 is Mars,	7 is Castor,
4 is the sun in eclipse,	8 is Venus.

In the immediate vicinity of the sun we have inserted at the points a, b, c, the three stars Delta, Theta, and Zeta Cancri. In our opinion the two stars b and c are the ones which might have been taken for two planets by the American observers. No doubt this hypothesis is a rather bold one, but then the hastiness, the difficulty, and the vagueness of the observation justify it.

Mr. Watson did not announce his observation of the second star till three weeks after the eclipse, and after he had revised in detail the conditions of his observation. Here is proof that this star had made a less impression on him than the first. But now the first is between the fourth and fifth magnitudes, and the second only between the fifth and sixth.

In any case, we can not but hold that the positions given are highly doubtful and can not be seriously assumed as a basis for calculating orbits, as has been done at the observatory.

Mr. Swift writes, "I have no doubt that one of the two stars was Theta Cancri, and the other the intra-Mercurial planet."

But if one of the two stars seen by Professor Watson is Theta Cancri, the other, whose difference from the former is, according to him, $18^m\ 57^s$ more to the west, and 13' more to the south, comes exceedingly near to Zeta Cancri, whose difference from Theta is $19^m\ 25^s$ also to the west, and 30' also to the south. If this star were a planet it would not have been possible to see it without at the same time seeing Zeta, which would have been quite near; but of Zeta our American astronomers do not speak.

Mr. Swift, however, answers this objection by saying that he saw simultaneously Theta and the Planet, and states the difference between them as follows:

	Right ascension	Declination
Star Theta,	$8^h\ 24^m\ 40^s$	$18^o\ 30'\ 20''$
Planet,	$8^h\ 26^m\ 40^s$	$18^o\ 30'\ 25''$

The two stars appeared red to him, shining at three degrees southwest of the sun, and presenting large disks of the fifth magnitude. He adds that he saw no other star, not even Delta, and that the distance between them was from 7' to 8' of arc. ("Nature," September 19, 1878.)

The two stars seen by Swift, then, are not the same two seen by Watson. The matter is a good deal confused.

Further, in observing the star Zeta Cancri, two stars could have been seen, instead of one, for Zeta is a double star. It is, indeed, a triple star, but a high power is required to see it triple, whereas a moderate power easily shows it double. But, inasmuch as only very low powers were employed, it is probable that no new cause of perplexity exists here.

To sum up: While it is possible that the American observers saw an intra-Mercurial planet, or even two, we can not, in view of the special difficulties of the situation, the confusion of the figures, and the negative observations of the other observers, concede it to be an absolute and incontestable fact that they saw even so much as one. The fact is not yet certain.

We would remark that no variable star is known to exist at that part of the heaven. Might it not be a comet? It did not present the characteristics of a comet.

Whatever the upshot may be, this discussion goes to show that in astronomy nothing is accepted without being first verified, and that this science is becoming more and more worthy of its reputation of being the exactest and the most absolute form of human knowledge.

This strange vicinage of the sun is unfortunate indeed in the annals of astronomy. Time and again have observers supposed that they saw planets passing before the sun; and, out of all the observations which have been made, there is not one which definitively settles the question.

The penultimate astronomical observation announced as appertaining to an intra-Mercurial planet was that of the German astronomer Weber, which was presented to Leverrier by Wolf, of Zurich, as having been made April 4, 1876, at Peckeloh. "A small, well-rounded disk of 12" of arc appeared suddenly during a partial clearness of the sky (eclaircie), which was turned to account for observing the sun. It was impossible to pursue the observation, owing to clouds." Wolf calculated that this observation accorded with two prior observations made in 1859 and in 1820, and fixed the period of the planet at forty-two days. Leverrier, too, seemed disposed to accept this observation, which gave occasion for a new memoir by the illustrious French astronomer, in which he brought together and compared all the observations of this kind. The stir which this made in the scientific periodicals is still remembered. But yet the little black disk observed by Weber was not a planet at all, but merely a sun-spot, round and without penumbra. It had been observed five hours earlier at the Observatory of Madrid and at the Observatory of Greenwich, and it was easy to prove that it was nothing but a simple sun-spot.

The fact that the spot was not to be seen the next day seemed to confirm the planetary hypothesis; but it was not sufficient, since there are ephemeral sun-spots, too. Nor is the roundness of the spot a distinctive character either. The proper movement then remains. Here we must note a circumstance which oftentimes must have caused illusions. When we observe the sun with a telescope not equatorially mounted, and whose support has not the two motions, vertical and azimuthal, as is commonly the case, the position of a sun-spot, by reason of the diurnal motion, is ever changing with relation to the vertical diameter of the disk. Even when an observer has had large experience, it is difficult for him to guard against the belief that the spot has changed its position on the disk.

This observation, then, had to be disallowed. But there remained others which were not to be discredited on the same grounds, and Leverrier, taking them to be more trustworthy, used them in calculating the orbit of the hypothetical planet. Different interpretations gave him five different orbits, with periods varying between twenty-four and fifty-one days. But he seems to have preferred that which gives a period of thirty-three days, and announced that on March 22, 1877, the planet in question might pass before the sun. Astronomers all over the world, with one accord, observed the sun on that day, to descry the transit, but the result was <u>nil</u>. No black point was to be seen.

Among the prior observations Leverrier accepts five as certain, viz.: Fritsche's, in 1802; Stark's in 1819; Cuppis's, in 1839; Sidebotham's, in 1849; Lescarbault's, in 1859; Lumnis's, in 1862. One of the best is no doubt that of Dr. Lescarbault, a country doctor, with a passion for astronomy, and who had vowed to the study of the heavens the time which was not spent in alleviating the wretchedness of earth.

This amateur astronomer, while observing the sun on March 26, 1859, from his humble house at Orgeres, discovered on its radiant disk a round, very black spot, which he was able to study for over an hour, and the proper motion of which he thus determined, no doubt taking account of the causes of error to which we have alluded. It was during this same year that Leverrier perceived the necessity of increasing by 38" the secular movement of Mercury's perhelion, and offered the hypothesis that a planet nearer to the sun than Mercury would account for the difference. Thus the observation made by my old and learned friend came as though on purpose to confirm the theory, just as earlier the telescopic discovery of Neptune had come to confirm so brilliantly the theoretical discovery of that distant planet.

Wellnigh twenty years have passed since 1859, and yet a fact which one might have supposed would be speedily confirmed, owing to the rapidity of the planet's revolution, and its no doubt frequent transits across the sun---this fact has received no confirmation. Yet search has been made for it throughout the whole region on both sides of the sun, if perchance it might be seen at the periods of its greatest distance. Perhaps not a day has passed for the last twenty years, but that the sun has been examined at one point or another of the globe, observed with the greatest care, sketched in all its details, even directly photographed.

The hypothesis of a single body comparable to Mercury, gravitating in close proximity to the sun, and on a plane probably little inclined to the solar equator, seems to us to be so open to objections as to be untenable. Still, the mathematical theory of universal attraction proves that there is a cause for the retardation observed in the motion of Mercury, and that this cause can not be found by augmenting the mass of Venus---a quantity now determined with great exactitude---but must be sought for in some disturbing mass between Mercury and the sun. But this mass may not be a planet worthy of the name of planet; it may consist of a great number of asteroids like the minute fragments which gravitate between Mars and Jupiter---asteroids so small that oftentimes they escape the notice of observers of the sun and of eclipses, though some of them may be large enough to be seen under certain rare conditions. This latter theory is the one which we adopt.

• Vulcan Denied But Observations Persist

THE CLOSING OF A FAMOUS ASTRONOMICAL PROBLEM

Campbell, W. W.; *Popular Science Monthly*, 74:494-503, 1909.

There is perhaps no more striking illustration of the power of scientific method than that relating to the discovery of Neptune in 1846. The planet Uranus, until then the outermost known member of our solar system, refused to follow the path computed for it by mathematical astronomers. With the progress of time the discrepancies between its predicted and observed positions grew constantly larger until, in the early eighteen-forties, the discordance amounted to fully 75 seconds of arc. This is a small angle---not more than one twenty-fifth the angular diameter of our moon---yet a very large angle to refined astronomy, for a discrepancy of two seconds would have been detected with ease. The opinion gradually developed that Uranus was drawn from its natural course by the attractions of an undiscovered planet still farther from the sun than itself. Adams in 1843 and Le Verrier in 1845, independently, and each without knowledge of the other's plans, attacked the then extremely difficult problem of determining the approximate orbit, mass and position of an undiscovered body whose attraction should produce the perturbations observed. Regrettable and avoidable delays occurred in searching for the planet after Adams's results were communicated to the astronomer royal, in October, 1845. Le Verrier's results were communicated to the Berlin Observatory in September, 1846, with the request that a search be made. The disturbing planet, later named

Neptune, was found on the first evening that it was looked for, less than one degree of arc from the position assigned by Le Verrier. If an energetic search had been made in England the year before, the planet would have been discovered within two degrees of the position assigned by Adams.

The above resume of this unsurpassed achievement of the human mind forms a natural prelude to the present article, as it was the immediate forerunner of another problem, famous for half a century, which has now been brought to a satisfactory conclusion.

The determination of the orbit of the planet Mercury gave great difficulty to its investigators, principally from two cases:

1. Being the innermost known planet in our system, remaining always near the sun, and usually lost to view in the sun's glare, fairly accurate observations of its positions could be secured only when the planet was near its greatest angular distances from the sun and on the rare occasions when the planet passed between us and the sun's disk. Consequently, observations of the highest accuracy were few in number; and,

2. There were large discrepancies between Mercury's predicted and observed positions, certainly not due to the attractions of any known members of our solar system.

Le Verrier, of Neptunian fame, undertook a systematic investigation of Mercury's orbit, making use of all available observations. His results were derived and published in 1859. His work established that there were peculiarities in the planet's orbital motion which could not be due to the attractions of known masses of matter. Chief among the peculiarities was a slow rotation of the orbit itself. It is best described as a forward motion of the orbit's perihelion amounting to 38 seconds of arc per century.

Le Verrier announced that the outstanding differences between prediction and observation could be produced and explained by the disturbing attractions of an undiscovered planet closer to the sun than Mercury and revolving around the sun in an orbit lying nearly in the plane of Mercury's orbit. The mass of (the quantity of matter in) the hypothetical planet would depend upon its distance from Mercury: if half way between Mercury and the sun, its mass would be two thirds that of Mercury; if further from Mercury, the necessary mass would be greater; if nearer, smaller. A group or "ring" of small planets, instead of one large planet, would serve equally well, provided the total mass of the planetoids were of the same order of magnitude. Le Verrier did not say that such an undiscovered planet or ring of planetoids did exist, but simply that it would account for the observed anomalies. The accuracy of his computations, published in detail, could not be questioned. The recognition of his masterly skill, and the memory of his entirely similar discovery of Neptune, assisted in convincing astronomers quite generally that a planet or group of planets existed. The discovery of the disturbing mass became at once a noted problem.

A body traveling around the sun in a circular orbit whose radius is only one half Mercury's average solar distance would never be more than 12° from the sun as viewed by terrestrial observers. A search for it by ordinary methods would accordingly be fruitless. A body large enough to shine brilliantly on a dark-sky background would be hopelessly lost in the bright sky near the sun. Mercury itself, though running out between 20° and 30° from the sun every few weeks, is seldom seen by any save astronomers; and they know where to look for it in the twilight sky.

Two special methods of discovery were applicable: (1) To detect the planet projected upon the sun's disk when its orbital motion carried it between us and the sun; (2) to search for it when the sky background was darkened at the time of a total solar eclipse.

Needless to say, a crop of discoverers by the first method grew up without delay. The observer of greatest note was Lescarbault, a rural physician of France. Immediately following the publication of Le Verrier's conclusions, Lescarbault announced that he had observed the transit of an unknown planet across the sun's disk several months earlier. Le Verrier journeyed to Lescarbault's home, investigated all the circumstances of the observation, weighed the evidence and concluded that a real planet had been seen. In fact, so convinced of its reality were many scientific men that the name Vulcan was given to it. Older and later reported observations of the same character, to the number of twenty, were collected by Le Verrier, and those which seemed to be in harmony with each other were made the basis of an orbit. Vulcan was found to be about one third Mercury's distance from the sun, revolving once around the sun in between nineteen and twenty days. In some of the text-books on astronomy appearing in the sixties and seventies, Vulcan was assigned a place in the solar system as conspicuous and as secure as that of Mercury itself.

Now it is probable that every one of the twenty observations referred to was erroneous, though made in good faith. In essentially every case the observer was inexperienced, and used a telescope of insufficient power, or one unprovided with measuring apparatus suitable for determining whether or not the subject observed was in motion across the sun's disk. Even the observation of Lescarbault was in doubt when it later transpired that a Brazilian observer of considerable professional experience was at the same hour studying the region of the sun in question and saw only uniform normal solar surface. The situation was not without its humorous side. For example, a Mississippi Valley weather prophet who saw Vulcan crossing the sun's disk, said it was about "as large as a new [sic] silver half dollar"! Many of the observations no doubt referred to small sun spots which, with small telescopes, would look round.

Vulcan was searched for by visual observers at the principal eclipses of the sixties, seventies and eighties. Two noted astronomers at the eclipse of 1878, Watson and Swift, believed that they saw two new planets near the sun. However, the two seen by Watson did not agree with those seen by Swift, and still other astronomers at the same eclipse saw no strange bodies in the same regions. As the assigned locations depended upon the hasty readings of graduated circles, in which one can so easily make errors, in the press and excitement of eclipse conditions, the astronomical world quickly, and no doubt correctly, concluded that the objects seen were well-known neighboring stars.

The perfecting of dry-plate photography gave renewed interest to the search for Vulcan, both when passing over the solar surface and at times of eclipse. Although the sun has been photographed almost daily during the past twenty years, at one observatory or another, no experienced observer has seriously claimed that his plates recorded an unknown planet crossing the sun. Neither were eclipse searches more successful: the well-known bright stars lying nearly in the direction of the sun were photographed, but no strange bodies. Curiously enough, the optical principles governing the efficiency of cameras in this search were overlooked for many years, and faint objects near the sun---say stars fainter than the fourth magnitude---were not observable, because their images, though formed on the photographic plates, were overwhelmed and buried from sight in the general darkening of the photograph by the bright-sky background. It was not until 1900 that the elements of the problem of photographing faint bodies near the sun were comprehended. While preparing for the eclipse of that year, three astronomers, Professor W. H. Pickering, of Harvard College Observatory, and Messrs. Perrine and Campbell, of the Lick Observatory, independently arrived at the same simple conclusion that the focal lengths of the intramercurial-search cameras should be relatively long, on order to reduce the intensity of the sky exposure on the plates without reducing the intensity of the star images, and thus let the latter be seen on the negative. The principles involved are so simple as hardly to call for elucidation.

. . . .

Professor Pickering, of Harvard, and Mr. Abbot, of the Smithsonian Institution, used such cameras at the total solar eclipse of 1900. The latter observer was favored with good conditions, in North Carolina, and he secured one photograph of a considerable area surrounding the eclipsed sun. Quite a number of the stars known to exist in this region were photographed; but in the absence of a duplicate photograph of the same region, he could not decide whether certain apparent images on the plate were due to unknown planets, or were defects such as always exist in photographic films.

At the eclipse of 1901, in Sumatra, Mr. Abbot, of the Smithsonian Expedition, and Mr. Perrine, in charge of the Crocker Expedition from the Lick Observatory, were prepared, with four cameras each, to secure duplicate photographs covering a large area extending east and west from the sun. Conditions were unfortunately against the success of Mr. Abbot's plans, but thin clouds at the time of the eclipse let 25 per cent. of the light come through to Mr. Perrine's photographic plates. The area covered in duplicate was $6^{\circ} \times 38^{\circ}$, extending along the direction of the sun's equator, with the sun in the center of the region. The plates recorded 170 well-known stars; and all apparent images not of ordinary stars were proved by the duplicate plates to be defects in the films. In two thirds of the area stars down to the eighth magnitude and many fainter ones were recorded; and in one third the area, covered with thicker clouds, stars were recorded down to the fifth and sixth magnitudes.

. . . .

The eclipse of 1908 in the South Seas was utilized by the Crocker Expedition to cover a region extending east and west along the sun's equator with duplicate exposures. Notwithstanding interference from rain and clouds at the beginning of totality, clear sky prevailed during the last two thirds of the four critical minutes. Dr. Perrine finds more than 500 images of well-known stars on the plates, and no images of unknown bodies. Stars are recorded down to nearly the ninth visual magnitude.

....

Inasmuch as planets, shining by reflected light, do not act upon photographic plates so strongly as stars of the same visual magnitude, we may say that exposures which recorded stars down to the ninth magnitude should have recorded planets down to the eighth. From the known brightness, distance from the sun and approximate diameter of a few of the asteroids revolving in space between Mars's and Jupiter's orbits, Dr. Perrine has computed that an average eighth-magnitude intramercurial planet could scarcely be larger than thirty miles in diameter and that roughly a million such bodies, of great density, would be required to supply the disturbing effect observed in Mercury's orbit.

Taking all these points into consideration, I think we may say that the investigations by Perrine, forming a part of the work of the Crocker Eclipse Expeditions from Lick Observatory, have brought the observational side of the Intramercurial Problem, famous for a half century, definitely to a close. It is not contended that no such planets will be discovered in the future; in fact, it would not be surprising, nor in opposition to the opinions here expressed, if several such bodies should be found; but it is confidently believed that any such bodies would fail hopelessly to supply the great mass of material demanded by Le Verrier's theory, as Perrine pointed out in discussing the Sumatra observations of 1901.

....

NOTE ON A SOLAR OBSERVATION

Markwick, E. E.; *British Astronomical Association, Journal,* 27:188-189, 1917.

I am rather interested in a spot observed on the Sun on April 5. On that date at $6^h\ 10^m$ p.m., G.M.T., using a hand draw telescope with object-glass 1.5 inches in diameter, and low power microscope inverting eye-piece magnifying about nine times, or perhaps a trifle more, the telescope being mounted on a small pillar and claw stand, a small spot was seen, not quite half way from the centre to the vertex, as shown on the accompanying diagram. The orientation of this is only approximately correct. As far as I could make out this spot was absolutely solitary on the disc; I could not trace any outliers or dots in its vicinity. It was very small, greyish, not quite black, and, so far as could be judged, circular. With 43 years' experience of Sun-spots I have always found that even a small solitary dot, such as this, generally gives some evidence, such as a slight irregularity in shape, or the presence of faculae, or other dots in the vicinity, that it is part of the photosphere and not a foreign body seen projected on the disc. But, in this case, possibly owing to my present very limited optical resources, I gained the impression that this spot might have been something other than a Sun-spot. It was watched for some time, and regarded as a min. vis. for the telescope used, in the way of a dark spot on a bright background. I may say that the definition on the Sun with this small glass is wonderfully sharp and clear.

On the previous day, April 4, at 5^h 30^m p.m. with the same glass, after a careful survey, there was not the slightest sign of a spot on the solar disc; and, again, on April 6, at 9^h 34^m a.m., there was the same state of things as on the 4th, not a sign of a spot. The spot described above must therefore (if it belonged to the solar disc) have burst out and died away between the dates named.

By the courtesy of Mr. E. W. Maunder, I am able to state that there was no sign of a spot on a photograph of the Sun taken at Greenwich on April 5, at 11^h 17^m, or on another taken at 11^h 20^m. This still further limits the period of visibility of the spot.

My chief object in reporting this observation is to ascertain whether any of our members observed the spot in question, and so definitely settle the matter as to its nature. I can quite imagine that in the early days of the telescope, and much later even, a spot of this nature might well have been taken for an intra-mercurial planet, or other body, between us and the Sun. The time available for its observation in my case was too brief to enable any change in position to be noticed. Finally, I may add that the possibility of its being a speck of dust or dirt in the eye-piece is ruled out by the fact that as light clouds passed over the Sun's disc, the spot faded away, to reappear again when he shone clear once more. Also the spot kept its place on the disc, however much the latter shifted, or was shifted in the field of view of the telescope. At the time of observation there were horizontal strata of cirrus cloud about, with clear gaps between.

THE MYTHICAL PLANET VULCAN

Anonymous; *Science*, 75:sup 8-9, June 3, 1932.

A new nail has been driven into the coffin of the mythical planet Vulcan by Dr. H. von Kluber, of the Astrophysical Observatory at Potsdam. Vulcan was once supposed to revolve around the sun in an orbit within that of Mercury. Such a body would never be seen in the night sky, but would be visible either when it passes in front of the sun, or during a total eclipse, when the bright solar disc is temporarily hidden by the moon, and faint objects in the same part of the sky are made visible. For many years observations to detect it were made at eclipses, without success. It had been quite definitely decided that no such planet exists.

In May, 1929, Professor Erwin Freundlich, director of the Einstein Tower at Potsdam, photographed the eclipse of the sun visible in Sumatra. Though made for other purposes, these plates showed such a profusion of star images that Dr. von Kluber decided to examine them carefully to make sure that Vulcan was not concealed there. For purposes of comparison photographs of the same part of the sky, but without the sun, were made six months later at Potsdam, with the same telescope. If a planet were present, it would betray itself by its motion between the time of the two exposures.

The search proves conclusively that there is no planet as bright as the ninth magnitude, considerably below naked-eye visibility, up to a distance of 40 minutes from the sun. This is approximately equal to one and a third solar diameters. Closer than this the bright solar corona, visible at eclipse time, might have made faint objects invisible, but even there, believes Dr. von Kluber, a planet as bright as the seventh magnitude, also too faint to be seen with the naked eye, would have been found. Thus, he has decided, if there is a planet closer to the sun than Mercury, it is extremely small.

EVIDENCE FOR INTRAMERCURIAL PLANETARY OBJECTS OR COMETS DISCOVERED DURING SOLAR ECLIPSES

Courten, Henry C.; *COSPAR Paper,* 1971.

Abstract. A number of objects which appear to be in heliocentric orbit have been discovered during several total solar eclipses. The small orbital eccentricities derived for three of the objects observed during the March 1970 eclipse suggest that they are planetary bodies. The remaining objects lend themselves more readily to cometary origin. If the existence of a new planet or asteroid belt is assumed, it appears reasonable to speculate that the projected distance of the object (or objects) is about 0.1 AU from the sun. The object is probably quite small (less than 800 km in diameter), thereby not seriously challenging the Einsteinian perturbation of Mercury's orbit. The orbital eccentricity is very low, and the visual albedo is less than 0.1. An equilibrium temperature of about 1300 K seems likely.

A TENTH PLANET?

Anonymous; *Chemistry,* 44:23-24, December 1971.

Another possibility is that a tenth planet might travel in an orbit between Mercury and the sun. This was first proposed over 100 years ago by French astronomer Urbain Jean Joseph Leverrier, who predicted the discovery of Neptune. He based his prediction of the tenth planet on irregularities in Mercury's orbit and named the unknown planet Vulcan after the Roman god of fire because of its nearness to the sun.

From time to time, astronomers sighted objects thought to be Vulcan but sightings were never confirmed and, after 1910 when Einstein accounted for Mercury's irregular orbit with his general theory of relativity, most astronomers abandoned the search for Vulcan. Recently, however, the Vulcan theory was revived, partially on reports made by Henry C. Courten, Dowling College, and Grumman Aerospace

Corp., Long Island, N. Y. On photographs made in Mexico during the 1970 eclipse of the sun, Courten and associates detected a number of mysterious particles which appear to be in orbits very close to the Sun (Miami Herald, June 15, 1970).

All the objects are very faint, and there is always the danger of detecting artifacts in the photographic plate. However, Courten feels that at least seven of the objects are real. His conclusion is based on a sensitive computer analysis to crosscheck the positions of the objects on separate plates. Also, some of the objects were confirmed by another observer in North Carolina, and one was confirmed also by a third observer in Virginia.

Courten is not sure what the objects are, but three possibilities exist: They could be comets, small planetoids, or galactic debris which is being constantly swept up by the sun as it travels through the galaxy and which gradually spirals into the sun.

Chapter 3
MERCURY

INTRODUCTION

Mercury is so small and close to the sun that details are elusive through the telescope. Nevertheless, a rather interesting repertoire of oddities has accumulated over the last two centuries. There are transient bright spots, terminator irregularities, brightness variations of the dark side, and even rare hints of rings. Close-up photos from space probes reveal a cratered, moon-like planet with little, if any, atmosphere. Therefore, the spots, rings, and terminator irregularities may not have objective existence. Luminescence of the dark side, however, may be excited by solar radiation, as seems to be the case with the moon.

Mercury's craters are beyond the resolving power of terrestrial telescope, although in periods of good seeing some indistinct surface markings can be seen. A few observers, such as Percival Lowell, have claimed to have seen linear surface markings, bringing to mind the Venusian spoke system and Martian canals. Some subjective factor may be at work here.

A final peculiarity involves Mercury's poor fit with Bode's Law. Some astronomers have speculated that this orbital anomaly is related to oddities of the earth-moon system, and that both problems would be solved if the moon were once a planet occupying the space between Mercury and Venus. A close encounter with Mercury might have ejected the planet (moon-to-be) from its orbit, leading to its capture by earth.

• Spots, Rings, Terminator Anomalies, Luminescence

THE BLUNTED CUSP EFFECT AND TERMINATOR IRREGULARITIES

Cruikshank, Dale P.; *Strolling Astronomer*, 17:129-132, 1963.

Observations of Blunted or Truncated Cusp Tips. For many years observers have reported that the south cusp of the crescent Mercury occasionally appears blunted as opposed to the usually sharp north tip. Schroeter noted this in 1800 and 1801; and, according to Antoniadi, the effect was confirmed by Burton, Noble, Franks, Trouvelot, Denning, and Antoniadi. The Jarry-Desloges observers (V. and G. Fournier) also noted this aspect from time to time, as well as small terminator deformations. Of this phenomenon, Schiaparelli says, "I must confess ....that I have always seen the entire southern cusp very well though its light was weaker (than the north); and only once (June 5, 1882) I found that cusp so little luminous, that from time to time, during the least distinct moments of vision, one could suppose it truncated. That unequal splendor of the polar regions is the real cause of the apparent truncation of the southern cusp..." ("Sulla Rotazione di Mercurio", Astronomische Nachrichten, 123, 2944, pp. 242-50, 1890). Because of the brightness diminution at the cusps approximately according to the Lambert cosine law of diffuse reflection, usually imperfect seeing conditions, and general low surface brightness of the image, one expects to find the cusp tips on the limit of visibility. As Schiaparelli points out, however, the apparent truncation of the southern cusp is due to a darker region in that vicinity; and it is a purpose of this paper to determine whether observations in the ALPO files are commensurate with such an interpretation.

Clearly, the planet is a complete sphere and observations of a blunted cusp simply represent inability to see the whole of what is really there. A dark region at the south cusp would produce the observed effect as Schiaparelli suggests; but of the planispheres extant, only those of Antoniadi, de l'Isle, and Wegner show anything of the sort in the proper place. The Dollfus photographic chart reproduced in Figure 1 [not reproduced] shows only a dusky continuum in this region, and Schiaparelli's own map shows both polar zones quite light and devoid of detail. We adopt the Dollfus map as generally the most reliable. This chart was made from composite photographs taken in 1942 and 1944 by Lyot and Camichel, and shows no particular dark region sufficient to produce the observed effect. The Fourniers often noted dark markings at both cusps but omitted them from their 1920 planisphere. The drawings of B. Lyot also show dark polar markings. On the basis of these results and the fact that other planispheres show dark polar areas, we are led to conclude that the Dollfus chart is inadequate at the poles. Such is not an unreasonable assumption in view of the limits of the photographic process. This opinion does not imply that the central and limb regions of the Dollfus planisphere are incomplete, though I suspect that they are. We must mention that all of the planispheres discussed above (except Wegner's) are drawn for mean librations. Probably, then, we must look to the libratory regions for the cause of the blunted cusp.

Fifteen drawings in the ALPO files (9% of all drawings made with $k < 0.5$) show a blunted south cusp tip. At least one of these may be omitted because of bad seeing. Of the remainder, eight were made by Gary Wegner, three by C. L. Johnson, two by C. J. Smith, and one by W. H. Haas. Haas' drawing shows the north cusp rounded to the same slight degree as the south and probably indicates imperfect observing conditions. One of Wegner's drawings is similar. Eight of the observations were made at an eastern apparition and six at a western so that no particular correlation is noted with respect to morning and evening apparitions. Figure 2 shows some representative views of the truncated cusps. These copies show only the limb and terminator; shadings and details are omitted.

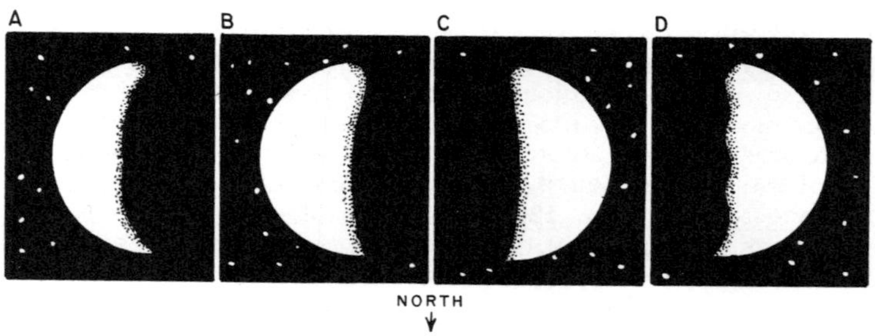

Mercury's blunted cusps

Since the cusp is not seen blunted in a large fraction of the observations on hand, we may assume that the dark region near the presumed south pole is a particular marking with definite boundaries. Then there should be a unique combination of latitude and longitude librations which puts the dark area in just the right place to give the observed illusion. Both librations were computed for the dates when blunted cusp observations were made, and the results are found in Table 1 [not reproduced] It will be noted that there is a strong preference for positive librations (east limb exposed) when Mercury is at eastern apparitions and for negative librations (west limb exposed) when at western apparitions. This result suggests that a large distinct dark area on the sun-averted hemisphere is responsible for the appearance of the blunted cusp, one extremity at the positively librated eastern hemisphere (eastern apparitions) and one at the negatively librated western hemisphere (western apparitions).

. . . .

THE LUMINOUS SPOT ON MERCURY IN TRANSIT

Jenkins, B. G.; *Royal Astronomical Society, Monthly Notices,* 38:337-340, 1878.

....

In the transit of 1799, which occurred in May, Schroter and Harding at Lilienthal, and Kohler at Dresden, saw a small luminous spot on the dark disk. The spot was not stationary, for Harding saw it change its position, and later in the day Schroter saw it sometimes on one part of the disk, sometimes on another. Other observers saw, not one, but two small spots of a greyish colour.

In the transit of May 1832 Professor Moll observed a spot, the periphery of which was not well defined, but was always situated in the same position, a little south of the centre of the planet, and proceding the centre.

Turning now to the November transits for further evidence of the luminous spot, we find that the transit of 1835 was not visible in this part of the world, and in the three previous ones of which we have some particulars, viz., those of 1730, 1780, and 1802, no mention is made of a luminous spot; but, as will be seen, it does not follow that it did not exist because it was not observed.

In November 1848 one spot was observed of a greyish colour, shading off on all sides from the centre.

In November 1861 Mr. J. W. Jeans saw a slight ashy light on the eastern limit or following the centre of the planet on its way to perihelion.

In November 1868 Mr. Huggins saw a point of light nearly in the centre which had no sensible diameter, but appeared as a luminous point. It was near the centre, a little south, and following the centre. It remained in the same position during the transit.

Mr. Browning, observing this transit, saw near the centre of the planet two bright grey spots close together, both a little south of the centre, one more conspicuous than the other.

So much for the luminous spot. Let us turn to the rings which have been observed round the planet during a transit.

In May 1707 the late Assistant of the then Astronomer Royal saw Mercury in transit encompassed by a thick haze or atmosphere.

In the transit of May 1753 a ring was observed round the planet; also in that of May 1786.

In May 1799 a dark or nebulous ring was observed, the tinge being of a violent hue and strongest near the planet. Schroter also saw the ring, the light of which he says was scarcely brighter than the surface of the Sun, but of another colour. Harding did not see it at first, but he did afterwards when he used a lighter shade.

In May 1832 the planet was seen at the Royal Observatory surrounded by a dusky tinge. In this transit also the ring has been described as of "a violet hue, the colour being strongest near the planet."

This is the sum of our knowledge about the rings round the planet in the May transits, and from it the natural inference appears to be that when the planet is in transit near its furthest distance from the Sun, it is generally surrounded by a dark nebulous ring.

The first November transit in which we have mention of a ring is that of 1736. Professor Moll says Plantade saw the reverse of what he saw in May 1832; so that the ring must have been a bright one.

In 1789 and 1802 rings---probably bright ones---were observed. The transit of 1835 was not visible in Europe; that of 1848 was, however, and many observers saw a bright ring, although Mr. Breen says he did not see any.

In November 1868 Mr. Huggins saw the planet in transit surrounded by an aureola of light, a little brighter than that of the Sun's disk.

Here the weight of evidence appears to be in favour of a bright ring round the planet when near the Sun, which would probably be the case. It should be stated, however, that Prosperin calls the ring of May 1786 a luminous one, and Ljungberg that of November 1802 a dark one; but it must be remembered that the ring of the transit in 1799, which there can be no doubt was dark, nebulous, and of a violet hue, is described by Schroter as scarcely brighter than the surface of the Sun, and of another colour.

Collecting these results and comparing them, it will, I believe, be found:

1st. That in the May transits, when Mercury is near its aphelion, the luminous spot is in advance of the planet, preceding the centre; in the November transits, when Mercury is near its perihelion, the luminous spot follows the planet.

2nd. The luminous spot has never been seen at the centre, but always south of it, and therefore cannot be due to diffraction, as Professor Powell suggests.

3rd. Sometimes in the same transit two spots have been seen close together, where shortly before only one was observed.

4th. In the May transits the rings round the planet were dark or nebulous and of a violet tinge; in the November transits they were bright.

5th. If we take the two transits which have received the most careful observation, the May transit in 1832, observed by Professor Moll, and the November transit in 1868, observed by Dr. Huggins, we find the contrast very great and very typical: in the one case a diffused spot preceding the centre, with dark ring surrounding the planet, in the other a sharply defined spot following the centre, with bright ring surrounding the planet.

6th. From all which I conclude that on May 6 next, when Mercury will be at almost its furthest distance from the Sun, the luminous spot will not be a point of light having no sensible diameter, but rather diffused, with a periphery not well defined, gradually sinking from a greyish-white to the dark colour of the disk, situated a little in advance of the centre and to the south of it, and that the planet will be surrounded by a dark nebulous ring, not a bright one.

VISIBILITY OF THE DARK SIDE OF MERCURY

Brenner, Leo; *British Astronomical Association, Journal,* 6:387, 1896.

Observing Mercury on the 18th May, between 22^h and 23^h G. M. T., I was astonished to see not only spots (as shown by the enclosed sketch) but even the dark side surrounded by an aureole, just like the appearance of Venus in July 1895. Fearing to be the victim of an optical illusion, I tried various eye-pieces (powers 146, 196, 242, 310, 410), changed the position of the planet in the field and shook the telescope, but both phenomena remained unchanged (respectively dancing with the illuminated disk in the same manner), so that there remained no doubt. Besides, after having made the drawing (at 22-1/2^h) I called Mrs. Manora, who believed for the first moment that it was Venus, as the appearance was so similar. She pronounced the dark side and the aureole to be very conspicuous objects, saying that she saw them at the first look, whilst she saw the spots on the illuminated disk later. The dark side was darker than the sky, just as I (with one single exception) have always found it in the case of Venus, so that I share now M. Flammarion's views on the explanation of this strange phenomenon. In every case I recommend observations of this kind to our Members, for, it may be, that it is just the present moment that is particularly favourable for the visibility of the dark side of Mercury. I will continue these observations, together with those of Mars, every fine forenoon. It seems strange that nobody else has made a similar observation up to date.

POSSIBLE LUMINESCENCE EFFECTS ON MERCURY

Cruikshank, Dale P.; *Nature*, 209:701, 1966.

A growing body of evidence leads to the conclusion that luminescence effects occur on the lunar surface. Large intensity fluctuations on the Moon have been observed photographically by Kopal and Rackham,[1] spectroscopically by Spinrad[2], and photometrically and polarimetrically by Gehrels, et al.[3] If the luminescence phenomena are induced by the solar wind (see Nash[4]), we might do well to look to other bodies in the solar system, which are similarly unprotected by an atmosphere, for evidence of luminescence.

Fluctuations in the intensity of portions of the surface of the planet Mercury were recorded by Antoniadi[5-7], who regarded the apparent obscuration of dark areas as the result of variable cloud cover. From his extensive observations with the 33-in. Meudon refractor, Antoniadi concluded that the density and extent of the 'clouds' varied with the changing distance of Mercury from the Sun, probably reaching maximum at perihelion and minimum after aphelion.

The newly derived rotation period of Mercury reported by Pettengill and Dyce[8] casts some doubt on the long-term changes reported by Antoniadi because his conclusions were based on the assumed 88-day period. Antoniadi[5] also noted, however, that changes occur on Mercury in a few tens of hours, and these cannot be simply explained in terms of the shorter rotation period. Other observers have also reported short-period changes on Mercury[9].

In the interval 1958-64, I made an extensive series of observations of Mercury with telescopes up to 82 in. in aperture. Visual studies of Mercury are difficult owing to the proximity of the planet to the Sun, but my observations generally confirm that variations in the intensity of certain of Mercury's markings can occur in a matter of several hours or days.

Fig. 1 shows four drawings of Mercury made in a nine-day period in 1963. Drawing A was made with a 36-in. reflector and the remainder with a 12-in. reflector, all in fairly good seeing conditions. Between the first and last drawings in Fig. 1, a darkening at the south cusp is apparent. Later drawings made as the crescent waned show the south cusp darkening further until it was the darkest region on the planet. The apparent change in the interval of three days between A and B may be related in part to the higher resolving power of the 36-in. telescope used for A. [Fig. 1 omitted]

The great similarities between the Moon and Mercury that have been pointed out by many investigators and the probable luminescence phenomena on the Moon may make it possible to interpret changes on Mercury as the effects of luminescence of the surface materials.

1 Kopal, Z., and Rackham, T., Icarus, 2, 481 (1963).
2 Spinrad, H., Icarus, 3, 500 (1964).
3 Gehreis, T., Coffeen, T., and Owings, D., Astron. J., 69, 826 (1964).
4 Nash, D. B., Trans. Amer. Geophys. Union, 48, 131 (1965) (abstract).
5 Antoniadi, E. M., La Planete Mercure et la Rotation des Satellites (Gauthier-Villars, Paris, 1934).
6 Antoniadi, E. M., J. Brit. Astron. Assoc., 45, 256 (1935).
7 Antoniadi, E. M., J. Brit. Astron. Assoc., 45, 301 (1935).
8 Pettengill, G. H., and Dyce, R. B., paper presented at American Geophysical Union, April, 1965 (see Sky and Telescope, 29, 339; 1965).
9 McEwen, H., J. Brit. Astron. Assoc., 45, 240 (1935).

MERCURY: ANOMALOUS ABSENCE FROM THE 3.4-MILLIMETER RADIO EMISSION OF VARIATION WITH PHASE

Epstein, E. Eugene; *Science*, 151:445-447, 1966.

Abstract. Radio observations of Mercury at 3.4 millimeters from July to October 1965 showed, contrary to expectation, brightness temperatures of only about 200°K, even when major fractions of Mercury's illuminated hemisphere were observed. There was no significant variation with phase.

The dark-side brightness temperature of 220° ± 35°K recorded at 3.4 mm (88 Ghz) during the April 1965 inferior conjunction of Mercury, and the brightness temperatures recorded at longer wavelengths, including the preliminary results at 8 mm, led us to expect great variation with phase in the 3-mm emission of Mercury. This expectation was based on the assumption that Mercury's surface layers behave as do those of Moon. [If Moon were moved to Mercury's distance from Sun, its 3-mm emission would show variation with phase of approximately ± 200°K about a mean temperature of $\approx$ 350°K.]

During observations of Mercury from 16 July through 17 October

1965 at 3.4 mm with the 15-foot (4.57-m) antenna of the Space Radio Systems Facility of Aerospace Corporation, no significant variation with phase was recorded, even though the observations covered almost a complex revolution of Mercury.

. . . .

The phase curve of Mercury appears in Fig. 2 [not reproduced]. The most significant features are the brightness temperature of only ≈ 200°K that was recorded when major fractions of the illuminated hemisphere were visible and the apparent absence of any strong variation with phase. Both features were completely unexpected. Brightness temperatures of ≈ 500°K were recorded on 2 days when least expected, that is, near inferior conjunction. None of the parameters of observation were unusual on these 2 days, 5 and 20 August. Subsequent observations of the positions of Mercury on 5 and 20 August indicated no time-constant background radio sources that could have caused spuriously high Mercury temperatures. Moreover, subsequent observations in a range of directions and over the same range of angular separations from Sun (≈ 8°) as that of Mercury on 20 August indicated no spurious enhancements of signal caused by Sun. No similar check was made for 5 August because Mercury was then ≈ 19° from Sun.

I can explain failure to record temperatures equal to the expected values (~ 500°K), well away from the time of inferior conjunction, only by making the unlikely assumption that I had unusually large (~ 2') undetected pointing errors, and that these errors decreased near inferior conjunction by just the right amount to cancel out any variation with phase. However, observations of Jupiter and Venus, made with the same procedures and on the same days as many of the Mercury observations, tend to disprove this explanation. The Venus and Jupiter brightness temperatures are internally consistent and are in good agreement with the findings of many other observers, thereby indicating absence of any large systematic errors in my equipment and observational procedures.

One may explain these anomalous results by suggesting that the emissivity of the surface layers of Mercury varies inversely with physical temperature. But, if this were so, the reflectivity would so increase with temperature that the longer-wavelength emission (arising from greater depths than the 3-mm emission) would not yield the high brightness temperatures actually observed away from inferior conjunction. I know no material that exhibits low emissivity at 3 mm only, but perhaps droplets about 3 mm in diameter of molten metal (that thus yield low emissivity and only at radio wavelengths ≲ 3 mm) develop in the surface layers on the hot side, and then on the cold side freeze into a material with normal emissivity; if so, observations 1 and 2 mm also should give an anomalous phase curve. It is difficult to imagine an atmosphere that would strongly absorb the 3-mm emission from the surface, but not the infrared emission, and also remain cold when the surface was hot. Independent confirmation of these results at 3 mm and at nearby wavelengths is obviously desirable.

• Surface Markings and Symmetry

MARKINGS ON MERCURY AND VENUS

Wesley, W. H.; *Knowledge*, 4:228, 1907.

In 1896 Professor Lowell published in the Monthly Notices (vol. lvii., p. 148) a series of drawings of Mercury, showing extremely curious markings on the planet, which he speaks of as much more distinct than the Martian canals, and, in fact, quite easy to see. They appear to have been observed when the planet's surface was not more than half illuminated, and the drawings show them as linear in character, though otherwise they bear little resemblance to the canals of Mars. I give as an example Professor Lowell's drawing of September 23, 1896, in which the markings appear mostly as stripes projecting from the terminator or running parallel with it. There is also a slight shading off of the cusps.

Professor Lowell promised further communications on this planet, but I cannot find any later observations, his work having been recently mostly devoted to Mars.

He also published in the same number of the Monthly Notices some drawings of Venus, showing a number of singular markings like the spokes of a wheel, but in 1902 he wrote in the Astronomische Nachrichten (No. 3, 823) that he had found reason to believe that these markings were due to an optical illusion. I do not know whether he still considers the markings on Mercury as objective, but those which he drew in 1896 do not seem to me at all like the canals of Mars, and, as far as I know, they have not been confirmed by other observers.

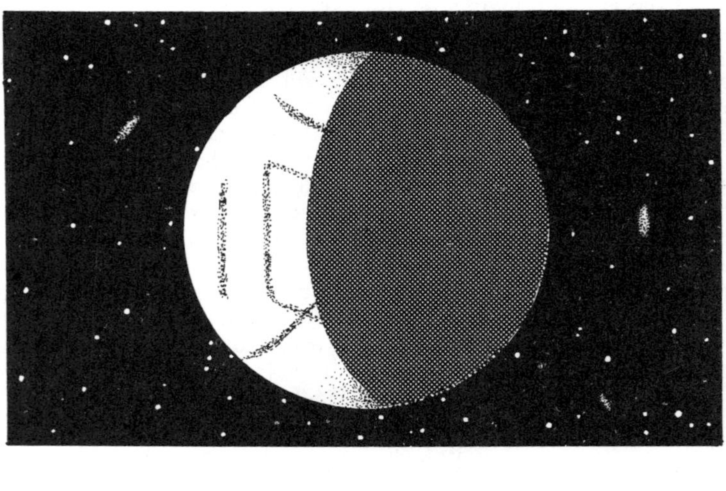

Surface markings seen on Mercury through the telescope

MERCURY: MORE SURPRISES IN THE SECOND ASSESSMENT

Metz, William D.; *Science*, 185:132, 1974.

Even more striking, Mercury, like the moon and Mars, appears to have evolved asymmetrically. Rough and heavily cratered crust, thought to be the primordial surface, seems to cover half the planet, while smoother plains seem to cover the rest. Why three of the five bodies studied among the inner planets have such modal asymmetry is perhaps the greatest puzzle of all.

....

• Mercury's Magnetic Field

MERCURY: MORE SURPRISES IN THE SECOND ASSESSMENT

Metz, William D.; *Science*, 185:132, 1974.

The Mariner 10 mission to Mercury produced dazzlingly clear pictures of what seemed from the earth only a fuzzy sphere. The early assessments---that Mercury is moon-like in appearance but has a trace atmosphere and evidence of an intrinsic magnetic field---have been substantiated. Now, after more study, the Mariner 10 investigators are able to quantify their characterizations of the innermost planet. They are finding that the magnetic field does not seem centered on the planet, that high energy electrons and protons are accelerated in some cyclic fashion that is still mysterious, and that Mercury underwent a period of global compression very early in its history.

Because Mercury rotates slowly (once in 58.6 days) and emits no radio emissions that can be detected from the earth, the early evidence for a magnetic field was surprising. The data still do not force one to conclude that there is an intrinsic magnetic field, but the case is strong. Mariner 10 found that the solar wind passing Mercury formed a detached bow shock. If it is not the result of a complicated process that induces a magnetic field around the planet, then Mercury has a significant intrinsic field. That field is apparently not tilted more than 10° away from the pole, but it seems to be offset by 47 percent of the radius of the planet. Perhaps it is the remnant of an extinct dynamo.

Since no magnetic field was expected at Mercury, particles were not expected to be accelerated to high energies by the interaction of the solar wind with the planet. But electrons and protons with energies above 100 kev were found. Furthermore, the large fluxes of electrons showed 6-second fluctuations that were at times accompanied by similar fluctuations of protons. Such well-defined structure was certainly not expected in the streams of high energy particles, and it will probably be quite difficult to find a physical mechanism that will accelerate particles with opposite charges simultaneously.

. . . .

MAGNETIC FIELD OBSERVATIONS NEAR MERCURY. . . . .

Ness, N. F., et al; *Science*, 185:151-160, 1974.

Abstract. Results are presented from a preliminary analysis of data obtained near Mercury on 29 March 1974 by the NASA-GSFC magnetic field experiment on Mariner 10. Rather unexpectedly, a very well-developed, detached bow shock wave, which develops as the super-Alfvenic solar wind interacts with the planet, has been observed. In addition, a magnetosphere-like region, with maximum field strength of 98 gammas at closest approach (704 kilometers altitude), has been observed, contained within boundaries similar to the terrestrial magnetopause. The obstacle deflecting the solar wind flow is global in size, but the origin of the enhanced magnetic field has not yet been uniquely established. The field may be intrinsic to the planet and distorted by interaction with the solar wind. It may also be associated with a complex induction process whereby the planetary interior-atmosphere-ionosphere interacts with the solar wind flow to generate the observed field by a dynamo action. The complete body of data favors the preliminary conclusion that Mercury has an intrinsic magnetic field. If this is correct, it represents a major scientific discovery in planetary magnetism and will have considerable impact on studies of the origin of the solar system.

• Strange Emissions and a Nonexistent Moon

MERCURY'S MOON THAT WASN'T

Anonymous; *New Scientist*, 63:602, 1974.

Now it can---or at any rate, will---be told: the story of how Mercury inadvertently and very temporarily acquired a moon thanks to an injudicious news leak; and how in the process some imaginary scales fell from the eyes of astronomers everywhere. The story comes by courtesy of Michael McElroy, professor at Harvard University's Center for Earth and Planetary Physics and one of the principals in the unfolding drama.

McElroy was a member of the scientific team that had a couple of instruments capable of observing in the extreme ultra-violet carried aboard Mariner 10, the unmanned spacecraft that flew past Mercury on 29 March. The instruments were there to measure any atmosphere the planet might have---and indeed they did so, revealing an extremely tenuous atmosphere consisting mainly of helium. But two days before the Friday morning flyby, one of the instruments began registering bright emissions at short uv wavelengths that had, in McElroy's words, "no right to be there", while the spacecraft was pointing towards the dark side of the planet. The next day, they were gone.

The Friday morning was occupied with the serious stuff of atmospheric observations; and then after the flyby, when the instruments were looking back at Mercury, the mysterious emissions---at wavelengths shorter than 1000 angstroms---were seen again. Moreover, the "object"---if that was what McElroy were indeed seeing---detached itself from the planet. The astronomers' first thought was that they were looking at a star; but they had of course seen it while looking in two quite different directions, and every astronomer knows that very short ultra-violet radiation can penetrate only a little way through the interstellar medium, suggesting that the object must be close. An attractive alternative to a star was that the object was a moon of Mercury that no-one had dreamt existed.

A hectic Friday night was spent at the Jet Propulsion Laboratory in Pasadena without the benefit of computers trying to work out the object's motion with respect to Mercury. When a figure of four kilometres a second began to come out, a speed consistent with that of a moon, the excitement mounted and the JPL managers were called in. They turned the then-dying spacecraft over full time to the uv team; and then everyone started worrying about a press conference scheduled for later that Saturday morning. Should the press be told of the astronomers' suspicions? Or should they keep quiet for a few more hours, watching for the object to turn back around the planet as it would if it were indeed a moon, and so settling the matter for sure?

In the event it turned out that the press, baying for news, had already been tipped off. So with all the usual words of scientific caution, McElroy and his colleagues told what they knew. Some papers---the bigger, more respectable ones---played it straight; many others ran excited stories about Mercury's new moon. And the "moon" itself? It headed straight on out from Mercury, and was eventually identified as a hot star, Crater 31. What the original emissions were from, the ones spotted on the approach to the planet, remains a mystery.

So ends the story of Mercury's moon. But so may begin a new chapter in astronomy, for the penetration of ultra-violet radiation from Crater 31 to Mariner 10 belies the belief that hydrogen in the interstellar medium absorbs these wavelengths and so closes them to observational astronomy. Already, for instance, the nearby Gum nebula has turned out to be quite a strong emitter in the extreme uv, and spreads across 140° of the night sky at 540 angstroms. Mercury may not have a moon, but astronomers may have a new window through which to observe the heavens.

• Mercury's Orbital Anomalies

THE MOON MAY BE A FORMER PLANET

Bailey, J. Martyn; *Nature*, 223:251-252, 1969.

There are regularities in the orbits of the planets which are not predicted by the laws of Kepler and Newton and which are related to the mechanism of formation of the solar system.

The orbital velocities of the planets lie on a smooth curve (Fig. 1). Beginning at about orbit 7 (Saturn) and proceeding inwards through Jupiter, the asteroids, Mars, Earth and Venus, the orbital velocities lie on a straight line. The point A corresponding to the orbital velocity of Mercury at orbit 1, however, lies far above this line. The orbit of Mercury, in addition, has an unusually large eccentricity of 0.206.

The fit is better if Mercury is assigned to orbit 0 (point B in Fig. 1). This implies, however, that a planet is missing from orbit 1. The predicted period $\underline{P}$ and the orbital radius $\underline{R}$ for such a planet is 0.41 yr and 0.55 AU respectively, calculated from the value of $\underline{R/P}$ obtained from the orbital velocity curve together with Kepler's relationship $P^2 = R^3$.

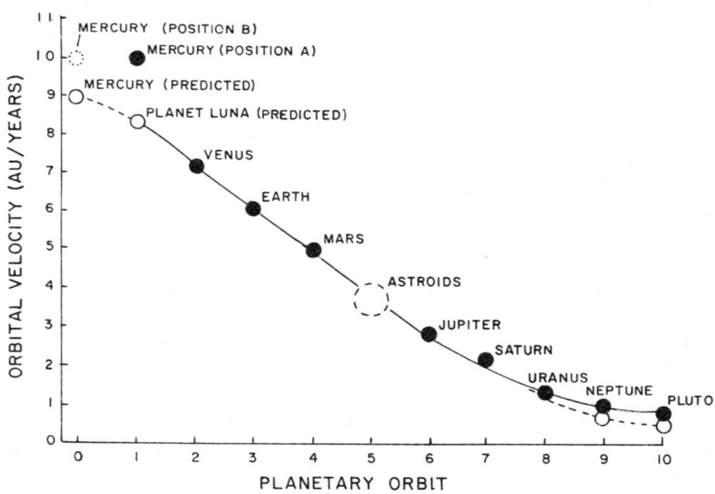

Orbital velocities of the planets

The Bode-Titius law, when written in the form $\underline{R} = 0.4 + 0.3(2)^{n-2}$ where $\underline{R}$ is the radius and $\underline{n}$ is the number of the planetary orbit, correctly predicts the orbits of the first eight planets except for Mercury. Used in this form when $\underline{n} = 1$, the law also predicts a planet at radius 0.55 AU. The corresponding value for Mercury ($\underline{n} = 0$) is then 0.475 AU.

The periods and orbital radii calculated for Mercury and the missing planet ("planet Luna") are given in Table 1 with those for Venus and Earth. The orbit of the missing planet is unstable, however. The predicted period 0.408 yr) is 5/4 times that calculated for Mercury (0.327 yr) and 2/3 of that of Venus (0.617 yr). There thus exists a condition of double resonance between Luna, Mercury and Venus, which would tend to induce eccentricity in the orbit of Luna. (The almost circular present orbit of Venus ($\underline{e} = 0.006$) may reflect transfer of "eccentricity" to Luna during this resonance phase.) Because of the relative closeness in the orbits of Luna and Mercury (7,000,000 miles), close encounters with Mercury may have occurred.

....

The capture of Planet Luna would explain a number of anomalies in the Earth-Moon system. The closeness of the two bodies and the relatively large mass of the satellite are unlike those of other planetary satellites. In the Jupiter system, for example, the mass ratio is of the order 1:10,000. The Earth-Moon system is best described as a double planet. But the present rates of tidal friction, if extended into the past, predict a maximum age for the system which is less than half that of the Earth itself. This has been interpreted by many, therefore, to mean that the Moon was derived relatively recently, either by fission from the Earth or by capture. The much greater inclination (23-1/2°) of the Earth's spin axis to the plane of the ecliptic than that of the orbit of the Moon (5°) supports origin by capture rather than by fission. It has also been noted that origin by fission would require that the primitive Earth had an unusually high rotation rate compared with the other planets.

The capture of Planet Luna in a direct orbit by the process described can be compatible with both the observed rotational angular momentum of the Earth-Moon system, and the time-scale proposed on the basis of tidal friction rates. The orbit of the former Planet Luna and the pre-encounter orbit for Mercury are in accord with the Bode-Titius law. The hypothesis accounts for the deviation in the present orbital velocity of Mercury, and predicts an eccentricity for the orbit of Mercury which is close to the observed value. It is concluded therefore that the Moon may be the former Planet Luna which once occupied an orbit between those of Mercury and Venus at a mean distance of 51,000,000 miles from the Sun.

A DYNAMICAL INVESTIGATION OF THE CONJECTURE THAT MERCURY IS AN ESCAPED SATELLITE OF VENUS

Van Flandern, Thomas C., and Harrington, Robert S.; *Icarus*, 28:435-440, 1976.

Abstract. The possibility that Mercury might once have been a satellite of Venus, suggested by a number of anomalies, is investigated by a series of numerical computer experiments. Tidal interaction between Mercury and Venus would result in the escape of Mercury into a solar orbit. Only two escape orbits are possible, one exterior and one interior to the Venus orbit. For the interior orbit, subsequent encounters are sufficiently distant to avoid recapture or large perturbations. The perihelion distance of Mercury tends to decrease, while the orientation of perihelion librates for the first few thousand revolutions. If dynamical evolution or nonconservative forces were large enough in the early solar system, the present semimajor axes could have resulted. The theoretical minimum quadrupole moment of the inclined rotating Sun would rotate the orbital planes out of coplanarity. Secular perturbations by the other planets would evolve the eccentricity and inclination of Mercury's orbit through a range of possible configurations, including the present orbit. Thus the conjecture that Mercury is an escaped satellite of Venus remains viable, and is rendered more attractive by our failure to disprove it dynamically.

TARDY MERCURY HURRIED

Hoffleit, Dorrit; *Sky and Telescope*, 1:17, August 1942.

In the summer of 1940, the U.S. Naval Observatory issued a general request for observations of the transit of Mercury which was to take place on November 11-12th of that year. Nearly 200 observers, may of them amateurs, distributed between terrestrial longitudes 175° E. and 157° W., responded. The results of their observations are discussed by G. M. Clemence and G. C. Whittaker in a recent Publication of the U.S. Naval Observatory. They find that the transit happened 36 seconds later than expected and lasted 18 seconds less. The tardiness is assumed to be caused by irregularities in the rotation of the earth, which lead to errors in the reckoning of astronomical time. The accuracy of the observed time deviation in this instance could be checked by observations of the moon, which yielded a value of the "fluctuation" of 35.3 seconds, in good agreement with the Mercury transit value.

The shortness of the duration of the transit is not wholly accounted for. Observational difficulties may readily account for part of Mercury's apparent haste, but the discordance between observation and theory is too large to be entirely attributed to the cause.

Chapter 4

VENUS

INTRODUCTION

The crescent of Venus hanging bright in the evening sky is like no other solar system planet. The planet's thick, hot atmosphere cloaks a surface seen only fleetingly by space probes and, perhaps, by a rare telescopic observer. Despite its atmospheric shield, Venus presents many enigmas. Most of these mysteries seem to involve optical and subjective factors: the Ashen Light, the Annular Phase, the Horns or Cusp Caps, the Radial Spoke System, and other phenomena seen---or thought to be seen---through the telescope. Then there was Neith, the lost satellite of Venus, seen by several astronomers during the last century. What was this transitory companion of our sister planet? The accepted value for the planet's period of rotation has also had a curious history. It has been set at about 24 hours, 225 days, and more recently, 237 days. This uncertainty reflects the near-invisibility of surface features. The current 237-day period puts the axial rotation of Venus in synchronism with the earth's orbital motion around the sun. Whenever the two planets are closest, Venus always points the same side toward us. (Why?)

THE FIGURE OF VENUS

Both Venus and Mercury go through moon-like phases, as befits planets closer to the sun than the earth. Venus, however, does not present the telescope user with a uniform, mathematically precise phase. The horns are often much brighter than the rest of the crescent. Sometimes, the horns project farther than the laws of optics say they should. Rarely, the horns extend all the way around the planet, producing a ring of light or "annular phase." On occasion, the dark portion of the planet seems to glow unnaturally. This is the so-called ashen light. Explanations fall into three categories:

1. The atmosphere of Venus is hot and dense, providing conditions for unusual optical refraction of sunlight.
2. Electrical processes are at work in the atmosphere. Visible effects, perhaps akin to the terrestrial auroras, are possible.
3. Some of the purported effects are merely subjective.

• Cusp Caps and Horns

A REVIEW OF SOME ALPO VENUS STUDIES

Cruikshank, Dale P.; *Strolling Astronomer*, 17:202-208, 1963.

5. Cusp Caps. Observers of Venus often note that one or both cusps are abnormally bright compared to the remainder of the disk. James Bartlett (Str. A., 12, 4-6, 1958, p. 43) contributed a very worthwhile paper on his statistical analysis of his own observations, those of Owen Ranck, and those of a group of ALPO observers lumped together. For the present time we will call these anomalous brightenings "cusp caps", though this term suggests a physical interpretation that is unproved. Bartlett showed that one or both cusp caps were visible in 54% of the observations (477 out of 830 observations), that the south cap alone was visible 11% of the time, the north 7%, and both caps simultaneously 35%. These figures correspond closely to those obtained when Bartlett's 221 observations or Ranck's 158 observations are considered alone. I believe that we can attach considerable significance to these values, particularly because both Bartlett and Ranck had such a long and continuous series of observations. This is prerequisite to a statistical study of features like the cusp caps.

Bartlett also found that the cusp caps appeared at virtually all phases of the planet, from which he concluded that seasonal effects on the planet are not important in governing the appearance of the caps, since all positions in the orbit were represented by cusp cap observations. This last statement does not follow, however, since phase is not uniquely related to Venus' heliocentric longitude alone, but also to the difference in heliocentric longitudes of Venus and the Earth.

Bartlett's work indicates that the south cap was most often the larger of the two. Some reports of details seen in the cusp caps were mentioned, but are probably not reliable enough for useful analysis. Dusky cusps were also discussed, but these are reported far less often than bright cusps.

The very fact that bright cusps occur indicates that the Lambert Law of diffuse reflection from a sphere is not strictly applicable to Venus, as we also saw in section 3.

Bartlett appears to favor the interpretation that bright cusp caps represent the actual poles of rotation of the planet, and that the inclination of the polar axis to the orbit plane must be small since the caps were seen at a wide variety of heliocentric longitudes. This does not follow, since we can only say (on the basis of a "polar cap" sort of interpretation) that the angle of inclination of the axis to the orbit plane is less than the angular extent of the larger cusp. This is the case with Mars (and with the Earth, correcting for the asymmetry of the caps). In fact, it would be of some interest for ALPO observers to make a continuous and consistently good series of observations of the angular extent of the cusp caps so that a statistical analysis could be made to determine the best value of this quantity. While it is not totally fair to judge Bartlett's 1958 interpretation in the light of 1962 observations, the recent infrared studies of Venus reported in many recent journals make it clear that the surface temperature of Venus precludes any sort of ices which could form the bright polar caps as required by Bartlett. P. Moore dismissed this idea some time ago (J. B. A. A., 65, 6, p. 232. 1955); and most atronomers, knowing that the surface temperatures on Venus are very appreciable by inference from atmospheric temperatures, rejected the idea of ice polar caps many years ago.

C. M. Pither (J. B. A. A., 73, 5, p. 197, 1963) has suggested on the basis of the frequent correspondence of cusp caps to the time of solar flares in 1961 that the caps are auroral in nature. The examples Pither gave, however, correspond to flares of class 1 or 1+ , which are not prone to produce terrestrial auroral activity. Furthermore, some of the flares that Pither cites were invisible from Venus at the time of the cusp cap observations (see Cruikshank, J. B. A. A., in press). While statistics such as Pither proposes are some interest and should be examined for ALPO and individual observers' records, Pither's own are not convincing to me. In fairness, we must admit that it is possible that auroral mechanisms in the Venus atmosphere may be triggered by flares which we ordinarily consider unimportant (1- to 2); the geometrical factor remains that flares must ordinarily occur within about 45° of the center of the solar disk as seen from the planet before intense magnetic storms are induced. Further, Venus does not seem to have an intense magnetic field by which to concentrate auroral activity toward the poles (Mariner II observations).

THE ORIGIN OF THE CYTHEREAN CUSP CAPS

Pither, C. M.; *British Astronomical Association, Journal*, 73:197-199, 1963.

Introduction. It is a well known fact that the Earth's aurorae have their origin on the Sun. Every once in a while there are tremendous outbursts on the photosphere of the Sun; these are called solar flares and it is the ionized particles from such flares which are directed to the Earth's magnetic poles, giving auroral displays. Venus is nearer the Sun than is the Earth, and on good grounds it is believed that Venus has a much stronger magnetic field than the Earth. Thus with a stronger magnetic field and closer proximity to the Sun, it is quite possible that aurorae occur in the Cytherean atmosphere. Active observers, and those who make systematic studies of Venus are well aware of the bright areas called cusp caps, which occur at the cusps of the planet, and it is hoped to show in this report, and others that will follow, that these cusp caps might well be a type of aurora, and that these too have their origin in the Sun.

. . . .

Report. There is an interval of some twenty to forty hours before a stream of particles leaving the Sun's photosphere arrives at the Earth. The distance of Venus from the Sun is approximately 67 million miles, this being some two-thirds the distance of the Earth from the Sun; thus ionized particles from a solar flare should arrive in the region of Venus after a delay of some fourteen to twenty-eight hours.

Not all the observations of cusp caps the writer received corresponded with flares, but by far the larger percentage did. For this report the observations are placed into one of two groups. The first, Group P, are the observations that were preceded by a flare. The second, Group N-P, contains the observations that did not appear to be connected with a flare. Summarising, the figures for each group are: Group P.---42. Group N-P.---9.

The total number of observations was 51; thus, as one can see, the proportion of observations falling in Group P is quite high.

The writer's original data sheets numbered some twenty sheets without an explanation of the report, and so would make this paper far too long; therefore only the basic details of the data sheets are given here.

Two scales are used in this report to describe the appearance of the cusp caps. The first is used to describe the intensity of the cap, i.e. its apparent brightness; o for bright caps, 1 for medium bright caps, and 2 for dull caps. The other scale indicates the size of the cap; L for large caps, M for medium, and S for small caps.

Group P Observations

Flares	Cusp Caps	Intensity	Size	Observer
1961 June 1^d 11^{h}05^m U.T.	June 2^d 06^h 30^m U.T.			
	N. & S. caps	o	L	J.H.R.
	2^d 04^h 00^m U.T.			
	N. cap	-	L	G.H.S.
1961 June 5^d 09^h 55^m U.T.	June 6^d 04^h 31^m U.T.			
	N. cap	o	M	C.M.P.
	6^d 06^h 40^m U.T.			
	N. & S. caps	o	M	J.H.R.
1961 June 18^d 07^h 42^m U.T.	June 19^d 04^h 04^m U.T.			
	S. cap	2	S	G.H.S.
	19^d 04^h 21^m U.T.			
	N. cap	o	L	C.M.P.
	19^d 07^h 10^m U.T.			
	N. cap	1	M	C.R.
1961 June 22^d 09^h 07^m U.T.	June 23^d 04^h 04^m U.T.			
	N. cap	-	S	G.H.S.
	S. cap	-	M	
	23^d 04^h 17^m U.T.			
	N. cap	o	S	C.M.P.

The writer believes it worth while to return to the cusp caps recorded on June 19. J. Hedley Robinson and the writer both agree that changes in the cusps of Venus can occur over fairly short periods of time; in the instance quoted there seems definitely to have been a change. On 1961 June 19, G. H. Smith was observing Venus at 04^h 04^m U.T., and recorded a small S. cap of intensity 2. At 04^h 21^m U.T., C. M. Pither (the writer) was observing; the S. cap had by then faded away, and in its place was a N. cap; this being large and bright, intensity o. By 07^h

10^m U.T., C. Raeburn was observing; he, like C.M.P., recorded a N. cap, but by now both the size and intensity had dropped and Raeburn recorded the cap as medium of intensity 1.

Group N-P Observations. On 1961 May 13, C.M.P. and G.H.S. were both observing; C.M.P. at $05^h\ 26^m$ U.T. and G.H.S. at $04^h\ 40^m$ U.T. On May 12 solar flares had been reported, occuring at such times that cusp caps should have been detected on Venus; but neither observer recorded caps. It could be that local conditions on Venus prevented such detection, or that the flare particles did not reach the vicinity of the planet.

Conclusion. During 1961 May and June, as can be seen, there are far more cases of correlation between cusp caps and solar flares than there are to the contrary. The writer believes that here is a good case for stating that the Cytherean cusp caps may be a type of aurora, which are the delayed effects produced in the atmosphere of Venus by solar flares. But at the same time it is clear that this report is not conclusive, and that further observations are essential before any firm conclusions can be drawn.

A REVIEW OF SOME ALPO VENUS STUDIES

Cruikshank, Dale P.; *Strolling Astronomer*, 17:202-208, 1963.

6. Cusp Bands. In addition to bright cusp caps, observers often report dusky bands or collars near the cusps which define or border the caps. P. Moore, (The Planet Venus, 1st ed., Macmillan, 1958, p. 53) suggests that the bands are simple contrast effects of the bright cusps, saying that his own observations "indicate that the collars can only be seen when the caps are unusually prominent". I suspect that many observers would disagree that they are seen only in the presence of bright cusps; I do. Because I have seen such dusky streaks with and without the adjacent bright cusp caps, I am led to believe that they are not different from the ordinary dusky bands which are often roughly perpendicular to the terminator. At the apparent poles, however, these bands are highly foreshortened, and this accounts for the usually narrow and rather dark appearance. This also may tell us why cusp bands or collars are often seen when no other features are visible (see Figure 16, Str. A., 16, 7-8, p. 179, 1962). Broad bands on the limit of visibility at the center of the disk will be darker and more easily visible near the poles, assuming that they are near the top of the visible portion of the atmosphere. That the dark bands lie at such a level is consistent with the fact that they are most prominent in the bluest regions of the spectrum. Observers might keep this in mind in planning individual programs of visual observations.

• The Annular Phase of Venus

THE RING OF LIGHT SURROUNDING VENUS

Keeler, J. E.; *Sidereal Messenger*, 1:292-294, 1883.

At the request of Dr. Hastings of the Johns Hopkins University, I send to the Messenger the following account of a singular appearance presented by the ring of light surrounding the planet Venus between the times of first and second contacts, which I observed during the transit of Dec. 5th, but for which I was unable to afford an explanation.

I used a telescope of 2-1/4 inches aperture and 4 feet focus, mounted on an equatorial stand in the open air, and having a positive eye-piece with power of 70, provided with a dark glass of a somewhat greenish tint. A mean time chronometer was placed in a convenient position for noting the time. As soon as first contact occurred, at $20^h\ 44^m\ 30^s$, I looked, for indications of the atmospheric ring around the planet, having been requested by Professor Langley to pay particular attention to the physical phenomena of the transit, but for the first three or four minutes could see nothing of it. Shortly afterwards, however, I caught a feeble glimmer of light, almost star-like in appearance, on the limb of the planet farthest from the Sun, which at $20^h\ 49^m$ presented the appearance of a curved streak of very faint silvery light extending for a short distance along the margin of the unimmersed portion of the planet's disc. The brightest part of this luminous arc was not directly opposite the Sun, but was situated about 20° to the west of a line joining the centres of the Sun and Venus. At the same time little horns of light, due perhaps to an optical illusion, appeared to rise from the cusps of the Sun at the margin of the planet, like the elevated rim of fluid which surrounds an immersed body through capillary action.

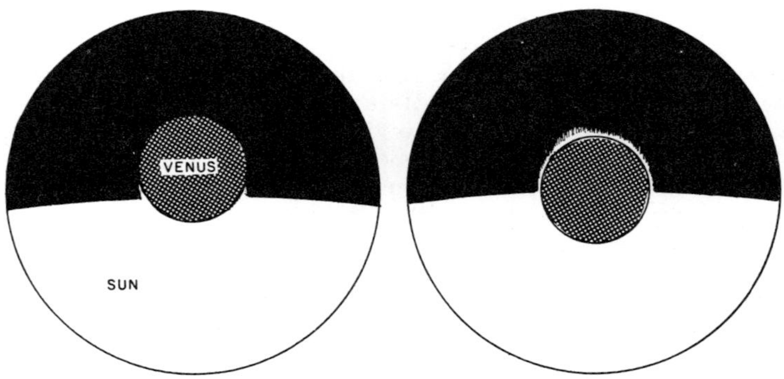

Annular phase of Venus. Left: 4.5 minutes after first contact. Right: 9.5 minutes after.

As the planet advanced, the arc of light gradually extended and brightened, until at $20^h\ 54^m$, or nearly half way between first and second contacts, the unimmersed portion was completely surrounded by a luminous ring. The light at the place on the margin when it was first noticed, however, much exceeded in brilliancy that of the adjacent portions, the brightest part extending along perhaps 30° of the planet's circumference, and on the western side the luminosity was more evident than on the eastern, when it was as yet barely discernible. The juncture of the luminous arc, first observed with the western cusp of the Sun, to which it lay nearest, occurred before the eastern edge became visible. The marginal patch of light now presented the appearance of a local brightening of a continuous ring of light surrounding the planet, and according to my impression at the time, lay without its contour, although thin clouds which had begun to gather, causing the image in the telescope to "boil," rendered a definite conclusion difficult. An independent drawing by Mr. Brashear with a reflector of about six-inch aperture, represents this spot of light as extending within the planet. At $20^h\ 57.5^m$ the appearance was still marked, and the ring of light quite brilliant all around the planet. After this my attention was withdrawn from it in preparing to observe the second contact through the fast thickening clouds. At emersion the sky was completely overcast, and observation was impossible.

The phenomenon I have described was also observed by Professor Langley with the large equatorial of the Observatory, temporarily reduced to six inches aperture, and by Mr. Brashear, of Pittsburg, with a Newtonian reflector. Prof. Langley estimated the centre of the brightest portion to be 30° from the line passing through the centres, toward the west, giving a position angle 10° greater than that recorded by me at the time. As the position angle of the planet on the Sun when half immersed was about 148°, the mean of these two independent estimates gives a position angle of 173° for the brightest part of the spot on the planet.

The appearance of Venus after the spot was first observed, and at $20^h\ 54^m$, when nearly half immersed, is represented in the accompanying drawing.

ANNULAR PHASE OF VENUS

Gibbs, Lewis R.; *Science*, 16:303-304, 1890.

An opportunity of observing an unusual, if not remarkable, phenomenon will soon occur; and I wish to call the attention of astronomers to it, as another opportunity will not present itself until after the lapse of eight years. This phenomenon may be conveniently called the annular phase of the planet Venus, though it be produced not be reflected light only, as in the ordinary phases of the moon, but partly also by the refracted light of the sun, which has passed through the planet's atmosphere. This phase I unexpectedly witnessed twenty-four years ago under the following circumstances:---

I desired to observe the prolongations of the cusps of the crescent of light, mentioned by several writers, and which I afterwards found had been observed by Madler in May, 1849, and used by him to obtain the amount of refraction in the atmosphere of Venus; but I had not then read his paper on the subject, and was unacquainted with his formulae.

It was well known, that, if Venus and the earth at any time occupied certain relative positions in their orbits, they would return very nearly to the same points, after an interval of eight years less two and a half days. It was also well known that Venus would transit the northern part of the sun during the forenoon of the 9th of December, 1874 (civil day at Greenwich), and would transit the southern part eight years less two and a half days later, or during the afternoon hours of the 6th of December, 1882. It was therefore evident that it would pass north of the sun, and very near it, eight years less two and a half days before the first of these transits, and would approach nearest to the sun about 2 p. m. (Greenwich time) on the 11th of December, 1866, least distance of centres being about 38' of arc. I therefore prepared to observe the planet on the forenoon of that day.

My observations were made in the open air, on the grounds of the College of Charleston, with a telescope presented to the college many years ago by William Lucas, Esq. This telescope is a refractor by Troughton & Simms, 5 feet focal length, 3-3/4 inches aperture, eye pieces used magnifying 70 and 120 diameters. I so placed my telescope that the apex of the north gable of the library building, 23 yards distant, screened its object glass from the rays of the sun; and the planet was easily found and distinctly seen above the roof of the library, least distance of nearest limbs about 22'. To my surprise, even atonishment, I saw not merely two cusps prolonged, but the whole circumference completely enlightened, the disk of the planet surrounded by a ring of light, broadest on the side nearest to the sun, narrower but quite bright on the opposite side. To have additional testimony to this fact, I immediately called to witness it Messrs. E. T. Frost and W. St. J. Jervey, two students in my astronomical class. They at once recognized the illuminated circumference, and said that it resembled in form the annular eclipse of the sun in October, 1865, which they had seen in this city in the preceding year. As said above, I was at this time unacquainted with Madler's observations and formulae, and, not having seen any intimation of the possibility of such a phenomenon, it took me wholly by surprise. I continued to watch the planet from 9 to 11 a. m., when the library building ceased to be available as a screen. This interval includes the instant of nearest approach of centres, which occurred about 9:30 a. m., Charleston mean time.

As far as I can learn, the only other persons who saw the phenomenon at that time were Professor C. S. Lyman of New Haven, Conn., and a few of his friends. In his equatorial of 9-inch aperture he saw the annulus or ring on the 10th completely formed; but the line of light on the side farthest from the sun was slender, faint, and only seen by glimpses. He saw it again on the 12th, but did not attempt to observe it on the 11th, the day of conjunction, when I saw it as a brilliant ring of light. He doubtless would have succeeded perfectly if he had abandoned the equatorial, which could not be screened, and used a more portable telescope, with some building as a screen.

In 1874 I watched the planet at intervals from the 30th of November to the 12th of December, the transit taking place on the night of the 8th and 9th, Charleston civil time. On the 2d of December I saw for the first time during this interval the distinct prolongation of the cusps, and watched their increase from day to day until the 8th, making eye-estimates of the number of degrees in the enlightened portion of the circumference, as I had not efficient means for making micrometer observations. On the 8th and the 9th I fully expected again to see the annular phase, but failed entirely to find the planet on both days. There were no clouds, at least not sufficient to entirely prevent observations, but there was a dense haze, and the region near the sun was strongly illuminated.

At this transit Mr. Lyman was more successful than myself, making good micrometer observations of the enlightened portion of the circumference, and seeing distinctly the illuminated ring on the 8th, the day before the transit. On the 9th he was, like myself, wholly unsuccessful in finding the planet, but on the following days continued his micrometer measures. The results of these observations he published in the American Journal of Science and Arts for January, 1875, with the amount of refraction in the atmosphere of Venus deduced from his observations, and also Madler's formulae by which it was deduced.

....

VENUS AS A LUMINOUS RING

Anonymous; *Knowledge*, 4:205, 1907.

Messrs. H. N. Russell and Z. Daniel describe their observations of this rare and interesting phenomenon, made at the last favourable conjunction in 1906, at the Princeton University Observatory, using the five-inch finder of the 23-inch refractor. On November 29, at 5h. 7m., G. M. T., the planet being only about 1° 49' from the sun's centre, it was possible when the air was steady to see the complete outline of the planet. On the side nearest the sun it was bright and easily visible, but on the opposite side it was very faint, and could be seen only for a few seconds at a time. When the complete circle was seen, the space within it always appeared a shade darker than that without, but it is possible that this was a subjective effect, as it was not noticed when the fainter part of the ring disappeared through bad seeing. The only other peculiarity noticed was a bright spot suspected several times in the bright part of the ring. From measurements of the extent of the crescent with a filar micrometer the length of the twilight arc in the atmosphere of Venus was determined to be about 60', which agrees fairly well with the value of 70' found from the observations made in 1898.

The ring phase of Venus may perhaps be visible again in 1914, if the atmospheric conditions are very favourable; after then there will be no further opportunity until 1972.

• The Ashen Light

ON THE VISIBILITY OF THE DARK SIDE OF VENUS

Schafarik, A.; *Report of the British Association, 1873*, 404–408.

It is well known that the unillumined side of the planet Venus has been sometimes seen shining with a faint grey light, like the dark side of the moon when illumined by the earth.

Schroter in 1806 thought he had made for the first time this remarkable observation; but it was found afterwards that Harding had made it almost simultaneously, and Olbers pointed out an old observation made by A. Mayer at Gryphiswald in 1759. Arago found a still older observation of the same kind made by Derham at a date not fixed, but certainly anterior to 1729, the date of publication of the French edition of his 'Astrotheology.'

Nevertheless this phenomenon is stated in the best text-books of astronomy to be one of the utmost rarity. Madler knows only two observers of it, the profoundly learned Humboldt only three, Arago only five; and even recently Dr. Winnecke, of Karlsruhe, believed that he was the only witness of that phenomenon in daylight since the time of A. Mayer; but under these particular circumstances it has been seen by eleven observers, and by five of them more than once.

It was known to me for a long time that there were on record far more observations of this phenomenon than is ordinarily supposed; and when, some years ago, I happened to be a witness of it myself, I undertook to collect all existing observations of it.

This I have now done; and as I have succeeded in collecting the surprising number of twenty-two observations, many of them repeated more than once, a short account of what I have found will perhaps be not uninteresting to astronomers.

1. The first observation recorded is that of William Derham, Canon of Windsor, referred to in his 'Astrotheology' as made in the perigeum of Venus, probably in bright twilight, when he saw the dark side of the planet shining with a dim reddish light. Arago, who mentions this observation, quotes from a French translation published in 1729. It would be interesting to know if this observation is found also in the first English edition published in 1714.

2. The second in order was Christian Kirch, first astronomer of the Royal Academy of Sciences at Berlin. He saw the phenomenon twice (June 7, 1721, and March 8, 1726), both times with moderate optical power and in bright twilight. He remarked that the bright crescent was apparently a part of a larger sphere than the faintly shining dark side. (Astronomische Nachrichten, No. 1586, vol. lxvii, p. 27.)

3. Third came Andreas Mayer, Professor of Mathematics in the Gryphiswald University, who, on October 20, 1759, observed Venus, culminating only 10° from the sun, with an unachromatic transit-instrument of only 1-1/2-inch aperture, and saw the whole disk "like the crescent moon which reflects the light of the earth." (Observationes Veneris Gryphiswaldenses, 1762, p. 19. Schroter, Beobachtungen des grossen Cometen von 1807, Appendix, p. 74.)

4. The fourth witness is Sir William Herschel, who about 1790 several times saw a part of the limb of the dark side in a faint light. Neither date nor time of day is given. (On the planet Venus, Philosophical Transactions for 1793.)

5. Count Friedrich Hahn, of Remplin, Mecklenburg, saw the phenomenon unusually well and often during the spring and summer of 1793, in twilight as well as in daylight. He employed excellent instruments, and gives a very detailed description of what he saw; also two sketches. No other observer seems to have seen the phenomenon so often and so well. (Berliner astronomisches Jahrbuch fur 1793, p. 188.)

6. The venerable old selenographer Schroter saw the phenomenon only once, February 14, 1806, in faint twilight, with an excellent telescope, and gives a very accurate description and sketch of it. He remarked an important feature in the phenomenon: the limb of the dark hemisphere was brighter than its central part. (Berliner astronomisches Jahrbuch fur 1809, p. 164, and Beobachtungen des grossen Cometen von 1807, Appendix, p. 66.)

7. Simultaneously with Schroter, and independently of him, C. L. Harding, at Gottingen, succeeded in observing the dark side of Venus on three different evenings---January 24, February 28, and March 1, 1806. On the second of these days the light was reddish grey, and on all of them the phenomenon was seen with the utmost sharpness and distinctness. (Berliner Jahrbuch fur 1809, p. 169.)

8. The well-known observer of the sun J. W. Pastorff, at Buchholz in Prussia, saw the phenomenon (as he reports) many times so distinctly that he could distinguish bright and dark patches in the faint grey light. Only one date and a corresponding drawing are given, February 10, 1822, at 5 p.m., when the breadth of the crescent was 0.23 diameter of the whole disk. (Berliner Jahrbuch fur 1825, p. 235.)

9. June 8, 1825, at 4 a.m., almost in full daylight, the phenomenon was witnessed by Gruithuisen at Munich. No particulars given. (Astronomisches Jahrbuch fur 1842, herausgegeben von Gruithuisen, p. 158.)

10. The next observation was made by M. Guthrie, near Bervie, N.B. (Great Britain), during the inferior conjunction in December 1842. Mr. Guthrie saw a narrow fringe of light around the whole disk of the planet. (Monthly Notices of the Roy. Astr. Soc., vol. xiv. p. 169.)

11. G. A. Jahn, at Leipzie, saw the dark side of Venus on September 27 and 28, 1855, at 11 a.m., in broad daylight. (John's Unterhaltungen im Gebiete der Astronomie, vol. ix. p. 320.)

12. Mr. Berry, of Liverpool, saw the phenomenon on the evening of January 14, 1862. (Month. Not. vol. xxii. p. 158.)

13. Mr. C. L. Prince, of Uckfield, observed Venus almost daily during her inferior conjunction between Sept. 23rd and 30th, 1863, in bright daylight, and could trace on every day the whole disk, or at least a faint fringe of light around the edge. (Month. Not. vol. xxiv. p. 25.)

14. Mr. W. Engelmann, of the Leipzie Observatory, saw the phenomenon repeatedly---most advantageously, as it seems, on April 20, 1865, immediately after sunset. The dark side was greenish grey, a little brighter than the sky. (Astron. Nachr. No. 1526, vol. lxiv. p. 223.)

15. During the inferior conjunction of 1867 Venus was well observed by Professor C. S. Lyman, of Yale College, Newhaven, U. S. The extension of the crescent over more than 180° was seen during a

period of eleven days: on 10th and 12th of December the thin bright crescent formed an unbroken ring; on the day of conjunction (11th December) the close proximity of the sun permitted no observation. (American Journal of Science, 2nd series, vol. xliii. p. 129.)

16. Mr. Th. Petty, of Deddington, near Oxford, saw the dark side of Venus on May 23 and June 9, 1868, probably during twilight. (Astronomical Register, No. 68, p. 181.)

17. In the same year I was observing Venus attentively for some months, chiefly in broad daylight, with a small but good achromatic. I saw spots on different occasions; and on July 4, 1868, at 1 p.m., I could see traces of the dark disk, though unsteadiness of the air and insifficient optical power prevented me from becoming certain of what I saw.

18. On February 5, 1870, the dark side of the planet, then near inferior conjunction, was seen (in daylight, I suppose) by Mr. R. Langdon, of Silverton, Devonshire. (Month. Not. vol. xxxii. p. 307; Astron. Reg. No. 115, p. 163, where the year is erroneously stated to be 1872.)

19. Captain W. Noble, of Leyton, Essex, saw the dark part of Venus very distinctly on February 22, 1870, only twenty-four hours before conjunction, in close proximity to the sun. In a later communication, Captain Noble adds that he saw the dark side always darker than the surrounding sky, and that he rarely failed to see it whenever Venus was in or near inferior conjunction. (Month. Not. vol. xxx. p. 152; Astron. Reg. No. 88, p. 74, and No. 130, p. 258.)

20. At the meeting of the Royal Astronomical Society, March 11, 1870, Mr. Browning stated that, without any special contrivance, he could see all the globe of the planet in his 12-inch speculum---perhaps on twenty different evenings, as Mr. Browning told me orally, and always in bright twilight. The unillumined side appeared darker than the sky around it. (Ast. Reg. No. 88, p. 74, and No. 131, p. 281.)

21. On August 9, 1870, I was regarding Venus in bright sunshine at 11 a.m., when a lady who was with me at that time immediately perceived the whole disk of the planet. I showed to her Schroter's drawing, which she declared to be in perfect accordance with what she saw in the telescope. I fancied only at moments that I saw a faint line of light all round the greyish disk. Illumination unusually large (0.35); air much disturbed at the time.

22. Dr. A. Winnecke, of Karlsruhe, saw the phenomenon twice, on September 25, 1871, at noon, and November 6, 1871, at 5 a.m. (Astron. Nachr. No. 1863, and No. 1866, vol. lxxviii, pp. 236 & 287.)

On the day subsequent to Dr. Winnecke's first observation, September 26, Captain Noble could not make out the dark hemisphere so well seen by him a year before that time, but he adds that the sky was not clear. (Month. Not. vol. xxxii. p. 17.)

From the above conspectus it appears that the unillumined side of Venus has been seen by 22 different observers:---

In twilight by 13 (once by 4, many times by 9).
In daylight by 11 (once by 6, many times by 5).

4 observers saw a faint line of light encircling the dark disk, 19 of them saw the disk itself. Of the 22 cases reported, 12 have been observed during the last eleven years, say one per year; and I am disposed to think that the phenomenon is a normal one, and that with sufficient optical power and attention under a favourable sky it is to be seen at every

inferior conjunction, though I would by no means advance that it is constantly visible, which would be a statement directly opposed to facts.

For the explanation of this remarkable phenomenon the following causes have been suggested:---

1. Phosphorescence.---This was the idea of Sir William Herschel, Harding, and partly of Schroter. It does not appear clearly whether they understood the word in its modern sense, meaning substances which absorb sunlight and emit it in darkness without being chemically changed, or whether they included under that name, like all the elder physicists, slow combustions also, like that of phosphorus and rotten wood, which in modern terminology do not belong to true phosphorescence. In both cases it is difficult to imagine the whole surface of the planet to be covered with such substances as sulphide of strontium, diamond, phosphorus, or rotten wood.

2. Auroral phenomena.---This was partly Schroter's idea; it is supported by a most extraordinary observation of Madler, who, during the whole evening of April 7, 1833, saw Venus surrounded by long bright immovable rays. Professor Zollner, of Leipzie, strongly advocates this idea, and trusts that the spectroscope will reveal bright lines in the grey light of the unillumined hemisphere of Venus.

3. Proper light.---An explication upheld by Pastorff, who supposed the atmosphere of the planet to be large and self-luminous. Possibly also the planet might still be incandescent, as is supposed to be the case of Jupiter by Mr. Nasmyth; but on this supposition the secondary light should be always visible, which is positively not the case.

4. The light of the Earth.---This, as seen from Venus, far exceeds the greatest brightness of Venus as seen by us; and according to the calculation of Dr. Rheinauer, of Munich (Grundzuge der Photometrie, 1861, pp. 58-77), the grey light of Venus, if resulting from this cause, should equal a star of the 14th magnitude. That this explanation is insufficient is so clear as to need no further proof.

5. Negative visibility, as it is called by Arago, or projection on the coronal light of the sun, as suggests Mr. Lynn (Astr. Reg. No. 109, p. 12) and, if I am right, Mr. Noble (Month, Not. vol. xxxii. p. 17). This explanation suits only those cases in which the unillumined side of the planet was seen darker than the surrounding sky (Messrs. Browning and Noble), but not those of the majority of observers, who make it brighter than the sky.

6. Accidental combustion and other illumining processes.---Gruithuisen suggests large luxuriant forests set on fire, an idea by no means absurd in itself; but, indulging in the fantastic cast of his mind, he brings it in connexion with general religious festivals of the inhabitants of Venus, a speculation in which it is not quite easy to follow the famous Munich selenographer. Immense prairies and jungles would do still better; but even these will hardly suffice for so frequent and general a phenomenon.

I will suggest another explanation, without laying too much stress on it, though perhaps it is not a mere fancy. The intense brightness of Venus, and particularly the dazzling splendour of her bright limb, is deemed by the late G. P. Bond and by Professor Zollner, a competent authority in photometric matters, not to be explicable without assuming specular reflection on the surface of the planet. This Professor Zollner supposes to be done by a general covering of water; and indeed if the faint grey spots of Venus, delineated in 1726 by Bianchini and rediscovered by Vico in 1838, are land, then nine tenths at least of the surface of Venus are covered by sea. Should Venus be in a geologically less advanced state, viz. less cooled than our globe, a supposition rendered not improbable by her considerable size and her nearness to the sun, then the present condition of Venus would be analogous to that of the earth in the Jurassic period, when large isolated islands were bathed by immense seas, blood-warm, and teeming with an abundance of animal life difficult to be conceived.

The intensity of the phosphorescence of the sea, shown not unfrequently by our tropical seas, gives us some idea of the intensity which this magnificent phenomenon could acquire under such unusual circumstances; and it is, I think, not unreasonable to expect that such a phosphorescence could be seen even at planetary distances. It would explain the fact that the edge of the dark hemisphere of Venus is seen brighter than its central part; for it is demonstrable by calculation and confirmed by observation (as in the case of the sea near the horizon, or the edge of the full moon), that a rough surface emitting diffused light is seen the brighter the more obliquely it is regarded.

It is satisfactory to think that my suggestion can be put to the test of physical inquiry. M. Pasteur found the spectrum of cucuyos (tropical phosphorescent beetles) a continuous one; and, according to Mr. Piazzi Smyth, the same holds good for the phosphorescent animalculae of the sea (Month. Not. vol. xxxii. p. 277), so that the spectroscope will be able to decide between Professor Zollner's hypothesis and mine.

Since the foregoing note was read before the British Association, Dr. H. Vogel has published observations of Venus with the large refractor of Baron Bulow (Beobachtungen auf der Sternwarte zu Bothkamp, Heft 2, pp. 118-132). He saw the secondary light of Venus on seven mornings between October 15 and November 12, 1871, in bright twilight. The light was yellowish, faint, brighter near the terminator, fading away on the other side, and never extended over more than 30° of arc on Venus. On five other mornings nothing was seen.

VISIBILITY OF THE DARK SIDE OF VENUS

Sale, Robert Killip; *British Astronomical Association, Journal*, 6:33, 1895.

This phenomenon, called by Webb "The phosphorescence of the dark side," is often questioned, but as it fell to my lot to have a very distinct view of it this year, I send you the simple statement of the facts.

On July 12th, there were staying in my house my wife's niece and her husband; and simply to demonstrate to them the possibility of picking up a planet in full daylight, I placed my little equatorial refractor on its iron pillar, and directed it to Venus. The planet's declination was $+10^{o}$ 4', and its meridian passage for my longitude, $3^h\ 18^m\ 8^s$. As it was then 4.20 in the afternoon Venus was well placed for observation. Without a thought of seeing anything extraordinary, I took a hasty glance, to be quite sure that the object was on the field, and was startled to see, not simply the illuminated portion, but the whole of the disk. As my friends, however, were not at all acquainted with astronomical matters, I made no remark, but simply called the lady to the eyepiece. To my surprise and delight, she almost immediately exclaimed, "Why, uncle, it is just like the moon is sometimes, "when they say the old moon is in the young moon's arms!" I then asked her husband if he could see anything like what she had described, and though at first he could not be sure of it, he afterwards affirmed that it was perfectly clear to him. Then for the best part of an hour we watched with increasing certainty and interest the very rare but unquestionable appearance. The sky was perfectly clear, and the air very steady. The most remarkable feature of the case is that Venus was more than two months off inferior conjunction. Nearly all the references I have seen to the phenomenon assume that it is only possible when the crescent is narrow, and the fact that on this occasion nearly half the disk was illuminated, may make against the acceptance of his story. But I was not deceived, and the independent and spontaneous testimony of my niece is sufficient corroboration.

A REVIEW OF SOME ALPO VENUS STUDIES

Cruikshank, Dale P.; *Strolling Astronomer*, 17:202-208, 1963.

2. Ashen Light. What about the ashen light? Too many good observers have seen the illumination of the night hemisphere of Venus for us to write it oft as an optical illusion. N. Kozyrev in the Soviet Union has spectra of the emission which suggest auroral activity involving the excitation of molecular nitrogen (Publ. Crimean Astrophys. Obs. 12, p. 169). I have seen the ashen light only once---when Venus was two hours from inferior conjunction. Some ALPO observers report the ashen light at almost any phase, sometimes as great as k = 0.82. I strongly doubt the reliability of the observations.

If the ashen light is auroral in character, then we should expect it to be most conspicuous and frequent during periods of intense solar activity. Further, since the Mariner II space probe observations indicate absence of an intense dipolar magnetic field associated with Venus, we might expect auroral activity to be more uniformly distributed around the planet as compared to the strong polar concentration observed on Earth. ALPO records are generally consistent with this interpretation, which some may consider ad hoc. The ALPO records should be compared to solar activity fluctuations to see if any degree of correspondence exists.

The above remarks pertain only to observations of the night hemisphere's appearing brighter than the background sky. Sometimes we hear reports of the night hemisphere's appearing darker than the background sky. Clearly, this difference is physically impossible, regardless of whether Venus is observed in daytime or at night; the planet can never subtract light from the sky but can only enhance it. James Bartlett's explanation of the appearance of the night hemisphere (Str. A., 16, 1-2, 1962, p. 10) at large phases as the result of the planet's being silhouetted against the Zodiacal Light and/or a coronal streamer from the sun is impossible as a passing glance at the geometry of the situation will show, and on the basis of what is known of the Zodiacal Light and the corona. Anyone can, however, easily demonstrate to himself how these misleading observations come about. If one looks at Venus when k is less than about 0.6 with very low magnification (about 5 or 10 per inch of aperture), it is at once seen that the portion of the night hemisphere adjacent to the terminator appears darker than the sky. The conclusion is obvious: the steep contrast gradient from planet to sky "darkens" the portion of the planet in question. Indeed, many sketches in the ALPO files show that just a fraction of the night hemisphere is so darkened---that portion nearest the terminator.

VENUS LIGHT CONFIRMED

Anonymous; *Astronomy*, 5:65, April 1977.

Recent results of Soviet and American space exploration have confirmed a phenomenon on Venus reported by generations of professional and amateur visual observers. Known as the ashen light, this phenomenon is a dull glow or coloration on the dark side of Venus, reported on rare occasions.

During the 1976 flights of the Russian probes Veneras 9 and 10, Soviet scientists detected a glow from the dark side, consisting of a series of spectral emission bands. The glow is strongest 56 miles (90 kilometers) above the Venusian surface.

At the January 1977 meeting of planetary scientists in Honolulu, University of Colorado scientists C. A. Barth and G. M. Lawrence reported identification of the emission bands. The bands are caused by glow created from the combination of oxygen atoms in Venus' high atmosphere. The oxygen atoms probably arise from breakup of oxygen molecules (O2) due to action of sunlight on Venus' sunlit side. Circulation of the atmosphere, taking about four days, brings the atoms around to the dark side where they recombine and emit the glow, which is spread through the visual part of the spectrum in at least 10 distinct wavelengths or colors. The strongest is at wavelength 4792 Angstroms.

The glow is strongest on the dark side near the evening terminator with a decrease toward the dawn terminator, but the glow is also patchy and variable. Variations in numbers of oxygen molecules brought to upper atmosphere levels by vertical currents may cause fluctuations in the total brightness from day to day. This agrees with old visual observations of ashen light, which indicate that the phenomenon may be visible on a few days but absent on most other days.

The maximum intensities observed by the Russians are strong enough to be seen by visual observers on Earth, confirming that the phenomenon could have been recorded by visual observers. The whole phenomenon is believed to be similar to the airglow of Earth, a faint light in our night sky that can be detected by instruments, which varies somewhat from place to place and time to time.

• The Phase Anomaly or Dichotomy

THE PHASE ANOMALY OF VENUS

Warner, Brian; *British Astronomical Association, Journal*, 73:65–70, 1963.

Observations of the planet Venus made at the time of theoretical dichotomy usually show that in reality more than fifty per cent of the whole disc is visible. The way in which a visual observer judges the time of dichotomy is to note when the terminator becomes straight along the whole of its length. It often turns out that at the time of theoretical half-phase the terminator is linear near the centre of the disc but the planet shows horns at the poles. Figure 1 shows four outlines of Venus taken from drawings made by members of the Mercury and Venus Section, all of which were made at times of theoretical dichtomy and which demonstrate the non-linear nature of the terminator at this time. In general there is a time difference of about four to six days between times of theoretical and observed dichotomy. This phase anomaly has been termed 'Schroeter's Effect' by Moore.

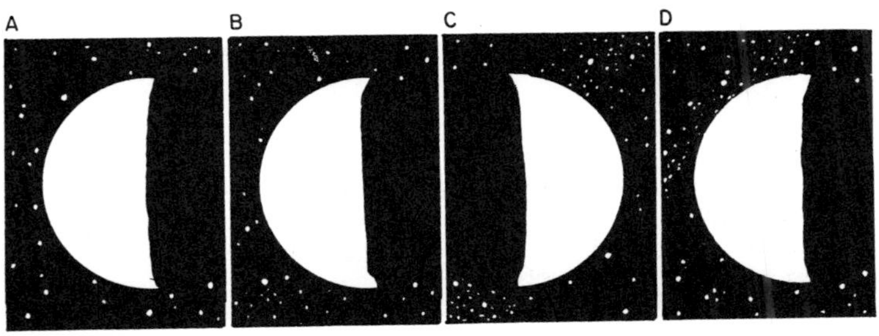

Four observations of the phase anomaly of Venus drawn at theoretical dichotomy (half phase)

The observational situation is, however, much more complex than Schroeter's Effect alone. It is obvious that a difference in observed and calculated phase at the time of dichotomy must be accompanied by incorrect apparent phases at other times, near dichotomy, but it was not until a book by Bronshten recently appeared that we knew just how large the (O-C)s are for phases other than 0.5. The results quoted in Bronshten's book are from measurements made by N. N. Michelson and V. N. Petrov. These results are depicted schematically in Fig. 2. Unfortunately only observational drawings were used by Michelson and Petrov, it would of course be preferable to have a long series of direct micrometric measures of the phase of Venus. I have measured a large number of drawings myself and arrive at identical conclusions to those shown in Figure 2, but with some additional points to be discussed later in this paper. It can be seen from Fig. 2 that the (O-C) phase can be as large as ± 0.10 for phases near 0.1 and 0.9.

I now propose to discuss in some detail the observed effects at the time of dichotomy and their possible theoretical interpretation. It is fair to say that a recent paper by Strong has stimulated this discussion.

It is quite evident that if Venus were a solid body devoid of atmosphere then an effect like that shown in Figure 2 could not possibly arise by physical means. The explanation of Schroeter's Effect must therefore lie in the detailed properties of the Cytherean atmosphere. In the interpretation of the prolongation of the cusps of Venus near the time of inferior conjunction Edson has postulated the existence of a thin scattering layer lying above the main cloud layer of Venus. Strong has suggested that this layer can qualitatively explain Schroeter's Effect. Contrary to Strong's statement at least two previous explanations of Schroeter's Effect have been given. The first was by Clayden in 1909. Clayden was well aware of 'the acceleration of phase in eastern elongation' and 'the retardation of phase in the western elongation'. He proposed an explanation based on the non-uniformity of the Cytherean clouds near the terminator. Also McEwan has proposed a refractional theory which cannot now be taken seriously.

....

ON THE PHASE ANOMALY OF VENUS

Brinton, Henry, and Moore, Patrick; *British Astronomical Association, Journal,* 73:119-120, 1963.

The 'Schroter Effect' or phase anomaly of Venus is well-known, and has been discussed in all the recent Section Reports. The present note concerns some rather puzzling observations made by the writers. Moore wishes to make it clear that he is writing as an observer, and not in his capacity as Director of the Section.

Venus was interesting in 1962 when, during July and August, it became clear that the phase was decidedly less than theoretical. Of the present writers, Brinton used his 12-inch reflector, together with one observation with the 18-inch refractor at the Pino Torinese Observatory in Italy, while Moore used his 8-1/2-inch reflector and, on occasions, Brinton's 12-inch. During observations made after dichotomy, it was noticed that there was an increased difficulty in determining phase.

Upon examination, it appeared that this was due to a discrepancy between the visible and theoretical shapes of the telescopic image of the planet. Being close to dichotomy, the image should have been almost exactly a semi-circle. In fact, it appeared as a segment bounded by a chord smaller than a diameter. Such an appearance could only result from viewing a sphere which is less than half illuminated, which seemed, on the fact of it, to be absurd.

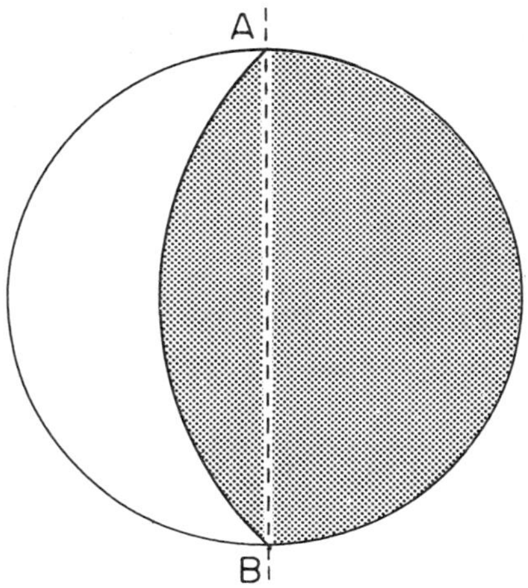

At theoretical dichotomy, the sun should illumine exactly half the disc of Venus.

In the absence of an immediately available micrometer we made a search of photographs of Venus. Many of these showed the effect to a marked degree. On the other hand photographs of the Moon and Mercury sometimes showed a similar, though lesser, effect. This suggested that photographic evidence in unreliable.

By way of a test, we experimented on September 1, with untrained, and therefore unprejudiced, observers, selected at random. Some of them were strangers, collected, to their astonishment, from the beach at Selsey. None of these observers had any idea of what they were going to see; but were merely asked to look through the eyepiece and draw what they saw. Making due allowance for variations in drawing ability, there was a clear tendency to see the effect observed by us.

We had noticed also that a very low power made the phase apparently much smaller than it appeared with a higher power. Our random observers reproduced this effect dramatically; x50, they drew the planet as a thin crescent whereas x130 they saw it as anything up to a half circle. This peculiarity can be accounted for on the assumption that the effect we have described is valid. In very low power, the eye tends to notice the ratio of height to breadth, the image being too small to judge the shape of the terminator. With a large image, the slope of the terminator catches the eye at once and determines one's estimate of phase.

Since these observations were made, members of the section have been asked to keep a watch for the effect and make measurements when possible. Preliminary results suggest that the effect is fairly common;

but certainly not permanent. They also suggest that it may be exaggerated in blue light and reduced in red.

The data at present available are insufficient to enable any firm conclusions to be drawn, but various lines of thought seem to be worth consideration. If the high albedo of Venus is due to lofty translucent clouds against a background of darker clouds at a lower level, there might be a tendency for increased reflectivity at the apparent poles, so that the cusps themselves might be accounted for in this way; and moreover the effects might well be extremely variable, and susceptible to changes in the conditions over Venus. However, speculation is rather profitless until more evidence is forthcoming, and the present note is designed merely to draw attention to an effect which is---if confirmed---both interesting and puzzling.

ON THE VARIATION OF THE PHASE OF VENUS FROM THEORY

Rushton, Minick; *Strolling Astronomer*, 15:49-52, 1961.

The difference between the times of theoretical and observed dichotomy of Venus has long been an established fact. Yet it comes as a surprise to many to learn that this deviation from prediction applies not only to dichotomy, but to all phases of Venus.

In August, 1958, a short article appeared in Sky and Telescope which mentioned the work of two Russian amateurs, Michelson and Petrov, on this subject.

. . . .

These observations indicate a definite pattern in which the crescent appears larger than expected, while the reverse is true for the gibbous phase. The first question to arise concerning these observations is that of their accuracy. Since the measurements were made from drawings, the question of "personal equation" must be carefully considered. It can be answered in part by the comparison of their work with that of other observers. However, accurate measurements of this phenomenon for extended periods of time are somewhat rare. Henry McEwen, a past Section Director of the British Astronomical Association, made over 900 measurements of the phase variation using a Slade micrometer mounted on a 5-inch altazimuth telescope. The eyepieces employed were a Ramsden of 156 diameters and a Zeiss orthoscopic of 180 diameters. All measurements were made in daylight in order to reduce the effects of irradiation. A discussion of the observations made during the period 1919-1927 appeared in the Journal of the British Astronomical Association. Figure 40 shows the results of McEwen's work. Many observers claim that the retardation of the phase is greater in western elongations (or apparitions) than its acceleration in eastern ones. The graph for the western elongation of 1927 contradicts this belief, for it in no way differs radically from the graphs of the other elongations. However, it must be stressed that the graph represents only twenty-five observations during one elongation. These curves in Figure 40 indicate a relatively constant rate of change during the gibbous phase

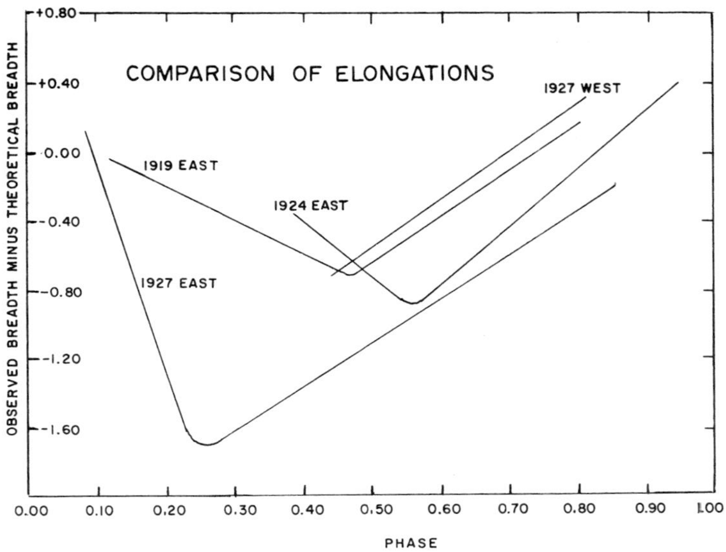

Graphical comparison of the observed phase of Venus with theoretical phase according to the micrometric measurements of H. McEwen

of the planet with the maximum deviation occurring near the time of dichotomy, although the evidence is insufficient to give more than a strong indication of this condition.

It is, of course, obvious that McEwen's results differ radically from those of Michelson and Petrov. While the two sets of results agree approximately as to the degree of the variation, they disagree entirely on its distribution. McEwen shows the planet as being smaller than its theoretical size, except near conjunctions. Michelson and Petrov show Venus as larger in the crescent phase and smaller in the gibbous phase than theory predicts. Since McEwen's measurements were made at the telescope, and not from drawings, his results may be considered the more reliable of the two sets.

The variation between the times of observed and theoretical dichotomy, although only a small part of the more general problem, is the best observed part of the deviation from theory. Apparent dichotomy is invariably late in western elongations (apparitions) and early in eastern ones, but the amount of the deviation from theory is highly variable. However, the average time lag is about 3-1/2-4 days.

The cause of the phase-error can only be guessed at, for the present state of knowledge concerning Venus offers little to explain it. Irradiation and various other factors affecting micrometric observations would tend generally to increase the apparent diameter, not to cause the results which McEwen obtained. One implication of this deficiency of illumination is the stoppage of light in the Venusian atmosphere by some cause other than absorption. Arthur Clayden suggested a sloping cloud surface along the terminator as the answer, but this would require a greater temperature difference between the day and night hemispheres than actually exists.

Many factors---seeing conditions, nearness of dusky markings, aperture, individual visual acuity, depth of the terminator shading, and the sky background---affect estimates of the phase, but their importance is not fully known. A study of the effect of these factors on the apparent terminator is vital to the accurate evaluation of micrometric measurements; but, as yet, this study has not been made. Relatively little work has been done on the phase-error of Venus, except for the short period near dichotomy. The major need is for frequent micrometric observations of the planet's apparent equatorial diameter over extended periods to time. These measurements, which should be made in daylight under standardized conditions, would contribute much to our knowledge of Venus. The previous work gives strong indications of a definite phase-error, but many questions exist as to its nature and degree. Much still remains to be done, yet real progress has been made.

• Venus' Bright Limb Band

A REVIEW OF SOME ALPO VENUS STUDIES

Cruikshank, Dale P.; *Strolling Astronomer*, 17:202-208, 1963.

3. Limb Band. What is the nature of the bright limb band of Venus? James Bartlett (Str. A., 15, 7-8, p. 133, 1961) recently discussed his observations of a bright band along the limb of the planet. He found that the band developed shortly after superior conjunction, was most conspicuous around dichotomy, and then diminished as the crescent became more narrow. Bartlett also reported a dusky inner band which was sometimes seen adjacent to, and parallel to, the bright limb band. He notes that this dusky band was usually brownish in color. He interpreted these bright and dark bands as a high cloud bank (or a propinquitously united group of several cloud banks) and the shadow cast by the clouds, respectively. The brownish tint was attributed to a scattering effect in the Venusian atmosphere.

Joseph Eyer (Str. A., 15, 9-10, p. 148, 1961) criticized Bartlett's article, saying that while the bright limb band at dichotomy is reasonable (assuming that scattering is maximum at an angle of 90° to the incoming lignt) the dark band cannot be a shadow of a cloud ridge. Even though the Rayleigh scattering function (which Eyer mentions) is strongly forward and rearward thrown and not at 90°, Eyer correctly showed that on the basis of Rayleigh scattering, the dusky band should be blue and not brown or red. Unfortunately, Eyer neglects the fact that Rayleigh scattering does not hold in the upper atmosphere of Venus, except perhaps in the very short ($\lambda <$ 3600 Å) wavelengths (T. Gehrels and T. Teska, Applied Optics, 2, p. 67, 1963). That Rayleigh scattering does not occur has been known for over half a century (see, for a discussion relating to this, Russell, Ap. J., 9, p. 294, 1899).

We now question the point of whether there does indeed exist a bright limb band on Venus. My own visual observations and those of most others indicate that the limb of Venus is far brighter than the terminator, as one would expect from the cosine dependence of brightness distribution across a diffuse reflecting sphere as given by Lambert in the eighteenth century. But, as with other complex phenomena where the physical senses of humans are insufficient to distinguish the true picture amid the subleties presented, we must turn to physical (as opposed to visual) observations. In the surface photometry of planets, Venus in particular, significant strides have been made. Among the important papers relating to this topic, V. I. Yezerskii's work "Photographic Photometry of Venus" (Trans. Gor'ki Astr. Obs., Kharkov State Univ., 12, pp. 73-165, 1957) is fundamental. Yezerskii's photometry done in 1932, 1951, 1953, and 1954, with phase angles from 26° to 139° and in four spectral regions, shows that the position of the brightness maximum along the intensity equator (the equator reckoned from the line of cusps) and in parallel directions, is located near the portion of the disk where the angle of incident sunlight equals the angle of "reflection". Clearly then, the limb band as Bartlett describes it is not a real feature of the planet. If the Venus atmosphere had an orthotropic scattering indicatrix as Eyer suggests, the brightness would be proportional to the cosine of the angle of incidence; and its maximum would then occur at the subsolar point. This is not observed.

It is probable that the actual brightness distribution across the Venus disk is a result of the combined effects of reflection and anisotropic scattering. It should be added that Yezerskii's observations confirm earlier photometry by R. S. Richardson (P. A. S. P., 67, p. 304, 1955) and N. P. Barabashev (Pub. Kharkov. Astr. Obs., No. 2, pp. 3-10, 1928).

While it is true that the eye perceives a bright limb on Venus, I contend on the basis of the observations reported above that this is a pitfall of contrast effects to which the eye cannot help but succumb. Fortunately, then, we need no enormous cloud banks, orthotropic scattering (which is hard to come by), or suchlike to explain our drawings of this phenomenon.

POSSIBLE MARKINGS ON VENUS

The Martian canal controversy stimulated hundreds of articles and even a few books down the years, but few have heard of the canals of Venus. When seeing is good, Venus seems to exhibit indistinct or "dusky" markings often arranged in a spoke-like pattern. These streaks may be atmospheric manifestations, for one can rather easily conceive convection cells in the planet's thick, hot atmosphere creating huge ridges visible through the earth's telescopes. On the other hand, large surface structures or volcanic activity might also disturb the atmosphere enough to be detectable from earth. Another line of thought holds that the spokes or canals are merely optical illusions, as they may be with some Martian canals. The mystery then moves into the category of psychology; why does the eye-mind combination see linear markings and patterns on difficult telescopic targets, such as Mars, Venus, and possibly Mercury?

• Vague Tracery and Radial Spokes

ILLUSIONS

Antoniadi, E. M.; *British Astronomical Association, Journal*, 8:94-95, 1897.

I was much amused to find in the account of our Meeting of October 27, that Mr. Griffiths thought it was pretty conclusive from my reasoning that the markings I had seen on Venus were illusory. As Mr. Griffiths did not even know what my markings were, I ought to pay no attention to the matter. But as such a course might lead to misconceptions, I regret to find myself under the obligation of overthrowing, in a few words, Mr. Griffiths' arguments.

I observed the planet Venus quite unbiassed by any preconceived ideas, and a subsequent comparison of my drawings with those of the most trustworthy observers has established that the markings seen by me are shown in part or entirely by Cassini, Bianchini, Schroeter, W. Herschel, Dr. Terby, Ormesher, Seabroke, Worthington, Niesten, Denning, Loschardt, Dr. Barnard, and, last but not least, by Prof. Schiaparelli himself.

As the subjoined diagram will show, no one is less entitled to speak of other people's illusions than Mr. Griffiths. This drawing was made after the publication of the "Flagstaff Observations," and, in agreement with them, we find canals, pilulae, and other black knots in considerable abundance. We could scarcely expect something sur-

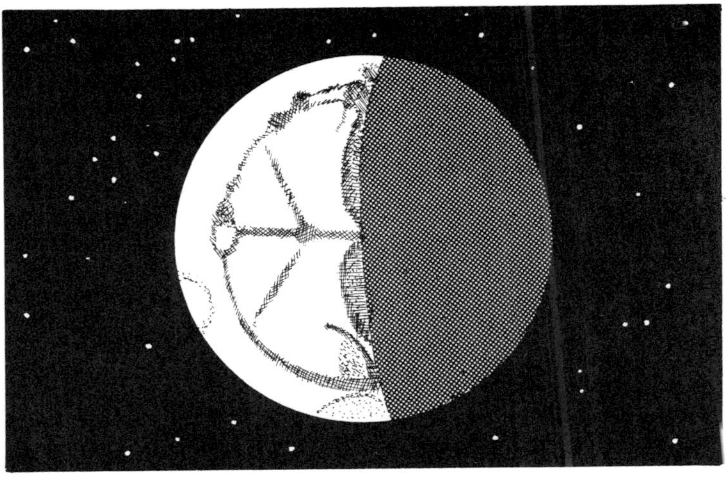

Venus markings sketched by Douglass

passing this unique production were we not to have Mr. Griffiths' own words in finding "very little geometrical symmetry in the markings of Venus," a few weeks later.

Under such circumstances Mr. Griffiths would have done better to leave this planet, at least, alone. For he was scarcely well inspired the day he decided to turn his attention on Venus, and expose himself to more than one deception.

THE MARKINGS ON VENUS

Douglass, A. E.; *Royal Astronomical Society, Monthly Notices*, 58: 382-385, 1898.

The reading public has been recently addressed on the subject of the markings on Venus in various attempts to show that the discoveries made at this observatory are unworthy of credit. No matter how futile such criticism must prove to be in the long run, some persons will be influenced by it if we do not from time to time make some rejoinder, or give out some statement which will show our continued activity in this line of work, our undiminished confidence in the results obtained, and our answering attitude towards adverse opinion.

In the last six years many thousands of hours have been spent by us at telescopes of 13, 18, and 24 inches aperture and their smaller finders, when the seeing was sufficiently good for profitable work on the finest known planetary detail. Expressed in standard terms, the seeing was practically always such that in a 6-inch aperture the spurious disc of the interference pattern was well defined, and a very large part of the time the rings of the same pattern were unbroken. I consider that any astronomer who cannot say the same for the seeing during his hours of work, and whose hours of work do not reach a commendable number, has no right to criticise our results; for he lacks the experience by which alone he becomes capable of judging.

Under proper conditions of air and aperture the markings on *Venus* are absolutely certain. Under proper conditions they are to me about as easy or difficult to see as the irregularities on the terminator of the Moon when it is near the first quarter, viewed by the naked eye. I have on a few occasions seen a large projection perfectly distinct. So it is with *Venus*. At the best seeing the markings are visible at the first glance.

To say that no markings save M. Antoniadi's symmetrical shadings of atmospheric contrast exist, or that the detail seen here is due to pressure on our objective, or to defective densities in the eye-piece, or to our own eyes, or to the imaginings of our brains; or, most ridiculous of all, to our looking all day at some map and then seeing it on the planet, is to offer suggestions too absurd to be taken seriously.

We use the telescope in both positions, normal and reversed: that shows that the markings are not in the lens. We use different eye-pieces and twist them in varying position angles: that shows that the markings are not there. We sit in different positions, so the markings are not there. We sit in different positions, so the markings cannot be in our eyes; and different persons in perfect independence find the same detail, so it is not a mental phenomenon.

....

The first reason why other observers have not seen these markings is bad atmosphere. When I began observing the third satellite of *Jupiter*, for days, even weeks, I drew nothing but hazy indefinite markings or belts, such things as M. Antoniadi describes as appearing to him on *Venus*. But one night after making several drawings of that character the seeing suddenly became superb, the curtain rose as it were, and I saw sharp distinct black lines about which uncertainty was impossible. The very same thing happened on the fourth satellite four days later. I had been drawing the same indefinable shadings, when one night the seeing improved, the curtain again rose, and I perceived sharp definite lines. After once thoroughly understanding the character of the object sought, I could see them and profitably study them under conditions of seeing formerly prohibitory.

The experience on *Venus* has been similar. On the day succeeding my first good view, I spent nearly the whole afternoon without catching a single certain glimpse. Suddenly the seeing improved for an instant, and I saw the same markings unmistakably. If it had not been for that glimpse, I would have gone away perfectly ready to believe that no markings existed. I am not surprised that other astronomers doubt them.

The second reason why some other observers have not seen them has been the fault of using too large an aperture. Six years ago I discovered "air waves," and over four years ago I explained theoretically why reducing the aperture is often beneficial. All this has been published in full elsewhere (*Am. Met. Jour.* 1895 and *Pop. Ast.* 1897).

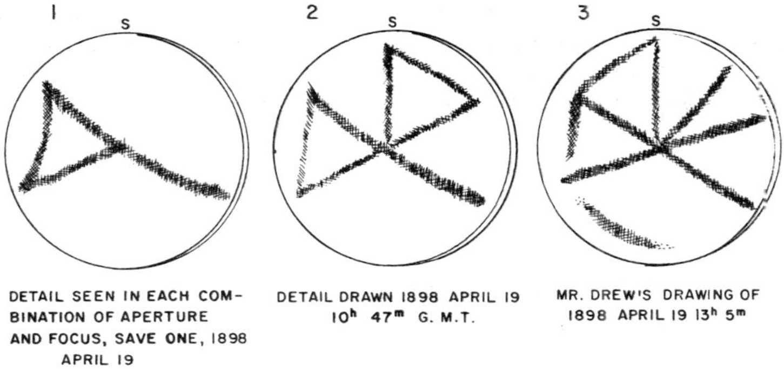

Antoniadi's drawing of markings on Venus

I decided long since that in planetary work the greatest efficiency is obtained with the smallest aperture which supplies the required illumination. There is a limit to this, however. An inch and a half lens shows the markings on Venus nicely, but they are not so well defined as in a lens of three inches, which in our atmosphere is a very satisfactory size to us. When the seeing is very bad an aperture of less than three inches will become necessary.

A third cause of failure is the effect of heating of the lens and tube by the Sun's rays. For this reason I have found it sometimes advantageous to use the small finder, which is far within the dome and well shaded.

A fourth cause comes from the air within the dome being colder than that without. This is likely to harm the seeing. If the interior is warmer than the exterior, it will certainly harm the seeing. In fact my latest experiments show that any dome at all is harmful. A sunshade surrounding the tube would be better both night and day.

A fifth cause of failure, and by no means the least important, is the lack of continuity of observations and the lack of a first good view. By the first I mean fair or good observations made many days in succession. For instance if the seeing is only fair it requires the work of several nights in succession, without intervals, to identify with certainty the longitude presented by a satellite of Jupiter. By a "first good view" I mean the necessity of one first-class observation before one understands what is sought. After that view the observer can obtain valuable results under conditions in which formerly he failed completely from ignorance of what he was after. It is the same in observing the Gegenschein. I have taught many persons to observe it, and I find that teaching consists in getting them to see it well once. After that they can be trusted to pick it out with very small liability to error. This, of course, is most true in atmospheres unclouded by smoke and unlighted by electricity.

No matter how difficult to obtain, a just hearing is our right. No one is entitled to cry out against us until he can show that his atmosphere is approximately as good as the one through which Mr. Lowell discovered these markings. Let our dubious friends, who attempt to show what we as well as they are deluded, devote a portion of their valuable time to work at the telescope under better atmospheric conditions, and no one will misunderstand the silence which will follow.

SPOKE-LIKE MARKINGS ON VENUS

Seagrave, Frank E.; *Popular Astronomy*, 27:406, 1919.

On the afternoon of April 23 at 3^h 5^m summer time I happened to turn the 8-1/4 inch to Venus, which was less than half an hour east of the meridian and in high northern declination (23°), and I was greatly surprised to find the "seeing" excellent. During the past forty or forty-five years I have never seen Venus so sharp and well defined---the most difficult of all the planets to get a satisfactory view of. For fully ten minutes the planet was perfectly steady and sharply defined and I am very sure that I could see faint traces of the spoke-like markings near the center of the disk. The markings were very much like those seen and described many years ago by the late Dr. Percival Lowell.

THE RADIAL MARKINGS OF VENUS AND THEIR MODERN RESURRECTION

Bartlett, James C., Jr.; *Strolling Astronomer*, 9:2-8, 1955.

In the year 1898 the astronomical world was startled anew by a strange communication from that famous observatory at Flagstaff, Arizona, whence had already come the highly controversial report of a Martian irrigation system. The dreadful uproar which resulted over the interpretation placed by Lowell on the canali discovered by Schiaparelli was in full swing; and now came the hardly less upsetting announcement that Venus too exhibited a well-marked system of straight lines... "in opposition to all previous observers", as Young commented. The reception accorded this novel report was if anything even more hostile than that accorded the Martian communiques. Lowell's work on Venus was met with general disbelief and even derision, and many were not slow in pointing out that the discovery of linear markings on practically every celestial body appeared to be a Lowellian specialty. As we shall see a little later such a charge was not only ill-tempered but ill-founded, for the other non-Martian linear markings, e.g. those on Jupiter, are today universally admitted (the Jovian lines we know as festoons) while Schiaparelli drew the markings of Mercury largely as streaks as to a lesser extent did Antoniadi. It remains then to see what confirmation can be offered for the streaks of Venus.

Lowell began his observations of Venus in 1896 with the discovery of "fingerlike streaks pointing in from the terminator". These streaks all appeared to meet in a kind of hub at or near the center of the disc, giving to the planet the strange appearance of a wheel. Although his critics chose to overlook Lowell's own evaluation of these markings, it is quite clear from his writings that he did not at any time claim any similarity, either in appearance or in origin, of the Venusian streaks to the Martian canals. The Martian lines he rightly regarded as being true surface features, whatever one may think of his explanation for them; but he specifically interpreted the streaks of Venus to be wholly of atmospheric origin.

His theory of the nature of the Venusian "spoke system" was grounded in his belief that the planet, like Mercury, turned one side constantly to the sun. This being so, it followed that there must be a point on the surface, the subsolar point, at which the heating effect of the sun would be greatest. This point would account for the "hub" of the system, which he described as representing "a funnel-like rise of hot stagnant air creating a partial vacuum, which would be filled by draughts of cold air from the night side coming in from all sides of the periphery, thus giving rise to a spoke system". Since he also believed that all water from the sunlit hemisphere had been evaporated and transferred by atmospheric currents to the arctic night side and there deposited as a universal glacier, the reason for the visibility of these assumed currents in the planet's atmosphere offered some difficulty, a difficulty which Lowell recognized but apparently never fully resolved.

. . . .

So much for the integrity of Lowell's belief. But what we wish to know is if Lowell's general work on Venus indicated anything more substantial than belief alone. One way of determining this is to compare his generalized picture of Venus to what is familiar today. Lowell divided the Venusian markings into various classes, among which he particularly mentioned "nicks" in the terminator and "the collar around the south pole". Terminator indentations, of course, are old stories; but in the "collar" we immediately recognize a feature familiar to every acute A. L. P. O. observer of Venus. Lowell's "collar" is simply the dark cusp band usually found as the northern border of the bright south cusp cap, which may indeed mark the actual south pole. Lowell evidently was quite satisfied that it did. There is a subtle point here which should be noticed. Lowell's work on Mars so overshadowed his work on Venus that the average modern observer is scarcely aware of the latter. Hence it is extremely unlikely that such an observer is influenced by suggestion when recording such a conspicuous feature as the south cusp band, which Lowell figured as far back as 1897. Moreover the cusp band is frequently reported by observers who fail to see anything else on the disc. We seem forced to admit that when Lowell described his "collar", <u>that</u>, at least, was not an illusion but an objective reality which remains visible today. Associated also with the "collar' were two dusky spots, which Lowell called Astoreth and Ashera, like large dark beads strung upon the narrow line of the cusp band. These too have been seen in our times. So too have the "notches" which Lowell described at both cusps. The general picture of Venus which Lowell gave us in the 90's is one that is easily recognized today by every close student of the planet, including those who have never heard of his Venusian work. We must concede therefore that in the main he was factual and correct with respect to the markings of Venus. Must we also believe that with respect to the streak markings he was entirely deluded?

We might believe it if, as Young alleged, Lowell's streaks were in fact entirely opposed to all previous work, but a little inquiry reveals that linear markings on Venus have a respectable and fairly long history. As long ago as 1801 Schroter discovered "a dim oblique dusky streak, like one on Mercury" while Gruithuisen "perceived repeatedly long vertical shades". Linear markings were also recorded by Fournier, Perrotin, Mascari, and others. In 1899 Schiaparelli recorded a long streak beginning at the south cusp band, which he used to measure the rotation, and streak markings were recorded by Phocas in 1939-45.

Lowell, therefore, was neither the first nor the only observer to record linear markings on Venus despite Young's crisp remark. True, Espin pointed out that the streak markings recorded by Gruithuisen and Fournier were altogether different in arrangement from those depicted by Lowell; and in this limited sense Young's statement may have been justified....but only in this sense. For we are not concerned with absolute identity of configuration, but only with the fact that linear markings on Venus were neither discovered nor invented by Percival Lowell. Exact identity is not important. When two observers can draw the same lunar crater so differently that it is hard to believe that they were observing the same formation, we need not be surprised that anything so fugitive as the streaks of Venus might not appear identical to different observers. It would be rather surprising if they did. Whether the _system_ presented by the Venusian streaks is precisely as Lowell figured it may be open to question. Neither Antoniadi nor Barnard was able to see the Martian canals as they were seen at Flagstaff, nor was Hale at Mt. Wilson. Few today however would care to deny that there does exist on Mars some kind of a linear network which on the whole rather resembles the drawings by Lowell. The resemblance may not there _is_ a resemblance. Hence the important thing is not that others' observations of linear markings on Venus do not agree exactly with Lowell's, but that linear markings were observed both before and since Lowell's time. We must therefore suppose that they correspond in some way to something real.

I first saw the streak markings of Venus on March 16, 1927, with the 6-inch refractor of the Maryland Academy of Sciences; and again in April of the same year with both the 6 and 10-inch refractors of the Academy. I did not again see them until the autumn of 1944, and then with my own 3-inch refractor (the long interval signifies nothing but neglect of Venus in favor of other pursuits). Since then I have repeatedly observed them from time to time, though I have never seen the entire "spoke system" as reported by Lowell.

They have been seen by others. O. C. Ranck rather consistently records them and they were well seen by Avigliano in 1954. It is instructive to note that Ranck's interpretation differs from Avigliano's in much the same way that Lowell's Martian canals differed from those of Antoniadi. Of course the region drawn by Ranck may not be the same as the one drawn by Avigliano. Our ignorance of the rotation is both complete and sublime.

But there is one modern observer whose work deserves special mention, because it practically reproduces the famous sketches made at Flagstaff so long ago. Mr. Richard M. Baum, the very gifted English observer, not only sees a wealth of streak detail on Venus but he sees it essentially as Lowell saw it.

....

So I read the riddle. I think too that this much can be regarded as fairly well established:

1. Linear markings do exist upon Venus
2. Indefinite dusky areas also exist, most of them probably arising only through slight differences in color and albedo compared to the surrounding area of the disc.
3. At least some examples of both classes are permanent markings.
4. The color sensitivity of the observer taken with the colors of the markings will determine which class of markings he will see predominately.
5. The validity of the above hypothesis is susceptible to test by color filters.

....

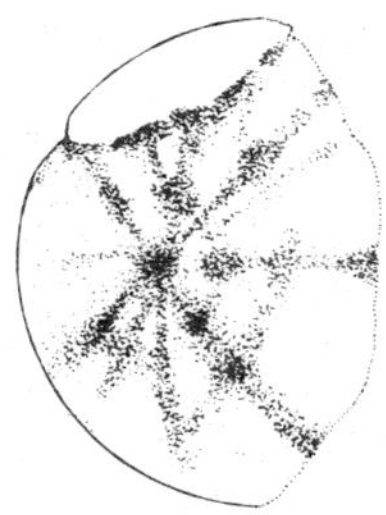

Radial spokes on Venus, as drawn by R. M. Baum, April 26, 1951

THE RADIAL MARKINGS OF VENUS: ANOTHER POINT OF VIEW

Moore, Patrick; *Strolling Astronomer*, 9:50-54, 1955.

Dr. J. C. Bartlett's paper in the January-February "Strolling Astronomer" raises a number of most interesting points, and with the full knowledge that I am treading on dangerous ground I propose to challenge him. Dr. Bartlett and I have corresponded for years upon a variety of topics, and it is scarcely necessary to add that I have the most profound respect for him personally and as an observer, so that I hope he will show me where I am wrong!

The first point seems to hinge upon Lowell's observations. Nobody in their senses would suppose that Lowell himself was anything but utterly honest and sincere, and any alternative suggestions can be treated with the contempt that they deserve. Few men sacrificed more in the cause of astronomy than Lowell did. But I cannot agree that his observations of Venus are accurate. He once described streaks as being "like steel engravings", and in his own words:

"The markings themselves are long and narrow; but unlike the finer markings on Mars, they have the appearance of being natural, not artificial. They are not only permanent, but permanently visible whenever our own atmospheric conditions are not so poor as to obliterate all details on the disk. They are thus evidently not cloud-hidden at any time ....The markings, which are of a straw-coloured grey, bear the look of being ground or rock, and it is presumable from this that we see simply barren rock or sand weathered by aeons of exposure to the Sun. The markings are perfectly distinct and unmistakable, and conclusive as to the planet's period of rotation."

It will be remembered that Lowell supported the 225-day rotation period. His observations were "conclusive" in this direction; yet the 225-day period has now been shown to be wrong without any shadow of a doubt. Here is an immediate and significant inaccuracy. But Lowell did more; he rejected the whole idea of an all-concealing Cytherean atmosphere, and went back to the ancient theory of Bianchini that the markings were permanent features of a solid surface.

In fact, his drawings and descriptions of Venus are so utterly at variance with those of others using comparable apertures that we have little choice but to reject them. His map of Venus is indeed curious. From a dark patch, Eres, a sort of focal centre, he drew well-defined strips which he named Adonis Regie, Æneas Regie and so on, while there were other less conspicuous streak centres---the whole forming a network. So far as I know, this network has been seen since only by observers using apertures of 12 inches or less. Antoniadi never saw it; nor did Barnard, and Barnard, who worked with the Lick 36-inch (and, moreover, found no necessity to reduce the aperture, as Lowell was usually forced to do with his 24-inch!), wrote: The planet was beautifully defined.....Nothing was seen of the singular system of narrow dark lines shown in recent years....Every effort was made to show them, by reduction in aperture and by the use of solar screens and magnifying powers. Previous attempts with the 12-inch here also failed."

So much for Lowell's map. We now come to the recent work by some observers, in particular by Richard M. Baum at Chester. I know Baum to be a splendid observer and a magnificent draftsman, and as he is also a very old friend of mine I feel I can say what I like! We have collaborated in lunar work, and aperture for aperture he invariably sees more than I do, quite apart from his infinitely greater skill at recording detail. The one object upon which we disagree is Venus. He sees a radial spoke system; I cannot.

The strange thing is that Baum has recorded the spokes with a 3-inch refractor, which is considerably smaller than the finder on the Lick telescope used by Barnard. Mr. Ranck's refractor is precisely the same size as this finder. I have used various telescopes up to and including a 15-1/4-inch reflector, but most of my work has been done with my own 12-1/2-inch reflector. My greater telescopic aperture must inevitably compensate for Baum's superior skill as an observer, but my only reason for mentioning my work is to point to a simple but significant experiment.

It is often said that a small telescope will show more on Venus than a large one, because of the reduced glare. This seems to me to be a complete fallacy (provided that Venus is observed against a light sky, of course). I have a 3-inch refractor of my own, and on a number of occasions I have drawn Venus with it before turning to my 12-1/2-inch. Invariably I see more with the larger telescope, but the markings appear more diffuse. The small aperture naturally sharpens them up, simply because of the lack of resolving power.

Venus is the most difficult problem observationally of all the planets, and we must be doubly wary. It is my contention that the "spoke system" of Lowell is merely an optical effect, seen by modern observers simply because of the inadequacy of their equipment. I do not believe that any observer, however skilful, can see with a 3-inch refractor details which the great Barnard and Antoniadi missed with the giant telescopes of Lick and Meudon respectively.

I do not dispute that some "streaky" features may exist, and some photographs support this. Differences in eyesight may also play a part, and a simple experiment once conducted by Baum and myself showed that his eyes are much more sensitive than mine with regard to short wavelengths. But streaks are different from radial spokes, and in any case there seems no basis for the contention that the streaks are even semi-permanent. Once we support the idea of permanency we must go back to the 225-day rotation period, which has been shown conclusively to be false.

I now bow my head to await the storm! [For the storm, see Richard M. Baum; Strolling Astronomer, 9:82-91, 1955.]

A REVIEW OF SOME ALPO VENUS STUDIES

Cruikshank, Dale P.; *Strolling Astronomer*, 17:202-208, 1963.

1. Dusky Markings. Do dusky markings exist on the visible cloud layer of Venus? Most observers say that they do, and I agree. In 1928, Frank Ross (Ap. J., 68, p. 57) reported his photographs of them in the ultraviolet at Mt. Wilson; and a number of others (G. Kuiper, R. S. Richardson, B. Smith, W. K. Hartmann, F. J. Eastman, B. Lyot, A. Dollfus, H. Camichel, etc.) have likewise photographed them, often with small instruments. W. K. Hartmann (Str. A., 16, 7-8, p. 171, 1962) showed that bands visible on ultraviolet photographs seldom correspond to those shown on drawings made at the same time. It is also clear from his report that markings on simultaneous drawings by different observers do not agree well. The implication is obvious; the bands and shadings are on the threshold of visibility at the contrast level and spectral response region in which the eye works, and observers cannot be absolutely objective in recording them on a sketch.

Clearly, these remarks apply to dusky shadings which are real. A wide variety of markings on the sketches in the section files are not in this category and appear to be results of contrast effects encountered when observing the bright image of Venus against a moderately dark, or worse, a very dark sky. Some would explain all shadings seen visually as contrast effects, but I don't believe it. My own observations are always made in daylight, and I can distinguish shadings about half the time. Further, very recent photographs in visual blue light show markings on Venus now and then and generally do not confirm sketches made by two persons with the same instrument at the same time (in an article to be published by W. K. Hartmann).

4. Radial Spoke System. In 1955 and 1956 there occurred in this journal a lively debate on the matter of a system of radial dusky markings with their center at the sub-solar point. J. C. Bartlett began (Str. A., 9, 1-2, p. 2, 1955) by reviewing old observations by Percival Lowell, more recent confirming observations by R. M. Baum, and to some degree, those of O. C. Ranck, D. Avigliano, and Bartlett himself. He concluded that both hard linear markings and dusky streaked markings exist, and that each has some degree of permanency on the disk of Venus. Patrick Moore countered (Str. A., 9, 5-6, p. 50, 1955) Bartlett's remarks by doubting the general authenticity of spoke systems, writing them off as illusions. He cited the observations of E. E. Barnard and E. M. Antoniadi with large telescopes, and U. V. photographic observations of streaks in support of his thesis. R. M. Baum (Str. A., 9, 7-8, p. 82, 1955) proceeded to correct "errors" in the Bartlett paper, flail Moore with some rather frail arguments, re-inflate Lowell's observations, and outline evidence "proving" the reality of the spoke system. In my view, experiments with artificial planetary disks, color sensitivity of the eye, and such prove nothing of the kind. Moore came forward again (Str. A., 9, 9-10, p. 112, 1955) to restate his views on the superiority of large versus small apertures, question again the observations of Lowell, and cite results of his own artificial disk experiments. Bartlett returned (Str. A., 10, 9-10, p. 102, 1956) to reemphasize the value of Baum's independent "recovery" of the Lowellian spoke system, and to assert its general reality.

Generally, I concur with Moore in that large apertures are better than small ones (given good seeing and high optical quality), and that the observations of Barnard and Antoniadi must carry considerable weight (and in the use of the word "Venusian"). With respect to Lowell's observations, to say that his sketches and maps of Venus, Mercury, and the Jovian satellites are highly suspect is a gross understatement.

While it would not be proper for me to enter the argument actively at this late date, it is fitting, I believe, to re-examine the so-called spoke system in the light of recent observations and on theoretical grounds. A. Dollfus has made important planetary observations under optimum conditions with large apertures in recent years. He reports that the dusky markings on Venus generally have the pattern of a radial system with the center at the subsolar point (L'Astronomie, 69, p. 413, 1955, or in English, see his chapter in Planets and Satellites, Kuiper and Middlehurst, ed., Chicago, 1962, pp. 534-571). Dollfus finds that the general appearance of the markings varies greatly from day to day but that the radial pattern or portions of it recur and that this represents the undisturbed condition of the markings observed. He offers a general map of the planet showing these permanent features. In many respects, this pattern reported by Dollfus is consistent with a model of the circulation of the Venus atmosphere proposed by Yale Mintz (Plan. and Space Sci., 5, p. 141, 1961). The Mintz model is based on convection in the planet's atmosphere (assuming a very slow rotation) with the principal circulation from the subsolar to the antisolar point.

Klaus Brasch (Str. A., 15, 9-10, p. 156, 1961) published a map of Venus made from composite sketches of the planet by members of the Montreal Centre of the RASC. He assumed, as did Dollfus, that the rotation period is long and that certain markings are semipermanent. While I do not agree in detail, Brasch's approach is clean and praiseworthy, and generally indicative of the good work of which the ALPO is capable.

In contrast with visual observations of the radial pattern of dusky features (and hard, linear features) we have the well-known ultraviolet photographs of Ross, Kuiper, etc. The markings shown in the ultraviolet (and more recently in the visual blue---W. K. Hartmann, unpublished---) do not correspond to a radial pattern.

While it is possible that visual observers record activity in a cloud level greatly different from that photographed, and while the Mintz model gives some theoretical reason for a radial pattern, I personally believe that it is not a major feature of the visible cloud layer of Venus the Dollfus and Brasch observations notwithstanding.

TRANSIENT VENUSIAN PHENOMENA

Transient visual phenomena are now admitted to be rather common on the moon; and Mars also possesses its repertoire of flashes and bright spots. Venus, though, is cloud-covered and transient events originating on the surface would pass unseen. Yet, bright pinpoints of light have been seen on Venus. Bright irregularities near the terminator dividing the bright and dark portions of the planet are not rare. On the dark portion of the planet, in particular, the sun cannot be a factor, as it might near the terminator. Electrical and/or chemical processes in the atmosphere can be postulated, but the nature of these phenomena can only be guessed at.

• Spots and Lights

WHITE SPOTS ON VENUS

Trouvelot, L.; *Observatory*, 3:416-417, 1880.

From Nov. 13, 1877, till Feb. 7, 1878, two remarkable white spots, strongly reminding me of those seen on Mars, have been observed on the opposite limbs, near the extremity of the cusps. The southern spot, which always appeared the brightest, became very prominent from Jan. 16, 1878, till Feb. 5, and appeared then to be composed of a multitude of bright peaks forming on its northern border a row of brilliant star-like dots of light. After the inferior conjunction, which occurred a few days later, the white spots were no more visible. This phenomenon has been independently seen by Mr. Seagrave of Providence, with his 8-in refractor.

OBSERVATIONS OF THE PLANET VENUS WITH A 6-INCH SILVERED GLASS REFLECTOR

Langdon, R.; *Royal Astronomical Society, Monthly Notices*, 33:500, 1873.

On the 2nd of January, 1873, there was a cloudy mark, of a semi-circular shape, extending nearly across the disk, and a dark spot in the centre; the illuminated disk itself was singularly egg-shaped. Bad weather prevented me from constantly observing this planet as I should like to have done, but on the 17th April, at 8 o'clock, p.m., I was viewing the planet with one of Mr. Browning's excellent achromatic eye-pieces, when I saw two exceedingly bright spots on the crescent---one close to the terminator, towards the eastern horn, and the other in the centre of the crescent. These spots appeared like two drops of dew; they were glistening in such a manner as to cause the surrounding parts of the bright crescent to appear dull by contrast. Cloudy weather prevented me seeing the planet again until the 19th, when the spots had disappeared, but the planet on this occasion was seen through the aurora, and the irregular and uneven appearance of the terminator was most beautifully depicted. The whole body of the planet also was distinctly visible.

ON THE OBSERVED APPEARANCE OF A REMARKABLE LIGHT SPOT ON THE NIGHT SIDE OF VENUS

Baum, Richard M.; *Strolling Astronomer*, 10:30-32, 1956.

In his valued reports on the activities of the Venus Section, Dr. James C. Bartlett, Jr. has from time to time had occasion to refer to a most peculiar phenomenon of the planet---the "star-spots", minute stellar-like brilliants flashing white in contrast to their duller surrounds. Though attention has only quite recently been centered on these objects, they are not in any way "new" to astronomy. Some 150 years ago the famous German planetary and lunar worker J. H. Schroeter observed then, in one particular instance of celebrity interpreting what he saw as a 25-mile high mountain! Towards the end of the nineteenth century the eminent French astronomer, E. L. Trouvelot, frequently saw them, especially depicting them in great numbers around and within the north and south polar zone. From his valuable memoir on Venus, published in 1892, we gather that he regarded these curious features as the snow-capped summits of lofty peaks protruding up through the dense cloud layers of the planet's atmosphere. Early in the recent century another distinguished observer, E. E. Barnard, not only recorded them, but subscribed to Trouvelot's hypothesis and as a consequence attempted to employ the "star-spots" as a means to determine the planet's controversial period of rotation. Needless to say he did not succeed. More recently Dr. Bartlett, Jr. has given the matter some thought and together with T. A. Cragg, another leading observer of the planet, seems inclined to follow both Trouvelot and Barnard in interpreting the manifestations. Over the period 1947-1954 I myself was able to see them. According to the late H. McEwen, the venerable British authority on Venus, all the so-called classic observers reported, at one time or another, having seen these "star-spots".

However it is not the object of this note to discuss the nature of the "star-spots" but rather to introduce them in order to differentiate their appearance from that of a most remarkable spot which I saw in March 1953, which at first sight seemed very similar to the accepted and known spots, but which upon reflection could not despite this facade be associated in any way with the "star-spots". This fact will emerge from a brief discussion of the general whereabouts of the latter, as we shall presently see.

Though there does seem to be a certain amount of individuality about the behaviour of each observed star-spot---some remaining as hard steely points of light, others which appear to scintillate, and yet others that exhibit strong spectral effects---all have one property in common, all have been seen on the illuminated side of the planet. When, however, they have been noted off the sunlit side they have always been seen a little way inside the terminator, on the night side. They have never been seen off the disk of the planet. From these basic facts it would appear that (a) these spots rely on the sun for their visibility, i.e. they are not artificial as some have been pleased to announce, (b) are the astronomical result of some comparatively rare phenomena of the apparent surface, and as a consequence (c) may be either towering cloud masses, which would seem to be the case with those noted off the terminator, or reflections from some cloud-bank so placed with respect to the sun as to reflect the incident light in the observed manner.

Though this explanation would seem to account for the star-spots as such, it does not do the same for the observation now to be related. Consequently I conclude that the nature of this unique feature is hidden entirely in the general ignorance that characterises our knowledge of Venus.

The observation under review was made in 1953 on March the 17th., at 19^h and continued until $19^h\ 40^m$. U. T. Seeing conditions at the time were estimated to be 4 (scale of 10). Two instruments were used, a 6.5-inch reflector (f/7.4), 216X, and a fine 3-inch refractor by Cooke (f/12), 120X.

Due to overcast skies it had not been possible to commence observations earlier than 19^h, so that when the cloud thinned and cleared it was to reveal a rapidly darkening sky, a condition not conducive to accurate observations of Venus, through reason of the overpowering glare from the planet. However within the hour haze began to form and with this acting as filter it was possible to make an observation and even to attempt a drawing. Happily seeing conditions which before had been classed as 0.5 rose to 4. Apart from some irregularities along the southern limb the most interesting part of this observation was the visibility of the dark side---a manifestation of that most curious phenomenon, the "ashen light".

On this occasion the dark side was seen to be brighter than the sky, as is the case when the planet is projected against a dark field, and to be veritably glowing with a strong phosphorescent glow of a deep ruddy hue. Upon closer inspection the whole of the dark hemisphere was found to be not uniform in hue but actually mottled over with minute nebulous granules. These were seen better by averted vision.

It was whilst studying this strange structure, that my attention was deflected by what appeared to be a speck of light of about the 8th magnitude situated in the southwest quadrant of the disk, i.e., in the southern part of the dark regions. The position angle of this feature measured round from the north point through west (in this case the right hand limb, taking west as right and the bottom as north, as with the Earth) was 171°. Its estimated distance from the terminator taken in angular measure from the drawing made was about 49°, where the disk is imagined flat. Nothing certain was seen by looking directly at the spot but by averted vision it was seen with certainty---additionally I could not be sure but that it was scintillating, which could have been pure fancy though.

Thinking that perhaps the object was due to an optical fault in the reflector, which was the instrument in use at the time, I switched over to the aforementioned refractor. A much sharper image was obtained and the dark side shone out with greater clarity. The fine mottling was also more distinctly perceived. Above all however out shone the stellar speck. Thus having satisfied myself of its objectivity, I at once returned to the reflector and continued to observe until $19^{h}\ 40^{m}$, after which time the planet could no longer be followed, owing to an irregular skyline---trees and distant hills. At any rate it is doubtful whether anything further could have been done, seeing was dropping and there was a good deal of interference from mist along the horizon.

From the time observations commenced and the spot was noted, no observable change took place in it, either in its light or in its position. Owing to the fact of the planet's descent into the horizon mist bank the spot did undergo a fading, but this was not inherent as it was shared by the disk as a whole.

Quite naturally this strange experience enthralled me, and upon the next fine occasion I again looked for the mysterious speck, but to no avail. Later searches and a thorough combing through of my Venus books have revealed that this spot record stands as a unique observation.

Unfortunately the most critical part of this study stands as unconfirmed. As for the viability of the dark side, its reddish tint, and the fine mottling, complete accord has been found with an observation made at the same time as I was observing by an independent worker two hundred miles or so away. For on this same evening at the same hour as

I was so engaged, Mr. M. B. B. Heath, F.R.A.S., using his fine 10-1/4-inch reflector was similarly so. He described exactly the same appearance as confronted me, a notable point in view of the wide difference of instruments used and also the accepted difficulty of the subject, but failed to detect even the slightest trace of the light speck. In later correspondence Heath considered that differences in our respective seeing conditions at the time could have accounted for this, also the fact that he did not continue to observe after 19^h, U.T.

To attempt an explanation of this report is to me fruitless; we know too little about the cytherean problem as a whole to consider an insignificant observation as this essentially is. Hence, apart from making one point I will go no further. The object's great distance from the terminator obviously precludes its having been sunlit, unless of course it was situated away from Venus out of the planet's shadow, and so placed in the line of vision as to appear in the observed position. If not illuminated by the sun and if a surface phenomenon, what then?

POSSIBLE ATMOSPHERIC EFFECTS

VENUS: ABNORMAL REFRACTION AND TRIANGULAR RAY

Gray, J. H.; *Marine Observer*, 32:122, 1962.

s.s. Gardenia. Captain J. H. Gray. Port Arthur (Ont.) to Ipswich. Observer, Mr. J. W. Spence, 2nd Officer.

18th September 1961. The planet Venus was observed rising at 0645 GMT and while low on the horizon it changed at regular intervals of 1 sec. from blood red to white; it actually resembled a shore flashing light. The colour changes continued until an elevation of about 4° was reached. When the planet attained this altitude, a fairly bright triangular ray of light was observed to move from it down to the horizon. This persisted for about 2 min. and then faded out; no more flashing was seen, the planet shining with a white light which flickered slightly for a short time. Air temp. 47^oF; visibility exceptionally good; sky cloudless; wind SW'ly, force 4.

Position of ship: 52° 15'N, 51° 55'W.

Note. Scintillation is due to irregularities in the density of the air in the lower layers of the atmosphere, causing bending of the rays of light from the planet, so that they travel along slightly different paths to the observer's eye. In the case of a planet, scintillation occurs when the changes in direction of its light rays are of the same order of size as its apparent diameter. We are at a loss to account for the triangular ray referred to in the observation.

THE MAEDLER PHENOMENON

Baum, Richard; *Strolling Astronomer*, 27:118-119, 1978.

One of the strangest observations ever recorded of Venus was made by the renowned German astronomer Johann Heinrich Maedler with a four-inch refracting telescope on April 7, 1833. At the time Venus stood east of the Sun and was well placed for observation. In his Beitrage (1841) Maedler tells us how on that evening numerous brushes of light were seen to emanate from the illuminated limb of the planet, then a crescent, and to diverge in a sunward direction. These brushes or rays were roughly fan-shaped, and invested Venus with the look of a broad multi-tailed comet. According to the sketch which accompanies Maedler's account (Figure 59), the brushes originated at the extremities of the crescent and extended along the limb to affect an arc of between 50° and 60° from the tip of each cusp. No brushes are depicted in the segment between, i.e., the center of the limb. Maedler watched for some considerable time and tested the appearance for illusion by rotation of the ocular and by change of magnification; but although he attempted no explanation, he concluded that the appearance was at least real.

The Maedler Phenomenon

"This is certainly possible," opined the Rev. T. W. Webb, "but," adding a note of caution, "it is an instructive instance of the oversights which may be incidental even to great philosophers, that it never occurred to him to try another telescope."

Here is a truly astonishing phenomenon. Uniquely different in its characteristics, gross aberration is immediately suspected. Venus is a difficult subject when studied with the telescope. On a dark sky its glare baffles and tricks even the most practiced eye. And yet it is hard to believe that Maedler, the eminent principal author of the first adequate description of the lunar topography (1837), could have been so deceived. He was a born observer, skilled, competent, and experienced. True, his phenomenon is bizarre but is that a reason to deny it reality? In fact, was Maedler mistaken? Were his mysterious coruscations simply the result of glare? Of dirty optics? Perhaps a greasy optical surface? If not, to what can they be ascribed?

Illusion, of course, cannot be ruled out. However, a work by John Gadbury, Nauticum Astrologicum, or the Astrological Seaman, published posthumously in 1710, suggests otherwise. Gadbury described himself as "a student of physick and astrology," and is mentioned in Aubrey's Brief Lives, and by Augustus De Morgan in his A Budget of Paradoxes (1872), who speaks of him as "a well informed astronomer" and the author of a treatise on comets (1665). In 1655 Gadbury published the first of a long series of annual Ephemerides, and the following year his Emendation of Hartgil's Astronomical Tables; in addition, he wrote prolifically on astrological subjects.

But Gadbury was more than a writer. He was a keen and careful watcher of natural phenomena. Between 1668, November and 1689, December 31 he kept a diary which, to quote his words, was "design'd for the service of philosophers, physicians, astrologers and all other faithful observers of the various wonders that are to be found in God's creation." It was, he further states, his own unaided work and a labor of love. Though little known, the diary is a valuable record of late 17th century London weather, and a fascinating register of optical phenomena seen from London. At that time Gadbury lived in Brick Court, College Street, Westminster, not far from the Abbey. The diary forms a supplement to the Nauticum Astrologicum, the manuscript of which was completed in 1698, and appears between pages 129 and 245. It is complete for the period stated.

Unlike his contemporary chroniclers, Gadbury did not confine himself to daytime events, but as he quaintly remarks, caused himself "the pains of nocturnal watchings." Thus many entries refer to coronae around Venus and Jupiter, and occasionally around Mars and some of the brighter stars. Saturn is also mentioned in this connection, and lunar haloes and coronae are frequently logged. Double and single rainbows were faithfully recorded: on 1682, June 8 a lunar bow was visible from Westminster at 23 hrs. But it is the observations of 1686, January 29 and February 21 (Old Style) which attract attention. On these dates Gadbury noted that Venus appeared "like a comet." The resemblance to the Maedler observation is striking.

To explain the Gadbury phenomenon, it is necessary to infer that the rays of Venus were reflected from a cloud of ice-flakes floating with their bases slightly inclined relative to the horizontal position. The rising or setting Sun is occasionally surmounted by such a luminous brush, and this is formed in precisely the same way. Sometimes the appearance is visible when the Sun is below the horizon. Minnaert also informs us that brushes or pillars have been seen beneath the Sun, although of much shorter length. Oblique or distorted pillars are not unknown. Some on record apparently deviated as much as 20° from the vertical. Minnaert believes this deviation is caused by air currents wafting the ice crystals into a slanted direction.

We do not know what happened back in 1833. There are many gaps and uncertainties in our knowledge. Maedler expressed no simile, but his drawing and description eloquently favor comparison with Gadbury. Putting aside the fact that the former observed telescopically, it is conceivable that his phenomenon is analogous to that of 1686, and thus belongs to the regime of atmospheric optics, being in all probability an exotic form of light pillar classified by Minnaert as oblique or distorted. It would be useful if observers would alert themselves to a possible recurrence of this interesting, if rare, phenomenon, especially telescopic observers of Venus. In the event a full description should be made, including details of weather conditions and the type of cloud involved.

ROTATIONAL AND ORBITAL ANOMALIES

The vague features seen on the surface of Venus were employed by astronomers in the past to estimate the planet's period of rotation. They obtained wildly different results and the "accepted" figure has changed many times. Today's accepted period comes from radar measurements, which supposedly penetrate clear to the surface of the planet. Another modern accomplishment is the determination that the upper atmosphere of Venus, or at least some of its visible features, sweep around the planet at high speeds. No wonder the early astronomers could not agree upon a rate of axial rotation.

Not well publicized is the anomalous advance of the perihelion of Venus. How well does the Special Theory of Relativity account for this observation?

Finally, the legends and wild tales of Venus' strange antics in the past are irresistable. Proponents of celestial catastrophism make much of such stories, but we have found only one such item in the "reputable" astronomical literature.

• Period of Rotation

NOTES ON THE ROTATION PERIOD OF VENUS

Antoniadi, E. M.; *Royal Astronomical Society, Monthly Notices*, 58: 313-320, 1898.

The question of the rotation period of Venus is one to which the reply has eluded many a distinguished astronomer. Nor is the reason of the failures far to seek. The determination is in itself impossible by ordinary visual methods on account of the formidable diffusion of light by the planet's atmosphere, and until the day arrives when the Doppler-Fizeau principle can be successfully applied to the limb of Venus, we may justly display scepticism as to the value of current rotation periods.

The first attempt to determine the rotation period was made by Dominique Cassini in 1666-1667. Cassini considered the spots to be animated by a motion either of libration or rotation, having a period of less than 24 *hours*, round an axis almost coinciding with the plane of the planet's orbit.

Bianchini, who affirms that he had seen *Venus* rotate (1726-1728), fixed the rotation period as 24 *days* 8 *hours*. The axis, he considered, formed an angle of only 15° with the orbital plane.

Influenced by Jacques Cassini's ideas (as shown by Schiaparelli), the fanciful Schroeter found a period of 23 *hours* 20 *minutes* 59.04 *seconds*, while not less biassed must have been Fritsch, in giving, from *four* days' observation, the period as 23 *hours* 22 *minutes*.

Never did scientists, however, indulge more freely in illusions than when the Jesuits of the Roman College undertook, in 1839, under De Vico's direction, the determination of the planet's rotation period. Eleven thousand micrometrical measures of spots taken by Palomba, with unprecedented zeal, gave the value of $23^h\ 21^m\ 21^s 9345$.

The honour of demolishing these fanciful results belongs to Professor Schiaparelli, who, in a masterly discussion of all the observations, concluded that "the rotation of $23^h\ 21^m$ or $23^h\ 22^m$ of Jacques Cassini, Schroeter and De Vico, is the result of a series of paralogisms and vicious circles."

Taking as a base the immobility of some spots relatively to the terminator, the illustrious Director of the Milan Observatory concluded that *Venus* constantly presents the same face to the Sun, or, in other words, that the rotation period coincided with the 224.7 days revolution round that luminary. Professor Schiaparelli's observations extend from 1877 December 8 to 1878 January 6. The fixed spots forming the ground work of the 224.7 days rotation may be described as follows:

(1) An elongated dusky shading *hm* trending obliquely from the south cusp along the path of a great circle of the dichotomous or crescent phase;

(2) A bright spot *h* near the cusp, bounded on the left by

(3) A slight diffused shading *p*.

Professor Schiaparelli notes that the spot *h* was seen by Professor Holden on 1877 December 15; that something like *h* and *p* was observed by Gruithuisen in 1813 and 1814, and by Vogel and Lohse in 1871.

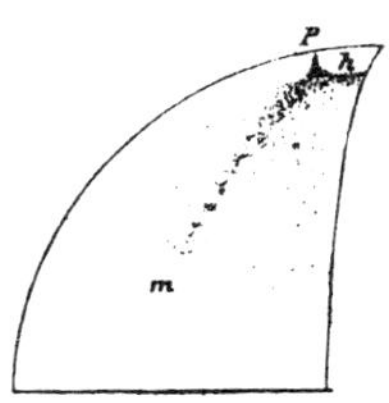

. . . .

Scarcely had Professor Schiaparelli announced his results, than M. Perrotin, starting from the apparent immobility of a vertical shading with regard to the terminator, concluded that *Venus* rotates in a period comprised between 195 and 225 days. M. Perrotin's drawings are of very great value. But we should point out here that he confirms Professor Schiaparelli by using a spot which, in spite of its immobility, was rejected by the great Italian himself in his determination of the planet's rotation period. In fact, M. Perrotin's vertical shading is obviously identical with the spot seen by Schroeter in 1788, and of which Professor Schiaparelli said: "The general stability of its form, which lasted for more than two months, its constant parallelism with the terminator, its almost invariable distance from the latter suggest that we have to deal here with some phenomenon of the atmosphere of Venus, depending much more on the Sun, and consequently on the terminator circle, than on any axis of rotation whatsoever."

In 1895 M. Leo Brenner announced that he had discovered the rotation period of Venus, which he fixed, with an accuracy equalled only by De Vico's, at $23^h\ 57^m\ 7^s.5459$.

Meantime more than one observer was engaged in confirming the long rotation. The last observations reach us from Flagstaff.

At variance with all his predecessors and successors, Mr. Percival Lowell finds the markings of Venus "as distinct really as those of the Moon. . . . The period of rotation coincides with the period of revolution. This planet is a desert."

. . . .

RESONANCE ROTATION OF VENUS

Shapiro, Irwin I.; *Science*, 157:423-425, 1967.

Abstract. Combination of two types of radar data shows the orbital plane and equator of Venus to be included by less than 2 degrees, and the sidereal rotation period to be 243.09 ± 0.18 days (retrograde)---remarkably close to the 243.16-day period for which the spin would be in resonance with the relative orbital motions of Earth and Venus. In this resonance, Venus would make, on average, four axial rotations as seen by an Earth observer between successive close approaches of the two planets. Estimates of the instantaneous spin period, accurate within about 0.01 day, would provide important information on the difference of Venus's equatorial moments of inertia, on their orientation, and on the magnitude of the tidal torque exerted on Venus by the sun.

ATMOSPHERIC TIDES AND THE RESONANT ROTATION OF VENUS

Gold, Thomas, and Soter, Steven; *Icarus*, 11:356-366, 1969.

The observed spin-orbit resonance of Venus, whereby the same side of Venus faces the Earth at each inferior conjunction, cannot be explained adequately by gravitational interaction with the Earth alone. The expected solar tidal drag on the solid body of Venus would easily overwhelm the Earth's couple upon any reasonable permanent deformation of Venus. If there exists, however, a solar atmospheric tide, partly thermally induced and similar to that known on the Earth, its torque may counteract that due to the solar solid body tide at a particular rotation period. The small interaction with the Earth is then sufficient to lock the period to one of the resonances in the vicinity of that angular velocity.

SUPER-ROTATING ATMOSPHERE OF VENUS

Hunt, Harry E.; *Nature*, 266:15-16, 1977.

The atmosphere of Venus is huge, composed mainly of carbon dioxide with ubiquitous clouds obscuring the surface which is characterised by temperatures of about 730 K and pressures of nearly 100 atmospheres. Centuries of visual observation provided little more than frustration to the early astronomers with the regular sight of a yellow, featureless cloud deck, which extends to nearly 100 km above the planetary surface, in complete contrast to Earth which averages about 50% cloud cover. But planetary observations in ultraviolet light have provided the most exciting and unexpected result which still remains today as a major unsolved problem in Venusian meteorology. The upper atmosphere in the neighbourhood of the cloud tops super-rotates. Features in this region move in a retrograde manner with a rotational period of only 4 d which is incredibly short when compared with the planet's own solid-body retrograde rotation of 243 d. So Venus possesses the largest ratio between the atmospheric and planetary rotation rates of any body in our Solar System.

Initial evidence of this atmospheric rotation came from the apparent motion of dark features seen in ultraviolet photographs, characterised by Y and ψ shapes centred on the equator (Boyer & Camichel Annuls Astrophys. 24, 531; 1961). But is it really a mass motion that is observed? A wave phenomenon propagating around the planet with a phase velocity corresponding to a 4-D period could produce a similar effect. Although Young (Icarus 24, 1: 1975) considered the super-rotation an illusion, a travelling wave-like disturbance driven by the ultraviolet albedo differences, spectroscopic measurements of winds by Traub and Carleton (J. atmos. Sci. 32, 1045; 1975) and the results of the Venera 8 Doppler drift experiment (Marov et al. J. atmos. Sci. 30, 1210; 1973) provide strong evidence that the visible parts of the Venusian atmosphere really are moving rapidly with respect to the planetary surface. The measured wind velocity is large enough to account for most, if not all, of the 100 m s^{-1} apparent equatorial motion seen in the ultraviolet markings.

But how persistent are the observed features? According to Boyer and Guerin (Icarus 11, 338; 1966) the Y feature persists for decades at essentially the same longitude in a coordinate system rotating with a period of 4.067 d. On the other hand, Beebe (Icarus 17, 602; 1972) and Scott and Reese (Icarus 17, 589; 1972) thought that owing to the selective nature of the ground-based observations, the features may actually appear at random longitudes and dissipate after 8-16 d. The high resolution images taken during the Mariner 10 flyby are valuable observations in the study of the lifetime of these features, since Dollfus (J. atmos. Sci. 32, 1060; 1975) has shown them to be representative of the more extensive time series of Earth-based measurements. Belton et al. (J. atmos. Sci. 33, 1363; 1976) found that the large scale markings ($\sim$1,000 km) situated between $\pm$ 45° latitude have lifetimes greater than 4 d and more with an apparent angular motion at the equator. Smaller scale markings (100-500 km) were found to have lifetimes between 1.5 and 4 d. These observations suggested to Belton et al. that the Y and polar ring features seen in the images are both visible manifestations of propagating waves.

. . . .

Why is meteorology of the upper atmosphere of Venus so different from that of Earth? The super-rotation requires an efficient high altitude thermal forcing, but for the Earth, the largest amount of available radiative energy is that absorbed at the surface. Could other planetary atmospheres exhibit this excess rotation? Leovy (J. atmos. Sci. 30, 1218; 1973) suggests that any atmosphere where heating tends to produce an equatorial thermal bulge, but which cannot develop instabilities because of slow rotation or large damping, must develop an excess rotation in the same sense as the planetary rotation at a high level. The super-rotation of the Earth's thermosphere reported by King-Hele et al. (Planet. Space Sci. 18, 1433; 1970) may be another example.

• Orbital Anomalies of Venus

THE MOTION OF THE PERIHELION OF MERCURY

Jeffreys, Harold; *Nature,* 101:103, 1918.

Many recent writers on this subject have treated the discordance in the motion of the perihelion of Mercury as if it were the only unexplained perturbation in the solar system. Yet there is an unexplained advance of the node of Venus of the same order of magnitude, the motions in a century being 43" and 10" respectively. The latter estimate is admittedly subject to greater uncertainty, but it is 3.5 times its mean error, and the probability that so large a discrepancy is accidental is only about 0.0004. Now, whatever may be the effect of departure from simple Newtonian dynamics, it cannot alter the plane of an orbit, which can be done only by the attraction of other matter, or to a negligible extent by a moving resisting medium. It is found that a distribution of gravitating matter that would represent the motion of the node of Venus would necessarily account also for the whole of the discrepancy in the perihelion of Mercury, so that departures from Newtonian dynamics to explain the latter make the former impossible to account for. It is, of course, possible that the excess motion of the node of Venus may be due to errors of observation, but the probability against this is about 2500 to 1, and it must be admitted that any theory with such an a priori probability against it is open to very grave suspicion.

VARRO'S STORY OF THE ANOMALOUS TRACK AND FIGURE OF VENUS

Anonymous; *Nature*, 20:351, 1879.

In a resume of the recent progress of astronomy contributed to an American work by Prof. Holden, of Washington, we note a reference to a communication made by M. Boutigny to the Academy of Sciences at Paris in December, 1877, calling attention to a passage in Varro, which describes the planet Venus, as having about the year B. C. 1831 (not B. C. 31, as misprinted in Prof. Holden's Report) "Changed its diameter, colour, figure, and course." M. Boutigny had probably overlooked the circumstance that this story of Varro's had been brought into notice long before, in a French work of astronomical authority, the "Cometographie" of Pingre, who, believing that the fable was originated by some celestial phenomenon, considered it was most probably due to the appearance of a bright comet. Pingre thus gives the fragment, preserved by St. Augustin:---"There was seen, says Varro, a surprising prodigy in the heavens, with regard to the brilliant star Venus, which Plautus and Homer call, each in his own language---the evening star. Castor affirms that this fine star changed its colour, size, figure, and track, which had never occurred before, and which has not occurred since. Adrastus, of Cyzicus, and Dion the Neapolitan, refer this great prodigy to the reign of Ogyges." Pingre's explanation will be found in "Cometographie," t. i. p. 247; the epoch he assigns for the phenomenon is "vers 1770."

The story has no astronomical importance, and is only noticed here from its revival so recently, as mentioned above.

PUZZLING PHYSICAL PARAMETERS

THE UNVEILING OF VENUS: HOT AND STIFLING

Anonymous; *Science News*, 109:388-389, 1976.

Another mystery in the Venera 9 pictures is the apparent shadows cast by the rocks. Avduevsky points out that as the lander descended it took continual measurements of the illumination from all sides. It recorded the sort of diffuse light expected under a cloud cover. "Then it landed, and all of a sudden these shadow." If they are shadows, they would indicate a directed light source in the Venus atmosphere, possibly a rift in the clouds or something more exotic. A facetious suggestion that got laughter all around was a Venusian standing over the lander with a floodlight. Marov points out that the black spots may not be shadows but depressions or differences in coloration of the surface. But then, why are they all on the same side of the lander and of the rocks?

The Venus clouds turn out to be more tenuous than anybody had thought. Nephelometer measurements show them to be made of aerosols or droplets about one micron in size that appear to be something like sulfuric acid. Their concentration is about one or two parts per million of the carbon dioxide that is the major atmospheric constituent. So they are actually more like a haze than heavy clouds. As a result, the surface illumination is brighter than anyone expected, and photography is much easier there. With the sun at a 30° angle from the zenith, the light flux at the surface is about 100 watts per square meter, an illuminacy of about 14,000 lux. "This value corresponds to the illuminacy at the terrestrial mean latitude in the daytime with overcast clouds," Keldysh writes.

....

WHY DID MARINER II FIND VENUS NON-MAGNETIC?

Anonymous; *New Scientist*, 17:10, 1963

Scientific speculation is aroused by the discovery of the American space probe, Mariner II, that the planet Venus has a negligibly small magnetic field. The announcement of this first revelation by Mariner was made last week by P. J. Coleman of the University of California at the annual meeting of the American Association for the Advancement of Science, held in Philadelphia.

There seems to be no reason to doubt that the spacecraft's magnetometer was working perfectly well both throughout its long journey to the planet and during the more crucial period of "fly-by". Values obtained for the interplanetary magnetic field tallied with those from earlier space experiments, notably the measurements made by Pioneer

V in 1960. In hard figures, what Mariner II found was that, at its closest approach to Venus---some 21,594 miles---there was no significant change in the interplanetary field greater than five gammas, one gamma being a hundred-thousandth of a gauss, or about a fifty-thousandth of the value of the surface field at the Earth's poles.

The discovery could mean one of a number of things. While no one knows for certain just why the Earth has a magnetic field at all, the widely accepted view at present is that it is the result of some kind of dynamo action in the Earth's core which is known from geophysical arguments to be an electrically conducting fluid. The conditions of pressure and temperature that generate such a liquid core are dependent, according to modern theories, upon the size of a planetary body. Moreover, the strength of the fluid dynamo and its associated magnetic field must depend on the rate at which the planet rotates.

Venus is approximately the same size as the Earth, and it is quite reasonable to assume that it may similarly possess a conducting core liquified by the long term build-up of radioactive heat, and by the enormous pressure inside it. Hence, by comparison with terrestrial theories, the most likely reason why it has little or no magnetic field is that it does not rotate very fast. In fact, there has been considerable dispute in recent years among astronomers as to what the length of Venus's day really is---estimates vary as widely as 22 hours and 225 days. It is a difficult thing to measure because the planet is shrouded in a complete and permanent cloud blanket so that no surface features can be tracked.

....

A less likely explanation is that the field of Venus is in the middle of undergoing a reversal of direction. Evidence collected from studies of rock magnetism indicates that the Earth's field has reversed its direction about once every million years or so, and that its intensity fell to zero for short periods in between. Conceivably, Mariner caught Venus at just such a moment.

....

VENUS BREATHES IN STEADY FASHION

Anonymous; *New Scientist*, 58:72, 1973.

Astronomers are well enough acquainted with periodic variations in the light from the stars, but a variable planet is quite a different matter. However, the planet Venus shows regular changes in the spectrum of its atmosphere, according to four scientists at Caltech's Jet Propulsion Laboratory. The strengths of carbon dioxide lines in the Venusian atmosphere swing through a four-day cycle (Astrophysical Journal, vol 181, p L5).

Over 20 years ago Gerard Kuiper noted day-to-day fluctuations in the infrared spectrum of Venus, but no one has yet got to the bottom of the basic cause of these changes. In order to study the oscillations, A. T., L. G., and J. W. Young and J. T. Gergstrahl obtained spectra nightly during the autumn of last year. Their data on the carbon dioxide line show an unmistakable oscillation.

The observed variation is not exactly periodic, but more akin to a relaxation oscillation in which the amplitude builds up on successive cycles and then suddenly collapses. In order to produce the observed changes the cloud deck of Venus must be moving up and down by as much as one kilometre, simultaneously over the entire surface of the planet. Such a large atmospheric oscillation requires a high input of mechanical energy. This condition is difficult to account for in the case of a slowly rotating planet heated uniformly by the Sun's rays. Therefore the cycle variations point to some unexplained deep-seated property of the atmospheric dynamics.

NEITH, THE LOST SATELLITE OF VENUS

It is fashionable these days to discover new solar-system satellites and planetary rings. Yet, modern astronomers quickly dismiss the several century-old observations of a Venusian satellite. Certainly, these old sightings of Neith, as the satellite of Venus was called, were just as sound and convincing as the recent reports of a moon circling Pluto. Some astronomers even saw the satellite in its proper phase! Because Neith has not been seen in modern times it never existed, for temporary moonlets are not permitted. If something unusual was seen near Venus, it must have been a faint star or asteroid, as concluded by one of the astronomers quoted herein. Nevertheless, the observations of Neith, like those of Vulcan (Chapter 2), remain enigmatic, although eroded by time and accumulated disdain.

THE SATELLITE OF VENUS

Webb, T. W.; *Nature*, 14:193-195, 1876.

That something strongly resembling a satellite has been occasionally seen near Venus, especially about the middle of the last century, is beyond a doubt. It is equally certain, and familiar to all experienced observers, that reflected images, or technically "ghosts," may, under certain circumstances, be formed in the eye-piece of the telescope, and might be the means of causing deception: and the whole matter is reduced to the simple inquiry, whether all the recorded instances admit of this easy explanation; though, if they do not, it must be remembered that the existence of a satellite would not necessarily follow.

. . . .

The name of Cassini at the head of them at once commands attention, but there is nothing in his two observations in 1672 and 1686 that does not lend itself to Father Hell's hypothesis, excepting the care and experience of such an observer, who must have been familiar with every telescopic defect. The observation of Meier, which seems to have lain unnoticed in the Astron. Fahrbuch, 1788, till brought forward by Schorr, is on that account worthy of being cited in full. "1759, May 20, about 8h. 45m. 50s., I saw above Venus a little globe of far inferior brightness, about 1-1/2 diam. of Venus from herself. Future observations will show whether this little globe was an optical appearance or the satellite of Venus. The observation was made with a Gregorian telescope of thirty inches focus. It continued for half an hour, and the position of the little globe with regard to Venus remained the same, although the direction of the telescope had been changed." During so lengthened an observation it seems natural to suppose that the eye must have been repeatedly removed and replaced, which could not have occurred without the detection of an optical illusion.

In 1761, when the expected transit drew attention to Venus, Montaigne, at Limoges, was persuaded to undertake the inquiry, though he had little faith in the existence of the satellite, and was not greatly disposed to enter upon an examination in which so many great men had failed. However, on May 3 he saw a small crescent 20' from Venus; it is expressly stated that the observation was repeated several times, and that after all he was not certain if it was not a small star; which, with a power of between forty and fifty, was not surprising. The next evening and on the 7th and 11th it was again seen, rather more distant, and each time in an altered position, but with the same phase as its primary; and on the 7th it was seen, and even much more distinctly, when Venus was not in the field. The improbability is obvious of such persistency in an illusion so readily detected. The cause may indeed have lain in the object-glass; such telescopes have been known. Wargentin, at Stockholm in the same year, found that his instrument produced a deception from this cause; and the 6-inch Cauchoix achromatic at Rome showed minute comites to bright stars a little too frequently for the credit of those who trusted it. Montaigne's changed position angles may be thought to indicate this cause of error, as his 9-ft. refractor probably admitted of rotation in its bearings, but it is a singular coincidence that these changes should all have been in the direction of orbital revolution, and still more, in such proportions as to be reconcilable with Lambert's calculated period of about eleven days; and it is quite unintelligible that he should not have subsequently detected the fault in his telescope, as from his estimation of angles and distances he was evidently not a novice in observation. Three years later, in 1764, Rodkier, in Copenhagen, saw such an appearance on two evenings with a power of thirty-eight on a 9-1/2-ft. refractor; on the latter occasion with a second telescope also. There is little in this to contravene the Vienna theory, especially as this second telescope had a coloured meniscus eye-glass, and he failed in finding it with two other instruments: but it is more remarkable that on two evenings a week later the same telescope told the same tale to four different observers, one of whom was Horrebow, the Professor of Astronomy, and who, we are assured, satisfied themselves by several experiments before the second observation that it was not a deception. That the necessary conditions for its being such could have been maintained before so many eyes, is, notwithstanding its admitted pale and uncertain aspect, what could not possibly have been anticipated. But we have not yet done with this temporary outbreak, so to speak, of visibility. Before this month of March was ended, Montbarron at Auxerre, far removed from all possibility of communication, and with a very different kind of telescope, a Gregorian reflector of thirty-two inches, which of course was fixed as to its optical axis, perceived on three separate evenings, at different position-angles, something which, though it had no distinguishable phasis, was evidently not a star, and which he never could find again.

There remains still the observation of the celebrated optician Short. It is indeed chronologically misplaced here, but has been intentionally deferred as affording the strongest point in the whole affirmative evidence. As his own account is an interesting one, and has seldom, if ever, been reprinted, our readers may not be displeased to see it here as it stands in Phil. Trans. vol. xli.:-

"An Observation on the Planet Venus (with regard to her having a satellite), made by Mr. James Short, F.R.S., at sunrise, October 23, 1740.---Directing a reflecting telescope of 16.5 inches focus (with an apparatus to follow the diurnal motion) towards Venus, I perceived a small star pretty nigh her; upon which I took another telescope of the same focal distance, which magnified about fifty or sixty times, and which was fitted with a micrometer in order to measure its distance from Venus, and found its distance to be about 10° 2' 0" (sic). Finding Venus very distinct, and consequently the air very clear, I put on a magnifying power of 240 times, and to my great surprise found this star put on the same phasis with Venus. I tried another magnifying power of 140 times, and even then found the star under the same phasis. Its diameter seemed about a third, or somewhat less, of the diameter of Venus; its light was not so bright or vivid, but exceeding sharp and well defined. A line, passing through the centre of Venus and it, made an angle with the equator of about eighteen or twenty degrees. I saw it for the space of an hour several times that morning; but the light of the sun increasing, I lost it altogether about a quarter of an hour after eight. I have looked for it every clear morning since, but never had the good fortune to see it again. Cassini, in his Astronomy, mentions much such another observation. I likewise observed two darkish spots upon the body of Venus, for the air was exceeding clear and serene."

THE PROBLEMATICAL SATELLITE OF VENUS

Anonymous; *Observatory*, 7:222-226, 1884.

A very interesting contribution to the literature of this enigmatical body has recently been made by M. Houzeau (until lately director of the Royal Observatory of Brussels) in an article appearing in 'Ciel et Terre' for 1884, May 15. M. Houzeau is unwilling to believe that the cases in which a satellite was seen near Venus were all illusions, "for all these observations were made either by celebrated astronomers, such as Cominic Cassini, or at least by experienced observers," and it might be added that in the cases of the observations of Short and Roedkiaer, the object was seen with more than one telescope, and with several different eyepieces. Short, indeed, actually measured the distance of the "satellite" from Venus with a micrometer, whilst Roedkiaer's observations were confirmed on more than one occasion by the other astronomers of the Copenhagen Observatory. M. Houzeau cannot admit the body in question to be a satellite which only becomes visible under accidental circumstances, "first on account of the impossibility of properly representing the observed positions by an orbit described round Venus, and further because the mass of the planet deduced from the least defective attempts would be seven times the real amount."

Some years ago M. Houzeau had suggested that the problematical satellite might possibly be an intra-Mercurial planet. "Let a small planet, revolving within the orbit of Mercury, come on some occasion into so close an apparent approach to Venus as to be visible in the field of the telescope with it, and it would appear beside the larger disk of Venus as a body of smaller size, presenting almost the same phase as the great planet. This is precisely what has been observed."

"There was a method of deciding whether this explanation was admissible. An intra-Mercurial planet could not pass as far from the Sun as Venus does or even Mercury. It could then be seen near the former only at times, when, by its apparent motion, it was not very far from the Sun. If the observations were made sometimes at distances from the Sun in excess of the greatest distance of Mercury, it would be necessary to seek another hypothesis."

In order to test this theory, M. Houzeau therefore drew up the following table of seven of the best known and best attested instances in which the "satellite" of Venus is believed to have been seen:-

No. of the Observation.	Date.	Venus morning or evening star.	Heliocentric longitude of Venus	Elongation.	Geocentric latitude.	Distance from the Earth. Radius of Earth's orbit 1.
1.	1645, Nov. 15	E	309°	31°	-2°.0	1.37
2.	1672, Jan. 25	M	162	46	4.8	0.59
3.	1686, Aug. 28	M	59	38	-0.7	1.17
4.	1740, Oct. 23*	M	68	46	-0.5	0.60
5.	1761, May 7	E	207	34	5.4	0.45
6.	1764, March 4	E	59	30	-0.7	1.38
7.	1764, March 28	E	98	35	1.2	1.24

* Old style. The line should read therefore:-
4. 1740, Nov. 3 M 86 46 0°.7 0.64.

M. Houzeau points out that this table completely overthrows the suggestion he had made, since in every instance Venus was further from the Sun than an intra-Mercurial planet could possibly be.

The table, however, gives rise to another hypothesis. Expressing the dates in years and decimals of a year, in order better to ascertain the intervals between them, M. Houzeau obtains the following table:-

No. of the Observation	Date.	Interval.	Number of periods.	Length of period.
1.	1645.87	years.		years.
2.	1672.07	26.20	9	2.91
3.	1686.65	14.58	5	2.92
4.	1740.84	54.19	18	3.01
5.	1761.34	20.50	7	2.93
7.	1764.24	2.90	1	2.90
Total interval		118.37	40	2.96

* The table as given by M. Houzeau is here altered for the true date of Short's observation.

It will be observed that the successive intervals are multiples of very nearly the same period, which is almost exactly the length of the interval between the two last dates.

M. Houzeau asks, "Is this agreement, six times repeated, wholly the effect of chance? Undoubtedly it may be only accidental; but the probability of the opposite view is so great that it will not be without interest to examine what a like periodicity would indicate, supposing it were established. There are two bodies, the one relatively large, and the other described as of much smaller dimensions, which appear side by side at nearly constant intervals. There can be no question that it cannot be a true satellite, since the two are separated in the intervals. It follows from the observed facts that the path they follow brings them together at fixed intervals of time. These paths are near one another throughout their entire length, for conjunctions have been observed at different parts of the orbit of Venus, on this and on the further side of the Sun, and to the east and west of that luminary. These conditions can only be satisfied by imagining two orbits sensibly concentric and of radii which differ but very little.

"To save time I will give the supposed star, which has to be mentioned several times, a name. Any name will do: I choose Neith, the name of the mysterious goddess of Sais, whose veil no mortal raised. I will say then: Venus and Neith come again to apparent conuunction every 2.96 years, i.e. about 1080 days, in their concentric orbits. It is evident that Neith moves either quicker or slower than Venus, and that after having either gained or lost a revolution it is again at the same longitude as the latter."

The second hypothesis is shown to be the true one by reference to Table 1, which shows that Neith and Venus have been seen together when Venus is at greatest elongation from the Sun. Neith must then be a planet a little exterior to Venus, and M. Houzeau shows must have a period of about 283 days. But if the period of Neith were 281 days, then five revolutions of Venus would exactly equal four revolutions of Neith, and the influence of the former upon the latter, which in any case would have been great, would occasion very considerable perturbations.

. . . .

It should be noticed that there are several observations extant, besides those here employed by M. Houzeau. Altogether there are about thirty that have fair claims for consideration. Amongst these are two series made at Copenhagen, the one in 1761 and the other in 1764, and detailed in 'Copernicus,' Vol. ii. p. 164, by Dr. H. C. Schjellerup. The entire series of the 1761 observations commences on Feb. 10 with the first of La Grange's observations, or, if these be rejected, with Montagne's on May 3, and extends to Aug. 13, the date of Roedkiaer's last observation in that year. Again, in 1764 we have Roedkiaer's observations of March 3 and the following days, and Montbarron's of March 15, 28, and 29. It is impossible that the two planets moving at such different speeds should have appeared to continue in conjunction for so long a time, especially in 1764, when Neith, the more slowly moving body, must have been the further from the Earth.

It seems clear, therefore, that we cannot add Neith to our list of planets; yet the theory is noteworthy from its ingenuity and from its affording another example, in which the solar system appears prolific, of a purely accidental numerical coincidence, none the less accidental that the exclusion of Fontana's observation (so-called), which clearly arose wholly from the defects of his instrument, and the inclusion of Roedkiaer's observation of 1768 Jan. 4, would have rendered it less perfect. It is curious, too, that this is the second such coincidence to which the observations of the "satellite" of Venus have given rise, since Lambert was able to combine all the observations in an orbit, which only failed by demanding a mass for Venus much larger than it really possesses. It might, however, be possible to revise Lambert's elements, since the mean distance of the hypothetical satellite which he obtained appears to rest mainly on Montagne's observations; and these were not measures but estimations, and are very greatly in excess of the distances as shown in the diagram in Baudouin's 'Memoire.' To diminish Lambert's distance would, however, only remove one difficulty to create another, since it would make it more than ever incredible that the planet should have been scrutinized for 115 years, and watched during three transits, without revealing any trace of so close a companion if it had any real existence.

It seems impossible to suggest any new theory which shall be more successful than its predecessors have been. Uranus and the asteroids, all then undiscovered, might, it has been rather hastily conjectured, have one or other been in conjunction with Venus at these different dates. Uranus was indeed in the neighbourhood of Venus on 1764 March 4, and on that occasion alone, but not even then near enough to account for Roedkiaer's observation on that day, whilst the minor planets are too small and too dull for the theory with regard to them to have any plausibility. There would seem to remain only the vague and unsatisfactory resource of ascribing as many of the observations as possible to false images, of regarding others as observations of stars, and in the case of the remainder, where the observations were too minute and precise to allow of these hypotheses, to adopt Webb's suggestion of "atmospheric reflection or mirage." Perhaps, however, it is better to be content with things as they are, and to leave the "satellite" of Venus, at least for the present, as an unsolved "astronomical enigma."

THE SUPPOSED SATELLITE OF VENUS

Anonymous; *Observatory*, 10:363-364, 1887.

This long-standing astronomical enigma seems at last to have received a solution, at once simple and sufficient, at the hands of M. Paul Stroobant, who has recently presented a paper on the subject to the Brussels Royal Academy of Sciences. M. Stroobant had been studying the question after it had been revived again by M. Houzeau's ingenious theory (see 'Observatory,' Vol. vii. p. 222), published in 1884, and in the course of his inquiry he tried to find out what was the star spoken of in the observation of the Copenhagen astronomers made on Aug. 4, 1761, and was led to the unexpected conclusion that the "satellite" itself was none other than a fixed star. Following out the line of research thus suggested, he found that in the great majority of cases there was at least a probability that a star had been observed, and in several instances there was no reasonable doubt at all that this was what had happened. Thus Horrebow evidently had observed θ Librae on Jan. 3, 1878, and the motion he ascribed to the "satellite" was really due to Venus herself. So Roedkioer seems on different occasions to have observed X4 Orionis, v Geminorum, and m Tauri. These are all fairly bright stars, being of the fifth magnitude or brighter, and might be easily seen even when close to the planet; but there was no star of similar brilliancy in the neighbourhood of Venus at the times Short, Montaigne, La Grange, and Montbarron made their respective observations. In Montaigne's case three stars of $7^m.1$, $8^m.4$, and $7^m.8$ respectively, are very near the successive places of the "satellite," so near that but for their faintness, especially as Venus was then at her brightest, and the fact that Montaigne describes the "satellite" as crescent-shaped and 10" in diameter, there could be no hesitation in concluding that they were the actual objects seen. M. Stroobant has found, however, that, with a telescope of six inches aperture, it was possible to see a star even of the ninth magnitude quite near to Venus, so that the objection from the faintness of the stars is much weakened, whilst the apparent size of the disk may have been due to bad focussing. Most of the observations are in this way at least plausibly explained as having been really of fixed stars, and several pretty certainly. A few observations are rejected ---those of Fontana, for instance, which were manifestly due to the bad quality of his telescope. Scheuten, during the transit of June 6, 1761, evidently saw only a sun-spot; and the only series left unexplained is that made at Copenhagen in March 1764. It is possible that in one or two cases it may be discovered that one of the larger asteroids was sufficiently near to have been observed; but in the majority of cases there can be little doubt but that the object seen was a fixed star.

M. Stroobant has completed his paper in a very thorough way. He supplies in one table the details of all the observations of the "satellite" available, in a series of other tables the different elements of Venus at the times when the observation were made, and has given a series of diagrams illustrating the position of Venus amongst the stars at the times named. He has likewise given a brief abstract of the different theories which have been started at various times in order to account for the observations of the "satellite," and, more important still, since some were not very accessible, has reprinted the original observations themselves. Altogether, M. Stroobant's paper is the most important contribution that has yet been made to the literature of an interesting and perplexing subject; and there can be no reasonable doubt but that he has, to a considerable extent at least, given a satisfactory explanation of what has been a puzzle to astronomers for so many years.

Chapter 5

THE ENVIRONS OF EARTH

INTRODUCTION

The earth itself is an astronomical object only for astronauts and extraterrestrial life. Nevertheless, the astronomical observations of artificial satellites have revealed the apparent superrotation of the earth's upper atmosphere---a feature sufficiently curious to warrant including it here (Superrotation also occurs on Venus). There is also evidence that interplanetary debris has accumulated at two of the earth-moon Lagrangian points of stability, creating in effect dust-cloud satellites of earth. The history of other natural objects in apparent orbit about the earth ranges from the infamous moons of Waltemath to the puzzling modern data collected by Bagby that suggests ephemeral natural satellites of earth. This subject is obviously closely related to anomalous meteors (Chapter 7) and the so-called "enigmatic objects" (Chapter 13).

• Super-Rotation of the Upper Atmosphere

SUPER-ROTATION OF THE UPPER ATMOSPHERE

Hughes, David W.; *Nature*, 249:405-406, 1974.

Observations of small changes in the orbital inclinations of artificial satellites have shown that the Earth's upper atmosphere (at altitudes of 150-400 km) is rotating about 20-30% faster than the Earth itself. This phenomena has become known as super-rotation. Even though it was discovered by King-Hele in the early 1960s, the cause of this 100 m s^{-1} west to east wind is still a subject of considerable debate. The change in inclination because of super-rotation is only about 0.1^{o} during the satellite lifetime, and so spatial variations and short term effects are very difficult to detect. No clear evidence has been found of year to year variations, such as might be caused by the 11-yr-periodicity in solar activity, but over daily periods the wind seems to maximise between 2100 and 2400 local time.

Possible causes of super-rotation are:

- Horizontal pressure gradients may be set up in the atmosphere as a result of thermospheric heating by solar ionising radiation, and bombardment by energetic particles. The resulting global pressure distribution produces geostrophic and cross-isobaric winds which are probably controlled by ion drag (the motion of the F region ions borne along by the neutral winds is strongly effected by the geomagnetic field). A problem with this theory is that heating of the auroral zone, which takes place during magnetic disturbances, can reverse the mean meridional pressure gradient. This would cause a drastic change in the superrotation, a change which has not been observed.
- Super-rotation might be caused by ionospheric or magnetospheric motions which couple with the neutral atmosphere by collision processes. Meridional electric fields can move F region ions quite efficiently even in the presence of the geomagnetic field. An average field directed towards the equator could produce the super-rotation, and these fields have been observed in high northern latitudes during negative magnetic bays. The effect should, however, be enhanced during magnetic disturbances. This is contrary to observations.
- The moving flame effect. A heat source acting on a fluid produces convective motion. If the heat source moves (as, for example, the Sun does in a westerly direction with respect to the Earth's atmosphere) the convective cells tilt, producing horizontal and vertical motions which interact to produce a mean horizontal flow. The flow rate is a function of the frequency of the heating function, and of the ratio of the kinematic viscosity and thermal diffusivity of the fluid. Permissible values of such parameters in the case of the Earth's atmosphere unfortunately yield inappreciable amounts of super-rotation; a situation which compares unfavourably with the case of Venus where the upper atmosphere rotates 50-60 times faster than the planet mainly because the heating function has a period of 240 days (the Venusian surface rotation period) as opposed to 1 day on Earth.
- Atmospheric waves may impart momentum to the F region, in which they are dissipated. If transmission efficiency depends on the direction of wave propagation, a net motion can be produced. As wave velocity is an inverse function of density, an imperceptible mean velocity at low levels can produce the observed thermospheric velocity.
- The Scott effect may produce super-rotation: a small brass cylinder suspended in a cylindrical vessel containing a low pressure gas, experiences a torque when a vertical magnetic field is applied, and if a temperature gradient is set up between the cylinder and the vessel. This effect might act on the atmosphere.

At present it seems that none of these theories can explain the phenomena of super-rotation completely. A sixth theory has now been put forward by Ved Mitra (Physics Department, MNR Engineering College, Allahabad, India) in a recent edition of Planetary and Space Science (22, 559; 1974). In this case, super-rotation is caused by meteoroid deposition. The meteoroids, small interplanetary dust particles which mainly originate from decaying comets, and which move in elliptical orbits around the Sun, carry, on average, a greater amount of angular momentum than that corresponding to the Earth's orbit. When they interact with the Earth's atmosphere the angular momentum must be conserved and therefore the excess orbital angular momentum of the meteoroids becomes manifest as extra spin angular momentum in the atmospheric layers in which they are arrested.

....

Unfortunately, there are two problems with Mitra's argument. The first is minor and concerns the size of the incident particles. Recent work (for example, Hughes, Space Res., 14, 789; 1974) indicates that about 95% of the mass incident on the Earth is in the form of particles of mass 10^{-9}-10^{-5}g which have very low geocentric velocities and different orbits from those assumed by Mitra.

The second point concerns Mitra's assumption that "the effective deceleration of the velocity initially possessed by a meteoroid takes place in the 150-400 km region". This is contrary to visual and radar observations of the more massive meteoroids ($>10^{-5}$g) which have maximum ablation at heights of about 90 km. Either it must be assumed that the super-rotation is produced by the minor amount of ablation and momentum transfer that occurs at heights above 150 km, in which case Mitra's 34 tons a day influx must be increased accordingly, or a process of momentum transfer from the lower layers where the ablation occurs to the regions of super-rotation must be invoked; similar to that propounded for the transfer of atmospheric wave energy.

It seems therefore that the cause of super-rotation still remains a mystery and that the Earth's spinning upper atmosphere is still the happy hunting ground for new theories.

• Other Moons of the Earth

THE SECOND MOON TO EARTH

Anonymous; *Popular Astronomy*, 6:253, 1898.

So good a paper as the Popular Science News has nearly a column article, in regard to a second Moon to the Earth. That which will interest astronomers about this second Moon is the alleged fact that its orbit has been computed. Its distance from Earth is 640,000 miles; its diameter is 435 miles. It is faint generally, and can be seen only by the aid of a large telescope. "Sometimes," says Dr. Waltemath "it shines at night like a Sun," and he thinks Lieut. Greely, when in Greenland in 1881, saw this little Moon, and by mistake regarded it as the

Sun, though it was ten days after the time when our Sun should have been invisible, and this fact was what puzzled Lieut. Greely. But Dr. Waltemath makes all plain when he says that Lieut. Greely saw this little Moon and mistook it for our Sun out of place.

That is a very good one for Lieut. Greely. It is the best joke of the season surely! It is not probable, at all, that anyone having any knowledge of the elements of astronomy would be mislead by such statements.

THE MANY MOONS OF DR. WALTEMATH

Ashbrook, Joseph; *Sky and Telescope*, 28:218, 1964.

Compiler's Digest. Dr. G. Waltemath, of Hamburg, announced in the late 1890s that he had discovered not only a second moon of the earth but also a whole system of midget moons. The "second" moon, Waltemath claimed was 435 miles in diameter and orbited at a distance of 640,000 miles from earth. It shone like a small sun at night on occasion. He suggested that the bright object seen in Greenland in 1881 by a Lt. Greely was actually this second moon. Public interest was aroused when Waltemath predicted his second moon would pass in front of the sun on the 2nd, 3rd, or 4th of February, 1898. Strangely enough, something of this sort was seen by Martin Brendel, in Greifswald, on February 4, and reported in Astronomische Nachrichten. However, experienced astronomers elsewhere were also watching the sun on that fateful day and saw nothing. Waltemath was contemptuous of conflicting observations and opinions. He even published a series of Astronomical Reports, subtitled Organ of the Union for Investigation of the Dark Moon of the Earth. Waltemath was clearly an eccentric, and his observations are suspect.

Ashbrook introduced Waltemath's story by referring to a note in the German astronomical magazine Die Sterne, in which a German amateur astronomer named W. Spill is described as seeing a second earth satellite cross our first moon's disc on May 24, 1926.

NEW NATURAL SATELLITES OF THE EARTH

Anonymous; *Sky and Telescope*, 22:10, 1961.

The discovery of two faint, cloudlike objects circling the earth at the same distance as the moon has been reported by the Polish astronomer K. Kordylewski, at Krakow Observatory.

His find is the result of many years of searching, based upon the idea that the most likely locations for additional natural satellites are the Lagrangian points of the Earth-Moon system. These places have the property that small bodies located near them will persist in stable orbital motion. Two of these points, known as L_4 and L_5, are at the same distance from the earth as the moon, the first 60 degrees ahead of the moon in its orbit, the other 60 degrees behind. In the Sun-Jupiter system, the corresponding points are each the center of a group of Trojan asteroids.

. . . .

[See also Sky and Telescope, 22:63, 1961.]

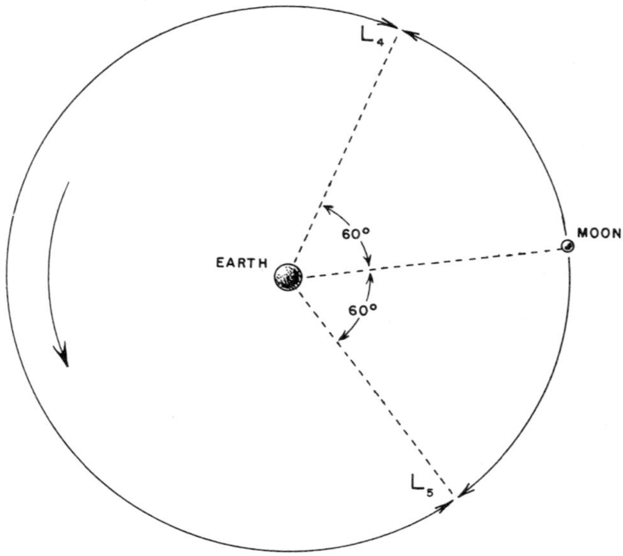

The L4 and L5 Lagrangian points form equilateral triangles with the earth and moon. Kordylewski's clouds were found near L5.

EVIDENCE OF AN EPHEMERAL EARTH SATELLITE

Bagby, John P.; *Nature*, 211:285, 1966.

The possibility that the Earth could acquire natural satellites from the debris of large meteoritic impacts on the Moon has been discussed by O'Keefe. A possible capture mechanism for bodies into temporary Earth orbit has been investigated by Baker. On February 9, 1913, a widely observed, long duration procession of meteors and fireballs gave rise to the suggestion by Chant, Mebane, and others that they actually were such ephemeral Earth satellites which were finally consumed at perigee. More recently, a long, extremely narrow field of both craters and meteorites in Argentina was investigated by Kohman et al. and they suggested that these were the end-result of a captured Earth satellite finally spiralling into the denser atmosphere.

During the mid-fifties I analysed evidence of two swarms of such objects in retrograde orbits. However, it has never been conclusively proved that such an ephemeral satellite, with a perigee in the upper atmosphere, has been observed while in orbit and before its final revolutions. I now suggest that the Earth has had such a satellite and that it was observed at least eight times during the period 1956-65. Fortunately, these eight known observations were recorded with sufficient accuracy to permit the determination of an orbit. These observations are listed in Table 1.

The first three observations were accidentally made by Metcalfe and forwarded to me as possible evidence of an inner planet. The brilliance of the objects to the naked eye caused most persons contacted to doubt that they were astronomical despite the fact that this was confirmed with optical aids. However, when Kayser and I made the fourth observation (also with optical aid) I computed an approximate orbit, assuming all four observations to be of an Earth satellite.

Table 1

Long.	Lat.	Date	Time	Azimuth	Elevation	Magnitude	Observer
87.7 W.	39.6 N.	Nov. 17	01:20	248	20		E. Metcalfe
		1956	01:33	242	0		E. Metcalfe
87.7 W.	39.6 N.	May 24	01:20	291	20		E. Metcalfe
		1957	01:37	291	0		E. Metcalfe
87.7 W.	39.6 N.	Nov. 10	23:30	287	40		E. Metcalfe
		1957	23:58	287	0		E. Metcalfe
118.5 W.	34.0 N.	Dec. 9	03:32	233	19		B. Kayser
		1964	03:35:40	260	3.5	3	J. P. Bagby
118.5 W.	34.0 N.	Dec. 29	02:20:39	250	12.5	2	J. P. Bagby
		1964	02:22:24	280	5	2	J. P. Bagby
118.5 W.	34.0 N.	Jan. 10	02:16:48	278.5	5.5	3	J. P. Bagby
		1965	02:16:56	280	5	3	J. P. Bagby
118.5 W.	34.0 N.	Oct. 25 1965	02:08	231	23	2	R. M. Hartmann
			02:11	281	4.5	2	J. P. Bagby
118.5 W.	34.0 N.	Dec. 14	02:57:09	303.5	30	3.5	J. P. Bagby
		1965	03:00:10	296	13	3.5	J. P. Bagby

Continued surveillance over the next 60 days during every possible evening twilight period resulted in the fifth and sixth observations being made. The fifth observation was made with optical aid, but the sixth was by naked eye only. These further observations made it possible to improve the determination of the orbits, and osculating orbits were distributed to about twenty individuals and astronomical agencies.

During most of 1965, no one else reported to me that they had seen the object. Then Hartmann and I made the seventh observation, and this further extended the orbit description. It also led to another period of evening twilight surveillance, which resulted in an eighth observation being made and further orbital refinement. These new results on orbits were recently distributed to thirty individuals and astronomical agencies, but so far no one else has reported an observation to me. It is probable that the object decayed in the Earth's atmosphere during January 1966.

....

RADIO ANOMALIES ASSOCIATED WITH AN EPHEMERAL SATELLITE STILL IN ORBIT

Bagby, John P.; *Nature*, 215:1050-1052, 1967.

Attention was recently directed in Nature to some radio reception anomalies observed in 1962, 1963 and 1965 at several different locations receiving the VLF stations GBR (Rugby) on 16.0 Kc/s and NBA (Panama Canal Zone) on 18.0 Kc/s. These anomalies consisted of an interfering signal just off the nominal transmitted frequency by some 10^{-7} parts, sometimes above the nominal value and sometimes below it. The interfering signal was often as strong as or stronger than the GBR or NBA transmissions, depending on the location of the receiving station. Two of the possible explanations for these anomalies have been discussed---nearby unknown transmitters operating on the same frequencies, and an alternate antipodal transmission path. I have studied the Nature report and others and offer some independent evidence of high frequency radio reception anomalies and a third possible explanation.

Early in January 1961 I listened to the high frequency time signals of CHU (Ottawa, Canada) on 7.33 Mc/s at Pacific Palisades, California. Occasionally, for periods of 30 min and more the signal, usually weak at my location, was strong enough not to require an external antenna. To examine this phenomenon further, I monitored CHU on 7.33 Mc/s almost continuously each evening during February and March 1962. There were many brief periods when the signal-to-noise ratio became very large, sometimes reaching 18 dB.

At first, I assumed these enhancements would correspond with the onset of evening twilight at a skip zone, but my early 1962 records clearly demonstrated that this was not the case for more than 80 per cent of the enhancements. CHU was monitored again in early 1963 between January 20 and March 3. Once more I found a definite pattern of reception enhancements, but it differed from that of early 1962. I tried unsuccessfully to correlate the times of enhancements with overhead passages of ECHO I and other artificial satellites.

During November and December 1965 I made reception measurements of CHU on 7.33 Mc/s and again found periodic enhancements. More than 90 per cent of them had a very definite pattern which differed from the previous patterns. Additional data for CHU were taken during December 1966 and January 1967 by B. A. Bagby and myself at Pacific Palisades. Again there was a very definite pattern to most of the enhancements, which differed from all preceding ones. It was obvious that the time between enhancement periods was different at each succeeding epoch, and that the daily advance was also changing. If these enhancements were due to reflexions from the passage of a particular Earth satellite through the upper atmosphere, then that satellite had a changing orbital period. I tried to correlate the various reception enhancements with the osculating orbit of the natural Earth satellite previously discussed in Nature, but because I was unable to account for all the various radio enhancement periods as being due to the natural satellite, I put the task aside temporarily.

I have re-evaluated the CHU radio data, and have found that the source of the earlier difficulty was the very elliptical orbit of the natural object. I had assumed that the apparent period deduced from the radio anomalies would always be the simple reciprocal addition of the object's sidereal period and that of the Earth's rotation, which is true only for a circular orbit. Allowing for the orbital situation at each radio epoch, I related the CHU radio enhancement records on 7.33 Mc/s in Table 1 to the osculating orbit of the natural satellite on each occasion.

I recently attempted to find a correlation between the GBR and NBA radio reception anomalies and those found for CHU, and tentatively concluded that the VLF radio anomalies could also be caused by the natural satellite. This would be confined to radio reflexions taking place near the object's perigee. Here the charged cloud associated with the passage of the body through the upper atmosphere would be large enough to reflect 16 and 18 Kc/s signals. This would temporarily provide an alternate transmission path similar to, but shorter than, that suggested by Isted as the cause of the VLF interferences. I computed the geocentric latitude and longitude of the perigee of this satellite on the various occasions of VLF anomalies and found that exceptionally favourable conditions were present during both the May-June 1962 and November-December 1965 periods. Fair conditions were present during the April 1963 period. For June 11, 1962, the perigee dip occurred near 14° S., 167°W. at 0300 U.T. For November 13, 1965, the perigee dip was near 30°N., 149°E. at 0800 U.T. These positions are approximate to the estimated locations of the postulated second transmitter in each case. The frequency shifts observed ($\sim 10^{-7}$ parts) on 16 and 18 Kc/s for the secondary VLF signals could have been due to the satellite's approaching or recessional velocity relative to both the receiving and transmitting stations. The rarely occurring maximum possible displacement, due to this Doppler effect, would be $\Delta f/f = 5.4 \times 10^{-5}$. More often this value would be much lower, and would have varied slightly during the brief enhancement period, as was observed.

Allowing for differences in the orbital situation, the GBR records of November 1965 are almost replicas of the CHU records for that same period, with reference to time of day and periodicity. The 1962 and

1963 VLF data and the 1961 HF data are too meagre for such a positive comparison.

In order to compare the radio anomalies with the orbit of the natural satellite beyond the period of time covered in my earlier paper, I made more optical observations. With the aid of the indirect CHU radio evidence, this was accomplished photographically near the object's apogee on both January 18 and 20, 1967. Because the CHU radio experiment was also being run during January 1967 the two sets of data could be directly compared. The successful pairs of plates (out of sixty pairs taken) were each exposed shortly after a radio enhancement period had begun at Pacific Palisades.

With the recent optical observations, orbital elements have been derived to cover the time period since October 25, 1965. The solution was quite unique, converging very rapidly during the trial and error computational process. The slowly changing inclination is not without precedent. This effect, due to the upper atmosphere near perigee, has been discussed at length by Nigam and others.

. . . .

TERRESTRIAL SATELLITES: SOME DIRECT AND INDIRECT EVIDENCE

Bagby, John P.; *Icarus,* 10:1-10, 1969.

Introduction. Several investigators have attempted to show theoretically how the Earth might capture small natural satellites. Some workers have conducted searches for natural satellites. Others have suggested that such objects exist or have even been observed. The existence of such objects has never been generally accepted, however, except for the Kordylewski cloud satellites believed to reside at the two LaGrange points in the Moon's orbit (60 degrees ahead of and behind the Moon). Research that I have been engaged in since 1956 now suggests that there may be several substantially large natural objects orbiting the Earth in similar orbits. These all appear to be fragments of a larger parent satellite which broke up about mid-December 1955.

Direct Observations. Originally, the orbits of the suspected natural satellites were computed from the first few direct observations alone. These first observations are summarized in Table 1. A simple and direct approach to orbital solution from a minimum number of observations for Earth satellites has been proposed by Briggs and Slowey (1959). The particular orbital computation method that I use, however, is an original graphical and analytical trial-and-error solution previously discussed for circular Earth orbits. It has been further refined for elliptical Earth orbits as a result of studying Kozai and Baker and Makemson. At first, there were several different possible orbits for the bodies. As more observations were accumulated, the number of possible orbits was finally reduced to three particular sizes and shapes with minor differences in orientation between the orbits of the individual bodies.

A thorough photographic and visual search for the first of these three orbit sizes and shapes was made by myself with the aid of others

Table 1. Direct Observations of Proposed Natural Satellites[a]

Observation:	I	II	III	IV	V	VI
Year	1947	1952	1956	1956	1957	1957
Date, UT	Sep 23 1	Jan 21	Jan 31	Jan 31	May 4	May 4
Universal Time (hr)	06.0 1.0	15.78	11.63	11.73	03.83	03.83
Latitude, N (degrees)	47.58	33.96	42.33	42.33	42.03	42.03
Longitude, W (degrees)	122.33	118.35	83.12	83.12	87.75	87.75
Right ascension (degrees)	304.5	302.76	180.51	180.56	194.3	197.5
Declination (degrees)	-24.7	-20.04	-5.55	-5.57	17.5	14.3
Position angle (degrees)	No record	57.8 5	45 8	45 8	Ambiguous	Ambiguous
Rate of motion (deg/min)	2.0 1.0	0.5 0.1	1.6 0.3	1.6 0.3	0.9 0.1	0.9 0.1
No. of objects	1	1	6	4	4	4
Aperture of instrument (cm)	8.9	10	15.2	15.2	8.5	8.4
Observer	J. M. Hammond	T. A. Cragg	D. Craig	D. Craig	J. P. Bagby	J. P. Bagby

a The position angle (of travel) is the stellar direction in which the objects appeared to be moving, where stellar "north" sould be 0 degrees and stellar "east" would be 90 degrees, etc.

during the period 1957 to 1963, and the results were decisively negative. The entire periphery of this first orbit was covered three times over. The second possible orbit is theoretically and practically implausible. This leaves only the third orbit as a probable solution. Before a search for this orbit could be organized, new information came from an unexpected source, discussed below, which not only indicated the proper orbit size and shape but made it possible to further refine the orbital elements of the individual objects. I now propose that the first two observations listed in Table 1 are of a parent body, that the third and fifth observations and the fourth and sixth observations are of two distinct groups (A and B, respectively) of the objects, with four or more members in each group. Finally, I suggest that the new information discussed below indicates the existence of further subgroups (C through F) consisting of one or more members each. When projected backward in time, the orientation of the orbits of groups A and B, as derived from the data in Table 1, converge on about December 18, 1955. It is assumed that this is the date when the parent body broke up.

Indirect Evidence. From an analysis of the considerable amount of published data on the orbits of artificial satellites, I have recently found that many sudden anomalies have been occurring in the orbital elements of a large number of the artificial bodies. These changes are often quite drastic and of a semipermanent nature. To my knowledge, they have not been previously noted or discussed. Inclinations have changed by up to tenths of a degree, apogees and perigees have fallen and increased from tens to hundreds of kilometers, and the lines of nodes and apsides have changed their orientation by up to ~ 50% of the normal rate. These anomalies are most easily seen by plotting the published tabular data. A typical example is shown in Fig. 1 for the orbit of Telstar 1. The curious thing about such sudden anomalies is that so far as is known over 90% have taken place at epochs when one or more of the orbits of the proposed natural satellite subgroups was intersecting the orbits of the artificial bodies affected. Under these conditions, one or more close passages of natural and artificial bodies could take place. This would be possible if the natural and artificial satellites in their own orbits were to reach the common intersection point at the same time.

When the orbit periods are commensurate with each other in simple ratios, such as 1:3, 1:2, 3:8, 5:7, etc., several close passages could take place within a short period of time. If an accelerative coupling is possible between natural and artificial objects, then for commensurate orbits a very serious perturbation could result in the orbit of the less massive artificial body. Such a resonant effect is well known in the case of the gaps in Saturn's rings that are caused by its largest moons (Baker, 1960, p. 25; Russell et al., 1945, p. 391). It is also known in the case of the gaps in the asteroid belt caused by the planet Jupiter (Brandt and Hodge, 1964, p. 287; Watson, 1956, pp. 18-19). In both of these cases, the perturbing force is gravitational coupling. In the present case, the perturbing force may be gravitational, but it is more likely to be electrostatic or electromagnetic, as discussed further below. These sudden perturbations of the many artificial satellites affected appear to be unaccountable from a consideration of the generally accepted perturbing conditions in near-Earth space.

....

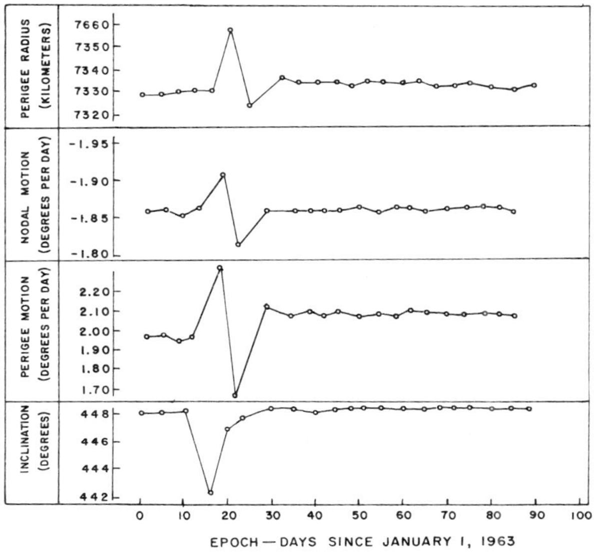

Behavior of some Telstar orbital parameters in early 1963

BAGBY'S PHANTOM MOONLETS

Meeus, Jean; *Icarus,* 19:547-549, 1973.

Bagby (1969) published a paper entitled, "Terrestrial Satellites: Some Direct and Indirect Evidence." The summary of that paper reads as follows:

> Several telescopic and photographic observations have been accumulated since 1947 which suggest that the Earth has at least ten close natural moonlets which broke off from a larger parent body in late 1955. A series of puzzling disturbances in the orbits of many artificial satellites now appear to be explainable as due to perturbations by these natural bodies.

Is this true? If the objects supposedly found in orbit around the Earth by Bagby are real, then this is a very important discovery. But there have been no confirmatory reports. When we look a little more closely at the article in question, we find that the author deduced almost all the "puzzling disturbances in the orbits of many artificial satellites" from a study of the semimonthly Goddard Satellite Situation Report. He seems not to have known that the values given in this publication are approximate only, and sometimes grossly in error, so that they may not be used in error, so that they may not be used for any precise scientific analysis.

. . . .

This is the nature of the "perturbations" considered by Bagby. They are completely illusory, and so are the moonlets needed to explain them.

Now, what about the "observations accumulated since 1947?" The stated observations are as follows.

On 1947 September 23± 1, an object was seen by J. M. Hammond; the motion is given as 2.0± 1.0 degree/min; the position angle of motion is not given. Note the date is uncertain.

On 1952 January 21, an object was seen by T. A. Gragg; the rate of motion was 0.5± 0.1 degree/min.

On 1956 January 31, six objects were seen by D. Craig; the rate of motion was given as 1.6± 0.3 degree/min.

On 1957 May 4, four objects were seen by Bagby himself; the rate of motion was 0.9± 0.1 degree/min, and the position angle was "ambiguous."

On 1968 March 27, an object of about magnitude 7 was observed photographically. Two positions five degrees apart were measured; the object took seven minutes to cover this distance.

On 1968 April 21, an object of magnitude 6 to 7 was observed photographically; three positions over an arc of 12.5 degrees were measured. This distance was covered in 14 minutes.

This is a rather meagre harvest. In the first four cases, the angular velocity is uncertain by 11 to 50%, so no reliable orbit can be calculated for the observed objects. In the last two cases, the angular velocity is more accurately evaluated, but nevertheless only a small arc has been observed. This is wholly insufficient to calculate an orbit, except if we make the assumption that the orbit is exactly circular.

And yet Bagby finds (by what method?) elliptical orbits for all the observed objects. He gives a semi-major axis of 14 065 km and an eccentricity of about 0.498. This yields a perigee and an apogee height of 680 and 14 700 km, respectively. It is not difficult to calculate that the corresponding topocentric angular velocity is about 46 degrees/min near perigee (when the object passes through the zenith), slowing down to 0.7 degree/min near apogee. The latter value is very near the speed of the objects seen in March and April 1968, so these bodies---assuming Bagby's orbits to be correct---must have been near the apogee.

But then we arrive at another major enigma. If the objects were of magnitude 7 as seen from a distance of 14 700 km, then they must be about 460 times as bright when near perigee. This corresponds to a difference of 6 magnitudes; near perigee, the objects stated by Bagby must be of magnitude 1. Why, then, are they not observed daily by thousands of persons?

To explain the so-called perturbations of "many artificial satellites" there must have been many very close approaches between these objects and Bagby's moonlets. But even when the least separation is only 30 metres centre to centre (and this is a near-collision whose probability is almost zero), the gravitational interaction is only about 6×10^{-7} of the attraction by the Earth; this value is given by Bagby himself. Since this is much too low to explain the "perturbations," Bagby invokes another mechanism: both the natural and the artificial satellites are electrostatically charged, so there is a repulsive force between the bodies. This seems a desperate hypothesis.

Bagby's story on the Earth moonlets thus appears to be groundless. There are no "puzzling disturbances" in the orbits of artificial satellites, and the objects cited by Bagby do not exist.

Chapter 6

THE MOON

INTRODUCTION

Down the centuries, astronomers have lavished more effort on the moon than any other astronomical object. This is natural, for the moon is close-by, intriguing, and a spectacular object on a clear night. We know a lot about the moon by virtue of this attention; but the same scrutiny has turned up a wonderful array of lunar anomalies and curiosities.

One class of phenomena apparently involves the atmospheres of the earth and moon and, perhaps, even the mind of the beholder. Included here are the apparent projection of stars onto the moon's limb and the strange dark patches sometimes observed under the moon. During eclipses, such phenomena seem to be multiplied; and we have the "dark" eclipses and peculiar lunar light rays. The lunar surface itself is the source of a wide variety of anomalies---permanent, regularly occurring, and transient. Diligent observers, for example, have often remarked on the weird lighting effects in some craters. These ghostly changes are overshadowed by brilliant lights and subdued patches of color that come and go. The moon is certainly far from being a dead planet where nothing ever happens.

The landings of the Apollo astronauts provided many clues and even some answers to long-standing lunar mysteries. On the whole, though, the more detailed scrutiny of the moon's surface and analysis of returned samples created more enigmas than solutions. We now have mascons, magcons, glazed rocks, radioactive hot spots, and much more to chew on. That unmatched celestial apparition called the moon is as much a puzzle and a challenge today as it was to the Stonehengers, although in different ways.

OPTICAL (?) PHENOMENA IN LUNAR OBSERVATION

The intervening space between the terrestrial observer and the lunar surface is occupied by the earth's unsettled atmosphere---a notorious distorter of images---and the moon's slight residual film of gases. The anomalies reported in this section probably owe their origins to these gases and the optical phenomena of refraction, diffraction, and interference. Indeed, this section is paralleled in several of the chapters on the planets; Venus, in particular. On both moon and Venus, there are irregular terminator and "ring-of-light" situations. Venus has a dense atmosphere where such optical events are to be expected. But what is happening on the moon? The occasional projection of stars being occulted onto the lunar limb just heightens the mystery.

Astronomical records also serve up a tantalizing menu of aurora-like phenomena, the so-called lunar zodiacal light, and strange triangular shadow-like effects. While these may not have science-shaking explanations, they should not be buried unsolved in the literature.

• Image Doubling

DOUBLING OF LUNAR DETAIL

Sartory, P. K.; *Strolling Astronomer,* 19:143-144, 1965.

Mr. P. K. Sartory at Mellow End, Seale, Farnham, Surrey, England on April 13, 1966 wrote in part as follows: "Whilst observing the moon one night during January 1966 with a very fine 5-inch Cooke refractor, doubling of the image was seen. The night was clear and seeing very good. It was at first suspected that tube currents were the cause of the effect, but on checking with a 4-inch Maksutov telescope the same effect was noted. The images were quite sharp and steady. Seven photographs were obtained, and all showed the effect quite clearly. It seems, therefore, that the effect is produced outside the instrument since two different telescopes were tested. The doubling of the image persisted for about 1-1/2 hours. A colleague of mine observing about 50 miles away from my station also observed the effect."

If we are to suppose that Wend, Sartory, and Delano have all observed the same effect, the differences in their descriptions then being regarded as ones of degree only, the explanation cannot be sought in the physics of Jupiter or satellite Io but must presumably be in the optical system, or perhaps in the atmosphere near the station of observation. The great scarcity of known observations of image doubling would suggest some highly unusual condition. The existence of the effect at two stations 50 miles apart may be troublesome from this point of view. Perhaps other readers will have something worthwhile to add to this discussion.

• Dark Triangular Patches

DARK TRIANGULAR PATCH UNDER MOON

Chivers, G. F.; *Marine Observer*, 31:188, 1961.

m. v. Silverdale. Captain G. F. Chivers, M. B. E. Bandar Mashur to Suez. Observers, the Master and Mr. M. L. Brisley, 3rd Officer.

21st December 1960. At 1800 GMT (2100 LMT) a dark triangular area was seen directly under the moon---then in its first quarter and at an elevation of about 6°---which extended from the horizon up to an angle of approximately 10°. The sky was cloudless and the visibility excellent, the horizon line being well defined. The observers found, on looking away from the moon and the dark patch for a few seconds, that when they looked back again the shadowy area was not seen for an appreciable interval. This agrees with the comments in Note 4 accompanying the report by m. v. Dartmoor on p. 193 of the October 1960 issue of The Marine Observer. Air temp. 78°F, wet bulb 74°, sea 76°. Wind SW, force 2. Visibility excellent.

Position of ship: 16° 54'N, 40° 54'E.

DARK TRIANGULAR PATCH UNDER MOON

Roberts, J. O.; *Marine Observer*, 30:193, 1960.

M. V. Dartmoor. Captain J. O. Roberts. Cadiz to Port Everglades. Observer, Mr. W. G. Shelton, 3rd officer.

15th March, 1959. Whilst approaching the Providence N. E. channel a dark patch was observed directly under the moon. The night was mainly clear, with occasional scattered rain showers and the moon three-quarters full. For the main part the horizon was clear cut, emphasised by what appeared to a light haze, although visibility was good. Directly under the moon the shading was reversed, the sea being lit up by the moon's reflection and the sky very dark. The shaded patch under the moon was directly over Abaco Island and seemed to rise to the moon, tapering away until lost in the night sky.

Position of ship: 25° 50'N., 76° 40'W.

Note 1. The appearance of a triangle of darkness between the moon and the horizon, contrasting with the triangle of light under the moon which extends from the ship to the horizon, is rare but not unknown. No satisfactory explanation can be given but the effect is thought to be physiological, due to contrast, rather than true. As long ago as 1856 an observer noted that the dark triangle disappeared when he screened off the moon and the lower triangle of light. And when, after turning round, he suddenly looked again, the illusion only re-appeared after a few seconds.

• Lunar Zodiacal Light and Auroras (?)

THE MOON'S ZODIACAL LIGHT

Trouvelot, L.; *American Journal of Science,* 3:15:88-89, 1878.

During the evening of April 3, 1874, the "Zodiacal Light" was particularly brilliant; especially close to the horizon, where it appeared as a segment of a circle, having an irregular wavy outline, giving it a vague resemblance to the beams of a faint aurora. Although the sky was clear, it was found impossible to observe with the telescope on that night, on account of the great disturbance of the atmosphere. At $9^h\ 45^m$ the declination needle indicated a very strong magnetic perturbation in Cambridge, oscillating through an angle of 3° 22'. However, no aurora was visible at this time, although the phenomena usually attending them were manifested during the evening by the tremulous appearance of the telescopic images.

While going home, I remarked in the east a strange conical light rising obliquely from the top of the roof of a building, behind which the moon, then about 15° or 20° above the horizon, was concealed from view. By going away from the building, the conical light, which closely resembled the tail of a comet, became brighter and brighter as it approached the moon, upon the western limb of which it rested. The base was at least as wide as the diameter of the moon; but it extended beyond, on each side, by a fainter light, which gradually vanished in the sky. The extension of this luminous appendage I estimated to be equal to eight or ten times the moon's diameter. It was not readily visible when the moon was in sight, as the brilliant light of our satellite overpowered its dim brilliancy. The axis of this appendage was found to be coincident, or nearly so, with the ecliptic; and its line prolonged in the west passed a little to the north of Jupiter.

The phenomenon had been observed for about fifteen minutes, when it gradually faded away until it almost totally disappeared five minutes later, although the sky was clear. A quarter of an hour after, the sky was overcast with dense vapors, which continued for nearly an hour.

At $11^h\ 0^m$ the sky had cleared up, and the moon shone brightly. The luminous appendage was still visible, and even appeared more brilliant than before. In order to ascertain whether this appendage was visible only on one side of the moon, or if it was seen on the other side, I went under the piazza of my house, and placed myself in such a position as to have the moon concealed by its upper part, the sky below being visible. As I expected, a similar appendage was observed on the eastern side of the moon, exactly opposite the western one; the axis of both wings, passing through the moon's center, being in the plane of the ecliptic.

Although at this moment no auroral light was seen in the north, yet, up in the zenith, there were evident signs of it, as luminous vapors assembled there and rapidly dissolved, arranging themselves into bands radiating from a center after the manner of the crown of bright auroras. At $11^h\ 20^m$, all traces of the luminous vapors in the zenith had vanished; and at the same time the appendages of the moon were almost totally invisible, although the sky remained clear.

The fact that the zodiacal light had been unusually brilliant during this evening, and that the two luminous appendages of the moon resembled it in shape and appearance, and were situated in the same plane, seems to indicate that the two phenomena are of the same order; while the magnetic perturbation and the auroral phenomena connected with the variation of brightness observed in the moon's appendages would seem to indicate some kind of connection between the zodiacal light and the aurora. The result of my observations of the zodiacal light and the aurora during the last seven years also seems to indicate some such connection, as when the zodiacal light was observed to be particularly bright, it has generally been followed by auroral phenomena. But only a long series of observations in this direction can solve the problem.

MOON'S ZODIACAL LIGHT

Holden, E. S.; *American Journal of Science,* 3:15:231, 1878.

In your Journal for February, 1878, p. 88, is a note by M. Trouvelot, on "the Moon's Zodiacal Light," in which he describes a conical luminous appendage about 4-1/2° long, extended on both sides of the moon, which was seen by him April 3, 1874.

In this connection an observation of a similar phenomenon by Messier, in the Memoires de l'Academie Royale des Sciences, 1771, p. 434, is noteworthy. In this memoir, Messier gives a rough woodcut of its appearance, from which its length on each side of the moon is shown to be about 2-1/2°.

The condition of the sky, as described by Trouvelot and Messier, appears to have been the same.

In the Comptes Rendus, July 2, 1877, p. 44, M. Hugo describes a similar phenomenon which he saw above the lunar disc, about 4° in length, and in this case, also, the sky appears to have been similarly affected. These are the only cases known to me.

AN AURORAL GLOW ON THE MOON

Haywood, John; *Sidereal Messenger,* 3:121, 1884.

Last fall I was led to inspect the Moon with more than usual care. On November 4, 1883, I saw on the dark part of the Moon a misty light different from that to be ascribed to the feeble illumination due to light reflected from the Earth. This appearance was seen repeatedly by myself and others in November and December. The weather in January and February was not favorable to observations. But on March 29th last and 30th, the phenomenon was distinctly seen by myself and others. One who had seen it last fall, on looking at it on the 29th, remarked at once that it appeared on a different part of the disc, and, on the 30th, he noticed that the light had narrowed down.

This accumulation of light shifts its place upon the disc; but generally is seen brightest along the dark limb of the Moon. The appearance is that of the light of the Aurora Borealis on the Earth, as it might show to one looking at it from the Moon. I suggest this explanation, therefore: that it is a light of the same nature as the Aurora Borealis, and produced by the same cause. So far as my observations go, this appearance is to be seen before the first quarter, and after the 3d quarter. At the time of quadrature, I have seen a line of light along the dark limb. I have been unable to see anything of it while the Moon is gibbous.

My telescope is a three-feet achromatic, two and one-fourth inches aperture, with a magnifying power of about sixty diameters. The eyepiece is negative. It is likely such diffused light is seen better with a low power.

LUNAR RAY

Bone, D. W.; *Marine Observer,* 11:49, 1934.

The following is an extract from the Meteorological Record of S. S. Transylvania, Captain D. W. Bone, New York to Moville. Observers, Messrs. C. B. Steuart, 1st officer, and A. Middleton, 3rd officer.

On 2nd May, 1933, at 0230 G. M. T., an orange-coloured vertical ray was observed projecting from the moon's upper limb at an altitude of about 6 degrees. This ray was very pronounced until it was covered by cloud at 0245 G.M.T. and was not seen again. Weather at time was very clear with sky 4/5 covered by Cumulus and Fracto-Cumulus cloud. Auroral displays had been noted to northward about four hours previously.

Position of ship: Latitude 53° 40' N. Longitude 25° 12' W.

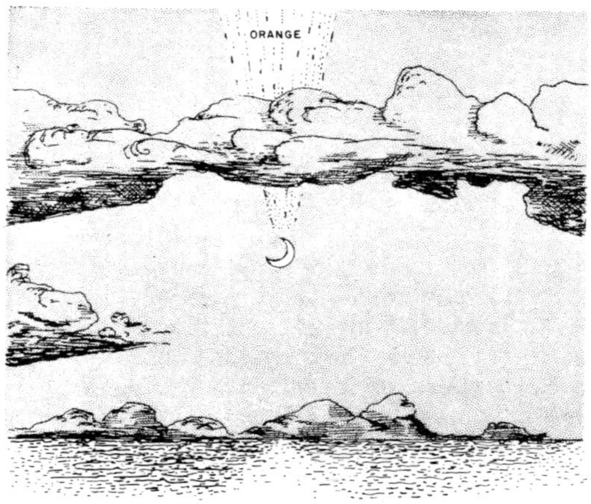

Lunar "ray" seen at sea

LUNAR PHENOMENON

Loftus, A. J.; *Nature,* 12:495, 1875.

I have pleasure in forwarding a brief account of facts relating to two very remarkable protuberances which were observed on the moon's disc in the Gulf of Siam, by Mr. E. C. Davidson, Telegraphic Engineer, and myself.

H. S. M.'s guard-ship Coronation (Champon Bay), July 13 (civil time), in lat. 10° 27' 40" N. and long. 99° 15' E., at midnight, the moon bore S. W. by W. magnetic, and its altitude was about 20°, when a prominent projection was seen with the naked eye on the moon's upper limb. The best glasses on board were soon brought to bear upon it, and the enclosed sketches (with due regard to proportion) were carefully made on the spot.

The protuberance, in colour, was similar to that of the moon.

On July 14, at 8 p.m., the moon was observed perfectly clear, but without a vestige left of the protuberance of the previous night. At this hour, however, a small one was noticed in a different position of the limb.

This also had disappeared before the moon rose on the evening of the 15th inst., when it finally presented its usual unbroken appearance.

A CURIOUS APPEARANCE OF THE MOON

Anonymous; *Scientific American*, 46:49, 1882.

A singular appearance of the moon was observed by several residents of Lebanon, Conn., on the evening of July 3. The moon, almost full, was about three-quarters of an hour high. An observer says: "Two pyramidal luminous protuberances appeared on the moon's upper limb. They were not large, but gave the moon a look strikingly like that of a horned owl or the head of an English bull terrier. These points were a little darker than the rest of the moon's face. They slowly faded away a few moments after their appearance, the one on the right and southeasterly quarter disappearing first. About three minutes after their disappearance two black triangular notches were seen on the edge of the lower half of the moon. These points gradually moved toward each other along the moon's edge, and seemed to be cutting off or obliterating nearly a quarter of its surface, until they finally met, when the moon's face instantly assumed its normal appearance. When the notches were nearing each other the part of the moon seen between them was in the form of a dove's tail."

PECULIAR PHENOMENON ON THE MOON

Harris, Frank B.; *Popular Astronomy*, 20:398-399, 1912.

I deem it a debt to the astronomical world to relate certain phenomena seen by myself on the evening of January 27 (Saturday) 1912. During the time mentioned I had been sighting the moon at half hour intervals. About 10:30 Eastern time I was surprised to see the left cusp showing the presence of an intensely black body about 250 miles long and fifty wide, allowing 2000 miles from tip of cusp to cusp. The appearance was fully as black comparatively as marks on this paper, and in shape like a crow poised.

Of course dark places are here and there on the lunar surface, but not like this. Not to be tedious I will say that every effort was made to eliminate any error of vision or other mistake. I tried at once to phone some Philadelphia astronomer but failed.

At that time I was unaware of Professor Doolittle of the University of Pennsylvania. The object was kept on view till 2 a.m. As the thermometer was at zero I turned in. This object resembled a crow poised as near as anything. Clouds prevented another view till Tuesday following. By that time or sooner it had disappeared, but some sort of disturbance seemed to be still at work. From the time of Pliny such appearances have been peculiar to volcanic eruptions and such seems very probable in this case. I was very sorry to fail in getting in touch with any one qualified to pronounce upon this with authority. The moon is very tricky and it is very unlikely that anything of this character will be seen in many years or hundreds of years even. I cannot but think that a very interesting and curious phenomenon happened.

ABNORMAL APPEARANCE OF THE LUNAR CRESCENT

Hopkins, B. J.; *Observatory,* 6:247-248, 1883.

Observing the Moon at $18^h\ 15^m$ Nov. 8, 1882, its age being 27.9 days, I was surprised to find that the "line" joining the cusps did not pass as usual through the Moon's centre, but a little to the E. of it; in other words, the exterior circumference of the crescent was less than a semicircle.

It is well known that both Schroter and Gruithuisen frequently traced the "horns" of the crescent for some little distance beyond a semicircle; but that the "cusps" should be observed not extending so far as to form the semicircle, is, so far as I am aware, one of those "things not generally known," and one, too, that cannot fail to be of interest to astronomers in general and selenographers in particular.

The foregoing observation having been made solely with the unaided eye, and not being able at the time to telescopically observe the Moon, it occurred to me that it was possible that this abnormal appearance of the lunar crescent was due to the "cusps" being such fine lines of light that the Moon's proximity to the Sun, combined with the brightness of the morning twilight, rendered them invisible to the naked eye. But on observing the Moon at 19^h on January 5th of the present year, its age being 26.5 days, I was exceedingly gratified to find the Moon presenting the same appearance as on the former occasion, the only difference being that the peculiar form of the crescent was more conspicuous.

That this was the true shape of the crescent, and not caused by the brightness of the twilight, I proved by viewing the Moon with a small terrestrial telescope of 1-1/2-inch aperture and power 25, with which the abnormal appearance was rendered most striking.

• Projection of Stars on the Lunar Disc

NOTE ON A PHENOMENON SEEN IN THE OCCULTATION OF A STAR AT THE MOON'S BRIGHT LIMB

Christie, W. H. M.; *Royal Astronomical Society, Monthly Notices,* 39:198, 1879.

At the disappearance of 17 Tauri at the Moon's bright limb, on November 10, 1878, a curious phenomenon was presented, which appears to throw some light on the apparent projection of a star on the Moon. I observed the occultation, under favourable atmospheric conditions, with the full aperture (nearly 13 inches) of the Great Equatoreal of the Royal Observatory, Greenwich, using a negative eyepiece with a power of 310, and my eye was much dazzled by the intense light of the Moon, only 19^h past the full.

The star is of 4 magnitude, and as it appeared to approach the Moon's limb, I fixed my attention on it, being afraid of losing it in the overpowering light of the Moon. I watched it steadily till it came up to the line of the limb, expecting it to disappear at a slight notch in the limb; but instead of that, the Moon's limb, to my surprise, seemed to recede for some 3 or 4 seconds of time, and the star disappeared gradually in a sort of luminous haze, through which it was seen with more and more difficulty as it advanced. At the instant of disappearance the star was seen apparently perfectly bisected by the limb, if limb it could be called, that is, completely shorn of its rays and half the disk on the one side (towards the Moon), and intact on the other.

This observation appears to accord perfectly with the explanation of the phenomenon of projection which the Astronomer Royal has given in vol. xxviii. of the Memoirs. The star was not bright enough to be seen distinctly projected on the Moon's disk, but was yet not so faint as to be overpowered by the irradiation at the limb. A spurious limb being formed, as the Astronomer Royal has explained, by the superposition of diffraction images, the intensity of light would increase gradually from the spurious to the true limb. At the former the star's light overpowered that of the Moon; at the latter the Moon's light overpowered that of the star. The circumstance that the diffraction image of the star (disk and rays) was cut off completely on the side of the Moon at the instant of disappearance is in perfect accordance with the Astronomer Royal's explanation.

. . . .

EVIDENCE OF AN ATMOSPHERE ON THE MOON

Gheury, M. E. J.; *Observatory,* 36:268-269, 1913.

On April 22 I was waiting for the reappearance of ϒ Scorpii (mag. 3.1) with a 3-1/2-inch refractor by Wray. My method of observing is to start a stop-watch immediately the reappearance takes place, estimating the interval elapsing between it and the "click" heard when the stop-watch mechanism is released, and to compare the watch with a chronometer directly afterwards. It is important to note that this leaves me ignorant of the exact time of reappearance, which is therefore more or less unexpected. It is also important to note that there was a veil of very thin clouds on the Moon, and I expected, therefore, the star to be dimmed by it. As I was watching, I saw suddenly a faint star appear in the field at the spot where the reappearance was going to take place. It was much fainter than I expected, and it was not until about a second afterwards that, being certain I was not mistaken, I started the stop-watch. Just as I was doing so, however, a much brighter star suddenly appeared exactly where the faint star was, the latter being no more visible.

The cloud veil was too thin to explain a sudden variation of brightness by the interposition of a variable thickness of vapour. (The crater of Langrenus, opposite which the reappearance occurred, was visible the whole time with all its details.) Everything happened as if a faint image of the star had preceded the reappearance of the star itself, the latter taking place about a second after the appearance of the image.

Scorpii is not a close double, so that the only explanation is refraction by a lunar atmosphere. It is possible that, exceptionally, a star reappears in an indentation of the limb corresponding to a deep valley along the line of sight, in which the atmosphere is abnormally dense. This might have happened on the night in question, hence the marked features of the phenomenon.

PECULIAR STAR OCCULTATION

Parkhurst, J. A.; *Sidereal Messenger,* 10:252, 1891.

On the evening of April 11, I observed the immersion of a star in Taurus, 6.5 or 7 magnitude, 60° from the north point, 10° from the vertex. The star disappeared at 8^h 10^m 1^s, Central standard time; re-appeared and disappeared again about one second after the first disappearance. The moon's altitude was 20°, the definition in the vicinity of the moon was unsteady; the details of the moon would be very distinct for an instant, then blurred. It would thus seem that the action of the earth's atmosphere would account for the above double disappearance.

THE PROJECTION OF A STAR UPON THE MOON'S BRIGHT LIMB

Waterfield, R. L.; *British Astronomical Association, Journal,* 33:250-255, 1923.

Part I. The Observation. The occultation of Aldebaran on 1922 November 6 was observed with a 3-1/4-inch refractor x 200 under the most favourable conditions of seeing and transparency. I had expected to see the star disappear at the instant at which its spurious system was precisely bisected by the Moon's limb. Instead, however, the star appeared to pass completely on to the Moon's surface and only disappeared when its first diffraction ring was completely within and, as far as I could judge, just past internal contact with the bright limb. The disappearance then took place instantaneously without any previous diminution in brightness. Immediately afterwards the focus was re-examined on the Moon's limb and found to be perfect. Since the radius of the first diffraction ring in a 3-1/4-inch telescope is 1".1, the projection observed was probably about 1".2 on the celestial sphere, or, after being magnified x 200, 4'.0 at the eye. It is interesting to note that the Rev. T. E. R. Phillips observing with a 12-1/4-inch reflector also, at that occultation, observed a projection of the spurious disc and first diffraction ring.

....

OPTICAL ILLUSION DURING THE OCCULTATION OF MARS

Thackeray, A. D.; *British Astronomical Association, Journal,* 48:126-127, 1938.

I have received the following observation of the recent occultation of Mars from Colonel C. B. Thackeray, which may be of interest to members of the Association:

"On July 17, while walking down the King's Road, Chelsea, with two others (unaware of the predicted occultation of Mars), we saw the Moon above the house-tops, with a planet in close proximity. We walked to the Embankment, to get a clear view, and watched it for some time with the naked eye. It was evident that an occultation was about to take place. But instead of disappearing at the dark limb of the Moon, the planet appeared to enter the dark field of the Moon's surface and to continue to travel across it, until it vanished on reaching the illuminated age of the Moon. We discussed this mysterious phenomenon as it occurred. It lasted several minutes. We could only account for it by some optical illusion, which we presumed would be a matter of common knowledge and frequent occurrence, to be easily explained by astronomers. Is this a common phenomenon, and how is it to be accounted for? There is no doubt whatever about the apparent optical effect produced. It was obvious to the three of us, and I have no doubt would have been corroborated by any others witnessing the occultation at the same time from the same place."

I have also heard of two other observations of the same optical illusion. Dr. A. Beer, at Cambridge, observing with the naked eye, noticed it when a telescopic observer could see that the occultation had not begun to take place. The explanation must be bound up with the difficulty of extrapolating the dark semicircle of the Moon. The resolving power of the naked eye is from 1' to 2', or about one-fifteenth of the lunar diameter, and any effects such as irradiation tending to enlarge the naked-eye view of the Moon and planet will help to create the illusion. It would be of interest to know whether previous observations of the illusion have been recorded. One would expect that the phase of the Moon, and possibly the altitude in the sky, might be factors of importance.

OPTICAL ILLUSION DURING THE OCCULTATION OF MARS

de N., E.; *British Astronomical Association, Journal,* 48:179-180, 1938.

The extraordinary phenomenon referred to in The Journal (48, 126) with reference to the above will appear still more peculiar in view of the following alternative experience. On the evening of the occultation in question, a well-known Tennis Club in the N. W. district of London was holding its annual open-air dance and reunion on the lawns of the club. I was there and explained the pending occultation to some dozens of the members present, all of whom watched the occurrence with interest--- but none of them saw any symptom of the optical illusion experienced by Col. C. B. Thackeray and his friends. The distance of this club from the Embankment is only about 8 miles, so that the "unanimity" of the two parties in agreeing entirely to differ as to what they observed is certainly rather intriguing. Viewed from this part of London, the naked-eye occultation proceeded according to anticipations, and was similarly observed by many of the party through a small telescope, which I had taken along for the purpose. The interest aroused in these non-astronomically minded people was very keen, and if anyone had noted the illusion under consideration, I feel sure comment would have been made.

POST-OCCULTATION RECEPTION OF LUNAR SHIP AMERICA RADIO TRANSMISSION

Salisbury, W. W., Fernald, D. L.; *Nature,* 242:601, 1973.

The reception of radio signals from the orbiting lunar spaceship America after its occultation behind the lunar limb is a confirmation of results reported for the Apollo 15 ship Endeavour. Similar observations arranged with the lunar command module during the Apollo 16 mission were unsuccessful because transmissions from the command module did not occur while the Moon was above our horizon.

The transmitter used for communication between the lunar command module and the lunar landing party was turned on almost continuously during a period of lunar orbiting between 1755 GMT, December 14, and 0400 GMT, December 15, 1972. The frequency of this transmission was 259.7 MHz with an audio signal at times impressed on the carrier. This transmission was intercepted by a special receiver attached to the 150-foot radio telescope of the US Air Force Cambridge Research Laboratory. This instrument is located at Sagamore Hill near Hamilton, Massachusetts, USA (latitude 42° 37' 55.6" N; longitude 70° 49' 04.5" W.)

The spacecraft transmitted during most of the time the Moon was above the local horizon. Post-occultation signals were received when the Moon was near the local zenith. This high lunar altitude facilitated the reception and the signal strengths were much greater than on previous flights.

Occultation of the signal occurred at 0143:36 GMT, December 15, 1972. The signal did not immediately go to a zero level; it seemed to go to zero at approximately 0144:02 GMT and reappeared at 0144:12 GMT and reappeared at 0144:12 GMT, persisted for 52 s, and finally disappeared at 0145:04 GMT. This behaviour indicates a dependence of the post-occultation signal on the exact topography of the lunar surface beneath the spacecraft.

Because the altitude of the Moon was higher than on previous flights, our signal strength was correspondingly greater, and we observed the Fresnel patterns that characteristically accompany occultation. When the signal reappeared, its level was approximately 20 db above noise (as shown by digital voltage recordings). The AM and FM magnetic tapes and digital sampling of receiver output voltage provide alternative records, and an oscilloscope gave visual and photographic reception.

The height of the America's orbit above the lunar surface was about 96 km at the time of occultation and was reasonably uniform thereafter. The occultation of the Moon ship occurred during 154° of its orbit relative to the lunar centre. During the elapsed time between the occultation and the disappearance of the signal, the ship traversed about 134 km of the lunar surface. The actual path of the signal was about 6 km less than this figure because the orbit of the ship was inclined to the line of sight to the Earth.

The surface distance traversed between the time of occultation and the time of disappearance of the signal projected on the signal path to the Earth is about 97 km. The effects of libration were insignificant.

Other results from this experiment, such as the possible significance of signals received during the other two occultations, are awaiting further reduction of the data.

Several possibilities for explaining such results have been considered, including refracted waves through the limb of the Moon or lunar surface waves. The immediate continuation of the signal after occultation favours the surface-wave explanation.

. . . .

• Young Moon Seen as a Circle

THE YOUNG MOON SEEN AS A CIRCLE

Stevenson, E. A.; *British Astronomical Association, Journal,* 28:223, 1918.

The phenomenon of the young Moon seen as a circle seems to have claimed little attention on the part of observers. This beautiful but not frequently seen circle appears to be confounded by many with the familiar and much-observed earthshine.

In a letter to the English Mechanic of 10th April 1914, the Rev. A. C. Henderson gave some notes of observations made by him, and expressed the opinion that the circle of silver light is caused by the lunar atmosphere refracting the light of the Sun and producing twilight round the Moon's disc. I have not come across any other attempt to explain the phenomenon.

In the course of a discussion on the subject, M. Camille Flammarion referred to a similar circle on Venus, and stated that when in inferior conjunction that planet shows a luminous ring (L'Astronomie, Juin 1916, p. 209). In a subsequent issue of the same journal, M. Le Baron Van Dedem communicated observations made by him on this striking phenomenon.

My observations have been made when the Moon was about two days old. The ring of silver light was continued from the crescent and was of almost electric whiteness. The dates of my observations were:--- January 17th, 1915; February 16th, 1915; January 7th, 1916, from 5.45 to 6.5 p.m.; February 6th, 1916, from 6 to 6.5 p.m.; December 26th, 1916, from 4.25 to 4.40 p.m.; December 17th, 1917, from 4.55 to 5 p.m. (the ring on this occasion was not so distinct as observed on a younger Moon). March 14th, 1918, 7 to 7.3 p.m. On the 16th February 1915, earthshine was present when the circle was observed and also on the following evening when no trace of the circle was visible. On February 6th, 1916, earthshine was seen at 6.50 p.m., after the ring of light had gone. The circle, as observed by me, has, although complete, never been of uniform brilliance. Its whiteness is in striking contrast to the colour of earthshine, which is of a pink or ashy hue.

LUNAR CRESCENT AND BRIGHT RIM

Wilkins, H. Percy; *British Astronomical Association, Journal,* 56:91-92, 1946.

When the Moon is a crescent, either before or after New, a remarkable appearance is often witnessed: in addition to the bright crescent the dark portion of the lunar globe is illuminated by the earthshine, while the dark limb seems fringed with light. To the naked eye the limb appears to be bordered externally with a narrow but inconspicuous fringe of light which disappears on telescopic examination, the limb itself then appearing brighter than the remainder of the earthlit portion of the disk.

This phenomenon has been remarked for a long time and has been ascribed to contrast---a lunar atmosphere or illusion.

It was shown by Proctor that if our satellite possessed an atmosphere analogous to ours, the refractive effect would be to surround the Moon with a ring of starlight similar to what he declared the Earth must exhibit when viewed from the Moon. It is, however, definitely established that a lunar atmosphere of such density does not exist; our satellite is an almost airless globe, and any atmosphere which may exist is of such tenuity as to be totally incapable of refracting starlight.

It is a matter of common observation that from New to shortly before Full the western limb appears brighter than the rest of the surface, while from Full to the time when the Moon disappears in the solar rays the eastern limb is the brighter. The limb always appears brighter than the interior portion irrespective of phase, and is undoubtedly a factor in the light-fringed appearance.

The limb is of a mountainous nature, except at one or two points the dark plains or seas do not reach the limb, and lunar elevations are, in general, brighter than the low-lying areas. This tends to make the limb bright, enhanced by the large portions of the northern, western and eastern parts which are occupied by the dark plains. Another factor is shadow concealment.

When the Moon is a crescent, shadows are very evident near the terminator; formations near the limb also cast shadows, although to a lesser extent owing to greater solar altitude; but from our viewpoint the elevations tend to conceal the shadows as well as the low-lying portions, which are often of a more or less dark tint. What to an observer directly above it would appear to be a circular mountain-ringed plain more or less shadow filled, we see as a narrow ellipse; if the walls are of considerable altitude the interior may be entirely concealed.

At Full the central regions alone are shadow free, but to us the elevations conceal their shadows; except that at ordinary Full Moon, that is when not eclipsed, the Moon's latitude enables us to see a strip of the darkened portion, and objects near one of the poles exhibit shadow. In a similar manner near New we may see a narrow strip of the illuminated hemisphere, so that the cusp is prolonged, and a portion of the limb which would otherwise be dark is seen bright by direct sunshine.

Even at Full the limb appears the brightest part, because the depressions are partially or even wholly concealed by the elevations; which is not the case with the more central portions, where the low-lying and often dark portions can be seen. Parts of the limb are occupied by very lofty mountain ranges, such as the Leibnitz, Doerfel and Rook mountains on the south, and the lower but still lofty mountains on the north and west.

Near New we see the dark portion by the reflection of a reflection, the earthshine, and the same effect is produced. To the distribution of light and dark areas on the linear surface, the fact that low-lying portions are usually dark and the elevated nature of the limb regions must be ascribed the bright-rimmed lunar disk, helped in the case of the bright sunlit crescent by shadow concealment at the limb.

The young earthlit Moon consists of a bright crescent illuminated by direct sunlight and the rest of the disk seen by reflected earthlight. The limb of the crescent appears still brighter owing to shadow concealment, and irradiation causes the bright crescent to appear to be a portion of a sphere of greater radius than the darkened area. The limb of this dark but earthlit surface appears brighter than the rest owing to such large portions of the inner regions being occupied by the in themselves dark plains and the concealment of the low-lying and usually dark areas near the limb. Irradiation and contrast with the dark sky cause an apparent enlargement similar to, but less pronounced than, the bright crescent, and the naked-eye effect is that of a border of light around the dark limb, sometimes intensified at one of the cusps by a narrow strip of directly sunlit elevations. With the telescope the external bright border seen with the naked eye no longer appears, but the true appearance is observed of a bright limb and the darker patchy interior.

Occasionally in addition to the bright limb two or three light patches are observed close to the limb in the earthshine. These have been seen with the naked eye, and may be due to fluorescence as well as to the brilliancy of the earthlight at the time. The actual amount of light reflected by the Earth varies according to the hemisphere turned towards the Moon: it is greater when the great land masses of Asia, Africa or

Europe are so situated than when the Pacific Ocean occupies the terrestrial disk seen from the Moon.

The light-fringed young Moon is a beautiful phenomenon, best seen after sunset in spring or before sunrise in autumn. Whenever seen, the appearance should be carefully noted. Any light-spots should be mapped when further interesting features may be recorded, and is commended to those lovers of nature who observe with the naked-eye or possess only small instruments.

THE LUNAR LIMB BAND

MacDonald, T. L.; *British Astronomical Association, Journal,* 55:63-64, 1945.

On several occasions during the occultation of a planet by the Moon a faint shading or line has been seen crossing the planet's disk parallel to the Moon's limb. Two such observations were recorded in America in 1944. On the first occasion (Jan. 13) the observation was confirmed in several places. W. H. Haas appeals for further observations of this phenomenon (Journal, R. A. S., Canada, 38, 351, 1944 Oct.). There appears to be slight photographic evidence for the appearance, which suggests it may be optical rather than psychological. It is, however, suggested that it may be evidence of an absorbing lunar atmosphere. Haas considers its width perhaps 4"; in the opinion of the present writer such an atmosphere would influence occultations of stars, but the matter deserves to be cleared up further. It is hoped that a search will be made for it at future occultations of planets. Other references are: Sky and Telescope, 3, Nos. 3 and 9, 1944, and recent issues of Texas Observers' Bulletin.

LUNAR LIMB BRIGHTENING

Johnson, Craig L.; *Strolling Astronomer,* 11:118, 1957.

Mr. Craig L. Johnson writes that on July 31, 1957 at 2^h 24^m, U. T., using a 4-inch reflector at 91X, he observed a slight ring of light reaching around the north limb of the moon. The ring was just barely brighter than the earthshine and was about 1,000 miles long. Mr. Johnson expresses confidence that the appearance was not an illusion and that it was not wholly due to contrast. The moon's age was 3.9 days, and the seeing was perfect (10 on a scale of 10, while moisture was literally running down the telescope tube). This lunar limb light is nothing new but has been recorded as long ago as Schroeter's time. It has been variously imputed to a lunar atmosphere, to optical effects, and to mountains on the limb. Perhaps some readers would enjoy investigating this facet of lunar affairs.

• Lunar Sunset Afterglows

THE LUNAR SUNSET PHENOMENON

Allen, L. H.; *Surveyor Program Results,* NASA SP-184, Government Printing Office, Washington, 1969, pp. 413-419.

Lunar Observations. Television pictures, taken by several Surveyors a few minutes after local lunar sunset, showed the lunar western horizon highlighted by the Sun as a thin, bright, jagged, discontinuous line. The Surveyor VII lunar sunset pictures probably are the most interesting in that this line was jagged, its brightest parts shifted with time, and its broken parts changed positions with time.

The brightest parts of this line have sometimes erroneously been called "beads" because they are remotely similar in appearance to "Bailey's beads." The discontinuous portions or the line are gaps and appear to be the result of lunar-surface features that lie beyond (farther west), casting shadows on the local horizon. As a general rule, the gaps appeared to grow larger with time during the Surveyor VI and VII missions, as the Sun moved farther below the horizon.

Because Surveyor VII was at a southern lunar latitude, the gaps appeared to shift slightly to the north as a function of time and by different amounts, which are assumed to be a function of the distance between the individual occulting objects and the point on the horizon being occulted. The brightest parts of the line shifted slowly to the south as a function of time and as a function of Surveyor VII spacecraft's southern latitude on the lunar surface. The lateral shifting

Illumination along western horizon about 15 minutes after lunar sunset as seen from a Surveyor spacecraft

effects were noticeable only on Surveyor VII, because it operated much farther from the lunar equator than any of the other spacecraft.

....

LUNAR THEORY AND PROCESSES: POST-SUNSET HORIZON "AFTERGLOW"

Gault, D. E., et al; *Icarus,* 12:230-232, 1970.

Observations of the western horizon shortly after sunset during the Surveyor VII mission revealed, along the crest of the horizon, a bright line of light similar to that previously reported for the Surveyor V and VI missions. Though not sufficiently well defined to be recognized at the time, the phenomenon also occurred during the Surveyor I mission. Although no sunset observations were made on Surveyor III, it appears that this postsunset phenomenon along the western horizon (and probably the eastern horizon at sunrise) is not an unusual event, but occurs regularly as the natural consequence of some aspect of the lunar environment.

The light was observed for periods of time up to about 2 hours after sunset. The center of the solar disk, therefore, is about 1.25 deg below the horizon when the "afterglow" either stops or the intensity falls below the limits of detection. Pictures of the light from the Surveyor VII mission are shown in Figs. 1 and 2 when the Sun was centered approximately 0.4 and 1.0 deg, respectively, below the horizon. In Fig. 1, the light intensity permitted normal shutter operation (exposure time, 0.15 sec); the bright line appears to extend only about 2 deg along, and 1/8 deg above, the horizon. The light intensity decreased rapidly; about 1-1/2 hour later, a nominal 1.2-second exposure showed a faint line of illumination extending at least 4 deg along the horizon. A 40-second exposure taken about 2 hours, 40 minutes after sunset, showed no edge of light along the horizon. This last picture (illumination provided by light backscattered from the ridges east of the spacecraft, and by earthlight) provides a valuable comparison of the rocks and horizon geometry with the shape of the bright regions in Figs. 1 and 2. A particularly striking facet of the phenomenon is the "mapping," or shadows, in the edge of light, apparently caused by the rocks extending along and above the lunar horizon line. [Figs. omitted]

Although no complete explanation can be offered at this time, the relative intensities of the light on Surveyors VI and VII suggest that scattering by small particles above the lunar surface is not the mechanism for the phenomenon. This conclusion is drawn from the fact that, while the intensity of the bright line appears to be greater for Surveyor VII than for Surveyors V or VI, the distance to the horizon and the path length of the light immediately above and along the surface is probably shorter. For equal spatial density of the particles above the surface, the longer path length, contrary to observations, should have produced a pattern of greater brightness. Alternatively, diffraction by small particles on the lunar surface, as discussed by O'Keefe et al. (1968) may provide a mechanism for producing the phenomenon; however, further study is required before any explanation is considered firm.

LUNAR TERMINATOR PHENOMENON

Anonymous; *Sky and Telescope,* 46:146, 1973.

A patchy glow has been discovered on the moon's horizon at dusk that is believed to be caused by individual dust clouds being continually formed above the hills, crater rims, and rocks which remain sunlit while the shadow of night advances over the lower surroundings.

J. J. Rennilson (California Institute of Technology) and D. R. Criswell (Lunar Sciences Institute, Houston) point out that the glow was originally reported by Surveyors 5, 6, and 7, and possibly also Surveyor 1, between 1966 and 1968. The Russian lunar rover Lunokhod 2 probably detected the glow in early 1973. Since the sites involved were well separated, the elusive phenomenon is presumably moonwide.

The leading edge of the 20-mile-wide pattern of cloud patches, which extends from pole to pole at the evening terminator, travels steadily westward at about nine miles an hour, the same rate as the terminator moves. The clouds are present over at least 0.1 percent of the terminator. The glow may also be a sunrise phenomenon, but the required observations at dawn on the moon have been impossible to obtain to date.

Dr. Criswell theorizes that the glow is produced by dust levitated above the surface by electrostatic fields. Grains at least 10 microns in size can be raised in this way, as the dust is churned up by fields which reach 500 to 1,500 volts of potential difference across sharp shadows. According to the theory, the fields are caused by solar X-rays ionizing atoms on the sunlit surfaces of rocks, so that they become positively charged relative to shadowed parts. But as the foot-high dust clouds get 20 miles or so into the night side of the terminator, the electrostatic fields become neutralized and the dust grains fall.

If this interpretation is correct, some 20 million tons of lunar surface dust may be churned up each year. And since the process would have been occurring on the moon for at least the past 3-1/2 billion years, Drs. Rennilson and Criswell suggest it has been a major process in mixing and spreading surface dust.

Their findings were reported at the recent meetings of COSPAR (Committee on Space Research) at Konstantz, Germany, in a paper read by Richard Shorthill of the Boeing Co.

ECLIPSE PHENOMENA

The best known eclipse phenomena are Bailey's Beads and the eerie shadow bands that race across the ground and up the sides of buildings. Since the latter are seemingly the result of the terrestrial atmosphere, they are assigned to the companion handbook, Handbook of Unusual Natural Phenomena. The former, Bailey's Beads, are generally ascribed to be sunlight shining between gaps in the lunar mountains. The data presented herein casts doubt upon this simplistic explanation.

As the earth's shadow passes across the face of the moon during an eclipse, it may become flattened and otherwise distorted. Color anomalies may arise also. Neither phenomenon is understood. When, finally, the moon is completely in the earth's shadow, it may remain fairly bright or become so dark as to be invisible. The brightness of the eclipsed moon may be related to solar activity and the consequent luminescence of surface materials under the influence of solar wind.

• Dark and Bright Eclipses

REMARKABLE APPEARANCES DURING THE TOTAL ECLIPSE OF THE MOON ON MARCH 19, 1848

Forster, Mr.; *Royal Astronomical Society, Monthly Notices,* 8:132-133, 1848.

"I wish to call your attention to the fact which I have clearly ascertained, that during the whole of the late lunar eclipse of March 19, the shaded surface presented a luminosity quite unusual, probably about three times the intensity of the mean illumination of an eclipsed lunar disc. The light was of a deep red colour. During the totality of the eclipse, the light and dark places on the face of the moon could be almost as well made out as in an ordinary, dull moonlight night, and the deep red colour, when the sky was clearest, was very remarkable from the contrasted whiteness of the stars. My observations were made with different telescopes, but all presented the same appearance, and the remarkable luminosity struck every one. The British consul of Ghent, who did not know that there was an eclipse, wrote to me for an explanation of the blood-red colour of the moon at 9 o'clock.

"The sky was of unusual brilliancy, as often occurs between showers; there was a bright aurora in the north, and a most magnificent meteor descended obliquely towards the north-west horizon about the time of the central eclipse. The western margin of the disc presented a rough, uneven appearance at this time. What would be the effect on the dark surface of the moon of extensive aurora borealis on our earth?"

Mr. Walkey, who observed the eclipse at Clyst-St. Lawrence, near Collumpton, says the appearances were as usual till 20 minutes to 9. "At that period, and for the space of the next hour, instead of an eclipse, or the umbra of the earth being the cause of the total obscurity of the moon, the whole phase of that body became very quickly and most beautifully illuminated, and assumed the appearance of the glowing heat of fire from the furnace, rather tinged with a deep red. The above description I gave to the editor of one of the Exeter papers, and some one has attempted to solve the peculiarity of the appearance by speaking of the umbra and penumbra, which might have been the cause of it. But such a solution has nothing to do with the appearance, the whole disk of the moon being as perfect with light as if there had been no eclipse whatever.

"Having spoken of this appearance, I was informed by one or two individuals that they had seen, between twenty minutes to nine two individuals that they had seen, between twenty minutes to nine and twenty minutes to ten, a very luminous appearance of the aurora borealis. Now, it strikes me that the light reflected from this northern effulgence might have caused the luminous appearance of the moon in this part of the country at the time when it was under the perfect umbra of the earth in other portions of England.

"Many more than threescore years have passed with myself, and during that period I have several times beheld an eclipse of the moon, but never before did my eyes behold the moon positively giving good light from its disk during a total eclipse. The phases of the moon perfectly corresponded with the authorised diagram up to the period of $8^h\ 40^m$ and after $9^h\ 40^m$ to the end of the eclipse."

ABNORMAL OBSCURITY OF THE MOON IN THE LATE ECLIPSE

Johnson, S. J.; *Royal Astronomical Society, Monthly Notices,* 45:43-44, 1884.

On the evening of October 4 there was a conspicuous return of the sunset after-glow that was common last winter. A peculiar state of the strata of our atmosphere might, therefore, indicate that something unusual was to be expected in the eclipse following.

At 3^h, penumbra barely perceptible with opera glass.

At $8^h\ 12^m$, very decided.

$8^h\ 18^m\ 5^s$, shadow had reached Grimaldi.

At $9^h\ 7^m$, for the first time, a very small portion of the eclipsed disk, extending only 4' or 5' inwards, could be discerned through the telescope, this appearing of a dark slate colour---the same tint as is usually observed at the commencement of an eclipse.

At $9^h\ 10^m$, the whole of the lunar circle began to be seen through the telescope, but without a trace of the ordinary coppery redness.

$9^h\ 29^m\ 7^s$, reappearance of small star (time by sextant).

$10^h\ 2^m$, middle of totality. To the naked eye nothing could be seen but a faint nebulous spot. That the obscurity of the Moon arose from lack of illumination, not from fog or cloud, was seen by the fact that stars of small magnitudes above and below the lunar disk shone as distinctly as on an ordinary dark night. The exact appearance of the Moon at this time would be described by quoting Kepler's words verbatim about the eclipse of June (not December) 1620. "Luna difficillime apparuit, emicuit tamen instar tenuissimae nubeculae, longe debilior quam via lactea, sine omni rubedine."

$10^h\ 8^m\ 21^s$, immersion of very faint star. Uncertain to 3 or 4 seconds.

$10^h\ 33^m\ 41^s$, star of magnitude 8-1/2 occulted. Appeared to linger for 4 seconds within the Moon's limb.

$10^h\ 39^m\ 28^s$, reappearance of a star.

$10^h\ 49^m\ 15^s$, sunlight breaking out.

The shadow left the Moon near the Mare Faecunditatis about 11.48-3/4, the penumbra being conspicuous to the naked eye at 11.51. Thus, while the Moon did not completely disappear during totality (except to one correspondent), the peculiar features of the eclipse were the complete invisibility of the eclipsed orb until nine-tenths were covered, also subsequent to the total phase after one-tenth was uncovered. In this respect it seems similar to the earliest instance of the kind given in modern times, that observed by Kepler in 1601, an eclipse of nine-tenths of the Moon's disk. His words are (Astronomiae pars optica), "Anno 1601 Decembri, tenuissimo cornu superstite, caliginosam tamen partem non vidi."

The usual explanation, that when the atmosphere is remarkably free from vapour the red rays would be absorbed, is hardly an adequate one.

(1.) Because the atmosphere was in an equal state of dryness on the occasion of the eclipse of July 12, 1870, when the Moon assumed the usual dull red or copper colour.

(2.) Because the last instance of the disappearance of the Moon was on June 10, 1816, when it could not be discerned even with telescopes, and this was one of the wettest summers in the century.

A REMARKABLE ECLIPSE OF THE MOON

Anonymous; *Sky and Telescope,* 27:142-146, 1964.

The total lunar eclipse of December 30, 1963, will take its place with about a dozen others since the year 1601 at which the moon was described as very dark or invisible. Many telescopic observers will long remember the dim and sooty "full" moon standing amid a rich field of stars.

More than 150 reports have been received by Sky and Telescope describing various aspects of the event. It is apparent that the moon's faintness at totality was not the only outstanding feature of the eclipse. At and after second contact, a brilliant blue band was seen at the western edge of the umbral shadow. So vivid was this fringe that some watchers thought the predicted time of second contact was five or more minutes off! Also, during totality a portion of the moon's disk toward the center of the umbra glowed with a reddish tint.

. . . .

One of the most remarkable happenings during this eclipse was the reddish glow on the northernmost part of the moon, toward the center of the umbra, where the shadow would be expected to be darkest. It seems to have been noticed shortly before second contact, as soon as light from the moon's uneclipsed portion had been sufficiently diminished.

Craig L. Johnson of Boulder, Colorado, was one of the first to see this large red patch. When the shadow had covered about 90 percent of the disk, the northern limb appeared reddish and the eastern limb bluish in 7 x 50 binoculars. He was able to keep the red glow in sight until 11:06 UT, one minute before mid-eclipse. Eric J. Lerner, who observed with binoculars at Baltimore, Maryland, watched the glow on the northeast limb from 10:23 UT (four minutes before second contact) to 10:45.

According to Lawrence B. Nadeau, at Holden, Massachusetts, the glow could be seen with the unaided eye. A typical description is that by Bremerton, Washington, amateur James Kousbaugh: "A few minutes after second contact, the glow was slightly brighter than the rest of the moon, just as if the shadow had missed a small portion of the northeast corner."

Clark Chapman, Buffalo, New York, records his impressions of the red glow and blue band after second contact: "At 10:31 UT, the southwest limb remained bluish while the northwest limb was a ruddy color. As the bluish limb faded, the two regions became more nearly equal in prominence. At about 10:33.5 the bluish limb seemed brighter but less strongly colored than the red one. The whole moon appeared colorless at 10:44."

. . . .

The blue fringe was noticed at 9:30 UT by Walter H. Haas, who judged its breadth to be 10 seconds, as seen in a 12-1/2-inch telescope. This blue rim became a striking sight as totality began. Rev. Kenneth J. Delano of New Bedford, Massachusetts, saw it as about one minute of arc wide at 9:30 and five minutes at 10:05. At the latter time, the inner two-fifths of the border seemed much darker than the rest.

About 10 to 15 minutes before totality, George Lovi began to notice the umbra's blue edge. This Lakewood, New Jersey, amateur states, "It became a striking and beautiful phenomenon just before the moon slipped completely into the umbra. I estimated that this zone was five to six minutes of arc across. It made an eerie sight during the first few moments of totality---I had never seen an eclipsed moon with a blue fringe."

At 10:33 UT the blue rim was seen as a long thin line by Mr. Woods, but seven minutes later it was gone. Richard F. Cressman of Allentown, Pennsylvania, timed second contact at 10:30:11 UT and the virtual disappearance of the blue band at 10:34:47, only a slight coppery tone remaining in its place. In Tokyo, Japan, T. Hara and H. Tanaka reported losing sight of the "pale blue light" at 10:34:20.

After totality, as the moon was emerging from the shadow, Mr. Haas again saw the fringe, perhaps 20 seconds of arc wide at 12:25. Robert Lozar, LaGrange Park, Illinois, also recorded the blue light at about that time.

WHY WAS LAST DECEMBER'S LUNAR ECLIPSE SO DARK?

Brooks, Edward M.; *Sky and Telescope,* 27:346-348, 1964.

The virtual disappearance of the moon during the eclipse of December 30, 1963, was a rare phenomenon that had not happened since the two total eclipses of 1913. It is worthwhile to ascertain the conditions in the earth's atmosphere that made last winter's event so unusual.

At a normal eclipse, the moon inside the earths umbral shadow is seen by light that has traversed our atmosphere twice, once on its way from the sun, when the light is refracted into the umbra, and again on its way back from the moon to the observer. We can distinguish between the effects of these two traverses.

For example, on September 25, 1950, the totally eclipsed moon looked purplish to observers in the northeastern United States. But the sun and moon had exhibited blue and other abnormal colors in Canada and the eastern United States during the two days before eclipse, and for two days following it in Europe. Outside of these regions the eclipse appeared normal. Harry Wexler, writing in Weatherwise for December, 1950, attributed the unusual phenomena to forest fires in western Canada, from which the smoke particles were of such a size as to scatter more red than blue light out of the eclipsed moon's light. Since the abnormal colors were not seen by all observers, the effect could only have occurred during the second passage through the earth's atmosphere.

Last December the situation was entirely different. The eclipsed moon was abnormally dark to everybody, no matter where they were located. Evidently the moon received less sunlight than would otherwise have been expected, due to obstruction by the earth's atmosphere on the light's first passage through it.

Both a dust layer and clouds (of ice or water) have been proposed as major causes of this obscuration along the earth's limb (as seen from the moon).

In a list compiled by F. Link of 11 very dark eclipses observed between 1601 and 1913, at least eight occurred within a year or two after a major volcanic eruption, suggesting that the obscuring matter was volcanic dust in the upper air. After the famous eruption of Krakatoa in August, 1883, dust was traced by many brilliantly colored sunsets around the world, and a dark lunar eclipse took place on October 4, 1884.

On the other hand, as long ago as Kepler's time, variations in brightness among eclipses of the moon were regarded as related to atmospheric moisture along the sunrise-sunset circle around the earth at the time of eclipse. When, for an observer on the moon, the sun is centrally behind the earth, the latter appears to be surrounded by a

Dark Lunar Eclipses*

No.	Date	Degree	Observers	Observations and volcanic explosions
1	1601 XII 9	11	Kepler	Eclipsed part invisible
2	1620 VI 15	18	Kepler	Moon invisible, stars visible
3	1620 XII 9	19	Cysat at Ingolstadt	Moon invisible
4	1642 IV 14	19	Many observers	Very dark or invisible. Explosion Avoe 1641 I, 1 km^3 ashes
5	1761 V 18	18	Many observers	Very dark or invisible. Explosion Jorullo 1759 IX 28??
6	1816 VI 16	15	Lee at London, Eule at Dresden	Moon invisible. Explosion Tambora 1815, spring, 150 km^3 ashes!!
7	1884 X 4	18	Many observers	Very dark and colorless. Explosion Krakatoa 1883 VIII, 18 km^3 ashes
8	1902 X 16	18	Barnard and others	Very dark. Explosions 1902 V Mt. Pelee and St. Vincent, 1 km^3; Santa Maria 1902 X 5 km^3
9	1903 IV 11	12	Many observers	Very dark. Explosions as above
10	1913 III 22	19	Many observers	Very dark and colorless. Explosion Mt. Katmai 1912 VI-X, 21 km^3
11	1913 IX 15	17	Many observers	

* F. Link's list of 11 unusually dark eclipses has been reproduced from "Advances in Astronomy and Astrophysics," Vol. 2 (Academic Press, 1963).

bright glow of sunlight. The reddish color of the eclipsed moon results from blue light having been scattered out of the sunlight by very small air particles, leaving the usual red sunset color to be transmitted through clear regions on the earth's limb.

The effect of the atmosphere is great because of the long path of sunlight passing through it at tangential incidence. Thus the brightness of a lunar eclipse may be a sensitive indicator of atmospheric conditions along the sunrise and sunset line. In fact, the accuracy with which an eclipse can measure these conditions is greater than that of the opposite procedure---predicting the moon's brightness from weather observations along the earth's terminator. In this way, a lunar eclipse can be useful to meteorologists.

....

It had been predicted in Sky and Telescope last December (page 326) that the eclipse might be unusual on account of volcanic dust from eruptions of Mount Agung on Bali earlier in the year. This was the case. The dust encircled the Southern Hemisphere, as indicated by unusual optical phenomena and colors of the sun and moon observed in Australia, South America, and Africa.

From the diagram [not reproduced] we infer that volcanic dust over the western part of Australia and the ocean south of it could account for the bright blue fringe seen on the moon at second contact. A similar effect at third contact (see upper left edge of moon as pictured on the front cover of Sky and Telescope for March) indicates volcanic dust at about the same latitudes in the eastern Pacific Ocean just west of South America. The blue color would require a greater scattering of red light out of the sunlight by large dust particles, as with the Canadian forest fire smoke in 1950. I have observed evidence of this preferential scattering in brilliant red skies after sunsets in Massachusetts following many clear days in November and December, 1963; also, some in February, 1964. Yet on these same occasions the sun itself did not look red as it set in a clear sky.

The faint reddish area seen just west of the moon's north pole after second contact was close enough to the umbra center to indicate relative absence of cloudiness in corresponding terrestrial regions. It was due to clear areas west of Sumatra in western Indonesia that were reported by a Tiros weather satellite. Again, a similar red region photographed on the moon's north rim before third contact indicates few clouds off South America, an inference also supported by Tiros observations.

The extreme darkness of the eclipse is evidence of the prevalence of volcanic dust and cloudy skies along the earth's limb all the way from western Australia, the extreme eastern south Indian Ocean, to the southernmost and southeastern Pacific Ocean.

. . . .

LUNAR ECLIPSES AND DANJON'S LAW

Hughes, David W.,; *Nature,* 253:503, 1975.

In 1920 Danjon, a French astronomer, noticed a relationship between the phase of solar activity and the brightness of a lunar eclipse (C. r. Acad. Sci. Paris, 171, 1127; 1920). This relationship is now known as Danjon's law and can be stated as follows: in the two years immediately after a solar activity minimum the shadow of the Earth during a lunar eclipse is very dark and has little colour. As the solar activity moves away from minimum the eclipsed Moon becomes brighter and redder until, during the seventh and eighth year after solar minimum the eclipsed Moon is at its brightest and is red, copper coloured or orange. The brightness curve then falls away very sharply to its minimum value. The maximum phase of the solar cycle passes unnoticed whereas the minimum phase is indicated by a sudden and considerable diminution in brightness and colour, this change forming a discontinuity.

To obtain this law Danjon had analysed eclipse records dating back to the time of Tycho Brahe. He subsequently used the law to estimate the times of solar minimum during the 17th, 18th, and 19th centuries, finding the general relationship that minima occurred in the years $1584.8 + 10.87n$ where n is an integer (C. r. Acad. Sci. Paris, 171, 1207; 1920). Notice that the length of the solar cycle, 10.87 years as derived from lunar eclipses, is less than the currently used value of 11.2 years derived from sun spots.

Link, of the Institut d' Astrophysique, Paris, has reconsidered the underlying causes of the variations of lunar eclipse brightness described by Danjon's Law in a recent edition of The Moon (11, 137; 1974). He finds that the phenomenon behind the law is not the absorption of the light in the terrestrial atmosphere, as originally assumed by Danjon, but some light source other than the normal refracted solar illumination. The problem is to find this source. Vassy (J. Sci. Met. 8, 1' 1958) hypothesised that the cause is a scattering layer in the Earth's upper atmosphere produced by aerosols. The spatial density of these aerosols could be affected by the corpuscular radiation flux from the Sun and the variation in density caused by the sudden change in the heliographic latitudes of the source of this radiation after solar minimum. Link examines the possible implication of this hypothesis on the twilight and daytime sky as observed from Earth. Looking at two extremes of shadow density he finds the eclipse of January 18, 1954 to be a thousand times less luminous than that of September 26, 1950. To explain the bright eclipse by aerosol scattering requires an aerosol density such that the twilight brightness would be a hundred times more than observed. The luminance of the daylight sky too is far from that needed to explain Danjon's Law.

Another source of additional light is the luminescence of lunar surface materials. This could be caused by corpuscular radiation striking the Moon during the eclipse or it could be an afterglow from the period preceding the eclipse. But recent laboratory examinations of lunar samples showed that the luminescence is much too low to explain the eclipse observations.

Two other facts must be borne in mind: first Fisher (Smithson, Misc. Coll., 76, No 9; 1924) found winter eclipses to be brighter than those occurring at other seasons and second, exceptions to Danjon's Law have occurred when very dark eclipses were seen after the eruption of the volcanoes Krafatoa (1883) and Katmai, Alaska (1913).

But we must return to Link's final conclusions: that a scattering layer in the upper atmosphere, or luminescence of the lunar surface---both caused by solar corpuscular radiation---describe qualitatively the general features of Danjon's law but are quantitatively completely unsatisfactory. So the cause of the relationship remains unknown.

• Distortions of the Earth's Shadow

THE LUNAR ECLIPSE OF OCTOBER 4

Noble, William; *Knowledge,* 6:325, 1884.

That well-known Belgian astronomer, my friend M. Ad. de Boe, of Antwerp, sends me an account of so remarkable and interesting an observation of his own of the eclipse of Saturday week that I am induced to hope that you will find space for a precis as short as I can make of it.

Briefly, then, at the stage of the eclipse represented in the annexed slightly exaggerated diagram the arc of the earth's shadow was quite decidedly peaked at a instead of being rigidly circular. "Il n' y a (says M. de Boe) dans cette observation aucune illusion. Deux personnes presentes la constaterent immediatement." Now, at the time of the phase indicated, the moon was on the horizon of the Cordilleras, while when she was half eclipsed she was similarly placed as regards part of the Pacific Ocean. At this time, however, all traces of the peaked appearance had absolutely vanished, and the periphery of the shadow was sensibly circular. Can the curious deformation figured have been the shadow of the Cordilleras, or had it its origin in the form of the lunar surface?

From all my friends abroad and at home I hear of the curious phosphorescent green tint of the earth's shadow during totality, a phenomenon which struck me forcibly in observing the eclipse myself.

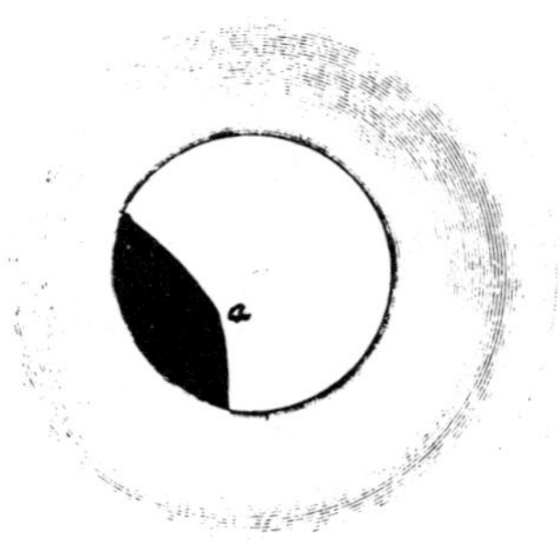

THE LUNAR ECLIPSE OF AUGUST 3

H., H.; *Nature,* 36:367, 1887.

It would be interesting to know if the following phenomenon was observed at other places. At 9. 30 p. m., local time, at Hamburg, a small cumulus cloud was observed a little distance below the moon, and the darkened part of the lunar surface was taken to be part of the cloud,

from its upper edge being flattened. Ten minutes later the cloud had passed away, but the flattened appearance on the moon remained, and it was evident that the earth's shadow was distorted, as seen in the annexed sketch. Several persons noted the peculiarity, which was visible until about 10.30 p.m. in a very clear sky.

THE LUNAR ECLIPSE OF AUGUST 3

C., M., and Malet, H. P.; *Nature,* 36:413, 1887.

I observe the account given by "H. H." (p. 367) of the eclipse of the moon as seen at Hamburg on August 3. Here the appearance was certainly unusual; at least I never saw anything like it. The shadow cast on the moon (with a perfectly cloudless sky) was irregular and jagged. I at first thought it was a cloud, but, on looking repeatedly at intervals, I continued to observe the same appearance; allowance being made for the progress of the eclipse. I was prevented by circumstances from watching continuously, but observed it at a little before 9, and again repeatedly between 9 and 10. (M. C.)

As seen from Killin, on Loch Tay, the shadow on the moon had no form similar to that given by "H. H.," in your issue of August 18 (p. 367); the sky was clear, and it seems possible that the clouds caused the straight lines shown in the diagram. (H. P. Malet)

• Light Rays during Solar and Lunar Eclipses

ON CERTAIN PHENOMENA SEEN DURING ECLIPSES OF THE SUN

Brett, John; *Royal Astronomical Society, Monthly Notices,* 35:14-16, 1874.

The eclipse of the Sun of the 10th of October last was observed in the Island of Guernsey under tolerably favourable circumstances. The wind was from the south-west, and the view only occasionally interrupted by clouds. Definition remarkably steady.

At 10^h 15^m G. M. T. the limb of the Moon, towards the following cusp, presented a series of prominent mountains, amongst which was one particularly deep indentation or notch.

The phenomenon to which this paper refers occurred at the time this particular notch arrived at the cusp and passed off the Sun's disk.

It consisted of a minute ray or brush of light, proceeding from the cusp of the Sun, and intruding upon the Moon's limb; the appearance in this instance being exceedingly transient, and disappearing within two seconds of time.

The rapidity of the intrusion was equal to that of the Moon's motion, so that it may be more correctly described as light dragged away from the Sun's cusp than light projected upon the Moon's disk.

The cusp was, of course, blunted by the notch or lunar valley, and it was at this moment that the encroaching light advanced over the limb.

The character of the ray might be called flickering and sparkling: an appearance such as might be caused by the successive illumination and obscuration of minute excrescences rapidly passing away from the parallel of the illuminating ray.

In estimating the extent to which the encroaching light advanced over the limb, the only available standard was the apparent depth of the lunar valley in which it occurred. I should say it advanced to a distance about equal to twice the height of the adjacent mountain from the bottom of the valley. From this rough estimate it will be seen that the phenomenon was exceedingly minute and subtle, and one which none but very practised observers would be likely to notice at all.

I, however, had the advantage of having seen a similar phenomenon before, under conditions which rendered it far more conspicuous; that is to say, in the eclipse of 1870, December 22, when the Moon was advancing over the Sun's disk instead of receding from it.

On that occasion a similar emanation proceeded from both cusps of the Sun at the same time, and these were seen with great steadiness and perfection of definition, although the time of their duration unfortunately could not be estimated, as they were only seen through a chasm in a passing cloud.

They were then of far greater length than in the recent eclipse: (vide *Monthly Notices*, 1871, March.) The telescope used on that occasion was a reflector of 8-1/2-inches aperture, but in this instance the instrument was a refractor by Cooke, of only 3-inches aperture, but of a quality which probably cannot be surpassed. The power employed was

100, with first reflexion prism.

The interpretation of the above observations presents certain difficulties, particularly on account of the rays appearing only at the cusps; but I ask leave to express my own opinion upon it, as follows.

The encroaching light must be reflected from something. I assume that thing to be the Moon's surface.

But before the light could be so reflected, it is obvious that it must have been to some extent diverted from its originally straight course from the Sun to the spectator. To effect this, I assume the existence of a lunar atmosphere, refracting it sufficiently to cause it to impinge upon the more prominent of those excrescences which the perspective view of the spherical surface of the Moon necessarily makes to appear to crowd more and more thickly towards the profile or limb.

A very small density of the refracting medium would be sufficient for the purpose.

The spherical form of the Moon would of course account for the weakening of the light towards the end of the brush; and the receding motion for the sparkling character of it.

But there remains the question why the refracted light is not seen as a fringe of twilight bordering all that part of the Moon's limb projected upon the Sun's disk; and, until the recent eclipse, this difficulty seemed more abstruse than it is now.

The fact that the intruding ray was only seen on this occasion in one valley of the Moon, and that the deepest of those visible upon the limb, would appear to corroborate the hypothesis that all the Moon's atmosphere of appreciable density is to be found secluded in her deepest valleys.

There is, however, another circumstance not to be overlooked in this interpretation, and which, combined with the last consideration, leaves, I think, nothing further to be explained; viz, that the extent of lunar atmosphere traversed by the incident ray at the cusp subtends a greater visual angle than at any other point on the limb; so that, when a deep lunar valley does appear at that point, the effect of refraction is exhibited under the most favourable circumstances.

It will not surprise me if it can be shown that my theory is untenable; but as far as the observations go, I respectfully offer them for consideration as good observations, and such as may possibly throw some light on the question whether the Moon has any atmosphere or not.

AN ECLIPSE PHENOMENON

Tennant, J. F.; *Observatory*, 25:168-169, 1902.

The mention of an observation of a ray of light projecting from the cusps of the Moon in the solar eclipse as having been seen in 1870 by the late Mr. Brett, calls my attention to the same phenomenon (apparently) having been noticed in 1868 by Captain Branfill and Major Hearne under similar circumstances.

Mr. Brett appears to have seen the lines only near totality in 1870, but Captain Branfill saw them when three-quarters of the Sun were eclipsed, and I see from Mr. Ranyard's volume of the R. A. S. Memoirs that they have been seen when only half the solar disc had vanished. It would seem, then, that these lines must be frequently present in partial eclipses, yet I do not recollect ever having seen them noticed.

The phenomenon seems not to be visible continuously, but we can hardly conceive that the totality of the eclipse can have anything to do with it, and the search for it and its cause would give an interest to the observation of partial eclipses such as they have not now. Have no such appearances been seen in making the micrometric measures of the cusps at Greenwich and elsewhere?

FINGERS OF LIGHT DURING LUNAR ECLIPSE

Davies, T. H.; *Marine Observer,* 34:126, 1964.

m. v. Canopic. Captain T. H. Davies. Sydney to Aden. Observers, Mr. K. Newton, 2nd Officer and Mr. R. Pedlow, Jnr. 3rd Officer.

6th July 1963. Between 2035 and 2300 GMT a partial eclipse of the moon was observed. At maximum eclipse when three-quarters of the moon was in shadow, its surface still remained visible, and fingers of light were seen illuminating the upper section which was in shadow, the appearance being shown in the accompanying sketch.

Only limited observation was possible due to the presence of 7/8-8/8 Sc. The end of the eclipse was not observed due to the sky becoming overcast.

Position of ship: 18° 40'S, 92° 33'E.

Note. Mr. H. B. Ridley of the British Astronomical Association comments:

"I think that the explanation of the 'fingers of light' mentioned in the text and shown in the drawing is fairly straightforward. The earth's atmosphere refracts light from the sun into the shadow-cone, so that a lunar eclipse, although total, is not complete; the moon is still quite plainly visible even when wholly immersed in the earth's shadow. The atmosphere scatters blue light (hence the colour of the sky) but transmits red; therefore the faintly visible eclipsed moon has a coppery hue, though this does not normally appear until the whole disc is immersed. The 'continental' areas of the moon are much brighter than the flatter, darker maria or 'seas', and show up very plainly during eclipse.

It is fairly evident that what the observer saw in this case was the comparatively bright north polar region, partially illuminated even though it was in the earth's shadow.

Although there is nothing exceptional about this observation, the officer concerned was quite justified in remarking on the phenomenon, which might have escaped the notice of a more casual observer."

• Mysteries of Bailey's Beads

BAILEY'S BEADS

Swift, Lewis; *Sidereal Messenger,* 9:170-173, 1890.

This curious phenomenon, always accompanying total and annular eclipses of the sun, is, if carefully observed, one of the most striking of the many phenomena visible on such occasions, and as, to my mind, no satisfactory explanation has been advanced for its production, and, inasmuch as I have studied carefully at three different eclipse occurrences its formation, I have thought that a short paper on the subject might not be wholly uninteresting to the readers of The Sidereal Messenger.

When the moon, advancing eastwardly over the sun, has reduced the sun's eastern limb to a very narrow crescent, it suddenly breaks up into many luminous objects extending from end to end of the crescent, and bearing, sometimes, so strong a likeness to a string of beads, that they have received the name of "Bailey's Beads," because first accurately described by Mr. Francis Bailey in 1836, though seen long before by Halley and others. Some four or five seconds before the advancing moon uncovers the western limb of the sun, they are in like manner produced at that edge of the disk, and present nearly the same appearance as at the eastern limb.

For a long time, perhaps quite naturally, the cause assigned was the visibility of the sun through the interstices of the lunar mountains. This theory was, until recently, from its very simplicity, received with almost universal credence, though after three or even two careful observations of this phenomenon, I think no astronomer would endorse it, as the idea that twenty or thirty lunar mountains are so evenly spaced and so precisely alike as to so admit the passage of sunlight that they shall produce contours so alike and so symmetrical is not tenable.

Perhaps I cannot do better than to reproduce my descriptions of them as observed at Denver, Colorado, in 1878 (as reported to Professor Colbert and published by the Chicago Astronomical Society in that year,) and, also, as seen at Nelson, California, on January 1, 1889.

"Several seconds previous to the formation of the beads, I observed near each end of the solar crescent a phenomenon which I have never seen described in any of the books, and, though reminding me of the "Black Drop" which I saw at the last transit of Mercury, yet it was very different from the latter. Please allow me to describe it. Imagine a long and very narrow crescent cut in a door between two rooms, one dark, the other brilliantly lighted, the observer being in the dark one. The appearance was as if two concealed persons in the lighted room, one on each side of the crescent, were busy in protruding and then withdrawing a series of long, slim, opaque objects like writing pencils. Only twice did I see the images appear opposite each other, as is always the case with the black drop. These appearances continued for about two seconds and then, after ceasing for a second, recommenced farther from the end of the sun's limb, now reduced to a much narrower crescent, but the images were shorter and larger and more irregular in size. This lasted only about a second and again ceased as if waiting for a grand *denouement*, and then came the formation of Bailey's Beads, beginning simultaneously at each end of the crescent, now only a curved line of silver light, and regularly and quickly meeting at the center they seemed to give a succession of quivers and instantly vanished.

These luminous beads appeared to be square (trapezoidal) and from the center each way decreased in a regular geometrical ratio, and, could one-half of them have been superimposed upon the other, they would have agreed in curvature, size, distance, shape and brilliancy. When I take into consideration the exact uniformity of their formation as to time and the other similarities above noted, I cannot subscribe to the generally received opinion that they are the sun's light seen through the interstices of the lunar mountains. The intelligent reader will not be slow to conceive how serious and numerous are the objections against receiving as true so improbable a theory, especially as those formed at the beginning are precisely like those at the end of totality."

The following report to Professor E. S. Holden, now incorporated in Lick Observatory Eclipse volume, gives their appearance as seen from my station at Nelson, California, which was as unlike that of my former observation as two things can well be, I quote from a manuscript which, in its final revision, was slightly changed:

"*Bailey's Beads*." "There has been much speculative writing concerning this singular and, at times, beautiful phenomenon observed at every total and annular eclipse for, at least, a century. The usually assigned cause is the shining of the sun through depressions between lunar mountains, a theory which, when compared with observed facts, has no support. As I saw them in 1878, every one was trapezoidal in shape and differed in size only, being mere points at each end of the crescent, but increasing gradually and regularly towards its center. At the California eclipse they bore no resemblance to those seen at Denver, for, instead of being trapezium shaped they, at both second and third contacts, resembled a curved series of dotted i's or, more exactly the letter a (dot and dash) of the Morse telegraphic alphabet, thus: .-.-.- .-.-. Professor Charles A. Schott says of those seen in 1869, "I would particularly notice their great regularity in width, outline and distribution." Professor D. G. Eaton says, "Bailey's beads were very conspicuous. Just before totality the thin crescent was broken by dark lines shooting out from and perpendicular to the edge of the moon."

From the above quotations added to observations of my own, I am irresistibly led to the conclusion that they, like the color of the protuberances, are not seen alike by all, and that the true cause has never been arrived at. Were I, however, to hazard an opinion, it would be that they result from a phenomenon closely allied to, if not identical with, that causing the "Black Drop." It is also quite likely that diffraction may have much to do with it and possibly may be its sole cause. Were the usually assigned reason the true one, as there are mountains all round the moon, then, on the occasion of an annular eclipse, when the moon so nearly equals the sun in size as to render the annulus exceedingly narrow, the "Beads" should be seen all round the moon. Again, long before a total phase is reached, the ends of the crescent, all the time reduced to a hair-like fineness, ought to reveal them, which is not the case."

. . . .

BAILY'S BEADS

Hollis, H. P.; *British Astronomical Association, Journal,* 37:220, 1927.

May I put on record an experience that occurred to me on the occasion of the Solar Eclipse, seen Annular from Paris, on 1912 April 17. Observing with binoculars (but it is forgotten whether with or without a dark shade) as the eastern limb of the Moon approached the eastern limb of the Sun, the diminishing crescent was suddenly and unexpectedly crossed by thin lines, not parallel, nor radial but in all directions, though I do not think that any two crossed. The lines broadened, and the Moon's limb appeared as a toothed-wheel. The bright space between the cogs, naturally disappeared, but as the whole appearance passed so rapidly it is not possible to say how this occurred precisely. If this is the phenomenon known as Baily's Beads, as I believe it is, it is difficult to see how it is connected with the contour of the Moon's limb except by making some extreme assumptions. I hope to see these thin lines on June 29 next.

[The article's incorrect spelling of Bailey is retained.]

• Color Phenomena

TOTAL ECLIPSE OF THE MOON

Brown, G.; *Marine Observer,* 24:16-17, 1954.

S. S. Pacific Importer. Captain G. Brown. Cristobal to London. Observer, Mr. T. M. Sims, 3rd Officer.

29th-30th January, 1953, 2305 to 0140 G. M. T. The commencement of the eclipse was not observed owing to almost stationary Cu covering the moon. During totality a small white patch of light of low brilliancy moved round the north pole of the moon until that phase came to an end at 0030. From that time the white patch increased in area until the end of the eclipse at 0140. During the total phase the face of the moon appeared to be coloured in bands of blue, green, yellow and orange as in the sketch, and stars were visible with the unaided eye within 2° or 3° of the moon.

Position of ship at 0000 on 30th: 33° 22'N, 45° 20'W.

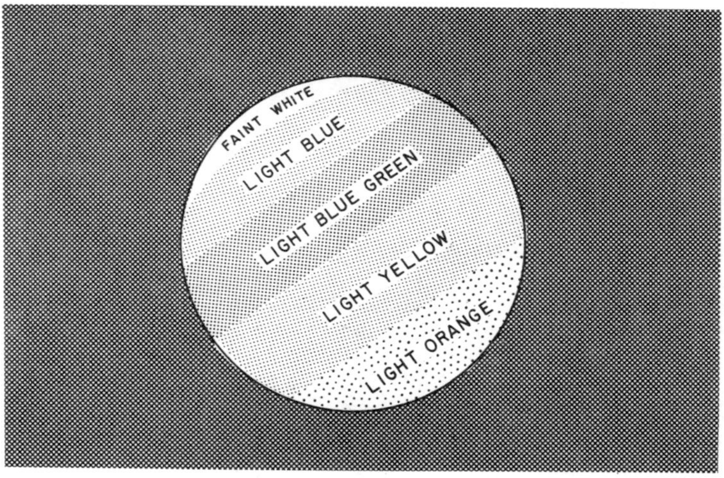

Note. The interesting part of an observation of a total eclipse of the moon is the degree of luminosity of the moon during the total phase and its colour. To an observer on the totally eclipsed moon, completely immersed in the shadow of the earth, the body of the earth would be quite invisible, since the sun is then behind it. The observer would, however, see a narrow luminous ring of reddish or orange colour, probably brighter in some parts than others, this being the section of the earth's atmosphere presented to him in profile and illuminated by the refraction of sunlight through the atmosphere. It is this light which gives the illumination of the moon which we see during the total phase.

The amount of light so transmitted to the moon and its precise colour varies to some extent on the occasion of different eclipses according to the degree of cloudiness and other meteorological conditions in the section of the atmosphere presented to the moon. The totally eclipsed moon usually remains sufficiently bright for the dark markings on its surface to be visible with the unaided eye. The normal colour is reddish, reddish-orange or coppery. On rare occasions in the past the totally eclipsed moon has almost, or even completely, disappeared from view. These "black eclipses", as they are called, are due to the presence of fine dust in sufficient quantity in the earth's atmosphere, due to great volcanic eruptions.

The bands of blue and green reported by S. S. Pacific Importer form a very unusual observation, as light of these colours would not normally be refracted by the earth's atmosphere.

Observations of this eclipse have been received from 18 other ships, but some of these were of a partial eclipse only. The colour cannot be properly judged in the partial phase owing to the proximity to the fully illuminated part. There are six of the total eclipse which give details of the colour as follows:

S. S. Empire Ken, Captain P. M. Burrell. 32° 20'N, 29° 30'E. Light brown, uniform over moon's surface except for a lighter arc in the northernmost part.

M. V. Inverbank, Captain R. A. Lorains. 00° 39'S, 59° 37'E. Bronze, tinged with bluish-white at the lower edge.

M. V. Kenilworth Castle, Captain T. H. Whatley. 24° 48'S, 35° 10'E. Dark reddish-brown, with faint pale arc at the lower limb.

S. S. Manistee, Captain R. W. Lundy, O. B. E., R. D., R. N. R. 04° 05'N, 02° 58'W. Dark orange.

S. S. Planter, Captain A. Robertson. 23° 20'N, 48° 11'W. Light orange-reddish, the moon's rim from N to SE being whiter.

M. V. Waipawa, Captain A. E. Warren. 15° 50'N, 77° 20'W. Dull orange.

THE COLORED LUNAR BAND

Swift, Lewis; *Popular Astronomy,* 3:269-270, 1896.

Speaking of the eclipse phenomena mentioned in last paragraph a few days ago, Dr. Swift says:

Starting from near the Moon's center a luminous band extended as far to the east as Aristarchus, passing very near if not exactly over that crater, from near which it bent sharply to the east, somewhat like in form to a huge figure 7, the up and down portion of the character following the curved outline of the lunar limb though distant from it by estimation one-fourth of the Moon's radius. In its curvature to the north it passed, as at the March eclipse, centrally over Plato and ended a short distance beyond it to the west. The band was of a pale drab color and of nearly uniform width throughout its entire length, and on both occasions I compared its width with that of Tycho and found the two equal, so that it must have been fifty miles wide.

PUZZLING LUNAR SURFACE FEATURES

Linne, which is presently a modest crater in the Mare Serenitatis, has fueled a long argument in the astronomical literature. Did it or did it not change from a prominent crater to an undistinguished one in the 1800s? The modern consensus is that the early observations and maps suggesting a prominent crater are in error. The Linne example is typical of several other controversies; viz., Vulcan, the moon of Venus, etc.; where older observations are swept into the trash heap when they are not in accord with modern data.

Whatever the truth in the Linne situation, some craters persistently display subtle changes in detail, shadow structure, and color. Much of this variation can be ascribed to changes in lighting, gas emanations, low-key volcanic processes, and other transient lunar phenomena. (See pp.

Two topological features bearing heavily on lunar (and possibly terrestrial) history are: (1) the asymmetrical distribution of lunar craters, and (2) the apparent erosion channels. The lunar basins facing the earth seem filled with basalt while those on the far side are not. Why the difference? And what kind of fluid scoured out the lunar "riverbeds", if that is what they are? In Chapter 9, the same questions are asked about Martian craters and channels. One should ask if the earth should not be included in this comparison of geological features by virtue of its huge ring structures and immense submarine canyons. (See the Strange Planet series of sourcebooks.)

• Crater Phenomena

ON THE OBSCURATION OF THE LUNAR CRATER "LINNE"

Birt, W. R.; *American Journal of Science,* 2:43:411-413, 1867.

The interesting phenomenon of a change in the appearance of the crater "Linne" was communicated to me by Herr Schmidt, the Director of the Observatory at Athens, an extract from whose letter is as follows:

"For some time past I find that a lunar crater situated in the plain of the Mare Serenitatis has been invisible. It is the crater which Madler named Linne, and is in the fourth section of Lohrmann, under the sign A. I have known this crater since 1841, and even at the full it has not been difficult to see. In October and November, 1866, at its epoch of maximum visibility, i.e., about the time of the rising of the sun on its horizon, this deep crater, whose diameter is 5.6 English miles, had completely disappeared, and in its place there was only a little whitish luminous cloud. Be so kind as to make some observations on this locality."

The earliest information respecting the crater I received from Mr. Buckingham, who favored me with a copy of a photograph taken by him on November 18, 1866. On this photograph the place of "Linne" is visible, but faint. I have during the last lunation received records of observations from the following gentlemen: Doctors Mann and Tietjen, and Messrs. Talmage, Webb, Slack, Grover, and Jones. On the 13th, when the terminator passed over the east boundary of the Mare Serenitatis, the place of "Linne" was seen by Messrs. Webb and Talmage; Mr. Webb's aperture 9-1/4-inch silvered glass reflector, and Mr. Talmage's 10-inch refractor of Mr. Barclay at Leyton. Mr. Webb described the appearance as an ill-defined whitishness on the site of "Linne." Mr. Talmage recorded "a dark circular cloud." The exact position of these appearances was carefully ascertained afterward and found to agree with the place of "Linne." Doctor Mann and myself at Leyton were prevented by a thin veil of cirrus seeing the "cloud" recorded by Mr. Talmage. With smaller apertures both Mr. Grover and myself were unable to detect the slightest trace of "Linne," while the small crater "Linne B" of Beer and Madler, and also Bessel, were very distinct with the shadows within them. On the following evening, December 14th, observations were made by Messrs. Webb, Slack, Grover, and Birt. A white spot was seen in the position of "Linne." Mr. Webb described it as the most conspicuous object on the east half of the Mare Serenitatis. Mr. Slack saw a whitish spot not remarkably bright, but could see no trace of a crater. Mr. Grover recorded "a tolerably defined roundish whitish speck," but he could not see the interior or margin of the crater, and "in this respect the spot showed very different from Bessel and other craters which were well seen." My own observations perfectly agree with the above. I estimated the light at 3°. On the 15th the spot was brighter, and I obtained the measures recorded below. On the 16th Messrs. Jones and Grover described the appearance as a white spot not over bright.

On the 20th Professor Foerster and Dr. Tietjen observed "Linne" with the Berlin refractor. The following is the translation of the letter which I received from Dr. Tietjen, dated Berlin, 21 December, 1866:

"On viewing the moon last night about 13^{h} M. T. Berlin with our refractor, in order to convince ourselves of the disappearance of the crater 'Linne,' Professor Foerster and myself perceived that crater very distinctly. If, therefore, an obscuration has taken place, on which certainly no doubt can exist, as it is affirmed by so competent an authority as Herr Schmidt of Athens, it has evidently now ceased."

Although Dr. Tietjen considers that the obscuration has ceased, it does not appear that either he or Professor Foerster has seen into the crater.

The whole of the observations are so accordant among themselves, and the measures appended so clearly indicate the white spot to be larger than the crater "Linne," as to leave no doubt that a change of some kind has taken place; and this conclusion appears to be supported by previous records which are here appended:

Date.	Authority.	Brightness.
1653,	Riccioli,	o
1788, Nov. 5,	Schroter,	0.5
1823, May 28,	Lohrmann,	7.0
1831, Dec. 12, 13,	Beer and Madler,	6.0
1858, Feb. 22,	De la Rue,	5.0
1865, Oct. 4,	"	5.0
	Rutherfurd,	6.0
1866, Nov. 18,	Buckingham,	2.0

The last four determinations of brightness are from photographs. There is some uncertainty in determining this element on the prints. Schroter, in plate IX of his Selenophotographische Fragmente, gives a large dark spot in the place of "Linne;" and the Rev. T. W. Webb informs me that "Linne" is not to be found on Russell's globe or maps, 1797, from which it may be inferred that the crater has previously been obscured.

. . . .

THE LINNE CONTROVERSY: A LOOK INTO THE PAST

Moore, Patrick; *British Astronomical Association, Journal,* 87:363-368, 1977.

Linne lies on the Mare Serenitatis, one of the most famous of the regular lunar seas. There are no really large craters on the Mare; Bessel is the most prominent of them. Naturally, the Mare was drawn and described by all the early observers. Schroter can tell us little (no doubt his main sketches perished with his observatory), but both Madler and Lohrmann are quite definite. I think it will be best to quote their descriptions. Bear in mind that the old German mile which they use is equal to about 4-1/2 English miles or 7 km.

First, Beer and Madler, in their classic work Der Mond, which was completed in 1838-39: "According to a measurement by Lohrmann and seven by ourselves, this crater lies at latitude 27° 47' 15" and longitude 11° 32' 28". It is 1.4 miles in diameter, brightness 6 degrees, borders not well defined at full moon." Lohrmann's account reads: "The second most conspicuous crater on the plain... it has a diameter of somewhat more than 1 mile, is very deep, and can be seen under all angles of illumination."

Nothing could be more positive, particularly as Madler added that "the deepness of the crater must be considerable, for I have found an interior shadow when the Sun has attained 30°. I have never seen a central mountain on the floor". Both Lohrmann and Madler actually used Linne as a reference point.

Oddly enough, the great map by Beer and Madler (and it really was a great map, even though the instrument used was a mere 95 mm Fraunhofer refractor) actually held selenography back to some extent. Unlike Schroter (and other observers, much more fanciful, such as Gruithuisen)

Beer and Madler regarded the Moon as inert, and contemporary astronomers tended to the view that since an accurate chart had been compiled there was really very little point in observing the lunar surface further. Neither Beer nor Madler did much more lunar work after 1839, and for the next quarter-century the only active selenographer of any standing was Julius Schmidt, yet another German, who went to Greece and carried out most of his work during his spell as Director of the Athens Observatory. Schmidt drew Linne in 1841 and 1843, and apparently agreed with the descriptions given by Madler and Lohrmann. Then came his observation of 1866 October 16, when he found that the old crater had disappeared, to be replaced by a small whitish patch. It was a startling discovery, but Schmidt had no doubt about it, and his announcement caused a major sensation. Telescopes were directed back to the Moon, and it was widely thought that a real change had occurred. Systematic lunar observation began once more, and has continued ever since.

. . . .

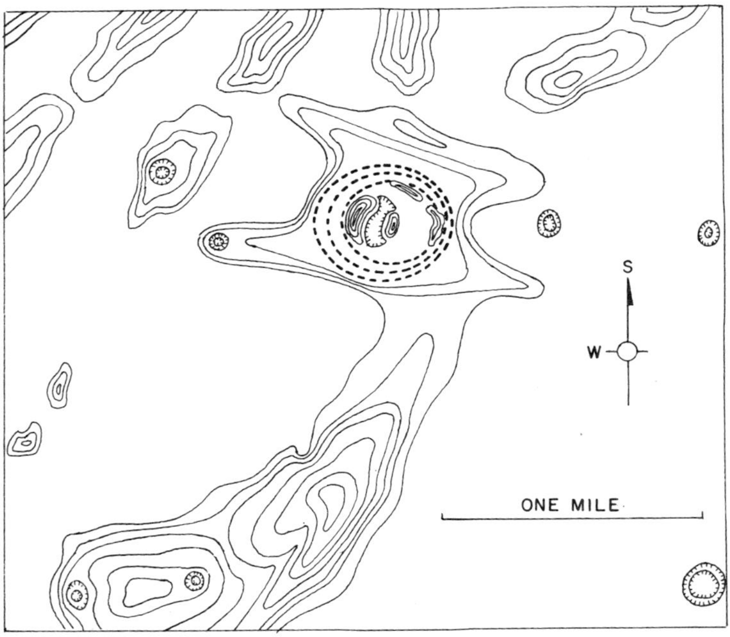

Topography in the vicinity of Linne in 1906 as drawn by Philipp Fauth. The present white spot is represented by dashed ellipses, within which are a craterlet and some small ridges.

Of the "three greatest selenographers" (Lohrmann, Beer and Madler) only Madler remained alive in 1866. He accepted the change as real, but he added a disquieting comment. He wrote: with the Bonn heliometer in 1867 May he "found it shaped exactly and with the same throw of shadow as I remember to have seen it in 1831. The event, of whatever nature it may have been, must have passed away without leaving any trace observable by me." In other words, to Madler, Linne looked the same in 1867 as it had done in 1831, which immediately weakens the evidence in favour of change. By no means everyone was satisfied, and there is an interesting comment by T. G. Elger, the first Director of the BAA Lunar Section, in 1895: "It is anything but an easy object to see well, as there is a want of definiteness about it under the best conditions, though the minute crater, the low ridges and the nebulous whiteness described by Schmidt and noted by Webb and others are traceable at the proper phase .... There are still many sceptics as regards actual change, despite the records of Lohrmann and Madler; but the evidence in favour of it seems to preponderate."

....

What, then, must we conclude? In my view, at least, we can at last reject the idea that any variation in Linne has occurred in telescopic times. No TLP events have been recorded in the area; it is not an active region. And perhaps the early results are not so surprising as has often been thought---and as I once thought myself. Schroter can tell us nothing; Beer and Madler used a telescope which was, by any standards, tiny; and when Schmidt drew Linne as a crater he was still young and inexperienced. Linne is a small object, and it is not hard to believe that observers of the years following 1866 were deceived into reporting changes which did not really occur. Unconscious prejudice may well have played a major role.

CHANGES ON THE MOON'S SURFACE

Anonymous; *Nature,* 66:40-41, 1902.

That the moon is a dead planet, devoid of water-vapour and air and consequently lacking any form of life, either of the animal or vegetable world has long been the belief of astronomers. New light upon the history of our satellite is, however, beginning to dawn, and it seems that the imagination of Mr. H. G. Wells, which illustrates so vividly the seasonal changes on the moon's surface and the appearance of vegetation of rapid growth, is supported by actual "results of observation," judging from an interesting article by Prof. William H. Pickering in the May number of the Century Magazine. Messrs. Pickering and Percival Lowell have during the last few years made numerous excellent observations on the planet Mars, and they have greatly increased our knowledge by accurately observing the surface markings and suggesting very plausible explanations of the phenomena observed. Such work was rendered possible by erecting an observatory in a locality where observing conditions were as near perfect as possible. Prof. Pickering has more recently turned his attention to an examination of the lunar surface, and the first results of this work have led him to some very definite and striking conclusions. The first of these is that there seems to be strong, if not fairly conclusive, evidence in favour of the idea that volcanic activity has not yet entirely ceased, and he quotes several instances in which small craters have disappeared while others have sprung up in different regions. The second, and perhaps more startling, announcement is that there is snow on the moon. He has observed that many craterlets are lined with a white substance which becomes very brilliant when illuminated by the sun, and a similar substance is found on the larger lunar craters and a few of the higher mountain peaks. The curious behaviour of these patches under different angles of illumination and their change of form have led him to suggest that an irregularly varying distribution of hoar frost may have something to do with the changes observed. The third remarkable deduction refers to the observations of "variable spots," which appear to be restricted between latitudes 55° north and 60° south; these spots are always associated with small craterlets or deep narrow clefts, and are often symmetrically arranged around the former. The alterations which these undergo have led him to seek the cause in the change in the nature of the reflecting surface, and the most simple explanation according to him is found in assuming that it is organic life resembling vegetation, but not necessarily identical with it. The new selenography consists, therefore, as Prof. Pickering remarks, "not, in mere mapping of cold dead rocks and isolated craters, but in a study of the daily alterations that take place in small selected regions, where we find real, living changes, changes that cannot be explained by shifting shadows or varying librations of the lunar surface." Prof. Pickering illustrates his article with numerous excellent and instructive drawings and photographs of portions of the lunar surface, and these give the reader a good idea of the changes referred to in the text.

CHANGE IN A LUNAR CRATER

Anonymous; *American Journal of Science,* 4:38:95, 1914.

Most, if not all, of the changes which have been observed in the moon and accepted as thoroughly verified are of a periodic nature. A marked case of a non-periodic change is recorded in a short article by William H. Pickering. The author says: "It is indeed the most marked non-periodic change that the writer has ever detected." The crater in question is named Eimmart and is situated at the northwest border of Mare Crisium, in latitude 24° and longitude 295°. This crater is about 40^{kms} in diameter. The general nature of the phenomenon observed is that while formerly, at each lunation, the crater apparently filled up gradually and overflowed with a white material, the source of which was a point at the foot of the northern interior slope, this change no longer occurs. The last "eruption" observed occurred in January, 1913. Only one observation was made in February and March, but less activity was then shown, while in April and May the phenomenon was so reduced in intensity as to be scarcely noticeable. Interesting drawings of the appearance of the crater during and after its maximum activity are reproduced in the paper, but the author does not discuss the geological or physical aspects of the results obtained with the Draper refractor.---Astron. Nachrichten, No. 4704, p. 414.

THE LUNAR VEGETATION

Pickering, W. H.; *British Astronomical Association, Journal,* 37:65-67, 1926.

Since the old question of lunar vegetation has again come up for discussion in the Journal of the B. A. A., I should like to point out once more that there are really two distinct questions involved. One is as to the facts. This is important. The other is as to their explanation, which is of much less consequence. Unfortunately, the interest of the latter, and the discussion which it has involved, has almost completely concealed and smothered the more important question. In denying the explanation, the rejectors have carelessly also denied the facts, without recognising that they are in a wholly incompetent position to do so. Some of them have even been so careless as to suggest that the darkening areas were shadows, not realising that shadows are most conspicuous under a low Sun on the Moon, and that they disappear completely when the Moon is full. The variable spots do the precise reverse of this. As has often previously been stated, they are inconspicuous, and in some cases invisible under a low Sun, but the central ones are strikingly conspicuous at full Moon.

In the case of Eratosthenes, to which I have called particular attention, they are perfectly obvious with a 5-inch refractor, and can even be seen with a 3-inch ("Popular Astronomy," 1919, 27, 579, Plates 44 and 45). Anyone can see them who will take the trouble to turn his telescope upon the Moon. During the three days including the full Moon, however, the novice may at first have a little trouble in the case of this crater, because the shadows have disappeared, and the changes are so marked that the crater itself is rather hard to recognise. The markings with their changing shapes are very easily photographed (Harvard "Annals," 53, Plates 2, 3, and 4. See also "Popular Astronomy," 1919, 27, 579, Plates 46 and 47). In the light of these photographs, for anyone to claim that these spots do not exist, or that they do not change their shapes, or that they are not visible at full Moon, is, to say the least, foolhardy.

While these spots are easily seen, and are interesting to watch, as their shapes change from night to night, yet the really interesting features are the much more delicate ones which are clearly visible only in a good climate. I have stressed them but little hitherto, beyond a few brief descriptions in "Popular Astronomy," because until those markings that are easily seen are generally accepted by planetary astronomers, it is not worth while to publish that which will not be believed. These delicate features---the so-called runs and plats---which are comparable in visibility to the canals of Mars, are small variable spots, some of which can be seen travelling slowly across country in a more or less unpredictable course, varying their positions from one lunation to the next. Their motions have been checked and confirmed by both Hamilton and Maggini, the former in "Popular Astronomy," 1924, 32, 69, and 237 (figure 5 inverted), the latter in "Popular Astronomy," 1925, 33, 629, Plates 31 and 32. When Harvard requested us to return the 11-inch Clark refractor, in order that it might be used in Cambridge for photographic purposes, all observations of the variable spots necessarily ceased; but now that we have acquired an excellent 12.5 in. Calver reflector, which shows many of the finer markings, we hope to take the matter up again and continue our investigations.

The variable spots, so far as I am concerned, have been under discussion for the past thirty years, but they were discovered originally by Maedler nearly a century ago. In spite of the large amount of material that has since been published in regard to them, the only two explanations that have been suggested are that they are due to shadows, and that they are due to vegetation. The spots themselves are very numerous, and occur in all the equatorial regions of the Moon, but are most noticeable where the soil is light, in the western hemisphere. They are not found in high latitudes, their extreme range lying between 55° and -60°. The most northern ones that are at all conspicuous consist of several small areas in the crater Atlas, latitude 46°. Schickard also contains some in about the same latitude south. There are three very conspicuous ones in Alphonsus, latitude -14°, described in Neison. Those in Eratosthenes are by far the most interesting, however, because they change their shapes with the age of the Moon more conspicuously than any others yet discovered. The search for others that exhibit similar changes, even if less marked, is suggested as an interesting study to the amateur.

. . . .

THREE RIDDLES OF PLATO

Carle, Jackson T.; *Sky and Telescope,* 14:221-223, 1955.

During the past century Plato, "the Great Black Lake," nestling at the northeast tip of the lunar Alps near the edge of Mare Imbrium, probably has been the most intensively observed feature on the moon. And many who have studied this great walled plain are reluctant to accept the general belief that the moon is a dead world where nothing ever happens. For it appears that subtle details on Plato's dark floor undergo changes which many experienced selenographers say cannot be explained by the varying illumination and libration.

. . . .

Plato is about 60 miles in diameter, with rim walls rising from 3,000 to 5,000 feet above the floor, and peaks up to 7,400 feet high.

If you first look at the Great Black Lake as the sun is rising there ---about a day and a half after first quarter---you will be struck by the extraordinary ruggedness of the west rim as revealed by its shadow cast upon the floor. With the sun low on the horizon viewed from Plato, even minor elevations cast long shadows, giving an exaggerated idea of the heights of the summits on the rim.

From our telescopic vantage point we see the shadow of the western ramparts extending completely across the 60-mile floor in jagged outline, and most of the interior of the crater is hidden in velvety blackness which knows no morning twilight. Even as we watch, the shadows recede across the floor with surprising speed. Within an hour the eastern half of the plain lies revealed in sunlight except for one needle-like black shadow still extending completely across Plato. Watch this shadow carefully as it quickly draws westward before the rising sun. Does its outline change? Does it grow broader or narrower? Does the shadow at any time present a hooked tip?

This shadow is cast by a solid, immovable mountain, and can change only from the increasing elevation of the sun and the nature of the surface across which the shadow falls. Herein lies one of the three riddles of Plato we shall discuss.

The testimony as to the conformation of the crater floor is surprisingly contradictory. In 1892-93 William H. Pickering studied Plato carefully at Harvard Observatory's station at Arequipa, Peru. His results, published in Vol. 32 of the Harvard Annals, report the floor of Plato as "extremely convex," much more so than the regular spherical curve of the moon. He also remarks upon irregularities and slopes in the floor.

The Mount Wilson picture on this page is one of the best photographs ever taken of Plato. In addition to five interior craterlets, the variegated pattern of light and dark splotches may be seen. The wedge-shaped light area at the upper right of the floor is the "sector." Under magnification, the original negative shows a series of parallel light and dark areas running from the upper rim across the floor. These look like rolling ridges with intervening hollows, extending from the south rim northward to the central part of the interior. On a number of occasions, I have seen a similar appearance, and so have other amateurs, among them David P. Barcroft and T. E. Howe. The latter's map of Plato, in the February, 1952, issue of The Strolling Astronomer, shows a ridge in the floor just south of the center. The ridge is one of two floor features mentioned in Barcroft's comments.

Twice I have noted dark oval patches soon after sunrise, which quickly disappeared; at other times, with similar illumination, no patches were seen.

On April 3, 1952, H. P. Wilkins and Patrick A. Moore observed Plato with the 33-inch Meudon refractor. The former reported that "the floor or plateau appeared remarkably uniform and level," but he does show some details in the accompanying drawing made that night. Similarly, in his book, A Guide to the Moon, Moore describes Plato from his view that night as "probably the most level spot on the Moon."

In Dr. Wilkins' drawing the shadow of the west wall extends about a fifth of the way across the floor. The long shadow of the southerly peak juts farther east, and is hooked to the south. Does this shadow appear hooked because of some strange conformation of the rim, or from falling upon an uneven surface? I have looked unsuccessfully for this hook at a number of sunrises, but this failure may mean that the appearance requires certain combinations of libration and sun direction. Is Plato's interior level, convex, rolling, or Irregular? Or perhaps, does it change? I don't know.

With good seeing, we may glimpse some of the tiny and difficult craterlets strewn over the darkish floor. In the apparently changing visibility and relative sizes of these elusive features over a period of years lies a second riddle posed by the Great Black Lake.

Since an intensive study of Plato was first organized by the British selenographer W. R. Birt and his co-workers in 1869, some 80 craterlets and spots have been charted. These have never all been seen by one observer or within a limited period of time. Some spots, having been seen easily, apparently disappeared---only to reappear years later and be reported as new discoveries.

The minute markings, according to E. Neison's The Moon (1876), include 10 craterlets, six doubtful craterlets, and 20 spots not represented as craterlets. T. G. Elger's authoritative The Moon (1895) states that there are 40 or more spots. "They are extremely delicate objects, which vary in visibility in a way that is clearly independent of libration or solar altitude."

Pickering mapped 71 craterlets and spots discovered by him and earlier observers, but was never able to see more than 39. He commented that large craterlets sometimes disappear.

An interesting comparison was made by T. L. MacDonald of three charts of Plato's spots, two by Dr. Wilkins for 1936-39 and 1941-42, and an independent map by Walter H. Haas for 1935-40. In the July, 1943, Journal of the British Astronomical Association, MacDonald noted that the contemporary Wilkins and Haas charts were definitely comparable, but the later Wilkins map differed from both. His cautious conclusion was, "To that extent there is quite definitely a case for variations."

In 1950, Dr. Wilkins was still undecided about the riddle of the vanishing craterlets, stating that the variations observed could be due to libration and poor seeing, but that the possibility of real changes warranted further observations.

The latest but certainly not the last word comes from Moore in his book, "The evidence is conclusive and we are bound to accept a certain amount of activity on the floor of Plato."

....

Plato's third riddle may confront us at any time without warning. Suppose you have been looking at this feature for several months. The half dozen most conspicuous craterlets and spots have become familiar and are almost always seen, except when poor seeing converts the floor into a shimmering, shapeless blob. Then one night, with Luna riding high in a cloudless sky, you cross your fingers, mutter the astronomer's prayer, unlimber the scope and turn it tentatively on the earth's companion, hoping that tonight you can shove the Barlow clear in, pulling the eyepiece way out for highest possible power.

If you're like me, you may want to approach Plato gradually, saving it as a main course after nibbling at other formations. Perhaps Copernicus is emerging after sunrise. It is a good feature for careful focusing and study of its terraces and central peaks. And to test the seeing you'll turn west to the chain of craterlets, which you find standing steady and clear. Then you look farther west to Archimedes and Aristillus. North down the rugged Alps and a side swing along the Alpine Valley, for perhaps tonight you can definitely find out whether that darkish spot nestled next to the southern cliff is a craterlet or a rock slide down the wall. It does look like a crater---very round. You test again on these two tiny craterlets just south of Plato---with a little focusing they are clear and steady---it's a good night.

Now to Plato. You look, and look again, and see nothing! Yes, the walls are clear and detailed, and that landslip on the east stands out. But the floor appears smooth, flat, and featureless, perhaps with a faint flickering glimmer of something where you know the comparitively prominent central craterlet should be. You see nothing, but this nothing is the something that has puzzled lunar observers for a century, the apparent obscuration of the floor of Plato at times when floor features should be readily visible.

. . . .

BANDED CRATERS

Robinson, L. J.; *British Astronomical Association, Journal,* 73:33-38, 1963.

In most general terms, a banded crater may be defined as a normal crater with one notable peculiarity---the floor and/or walls are scarred by darkish patches of material. These dusky markings may take various forms: from large irregular patches to very regular and linear ray-like areas. Nevertheless, this entire continuum may be organized into the following categories.

Class I. Aristarchus Type

This form of banded crater is probably the most intriguing, for the major bands arrange themselves radial to the centre in a spoke system. This point, while being within the crater, need not necessarily be the geometrical centre of same. The bands, themselves, do not meet at this origin but rather initiate from a dark ring which is quite concentric to the rim of the crater, see Fig. 1a. These craters are very bright; being the brightest crater-like objects on the lunar surface.

Class II. Pytheas Type

A crater carrying this designation may be said to be a Class I crater with the central ring filled with dark material. In most cases, this ring covers a very substantial portion of the crater floor. Short radial bands emerge from the periphery of this dark ring, see Figure 1b. Under a high Sun these craters give the appearance of a bright ring of light surrounding a dull area.

Class III. Birt Type

This third class is somewhat of a hybrid between the Aristarchus and the Pytheas types. The Birt type crater has rather long bands stemming from a non-central, dark nucleus; this nucleus is much smaller than in the case of a Class II crater, however, see Figure 1c. The interior walls of these craters, being very bright, are much the same as a Class I object.

Class IV. Messier Type

This class of craters is most singular, for a single east-west band is its earmark. Though these craters are not nearly as common as the above types, they are still characterized by bright interior walls, see

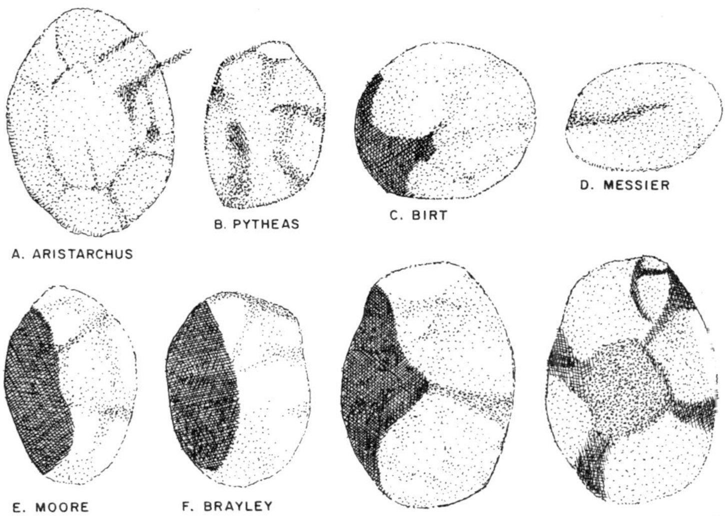

Banded lunar craters. (Fig. 1)

Figure 1d.

Class V. Moore Type

The craters forming this class have one or two dark bands which extend from a non-central dark area; this dark area covers approximately one-half of the crater's floor. This large dull area is confined to a single side of the crater. Due to the great extent of this dark material, these craters are noticeably less bright than those of, say, Class I, see Figure 1e.

Class VI. Anaxagoras Type

This final class of banded craters is much different from any of the aforementioned. Whilst the other craters have dark bands, these craters have bright bands. Up to date, only Anaxagoras, Ramsden, and Brayley have shown this feature (Brayley's bright bands being discovered by the author in 1960). Wilkins, however, states in The Moon that bright bands were recorded in Proclus by Thornton; to this writer's knowledge this observation has not been confirmed. Within these craters one may find dark bands in addition to light ones and, again, the high reflectivity of the interior walls persists, see Figure 1f.

The above classification should be considered only as a descriptive aid. It is very difficult to place some craters within the above scheme, for there are many intermediary cases. Furthermore, in addition to the basic classification difficulties, this author has found cyclic variations in the appearance of markings within particular craters; these changes occurring with varying angles of illumination, see Figure 1 (g and h). It is encouraging to note, however, the basic agreement between researchers in this field. Abineri and Lenham in their most valuable early work came to much the same basic conclusions as did this writer with respect to a general system of classification.

As to what constitutes the bands or the dark areas, nothing definite can be said. A single observation by the author found the bands of Aristarchus to be composed of numerous filaments arranged along the axis of the band. It is believed that these 'filaments' are rilles of very small size---this observation is given not as the origin of the dark nature of the bands, but merely as an explanation of the filaments.

Having defined a banded crater, it is now possible to view their common physical characteristics. A study by both visual and photographic means reveals the following:

1. These craters all have definite external walls which rise high above the surrounding plane.

2. The walls of banded craters brighten under high solar illumination. As proof of this high-Sun aspect, several excellent photographs were studied, and all craters which appeared to have bright walls were marked, and a catalogue was compiled. Once the moon had been fully studied between longitude $+60^{\circ} > \lambda > -60^{\circ}$ and latitude $+60^{\circ} > \beta > -60^{\circ}$, the positions of all catalogued bright craters were compared with the 188 known banded craters of the Abineri and Lenham Catalogue. The following results were obtained:

All of the 188 banded craters were confirmed except for those listed below:

1	Total number of unconfirmed banded craters	24 or 12.8 per cent
2	Unconfirmed because of proximity to limb	9 or 4.8 per cent
3	Unconfirmed because of insufficient position data	5 or 2.7 per cent
	Remaining unconfirmed	10 or 5.3 per cent

In light of this fine correlation, the author now argues that <u>all</u> banded craters appear bright under high solar illumination.

To continue with other physical similarities amongst bright craters:

3. The floors of these craters lie below the surrounding plane.

4. All bright craters, unless intruded upon by another crater, are quite circular.

5. The larger bright craters have central mountain masses.

6. The very bright bright craters are origins of ray systems.

With these comments we conclude our study of the obvious aspects of bright craters. Other papers will be concerned with the less obvious characteristics of bright craters: particular emphasis being placed upon the mutually consistent functions of bright craters and terrestrial craters of known meteoritic origin.

....

• Distribution of Lunar Craters

THE MYSTERY OF THE HEMISPHERES

Anonymous; *Science News,* 105:241, 1974.

A major surprise in the early days of lunar exploration was the discovery that the soft maria visible from earth were far more rare on the moon's farside, presumably because of some one-sided influence of the earth. Now refinements of Mariner 9 data show one hemisphere of Mars to be far rougher than the other, and Mariner 10 suggests the same asymmetry for Mercury. Data files grow, observes Bruce Murray of the California Institute of Technology, yet so does the mystery of hemispherical asymmetry. "We now know," he says, "a little less about the moon."

OLDEST AND LARGEST LUNAR BASIN?

Cadogan, P. H.; *Nature,* 250:315-316, 1974.

The nearside-farside asymmetry is evident from the dark mare material which fills the circular basins on the nearside. Although there are similar basins on the farside, they are apparently not filled with mare basalt. Laser altimeter data suggest that this may be a result of the greater thickness of low density crust on the farside. The surface of the farside is largely elevated from a sphere of radius 1,738 km, centred on the Moon's centre of gravity, and the nearside is correspondingly depressed. Although this may explain why Maria occur only on the nearside, the differences in crustal thickness are still unaccounted for. Either the Moon accreted heterogeneously, or else low density crust was, at some early stage of lunar evolution, transferred from the nearside to the farside. A possible mechanism for the transfer of vast quantities of crustal material could involve a very major impact on the nearside.

. . . .

DISTRIBUTION OF CRATERS ON THE LUNAR SURFACE

Fielder, Gilbert; *Royal Astronomical Society, Monthly Notices,* 129:351-361, 1965.

Summary. A fresh attack on the vital problem of the origin of the lunar craters has been made by analysing the surface distribution of craters of given diameter. The distribution shows a general clumpiness in both the lunarite (bright regions) and the lunabase (dark regions).

Craters between 30 and 40 km in diameter, situated in the lunarite, are non-randomly distributed at the 2 per cent level of significance. This argues against the impact theory.

It is found that the number-density of differently-sized craters is slightly greater in the following half of the Moon then in the preceding half. This result is shown to apply equally to the lunarite and, taken separately, to the lunabase, and again argues against the theory that the craters were produced exclusively by impact.

In assessing the origin of the craters on the basis of the observed frequencies and distribution of craters alone, it is concluded that the ratio of the number of impact craters to the number of endogenic craters is not very large. If only one theory is allowed, it must be that the craters are of internal origin.

COMMENTS ON 'DISTRIBUTION OF CRATERS ON THE LUNAR SURFACE'

Marcus, Allan; *Royal Astronomical Society, Monthly Notices,* 134:269-274, 1966.

Summary. In a recent study of the distribution of centres of lunar craters, Fielder found two apparent anomalies which he used as arguments against the meteoroidal impact hypothesis for the origin of lunar craters. First of all, the Poisson distribution gave a very bad fit to the number of crater centres in equiareal sectors of the continents and the maria, especially among craters smaller than 40 kilometres in diameter. The second difficulty was a systematic excess in the number density of small craters in the western (trailing) half of the Moon. We will show that these observations could also have been expected even under the impact hypothesis, since the numbers of small and of large craters in a finite region are negatively correlated. Available crater statistics therefore neither preclude nor establish the impact or volcanic hypothesis for the origin or craters.

• The Lunar Channels

WATER ON THE MOON

Urey, H. C.; *Nature,* 216:1094-1095, 1967.

The possibility that water has existed on the Moon for varying lengths of time, both in liquid and in solid form, and both beneath the surface and on the surface, has been widely discussed during the past 10 years. The subject has been discussed repeatedly at scientific meetings and has been received mostly with great scepticism. Evidence supporting this view has recently become quite overwhelming and, in fact, no communication seems necessary to point out the evidence from the Orbiter 4 and 5 pictures. Because many people are not aware of this evidence and suggest that the effects are caused by other liquids, that is, lava, dust-gas or possibly even vodka, a brief discussion of the evidence may be in order.

Gold pointed out that it was unreasonable to believe that all the smooth areas within the craters and between the craters were caused by lava flowing from deep in the Moon, and he suggested that dust produced by particle erosion was the origin of the material filling these craters and the maria. Because of the lack of dust river valleys, I disagreed with this and suggested that dust from the great collisions and temporary rains washed the dust into the low areas, and the water escaped into space or sank below the surface. Gilvarry postulated much more extensive oceans lasting for billions of years. Although many of Gilvarry's observations are reasonable, the long length of time assumed, with no evidence for mature river valleys, unfortunately led to definite non-acceptance of his ideas. Gold and Kopal have suggested that water has escaped from the lunar interior and formed permafrost similar to that found in the arctic regions of the Earth. Safronov and Rouskol argue that water could not have escaped slowly from the lunar interior and formed oceans because the escape rate of water from the Moon would be so great that only a very small atmosphere of water could have accumulated. This conclusion is correct, and sources of surface water, which have been suggested, have been of a catastrophic kind, that is, a degassing of the lunar surface and the colliding objects in the great collisions on the Moon, a great splash from the Earth during the period of accumulation of the Earth, comet head collisions, etc. I suggested that the lake-like areas near and on the walls of Alphonsus were indeed caused by water (see under Moon, Encyclopaedia Britannica, plate III, for pictures of these). Gold suggested that the great central mass in Alpetragius is a pingo*. I have argued that the carbonaceous chondrites may come from the Moon, for those of the type 1 variety, particularly, show that they were once immersed in water and that the Moon has certainly had liquid water on its surface.

Schroeter's valley, to the west of Aristarchus, has been known since the late eighteenth century, but terrestrial observations could not establish its origin. Orbiter 4 and 5 pictures show this valley in great detail as well as many neighbouring valleys. Schroeter's valley begins in a mountainous region in an enlarged area called the "Cobra Head", and extends in a generally westerly direction into a smooth maria area. Its greatest width is some 8 km and its depth some 600 or 700 m. Other smaller valleys with "cobra heads" are found north of Aristarchus and

---

* In regions of permafrost, that is, northern Canada and Siberia, great cones of ice covered with soil are formed by water and ice being extruded from the lower regions. They often look like small volcanoes with sloping sides and a depression like an irregular volcanic cone at the top. In some cases the ice has melted leaving a crater much like lunar craters but of much smaller diameter.

Prinz. Branching occurs in a few cases but mostly there are no tributaries and no deltas. Possibly there are small deposits at the maria ends, but they are much too small to account for the materials eroded from the valleys. South of Aristarchus and northerly from Marius is a valley extending to more than 160 km in a very smooth maria area. It decreases gradually and uniformly in width from the south-easterly end to the north-westerly end where it runs off the picture available to me (Orbiter 4, frame 150, high resolution). Near the larger end it is about 900 m wide and apparently goes to near zero at the other end. It is a very crooked rille and cannot possibly be a physical fracture. It has no tributaries. Similar rilles are found in many places, for example, within and near the Alpine valley and many other areas. They seem to be a general lunar feature.

But why are there no sediment deposits at either end of these valleys and no tributaries? (There are a very few examples of tributaries.) The walls seem to be formed by slumping, and north of Krieger there is a row of craters which seem to form an initial stage of such slumping. Water must have run below the surface and formed a tunnel which then caved in. But where did the excavated material go? If the maria are underlaid with ice, it is only necessary to melt the ice and let the water drain into the desert sands. (The explanation offered here was suggested by Gold.) But could it have been lava? Would lava flow in a narrow stream for hundreds of kilometres and disappear without a trace? I believe that the answer to both these questions is "No". But the details of these valleys are very varied.

While flying over northern Greenland on September 9, 1967, I saw crooked rille-like structures in the Greenland ice cap. Some of these looked like fractures but others, at least over limited regions, looked much like the crooked rilles on the Moon.

Krieger, north of Aristarchus, is an irregular crater with a recent smaller collision crater on the southern wall, a curious depression in the plane south of the crater and an irregular mass covering the north wall and an area north of the crater. There is a break in the western wall, and from this break an irregular gully or stream bed extends a short distance to the west. This seems to have been a surface stream, and a suggestion of a sediment deposit can be seen at its western terminus. This appears to be a surface stream bed quite different from the other rilles discussed. North of Prinz one of the small rilles of the Schroeter valley type crosses an elongated sunken area, transverse to this valley. It has a flat floor, and a small stream bed, similar to that coming from Krieger, crosses it. These two stream beds indicate that streams flowed on the surface and thus that an atmosphere had a pressure equal to and probably much greater than the vapour pressure of water at its melting point, that is 4.6 mm of mercury. Could Krieger be a pingo that has melted and collapsed? If so, it is a very large example as compared with terrestrial pingos. Krieger is 25 km in diameter. I have found a picture of a terrestrial pingo 300 m in diameter. But in a billion years, say, it may be that pingos get very much larger. Are there unmelted and uncollapsed pingo areas on the Moon? Probably some of these volcanoes which are referred to so confidently are pingos.

South of Aristarchus is a crater similar in size to Krieger (name unknown; it appears on Orbiter 4, picture frame No. 150, 1 of 3 high resolution). There is a break in the northern wall and there is a rather smooth pile of sediment outside the crater at this point. Some liquid drained out of the crater and left this material. Possibly this was lava, or mud, or a flow of dust and gas. In view of the overwhelming evidence for water as the agent producing the other valleys, I favour the view that effects seen in this crater as well as Krieger are caused by water.

In fact, as mentioned here, several serious students of the Moon concluded, or at least surmised and suggested, that water was present on the Moon at some time in the past. This evidence for lake-like structures has been presented repeatedly, each author discovering for himself the same evidence previously recorded by others. The suggestions were reasonable and the evidence valid. The fact that there seem to be no mature river valleys anywhere argues that these seas were present for only a short length of time. It is not possible to give the time at which they were present. Was it 4.5×10^9 yr ago when the Moon was captured by the Earth, and may the Moon have captured some water from the Earth at that time, or was it some 2×10^9 yr ago when the Moon's orbit changed from retrograde to direct and when the Moon was near the Earth? Or has the surface water appeared repeatedly? Thus water may have been acquired by the Moon early in its history; it escaped into space leaving water below the surface; the atmosphere was lost and low temperatures below the surface were established. Then a comet collided with the Moon, gave it an atmosphere, the temperature became warmer as on Earth, water melted and flowed out of the surface in springs. The atmosphere escaped and the water froze again until the next comet collision. The Safronov and Rouskol arguments seem to be valid and only a catastrophic origin for the water on the Moon seems possible.

Undoubtedly, someone will eventually measure and record these stream beds in great detail. This communication is to point out that the maria of the Moon are dried-up or frozen seas and that water has aided in forming the final features of the Moon. These conclusions in no way determine what the composition of the solid materials in the maria may be or whether this material had a volcanic or other type of origin.

LUNAR RIVERS

Lingenfelter, Richard E., et al; *Science,* 161:266-269, 1968.

Abstract. Mature meanders in lunar sinuous rills strongly suggests that the rills are features of surface erosion by water. Such erosion could occur under a pressurizing ice cover in the absence of a lunar atmosphere. Water, outgassed from the lunar interior and trapped beneath a layer of permafrost, could be released by a meteoritic impact and overflow the crater to form an ice-covered river. A sinuous rill could be eroded in about 100 years.

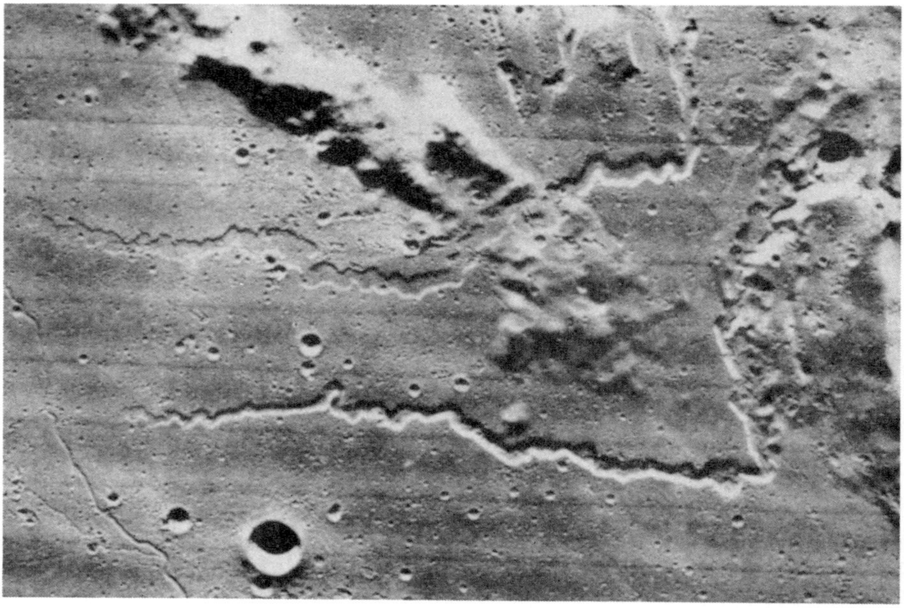

The sinuous rills Rima Prinz I and II. Width of field 67 km. (Courtesy NASA)

• Lunar Ray Puzzles

LUNAR RAYS

O'Keefe, John A.; *Astrophysical Journal,* 126:466, 1957.

It is well known that the ray structures observed on the lunar surface are conspicuous only at, or near, full moon or when the dark side of the moon is visible near new moon. The phenomenon is described by Proctor (1873) and is well shown in the photographic lunar atlas of Pickering (1903). According to Moore (1953), no satisfactory explanation has been offered. It is noteworthy that the rays cover most of the visible face of the moon; thus those at her western limb are more conspicuous at full moon, when the sun is on their horizon, than at first quarter, when the sun is in their zenith. The altitude of the earth seen from these points is approximately the same at both times. Thus it is an effect neither of the sun's altitude nor of the earth's altitude as seen from the moon. In fact, the rays are most clearly seen when the source of light and the observer are nearly in the same line. It follows that the matter of the rays must possess the property of retrodirecting the rays, i.e., returning them chiefly along the direction from which they came.

The most precise optics of this type has the form of the corner of a cube---a somewhat improbable form for the moon. Less precise reflectors can be produced by scattering transparent beads over a white surface. The mechanism here is simply that if the bead forms an image of the source on a white surface, then, when the bead is viewed from the direction of the source, nothing can be seen except the image of the source; and hence the surface appears much brighter than when viewed from any other angle. The gleaming of a cat's eye in the darkness is a familiar example. If the focus is not precisely on the screen, the concentration of light will be imperfect to that extent; and the cone into which the light will be returned will be broadened and its intensity correspondingly reduced. This mechanism is employed for motion-picture screens.

The essential feature here is some sort of lens whose focal length is as short as possible. The only plausible natural formations of such a kind would be some sort of small transparent spheres. The hypothesis is therefore proposed that the lunar rays consist of small transparent spheres.

LUNAR RAYS: THEIR FORMATION AND AGE

Giamboni, Louis A.; *Astrophysical Journal,* 130:324-335, 1959.

Abstract. The nature of the rays of Tycho and Copernicus suggests that they were laid down at a time when the angular velocity of the moon was markedly different from that observed today. The rays of Tycho indicate that the lunar sidereal period was between 0.5 and 6.8 days and that the poles were in approximately the same position then as they are today. These rays were created within 8 x 10^7 years of the time the moon was formed. The rays of Copernicus support these observations and indicate that the rotation of the moon at the time that these rays were formed was similar to the rotation which existed at the time that Tycho was formed.

Introduction. Several observers have noted that the rays in the vicinity of a lunar crater, if extended,do not always pass through the center of the crater. Many extended rays would be tangential to the crater; other rays would miss the crater entirely. Observers have also noted both single rays and pairs of rays. No consistent explanation of these phenomena has yet been advanced. This paper presents a possible explanation for both these phenomena in terms of the lunar sidereal rotation and the equations of motion for a body moving in a vacuum. The theory is developed for the rays of Tycho and Copernicus.

....

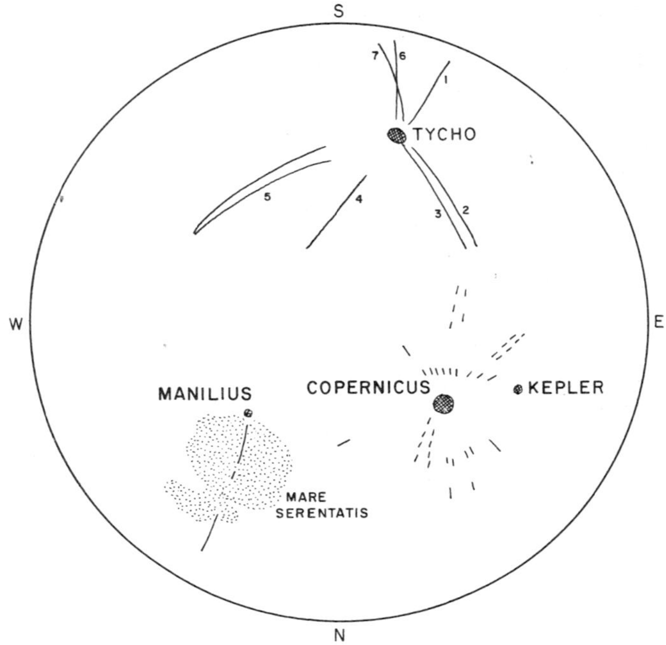

Selected rays from Tycho and Copernicus illustrating how some do not converge at the crater

• Peculiar Lunar Terrain

CURIOUS LUNAR FORMATION

Gaudibert, C.; *English Mechanic,* 18:638, 1874.

Perhaps you will find a corner for the inclosed sketch, which is the rough outlines of that formation which Mr. Webb tells us, p. 94 of "Celestial Objects," third edition, "attracted so much attention" in 1822; and not at that time only, but much later also, as the reader will gather in reading in that invaluable book the precise history and description of the object my sketch roughly represents. Doubtless drawings of this object exist, and it would be interesting to compare them with its present state. It seems almost certain that Gruithuisen found changes here in the disappearance of the east "ribs," and I could not see that "meridional wall," against which abut the west "ribs," with the exception of the north-west "rib," which seems to have been covered by a meridional

1874 view of Gruithuisen's peculiar lunar structure; the "lunar city" of some popular books

wall. At the extremity of the next "rib," instead of a wall I see a depression, and beyond a prolongation of the "rib." The third and fourth "ribs" have no prolongation, but the south one has just a depression, and then what seems to be a continuation of the "meridional" wall; so that we have the two extremities of this wall without the middle. It would be interesting also to know whether the east side is now in the same state as it was after Gruithuisen lost the "ribs," and also if the three craterlets north of this object have been observed. They seem of recent date. Observing this object with a power of 550, I saw its surface covered with minute hillocks, with a larger mound at the latitude of the second "rib." The terminator was passing through Stadius when I made my observation.

SURFACE LINEAMENTS DISPLAYED ON LUNAR ORBITER PICTURES

Fulmer, Charles V., and Roberts, Wayne A.; *Icarus,* 7:394-406, 1967.

A lunar surface fabric composed of near parallel topographic ridges, valleys, and similar cross structures was noted by many of the early selenographers (Beer, W. and Madler, J. H., 1837, Der Mond, Simon Schropp, Berlin). More recently, observers using larger instruments, with greater resolving power, have noted, during optimum seeing conditions, that the lunar surface is marked by numerous fine linear structures. It is not surprising to find similar parallel and near-conjugate sets of lineaments well displayed on some of the NASA, Lunar Orbiter photographs. The lineaments displayed on the Orbiter photographs are closely spaced, fine, parallel striations that are generally oriented in a NW-SE and NE-SW cancellate design. Locally a tertiary E-W or N-S lineament may also be developed, but it is frequently less distinct and quite often displays a braided appearance. The interspaces between lineaments of the same attitude are frequently higher or lower in elevation than the adjacent area. The difference suggests the possibility that differential movement has taken place along these preferred lines, creating a stepped or seriated appearance to the surface.

The cause of the lineaments is not known; however, on Earth, lineaments (straight, parallel) are generally the surface expression of planar discontinuities or breaks in the subsurface rocks such as fractures or joints, faults, shear zones, or stratigraphic layering. The intersection of these subsurface features with the surface frequently results in visually detectable lineaments. Where many such parallel lineaments occur, the simplest explanation is a fracture set or tilted stratigraphic layering. Most fractures are the result of complex forces other than uniaxial tension and a system of two or more fracture sets results. Thus, the simplest explanations of a set of moderate to steeply dipping lineaments are fractures, and this is presumed the origin of the sets noted in the lunar pictures. What appears to be layering related to rock texture or composition is exposed in pictures of many "far-side" and "near-side" craters such as Copernicus. Lineament azimuthal changes, with elevation differences or slope directional gradients occur for the intersection of a subsurface plane with the irregular planetary surfaces except for vertical planes. Thus the topography-lineament relationship can provide an estimate of the inclination of the subsurface plane. In many cases the lineaments appear to lie in vertical planes; however, apparent dips as low as 30 degrees from the horizontal have been observed where the angles of illumination and observation were favorable. Surface striations of the type noted here have also been found on the far-side photographs. Both here, and on the better known visible disc, one or sometimes both of the lineament directions are nearly parallel to the larger structural features that have been described elsewhere.

. . . .

PUZZLING LUNAR TERRAIN SNAPPED BY APOLLO 16

Anonymous; *New Scientist,* 55:127, 1972.

A camera aboard the orbiting command module of Apollo 16 captured this shot of part of a 100-km crater on the Moon's far side. It lies adjacent to the crater King, a major landslide in the walls of which can be seen on the left of the photo. The marked texture of the lunar surface here, with its pronounced linearity, awaits an explanation---it looks strikingly like the work of a giant glacier! Careful scrutiny of the boundaries of the smooth-floored, "ponded" feature in the centre reveals other, smaller, "ponds" at a variety of levels. [Photo omitted]

TRANSIENT LUNAR PHENOMENA (TLPS)

Thousands of TLPs have been recorded down the centuries; ranging from bright flashes of light to subdued ruby-red patches to peculiar obscurations veiling lunar features. Early astronomers had no doubt that they were viewing volcanoes in action and that glints of sunlight from snow-clad lunar peaks accounted for light flashes. This earth-like view seems naive after the manned lunar landings, which revealed anything but a terrestrial landscape and certainly no obvious volcanoes or snow. Nevertheless, TLPs kept on occurring.

A notoriously active spot on the moon is the crater Aristarchus. Glows and lights appear there often. Yet closeup photos from spacecraft show no vents or lava flows. Whatever causes TLPs leaves little evidence behind. The release of internal gases from the moon has been suggested as a cause of some TLPs, particularly the obscurations. The fact that TLPs seem more frequent when the moon is closest to earth gives some credence to the gas-venting theory. Conceivably the gases could fluoresce under the solar wind to create glow effects, but the brilliant flashes and rare lightning-like phenomena seem to transcend this explanation. Something is going on---but what?

• Lights and Bright Spots

OF THREE VOLCANOES IN THE MOON

Herschel, Wm.; *A Popular Display of the Wonders of Nature*. . ., C. C. Clarke, W. Tweedie, London, 1837, pp. 486-487.

April 19, 1787, I perceive three volcanoes in different places of the dark part of the new moon. Two of them are either already nearly extinct, or otherwise in a state of approaching eruption; which, perhaps, may be decided next lunation. The third shows an actual eruption of fire, or luminous matter.

April 20, 1787, the volcano burns with greater violence than last night. I believe its diameter cannot be less than 3", by comparing it with that of the Georgian planet; as Jupiter was near at hand, I turned the telescope to his third satellite, and estimated the diameter of the burning part of the volcano to be equal to at least twice that of the satellite. Hence we may compute that the shining or burning matter must be above three miles in diameter. It is of an irregular round figure, and very sharply defined on the edges. The other two volcanoes are much farther towards the centre of the moon, and resemble large, pretty faint nebulae, that are gradually much brighter in the middle; but no well defined luminous spot can be discerned in them. These three spots are plainly to be distinguished from the rest of the marks on the moon; for the reflection of the sun's rays from the earth is, in its present situation, sufficiently bright, with a ten-feet reflector, to show the moon's spots, even the darkest of them: nor did I perceive any similar phenomena last lunation, though I then viewed the same places with the same instrument.

The appearance of what I have called the actual fire or eruption of a volcano exactly resembled a small piece of burning charcoal, when it is covered by a very thin coat of white ashes, which frequently adhere to it when it has been some time ignited; and it had a degree of brightness, about as strong as that with which such a coal would be seen to glow in faint daylight. All the adjacent parts of the volcanic mountain seemed to be faintly illuminated by the eruption, and were gradually more obscure as they lay at a greater distance from the crater.

SCHROTER AND LUNAR TRANSIENT PHENOMENA

Bispham, K.; *British Astronomical Association, Journal,* 78:381, 1968.

A few days before reading the Lunar Section Report by P. Moore, printed in the February issue of the Journal, I had been reading and doing some abstracting from Schroter's Selenographische Fragmente, and came across the quotations which I append below in a translation. These are in connection with Moore's remarks on the rarity of flashes on the Moon.

"As I was observing the night side of the Moon on the 15th October, 1789 in the morning with the marking Plato, together with the Mare Imbrium in view, but absolutely nothing of the illuminated lunar hemisphere in the field, there arose, soon after 5 o'clock, in, or rather in front of the dark lunar disk, and just, in fact, in the centre of the Mare Imbrium, as far as I was able to report, after my sudden astonishment, and almost right in the middle of the field of telescope, quite suddenly and quickly, a bright burst of light, which was composed of many single, separate small sparks which had just such a white light as the illuminated day side of the Moon, and which all moved away together in a straight line northwards over the northern part of the Mare Imbrium and the other parts of the Moon's surface bordering this to the north, but from there onwards through the remaining empty field of the telescope.

"When this light shower had travelled half its distance, a further, similar burst of light arose to the south of it over just the same place or perhaps somewhat more to the east of where the first originated: the second was exactly the same in appearance as the first, composed of like, small, white sparks of light which flashed away in the same direction and in an exact parallel direction towards the north, as far as the edge of the telescopic field.

"Startling as this impression was also, which this distant light phenomenon made on me, nevertheless, I quickly recovered my wits, re-imagined this phenomenon repeatedly, and vividly after it had quite disappeared and estimated, using a pocket watch, the time it took each shower of light to travel from the position of origin to the edge of the telescope field as about 2 seconds, and thus the complete duration of this phenomenon as 4 seconds, in which time it was all over without leaving the faintest trace behind."

Of the multitude of changes reported in Schroter's book, this is surely one of the strangest.

LUNAR VOLCANOES

Anonymous; *American Journal of Science,* 1:5:176-177, 1822.

Dr. Olbers observed on the 5th of last February, the phenomenon which some philosophers have attributed to volcanos in the moon. He declared that he never perceived it more distinctly. The spot called Aristarchus, threw out a very vivid light, and appeared like a star of the 6th magnitude, placed on the north-east of the moon. The evening of the 6th unhappily was not so fine as that of the preceding day, and Dr. O. could not pursue his observations, but the English journals announce that Capt. Kater had made on the 7th of Feb. a report to the Royal Society of London in which he affirms that he had seen a lunar Volcano in actual eruption. Dr. Olbers thinks that the observations of Capt. K. coincide exactly with his own, but he differs from him with respect to the cause. He does not admit the existence of a volcano in the moon; he thinks that the phenomenon which Capt. K. regards as such, is produced by the reflection of the light cast by the earth on the open immense rocks of a smooth surface, situated on the part of the moon called Aristarchus. Should these rocks, says Dr. O. send back only a tenth part of the light which they receive from the earth, (our mirrors return one half of the incident light) the effect would be equal to a star of the sixth magnitude. It is in this way that Dr. Olbers accounts for our always seeing those spots in the same place, and also why they do not show themselves at each lunation. On the 6th of March, Dr. Olbers could distinctly see all the spots of the moon; Grimaldi, Copernicus, Kepler, Manidius, &c. Aristarchus, was particularly remarkable, but it was not so splendid as on the 5th of February.

The hypothesis of volcanoes in the moon is not modern, and at present it is almost rejected, and the explanation of Dr. Olbers is generally admitted. The spot Aristarchus, is plainly to be seen when the moon is illuminated by the sun, and hence it is natural that it should appear more luminous than the rest of the disc, when it is enlightened only by the earth. As to the variation of extent which is remarked commonly in the spots at the beginning of a lunation, the phenomena of refraction, produced by the position of the moon near the horizon, are sufficient to explain it without having recourse to Lunar Volcanoes.

LIGHTNING-LIKE PHENOMENA ON THE MOON

Giddings, N. J.; *Science,* 104:146, 1946.

The following observations were made some time ago and seem to me to be worthy of recording in Science. During the evening of 17 June 1931, I was working in the yard near our house at Riverside, California, and happened to glance at the moon. It was an unusually fine, clearly outlined new moon, and as I stood looking at it, suddenly some flashes of light streaked across the dark surface but definitely within the limits of the moon's outline. Since this was a phenomenon which I had never seen before, I continued to watch it and saw similar flashes streak across the moon again in a moment or two. Without mentioning what I had seen, I called my wife's attention to the new moon. She admired it. When I asked her to watch it closely to see if she noticed anything strange, she said: "Oh, yes, I see lightning on the moon," adding that this appeared to be confined to the moon. We watched it for some 20 or 30 minutes during which the phenomenon must have occurred at least six or seven times. The facts were recorded in my notebook as of approximately 7:40 p.m., 17 June 1931. At the time, I was inclined to attribute the phenomenon to some sort of sunlight reflection from mountain peaks on the moon or possibly some sort of electrical activity. I wrote the Mount Wilson Observatory regarding the phenomenon, and the reply very courteously discounted my observations. The observations were carefully made and carefully verified, and at the time I assumed that the phenomenon was probably something which astronomers were familiar.

LIGHTS IN THE MOON

Ogilvy, C. Stanley; *Popular Astronomy,* 57:229-233, 1949.

"Till clomb above the eastern bar
The horned Moon, with one bright star
Within the nether tip."

The usual reaction of the astronomer who encounters these famous lines from Coleridge is a smile and perhaps a charitable shrug. Yet the fact remains that in Coleridge's time the question of "lights in the moon" was still an open one; and even today we may find a background of plausible fact behind what seems at first to be only poetic fantasy.

There are three obvious interpretations of the anomalous star between the points of the crescent. (1) Coleridge simply made a blunder, through plain ignorance of astronomy. (2) He meant "almost between." (3) He was employing poetic license; anything could happen on the "spectral sea" being described by the Ancient Mariner.

John Livingston Lowes, whose "Road to Xanadu" is based on a most exhaustive study of sources, suggests that none of these three explanations is the correct one. Far from being an astronomical ignoramus, Coleridge was an ardent reader of the Philosophical Transactions of the Royal Society of London, the outstanding learned periodical of the day, as well as any other scientific literature he could lay his hands on. Lowes finds evidence in Coleridge's personal notebook that he had had occasion to refer to Vol. 5 of the Transactions on another matter. At the top of a page of that volume, where "it leaps to catch the attention of the most careless reader," is the statement that "in November 1668, a Star appear'd below the Body of the Moon within the Horns of it." This information was supplied by Cotton Mather, from Boston, where it was supposed to have been observed. It could not have actually been below the bright part of the moon, since the horns never point downward after dark; but the original reference from which Mather took his data says: "In November following [1668] appeared a Star between the horns of the moon in the midst." Mather's introduction of the spurious "below" is typical of the inaccuracies of early scientific reporting which bedevil the seeker of precise historical data.

We come now to much more authoritative sources, world-famous astronomers whose opinions Coleridge must surely be forgiven for having credited. In 1794, just three years before the "Ancient Mariner" was written, Nevil Maskelyne wrote in the Transactions "An Account of an Appearance of Light, like a Star, seen in the Dark Part of the Moon, on Friday the 7th of March, 1794." This account contains letters from two independent and widely separated observers, both of whom submitted sketches (in substantial agreement with each other), and one of whom---one William Wilkins by name---ascertained the time as about 8 p.m. by consulting a neighbor shortly after he had made the observation. This matter of the time is of considerable importance, as will be seen presently.

Wilkins wrote "I was, as it were, riveted to the spot where I stood, during the time it continued [about 5 minutes] and took every method I could imagine to convince myself that it was not an error of sight." He was something of an amateur astronomer, and had a reflecting telescope at home. He "had determined to return and use it on this occasion; for it was so fixed, I had scarcely a doubt but its appearance would continue; but almost at the same instant, and whilst I was looking, it totally disappeared." And in another letter, after further questioning by Maskelyne, Wilkins answered, "I am very certain of this spot appearing within the circumference of the moon's circle."

The other observer, Thomas Stretton, saw what he described as "a light like a star, and as large as a middle-sized star, but not so bright [!], in the dark part of the moon." Stretton was not sure of the time, but the evidence (based on his later estimate of the then position of the moon) brought it closer to 7 than to 8 o'clock.

We move swiftly to the denouement: on the same night as the observed phenomenon, a few minutes before 7 o'clock, Aldebaran was occulted by the moon, in a position quite close to where the "star in the moon" occurred. Maskelyne was of course aware of this, and admitted it to be "a singular coincidence of circumstance," yet stoutly defended the evidence of the 8 o'clock observation. He discounted the time element of the other observation, and was in fact easily able to persuade Stretton that he might have been in error. We cannot so lightly dismiss the coincidence of the occultation. The moon was about 5 days old---at precisely the phase when the optical illusion that the bright part of the disc is of larger radius than the dark part is most pronounced. Even an experienced observer is surprised at the tiny appearance of the dark disc which sometimes (very faintly at this phase) can be seen illuminated by earthshine. This conspiracy of nature would of course place a star about to be occulted well within the circumference of the apparent disc. The sudden and complete disappearance as reported by Wilkins fits exactly the description of an occultation. (Stretton lost the star by going indoors.) Finally, if the star seemed fainter than it should have (Aldebaran's magnitude is 1.06), that is easily explained by the proximity of the bright moon.

Maskelyne offers no explanation of the event. He remarks only that it was probably "of the same nature with that of the light seen of late years in the dark part of the moon by our ingenious and indefatigable astronomer, Dr. Herschel, with his powerful telescopes, and formerly by the celebrated Dominic Cassini."

This leads us to Sir William Herschel's account, which is most specific. He speaks with confidence of volcanos. This under the date April 19, 1787, 10h 36m sidereal time: "I perceive three volcanos in different places of the dark part of the new moon. Two of them are either already nearly extinct, or otherwise in a state of going to break out; which perhaps may be decided next lunation. The third shows an actual eruption of fire, or luminous matter. I measured the distance of the crater from the northern limb of the moon and found it 3' 57".3." No guesswork here. And now we have another curious coincidence. The careful sketch which Wilkins made in his notebook after observing the phenomenon of seven years later shows the bright spot, or star, about 4' from the northern limb!

"April 20, 1787, 10h 2m sidereal time. The volcano burns with greater violence than last night. I believe its diameter cannot be less than 3"..." He goes on to compare this with the one he had seen four years previously which, "though much brighter than that which is now burning, was not nearly so large...Seen in the telescope, it resembled a star of the fourth magnitude as it appears to the natural eye." Presumably, then, none of Herschel's lights were visible to the naked eye, if it required a telescope to bring them up to an apparent magnitude of 4.

Herschel subsequently revised his opinion, feeling by no means so sure of his volcanos in later years. Clerke states that he was "completely, though temporarily, deceived into the belief that he had witnessed volcanic outbursts," and on the same page attributes them to "the reflection of earthlight at a particular angle from certain bright summits." Modern astronomy texts seldom mention the matter, but Arago devoted a chapter to it, discussing Maskelyne's and Herschel's reports in a mildly reproving tone, much as if he were shaking a finger at these two distinguished gentlemen for having allowed themselves to be so deluded.

Haas, in the May, 1947, issue of this monthly, describes lights observed on the moon, but not of a durable character, and only specks even with a telescope.

Is there, then, any possible source of bright light in the dark of the moon which might be mistaken for a star? One occurs immediately to the modern astronomer, knowledge of which was not available in Coleridge's day: a wandering asteroid might, by great coincidence, line itself up between the earth and the moon for a brief interval. Hermes, for example, can come within 220,000 miles of the earth---20,000 miles nearer than the moon. But when Hermes was 485,000 miles away in 1938, its magnitude, according to Duncan, was 8, and its velocity was 5° an hour. Thus at its nearest possible distance Hermes might just be visible as a 6th magnitude object, but could not possibly be classified as a "bright star." It would in all probability move at an apparent velocity so high that it would cross the entire face of the moon in a matter of a few minutes. Moreover, it would, if it were in line with the crescent moon, be crescent itself, which would further greatly reduce its magnitude.

It is not inconceivable that at some time in the past another asteroid, large enough to appear as a bright star, may have found its way temporarily into the required position between earth and moon. But the possibility is exceedingly remote; and the behavior of Maskelyne's star does not, of course, confirm such a hypothesis for that particular instance.

If we still insist, then, on a realistic explanation of what the Ancient Mariner was supposed to have seen, we must fall back on the theory that, although a star was not actually between the horns of the moon, it seemed to be. Since it was a morning moon (it "clomb above the eastern bar,") the star must have been seen after emergence, and there was no danger that it would suddenly do a disappearing act, further to confound the already overwrought Mariner.

Perhaps the only remaining alternative is the answer given recently by an undergraduate to the familiar examination question, "Comment on Coleridge's line, 'The horned moon with one bright star within the nether tip'." The student wrote, "The horned moon refers to the crescent moon and the bright star refers to a speck in Coleridge's eye. If the star had been beyond the moon it would have been occulted, and if it had been nearer than the moon it would have burned Coleridge up." With this sage observation we might well bring our discussion to a close.

LIGHTS ON THE MOON

Vreeland, Frederick K.; *Popular Astronomy,* 57:354-355, 1949.

I have noted with interest Mr. Stanley Ogilvy's article on "Lights on the Moon," in your May issue, because of the bright spot that appeared in the umbra on the occasion of the recent total eclipse. I have looked for reports on this occurrence in the recent publications, but have found none. Surely it must have been observed by others.

We were observing with a 4-1/4-inch refractor and watching the three stars below the moon to see whether any of them would be occulted before the sunlight blotted them out. At the end of totality, when the crescent of light had covered about 1/16 the diameter of the moon, a bright spot of light appeared in the shadow, near the middle of the illuminated zone and about 1/8 diameter from the limb. It was star-like and white, in marked contrast with the red-orange color of the shadow. It was confirmed by three independent observers.

We watched it as the light approached it, and it remained unchanged until the light covered it. Then it was still visible, but appearing as a lighter spot of substantial area.

When the moon became illuminated we identified it as the high crater, Aristarchus, and it was apparent that something on the mountain must have caught a reflection from something in the illuminated zone.

Three features were conspicuous:

1. The star-like appearance of the gleam,
2. The white light, in contrast with the red-orange of the background umbra,
3. Its appearance only after sunlight had touched the moon.

Study of photographs shows high and very bright mountains in the illuminated zone, the highest and brightest being near the end of the illuminated crescent and about 250 miles from Aristarchus. There are other and nearer mountains, apparently lower, farther North. These would seem to be the source of light for the bright spot.

A little calculation indicates that for two mountains 200 miles apart to have line-of-sight visibility over the hump of a sphere the size of the moon, they should have an altitude of about 26,000 feet above the sphere. For a distance of 130 miles the altitudes should be over 10,000 feet. Since differences of altitude over 26,000 feet are known on the moon, and since the space between Aristarchus and the bright mountains looks like a plain and might be a depression, the reflection hypothesis seems not unreasonable.

A BLUISH BRIGHT SPOT NEAR OR ON THE MOON

Logue, Daniel A., Jr.; *Strolling Astronomer,* 8:146, 1954.

Mr. Daniel A. Logue, Jr., of Larchmont, Penna. has reported a curious observation on January 5, 1955, 1 hr., 0 mins.---1 hr., 30 mins., U.T., colongitude 44.°1. He says: "I saw a strange blue light above the surface of the moon where the night and day meet. I observed this light for more than 30 mins.; it did not move. It appeared like a star, in that rays of light came from it." Mr. Logue adds that he at first thought that the object was a star but later decided that it must be on the moon itself. He employed a 6-inch reflector at 340X in good seeing. An accompanying drawing places the blue spot near the rugged southeast limb of the moon, but the Editor is unable to identify the craters drawn. One conjecture is that Mr. Logue saw a very high lunar peak with its summit only in sunlight and detached from the illuminated regions, but the blue color is then left unexplained.

ANOTHER FLASHING LUNAR MOUNTAIN?

Anonymous; *Strolling Astronomer,* 10:20, 1956.

Mr. Robert M. Adams has called to our attention a curious lunar observation by Mr. Robert Miles of Woodland, Calif., an observation rather reminiscent of Mr. Brian Warner's article on pp. 130-131 of our November-December, 1955 issue. Mr. Miles says in part: "I noticed a flash of white light that caught my eye. At first I thought it could have been a lunar meteor. But it kept flashing on and off..... The light was very bright but changed its color to a very bright blue, like an arch light. It was brighter than the sunlit portions that I was looking at." Sketches indicate that the object in question lay on the night side of the terminator and perhaps about 100 miles east of the gap in the mountains on the east boundary of the Mare Crisium. Mr. Miles found the duration of visibility of the flashing light to be from $3^h\ 0^m$ to $4^h\ 30^m$, U.T. on January 17, 1956. The colongitude was then $320^o.5$ to $321^o.3$. These colongitudes seem inconsistent with the sketches, and the Editor suspects that the U.T. date was really January 16. It would be interesting to hear whether anyone else was observing this lunar area at this same time---or better still, was then photographing the moon.

OBSERVATION OF A VOLCANIC PROCESS ON THE MOON

Kozyrev, Nikolai A.; *Sky and Telescope,* 18:184-186, 1959.

In October and November, 1958 together with V. I. Ezerski of the Kharkov Observatory, I was conducting spectral investigations of Mars, using the 50-inch reflector of the Crimean Observatory of the Academy of Sciences of the Soviet Union. At that same time, I decided to obtain systematically some photometrically standardized spectrograms of lunar details, in particular of the crater Alphonsus.

During these observations, the slit of the spectrograph was always oriented east-west on the sky. The linear dispersion was 23 angstroms per millimeter in the vicinity of the hydrogen-gamma line, and the scale of lunar details about 10 seconds of arc per millimeter. The normal exposure on Kodak 103a-F emulsion was 10 to 30 minutes.

Nothing special was noticed on the spectrograms of Alphonsus up to the night of November 2-3, when three spectrograms were taken with the slit running through its central peak, as shown in the accompanying picture. While I was taking the first spectrogram, at 1^h Universal time, and guiding on the image of the central peak, the latter became strongly washed out and of an unusual reddish hue.

After taking this spectrogram, however, and in accordance with our program, we changed over to observe Mars, and the next spectrogram of Alphonsus was made from 3:00 to 3:30 UT, a 30-minute exposure. Only the central peak of this crater showed on the slit, and I was struck by its unusual brightness and whiteness at the time.

During the exposure I did not take my eye away from the guiding eyepiece, but suddenly I noticed that the brightness of the peak had fallen to its normal value. The spectrogram exposure was then immediately stopped and the following one was taken, from 3:30 to 3:40, with the same position of the slit.

I did not give serious thought to my visual impressions, believing that all the peculiarities I had noticed were caused by a change in the quality of the observing conditions. Therefore, it came somewhat as a surprise when development of the spectrograms showed that all the changes noted visually had in reality occurred on the central peak of Alphonsus.

On the first spectrogram (not reproduced), the central peak is considerably weakened in violet light compared with the neighboring details of the crater, a fact that was not observed on earlier spectra. Measurement of this photograph showed that the absorption varied inversely with the wave length, and the calculated general absorption turned out to be equal to 15 or 20 per cent in the visual region.

On the second spectrogram, this absorption is not noticeable, and an emission spectrum stands out, composed of a series of broad bands superimposed on the usual spectrum of the central peak. Below this second spectrogram is reproduced the 10-minute spectrum that was taken immediately afterward, showing the normal appearance of the crater. Therefore, the phenomenon of gas effusion lasted not longer than 2-1/2 hours and not less than half an hour.

On the following night, November 3-4, I obtained two more spectra of Alphonsus, but its condition continued to be normal. Then the moon entered the last quarter phase and this region of its surface was in shadow and unobservable.

These observations are interpreted as showing that on the morning of November 3, 1958, there occurred a volcanic phenomenon. First there was an ejection of dust---volcanic ash (appearing reddish in the guiding eyepiece)---and afterward an efflux of gas (causing the emission spectrum). The effusion of gas could come from magma rising to the lunar surface.

....

JAPANESE SAW PINK PATCH ON THE MOON

Anonymous; *New Scientist,* 22:334, 1964.

On two occasions late last year, astronomers at the Flagstaff Observatory, in Arizona, reported reddish patches on the Moon, in the region of the brilliant crater Aristarchus (New Scientist, Vol. 20, p. 367 and Vol. 21, p. 172). No photographs were taken, however, and the reports remained unconfirmed, though they are very well substantiated.

A fresh report of colour in the same region has now been received. It has been communicated by T. Sato, astronomer at the Rakurakuen Planetarium at Hiroshima in Japan, who is an experienced lunar observer. The observation was made not by Sato himself, but by seven young Japanese amateurs using the 10-in. reflector at Rakurakuen. On 28 December, 1963 (UT date) they were carrying out observations in preparation for the lunar eclipse on 30 December. Until 15h 05m UT nothing strange was seen, then at 15.55 one of the observers, Y. Yamada, noted a large, distinct pink patch covering the southern part of Aristarchus, which was soon confirmed by the other six observers. It gradually spread toward Herodotus until 16.26, when clouds prevented any further observation. The patch was not markedly brighter or darker than the adjacent area.

The position of the patch corresponds quite well with one of the patches seen at Flagstaff Observatory previously.

. . . .

ANOTHER LUNAR COLOR PHENOMENON

Anonymous; *Sky and Telescope,* 27:3, 1964.

As told on page 316 last month, three short-lived reddish spots were seen near the lunar crater Aristarchus on October 29, 1963, by James A. Greenacre and Edward Barr with the 24-inch refractor of Lowell Observatory. In a special search for similar happenings, they observed another colored spot in the same part of the moon on November 27, 1963.

This feature, like the others, seemed a light ruby red, according to Mr. Greenacre. It was larger than the previous ones, being about 12 miles long and 1-1/2 miles wide, on the rim of Aristarchus. The marking's coordinates were determined with the aid of the Orthographic Lunar Atlas; it extended from Xi= -.682 to -.685 and from Eta = + .391 to + .398.

The ruby flush was first seen at 5:30 p. m. Mountain standard time, and lasted until 6:45. This 1-1/4-hour duration was longer than for any of the earlier three. Four observers saw this spot in the 24-inch refractor, including John S. Hall and Fred Dungan. Dr. Hall, who is the director of Lowell Observatory, described the reddish glow as delicate.

. . . .

The possibility now emerges that small, transient reddish blotches on the moon are fairly frequent phenomena. However, adequate study may require highly experienced lunar observers with telescopes larger than 12 inches. It is perhaps significant that all four Lowell sightings were made less than two days after the Aristarchus neighborhood emerged into sunlight.

COLOR EVENTS ON THE MOON

Moore, Patrick; *Sky and Telescope,* 33:27, 1967.

Much has been heard recently of the Moon-Blink project, which originated in the United States. Basically, the moon is observed in rapid alternation through red and blue filters, so that a color phenomenon (almost always a red glow) will show up as a flickering spot. Here I propose to give a brief description of the British Moon-Blink program, with particular reference to observations of the crater Gassendi on April 30 and May 1, 1966.

The British Moon-Blink device is much simpler than the American, and the moon is viewed directly through an eyepiece instead of with an image tube. Such a system was devised some time ago by V. A. Firsoff. In early 1966, Peter Sartory of Farnham, Sussex, England, constructed a blink device consisting of a filter wheel carrying 120-degree sectors of red gelatin filter (Wratten 25), blue gelatin filter (Wratten 80B), and clear. These join without intervening obstructions. A knob and 1:1 gearing permit smooth manual rotation of the wheel, which is in a protective housing, and there is no appreciable shake. Sartory has described this apparatus in the Journal of the British Astronomical Association, 75, 98, 1966.

Such equipment has been constructed by a number of members of the BAA Lunar Section, which has numerous experienced observers. While the telescopes they use differ widely in aperture, most are reflectors of from 8 to 15 inches.

On April 30, 1966, at 21:30 Universal time, Sartory was using the device on his 8-1/2-inch reflector when he noted a marked "blink" on the outer southeast wall of Gassendi. The blink was still visible at 21:55, when Sartory left the telescope to telephone other members of the Lunar Section team, including me (at Armagh in Northern Ireland). Together with T. J. C. A. Moseley, another member of the Lunar Section, I had been observing with the 10-inch refractor at Armagh Observatory, but there had been a period of indifferent seeing. We received Sartory's message at 22:05 Universal time, and had the 10-inch in action again by 22:17.

Deliberately, Sartory had not told us the precise location of his lunar phenomenon, and at that time we did not have a blink device (we do now). Almost immediately, however, we detected a reddish patch outside Gassendi's southeast wall, subsequently found to be in the exact position of Sartory's blink. It was seen at 22:18, and lost finally at 22:34. Sattory had recommenced observing at 22:27, but the patch was fading, and he lost it two minutes later. At no time did he see any color, which indicates that the reddish hue was beyond the reach of his reflector even though detectable with the Armagh refractor.

Similarly alerted was P. Ringsdore, secretary of the Lunar Section, at Ewell, Surrey. Using an 8-1/2-inch reflector without a blink device, he saw no definite color, but a distinct obscuration in a position which was found, when the observations were compared later, to be exactly the same.

Sartory also suspected two blinks well outside the southeast wall of Gassendi, but these were not seen at Armagh or by Ringsdore. However, at 23:21 Moseley and I independently noted a larger, elusive red-

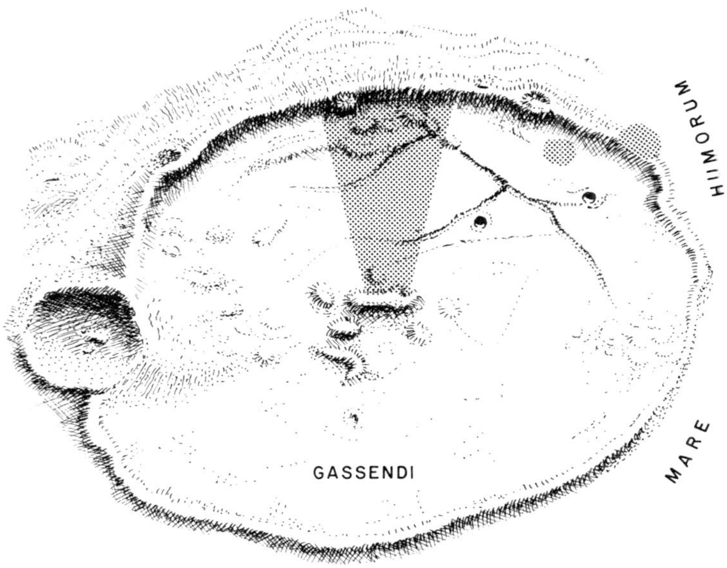

Transient lunar phenomena at Gassendi on April 30 and May 1, 1966. Shaded areas mark color phenomena. (South is up.)

dish patch over the east wall, which was gone by 23:25. Ringsdore also saw an obscuration from 23:20 until 23:28. Though his telescope was smaller, his sky conditions were extremely good. For the last observation at Armagh, Moseley and I used my own 12-1/2-inch reflector.

On the following night, Sartory began observing with his blink equipment under a good sky. He could not telephone Armagh, because of a service failure, and so his findings could not be compared with ours until much later. By now the floor of Gassendi was fully illuminated. No effects were seen in the positions of the previous night's activity, but a large and definite triangular blink was present on the floor, its apex almost touching the central mountains and its base near the east wall. At Armagh, Moseley and I were at first handicapped by drifting cloud, but when conditions improved we saw a triangular reddish "wedge" in what proved to be the same position. On May 2nd everything was normal.

This series of observations aroused much interest, and various other reports regarding other areas have come in since. These may be described as "two-way." For instance, between 0:50 and 1:20 on August 1, 1966, Moseley, P. G. Corvan, and I, using the Armagh 10-inch with a blink device, detected a blink outside the southwest wall of Aristarchus. Using the telephone link, we were able to have it confirmed by other observers both visually and with a blink device, although, as before, we refrained from giving details of its location.

There is no doubt that observations of this kind are well worth following up, and the Lunar Section now has a strong team of observers engaged in this work. Several Moon-Blink devices are in action, and more will be available in the near future.

One interesting finding is that there seem to be certain areas of the moon which will always give weak blinks, indicating permanent features and not transient phenomena. There is, for example, one of considerable size in Mare Frigoris, north-northeast of Plato. At the moment, systematic observations are being made with a view to charting these regions. of which there are more than might be expected.

Just what the precise nature of these phenomena is remains to be seen; in any event, the first step is to carry out a close study of the phenomena themselves. Extreme care is necessary, and one must be wary of jumping to conclusions. But the fact that observations have been made and confirmed by completely independent observers does seem significant.

TLPS AND SOLAR ACTIVITY, AND OTHER PHENOMENA

Botley, Cicely M.; *British Astronomical Association, Journal,* 86:342-343, 1976.

Before the discussion at the September meeting I had been investigating historic reports of possible TLPs, following a note by Mrs. Cameron in the Strolling Astronomer. A preliminary account has been published in the Lunar Section Circular, but for the benefit of non-members of the Section, I should like to say that I think I have found an accord between such reports and Schove's rich auroral periods, namely:

565-580	About 577 by Gregory of Tours observed a 'light on the Moon';
1095-1152	Total eclipse 1096 Aug. 6;
1520-1582	1540 Nov. 26 a "Starlike appearance on dark side" near Callipus;
1616-1640 or 1648	Aristarchus noted as "Red Mountain" by Hevelius;
1715-1746	"Lightning" observed on face of the Moon during a total solar eclipse;
1770-1788	Period of Herschel's "volcanoes", including coincidence between observation and great aurora 1787 April 18-19.

This list does not include the famous Elizabethan observation of 1587, but then an aurora was still being reported. Schove's minimum is 1587 but this is approximate and large spots often occur at that time. As for the New England observation of 1668, Link gives an aurora for Hungary in February 5.

WAS THE FORMATION OF A 20-KM-DIAMETER IMPACT CRATER ON THE MOON OBSERVED ON JUNE 18, 1178?

Hartung, Jack B.; *Meteoritics,* 11:187-194, 1976.

An interesting reference to an event which apparently occurred on the Moon on July 18, 1178 is given by Newton (1972, p. 690). "...the upper horn of the new moon seemed to split in two and a flame shot from it." A search for a more complete description of the event led to the following report written in the medieval chronicles by Gervase of Canterbury (Stubbs, 1879). The original Latin version is presented first; an English translation by Dr. Richmond Y. Hathorn, Classics Department, State University of New York at Stony Brook, follows.

[Latin text omitted.]

> In this year, on the Sunday before the Feast of St. John the Baptist, after sunset when the moon had first become visible a marvelous phenomenon was witnessed by some five or more men who were sitting there facing the moon. Now there was a bright new moon, and as usual in that phase its horns were tilted toward the east; and suddenly the upper horn split in two. From the mid-point of this division a flaming torch sprang up, spewing out, over a considerable distance, fire, hot coals, and sparks. Meanwhile the body of the moon which was below writhed, as it were, in anxiety, and, to put it in the words of those who reported it to me and saw it with their own eyes, the moon throbbed like a wounded snake. Afterwards it resumed its proper state. This phenomenon was repeated a dozen times or more, the flame assuming various twisting shapes at random and then returning to normal. Then after these transformations the moon from horn to horn, that is along its whole length, took on a blackish appearance. The present writer was given this report by men who saw it with their own eyes, and are prepared to stake their honor on an oath that they have made no addition or falsification in the above narrative.

The event reported in 1178 appears to be unique. Although a wide variety of short-duration events or transient lunar phenomena have been reported by observers of the Moon, to my knowledge no similar event of the magnitude and complexity observed has ever been described.

A variety of hypotheses may be advanced to explain the phenomena observed. The fact that five men were "prepared to stake their honor" leads me to believe that something was observed and that the report was not entirely the product of someone's imagination. The events described may have originated in the atmosphere of the Earth as a result of cloud layers or turbulence just above the western horizon or the entry of a meteroid along the line of sight to the Moon. I would prefer, however, to rely on the report which repeatedly refers to the Moon and never mentions clouds or the sky or anything not related to the Moon.

Such an event on the Moon must have had either an external (impact) or internal (volcanic) origin. From the report alone it may not be possible to explain the origin of the phenomena observed. However, with the report and our present understanding of the lunar surface and processes occurring in the Solar System, I would suggest the description is consistent with the occurrence of a very large impact on the lunar surface. "The upper horn (of a new Moon) split in two." A portion of the sunlit crescent visible at the earth was either obscured by the ejecta cone or cloud produced by the impact or darkened by the shadow of the ejecta. "A flaming torch sprang up, spewing out fire, hot coals, and sparks." Incandescent gases and solids or liquids were ejected. "The moon...writhed...and...throbbed like a wounded snake." Gases produced or released by the impact formed a temporary atmosphere which was in turbulent or non-uniform motion, thus causing the light from the moon to pass through variable amounts of material with variable indices of refraction.

Up to this point the phenomena described are reasonably consistent with what might be expected based on experience with laboratory impacts and study of large terrestrial impact craters. The remainder of the description is less easily interpreted, mainly because we are not familiar with the short-duration effects of a large impact on a planetary body. This report by at least five men almost 800 years ago probably represents the best observational data on such effects that exist.

"Afterwards it (the crescent moon) resumed its proper state. This phenomenon was repeated a dozen times or more, the flame assuming various twisting shapes at random and then returning to normal." Evidently, "the flame" continued throughout the writhing process, which I attribute to the large-scale motion of gas clouds. Because the duration of a "flame" consisting of incandescent gases or other material among the ejecta should be short compared to the time scale for the motion of large gas clouds, some uncertainty exists regarding the interpretation of "the flame." Instead of consisting of incandescent material, the appearance of a "flame" could be produced by sunlight reflected from dust particles moving as ejecta from the crater or suspended by a temporary gas cloud above it.

"Then...the moon from horn to horn, that is along its whole length, took on a blackish appearance." This observation is consistent with the presence around the entire moon of a temporary atmosphere in which a sufficient amount of dust was suspended to block a significant amount of light reflected from the moon's surface.

The most decisive test of the validity of the report and the impact hypothesis explaining it would be the identification of a surface feature on the Moon which corresponds to the event reported. The occurrence of the event at the "midpoint" of the "upper horn" of a "new Moon" establishes its location at a latitude near 45° north and longitude near 90° east.

The size of the event may be estimated based on the fact that the observations were made by human eyes without optical aids. If it is assumed that the ratio of the linear dimension of an object to the distance from which the object is just visible to the unaided eye is about 10^{-4}, then the smallest shadow on, or ejecta cloud above, the Moon that could have been observed at the Earth would have had a linear dimension of about 40 km. Based on this I estimated that the ejecta cloud and the shadow actually observed would have had linear dimensions in excess of about 100 km; the diameter of the resulting crater would exceed 10 km; and the bright rays associated with such a recently formed crater would extend at least 100 km from the crater.

Does a crater corresponding to the medieval observations and our interpretations and estimates actually exist on the Moon? The answer is yes! The criteria we used in search of the actual crater formed only 800 years ago are that the crater must have a:

1. selenographic latitude between 30° and 60° N.
2. selenographic longitude between 75° and 105° E.
3. diameter greater than 10 km.
4. surrounding pattern of prominent bright rays.

Using Lunar Orbiter and Apollo mission photography, we found an impressive 20-km-diameter crater at a selenographic latitude of 36° N. and longitude of 103° E. surrounded by very prominent bright rays extending for hundreds of km from the crater. We concluded that this crater is the one formed on June 18, 1178. The crater, Giordano Bruno, is shown in [the figure].

Lunar photograph looking northwest, showing the 20-km crater Giordano Bruno. (Courtesy Lunar Science Institute)

Not only does Giordano Bruno satisfy the above-stated criteria, but there is independent evidence to suggest it was among the most recently formed large craters on the Moon. The crater is "the center of a ray system rivaling that of Tycho", yet its rim diameter is only about one-fifth that of Tycho. The first photographs of Giordano Bruno, made in October 1959 during the Lunik III mission, showed a ray system so bright and impressive that the actual crater diameter was over-estimated by a factor of three. Based on a survey of large bright-rayed craters on the Moon, I have not found a single crater with a larger ray-length-to-crater-diameter ratio than that for Giordano Bruno. These observations support the idea that this crater may have been the most recently formed large crater on the Moon.

....

• Cloud-like Phenomena

THE MOON—MARE CRISIUM

Hardy, Jas. D.; *British Astronomical Association, Journal,* 7:139–141, 1897.

Now, at various times when studying the floor of M. Crisium, I have noticed waves of light and shade, so that it was difficult at times to see objects with which I was perfectly familiar. Also clouds have passed over the object I have been viewing. These clouds have been seen by other observers, and are mentioned in Neison. That they belong to the moon there can be no doubt, and I gradually come to the conclusion that vapour of some kind still existed on the moon. I had my surmises verified on one particularly fine night---the best I ever had---when I plainly saw a well-defined cloud pass over the object I was copying.

....

UNUSUAL LUNAR PHENOMENON

Goddard, A. V.; *Popular Astronomy,* 40:316–317, 1932.

I was observing the moon on the night of April 14, at about 10:30 Pacific Time and beheld a most interesting phenomenon. I was using a 16-inch glass and the atmosphere was very steady. A friend and I noticed the unusual absence of all white spots and markings in Plato. At 10:57 P. S. T. in long. 10° E and lat. 51°N a white spot made its appearance in less than a minute it had spread in a northeasterly direction until it almost reached the rim of the crater. This observation was verified by my friend as I rather doubted my own eyes.

White markings are almost always visible in Plato but this is the first time I have ever seen one appear suddenly. It appeared and moved like a cloud of steam but, considering the size of Plato and its rapid motion, this idea seems untenable.

A more practical explanation would be that some pecularity of the surface does not show until the rising sun reaches a certain angle and we happened to be observing at that particular instant.

The terminator had passed beyond Copernicus so it must have passed about 8 degrees beyond Plato.

I would be glad to hear the views of others on this occurrence.

CLOUD ON THE MOON

Jackson, J. G.; *English Mechanic,* 35:326, 1882.

Last evening (May 19th) on observing the moon's slender crescent, about two days old, I was struck with a very peculiar appearance on the westerly side of the Mare Crisium, just on, or immediately within, the dark of the 'terminator'.

It seemed a curved feathery mist or cloud lying just over the edge of the 'Mare', and against the spur or range of mountains bounding the westerly side of the great valley. It seemed to be divided longitudinally by a faint dark line, and looked not unlike a feather. It must have been more than 100 miles long, by 40 or 50 miles wide.

The definition was excellent, and I watched it for nearly an hour, showing it to several friends. In colour and appearance it was so strikingly different from the other illuminated parts, and so different from anything else I have ever seen on the moon, that I scarcely think it possible to be mistaken in its vapoury character.

• Lunar Luminescence

EXCITATION OF LUNAR LUMINESCENCE BY SOLAR ACTIVITY

Kopal, Zdenek, and Rackham, Thomas W.; *Icarus,* 2:481-500, 1963.

Abstract. An anomalous increase in surface brightness of the Moon in an area of over 60,000 km^2 around and north of the crater Kepler was observed to occur twice on the night of November 1-2, 1963, on eight photographs secured with the 24-inch refractor of the Observatory du Pie-du-Midi between 22.35-22.12 U.T. on November 1st, and 00.20-00.35 U.T. on November 2nd, on Kodak 1-F plates exposed through an interference filter of 45 A half-width centered on 6725 A. Control photographs taken through an interference filter of 95 A half-width centered on 5450 A failed to show any such effect. The enhancement in the red was not only observed to recur twice during the same night, but plates taken between 00.20-00.35 U.T. on November 2nd disclosed that the degree of enhancement (resulting nearly in a temporary doubling of surface brightness) increased appreciably within 15 minutes of observation.

This observed enhancement is interpreted to be the result of luminescence of the respective parts of lunar surface excited by solar activity. Two Class-1 flares were indeed observed to occur on an identical place of an otherwise calm Sun earlier that day at the Sacramento Peak and McMath-Hulbert Observatories at 13.58 and again at

15.55 U.T. The time interval between these flares suggests that the recurrence of the observed brightening of the Moon may have been due to surface luminescence stimulated by them after a transit time close to 8.5 hours---of sufficient duration to rule out solar X-rays or UV light as the exciting source, and to direct attention to corpuscular radiation. If so, this transit time would correspond to a particle velocity of 5000 km/sec, and the energy flux to a density of the order of 10^3 particles cm^3.

LUNAR LUMINESCENCE

Flamm, E. J., et al; *Nature,* 205:1301-1303, 1965.

In a recent article, Kopal has directed attention to a report by Sir William Herschel of 'volcanoes' in the neighbourhood of the lunar crater Aristarchus seen in April of 1787. Both he and Middlehurst note the similarity between Herschel's description and two recent reports by Greenacre of reddish luminescence near Aristarchus in October and November 1963. In Table 1 are listed 16 additional transient luminescent events in the vicinity of Aristarchus observed during the period 1783-1963. Included in the second column of Table 1 are the colour, the visual magnitude, and the dimensions of those luminescent spots for which these data were reported. Of the 19 observations listed, the first 16 were made 2-6 days after new Moon; that is, when Aristarchus was on the dark side of the lunar surface far from the terminator. The three most recent observations were made shortly after full Moon, with the crater in sunlight. In addition to these reports, the literature contains references to luminescent events in the vicinity of several other lunar features including Pitatus, Heraclides Promontory, Alpine Valley, Carlini, Alphonsus, and Kepler. All these events, except those involving Alphonsus and Kepler, were on the dark side of the Moon. The latter two events as well as the luminescence of Aristarchus observed by Kozyrev in 1961 were recorded photographically. Several of the remaining events were observed visually through different telescopes by two or more independent observers.

Table 1. Observations of Luminescence in the Vicinity of Aristarchus

Date of observation	Description	Observer
May 4, 1783	Red; fourth mag.; 3 in. arc	Herschel
April 19-20, 1787	Red; fourth mag.; 3 in. arc	Herschel
March-May, 1789		Bode
Feb. 4-7, 1821	Sixth-seventh mag.; 3 in.-4 in. arc	Kater Olbers
May 4-6, 1821	1 in. arc	Ward Baily
Jan. 27, 1822	Eighth Mag.	Struve

May 1, 1824	Ninth-tenth mag.	Gobel
April 22, 1825		Argelander
		Gobel
Dec. 25, 1832		Smyth
Dec. 22, 1835	Ninth-tenth mag.	Smyth
June 14-16, 1866	Reddish-yellow	Temple
April 9, 1867	Seventh mag.	Elger
May 7, 1867	Reddish-yellow	Temple
March 1-3, 1903		Rey
		Gheury
Feb. 22, 1931	Reddish-yellow	Joulia
March 30, 1933	White	Douillet
Nov. 26, 28, Dec. 3, 1961	Red and blue	Kozyrev
Oct. 29, 1963	Reddish-orange to light ruby	Greenacre
Nov. 27, 1963	Light ruby	Greenacre and Boyce

Kopal and Rackham have proposed that the luminescence observed by them in the vicinity of Kepler was excited by streams of corpuscular radiation emanating from a solar flare. Accordingly, Kopal suggests that the luminosity of Aristarchus seen in 1787 as well as in 1963 was likewise due to flare-accelerated particles directed to the crater by interplanetary magnetic fields. In support of this suggestion, he notes that May of 1787 was the peak of an unusually active solar cycle. It therefore seems worthwhile to consider the distribution of the events in Table 1 with respect to solar activity:

Yearly mean sunspot No.	No. of events
0- 30	15
30- 60	2
60- 90	0
90-120	1
120-150	1

Clearly, the 19 events are negatively correlated with solar activity. This fact alone does not necessarily eliminate solar flares as a cause of the luminescence, since major flares have occurred, although infrequently, very near solar minima.

. . . .

• Lunar Meteors (?)

DOES ANYTHING EVER HAPPEN ON THE MOON?

Haas, Walter H.; *Royal Astronomical Society of Canada, Journal,* 36:237-272, 317-328, 361-376, 397-408, 1942.

Possible Lunar Meteor. The observation about to be reported is so surprising that it is likely to impress many conservative minds as ridiculous, but I do not wish on that account to suppress anything that may advance knowledge of the moon.

On July 10, 1941, I was observing the nearly full moon with a 6-inch reflector at 96X, the seeing and the transparency both being 4. At $5^h 44^m$, while drawing the crater Hansteen, I saw a tiny luminous speck move across the moon's surface at a uniform rate. It appeared about one north-south diameter of Gassendi west of that crater and travelled almost due east until it vanished just short of Gassendi's west wall. The speck was far smaller than any of the central peaks of Gassendi, and it is thought that the angular diameter could not have exceeded 0''. 1. The brightness was constant along the whole path, and the stellar magnitude was estimated to be + 8. The duration was about one second.

Near $5^h 41^m$ I had seen a fainter speck somewhere south of Grimaldi but only by vision so indirect that little more would have been thought about the matter if the second speck had not seen appeared.

The end-points of the motion of the well-seen speck were strikingly definite; and we may accordingly exclude as the cause of the appearance any terrestrial object low in the atmosphere (e.g., the appearance any terrestrial object low in the atmosphere (e.g., thistle-down), for it would have moved all the way across the field of view. One thinks next of a telescopic shooting star very near its radiant. But since I have on no other occasion seen similar specks and know of no such observations by others, it is apparently very difficult to see shooting stars against the moon; and one is justified in giving some figures based on the assumption that the speck was a lunar meteor.

An angular diameter of 0''. 1 would mean a linear diameter of about 600 feet, a value similar to the observed sizes of terrestrial fireballs. As seen from the surface of the moon the stellar magnitude would have been 8 -16.6 = -9 (roughly). On Plate 14B in Pickering's Atlas, I measured the length of the projected path to be 63 miles, a value probably not in error by more than 10 miles. The corresponding velocity relative to the moon was at least 63 miles a second, the exact value depending upon the undeterminable orientation of the real path in space; but the velocity is in any case surprisingly large, the more so when we bear in mind that the meteorite, which moved eastward, overtook the moon. (pp. 397-398)

• Analyses of TLPs

LUNAR TRANSIENT PHENOMENA: TOPOGRAPHICAL DISTRIBUTION

Middlehurst, Barbara M., and Moore, Patrick A.; *Science,* 155:449-451, 1967.

Abstract. The sites named in nearly 400 reports of lunar transient phenomena fall into three classes: (i) sites peripheral to the maria, (ii) ray craters, and (iii) ring plains with dark or partially dark floors; none are known in the rugged highland area of the southeast (International Astronomical Union, 1964; classically southwest) quadrant. Permanent records are few; the sites where known are consistent with the visual records.

COMPARATIVE ANALYSES OF OBSERVATIONS OF LUNAR TRANSIENT PHENOMENA

Cameron, Winifred S.; *Icarus,* 16:339-387, 1972.

From the author's collection of more than 900 reports of Lunar Transient Phenomena (LTP) covering the period 1540-1970, 771 positive plus 112 negative observations (several times more than any previously published analyses) with sufficient ancillary data were analyzed for five hypotheses of causes. Approximately one third were of Aristarchus, divided almost equally between one observer, Bartlett, and all others. Treated as two groups they were divided into four categories: (1) gaseous, (2) reddish, (3) bluish, and (4) brightenings, and analyzed separately and combined with respect to the hypotheses. Six other individual LTP sites also were analyzed. The five hypotheses involved effects of tides, sunrise, low-angle illumination, Earth's magnetic tail, and solar particles. Novel analyses were made by [1] histograms with respect to (a) phase (ø) of anomalistic period, (b) Moon's age; [2] plot on a representative cycle of anomalistic orbits; [3] ratios of observed to expected percentages (O/E); [4] plot of distributions of (a) LTP sites (< 100), (b) dark, flat-floored craters, and (c) the author's (1967) ringdike candidates.

The histograms show: [1] caprice in analytical behavior between (a) the sites individually, even between near neighbors, and collectively, (b) the different categories, and (c) Bartlett and All Others, especially in ø: [2] any tidal correlations found were almost exclusively with perigee; [3] in lunar age, rather persistent correlations were found with feature sunrise and with the magnetic tail; [4] differences between Bartlett and All Others for the reddish phenomena, and similarities between the Bartlett bluish and All Others' brightenings suggest variant optical physiological responses, indicating higher blue-sensitivity and lower(?) red-sensitivity for Bartlett; [5] inverse relationships between Bartlett's gaseous and reddish phenomena in both tidal and age diagrams, and the same inverse relation between his reddish and All Others' suggest that Bartlett glimpsed true ground color in the absence of obscuring matter; [6] the sharp perigee and apogee correlations found by Burley and Middlehurst (1966) smeared out or disappeared in the much larger number of observations of all sites analyzed here. Most perigee or apogee correlations clustered more at times of greater orbital eccentricity than at lesser eccentricity, as shown by the orbital plot; [7] absent and present (especially the onset) phenomena correlated similarly, instead of oppositely as expected, and both were found to advance around the orbits on the orbital plot, which weakens any positive correlations found; [8] phenomena have been reported for all ages, from 0^d to 28^d with > 50% within ± 4^d of full Moon (< 1/4 near perigee).

The ratios (O/E revealed: [1] universely high correlations with sunrise, not all accountable by observational selection; [2] rather high correlation with Earth's magnetopause, but not the bow-shock front turbulence; [3] decline of perigee correlation with increasing number of observations, from (O/E) = 1.9 to 1.1; [4] little correlation with direct bombardment of solar flare particles but apparent enhancement by the magnetic tail when the Moon was within it at time of arrival of the particles in the Earth-Moon system.

Some facts indicate that some of the brightenings result from unknown atmospheric, instrumental, and geometric effects, but many puzzling aspects still remain.

Comparison of the distributions (nonrandom) of the LTP sites, dark craters, and the author's proposed ringdikes shows strong affinities for the mare edges for each, implying internal activity (probably degassing) as the source of most LTP.

Considering everything the data show, the author is skeptical that there are any decisive external (even tidal) influences on this internal activity released at LTP sites. [excerpt]

LUNAR PHYSICAL AND CHEMICAL ANOMALIES

The manned lunar landings harvested a rich lode of anomalous data generally unobtainable through the telescope. There are few connecting threads; so a simple listing is presented below to guide the reader. Even though the list is incomplete, it is manifest that the moon possesses mary inhomogenities, suggesting a history punctuated by energetic events.

- Glazed rocks, implying (perhaps) bursts of thermal radiation
- Lunar soils containing an ubiquitous micrometeorite component similar in composition to carbonaceous chondrites

-Lunar volatiles inferring cometary impacts
-Mass concentrations (mascons) existing below the surface
-Surprisingly high heat flows from the lunar interior
-Existence of infrared hot spots
-Radioactivity concentrations (radcons)
-Unexpectedly high magnetic anomalies (magcons)
-Curious seismic events, inferring a layered moon and correlations with both TLPs and orbital parameters

• Glazed Lunar Rocks

GOLD AND THE GLASSY CRATERS

Anonymous; *Nature*, 226:598, 1970.

One of the more awkward discoveries of the Apollo 11 mission is that small craters between 20 cm and 1.5 m across frequently contain lumps of soil coated on their top surfaces with glass, but Professor T. Gold's explanation of this phenomenon in terms of a solar flare powerful enough to cause glazing within the craters has not been immediately accepted (Science, 165, 1345; 1969). The problem is, of course, to invoke glazing within the craters and nowhere else, for there is no evidence that the glazed patches have anything to do with the glass beads which make up such a significant part of the lunar soil. This is why Gold hit upon solar heating as the mechanism, because calculations by D. Buhl, W. J. Welch and D. G. Rea in the Journal of Geophysical Research indicate that the temperature within the craters is likely to be 10 per cent or so higher than that of the surrounding lunar surface when the Sun is above the horizon (73, 5281; 1968). Thus Gold proposed a solar flare which would have been just powerful enough to start melting the soil within craters but to leave the surface intact.

Obviously people are finding the degree of selectiveness necessary for this process hard to accept, and three papers challenging, or at best expressing doubt, about Gold's hypothesis appear in a recent issue of Science. J. Green of the Douglas Advanced Research Laboratories, for example, who shows his hand at the start by admitting his belief that almost all the major surface features are volcanic in origin, finds it easier to believe that volcanic bombs are responsible both for the craters and the glassy patches which they contain (168, 608; 1970). Volcanic bombs are large blobs of molten lava protected by a solidified crust which are thrown out of volcanoes. On impact they could produce a crater and their molten contents are likely to freeze rapidly in the lunar environment to leave behind the glassy patches.

Then E. D. Dietz and P. J. Vergano of Owens-Illinois Inc. of Ohio, famous for its glass-ceramic telescope mirrors, point out that the appearance of the glassy patches on the Apollo 11 photographs do not match what might be expected for the glazing of basalt by solar heating (168, 609; 1970).

Finally W. R. Greenwood and G. Heiken of the Houston Manned Spacecraft Center suggest that the glass is lunar surface material fused by the impact of the meteorite producing the crater and which has fallen back after the explosion to land on the crater floor (168, 610; 1970).

But although Gold agrees that the solar flare hypothesis is not strong, he refuses to accept that the alternatives are any better (168, 611; 1970). The drawback to the meteorite hypothesis, for example, is that the explosions are violent enough to make it unlikely that material would fall back precisely to the center of the craters where the glassy patches are found. Gold also points out a general drawback to theories which have the glass formed at the same time as the craters. Because the glass appears to be barely eroded at all, the implication is that the craters have not been modified since their formation, and this is hard to square with what is known about the size distribution of meteorites.

Thus even Gold would seem to agree that the explanation of the phenomenon is still in doubt. Obviously everyone would like to get their hands on a sample of the glass, but unfortunately this is something that neither Apollo 11 nor 12 was able to provide.

GLAZED LUNAR ROCKS: ORIGIN BY IMPACT

Morgan, John W., et al; *Science,* 172:556-557, 1971.

Abstract. The glassy coating of lunar rock 12017 is enriched in 15 trace elements relative to the crystalline interior. It apparently consists chiefly of shock-melted rock, somewhat richer in rare earth elements and alkali metals than rock 12017 itself. The glass has been contaminated by about 0.5 percent carbonaceous-chondrite-like material or, alternatively, by a mixture of 0.06 to 0.3 percent fractionated meteoritic material and approximately 10 to 15 percent local soil. The glazing seems to represent molten material splashed from a nearby meteorite impact and not in situ melting by a sudden increase in solar luminosity.

• Lunar Chemistry and Catastrophism

MOON: POSSIBLE NATURE OF THE BODY THAT PRODUCED THE IMBRIAN BASIN....

Ganapathy, R., et al; *Science,* 175:55-59, 1972.

Abstract. Soils from the Apollo 14 site contain nearly three times as much meteoritic material as soils from the Apollo 11, Apollo 12, and Luna 16 sites. Part of this material consists of the ubiquitous micrometeorite component, of primitive (carbonaceous-chondrite-like) composition. The remainder, seen most conspicuously in coarse glass and norite fragments, has a decidedly fractionated composition, with volatile elements less than one-tenth as abundant as siderophiles. This material seems to be debris of the Cyprus-sized planetesimal that produced the Imbrian basin. Compositionally this planetesimal has no exact counterpart among known meteorite classes, though group IV A irons come close. It also resembles the initial composition of the earth as postulated by the two-component model. Apparently the Imbrian planetesimal was on Earth satellite swept up by the moon during tidal recession or capture, or an asteroid deflected by Mars into terrestrial space.

VOLATILE-RICH LUNAR SOIL: EVIDENCE OF POSSIBLE COMETARY IMPACT

Gibson, Everett K., Jr., and Moore, Gary W.; *Science,* 179:69-71, 1973.

Abstract. A subsurface Apollo 16 soil, 61221, is much richer in volatile compounds than soils from any other locations or sites as shown by thermal analysis-gas release measurements. A weight loss of 0.03 percent during the interval 175° to 350°C was associated with the release of water, carbon dioxide, methane, hydrogen cyanide, hydrogen, and minor amounts of hydrocarbons and other species. These volatile components may have been brought to this site by a comet, which may have formed North Ray crater.

• The Mascons

MASCONS: LUNAR MASS CONCENTRATIONS

Muller, P. M., and Sjogren, W. L.; *Science,* 161:680-684, 1968.

The Lunar Orbiter missions have provided both high quality photographs of the moon, and supplementary scientific information concerning the gravitational field of the moon. Previous investigators have concluded that the moon was gravitationally rougher than anticipated in the sense that comparatively high degree terms in the spherical harmonic expansion would be required for effective representation of the gravity field. This roughness of the moon has been of interest to the Apollo Project because of the resulting perturbations on the trajectory of the Apollo orbiting spacecraft. For these reasons, a new analysis has been done with the use of the accurate tracking data received here by the NASA Deep Space Network operated by Jet Propulsion Laboratory.

We now report that this new processing of the Lunar Orbiter data has produced unexpected results. A study of local accelerations on the spacecraft resulted in a gravipotential map of the lunar nearside which has revealed very large mass concentrations beneath the center of all five nearside ringed maria (Imbrium, Serenitatis, Crisium, Nectaris, and Humorum). In addition, they were observed in the area between Sinus Aestuum and Sinus Medii (presumably a newly discovered ancient ringed mare), and Mare Orientale. The Urey-Gilbert theory of lunar history has predicted such large-scale high-density mass concentrations below these maria, which, for convenience, we shall call mascons.

....

Even though we have computed both masses and depths for the largest mascons, nevertheless our present quantitative data require further refinements. One may easily compute approximate masses from an assumed depth, such as 50 km. The Mare Imbrium mascon yields numbers on the order of 20×10^{-6} lunar masses. A spherical nickel-iron object about 100 km in diameter would be a rough equivalent. This type of calculation gives a qualitative estimate for the large size of these objects.

The presence of large mascons under every ringed maria, excepting Iridum, and their relative absence elsewhere, suggests a relation between the two phenomena which may be similar to that suggested by Urey and Gilbert. Among questions that arise are these: Does each of these mascons represent an asteroidal-sized body which caused its associated mare by impact? If not simply the original impactor itself, by what processes were they formed in the lunar interior? Is the presence of these objects consistent with a molten lunar interior?

• Lunar Thermal Anomalies

NONUNIFORM COOLING OF THE ECLIPSED MOON....
Shorthill, R. W., and Saari, J. M.; *Science,* 150:210-212, 1965.

Abstract. Infrared scanning during a total eclipse has revealed hundreds of hot spots, many identified with craters smaller than the detector resolution. Areal corrections show that some of these features may have the thermal properties of bare rock. Correlation of thermal response with albedo and radar reflectivity shows discrepancies. There is a concentration of hot spots in Mare Tranquillitatis.

....

The identification of the thermal anomalies, 400 of which have now been cataloged, shows that most (~ 90 percent) are craters which are visually bright in some respect at full moon (ray craters or craters with bright interiors or rims). The remaining are associated with "white areas" at full moon (for example, in Deslandres), which may have a very small crater (for example, Linne or Posidonius γ), or with rilles, such as Rima Hadley.

....

The interpretation of these thermal features on the eclipsed moon is outside the scope of this report, but several puzzling aspects of the problem can be mentioned. First, although the hot spots are associated with bright features, conversely, there are bright features which are not thermally anomalous; thus there does not appear to be a direct relationship between visual brightness and the thermal response. For example, the ray craters Menelaus and Dionysius have the same brightness and appearance at full moon, but during the eclipse Menelaus was only 11°K warmer than its environs, whereas Dionysius was 44°K warmer. Further, the ray crater Euclides was only slightly warmer than its environs during totality, even though other craters similar in size and appearance gave large thermal responses. Second, the fact that ray craters show enhanced radar returns as well as anomalous eclipse cooling has been cited as showing that the surface of these features is denser or rougher, or both. However, a comparison of the radar reflectivity contours with our infrared eclipse data shows there is no simple correlation between the measurements. For instance, enhanced radar returns are found on Eratosthenes and Posidonius, but these craters do not show anomalous eclipse cooling. Also, the radar contours for the ray craters Copernicus and Tycho are decidedly different from their isotherms during totality. It is interesting to note that the uplands give a higher radar return than the maria; the reverse is generally true for the infrared eclipse data. Since the two experiments do not measure the same parameters, it may not really be surprising that there is some lack of correspondence, and, in fact, the differences between them may be as interesting as the similarities. Finally, the nonrandom distribution of hot spots (there is a concentration in Mare Tranquillitatis) can be understood in terms of a random impact origin of the craters if the original surface in which they were formed had local properties which in some way enhanced and prolonged the anomalous thermal behavior. On the other hand, this nonrandom distribution might not be surprising if the craters were formed by internal processes. In fact, internal processes may explain the anomalies observed on the "white areas," which give the appearance of being caused by a deposition of some material of higher albedo over areas large compared to their sources on the surface.

LUNAR ECLIPSE: INFRARED IMAGES AND AN ANOMALY OF POSSIBLE INTERNAL ORIGIN

Hunt, Graham R., et al; *Science,* 162:252-254, 1968.

Abstract. Infrared images of the lunar eclipse of 13 April 1968 were obtained and compared with infrared images of the 19 December 1964 eclipse. A similarity of apparent strength and distribution of most thermal anomalies on the maria is evident from inspection of these images, indicating that these features are not ephemeral. One new linear thermal anomaly was discovered, which is thermally enhanced during the lunar afternoon. Its close relation to a lunar crustal fracture line and other features of probable internal origin suggests that this anomaly may be of internal origin.

LUNAR RESEARCH: NO AGREEMENT ON EVOLUTIONARY MODELS

Hammond, Allen L.; *Science,* 175:868-870, 1972.

Measurements of the temperature gradient and the thermal conductivity in soil at the Apollo 15 site by M. Langseth, of the Lamont-Doherty Geological Observatory, and his colleagues indicate a heat flux from the moon of 3.3 microwatts per square centimeter (about half of the heat flux from the earth), a value that is considerably higher than was expected. The result is also higher than that predicted by thermal calculations based on geochemical models of the moon's interior, and, if sustained by further measurements on Apollo 16 and 17, it might lead to a revision in estimates of the amount of radioactive materials (whose decay is thought to be the heat source) present within the moon. [Excerpt]

• Concentrations of Radioactivity on the Moon

LUNAR RESEARCH: NO AGREEMENT ON EVOLUTIONARY MODELS

Hammond, Allen L.; *Science,* 175:868-870, 1972.

The region around the Imbrium basin, however, appears to have some unusual features. Observations of the gamma rays given off by radioactive materials showed much higher concentrations of uranium, thorium, and potassium in Mare Imbrium and in the neighboring lava flows of Oceanus Procellarum than elsewhere on the moon. The observations were made from lunar orbit, and the flight path of the Apollo 15 spacecraft allowed about 15 percent of the moon's surface to be mapped. The results of the experiment, conducted by a team headed by James Arnold, of the University of California at San Diego, indicate concentrations of about 10 parts per million of thorium in the Mare Imbrium soil, compared with 1 to ppm in the eastern part of the moon. The high concentrations of radioactive elements in the Imbrium region are similar to concentrations found in KREEP basalts, which were found in large quantities at the Apollo 14 landing site. This material has also been found, in smaller quantities, at all the landing sites, but why the Imbrium and Procellarum regions should be the overwhelming source of the radioactive elements on the moon's surface, as they appear to be, has been difficult for geochemists to explain. [Excerpt]

DETECTION OF RADON EMANATION FROM THE CRATER ARISTARCHUS. . . .

Gorenstein, Paul, and Bjorkholm, Paul; *Science,* 184:792, 1973.

Abstract. The alpha particle spectrometer aboard the Apollo 15 command/service module was designed to detect alpha particles from radon decay and to locate regions with unusual activity on the moon. A significant increase in radon-222 activity was detected from a region containing the crater Aristarchus. The result is interpreted as probably indicating internal activity at the site. By analogy with terrestrial processes, increased radon emanation may be associated with the emission of other volatiles.

• Local Magnetic Fields on the Moon

APOLLO 12 MAGNETOMETER: MEASUREMENT OF A STEADY MAGNETIC FIELD ON THE SURFACE OF THE MOON

Dyal, Palmer, et al; *Science,* 169:762-764, 1970.

Abstract. The Apollo 12 magnetometer has measured a steady magnetic field of 36 ± 5 gammas on the lunar surface. Surface gradient measurements and data from a lunar orbiting satellite indicate that this steady field is localized rather than global in its extent. These data suggest that the source is a large, magnetized body which acquired a field during an epoch in which the inducing field was much stronger than any that presently exists at the moon.

MAGNETIC PROPERTIES OF LUNAR SAMPLES

Strangway, D. W., et al; *Science,* 167:691-693, 1970.

Abstract. A breccia sample (10023) from the moon was found to have a strong and fairly stable remanent magnetization. If this sample was not magnetized by local fields in the spacecraft or in the lunar receiving laboratory, it must have been magnetized on the moon. This could have happened in a variety of ways, such as cooling through the Curie temperature, by continuous thermal cycling, or by impact, but all of these require the presence of a magnetic field. Such a field could have been of internal origin in the moon, or it could have been a residual effect from the earth's magnetic field at a time when the moon and the earth were much closer together. Thermomagnetic studies identify the presence of iron with about 1 percent nickel (igneous), iron with about 5 to 10 percent nickel (meteoritic), iron with about 33 percent or more nickel (meteoritic), and ilmenite.

CONCENTRATIONS OF MAGNETISM FOUND ON THE MOON

Anonymous; *Aviation Week,* 19, September 4, 1972.

Localized concentrations of higher magnetism or magcons dot parts of the surface of the moon similar to but more diverse and localized than the mass concentrations or mascons first detected by lunar orbiter.

Magnetometers on Apollo 14 and Apollo 16 traverses have detected local fields as high as 300 gammas, Prof. S. K. Runcorn of the University of Newcastle-Upon-Tyne said at the International Geological Congress here, but Explorer 35 generalized magnetic mapping of the moon shows the field is negligible---10-15 gammas.

Few of these magcons are found in the mares, and Prof. Runcorn said this is explainable if it is assumed the mare regions are fairly uniformly magnetized. In the highlands, where the distribution frequencies are higher, the crustal plates are more broken by craters, giving the magnetic field a chance to escape.

Apollo 15 and Apollo 16 subsatellite plots show the widespread character of lunar surface magnetism. To account for this and remanent magnetism in samples, a magnetic field on the order of 1,000 gammas at the time of extensive lunar melting 3.2-4.1 billion years ago was postulated by Prof. Runcorn.

"The question then arises," Prof. Runcorn said, "whether this 1,000-gamma field existing for about 1 billion years was external or internal, like the earth's." Although many assume the moon approached the earth in its early history and acquired its magnetism from the earth, Prof. Runcorn said tidal forces would have prevented an earth-moon close approach. His conclusion is that the lunar field originated internally.

Extensive gravitational differentiation in the early melting history of the moon is consistent with the sinking of heavy iron to the center to form a core---albeit a small one---to act as a dynamo to create the field. As the moon solidified and the liquid core disappeared, the field could have decayed away.

. . . .

LUNAR RESEARCH: NO AGREEMENT ON EVOLUTIONARY MODELS

Hammond, Allen L.; *Science*, 175:868-870, 1972.

Even more disturbing to many geochemists is evidence that the moon had a magnetic field, and hence presumably a molten iron core, throughout most of its first 1.5 billion years. Residual magnetism has been found in many of the lunar rock samples. In an experiment conducted by Paul Coleman and his colleagues at the University of California in Los Angeles, a magnetometer in a small satellite launched into lunar orbit by the Apollo 15 crew has now recorded measurable amounts of residual magnetism over much of the moon, an indication that the magnetized samples are not isolated phenomena. Coleman believes that the residual field is due to a magnetized crust, which would imply that the moon was immersed in strong magnetic field at the time that the crust was formed.

Neither the magnetic field associated with the solar nebula nor the dipole magnetic field of the earth, according to S. Runcorn of the University of Newcastle-upon-Tyne in England, could have magnetized the moon's crust because of the orientation of the field in the solar wind and the length of period during which the moon would have had to be very close to the earth. Hence, Runcorn concludes, no external field could have produced the residual magnetism, and the moon must have had an internal field from its early moments until some time later than 3.1

billion years ago (when the youngest mare samples were formed---thus the last date for which evidence of rocks formed in a magnetic field is available). As far as is known, planetary magnetic fields are generated by dynamos resulting from convection in a molten iron core. If the moon has a small iron core (an iron core larger than about 0.2 lunar radii is ruled out by the moon's known mass and amount of inertia) and the core was at one time molten, the temperature, at least in the moon's core, must have been much higher than projected in geochemical models of the moon's thermal history. Runcorn believes that the moon's deep interior was initially quite hot and that the convective process---which is a more effective means of transporting heat than the conduction that would occur in a solid core---cooled the moon rapidly to the point where convection could no longer take place, thus turning off the magnetic field.

The magnetic evidence therefore leads to a picture of the moon's evolution that conflicts with models based on geochemical considerations. It is still not known, for example, whether if the moon's core were molten during the early part of the moon's history, the outer layers could have remained cool enough so that only partial melting took place. [Excerpt]

• Lunar Seismic Phenomena

SEISMIC DATA FROM MAN-MADE IMPACTS ON THE MOON

Latham, G., et al; *Science,* 170:620-626, 1970.

Impacts on the lunar surface of the Apollo 12 Lunar Module (LM) ascent stage and the third stage of the Apollo 13 Saturn booster (S-IVB) generated seismic signals that were recorded by the seismometers installed on the moon by the Apollo 12 astronauts on 19 November 1969. The seismometers are part of the emplaced science station called ALSEP (Apollo Lunar Surface Experiments Package). Approximately 160 events of natural origin have been recorded by the seismometers during the first 7 months of operation. However, few criteria have emerged for estimating the time or location of these natural events from the study of their seismograms. Hence, the man-made impacts, whose time and location are well known from the National Aeronautics and Space Administration's tracking information, have emerged as extremely important tools for the seismic exploration of the moon.

. . . .

1) The lunar signals have extremely long duration. The LM impact signal was detectable for about 1 hour, and the S-IVB impact signal was detectable for more than 4 hours. A missile impact signal on earth would last only a few minutes at an equivalent distance. The lunar

signals build up gradually to a maximum and then decay very gradually. The LM impact signal reaches its maximum intensity approximately 7 minutes after its beginning, and the S-IVB impact signal reaches its maximum value after approximately 12 minutes. The smoothed envelopes of the signals reach maximum peak-to-peak ground displacement amplitudes of 9.5 and 75 nm for the LM and S-IVB, respectively.

2) The earth missile impact signal contains distinct phases corresponding to the arrival of various types of seismic waves. Such phases are less distinct in the lunar signals.

. . . .

The lunar seismic reverberation can be explained equally well as resulting from dispersion of surface waves or from scattering or, perhaps more likely, from a combination of these mechanisms. Scattering of surface waves implies the presence of heterogeneity in the outer shell of the moon on a scale from several hundred meters, or less, to several kilometers. Surface irregularities may contribute significantly to the scattering. The dispersion hypothesis requires the presence of a low-velocity outer zone. The presence of this zone to depths of several meters has been confirmed by measurements of seismic waves from sources associated with the LM and from Surveyor measurements. Very low absorption of seismic wave energy in the lunar material is inferred, independent of the assumed mode of propagation. This may be a consequence of the absence of fluids in the near-surface materials or low temperature or a combination of these factors.

. . . .

MOONQUAKES

Latham, G., et al; *Science,* 174:687-692, 1971.

Abstract. Although the average rate of seismic energy release within the moon appears to be far below that of the earth, over 100 events believed to be moonquakes have been recorded by the two seismic stations installed on the lunar surface during Apollo missions 12 and 14. With few exceptions, the moonquakes occur at monthly intervals near times of perigee and apogee and show correlations with the longer-term (7-month) lunar gravity variations. The repeating moonquakes are believed to occur at not less than 10 different locations. However, a single focal zone accounts for 80 percent of the total seismic energy detected. This active zone appears to be 600 kilometers south-southwest of the Apollo 12 and 14 sites and deep within the moon. Each focal zone must be small (less than 10 kilometers in linear dimension) and fixed in location over a 14-month period. Cumulative strain at each location is inferred. Thus, the moonquakes appear to be releasing internal strain of unknown origin, the release being triggered by tidal stresses.

. . . .

Since this report was written, a third seismic station has been installed near Hadley Rille as part of the Apollo 15 mission. Two moonquakes from the most active source (A_1 zone) were recorded by the three stations of the Apollo seismic network during the first two perigee periods following activation of the new station. The epicenter has been tentatively located at 21°S, 28°W, 600 km SSW of stations 12 and 14. The depth of the focus is about 800 km---deeper than any known earthquake. Unlike the earth, the moon must be sufficiently rigid at this depth to sustain appreciable shear stress. The remaining foci have not yet been located.

In addition to the repeating moonquakes, moonquake "swarms" have been discovered. During periods of swarm activity, events occur as frequently as one event every 2 hours over intervals lasting several days. The source of these swarms is unknown at present. The occurrence of swarms does not appear to be related to lunar tides, although present data are not sufficient to rule out this possibility. [Excerpt]

LUNAR DUSTSTORMS

Hughes, David W.; *Nature*, 254:481-482, 1975.

One piece of apparatus left on the lunar surface by the Apollo 17 astronauts was a three-axis microparticle detector. This was designed to study the cosmic dust environment at the lunar surface to determine the nature and extent of luna ejecta produced by meteorite impacts at other spots on the Moon, to measure possible increases in flux produced by Earth focusing and to look for possible interstellar particles. Berg, Richardson, Rhee and Auer (Goddard Space Flight Center, Greenbelt, Maryland) present preliminary results from this experiment in their recent letter to Geophysical Research Letters (1, 289; 1974). The sensors are on three sides of a cube, one looking up, the other two looking along the lunar surface, pointing 25° north of east and 25° south of west respectively. The cube stands on four legs and is about 20 cm above the level of the lunar soil. The mass threshold is typically 10^{-13}g. The apparatus is on for 76% of the time, being switched off at the height of lunar day. 1,117 particles have been so far recorded during 203 days of observation.

One fascinating result from the analysis is that the particle event rate increases when the terminator (the day-night dividing line) passes over the apparatus. This increase starts some 40 h before sunrise and ends about 30 h after, the event rate typically increasing by a factor of 100 over the 'non-terminator' rate. The sunset effect is less consistent in time of appearance, duration and quality but is definitely present. Berg et al. rule out the direct effect of solar electromagnetic radiation as a cause because the event rate goes up long before the Sun illuminates the lunar region under consideration. Particles emanating directly from the Sun can also be discounted because the phenomenon always ends when the Sun is still rising in the sky. Rigorous thermal testing of the apparatus prior to launch also rules out temperature effects. The authors conclude that they are observing lunar surface particles, moving westward at sunrise and eastward at sunset, just as if a duststorm was sweeping across the lunar surface. This storm is along the terminator line, the dust particle moving always in the direction away from the Sun.

This could be a perfect example of electrostatic particle levitation and subsequent horizontal soil transport. It has been shown that solar ultraviolet radiation will raise the surface of the Moon to a potential of 20 to 40 volts and will give rise to a layer of high-electron density and electric field intensity that extends several centimetres above the lunar surface. This field is of the order of 60 V m^{-1} and can easily support small positively charged dust particles. Moving them bodily from place to place is another problem which requires some circumstance such as deviation of the potential of dust grains from the median value, this leading to the build-up of repulsive and attractive forces.

Dust movement could also be due to variations in the electrical surface charge with position on the lunar surface. This idea was first put forward by Criswell (Geochim. Cosmochim. Acta., 3, 2671; 1972) who suggested that large differences in the electrical surface charge could build up across light-dark boundaries due to the highly resistive nature of the lunar surface. If this is the cause of the duststorms the horizontal variability of the electrical field as a function of solar zenith angle (or distance perpendicular to the terminator) could be inferred from the time variability of impact events obtain from the data of Berg et al. The electrical charge of the lunar surface, positive in the Sun and negative out of the Sun, is confirmed by the transport direction of the positively charged dust particles.

Confirmation of dust churning in the terminator zone has been obtained from Surveyor spacecraft observations of lunar horizon glow. Surveyors 5, 6 and 7 clearly detected a line of light along the lunar western horizon immediately following sunset. These observations have been discussed by Rennilson (California Institute of Technology, Pasadena) and Criswell (Lunar Science Institute, Houston) in a recent paper in The Moon (10, 121; 1974).

. . . .

METEOROID STORMS DETECTED ON THE MOON

Duennebier, F. K., et al; *Science,* 192:1000–1002, 1976.

Impacts of masses greater than about 50 g generate lunar seismic disturbances that are detected by the Apollo passive seismometer network. Thus, the moon acts as a sounding board for meteoroids and the seismic data yield information on the mass and spatial distribution of these large particles in space. In this report we describe time variations in the number of meteoroids hitting the moon. The large fluctuations observed suggest that some massive fragments are not randomly distributed in space, but are concentrated in clouds that sweep past the earth-moon system.

. . . .

It could be argued that these events represent cometary particles such as those which produce the well-known yearly showers. Cometary particles, however, are normally much smaller (less than 1 g) than the particles detected by the seismometers. Also the lunar events show no yearly pattern and, indeed, no significant periodicity. Thus, these events are classified as sporadics, or meteoroids unrelated to known groups. Hoffman et al. detected similar changes in meteoroid activity for masses less than 10^{-7}g. Their analysis of HEOS-2 satellite data suggests that some micrometeoroids (groups) have a lunar origin and others (swarms) a cometary origin.

The mass distribution also appears to vary. During the period of observation, 35 percent of the impacts observed were small by the criteria given earlier. During the three largest meteoroid storms, this percentage increased to about 58 percent. Thus the clouds appear to contain a larger percentage of small objects than does normal space.

The phase of the moon during high meteoroid activity is related to the possible orbit of the clouds. Both of the peaks late in 1974 occurred near new moon on consecutive months. Thus, the apparent motion of the cloud as viewed from the moon was toward the sun, since the seismic array is approximately centered on the front side of the moon. The peak in June 1975 occurred at full moon; thus, the apparent direction of motion was away from the sun. According to Dainty et al. most impacts are observed from these directions, implying orbits with aphelion distances between 2.8 and 5.0 A.U.

It should be stressed that a cloud of meteoroids, as mentioned here, is very diffuse and is observed only because the moon is such a large detector. Peak activity corresponds to an infall rate of only one every 3 days over an area 100 km in radius. Thus, these clouds do not appear as meteoroid showers visible from the earth.

....

APOLLO 12 SEISMIC EXPERIMENT LINKS RED LUNAR GLOW TO QUAKES

Strickland, Zack; *Aviation Week*, 57, August 10, 1970.

For hundreds of years astronomers have puzzled over a strange red glow which sometimes appears on the surface of the moon when it is nearest the earth.

The Apollo 12 lunar surface seismic experiment may have solved that ancient mystery, but at the same time created another concerning the peculiarly precise moonquakes which possibly trigger the coloration.

Signals from the Apollo 12 seismometers have been analyzed at the National Aeronautics and Space Administration's Manned Spacecraft Center showing the presence of moonquakes at the time the moon's red glow is most evident. Both events occur when the moon is at perigee about the earth, a period when tidal forces between the two is strongest.

Dr. Gary Latham of Lamont Dougherty Geophysical Observatory at Columbia University, the principal investigator of seismic activity on the moon, believes the two events can be correlated. His theory is that the tidal forces cause the surface of the moon to "pop", creating a moonquake and allowing gases trapped beneath the lunar surface to escape. These gases may be the orange-red glow which astronomers have watched for centuries.

. . . .

LUNAR RESEARCH: NO AGREEMENT ON EVOLUTIONARY MODELS

Hammond, Allen L.; *Science,* 175:868-870, 1972.

In addition to naturally occurring seismic events, six artificial events caused by the impact of spent Saturn rocket stages and lunar modules on the moon have been analyzed by a team headed by Nafi Toksoz, of the Massachusetts Institute of Technology. The times required for the seismic signals to travel through the upper layers of the moon, when combined with laboratory measurements of seismic velocities in lunar rocks, indicate that the moon has a crust about 65 km thick with sharp boundaries, at least in the Fra Mauro region. The upper 1 to 2 km appear to be composed of broken rocks and rock fragments. Below this, to a depth of about 25 km, the sound velocity appears to be similar to that in mare basalts. A second, deeper layer of the crust extending down to 65 km appears to be of different composition, with higher seismic velocities similar to those measured in plagioclase-rich materials. Seismic velocities below the crust, in what would correspond to the mantle of the moon, are still higher and seemingly do not correspond, according to Toksoz, to common rocks.

The existence of a layered crust and the apparent differences in composition in the two layers are evidence that the moon has undergone differentiation and that the layers were formed at different times. The seismic evidence thus appears to agree with the hypotheses, based on geochemical data, that a moonwide crust was formed from early melting of an aluminum-rich material and that this original crust was partially covered by the outpouring of lavas which formed the maria. [Excerpt]

POSSIBLE SIDEREAL PERIOD FOR THE SEISMIC LUNAR ACTIVITY

Sadeh, Dror; *Nature,* 240:139-140, 1972.

Striking similarities exist between the lunar seismic response of the Apollo 12 vertical seismometer as reported by Latham et al. and the response reported by Sadeh et al. in the preceding communication. The Apollo 12 station has been operating since November 1969, and moonquakes have been recorded for more than two years. The interesting feature of this seismic activity is its periodicity, of approximately 27 days. This is seen in the upper part of Fig. 1 [not reproduced], where seismic activity is plotted as a function of time. As the exact days of maximum activity are known, it can be correlated with known characteristic periods of the Moon. The anomalistic month or the time between two perigees is 27.554 days. The nodical month or the time between two nodes is 27.212 days. The sidereal period or the rotation of the Moon as seen by a fixed star is 27.321 days.

Effect of Pulsar CP1133. In Fig. 1b the dates of occurrence of the maximum activity are plotted with respect to the time of perigee. If the effect is correlated to perigee, the points should have been on a horizontal line. If the effect is due to an object outside the solar system, it should follow a sidereal lunar period. The line drawn corresponds to the lunar sidereal period and corresponds to the position in the sky of the pulsar CP1133; this pulsar is in the horizon for the location of Apollo 12, at the time of maximum activity.

The published data for 1970, as presented in Fig. 1, are not conclusive enough to distinguish between the sidereal period and the anomalous month. Unpublished data, supplied by the Lamont-Doherty group, for the 1971 events agree well with the sidereal period and rule out the anomalous month period. It can still be a nodical month period, but this period has only a geometrical and no physical meaning.

One other similarity between the lunar and Earth based experiment is the fact that when a star rises at the seismometer location, the seismic response is higher than when the star sets. This is seen in the upper part of Fig. 1 where between almost every 27 day peak there is a smaller peak and in Fig. 2 where the data obtained in Eilat are plotted. To obtain this plot, 10 days were superimposed to give a more significant analysis. It is not hard to explain this fact if the gravitational wave is polarized.

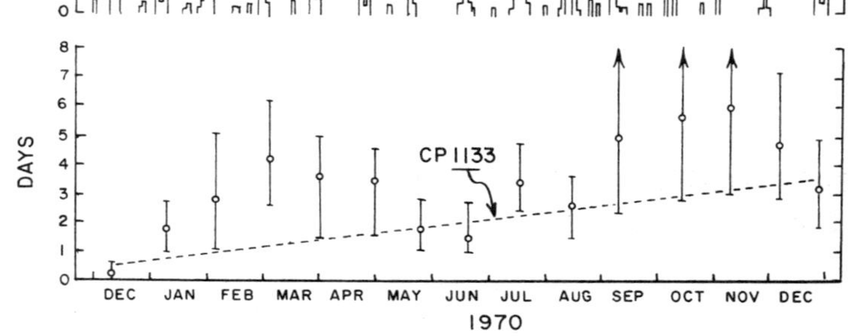

Possible correlation of moonquakes with pulsar CP1133

In conclusion, three of the predicted effects due to gravitational waves are present in the lunar seismic data, (1) A sidereal period (which even without reference to gravitational waves is an interesting observation). (2) It appears twice in a lunar day. (3) It corresponds to the position in the sky of a nearby pulsar. It has still to be proven that the period is the same as the pulsar's.

Further Proof. One other proof will be if the seismic activity for the Apollo 15 station peaks two days later than the Apollo 12 station. The individual events could be confusing because they can be genuine moonquakes. Gravitational waves should arrive at all the stations at the same time, but at a different intensity depending on the position of the star with respect to the station. One mechanism to explain why individual events are recorded can be that the gravitational waves trigger a genuine moonquake and, therefore, the integrated activity has a maximum at the right time while the individual quakes are random.

The gravitational wave interpretation of the lunar seismic response has the same drawback as the Earth based experiment; namely, the effect is orders of magnitude larger than the theory predicts.

LUNAR ORBITAL ANOMALIES

The moon has always been assumed a permanent companion of the earth in its journey about the sun, except, of course for a modest retreat ascribable to tidal friction. This permanency is now questioned by small discrepancies between theory and measurement that have cropped up. For example, when all factors are included, the earth-moon system may not be stable. Further, the moon's orbit seems to be expanding faster than it should; and some astronomers speculate that gravitational forces may be weakening. In essence, this section draws together a few observations that do not quite fit---small things but potentially far-reaching.

DISCREPANCIES BETWEEN RADAR DATA AND THE LUNAR EPHEMERIS

Smith, Carl R., et al; *Science*, 160:876-878, 1968.

Abstract. Precise measurements of the Doppler shift of radar waves reflected from Moon disclose unexpectedly large discrepancies---averaging about 0.6 centimeter per second---between the radial velocities and the predictions based on the Eckert-Brown lunar ephemeris. These residuals have a rapidly changing component corresponding to a relatively large, variable, and unexplained discrepancy in radial acceleration of about 10^{-4} centimeter per second, per second, in magnitude and about 1 day in period.

IS THE EARTH EXPANDING?

van Diggelen, J.; *Nature*, 262:675-676, 1976.

There have been many determinations of the acceleration of the Moon's mean longitude, mostly based on observations of occultations since 1680 and on records of ancient eclipses. All these results have been expressed in Ephemeris Time, based on the annual apparent motion of the Sun about the Earth. Since 1955 Atomic Time came into use based on the period of oscillation of the caesium atom which is, by definition, constant. A new analysis of lunar occultations observed from 1955 to 1974 using Atomic Time, by Van Flandern (unpublished) gives a value for the acceleration of the Moon's mean longitude differing from other determinations which used the Ephemeris Time scale. His new results differ from earlier work. The data reveal a difference between Ephemeris Time and Atomic Time of 25 s a century. Van Flandern suggests that the most probable explanation is a decrease in the gravitational constant, G, by 8 parts in 10^{11} per yr. Such a decrease would mean that the Earth is expanding somewhat and there is some geological evidence supporting this. We tried to check this conclusion using a method for determining the lengthening of the day verified by Wells in which the number of tiny lines of ridges on well preserved epitheca of corals in their annual growth increments are counted. We found no evidence for any expansion of the Earth in the past 5×10^8 yr.

More extensive results have been published by Panella et al. giving the number of days in the year for different geological periods. Their oldest corals are from the Cambrian, 520 Myr ago. The Table of Panella et al. giving the number of days, T_E, in the sidereal year has been extended by us to derive the length of the day, D, in hours and the angular rotational velocity, Ω, of the Earth. By plotting the time dependence of Ω (Fig. 1) and by assuming a linear decrease, which seems fully justified, we find a value of the time derivative

$$\dot{\Omega} = (-6.88 \pm 0.05) \times 10^{-22} \text{ rad s}^{-2}$$

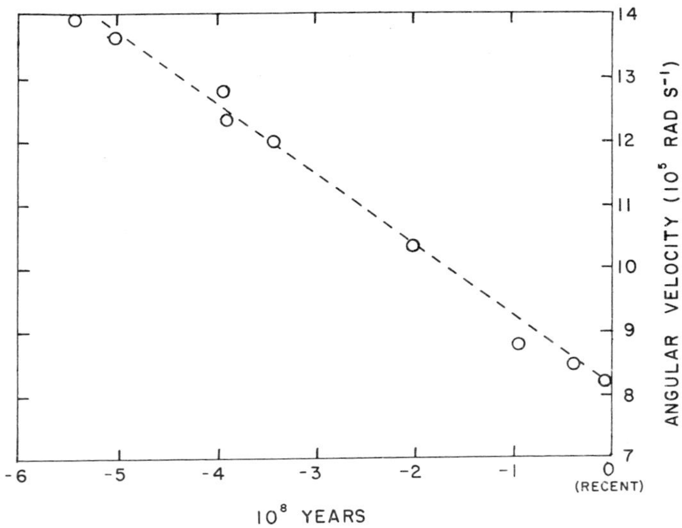

The rotational velocity of earth as derived from fossil corals over the past 540 million years. (Fig. 1)

From this it follows that

$$\dot{\Omega}/\Omega = -(29.91 \pm 0.05) \times 10^{-11}\ \text{yr}^{-1}.$$

The total angular momentum of the Earth-Moon system is given by

$$P = C\Omega + mvr + cw$$

where m is the mass of the Moon, v is the linear velocity of the Moon in its orbit, r is the distance of the Moon, w is the angular velocity of rotation of the Moon, and c is the moment of inertia of the Moon. As both $c \ll C$ and $w \ll \Omega$ we can ignore cw so that

$$P \simeq P_E + P_M$$

where $P_E = C\Omega$ and $P_M = mvr = m(GMr)^{1/2}$ where M is the mass of the Earth.

As a closed system cannot lose angular momentum, the Moon must gain the angular momentum lost by the spinning Earth. If we define the present angular momentum of the Earth as P_E 1, the present value for the Moon is $P_M = 4.964$, and $P = 5.964$. Therefore

$$P_E = 5.964 - 4.964(r/r_o)^{1/2} \qquad (1)$$

where r_o is the present distance of the Moon and we ignore the small variation of G.

Scrutton has studied other bands between the yearly expansions found on some fossil corals and he suggested that they might reflect monthly fluctuations in growth influenced by variations in the light of the Moon. Each band contained 30.6 d and the corals investigated lived 380 Myr ago in the Middle Devonian. From his 30.6 d we can calculate the length of the synodic month in hours for that time by using the length of the day, D. By using the number of days in the year we can then calculate the length of the sidereal month in the Middle Devonian and with the third law of Kepler we can easily derive from it the lunar distance r at that time.

The value of P_E in the Middle Devonian can now be found by substituting the value of r in equation (1), and from its value 6.53 x 10^{40} g cm^2 rad s^{-1}, we derive

$$\dot{P}_E = -1.76 \times 10^{40} \text{ per } 10^9 \text{ yr}$$

and

$$\dot{P}_E/P_E = (-30 \pm 3) \times 10^{-11} \text{ per yr}$$

As $P_E \simeq 0.4MR^2\Omega$, where R is the radius of the Earth, we may write

$$\frac{\dot{P}_E}{P_E} = 2\frac{\dot{R}}{R} + \frac{\dot{\Omega}}{\Omega}$$

The value of $\dot{\Omega}/\Omega$ derived from Fig. 1 is practically the same as the value of $\dot{P}_E/P_E$. The conclusion seems inevitable that

$$\dot{R}/R \approx 0$$

so that the Earth has not expanded much during the past 5 x 10^8 yr.

An expansion of the Earth has already been pointed out by Holmes and by Wesson in a discussion of continental drift and seafloor spreading. If G is decreasing, the Earth should expand because of the decreased weight of the surface layers. This idea has been developed by Van Flandern (unpublished) in his discussion of the decrease of G.

. . . .

TERRESTRIAL, INTERPLANETARY AND UNIVERSAL EXPANSION

Klepp, H. B.; *Nature*, 201:693, 1964.

A remarkably close agreement between the rate of increase of the Earth's radius and that of the universe, according to Hubble's law, was recently pointed out by MacDougall et al.

Extending this line of investigation to the recession of the Moon from the Earth, assuming the same expansion rate, Hubble's constant H= 100 km/s megaparsec^2. We then arrive at a recession rate for the Moon of 3.9 cm per annum, whereas the observed recession, found as far back as 1878 by G. H. Darwin, amounts to 13 cm per annum. This recession is usually ascribed to a tidal effect. Does perhaps part of this recession stem from the general expansion of the universe?

Both these 'local' applications of Hubble's law have already been pointed out recently by me. Hubble's law appears, in my theory, as an apparent, general expansion, due to a combined general contraction of the measuring rod and the time scale, in conjunction with a combined continuous creation of new matter and generation of gravitational fields.

EVIDENCE FOR WEAKENING GRAVITY

Anonymous; *Science News,* 105:237-238, 1974.

. . . .

Now Thomas Van Flandern of the U. S. Naval Observatory reports that he has discovered such evidence for a weakening of gravity. He told the meeting of the American Geophysical Union in Washington this week that the evidence comes from a study of the motions of the moon.

If gravity is weakening, the orbits of planets around the sun or of satellites around planets will expand, and the orbital period of these bodies will correspondingly increase. Some such expansion is provided by tidal forces in these systems, and the trick is to subtract out the tidal part and see if there is any left over.

Working with the calculations of two other observers, Van Flandern reports he has found there is an increase of four centimeters a year in the radius of the moon's orbit that is not accounted for by tidal action. "This is the first numerical result which appears to have as its most probable explanation that gravity is decreasing."

. . . .

ASTRONOMICAL EVIDENCE CONCERNING NON-GRAVITATIONAL FORCES IN THE EARTH-MOON SYSTEM

Newton, R. R.; *Astrophysics and Space Science,* 16:179-200, 1972.

Summary. The acceleration of the Moon due to non-gravitational forces has averaged about -42''/cy^2 over the past 2000 yr, and the acceleration of the Earth's spin has averaged about -28 parts in 10^9 per century over the same interval. The errors in these estimates probably do not exceed about 10% or so. In contrast, the present values of the accelerations are uncertain by a factor of more than 2. There is also a reported decrease in the obliquity of the ecliptic by about 1/4 s of arc per century that probably results from systematic errors in observations made in the 19th century. Non-gravitational effects upon ε of the size that we expect are too small to be shown by either ancient or modern data.

The parameter $\langle D'' \rangle$, which is a linear combination of the accelerations of the Earth and Moon, can be followed as a function of time with high confidence from about 700 BCE to the present. From its behavior, we are apparently forced to conclude that there was something like a 'square wave' in the non-gravitational forces that began about 700 CE and that lasted until about 1300 CE. During the time of this square wave, the accelerations apparently changed by factors of around 5.

We are seriously lacking in mechanisms to explain the non-gravitational forces. The only mechanism of tidal friction (the 'shallow seas' model) that has been evaluated quantitatively provides only about one-fourth of the necessary amount of friction, and it does not provide for much change with time within a period as short as historic times. Forces of non-tidal origin, which are of the same order as the tidal forces, may be due largely to core-mantle interactions. There are no quantitative theories of these interactions; there are only models whose parameters are uncertain within many orders of magnitude.

The reason for including this paper in this symposium is to answer this question: Can we infer the time scale of evolution of the Earth-Moon system from astronomical evidence? At present, we cannot do so with confidence, because the forces involved seem to have changed greatly within historic times, and we know of no mechanisms that can explain either the magnitude or the time behavior of the forces. Thus any attempt to extrapolate the forces from historic time into geologic time is fraught with uncertainty.

This is not a conclusion of despair. Our knowledge of the situation has increased rapidly during the past few years and should continue to increase rapidly. The answer to the question may be quite different a few years from now.

Since we need to extend research into new areas in order to understand the situation, a little speculation is in order. It is probable that we need mechanisms of tidal friction that provide more friction than the current 'shallow seas' mechanism and that provide for large changes within a period as short as 1000 yr. There is no reason to limit ourselves to only one important mechanism. Jeffreys (1968) suggests the energy dissipation in breaking waves, and their interaction with the tides, as a likely mechanism. He works out some of the necessary theory, but makes no quantitative estimates of the total dissipation. It does not seem likely that this contribution would change appreciably within historic times.

Earlier I suggested shelf ice around Antarctica as a possible source of friction. Such ice may be an efficient absorber of tidal energy, it is closely coupled to all the major oceans, and it could vary rapidly with time since it is probably a sensitive function of climate, which has changed appreciably within historic times.

There is another possibility. It was suggested in Section 2 that the interaction between tidal currents and non-tidal ones might provide more friction than tidal currents alone. Further, there is the possibility that non-tidal currents either reflect or cause changes in climate. Hence, not only is there the possibility that this interaction may provide more friction, but there is also the possibility that it may have changed rapidly with time even within historic times.

However, it is possible that tidal friction has remained constant during the past 2000 yr. If we conclude that this is the case, the data force us to conclude that the forces of non-tidal origin have changed greatly during that time. It seems harder to explain rapid changes in the non-tidal forces than in the tidal ones.

MOON, JUNE: GONE BALLOON?

Anonymous; *Science News,* 111:361, 1977.

The moon has been held responsible for inspiring everything from versification to the growth of plants. Suppose we didn't have it. Someday the earth may not, although by then perhaps we humans will have exterminated ourselves. Such is one of the possible conclusions of a study of the dynamics of the sun-earth-moon system by Victor Szebehely and R. McKenzie of the University of Texas at Austin (Astronomical Journal 82:303).

Earlier studies had concluded that the earth-sun-moon system is stable, but they had ignored certain of the smaller dynamic variables (the eccentricity of the earth's orbit and the effect of the moon's mass). Including these variables leads Szebehely and McKenzie to conclude that the system is in fact slightly unstable. Someday in the future the moon may come loose and become a planet in its own right. Alternately, the calculation may support a history in which the moon was originally part of the earth, then became detached and went into orbit and will eventually fall back.

Chapter 7

METEORS AND METEORITES

INTRODUCTION

The sun's environs are populated with bits of matter ranging in size from motes of dust to the major planets. The bits of cosmic debris at the small end of this distribution enter the scientific domain when they pass in front of the sun or moon as dark objects or fall into the earth's atmosphere to burn up as vivid streaks in the night sky. Some of the bigger meteors survive the conflagration and reach the earth as meteorites. Once in hand, these bits of presumably extraterrestrial matter can be analyzed for clues about the history and constitution of the solar system. So far, so good; but this simple scenario is marred by many observations and analyses that contradict our conventional view of the universe. We cannot ignore meteors and meteorites because, except for the Apollo moon rocks, they represent our only samples of matter beyond the atmosphere.

METEORS IN FLIGHT

Most meteors or falling stars appear as silent, far-away streaks of light that die out high in the atmosphere. The bigger meteors penetrate farther into the atmosphere with fiery displays and considerable noise. The anomalous meteors are those that are erratic in flight, that move too fast or two slowly, or are bunched in space and time in patterns we cannot explain. In particular the problem of stationary meteor radiants has perplexed many astronomers. Why should so many meteors streak into the atmosphere from the same points on the celestial sphere even though the earth moves? Some astronomers believe they know why, as detailed below; others still wonder. Lastly come the so-called "telescopic meteors"---objects bright or dark seen floating nonchalantly through the telescope's field of view, often in immense quantities.

Seeds and birds say most astronomers. And probably most telescopic meteors have such mundane origins but, as usual, there are deviants that undermine the most logical theories.

• Anomalous Meteors

A PECULIAR VARIETY OF METEORS

Denning, W. F.; *Royal Astronomical Society, Monthly Notices,* 45:408-409, 1884.

On the morning of April 20, at $1^h\ 36^m$ A.M., while watching the progress of the Lyrids I saw a remarkable meteor, projected apparently on the stars ε and δ of Serpens, as it rose upwards with great rapidity in the south. The meteor was about the third magnitude, but the singularity about it was its marvellous velocity and seeming nearness. It appeared to be in the air, a few yards distant, and I believe its path, extending (as it instantaneously impressed me) over some 16° on the background of the sky, must have been traversed in less than the twentieth part of a second. Of course there is a great difficulty in estimating such short intervals, but I feel confident the duration was even more transient than that assigned.

Now and then I have observed similar meteors before. They immediately strike one by their close proximity and enormous velocity. They are mere gleams of pale white light, which have little analogy to ordinary shooting stars, and suggest an electric origin, though I do not know whether the marvellous quickness with which they flash upon the eye is not to be held responsible for the sensation of nearness. They are somewhat rare, and one may watch through several whole nights without a single example, but, as far as my memory serves, I must have witnessed some scores of these meteoric flashes. I never register the paths; they are so rapid as to make but a vague impression on the retina, and the direction is necessarily much involved in doubt. As to the meteor of April 20, it ascended in Serpens from the western boundaries of Scorpio, and probably diverged from a radiant near the bow of Sagittarius. It was the only one I saw of the kind alluded to, or which could have belonged to a radiant so low in the S. E. amongst eighty-one meteors recorded during the nights of April 18, 19, and 20, 1885.

I have consulted several catalogues for notices of abnormal meteors such as the one described, but in most cases there is an absence of individual notes, and I have failed to gather much evidence of the kind required. But a most excellent instance, and I think the only one, is referred to in Col. Tupman's catalogue of nearly two thousand shooting stars observed by him while cruising in the Mediterranean during the years 1869-71. The particulars are:---

1870, January 9, $14^h\ 59^m$; mag. 3; path from $169^\circ + 20^\circ$ to $157^\circ - 10^\circ$; length 31°; duration 0.1 second. "An instantaneous flash; seemed to be in the air, quite near. Very curious."

It will be seen that Col. Tupman's description is very similar to my own. Though his meteor had a path of no less than 31°, he estimated the duration as only the tenth of a second.

On the whole I incline to the belief that meteors of this abnormal class give the idea of great nearness as the result of their astonishing speed, and that they will be invariably found directed from radiants close to the Earth's apex. Their appearance, however, is such as to vividly impress the observer as to their special character. It therefore seems desirable to mention the circumstances, so that workers in this department may record such further instances as they may notice, notwithstanding the uncertainty attached to the path directions of such very transient phenomena. It is just possible they may indicate a form of meteoric display essentially different to that commonly understood.

ERRATIC METEORS

Hopkins, B. J.; *Royal Astronomical Society, Monthly Notices,* 46:27-28, 1885.

As is well known to observers of meteors, the apparent paths of those bodies, as seen by us projected upon the celestial vault, are generally in the form of a curve, but sometimes they appear to travel in straight lines, while a comparative few have no appreciable path, being seen only as points of light, which are no sooner observed than they are gone.

There is, however, a class of meteors that I have occasionally observed, though never found described in the text books, which differs from those usually seen in that they travel in a zigzag or wavy path; from which circumstances and the rarity of their appearance I propose designating them "erratic" to distinguish them from ordinary meteors, which, it is perhaps needless for me to mention, they resemble in every other particular.

As I have said, these meteors are very rarely seen; I first observed one in the year 1879, and I have only noticed two since---one in 1881 and the other on September 9 of the present year. Details of these meteors I give in the following extracts from my note-books, which, together with the charts showing their course among the stars and the form of their paths, will make clear what I mean by erratic meteors.

Extracts from Note-books.

1879, Oct. 20. $10^h\ 49^m$ G. M. T. Observed a bright yellow meteor, which described a zigzag path with two bends, as shown in diagram. Appeared between α and β Cameli, disappeared near θ Persei.

1881, June 25. $12^h\ 4^m$ G. M. T. Observed a bright white meteor = ♃. Appeared just south of κ Draconis, disappeared near γ Ursae Majoris, after pursuing a wavy path.

1885, Sept. 9. 10^h 15^m G. M. T. Bright orange-coloured meteor = ♃ . Appeared near φ Persei, disappearing just north of κ Persei, after describing a zigzag path with three bends.

For the sake of clearness, the scale being so small, the bends in the paths, as shown on the charts, are greatly exaggerated; in no case did a bend extend over more than 20 minutes of arc.

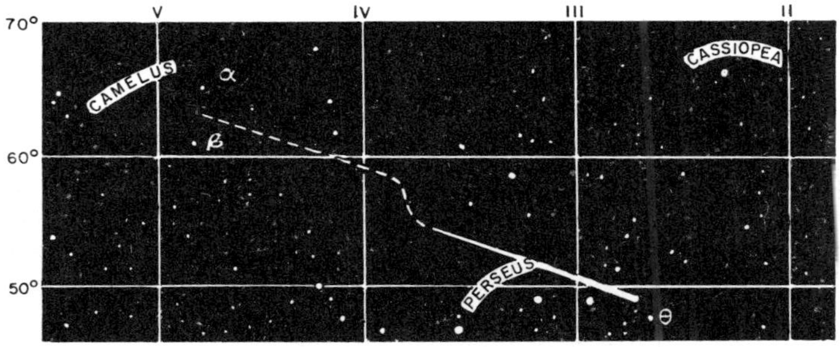

METEOR OF OCTOBER 20, 1879

NOTES ON AN ERRATIC METEOR

Hopkins, B. J.; *Royal Astronomical Society, Monthly Notices,* 47:73-74, 1886.

In November 1885 I had the honour of reading before the Society a Paper, containing a series of observations of mine of a class of meteors which from the apparent form of their paths I termed "erratic." I pointed out in that Paper that the appearance of these bodies was comparatively rare, and in proof of that statement I may here remark that, in a correspondence on the subject which I had with Mr. Denning, he informed me that though he had observed considerably over 1,300 meteors during the year 1885, only four of them described paths similar to those mentioned in my Paper above referred to.

On account of their being so rarely seen, I have thought that a few remarks upon one I had the good fortune to observe on December 4, 18 hrs., would not be without interest; particularly as the meteor presented not only the peculiarity of a wavy path, but also had that path broken in two. The meteor referred to was of a brilliant white colour, and equalled Jupiter in apparent magnitude; it made its appearance near ν Ursa Majoris, and disappeared between ι Draconis and γ Ursa Minoris. The first portion of its flight extended from the point of its appearance to just beyond ζ Ursa Majoris, where it made a slight curve, the convex side of the curve being towards that star; at this point the train of light following the meteor in its flight divided, and the remainder of its path was continued at a short distance (about 30') above the level---if I may use such a term---of thepath pursued by it at its first appearance, though parallel to it. I have on one or two previous occasions noticed breaks in the path of a meteor, but never took any particular notice of it, looking upon it as merely an optical illusion; or the phenomenon has been so slight that I have not been able to decide whether I had actually seen it, or if it was due to fancy. Such an appearance as I have tried to describe would be caused, for instance, if at the point of disappearance of a meteor another made its appearance, and so as it were continued the path of the first. But in this case, owing to the brightness and slow motion of the meteor (it was visible for two seconds), and to the fact that I was looking at the part of the heavens in which it appeared at the moment it did so, thus having it in view from first to last, I can most confidently affirm that it was not an illusion due to the cause just mentioned, or to the imagination.

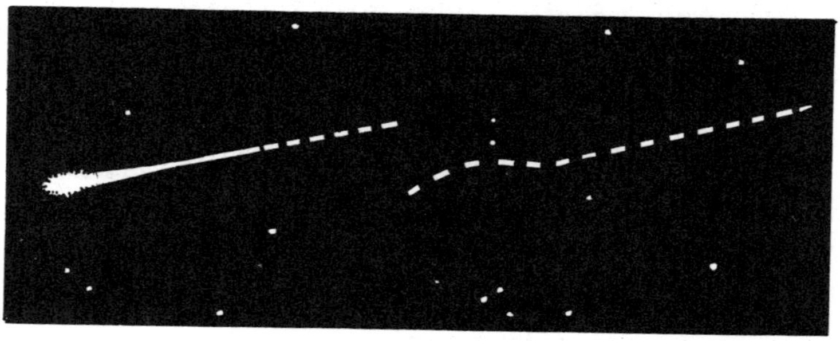

A suggestion put forward, I believe, by Mr. Ranyard, during the discussion which followed the reading of my former Paper, viz., that the irregular shape of some meteors is the reason of their describing curved paths, will probably account for the slight curve I observed in the path of this meteor. But it is not so easy to explain the break in the path. Mr. Denning has observed meteors with divided paths, and accounts for them by ascribing the appearance as due to variations in the light of the meteor; this meteor did not, however, vary in brightness, and differed from those observed by Mr. Denning in that the second portion of its path was not in a line with the first. Whatever caused the meteor to change its course seems also to have been the cause of the momentary disappearance of the meteor, and it is in hopes of some explanation of the phenomenon being forthcoming that I venture to bring this observation before the notice of the Society.

A METEOR TRANSITS THE SUN

Anonymous; *Nature*, 120:201, 1927.

Astr. Nach. No. 5505 contains a note by A. Stenzel of Hamburg, describing an observation of the transit of a round black object, some

18" in diameter, across the sun's disc from R. A. 120° (about) to 5° (about), on Mar. 15 at 15^h 25^m U. T. The time of passage was 6 seconds. The object was sufficiently distinct to make certain that it was not a distant bird. The observer notes that it is the first undoubted observation of the kind that he has made in twenty seven years of observing. A very similar transit across the moon was noted by Dr. W. H. Steavenson, a few years ago, and described in the B. A. A. Journal. The rate of apparent motion was about the same in each case. The slowness of the motion makes it probable that the distance, and therefore the size, of the object were considerable. Most meteors cross the telescopic field in a small fraction of a second.

• Telescopic Meteors

BRIGHT OBJECTS PASSING THE SUN

K., D. T.; *Astronomical Register,* 7:138–139, 1869.

When looking at Venus, just after noon on the 2nd of May, I saw a remarkable phenomenon. I call it a brilliant shower of meteors. These meteors, as I term them, were exactly like planetary bodies as seen in the telescope, varying in size and brightness, and some of them appearing larger than Jupiter. Many of the larger ones were fringed on one side; the fringe appearing somewhat bluish. They shot across the field of view for about an hour and twenty minutes in an almost continuous stream, 15 or 20 being seen together, and were still passing when I left the spectacle, which I could no longer stay to witness. The time occupied by a single meteor traversing the field was probably half a second. Their course was westward. I did not observe the way of the wind.

I find the subject mentioned in the Astronomical Register (vol. 5, pp. 157, 179, &c.,) where the meteors are said to be the seeds of dandelion highly magnified. But does not the date disprove such a supposition? A few specimens of dandelion in seed may perhaps be seen at the beginning of May, but I have only seen one plant this year bearing seed. The thistle and the poppy flower in June. I cannot help think this supposition altogether fanciful. Would the seeds of the dandelion or thistle, respectively, of the form of a parachute and a shuttlecock, by any magnifying power or process, present clear, round, planetary discs; and would they move in a straight line rapidly across the field of view?

An unscientific person to whom I showed the sight while it was going on, when told that the meteors were supposed to be floating seeds, said that they were just like shooting stars in the winter. Is not this a commonsense view of the matter? Why should they not be meteors of periodic recurrence, like the November ones?

DARK OBJECTS CROSSING THE SUN'S DISK

Anonymous; *Royal Astronomical Society, Monthly Notices,* 30:135-138, 1870.

An unusual phenomenon was noticed by Lieut. Herschel while observing the Sun at Bangalore, in India, on the 17th and 18th of October last, the following particulars of which are extracted from a letter to his brother, Prof. A. S. Herschel, dated Bangalore, October 20-25th, 1869. At about noon on the 17th, while preparing to observe the red prominences of the Sun, with an equatoreal refractor of 5 inches aperture armed with a spectroscope, Lieut. Herschel first threw the Sun's image on a sheet of white cardboard, placed as a screen, to obtain a general view of any spots which might be visible on its disk. Some dark shadows were soon noticed crossing the Sun, and afterwards some light streaks beyond its border. The first were attributed to birds, and the second to sparks inside the tube; but their frequency first, and then their uniformity of direction attracted consideration, as evidently indicating that an unusual phenomenon was in progress, and a few minutes' attention showed that what were dark shadows on the Sun were luminous moving images beyond its border. The possibility of the passage of a meteoric stream having here suggested itself, the Sun's image was sketched, and a pencil was drawn across it wherever a shadow passed. In ten minutes thirty or more lines were drawn, and their accordant direction proved that it was really a continuous stream.

The clock-work was now adjusted, and a friend's assistance was obtained to move the screen, so as to keep the Sun's image fixed upon it by means of the positions of images of conspicuous Sun spots; while, whenever a shadow appeared a ruler was placed, and a line was instantly drawn in the direction which it had taken. After about ten minutes the observers changed places, and thus secured two diagrams containing forty-four and thirty-eight lines respectively, and noted the N. and S. direction by shifting the image of a solar spot in declination. The apparent sizes of the shadows were defined by three marks upon the margins of the diagrams. A similar image of the Sun being next cast with the large finder of the equatoreal, it was found that the shadows crossed that as well. On diligently looking through the finder, and also through the main tube of the telescope, the images were at length visible there, not as shadows crossing the Sun's disk, but as ill-defined passing sparks near the Sun's border. They were thus seen and repeatedly examined by both of the observers until the Sun set.

At seven o'clock on the following morning the appearances were the same, the bodies still passing in a continuous stream. Fresh drawings were made; and it was found possible to obtain views of them in the spectroscope. Soon after noon on the 18th, the following principal features, or apparent characters of the bodies, were recorded. 1. Their direction is towards about 150° E. of North, but it is almost certain that there are two streams. 2. They are not very distant. The majority of them are completely out of focus when the Sun is in focus. When focus was adjusted on a passing cloud they all appeared much better defined. 3. They are brightest near the Sun, as well as most frequent; in spite of the overpowering tendency of the Sun's light they distinctly lose their brightness as they leave the Sun, and acquire it as they approach 4. They vary greatly in size and velocity, and in distinctness of definition. As a rule the smaller they are the slower they move, and the more distinct is their form; but there are exceptions. The slower ones can be followed up, by casting loose, and may be traced several degrees from the Sun. 5. Their motion is exceeding irregular; not to the extent, however, of *in the least degree* making their average direction and velocity uncertain, but only in comparison with that regularity which is to be expected in cosmical matter. One only out of hundreds (if it was one) was rectrograde in direction. Not unfrequently, however, their path is contorted or devious. On one occasion (the most marked instance that I can recollect) one entered the field slowly in the usual direction, but on reaching the centre seemed to meet a transverse current, with which it was swept away at right angles to its former course. On the whole the motion resembled that of floating particles, subject to the influence of a mingling of many currents. 6. Their number is anything short of infinity. Fifteen or twenty in a minute were repeatedly counted across the field of view, 45' in diameter. 7. Their form is very difficult to describe. For a long time the impression was as of a half-moon moving diameter forwards, or sometimes edgeways. Then the feeling was that there were large luminous snowflakes of various sizes, the smaller ones being almost stellar in their distinctness and brilliancy. But since I tried focussing on a distant cloud they seemed with one accord to take a tangible and real shape, a kind of double crescent with a bar across, and wings or phantom-like appendages accompanying, thus---

Whatever was their shape, it had reference to the direction of their motion, and not to the Sun. 8. Their spectrum is solar. Vivid flashes from top to bottom are seen as they cross the slit."

Later in the afternoon, having by continued effort managed to see them without reflecting eyepiece or dark-glass almost up to the very edge of the Sun, with a power of 55, Lieut. Herschel almost satisfied himself that that shape which he had seen so distinctly was a delusion, and that the real shape was a disk; when light clouds, passing over, at first interrupted a perfect view of the objects, but eventually solved the question of their nature. The particles streaming by in regular direction were intensely brilliant, and many of them moved so slowly as to take many seconds crossing the field of view.

At last one of the objects *paused*, *hovered*, and whisked off, and in that instant the observer writes that he saw---

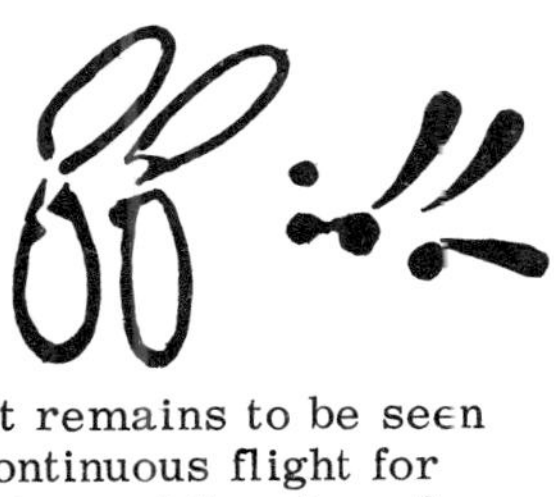

"There was no longer any doubt; they were locusts, or flies of some kind. The next morning (October 19th) they were still streaming by in hundreds in the same direction; but I paid little attention to them now, but put up the spectroscope to look at prominences. After a while they were passing the slit so frequently that I might have counted twenty or thirty in a minute. It remains to be seen if the appearances will continue. As it was, the continuous flight for two whole days, in such numbers, in the upper regions of the air, of beasts that left no stragglers is a wonder of natural history, if not of astronomy."

At the time when the above description was written, the *Homeward Mail* contained the news that countless locusts had descended upon certain parts of India. An appeal has also recently been made in the daily newspapers, stated that a famine has arisen at Jerusalem on account of the destruction of every green herb there by the devastations of innumerable locusts.

Among the appearances which may, perhaps, receive a partial explanation from the above observations of Lieut. Herschel, are some very similar phenomena recorded by the observers of the total solar eclipse on the 7-8th of August last at some of the stations in the United States of America. At Ottumwa, "about twenty-five minutes before totality, Prof. Zentmayer observed some bright objects on the ground glass, crossing from one cusp to the other of the solar crescent, as indicated by the accompanying cut by the lines from A to B; each object occupied

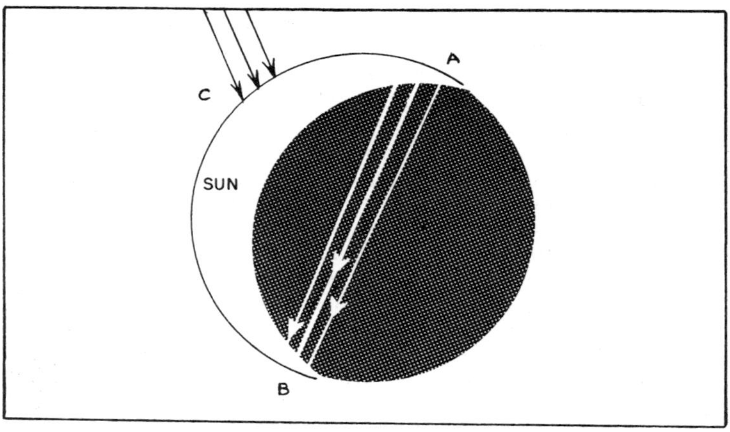

about two seconds in passing, and they all moved in right lines, nearly parallel, and in the same direction. These points were well defined, and whatever they might be, must (in order to produce such sharply defined images on the ground glass) have been several miles distant from the telescope. After calling Prof. Himes' attention to this phenomenon, and observing some eight or ten bodies in all, Mr. Zentmayer then noticed three others coming in from the limit of the field, and disappearing in the solar crescent, as shown at C; but not reappearing on the other side. It is worthy of note that the direction of motion of the three bodies last mentioned coincided with that of the wind blowing at the time, but that of the others did not; and they are thus, as also for other reasons, unlike the plant seed noticed some years ago by Mr. Dawes." Prof. Coffin also saw "meteoric bodies cross the telescope from east to west like bright flakes."

The eclipse having taken place very near the periodical date of the 10th of August, it is interesting to remark that at Vevay, Indiana, during the eclipse, several meteors were seen, at an altitude of about 45°, taking a westerly direction. On the other hand the bright objects seen near the Sun's disk by Prof. Zentmayer, and the bright flakes noticed by Prof. Coffin, were not impossibly caused by the distant passage between the observer and the Sun's disk of some winged tribe as the extraordinary flight of locusts, seen through the telescope by Lieut. Herschel, in India, in October last.

ON LUMINOUS MATTER IN THE ATMOSPHERE

Waldner, Henry; *Nature,* 5:304-305, 1872.

Much has lately been written and lectured on atoms, molecules, organic matter suspended in the air, effects of the light passing through the sky, abstracting its blue colour, and changing it into red. May I therefore be allowed to add some facts which I noticed during a long and careful observation of a hitherto almost unknown phenomenon to which my attention was drawn by chance.

Some years ago I had directed my excellent six-feet of Merz, Munich, towards the sun in order to draw the sun-spots in the camera-obscura. One day (April 27, 1863), when the sun had scarcely passed, and I was pushing the instrument to get its disc again in the field, I was astonished to perceive a mass of luminous little bodies, apparently coming from the sun, and passing altogether with great velocity towards the east. They brightened in a white and sparkling light, and were as numerous as stars; but as their velocity was much too great, and as they disappeared when I followed them to some distance from the sun, I was inclined to take them for little bodies floating in the atmosphere, and getting their light from the sun, an opinion which soon became stronger when I grew aware that I had to draw out the eye-piece some millimetres in order to get them quite clear and distinct. As I had never heard of the existence of any such bodies, I resolved to give notice to Dr. Wolf, Director of the Observatory at Zurich, who convinced himself of the strange phenomenon, and, encouraging me to persist in my investigations, told me that the late Sig. Capocci, on the Capodimonte Observatory at Naples, had mentioned these little bodies appearing to him under similar circumstances on May 11, 1845. Since that time Prof. Dr. Edward Heis, of Munster, Westphalia, in his "Wochenschrift fur Astronomie," 1869, March 24, also gave full corroboration to this fact. I therefore went on, and uniting the investigation to the daily labour of observing and drawing the sun spots, my arrangement of the camera-obscura improved and ensured these results as well. Convinced of the importance of the phenomenon, I resolved to direct my whole attention to it, and to examine it thoroughly. I decided to find out not only the distance, the size, the shape, the frequency, the velocity, and the nature of the light of these little bodies, but also to take notice of their daily direction by comparing it with the simultaneous direction of winds and clouds. I continued my observations during a period of three years.

As I mentioned above, I was obliged to draw out the eye-piece of the telescope in order to have the little objects more distinct. Now, everybody knows that the focal distance of any lens, or system of lenses, such as the telescope is, will differ according to whether the beams come from a more or less distant object. The little bodies did not appear distinct in the focus of the sun; I had to draw out the eye-piece; but if the focal distance was greater, their distance was smaller than that of the sun, and by means of a scale placed on the eye-piece, I soon obtained the result that these little bodies belong to our atmosphere, floating in a stratum of about 4,000 metres down to about 200 metres, the most numerous swarm passing almost always at a distance of not less than 500 metres. Here I remark that for my observations I had chosen the time of the sun being in, or about, the meridian, for then I was sure to have its light as strong, and the sky as clear as possible, while mostly preferring a magnifying power of only 48 diameters.

Taking the little bodies in the right focus, I was enabled not only to draw their shape, which I found very various, but also to measure their apparent diameter, which did not differ less, and depended much on distance, the nearer ones being larger, and, as I learned from the scale the accurate distance of every one, I calculated their diameter to vary from 10 to 59 millimetres, the average being 32 millimetres. Their shape was very various, too. The greater number were oblong, angular, resembling flakes, some few were orbicular, while some smaller ones were star-shaped, with transparent arms.

With respect to their frequency, I was surprised to find on certain days, especially in April and May, an incalculable number of little bodies in the field of the instrument, passing without interruption for hours. In general I found their number to be connected with the purity of the sky; and every day I noticed the average, the daily minimum occurring in the morning and evening hours, the maximum in the noon-tide hours; also the annual minimums in the summer and winter months, the chief maximum from April 20th to May 15th, the second, much lower maximum in August and September. I often saw their number increase soon after clouds had passed.

The velocity of the bodies, irregular in the lower strata, being about 2 metres in a second, became greater and more regular in the higher ones, where, for instance, at a distance of 3,000 metres, I found them to pass 8 metres during the same period, a rapidity agreeing closely with that of the cirri, which often passed at or above this distance. Whether far or near, all these little bodies glittered in a magnificent white light behind the sky, but as it retreated farther from the sun its blue colour became darker, the light of the bodies consequently diminished, and was more and more absorbed, when I followed them to some five or more degrees from the sun, in whose proximity they always brightened most, but passing over its disc, appeared to be rather dark, changing, however, suddenly into white when they emerged and entered the blue again. It became obvious that the little bodies I had before me were of small density, partly opaque, apparently of a white and reflecting surface, the edges of which were lit up by the sunbeams.

The course of the higher ones (at some 1,000 metres distance) being generally parallel, and their reciprocal velocity of about the same rate, I noticed much variety in the lower strata, where their flight was often of great inconstancy, changing their direction every moment, or falling, and second after second augmenting their focal distance, by the change of which, taken on the eye-piece scale, I learned that these bodies did not quite follow the law of gravitation, losing time; a fact not surprising to me, already convinced of their small consistency. In comparing the daily direction with the simultaneous course of winds and clouds, there was a remarkable conformity. Accepting the direction of the clouds to be the same as that of the wind in the stratum they pass through, a supposition not far from the truth, to which, of course, I was forced, having no weather-cock in such high regions, I found the direction of the little bodies and the clouds (in about the same stratum) to be (1) accurately the same in 31 per cent.; (2) differing not above 90 degrees in 49 per cent.; (3) differing not above 180 degrees in 67 per cent.; and (4) of quite opposite direction in only 1-1/2 per cent. This conformity is so evident that when the sky is cloudless, starting from the distance and direction of the ever-passing little bodies, one might easily learn the direction and perhaps the velocity of winds in the reciprocal strata, a fact of course of no little value to meteorologists and even mariners.

Taken altogether, these results could not but lead to the opinion that what I had to deal with were ice-crystals and flakes of snow. Here it may be recollected that already, in the seventeenth century, Marionette, the renowned discoverer of the law of gas-expansion, pointed out that parhelions and mock-moons are caused by ice-crystals floating in the sky; and indeed, if we consider the above results, we are forced to believe him. Firstly, we learned that these bodies belong to the atmosphere; we also found them in its lower strata. Their average size of 32 millimetres, their flake-like shape, their incalculable number, will also strongly convince us. But while the minimum during the winter months might seem rather unaccountable, the chief maximum occurring in April and May, it may be remarked that from September to March the sun, although in the meridian, does not light up so strongly the rather misty sky; and that many days the sun will not appear at all. Now, referring to the chief maximum, from about April 20 to May 15, is it not astonishing that it occurs on the very same days which, especially those of May, were at all times well known from their low temperature, and called in Germany "the Latins" (Pancratius, May 12; Servatius, May 13, &c), and were much feared by gardeners? But are the enormous masses of ice-crystals found in the atmosphere during these days the origin of its low temperature, or does the latter favour the formation of snow masses? I only mention the fact that, for instance, heat is absorbed when snow is melting, and would be happy to direct the attention of meteorologists in any country to this phenomenon, inviting contributions of facts and correspondence. Finally, the velocity of the bodies being the same as that of the clouds, their reflected magnificent white light, their regular courses in the higher regions where strong winds are generally blowing, their irregular or even falling movement and small density in the lower ones, and their very remarkable conformity of direction with simultaneously passing clouds, will give much support to my explanation.

REMARKABLE FLIGHT OF TELESCOPIC METEORS

Brooks, W. R.; *Observatory,* 7:49-50, 1884.

Prof. W. R. Brooks, who rediscovered Comet Pons last September, describes a remarkable observation, which he was fortunate enough to make whilst engaged in comet-seeking, in the following terms:---"Whilst sweeping on the evening of November 28th, it was my pleasure to observe a wonderful shower or flight of telescopic meteors, about ten degrees above the horizon and near the sunset-point. They were very small, none of them visible to the naked eye, most of them leaving a faint train, visible in the telescope for one or two seconds. The motion of most of them was to the northward, with an occasional group to the south of the Sun, moving southward. This observation occurring at the time when the unusual red-light phenomenon was at its height, the theory is suggested of a possible connection between that phenomenon and the passage of the Earth through a mass of meteoric matter more or less attenuated." Referring to the drawing given above, he adds, "As may readily be inferred, the wonderful sight is a difficult one to represent in a drawing, but I have endeavoured to give some idea of the appearance at its maximum stage. The instrument used was my nine-inch reflector with comet eyepiece giving a field of one and one-half degrees. The field shown in the drawing was a few degrees north of

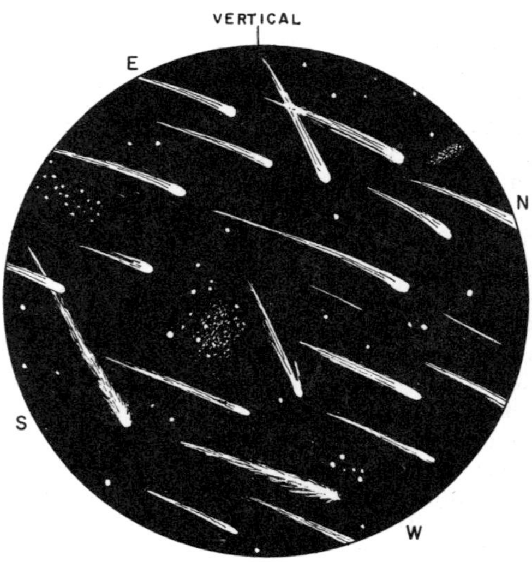

the sunset-point and about ten degrees above the horizon. The faithful comet-seeker frequently in a single night's work encounters numerous telescopic meteors singly, very rarely two at once; but this flight is quite unprecedented in my experience."

Mr. Barnard, well known as another industrious and successful comet-seeker, has noted similar phenomena during the period in which the remarkable red sunsets were most conspicuous. One of his observations is thus described in the 'Sidereal Messenger':---"So late as Dec. 15, he saw with the telescope small bright bodies close to the Sun. They were visible at the rate of five or six per minute, and were all moving to the north of east quite rapidly. Occasionally a larger body was seen to flash across the field, blurred by being out of focus. Generally they looked like little stars, many as bright as those of the first magnitude. Mr. Barnard could follow the slower-moving ones with the telescope for five or six degrees from the Sun, where they became faint and were lost. He was unable to detect any crossing the Sun; they seemed to be some distance from it, and required generally an increase of focal distance to be seen favourably. He thinks they are small particles drifting with the air-currents at considerable altitude."

ON DARK METEORS

Hopman, Frits; *British Astronomical Association, Journal,* 8:127-131, 1898.

It has already been stated in the "Journal" of the British Astronomical Association (Vol. VII., No. 2), that Mr. A. M. du Celliee Muller, at Nimeguen (Holland) was the first astronomer, who ever observed a dark meteor passing before the moon. Other scientific men, however, have claimed for themselves the priority in the discovery, and I have, therefore, thought it my duty in writing this paper to closely investigate the right of the respective claimants. It seems that five astronomers have, in different parts of the world, and independently of each other, observed dark meteors. Their names, arranged according to the date of their observation, are as follows:---

April 4, 1892. A. M. du C. Muller, Nimeguen.
July 22, 1896. Prof. W. R. Brooks, Geneva.
August 22, 1896. Gathman, America.
September 28, 1896. Ph. Fauth, Landstuhl.
February 16, 1897. O. Hoffmann, Buda Pesth.

Mr. A. M. du Celliee Muller gives in the Dutch periodical "De Natuur" (of the 15th of January 1893) the following account of his discovery:---

"On the 4th of April 1892, observing the moon with a Molteni telescope, I have seen phenomena, which according to Prof. H. G. van de Sande Bakhuyzen, have never been seen before by any astronomer. I observed little black specks or disks, which moved with great rapidity before the disk of the moon. The duration of these transits was about one-fifth of a second, and within 41 minutes seven transitions took place. When I observed similar phenomena again on the 2nd of December following, I came to the conclusion that the little specks observed by me were meteors, which, being outside our atmosphere, and consequently not incandescent, stood out black against the bright moon's disk."

Some information about Prof. Brooks's first observation of dark meteors is to be found in the "Scientific American" of the 1st of August 1896, and is to the following effect:---

A despatch from Geneva, New York, dated July 22, says: 'Prof. William R. Brooks, Director of Smyth Observatory, while observing the moon last night, made a most interesting and unique discovery. A dark round object was seen to pass rather slowly across the moon in horizontal direction.' Prof. Brooks believes that it was the passage of a dark meteor between the earth and the moon, far beyond the earth's atmosphere, so that it remained non-luminous. The observation is new in astronomical records.

We have already seen that this was not the case.

In September 1896. Mr. du Celiee Muller having meanwhile made a series of observations of dark (or as he styles them) "cosmic" meteors, published these observations in the "Mitteilungen der Vereinigung von Freunden der Astronomie and Kosmischen Physik," Jahrgang VI., Heft 8, whereupon in the next number (9) of that same periodical, Herr Ph. Fauth, at Landstuhl, Rheinpfalz, stated that he had himself seen the phenomena, which, for the first time, had been observed at Nimeguen. It seems to me though, both from Herr Fauth's own paper, and from his letter (to Mr. du Celliee Muller) that although he saw from their distance that the black specs were moving outside our atmospheric zone, he did not recognise them as meteors, and it was only in recording his observation of September 28, that he definitely named them as such; at any rate this was the first observation to that effect published by him.

Furthermore, as we have already stated, Mr. O. Hoffmann at Buda-Pesth announced that he had seen a dark meteor crossing the moon, but this occurrence, it will be remembered, took place five years after the published observations of the Dutch astronomer. Mr. du Celiee Muller's discovery was received both in Holland and abroad with a considerable amount of scepticism, which is the more remarkable as the two principal theories about meteors (those of Arago and Schiaparelli) assume the existence of numberous very small heavenly bodies, which pass through our solar system in all directions. And even Annexagoras (according to Theophrastus, Stob. Eclog. Phys., lib. I., p. 560), says, "Lower than the moon and between her and the earth there are other dark bodies, which may also produce moon eclipses." But some astronomers, and amongst them the late Prof. Krueger, of Kiel Observatory, were of opinion that Mr. Muller had seen birds passing before the moon, and indeed, not wholly without reason, for it is well known that the objects which Capocci and Nobile at Naples observed passing before the sun have been shown by the late Prof. Kaiser to have been birds. In our present instance, however, this was not the case, as will be shown hereafter Others suggested the black specks to have been plant seeds, while a third body of scientific men were of opinion that the phenomenon must have been caused by particles of dust in the eye of the observer. Subsequent observations have proved beyond all doubt that neither of the suggestions were in conformity with the truth; at the time, however, the false theories had to be refuted.

First of all, Mr. du Celliee Muller had his right eye carefully examined by a specialist, Dr. Nicolai, at Nimeguen, who stated in his testimonial that he found no dust particles in the eye. Later on the fact that Mr. Muller and Mr. van Dyk watched the same meteor passing before the sun, the image of which was projected on paper in the well-known method, showed as clearly as possible that in this case "mouches volantes" were out of the question.

On the 2nd of March, while a strong southerly wind was blowing and the clouds were drifting towards the east, one of these black spots was seen moving in a perfectly vertical direction. Therefore the object must have been moving above the regions where the upperwind was blowing. Again, at that great distance the object must have been of a considerable dimension and consequently could not have been a plant seed.

The explanation of the phenomenon by the hypothesis of birds has been refuted, amongst others, by Mr. Gouka, who, while observing the moon with a magnifying power of 200, saw a meteor which appeared as the very smallest speck. Now even the smallest bird must have been at a distance of 20,000 meters at least, to appear when seen with such magnifying power as a point without a visible diameter. This distance implies a velocity of 186 meters and a height of 12,000 meters above sea level!

So the identity of the dark meteors was proved. But now the question arose, how it was possible that a phenomenon of this kind could have remained undiscovered for so long, and indeed it seemed strange considering that the moon is closely watched year after year by so many astronomers in all parts of the globe. Mr. Th. E. J. Kramers, at Schiedam, suggested that the meteors might belong to a swarm which the earth had only recently captured and which was now rotating round her. Dr. G. J. D. Mounier, at Utrecht, remarked that since the great telescopes allow only a comparatively small portion of the moon or the sun to be surveyed, the chance of observing the transit of a dark meteor is small in the same proportion and many may pass unnoticed. Finally, Dr. Nyland is of opinion that the fact that dark meteors which formerly were never seen, are now observed in several countries, finds its cause in the turning of the lines of the nodes of the meteoric orbits, an explanation which Mr. Muller also adheres to.

. . . .

Finally I mention here a curious phenomenon, which some of the dark meteors reveal. Sometimes such a meteor appears before the disk of either sun or moon, and before it has reached the opposite edge suddenly disappears. The only explanation of the "fragmentary chords" (as Mr. Muller calls them) which has been given, is inserted in the paper on cosmic meteors, which appeared in "De Natuur." It is supposed that the meteor, while passing before the sun or moon, suddenly enters our atmosphere, becomes incandescent, and consequently ceases to be visible.

. . . .

DAYLIGHT METEORS

Coles, John; *Observatory,* 24:387, 1901.

Yesterday, 18th September, having let my sidereal watch run down, I set it to sidereal time in the usual manner, to test its accuracy, I went into my observatory and brought Arcturus into the centre of the field. This was at 2^h 15^m P. M., in bright sunshine. While I was looking at the star, a shower of meteors, which nearly occupied the whole field of vision, passed from west to east, some nearly as bright as Arcturus. I then set my driving-clock in motion, and thinking that possibly there might be something wrong with my eyesight, I called my wife, and on looking through the telescope she at once saw the meteors, but the principal shower had then passed. However, they continued to pass across the field at the rate of about twelve a minute, sometimes more, while the star remained in the centre of the field, all coming from east to west. I continued watching until 3 P. M., when the meteors had nearly ceased to appear.

I have never heard of a shower of meteors being observed in bright sunlight, but I am perfectly convinced that I am not mistaken, especially as my wife, who has very good eyesight, saw them quite plainly. My telescope, a 7-inch Newtonian, is in perfect adjustment, a Kelner eyepiece was used, and my G. M. T. is correct to within a second. As I have mentioned, Arcturus remained in the centre of the field after the driving-clock was in motion.

DAYLIGHT METEORS

Noble, William; *Observatory,* 24:418, 1901.

Like Mr. Coles, I have observed daylight meteors (?), and possibly my experience may enable him to account for the phenomenon witnessed by Mrs. Coles and himself on September 18.

Some years ago, when scrutinizing the Sun's limb---which at that particular instant I had suffered to travel just out of the field---I was startled to see a shower of luminous objects driving rapidly and nearly horizontally, which must, from the direction of their drift, have apparently crossed the Sun's disc immediately afterwards. They were notably bright; but as I was struck with the imperfection of the definition of them, and with the fuzziness of their tiny discs, it occurred to me to try to focus accurately upon them. On doing so, I found at once that the eyepiece had to be racked out, and on obtaining the true focus for my meteors (?) discovered that I was looking at hundreds of pieces of thistle-down driving on the wind, and brilliantly illuminated by the sunlight. Having once seen them clearly and sharply defined and satisfied myself as to their nature, I discontinued my watch, so am unable to say how long the shower lasted. Inasmuch as the thistle-seeds were probably less than 30 or 40 yards from the object-glass of my equatorial, and the Sun was more than 92,000,000 miles off, the necessity for a considerable change of focus will be at once apparent.

Should Mr. Coles witness a recurrence of his daylight shower, perhaps I may venture to suggest that he should repeat my experiment of focussing, not on Arcturus, but on the objects themselves.

TRANSIT OF A DARK BODY ACROSS THE MOON'S DISC

Steavenson, W. H.; *British Astronomical Association, Journal,* 31:107-108, 1920.

On 1920 Nov. 23, at about $11^h\ 0^m$, I was observing Plato with the 28-inch Equatorial of the Royal Observatory, using a power of 450. While I was watching the crater a small black object entered the field on the North side and passed nearly centrally across it in an upward direction---that is, towards the South. The field of view is about 6 minutes of arc in diameter and I should judge that the time taken by the object to cross it was between 2 and 3 seconds. The suddenness of the appearance and the rapidity of transit made it impossible to observe the object carefully enough to be certain of its exact shape, but I can say with confidence that its apparent diameter was of the order of one second of arc and that it was in perfect focus. At the time, the observation did not seem to me to be of much interest; as I had more than once seen the same sort of thing when observing with a small telescope and had always attributed the appearance to a distant bird or the fall of a small speck of dust across the diaphragm of the eyepiece. But it has since occurred to me that neither of these explanations is quite satisfactory in this case. The eyepiece was a positive one, and was focussed on the wires of the micrometer, of which it gave, of course, an erect image. It follows, therefore, that the object, if it were a speck of dust, must have been moving upwards in the plane of the wires, which seems highly improbable, especially as the body was moving with uniform velocity in a straight line. On the other hand, one cannot exclude altogether the possibility of internal currents in such a large tube.

But, if the object were not in the plane of the wires, it must have been outside the telescope altogether. In that case it is possible to get some idea of its distance, or at least of its minimum distance. The focus of the objective is 28 feet and the telescope was focussed on the Moon. Commander Ainslie has pointed out to me that, for an object 10 miles away the eyepiece would require a shift of about 1/6 of an inch.

Now it is certain that such a large change as this would put both Moon and wires completely out of focus. Indeed, half this amount would certainly do so. This being so, I think there is no doubt that the object, being in perfect focus, must have been at least twenty miles away, assuming it to have been outside the telescope.

The body, therefore, can hardly have been a bird. There remains the possibility that the object was a meteor. In that case it must have been travelling almost straight towards the telescope or its lateral motion would have been immensely greater. The position of the Moon was, roughly, R.A. $2^h\ 2^m$, Decl. + $12^o\ 48'$. Mr. Denning tells me that he has often found active radiants near this point during November, the nearest being one at $2^h\ 0^m\ +16^o$. Assuming the apparent diameter of the object to have been exactly one second of arc, it follows that, if twenty miles away, it must have been about 6 inches in diameter; but it certainly would not have been a dark object when so close to the Earth. If it was further away it must have been larger still, so that in any case the meteoric theory demands the appearance on that night of an exceptionally bright fire-ball such as would have attracted attention even in bright moonlight.

Alternative explanations may suggest themselves, such as a very large meteor moving laterally far outside the Earth's atmosphere, or even a small satellite of the Earth. Possibly no entirely satisfactory explanation could be founded on such meagre data, but it seems right that such an observation should be put on record rather than be disregarded altogether.

AN OBSERVATION OF UPPER ATMOSPHERE DEBRIS

Spiess, Eugene; *Strolling Astronomer,* 13:118, 1959.

About two years ago while pointing the 'scope toward the clear blue sky in order to measure the exit pupil with a micrometer, I was amazed to see what appeared to be white round luminous objects travel across the field of view at great speed. Thinking they were reflections or eye defects, I moved the 'scope; but this showed that they were real objects picked up by the objective lens. They would only be in the field of view from a fraction of a second to one second, and I was unable to focus on them. These objects intrigued me, and I started keeping a record of number and time. The record showed no pattern of any kind.

In order to try and resolve them, I mounted three refractors with the optical axes parallel to each other. One telescope I focused on the Sun (with suitable filters), the second I focused on a point closer, and the third was given a still closer focal setting. By moving from one eyepiece to the other quickly, I could determine which 'scope was the closer to being in focus. By experimentally focusing, I was able to resolve the objects. They turned out to be out-of-focus objects carried by the air streams at different altitudes. This evidently was the reason for the difficulty in trying to focus one 'scope on the objects seen for such a short time.

Anyone wishing to observe these objects should point the 'scope to the ecliptic from ten to twenty degrees east of the Sun against a clear sky. They can be seen either east or west of the Sun; but with the 'scope east of the Sun, there is no chance of the Sun's creeping into the field of view, with possible eye damage. You may not see any objects the first time you observe. They seem to appear sporadically in singles or in groups. Between two and three o'clock in the afternoon in a cloudless sky I have observed up to thirty-five in one hour. They are all shapes and sizes. One which I got in focus appeared to be an unrolled roll of toilet paper. I could observe it to be changing shape as it travelled across the field.

Perhaps if someone would calibrate his 'scope for known distances at different focal settings and measure the field of view, he could calculate the speed of the jet streams in the upper atmosphere. These objects are interesting to observe so take a look sometime.

• Stationary Meteor Radiants

THE LONG DURATIONS OF METEORIC RADIANT POINTS

Denning, W. F.; *Nature,* 31:463-465, 1885.

. . . .

The long visible duration of a large number of radiant points of shooting stars is, it must be confessed, a fact which defies satisfactory explanation. The ingenious theory which had attributed to meteor streams an identity with cometary orbits, required that the visibility of such streams should be of very brief character, though in the case of an abnormally wide system or of a shower directed from a point near the earth's apex the duration might be longer than usual, but the radiant point could not maintain a perfectly fixed position amongst the stars. This general view of the subject is, however, not accordant with the results of recent observations, for while there are undoubtedly some cometary showers which display all the peculiarities taught by theory, there are many other streams which continue visible for several months and retain a stationary position in the firmament. It is evident therefore that these streams are presented to us under totally different circumstances as regards orbit to the true planetary showers, and are amenable to conditions and laws which form a problem the solution of which is arrested by no ordinary difficulties.

The multiplicity of streams would naturally originate a false appearance of long duration in certain radiant points, but observations of very precise character would soon show that the point of radiation, as successively determined, differed considerably, being not, in fact, confined absolutely to the same point in the sky. But it is now proved that there are no differences, other than those introduced by small unavoidable errors of observation, in the centres from which shooting stars continue to fall during several months. Indeed, it seems a probable inference from the observations that some showers exist all the year round, though not visible during the epoch when they are very near to the sun.

That such long enduring radiants of meteors can have a community of origin and belong to physically associated streams in the same degree as the true cometary meteor showers is very difficult to understand. But the fixity of the radiant over so long an interval would yet seem to indicate some bond of close affinity existing between them. At any rate we have no reason to suppose that a large number of showers, distinct in themselves, can occur consecutively from the same points of the sky owing to a common peculiarity of grouping.

At intervals of six months the earth's motion in space is in exactly opposite directions, and yet these streams of meteors enter the atmosphere from the same apparent radiants. Evidently therefore the meteoric particles, which individually move in parallel flights, are travelling independent of solar attraction and are presented to us under a totally different aspect to the cometary showers the phenomena of which are clearly understood.

If meteoric streams of great width are encountered by the earth as the result of the sun's proper motion in space then it would appear that to give the phenomena of stationary radiants they must move with enormous velocities. This is not borne out by the observations, for the meteors of these long-enduring streams exhibit appearances similar to what is generally observed in the meteors from the cometary showers. The farther the radiant is removed from the earth's apex the slower become the motions of the meteors, they lose the streak-generating capacity, and their colour changes from white to yellow or red, indicating a lower degree of incandescence as the result of a less violent friction with the atmosphere. There are exceptions, however, for the meteors from some radiants retain a velocity much greater than that theoretically assigned.

There is a very pressing need for further observations specially directed to the visible trajectories of shooting stars. The apparent motions of the corpuscles belonging to a stream depend upon several conditions which are very liable to originate discordances. The particles near the radiants move slowly in short courses owing to foreshortening, and when the radiant is near the horizon the flights are longer and more gradual than when it has reached a considerable altitude. The Geminids of December, for instance, appear very slow in the early hours of the evening, but in the morning their swift, diving courses would lead the observer to attribute them to an entirely separate family were it not that the radiant occupies an identical place to that determined from the slower meteors recorded several hours before. Some showers also doubtless furnish meteors which become igneous at greater distances than others and more relatively slower than those belonging to streams formed of materials not so readily combustible. Moreover the specific gravity of the particles of different systems probably varies to some extent, and their individual forms may not always coincide, so that the effects of atmospheric resistance must necessarily introduce peculiarities in the observed flights.

The idea occurred to me that these long-enduring radiants must result from terrestrial meteor streams, i.e. streams revolving around the earth in an excentric orbit with perigee near the outer limits of the atmosphere. If streams of this character existed and were closing in upon the earth we should have the phenomena of stationary radiants. And the fact of their closing in upon us would be rendered possible on the assumption of a resisting medium (similar to that affecting the motion of Encke's comet), or that at each return to perigee the atoms encountered the tenuous outer region of the atmosphere, which, though not sufficiently dense to render them incandescent, would slightly diminish their velocity and thus bring about a contraction of the orbit. But there are difficulties to the adoption of such views, one of which is that the meteors from such streams would exhibit a consistency of velocity whatever the relative position of their radiants with regard to the earth's direction of motion, and this does not accord with the facts of observation.

The earth's atmosphere probably extends in a barely appreciable degree a much greater distance than ordinary estimates allow. The computed heights of certain meteors deduced from multiple observations, and the phenomena of minute, telescopic shooting stars, which are evidently far exterior to ordinary naked eye meteors, render this highly probable. The former are very numerous, though to what degree is only known to those who have been habitually engaged in sweeping the heavens with a telescope of low power and large field.

. . . .

THE STATIONARY RADIATION OF METEORS

Denning, W. F.; *Observatory,* 36:334-339, 1913.

More than 40 years ago Mr. R. P. Greg, of the Luminous Meteor Committee of the British Association, found that certain showers continued visible much longer than theoretical considerations warranted. In 'Monthly Notices,' xxxviii, p. 352, he says:---"In not a few cases the duration would seem to extend over two or three months without any special intermission." Schmidt, Heis, and others had also obtained results which seemed to support long duration, but I think that many of the radiants derived from observations made 40 to 70 years ago were very uncertain. Mr. Greg adopted a system of averaging bordering showers, and the errors of observation were such that it was impossible to say whether or not radiants were long continued in exactly the same positions.

Soon after meteoric observations had been entered upon at Bristol, I was struck with two things, viz., the motion of the Perseids and the recurrence of a great many radiants from the same apparent positions during several months. I thought at first that accidental coincidences would explain the matter, but on closer observation found that certain centres became again and again the focus of meteoric flights, while the space immediately surrounding them remained absolutely void. This would not occur in the case of successive though different showers, which would be seen scattered irregularly around a star and would not retain a similar place relatively to it.

I gave some particulars of my observations in 'Monthly Notices,' xlv. pp. 93-116. Nearly 30 years have passed since that paper was published, and from the materials gleaned during the interval I see no reason to alter my views. In fact, the conclusions formerly expressed have been very strikingly corroborated. I regard the persistency of certain radiant centres not only as an observational fact, but as a most remarkable and significant feature. It is not one easy to recognize in a convincing fashion because the showers are very feeble and a large number of accurate observations are required. In order that my ordinarily-determined radiants might have adequate support, I have obtained about 570 radiants by double and multiple observations of meteors, and further data by collecting records of stationary meteors. But I do not regard the latter as always affording reliable evidence of the exact place of a radiant, because so-called stationary meteors are often very short paths near the radiant but sometimes several degrees distant.

Various ideas have been put forward to explain stationary radiation, but no really satisfactory theory has yet been formulated. Greg, Proctor, Turner, A. S. Herschel, Callandreau, Bredichin, W. H. Pickering, and others have discussed the subject, and various astronomers have given a direct negative to the fact that fixed radiants do or can exist at all. It was pointed out many years ago that the meteors of a stream having parallel directions must constantly undergo changes in their apparent radiant from night to night, as exemplified in the case of the July-August Perseids, which move 1° diurnally to the E. N. E. A few astronomers, in fact, recognizing the insuperable difficulties in explaining stationary radiants, reject them as unproven and suspect them as brought about by many succeeding but distinct showers from directions forming the same apparent radiant during long intervals. But such sceptics wilfully neglect the very real and tangible evidence which supports the fixed radiation of meteors. It is a thing that lives indelibly in the sky, and though previous attempts to account for it are not generally accepted, no observations have succeeded or will succeed in obliterating it from the firmament.

In the future there may be some who will dispute its occurrence, but I think that when one or more really master hands at meteor-observing apply themselves sedulously to search out the behaviour of meteoric radiants, then the fixity of many positions must essentially be substantiated. It is not, however, every observer who will possess the natural aptitude for the work to attain the necessary accuracy, even if he has the patience and application to gather a great number of observations. Not three or four observers in fifty may be able to get radiant points to within 1° or $1\text{-}1/2^{\circ}$, but many will average errors of 3° to 5°, and not a few will even require wider limits than this. There is a very great difference in the capacity of individuals in this field of work.

Had I not observed with my own eyes the long duration of many showers, and obtained evidence of a character sufficiently striking to demonstrate it, I should not have accepted this feature, but would have considered it as arising from chance coincidences. I am not at all surprised at the scepticism still existing with regard to the fact.

Obviously there must occur amongst the almost innumerable swarms of meteoric systems which encounter the Earth many suggestive agreements in the radiant positions. In numerous cases it must be difficult to say where one shower begins and another ends. And there is the difficulty in determining the exact centres of the radiants; the probable and possible error is about $1\text{-}1/2^{\circ}$ for the most accurate observer, and it is sometimes difficult to say for certain that one radiant is the continuation of another, because they are often very tenuous and there is not identity, if there is close contiguity in the deduced positions. However, allowing for the small error of determination, the radiants come out the same. The significant point is that the positions are grouped very close to one point, while outlying areas show no radiation (see 'Monthly Notices,' xlv. pp. 106-112).

The meteors from the stationary radiants vary their velocity according to their direction relatively to the Earth. When the radiant is at or near the Earth's apex, the meteors are very swift and streaking. They become slower as their distance increases from the apex until, when the radiant six months later is at the antiapex, they move very slowly---in fact the velocities seem to range from about 45 to 8 miles per second, consistently with that of the well-known cometary streams of meteors. But there are some curious exceptions, and more critical observations are needed.

Mr. R. P. Greg tried to explain the long duration during a period of three months, and gave a diagram in B. A. Report, 1867, p. 413. In 'Monthly Notices,' xxxviii. p. 353, he says there must be very wide or diffused streams, and "these not improbably coincident with a considerable degree of parallelism in the direction of the orbits of the Earth and meteors at the time."

When the subject was brought before the R. A. S. in 1883 December, Mr. Ranyard and Mr. Proctor thought the only explanation must lie in meteors of very great velocity (600 or 800 miles a second), but this did not accord with observation.

Mr. Proctor gave it as his idea that certain hyperbolic comets and systems of enormously swift meteors might have been originally ejected from "giant suns like Sirius" ('Monthly Notices,' xlvi. p. 405). But, as Prof. Herschel remarked, had such high-velocity meteors produced "the phenomenon of meteor-streams with radiant-points sensibly fixed, and independent of the Earth's motion and position in its orbit, there can be no doubt that the enormous speed of such meteors would long ago have been detected."

Prof. Turner gave an ingenious explanation in 'Monthly Notices,' lix. pp. 140-151, and thought that the Earth's accumulating perturbation upon meteors passing very near it may have induced the fixed radiants, but this theory required the maintenance of equal velocities, which is not observed. Prof. Herschel also gave his views in the same number of the 'Monthly Notices.'

It is not proposed in the present paper to adduce more than a few examples of stationary radiation. A few typical instances may suffice. Thus there are two showers in Perseus, one between α and β Persei, the other near ϵ Persei, which I have pretty fully investigated. There are well-defined radiants at the places given on the dates mentioned:---

48°.0	43°.7.	61°.6	35°.7.
July 16	July 31	Aug. 10	Sept. 7
20	Aug. 4	11	8
23	7	22	9
24	10	25	15
25	13	Sept. 3	16
27	14	4	17
28	16	5	18
30		6	

Also in Sept., Oct., Nov., and Dec.

Also in Oct., Nov., and Dec.

[A much more extensive tabulation is omitted here.]

STATIONARY RADIANTS

Pickering, W. H.; *British Astronomical Association, Journal,* 25:389-391, 1915.

It is hard to see why there has always been so much trouble about understanding the general theory of the so-called stationary radiants. They must necessarily occur, and had they not already been discovered, could have been predicted from theory. It is true that they are not really stationary in a geometrical sense, but they are sufficiently so to deserve their name, their motion being very slow. The meteors should appear in the same place for several days, or even weeks, at a time, and a few months later may reappear in practically the same location. This is precisely what they have been observed to do. These radiants are by no means of rare occurrence; in fact, they are the rule rather than the exception.

Mr. Davidson discusses the subject in the April number of this Journal, and with what he says I quite agree. The geometrical principle is very simple, and may be explained in a few words. Meteors whose orbits are nearly perpendicular to that of the Earth, or which are but slightly inclined to it and moving in the opposite direction, cannot be stationary. Examples of these two cases are the Perseids and Leonids, which form perhaps the two best-known examples of shifting radiants. On the other hand, all meteoric showers lying near the plane of the ecliptic, with direct motion, must necessarily be stationary or nearly so, and these form the great bulk of all the showers known.

In the figure let S represent the Sun, the circle PB the Earth's orbit, and the three ellipses three meteoric orbits. Let us first consider the middle one. If at the point of its intersection with that of the Earth we construct a parallelogram whose lower side will represent the speed of the Earth, and its diagonal that of the meteor, then by the principle of the parallelogram of motions, the intersecting meteor would seem to be approaching us from the direction C. If we take A as at longitude 0°, then the longitude of the radiant of the meteor will be the angle CDA, or a little more than 90°.

Let us now consider the smallest ellipse, tangent to the Earth's orbit at B. Since the meteor striking the Earth at this point will be at its aphelion, it will be moving more slowly than the Earth, which will overtake it. The longitude of its radiant will therefore be 90°. Let us now consider the largest orbit. Since the meteor reaching the Earth at P will be at its perihelion, it will be moving faster than the Earth, and will overtake it. The longitude of its radiant will therefore be 90°.

In short, no matter where the meteor's orbit intersects that of the Earth, its radiant will have approximately the same longitude, 90° plus that of the aphelion of its orbit. Every meteor whose aphelion had the

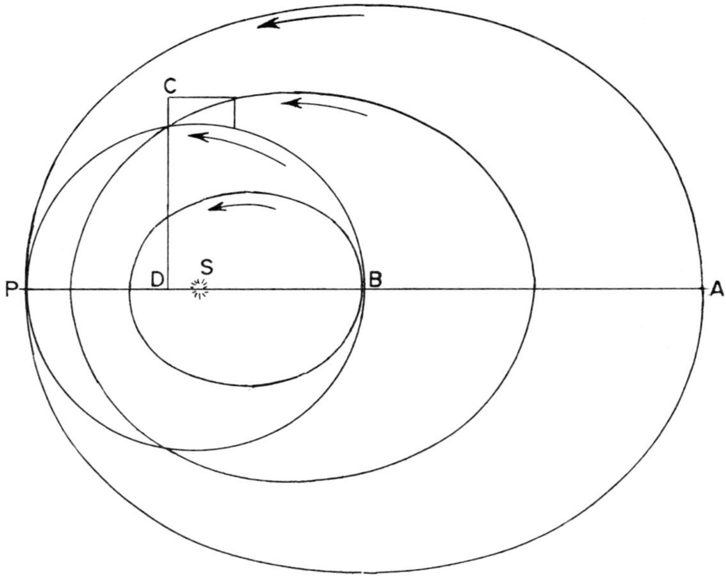

same location on the sphere will have the same radiant. It is true the coincidence is not quite exact, but the deviation in either direction amounts to only 16 in extreme cases, and the attraction of the Earth by perturbing the path of the meteor will nearly always tend to diminish this by several degrees. An increasing in the eccentricity of the orbit will make but little difference in the result, since by a curious coincidence the effect of the increase in speed attained will be practically neutralised by the change in the angle between the two orbits at their intersection.

Meteors intersecting the Earth's orbit at 140° from B will have stationary radiants for several weeks at a time. For meteors intersecting at 90° from B the motion of the radiant will be rather slow, but they will reappear from exactly the same radiant six months later.

. . . .

• Unusual Distributions of Meteors in Time and Space

METEORIC TIME-OF-FALL PATTERNS

Millard, Hugh T., and Brown, Harrison; *Icarus,* 2:137-151, 1963.

Abstract. The yearly and monthly time-of-fall patterns for all meteorite falls and for certain types of meteorites were examined for departures from randomness. The histogram of falls per decade contains at least three peaks and a sharp decrease after 1940 which is believed to reflect a decrease in the rate of influx of meteorites. The yearly fall pattern of the calcium-rich achondrites parallels the peaks in the falls per decade histogram very closely. The only monthly fall pattern which appears to depart greatly from randomness is that for the veined-olivine-hypersthene chondrites.

FEW LEONIDS EXPECTED

Anonymous; *Nature,* 224:639-640, 1969.

Meteor showers are caused by streams of dust particles released by the disintegration of comets, and the explanation of the periodicity in the activity of the Leonids is a concentration of particles at a particular point on the orbit. It must be a small concentration, however, because the magnificent shower of 1966, which was widely observed across the United States made little if any impression over Europe. It is also odd that the recurrences of 1899 and 1932 were weak compared with those in 1799, 1833, 1866 and 1966. At the time the accepted reason was that the stream had been perturbed by the planet Jupiter, but it is hard to see how the stream could have been perturbed back again by just the right amount. An alternative explanation, equally difficult to accept, is that in 1899 and in 1933 the peak of the shower occurred over uninhabited regions or was masked by bad weather conditions. Clearly the Leonid shower is nothing if not unpredictable, which is why meteor observers will be keeping watch in the early hours of November 17. (Excerpt)

THE CHANGING MICROMETEOROID INFLUX

Anonymous; *Nature,* 251:379-380, 1974.

The disparity between the flux of micrometeorites at 1 AU obtained from lunar microcrater counts and from satellite data has been known for many years. A typical example is shown on page 380, the satellite data coming from the work of Alexander (Baylor University, Waco, Texas), the data from the study of microcraters on the surface of lunar rocks and spherules coming from the work of Neukum and Schneider (Max-Planck-Institut fur Kernphysik, Heidelberg). In the 10^{-10} g region the two techniques give results which differ by a factor of 200 and it is of great importance to decide if this difference is due to some physical cause or is just an unresolved systematic error in one or both techniques. Fortunately the influx of meteoroids to the lunar surface is the same as that to the Earth's upper atmosphere (most of the satellite observations are made in near-Earth space), the circumterrestrial dust cloud, thought in the late 1960s to exist, now being discounted as a figment of incorrect callibration. But satellite-borne detectors do measure the present day influx, whereas lunar microcrater counts measure the mean influx over the last few tens of thousands of years.

To convert the crater population on a given lunar sample into an interplanetary dust flux the size and form of the craters have to be related to the mass of the causative meteoroid. One immediate problem is that crater size also depends on meteoroid velocity, density, and shape. Up to now the meteoroid impact velocity has been thought to be about 20 km s^{-1}. Two recent results indicate that it is much lower. Doppler shift observations of the solar spectral lines scattered by the solar system dust cloud give a geocentric velocity of about 5 km s^{-1} and this observation has recently been backed up by measurements of microcraters found in glass, steel and pyroxene surfaces exposed for 46 days during the Skylab mission. Nagel, Fechtig, Schneider and Neukum (Max-Planck-Institut, Heidelberg) reported these observations at the recent COSPAR Meeting (Sao Paulo, 1974). They compared the spall zones, smoothness of the central depression, fracture damage and general shape of the 47 observed craters (diameters $> 1\ \mu m$) with the form of craters produced by simulated micrometeorites, these particles having been accelerated using a Van der Graff accelerator to known velocities in the 1 to 20 km s^{-1} range. In many cases these characteristics indicated the relatively low impact velocity of about 7 km s^{-1}. So both the curves in the figure have to be moved to the right by factors varying with the velocity dependence of microphone response, penetration hole size and crater diameter. As these effects are functions of meteoroid momentum or kinetic energy the use of this new velocity determination is still not sufficient to bring the curves together.

Two other things are assumed when dealing with microcraters, first that the ratio of the diameters of the microcrater pit and the impacting particle n is 2 and second that the density of the particle is 3 g cm^{-3}. Both these assumptions need to be reconsidered; recent laboratory experiments have shown that n decreases with particle density and also that the mean particle density varies with mass. Both these factors make the interpretation of microcraters more difficult. Measurements of crater circularity have led Brownlee (University of Washington) et al. (4th Lunar Science Conference, Houston, 1973), to the conclusion that the majority of microparticles have equidimensional shapes and that shallow craters are produced by oblique incidence so the direction of the incident flux has also to be taken into account.

The main problem however lies in estimating the time the rocks

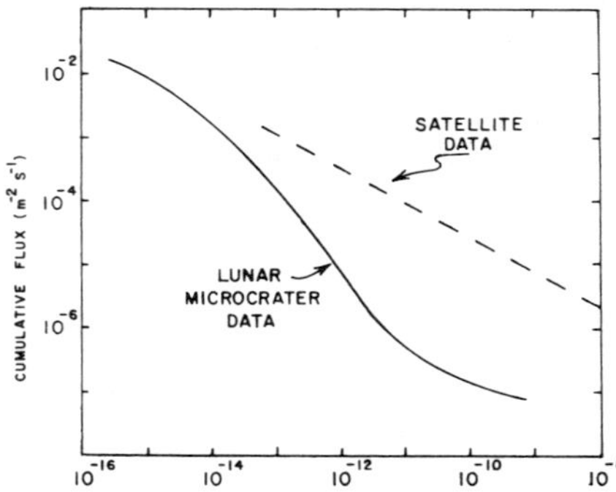

Number of micrometeoroids with mass greater than the given mass that impact on a square meter per second at 1 A.U. from the sun

under consideration have remained exposed on the lunar surface. One way of doing this is to count the tracks in the rock surface produced by retarding solar flare particles. Knowing the present day flare flux and assuming that solar activity has remained constant for about 10^6 years gives the exposure age. This second assumption is the stumbling block; the data of Neukum and Schneider could indicate that either a lower particle flux existed in the past or that the solar flare flux was higher then than now---both would produce the same result.

McDonnell, Ashworth, Flavill, Bateman and Jennison (University of Kent) overcome this problem by considering surfaces with an equilibrium distribution of craters. Laboratory simulation indicates that solar wind sputter erosion (0.043 A yr^{-1}) is the dominant factor determining crater lifetime. Knowing this erosion rate enables the microparticle influx to be derived from the equilibrium distribution. The authors conclude (COSPAR, Sao Paulo, Brazil) that the microparticle influx has increased by a factor of four in the last 10^5 years.

This would account for the discrepancy shown in the figure but creates the question of why this flux has increased.

METEOR RATES, VOLCANOES AND THE SOLAR CYCLE

Hughes, David W.; *Nature,* 252:191-192, 1974.

A world-wide increase in radar meteor echo rates occurred in 1963 and the cause of this has been puzzling scientists ever since. These echoes are obtained by reflecting short (100μs) pulses of radiation (in the tens of MHz region) from the ionised trains produced in the upper atmosphere (at heights of 80 to 100 km) by incoming meteoroid particles. The number of echoes seen in 1963 was a factor of 1.5 to 2 greater than in previous and subsequent years and this increase was recorded in Christchurch, New Zealand, Ottawa, Canada and Onsala, Sweden, so instrumental effects can be ruled out. The effect was periodic, in New Zealand occurring during the winter (June, July) and recurring on successive years with reduced magnitude. In the Northern Hemisphere the effect occurred initially in May 1963 but on following years recurred in the winter months. January-March. The Swedish data showed a rate increase for all echo duration. Canadian data however indicated that the anomalous increase was restricted to the smaller meteors. The rates were also proportionally higher throughout the day.

Kennewell and Ellyett of Newcastle University, New South Wales have reconsidered the possible causes of the rate increase in a recent article in Science (186, 355; 1974). They divide the possible causative agents into extraterrestrial and terrestrial.

Consider the extraterrestrial possibility first. There are no reasons why the influx of meteoroid dust to the Earth should not vary from year to year. For example certain meteor showers are distinctly periodic in their activity. The Leonids and Giacobinids being perfect examples. Showers are fed by the decay of the parent comet and it takes a considerable time for the new meteoroids to move around the comet orbit and form a complete, uniformly dense toroid of dust in space. The loop formation times of the Quadrantid and Perseid meteor streams are for example 330 and 4,000 years respectively. The activity of streams which are young and not completely formed is expected to vary with the periodicity of the original comet (33 and 6 years respectively for the Leonids and Giacobinids) the dust still being closely grouped around the comet. A considerable proportion of the sporadic meteoroid background ($\sim$30%) is made up of minor streams and therefore the mean periodicity of these streams might be reflected in the total influx rate. But the 1963 increase extended over a period of some months which is incompatible with the period of activity of most meteor showers, usually less than a month. Also the fact that the recurring annual variations after 1963 were 6 months out of phase in opposite hemispheres cannot be explained. And the form of the monthly mean diurnal variation remained unchanged ---this being something which would vary decidedly if a shower was active. Kennewell and Ellyett conclude that the rate in crease in 1963 was not caused by an increased particle influx.

Is the cause terrestrial? Did the properties of the upper atmosphere change in 1963, resulting in the improved detection of meteors? The two main atmospheric parameters which affect meteoroid ablation are density and scale height. Now a uniform increase in atmospheric density down to meteor heights would cause no increase in the number of meteor echoes since the entire meteor population would simply rise in height. This would change the average range of the echoes. McIntosh found a change in range of about 3 km in 160 km in 1963, a change which would have an insignificant effect on the rate. The initial radius of a meteor train varies as a function of height, also the reflected power from a train decreases exponentially with its radius. This produces a height ceiling effect, the echo height distribution being curtailed at the top. Smaller meteoroids which burn out at these great heights will be observed in greatly reduced numbers. But a simple increase in density just raises the height ceiling with the mean height, while the observed rate remains the same.

If the density gradient varies, as a result, for example, of a change in scale height the effect is more complex. First a larger train electron line density will result from a particular meteoroid thus allowing smaller meteoroids to come within the radar detection range. Second, a decrease in scale height will decrease the length of the train, making it more difficult to detect and will tend to cancel out the previous gain.

Kennewell and Ellyett propose that the meteor rate increased because the scale height around 90 km changed after the massive injection of volcanic dust into the upper atmosphere by the eruption of Mount Agung on the island of Bali (8°S, 155°E) in March 1963. This dust was mainly concentrated at a height of 20 km but also formed a minor layer in the mesosphere where, by absorbing solar radiation, it led to the heating of the immediate atmosphere, a lowering of density in the region and a subsequent increase in the atmospheric density gradient higher up. The lag between the southern and northern hemisphere rate increases is thought simply to be due to dust transport times. Kennewell and Ellyett also show that total atmospheric dust content measurements made in Australia correlate with the meteor rate and that the biennial oscillation of the equatorial stratospheric winds could inject small particles into the mesosphere on alternate years and produce a two-yearly variation in meteor rates.

A cautionary note must however be sounded because 1963 is not the only year of excessive meteor rates. As Ethiopian rainfall, thunderstorm activity and the quality of claret vary with the solar cycle why shouldn't the meteor rate? This question was answered in the affirmative by the Czechoslovak astronomer Bumba in 1948 (Bull. Astr. Inst. Czech., 1, 93; 1949). He calculated, among other things, the mean annual rate of visual meteors during the period 1844–1933 as a function of position in the solar cycle and his results are shown in the figure.

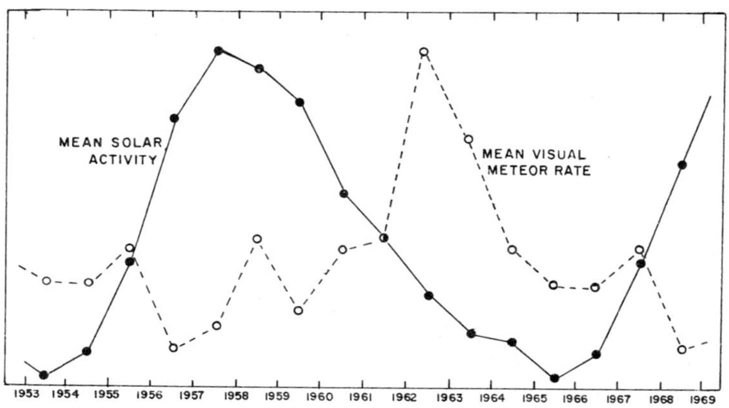

Apparent relation between solar activity and visual meteor rate

He also found that the ratio of bright to faint meteors and the mean observed geocentric velocity varies with the solar 11-year periodicity. There is an inverse correlation between solar activity and visual meteor rate, the rate maximising near solar minimum. This was also mentioned by Lindblad who found that the height of first appearances of meteors from a given shower remained around 110 km whereas the average end point rose from 85.2 km in 1956 to 96.0 km in 1963---a change of 11 km. Lindblad also found the 1953 rate considerably higher than the average.

If the increase around 1963 is just a solar cycle effect (a deduction supported by the work of Bumba and Lindblad) and not a one-off affair caused by a volcanic explosion (and it can be seen from the figure that 1963 coincides exactly with visual rate increases seen between 1844 and 1933) then similar increases should be found at other times of solar minimum, for example during 1972-73. So now is an excellent time for visual observers to peruse their records to see if this increase occurred. Also the experts in meteor ablation theory and upper atmosphere physics should try and work out why the solar cycle-induced variation in atmospheric scale height has such an effect on observed meteor rates. They could also investigate whether some unknown factor affecting meteor physics and ion chemistry varies with an 11-year periodicity. As the meteor rates are proportionally higher during the day perhaps the decrease in solar X-ray and ultraviolet flux at solar minimum have an effect.

METEOR SWARMS COLLIDING WITH THE MOON

Hughes, David W.; *Nature,* 262:175-176, 1976.

Meteoric dust particles are continually striking the Moon with velocities between 2.4 and 74 km s^{-1}. On impact they lose their considerable kinetic energy and excavate a crater thus producing a large quantity of hot ejector most of which subsequently falls back to the lunar surface but some of which exceeds the escape velocity and moves off into space. A lunar seismic disturbance is also produced. The latter can be detected by the passive seismic network deployed by the Apollo astronauts. This consists of four stations arranged around an approximately equalateral triangle with sides about 1,100 km long. The Apollo 15 seismometer makes the northern apex at Hadley-Appenines, Apollo 16 is in the east in the crater Descartes, and the west apex has Apollo 12 in Oceanus Procellarum and, 180 km away, Apollo 14 at Fra Mauro. Each station has three long period ($>$1 s) seismometers, sensitive to motion along three orthogonal directions, and a fourth short period ($<$1 s) seismometer sensitive to vertical motion. Each can detect a ground movement of about 0.05 nm.

Fortunately lunar scientists think that the seismic signals produced by meteoroid impacts can be distinguished from those caused by moonquakes. Impact events have sharper first signal rise times, indistinct shear wave arrivals and lower frequencies in the wave train. This is due to the fact that the impact waves must traverse the surface scattering zone twice between the impact point and the seismometer. Moonquakes, caused by tectonic strains within the lunar interior cross the zone only once so the energy in their seismic wave is less diffuse.

During the 924 days between January 1, 1973 and July 13, 1975 the seismic network detected 815 signals interpreted as being caused by meteoroid impacts. These have been statistically analysed by Duennebier, Nakamura, Latham and Dorman (Geophysics Laboratory, University of Texas) in their recent paper in Science (192, 1000; 1976). If an event is detected at only one seismometer it is regarded as "small". The signal amplitude is probably proportional to $E^{1/2}/r$ where E is the energy of the meteoroid and r is the distance between the sensor and the impact point, so most of the small events are caused by impacting particles with masses between 50 and 10^3 g. the larger events, detected at two or more stations, being caused by 5×10^3 to 5×10^4 g meteoroids. The authors find that a plot of impact rate as a function of time shows several sharp peaks in activity. Also, the frequency of intervals where there are no impacts is significantly more than expected---expected, that is, if the meteoroids are scattered randomly in space and the resultant impact rates follow a binomial distribution. Duennebier et al. conclude that some of the objects in the 50 to 50,000 g size range are not distributed randomly in space but occur in swarms. One of the more impressive swarms detected by the network was estimated to be about 2×10^7 km across and have a total mass between 10^{13} and 10^{14} g.

Could these peaks in the rate curve be caused by the Moon passing through a stream of particles produced by a decaying or decayed comet? Meteor showers caused by just such a process are common on Earth. The problem is that only about 10% of the mass in a meteor stream is made up of particles with individual masses over 50 g (according to Hughes, Space Res. XV, 565; 1975). A considerably smaller percentage of the sporadic Solar System dust cloud is made up of these large particles, however, so the meteor streams should stand out prominently above the background. Now 50 and 50,000 g particles impacting with the Earth's upper atmosphere at around 40 km s^{-1} produce visual meteors of magnitude about -4 and -12 respectively and an increase in flux of these very bright meteors and fireballs has been noticed during meteor showers, Meteor streams are always detected from Earth at the same solar longitude (about the same time each year) and Duennebier et al. note that the lunar events show no yearly pattern. The fact that the Moon oscillates by 3.8×10^5 km about the Earth's orbit does not help to get over this problem as meteor streams such as the Perseids and Geminids are 2.0×10^6 and 1.0×10^6 km wide. Both these streams are rich in fireballs.

It was also noticed that the mass distribution of the incident particles changed during the high rate events, the percentage of smaller objects being larger in the swarms than during the normal sporadic influx. Unfortunately this is exactly the opposite to what is observed with visual meteor showers, where the percentage of large particles increases.

Two of the maxima occurred on consecutive months, near new Moon, whereas the June, 1975 maxima occurred at full Moon. This contrasts with the timing of the moonquakes observed at the Apollo 17 site (Cooper and Kovach, Proc. Lunar Sci. Conf. 6th, 2863; 1975) which appear to be closely correlated with the 29-d lunar month and the 300 K temperature variation of the lunar surface layer that occurs between mid-day and midnight.

So these data reveal something of a mystery. Large moonquake rates result, according to the authors, from the impact with the Moon of swarms of particles with masses greater than 50 g. Should similar swarms be seen on Earth? The answer is no, because even the peak lunar activity only corresponds to an infall of one large meteoroid every 3 days over an area 100 km in radius. The converse though is puzzling. The annual meteor streams seen on Earth contain a significant flux of particles in the mass range supposedly detected on the Moon. This flux also exceeds the sporadic (non-shower) flux by nearly an order of magnitude. So why haven't these been observed in lunar seismic data? Using the influx rates for the large meteoroids which produce fireballs in the Earth's atmosphere, obtained from the American Prairie Network of all-sky cameras (McCrosky and Ceplecha in Meteorite Res., 600 (edit. by Milman, P.,) D. Reidel, Dordrecht, 1969), about 250 particles with masses greater than 50 g should hit the Moon each day. Even going to the upper limit (50,000 g), of Duennebier et al.'s mass range the rate is about 10 per day. This compares with the maximum rate of 12 events per day observed by the seismic network. If meteoroid impacts are being detected the major meteor showers should stand out clearly above the sporadic background. The fact that Duennebier et al. are not seeing these showers introduces the possibility that they are seeing something else.

ANALYSIS OF FALLEN METEORS

The overwhelming majority of fallen meteorites can be classified easily into three groups: stony, iron, or intermediate, and thence into many finer categories. But many do not fit into this pattern, such as the obviously impossible "sedimentary meteorites." Hydrous meteorites also pose problems in explanation. Could bits of the earth have been blown into space by huge volcanoes or as ejecta from the impacts of large meteorites? Such catastrophism embarrasses most scientists. Therefore, most deviant meteorites are classified as hoaxes and misidentifications. Meteorites can be dated in several ways. A few seem much younger than they should, indicating perhaps some flaw in the dating methods or in our currently accepted history of the solar system. Of all the puzzles found in the laboratory analysis of meteorites, the apparent presence of life forms or "organized elements" has created the most controversy. Here we present some historical gems from this century-old argument. It is not settled yet. It is even heating up with the recent speculations about the possible origin of life in space. It is, in truth, a classic kind of controversy seen often on the frontiers of science. The existence of the anomaly (the supposed life forms) is usually blamed upon terrestrial contamination, hoaxes, and misidentification. Of course, if life really is out there, there could be meteorites of sedimentary origin, perhaps containing fossils!

• Meteorites of Anomalous Chemical Compositions

FALL OF A METEORITE IN NORWOOD, MASSACHUSETTS

Very, Frank W.; *Science,* 31:143-144, 1910.

During the night between October 7 and 8, 1909, a meteoric stone fell to earth on the farm of Mr. W. P. Nickerson, of Norwood, Mass. The meteorite is a ham-shaped mass of very hard gray stony material, much corrugated on the surface, about two and one half feet long in its greatest dimension, one foot to nearly one and one half feet broad, and varying from one foot to one half foot in the third dimension. I estimated its volume as about 1.75 cubic feet, its weight as perhaps 275 pounds, and its density as not much over 2.5. The material has a flow structure, like that of an ancient lava which has solidified during flow, but is completely crystalline. It is, therefore, entirely different from any meteorite on record. The stone is about as hard as petrosilex, and has a slight salty odor. Laminae from 2 to 4 millimeters thick, perhaps on an average 5 to 10 mm. apart, disposed in a parallel order, project from the surface to the extent of several millimeters, resembling in this respect a much weathered piece of laminated felsite, except that there has been no chemical alteration of the superficial layer such as occurs in felsitic weathering. The laminae are distinctly parallel, their general direction transverse to the longer axis of the mass. The projections, although rounded, exhibit a remnant of crystalline form. They are in fact phenocrysts of plagioclase feldspar. Several small cavities, a few millimeters in diameter, are recognizable, but the greater part of the surface is without any pitting, other than that of the normal, and everywhere present, structural corrugation.

The bolide fell vertically through the bars of a gateway, breaking every bar and burying itself in the sand directly underneath to a depth of three feet. It was this fresh break which attracted the attention of one of the farmer's men in the early morning of Friday, October 8. The top of the stone was about six inches below the level of the surface in the interior of a cavity in the ground not much over a foot wide. The top of the stone was still appreciably warm the following morning at 7 a.m., according to Mr. Nickerson, and the bottom was decidedly warm ("hot" is the word used by the man who first felt it). A neighbor, Miss Stuart, of Westwood, in whose candor and honesty I have complete confidence, arrived at the spot just after the stone had been exhumed, handled its surface without gloves, and declares that it was so hot that she did not care to keep her hands on it very long. One of Mr. Nickerson's hired men independently told me the same. The moisture in the surrounding earth had been converted into steam which, in blowing off during its escape, had brushed off, and thus cleansed the lower surface of the meteorite---the surface of impact---which was cleaner than the upper surface, a fact which attracted the attention and surprise of the diggers who could not account for it. The sand had been so thoroughly dried that it sifted back into the hole as the stone was pried out, although the surrounding soil of the pasture was damp. The bolide passed through the bars so swiftly that the rather weak side supports were not injured. One hard wood bar was cut with a sharp fracture. Some smaller and weaker ones were more or less torn.

It seems to me probable that when a bolide succeeds in penetrating to the denser layers of the atmosphere at a very low angle, the upward elastic reaction of the air becomes so great that the meteorite rebounds, but if the angle of the path is a high one, atmospheric friction and impact retard the meteoric velocity to so great an extent that gravity gets the victory, and the last part of the meteor's fall is vertical. If this conclusion is correct, there should be some evidence that bolides which strike the ground fall more often than not in a vertical direction. I am not aware that such evidence has been sought, or especially noted. The present instance is so well authenticated, that it seems worth putting on record. Subsequent investigation has proved that the fall of the meteorite occurred at about quarter before seven o'clock on the evening of Thursday, October 7, as witnessed by several people in Norwood.

ON THE SO-CALLED NORWOOD, "METEORITE"

Hovey, Edmund Otis; *Science,* 31:298-299, 1910.

The issue of Science for January 28 contains an article by Professor Frank W. Very entitled "Fall of a Meteorite in Norwood, Massachusetts," descriptive of what he supposes to have been a meteoritic stone said to have fallen on the farm of Mr. W. P. Nickerson, of Norwood, Mass., during the night between October 7-8, 1909. On account of the specific character of the description and for fear that this may be successful in giving the "Norwood meteorite" a place in the literature, I feel that another opinion with regard to the character of the specimen should be placed on record.

I saw the newspaper account of this fall directly after its occurrence, and after correspondence with Mr. Nickerson took the first opportunity that presented itself to examine the specimen, which was then on exhibition in a "dime museum" in Boston. Mr. Nickerson himself met me there and showed me the stone. Professor Very's account of the appearance of the mass is sufficiently accurate, but his interpretation of it is entirely erroneous. As a matter of fact, the specimen is a characteristic glacial bowlder of a basic igneous dike rock, the matrix in which has been weathered so as to leave the characteristic large phenocrysts of plagioclase projecting from the surface. There is no surface indication whatever of flowage or of the skin which is characteristic of freshly fallen stony meteorites. I broke off a piece of the stone and examined the fresh fracture with the greatest care under a hand lens without finding any indication of the existence of metallic iron in the mass. Since reading Professor Very's article, I have had a thin section of my fragment made. Microscopic examination of this proves the rock to be ordinary labradorite-porphyry---a diagnosis which has been confirmed by Dr. H. S. Washington, who has called my attention to his description of this rock type from Essex County, Mass.

Mr. Nickerson told me about the broken bars of the gateway under which the mass was found and the other circumstances as related by Professor Very, but he added a statement with regard to a bright flash of light which he had noticed in the sky during the evening of October 7. His description, however, was only that of an unusually brilliant shooting star. A meteorite of the size of this specimen would surely have illuminated the region over many square miles with almost the light of day, judging from the reports of known meteorites which have been seen to fall, but no such occurrence was reported from Norwood. If the falling of a meteorite was the cause of the broken bars, the mass has not yet been found, or at any rate it was other than the specimen described by Professor Very and seen by me.

The circumstantial nature of the observations made by the several persons who had to do with digging up the "meteorite," as quoted in the article to which reference is made, are not as conclusive to me as they are to Professor Very, through scepticism engendered by the falsity of nearly all of the many reports that have come to my office during the past sixteen years in which people have described "meteorites" that they "had actually seen fall" at their feet or on the lawn in front of their houses, or in the road, or in some other very near-by place. On request, samples of some of these "meteorites" have been sent in, one of them proving to be a piece of fossiliferous limestone, another a bit of furnace slag, another a glacial bowlder of trap rock, another a glazed stone that had been used in the wall of a limekiln, another a glacial bowlder of quartzite covered with a film of limonite. The list might be extended almost indefinitely, but it is not worth while. In almost every case mentioned, the mass when found "was so hot that one could not bear his hand on it."

HYPOTHETICAL METEORITES OF SEDIMENTARY ORIGIN

Cross, Frank C.; *Popular Astronomy,* 55:96-102, 1947.

Abstract. In 1939, Dr. Assar Hadding, the Director of the Geological Institute in Lund, Sweden, described two specimens (one of limestone and one of sandstone) that he believed to be of cosmic origin. This paper briefly reviews Hadding's account and discusses three other specimens, found in the United States, that may deserve consideration in connection with the specimens found in Sweden.

One of the most momentous discoveries that could be made in the field of meteoritics would be the positive identification of a meteorite of sedimentary origin. It would offer the first tangible evidence of the existence, at least in times past, of conditions favorable to life beyond the confines of the Earth. If the specimen were composed of limestone containing distinguishable shell fragments, the proof of extraterrestrial life would be undebatable; it would prove that certain meteorites, if not all of them, are products of the disintegration of a planetary body (or possibly several such bodies) having a hydrosphere as well as a lithosphere.

In December, 1939, Dr. Assar Hadding, the Director of the Geological Institute at Lund, Sweden, presented a paper before the Swedish Physiographical Society, supporting the planetary origin of meteorites and describing two specimens---one of limestone and one of sandstone---that he believed had fallen from the sky. He related, with regard to the limestone specimen, that on Easter eve, April 11, 1925, "a beautiful meteor was observed moving towards the west across Ostergotland and the Baltic outside." The next day he was informed that the stone had fallen "near the farm of Bleckenstad, just south of Mjolby." Unable to go there himself, he immediately sent one of the assistants of the Institute, Dr. Sven Holgersson, who returned two days later, bringing a sample of the "meteorite" and a report of his investigation at the site where it was found. The fall had been observed by several persons. "One of these, farmer Oskar Gustafsson," Hadding stated, "had at once collected bits of the crumpled stone and given them to Holgersson. The ground around the place where the stone had fallen was carefully examined in search of other stones that had more similarity to known meteorites. No such were found, but only a small heap of fragments from a stone of exactly the same kind as the one found by Gustafsson. The fields around the place also were searched, and more fragments were found. They all lay in a narrow belt along the course that the stone was said to have taken.....

"Oskar Gustafsson, a respected and reliable man, whose words could not be doubted, had related that he had seen the bright body in the sky and that it was seen to fall. He was standing on the road leading up to his farm, and, in the field in front of him, some 50 meters from the road, two children were playing, his niece and nephew. Suddenly he saw the falling body sweeping over the heads of the children like a white ball and breaking against the ground. 'It looked like a newspaper that had been crumpled up into a ball. Somehow it fluttered open.' The children were scared out of their wits, the boy fell onto the ground and then rushed to his uncle crying, 'The Moon is falling down! We must go home!' Gustafsson had taken the children up to the house and then gone back in order to find out what had fallen down. He found the split-up white stone and knew at once that it was limestone. He collected some of the splinters, finding the whole thing rather curious. One thing he was convinced of. Nobody could have thrown the stone there, and, in the 25 years that he had owned the farm, he had not limed the land a single time. He was absolutely certain that the stone must have fallen from the sky."

After Holgersson returned to Lund, Hadding sought a supplementary report from Dr. Sven Zensen, of the Swedish Riksmuseum, who also had gone to Bleckenstad to investigate the fall. Apparently, Zensen found nothing either to strengthen or to discredit Holgersson's report, for Hadding observed that "there were no essential new points." All the fragments of the stone found by Gustafsson and Holgersson "were lying loosely on top of the crop and the grass in the field," Hadding stated. "They must have only just got there." In conclusion, he expressed the conviction that, "under the circumstances," the observations on the fall "could hardly have been better. The stone, or rather the splinters that were preserved, were, according to Gustafsson's account and Holgersson's investigations, the very stone that fell."

This narrative, which is quoted from an English translation published in Sweden, is followed by a description of the specimen: "a white, or slightly grayish-yellow, fine-grained mass, somewhat porous," composed of almost pure calcium carbonate. Small fragments of calcareous shell, undefinable in character, are scattered thru it. Among Swedish rocks, Hadding stated, he knew of none that bore a striking resemblance to the specimen. "It is impossible to mistake the splinters for any of them; nor can they be mistaken for burned or slaked lime from a sugar refinery, or any other product to be found in the country." The breakage of the stone, which reduced it to what Hadding called splinters, made impossible any satisfactory determination of its surface features. Some of the splinters, however, revealed a peculiar gloss, according to his statement, which he had never seen "in natural limestones or in burnt or otherwise prepared stones." He believed that the gloss may have resulted from atmospheric friction during the flight of the stone, which he designated as the "Bleckenstad meteorite."

The sandstone specimen received much less attention in Hadding's paper. "Several years ago," he stated, "I was sent, from a farm in South Sweden, a small sample of stone that was said to be a meteorite. A farmer had been walking with his wife and daughter in the garden one morning when suddenly a stone came spinning thru the air. Nobody had thrown it. It was crumpled against the garden path, but the fragments were collected and sent to me. It was loose sandstone. In spite of the assurance of the absolutely trustworthy finders that the stone was a meteorite, I presumed some kind of mistake, and the fragments were not preserved for further investigation. Alas! Others may have acted like me!"

I believe that the publication of reports such as Hadding's serves a valuable purpose, especially if the reporters use care in discriminating between credible evidence and presumption. Meteoritics is a science that can be advanced best thru a liberal exchange of information that may possibly establish the existence of meteorites of types now unknown. It should not be necessary for a researcher to withhold a description of a puzzling specimen until he feels prepared to commit himself to a positive identification of it. The decision to publish should be based upon a consideration of the value of the report to other researchers whose observations it may supplement.

Among the thousands of specimens (including many of unquestionably cosmic origin) that were sent to me by correspondents in all parts of the country, as a result of my numerous magazine articles on meteorites in the period between 1934 and 1941, were two specimens that may conceivably add at least a footnote to Dr. Hadding's observations. A third specimen, which will receive my attention first, was offered for my study several years ago by Dr. H. H. Nininger. It was discovered in northern New Mexico in the search for fragments of the Pasamonte Ranch fall of March 24, 1933, which it superficially resembles. This specimen has roughly the form of a sphere-quadrant with a radius of about 3 cm. Approximately two-thirds of the surface exhibits a thin fusion crust that may be assumed to have covered all of it before the stone was broken. Except for a narrow band of brownish-gray, which borders about 3 cm. of the broken surface, the crust is black, as a result of the inclusion of free carbon. The interior, revealed by the fracture, is a porous, dirty-gray limestone, composed of fragmentary shells. I have been unable to observe any of the shell forms that appear to me to be identifiable. It is the fusion crust that seems to me to make the specimen worthy of attention. I am aware, of course, that the presence of such a crust is not sufficient evidence to justify the identification of the specimen as a meteorite. On the other hand, the formation of a fusion crust on a specimen of limestone by processes not related to the fall of a meteorite is not an easy phenomenon to explain. Limestone does not fuse directly under heating. It is first reduced to CaO, which fuses at 2572° C., a temperature difficultly achieved by artificial means. Quartz, by contrast, fuses at temperatures from 1470° C. upward. It is difficult to understand how any small mass of limestone could be heated intensely enough to fuse a crust on it without reducing, or calcining, the entire specimen. Then, on exposure to atmospheric moisture, it would unite with H_2O to form calcium hydroxide, which would eventually revert to limestone by absorbing carbon dioxide from the air. Any limestone specimen, under ordinary circumstances, might be expected to disintegrate before these processes were completed. After calcination, if it remained intact, even a moderate disturbance would normally reduce it to a powder. The next hazard would be slaking, which would result from its first contact with water. These hazards would be circumvented, however, if the surface fusion were, in some manner, accomplished before the heat could penetrate into the interior of the specimen. Such heat is developed in the flight of a meteorite. It would seem, moreover, that the fusion must have been almost instantaneous to free the carbon from the carbonate and to involve it in the melt before it could escape as an oxide. An artificial origin of the fusion crust seems to me to be difficult to accept. Even if it could have been artificially produced, the discovery of the specimen in a barren wasteland in northern New Mexico would still be puzzling. Certainly it could not have been formed by volcanic action, and the possibility is hardly worth considering that it could have been formed by lightning.

The other two specimens---the two that were submitted to me directly by their finders---are composed of grayish, fine-grained sandstone. One was sent to me in 1940 by Mr. Leonard Fishel of Trevlac, Brown County, Indiana, who stated that he had found it in a pasture, remote from any source that might have suggested an artificial origin for the thin, greenish, glass crust that encompasses it. The specimen puzzled me. Eventually, however, my interest lagged; I returned it to Mr. Fishel and thought no more about it for several months. Then another specimen of remarkably similar appearance came to me from Miss Kathleen Auvil, of Montrose, Randolph County, West Virginia. Miss Auvil said that she had found the stone near the head of a small creek, "in a country very aptly described as the West Virginia hills." The site was 15 miles from Elkins, Randolph County, and 3 miles from the nearest village, and no traces of any pottery works or any other industry that might have accounted for the specimen were known to her anywhere in the vicinity. This discovery revived my interest and prompted me to request Mr. Fishel to submit his stone to me again for comparison and further examination. Both stones have now been in my possession for about 5 years.

Unfortunately, a considerable part of the Trevlac stone has been broken away and lost. A minor part of the breakage occurred before Mr. Fishel found the specimen; the rest was done by an experimenter who volunteered to help me ascertain the character of the stone and who apparently failed to recognize that he was destroying valuable evidence! I had submitted it to him because he had been recommended to me as an authority on ceramics and the fire-clay industry, whose opinion would be helpful in probing the possibility that the glassy crust on the specimen had been formed in a kiln. He was very doubtful that it could have been the product of any fire-clay operation. The portion of the Trevlac specimen that remains unbroken is roughly a quadrant of a sphere, or a triangular pyramid with one face broadly rounded. It has a maximum length of slightly more than 8 cm. Aside from the fusion crust, the surface features of the specimen offer no strong evidence either to support or to discount the supposition that it may have been subjected to atmospheric erosion. It exhibits no piezoglyphs and no flow lines, nor any pits or notable depressions of the type observable on many stones of cosmic origin; on the other hand, it can be fitted readily into the range of forms exhibited by stony meteorites that have not maintained a fixed orientation during their flight. The transparent crust of greenish glass that covers virtually the entire unbroken surface of the stone appears to average about 1/2 mm. in thickness, a measurement wholly compatible with the hypothesis that it may have been formed by the heat of atmospheric friction. A network of cracks in the glass indicates that it cooled rapidly after exposure to the heat that formed it, and in several places small areas of the surface appear to have sloughed off before the heating ceased. These areas show an incipient fusion such as may sometimes be observed on meteorites that have broken in flight. If the fusion of the stone proceeded far enough at any point to penetrate more than a fraction of a millimeter below its surface, no accumulations of glass remain to bear evidence of the process. Either the fusion was halted before it penetrated farther or the plastic material was removed almost as rapidly as it was formed. I am hesitant to say that the stone shows evidence that it has been sculptured to some extent by external fusion; yet the shallow, irregular undulations of the surface and the rounded edges definitely suggest such a possibility. The most remarkable feature of the specimen is, however, not the fusion crust. It is a crack that penetrated into the stone about 29 mm.,---that is, before my collaborator of the ceramics industry chipped away one face of it! Fortunately, the other face still remains, and probably enough of the crack is still accessible to examination to confirm my description of it as it originally appeared when Mr. Fishel submitted the specimen to me. On both faces of the crack the stone was completely covered by a film of glass which may still be observed on the face that remains unbroken---a film so minutely thin that it follows, in some degree, the contour of the sand grains beneath it. There is no checking or cracking of the glass. This feature, as well as the thinness of it, differentiates it from the glass that covers the exterior of the specimen. When the stone came to me, the deeper part of the crack contained a considerable amount of crushrock, or myolinite, fused barely to the point of making the particles adhere together and to the glassy faces. The evidence of crushing was unmistakable. An almost identical appearance may be obtained by placing a bit of sand between two iron plates and grinding them together. Still deeper in the stone and still observable, the crack contracts into a minute glassy vein that continues for slightly more than 1 cm. I believe that the evidence is conclusive that the two opposing films of glass were formed in the crack. The hypothesis is untenable, in my opinion, that the glass could have entered the crack from outside to coat the two faces of it without flowing together. An equal amount of molten glass introduced from the outside could have proceeded only a few millimeters before congealing, instead of spreading uniformly over the two faces in an almost microscopically thin film and penetrating to the deepest recesses of the crack to form a minute vein of glass. The presence of the crushrock seems to me to point unmistakably to the conclusion that the glass in the crack was formed by faulting under high pressure.

Small dark veins are very common in meteorites. Dr. Virgil E. Barnes, of the University of Texas, who has made a study of them, states that in composition they resemble the fusion crusts of the meteorites in which they appear. He believes that they were formed by faulting. In The University of Texas Publication No. 3945, he describes a meteorite found near Cuero, Texas, which exhibits cracks and veinlets that seem to correspond, in some degree, with the crack and vein in the Trevlac stone. "Surfaces of the crack," Dr. Barnes states, "have well-developed slickensides along them, indicating that they are fault surfaces... The filled portion of the cracks is probably a myolinite which is coherent enough to hold the meteorite together." Veins similar to the veins in meteorites have been found in terrestrial formations, Barnes states, but "only in metamorphosed rocks, and [they] must have been formed at depths of several miles below the surface and consequently under rather high pressures." No terrestrial sandstone is known, of course, to exist at such depths. As one possible origin for the veins in meteorites, he suggests that they may have been caused by the "disaggregation of a celestial body during a close approach or collision."

The sandstone that composes the Trevlac stone is extremely compact and has been partially metamorphosed by heat. It will even take a degree of polish with a fine emery cloth. The evidence is strong, however, that the partial metamorphosis did not result during the same application of heat that formed the fusion crust. There is no gradation in the degree of metamorphosis from directly under the fusion crust to the center of the specimen. It is fully uniform.

E. P. Henderson and Harry T. Davis have described a stony meteorite from Moore County, North Carolina, that reveals "very definite evidence of reheating" before it entered the atmosphere of the Earth. They do not hypothesize on the origin of the heat, except to comment that it was likely applied at the place where the meteorite originated rather than during a close approach to the Sun. An obvious possibility is that it may have been generated by the disintegration of a planetary body in some cosmic catastrophe. According to the authors of the description, the heat that caused the alteration in the specimen "was somewhere between 1090°....1510° C." A similar degree of heat would probably have caused the partial metamorphosis observed in the Trevlac stone.

The Montrose stone, while it shows a remarkable resemblance to the Trevlac stone, exhibits no feature of comparable interest to the crack in the other specimen. It is roughly a truncated, quadrilateral pyramid, with a maximum altitude of about 9 cm. and a base averaging about 5 x 6 cm. The glass crust that envelops most of it is greener and also slightly thicker than the crust of the Trevlac stone. Moreover, it shows less uniformity in various ways. Deeper shades of green in some places indicate some thickening of the glass, especially on one side, where the irregularity of the surface suggests considerable sculpturing as a result of material fused away. In two spots, more especially, the glass has blistered and broken. The principal departure from uniformity is, however, on another side where the fusion has resulted in forming a rather extensive area of slaggy glass, filled with gas cavities and unfused grains of sand. The interior of the stone, like the interior of the Trevlac stone, shows evidence of uniform heating. It is evident likewise that the heat that affected the interior of the stone was not the same heat that formed the fusion crust. If the heat that fused the surface had been continued long enough to penetrate to the center of the stone and to cause a partial alteration of the sand grains there, and if no force had been present to remove the molten material from the surface almost as fast as it fused, considerable accumulations of glass on the outside would surely have resulted. A peculiar feature of the interior of the stone is observable on the basal cross-section, which has been ground flat to facilitate examination, as a border of darker sand grains flanking two sides of the quadrilateral to a depth of about 1-1/2 cm. Toward the center of the stone several shadowy filaments diverge from the border, giving it the appearance of a spray. The border and the filaments attached to it exhibit no similarity to the concentric bands to be seen in concretionary formations.

Dr. Hadding makes the logical point that sedimentary meteorites, if they do exist, are bound to be extremely rare in comparison with meteorites of the heavier types. As the heavy magma of the Earth far outbulks the sedimentary and eruptive layer that envelops it, so would the heavy nuclear materials compose the greater part of any planet that may have disintegrated into cosmic rubble. Hadding reasonably remarks also that "we should consider a stone that looked like granite or some other known rock not as one of cosmic but as one of tellurian origin." It would be hard to identify as a meteorite and it would be even harder to tain its acceptance as one!

I wish to emphasize, in conclusion, that the purpose of my discussion has not been to argue that sedimentary meteorites do, of a certainty, exist; rather, it has been to set down certain observations that may conceivably be helpful to other researchers who may become interested in exploring the possibility of their existence. If sedimentary meteorites do exist, they should be grouped and named to distinguish them from other types of cosmic missiles already known. Even while they remain in the realm of speculation, some designation will facilitate discussion of them. I suggest, therefore, that the sandstone variety be called amathosites and that the limestone variety be known as calcarites.

HYDROUS MINERALS IN METEORITES

Hughes, David W.; *Nature*, 256:697, 1975

The water content of rocks, whether they be terrestrial or meteoritic, can provide an important clue to their place of origin. Meteoritic minerals are nearly always fresh and undecomposed---which contrasts with many of the water-containing products of weathering and erosion which occur in such profusion on Earth. By studying the chemical composition and paragenesis (the groupings of different types of minerals) of meteorites it is found that in the main they must have crystallised from a fiery melt which had an exceptionally low water content in contrast to the wet volcanic terrestrial melts which solidify to form igneous rocks. Only in type 1 and 2 carbonaceous chondrites is water reasonably abundant. These were formed at low temperatures and have not been subsequently heated above 350°C. They are thought to have compositions very close to that of the primordal dust which coalesced to form the Solar System and they contain many primary volatile components. These meteorites are thought to be parental to other types which are formed by complicated thermal evolution processes and geochemical depletion reactions.

Ashworth and Hutchison in this issue of Nature (page 714) report having found hydrous minerals in the meteorite Nakhla, a diopside-olivine achondrite which fell in Egypt in 1911 and in the meteorite Weston (a chondrite which fell near Weston, Connecticut in 1807). Both are meteoritic types in which water is rare.

The water in Nakhla and Weston is in the form of iddingsite, a mixture of hydrothermal alteration products of olivine ($(Mg, Fe)_2SiO_4$). The iddingsite is found as red-brown veins along cracks in the olivine. These veins were probably formed by shock deformation, and water present at this time percolated down the cracks and caused mineral alteration.

Ashworth and Hutchison conclude that magmatic water must have been present during the formation of this shocked olivine, and that the meteoritic mineral originated on a body in the Solar System that had a hydrous atmosphere. There is however still the possibility that the meteorites were 'dry' when they left the parent body and that they picked up the water later. The water could also have come from an admixture of some carbonaceous chondritic material in with the meteorite; this however has not been chemically identified. The interaction between solar wind ions and meteoritic minerals causing hydrogen implantation is discounted because lunar breccias, which have been exposed to the solar wind for aeons, are generally anhydrous. That leaves terrestrial weathering as the alternative source of water. Was the hydrous mineral produced by the hydrolysis of primary tawrencite ($FeCl_2$) by water vapour in the terrestrial atmosphere? As Weston contains an order of magnitude more water than Apollo 16 rocks this requires a very high initial content of chlorine. Both Nakhla and Weston were observed falls and were collected soon after hitting the Earth. However as Nininger (Out of the Sky, Dover, 1952) points out, we never see a truly unweathered meteorite, the atmosphere through which they fall being nearly saturated with water---to say nothing of the air in the vicinity of the museum shelf on which they have been sitting for over half a century.

So the mystery remains---did the meteorites come from a wet planet or have they picked up their water later? It would be heartening to think that Earth has not been the only wet place in the life of the Solar System.

THE CUMBERLAND FALLS METEORITE

Merrill, Geo. P.; *Science,* 50:90, 1919.

The stone described by Professor A. M. Miller In Science for June 6 of the present year, and of which the National Museum has secured the major portion, proves of exceptional interest. In fact, it is scarcely too much to say that it is one of the most remarkable falls yet reported on the American continent. The stone is a coarse enstatite breccia, closely compacted, showing evidences of compression while under a considerable load and other indications of its having formed a portion of a body of considerable size, even of planetary dimensions. The most striking macroscopical features aside from its brecciated structure are the occasional enclosures sometimes 4 or 5 cm. in diameter, of a dark, nearly black, chondritic stone. I do not recall another instance of so plain an admixture of stones of quite different type. Such a stone finds no exact position in the classification of Brezina. Following out the general plan, however, I have made a place for it among the achondrites and designated it a Whitleyite---a magnesia-rich stone brecciated in structure, consisting essentially of enstatite, poor in iron and carrying enclosures of a black chondrite. The results of further studies will be published elsewhere.

METEORITE WITH UNIQUE FEATURES

Nininger, H. H.; *Science,* 139:345-347, 1963.

A 1955-lb meteorite of unique structure has been recovered from a remote location in the Bondoc peninsula on southern Luzon Island in the Philippines.

The meteorite was located during our visit to Manila in 1958 through information obtained from the National Bureau of Mines in Manila. The

Bureau had received a small sample, but was not interested in pursuing an investigation. The specimen, together with meager information, was then turned over to us; since that time we have been working through a local field party for its recovery.

The first sample was a badly oxidized nickel-iron nodule weighing 684.1 grams with a specific gravity of 7 plus. Later samples from the field ranged from 3.26 to 6.35 in specific gravity with an average of 3.94. All local geologists and mining men called them "low-grade iron ore" and not meteoritic. It was felt that this was a very unusual meteorite and efforts to recover it were intensified. These were successful, and the meteorite arrived in Sedona, Arizona, on 10 August 1962.

A section measuring 71.2 by 43.8 cm was cut from near the small end of the generally oval-shaped mass. This cut showed 15 large nickel-iron chondrule-like nodules similar to those described by Brezina in the mesosiderite from Mincy, Mo. (1895). The Luzon specimen, however, is an aerolite, most of which is free of metallics. Those present are all located in an outer zone surrounding a roughly rectangular block of what appears to be pure crystalline enstatite measuring 22 by 15 cm. within which Carleton Moore and I could find no trace of metal. The zone surrounding the metallic chondrules, of which the largest measures 7 by 4 cm. seems on casual inspection, to be a mineralogical complex of enstatite, olivine, and other unidentified minerals.

A remarkable feature of this meteorite is its multiple magnetic polarity. More than 90 each of positive and negative poles have been located without investigating the under surface of the stone. Each pole exercises the compass needle much more energetically than the two poles found on the many meteorites that I have previously tested.

A sample from the surface of the stone was submitted to Edward Anders of the Enrico Fermi Institute of the University of Chicago for an estimate of terrestrial aging. His report is as follows:

"The Al^{26} content of Bondoc, as measured by y-y coincidence spectrometry, is 5.0 ± 1.7 disintegrations per minute per kilogram, compared to about 54 dpm/kg for an average chondrite. This low activity may be due to any of the following three causes, acting singly or in combination.

"1) Short cosmic-ray exposure age. If the pre-atmospheric mass of the meteorite was not much greater than its present mass, the low Al^{26} content may indicate an exposure age as short as 100,000 to 40,000 years.

"2) Long terrestrial age. Assuming negligible shielding and a long exposure age (greater than a few million years), the low Al^{26} content may be attributed to decay since the time of fall. The meteorite must then have fallen 2.5 ± 0.3 million years ago.

"3) Shielding. Assuming a long exposure age and a short terrestrial age, one can attribute the low Al^{26} content to shielding. If the recovered mass of Bondoc came from the center of the meteoroid, the pre-atmospheric mass must have been at least as great as 6 tons, and may have been greater if Bondoc was located closer to the surface."

A mineralogical analysis of the meteorite is now being made by C. Moore, curator of the Nininger Meteorite Collection at Arizona State University. It appears that this specimen must be classified in a new subclass of the aerolite group.

The meteorite will bear the name Bondoc, a station on the Bondoc peninsula that appears on the latest map of Luzon Island.

• Meteorites with Anomalous Isotopic Compositions and Ages

TWO METEORITES OF UNUSUALLY SHORT COSMIC-RAY EXPOSURE AGE

Anders, Edward; *Science,* 138:431–433, 1962.

Abstract. The chondrites Cold Bokkeveld and Farmington seem to have reached the earth less than 200,000 years after they left their parent body. Either these meteorites came from the moon, or, more probably, the collision life times of interplanetary objects are occasionally much shorter than has been assumed heretofore.

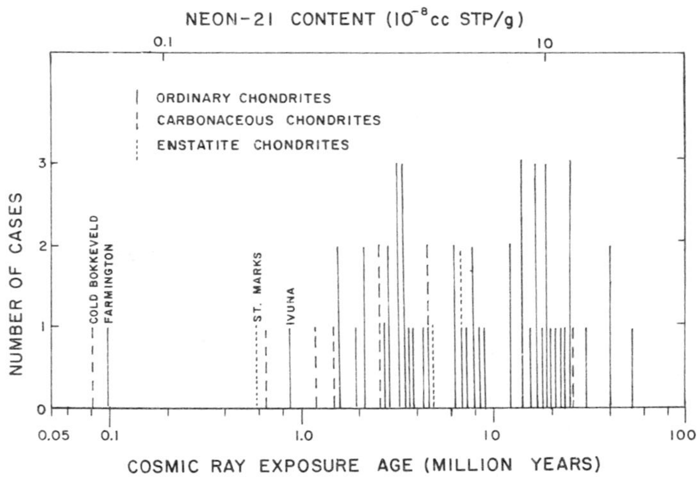

Histogram portraying the unexpectedly short cosmic-ray-exposure ages of some meteorites

"AGES" OF THE SIKHOTE ALIN IRON METEORITE

Fisher, David E.; *Science,* 139:752-753, 1963.

Abstract. Measurements of the potassium and argon content of the Sikhote Alin iron meteorite, by activation analysis, enable a potassium-argon "age" of 1.7 x 10^9 years to be calculated. Such an age is vastly different from all the ages previously measured for iron meteorites.

IRON METEORITES WITH LOW COSMIC RAY EXPOSURE AGES

Cobb, James C.; *Science,* 151:1524, 1966.

Abstract. Analysis of argon-38 and argon-39 produced by cosmic rays in four iron meteorites gives normal amounts of the radioactive product argon-39 and abnormally low amounts of stable argon-38. This indicates that these meteorites were exposed to cosmic rays for unusually short periods of time. These exposure times are one or two orders of magnitude shorter than those for the average iron meteorite, and they overlap the periods found for chondrites. It is suggested that perhaps 20 percent of the iron meteorites have similarly short exposure periods.

METEORIC EVIDENCE FOR SUPERHEAVY ELEMENTS

Anonymous; *New Scientist,* 53:583, 1972.

Meteorites are a valuable source of information on the physical conditions prevailing during the early history of the solar system. Particle tracks and trapped decay products are a fossil record of the radioactive nuclei which existed in the nebula from which the Sun condensed. One problem of particular interest to the nuclear physicists is the origin of xenon fission products in meteorites. Some of the xenon isotopes, due to the decay of plutonium, are found in most meteorites. But xenon-136 is a puzzle because it occurs only in meteorites which are rich in volatile elements.

According to Edward Anders of Chicago University and John Larimer of Arizona State University, the xenon-136 may come from the decay of a superheavy element. They base their prediction on the isotope's predilection for volatile-rich meteorites. This suggests, they say, that its parent nucleus must be a volatile element also.

Beyond the man-made elements lies an island of nuclear stability centred on atomic number 114. The superheavy elements from 112 to 119 may well be volatile since their lighter analogues in the periodic table are all volatile elements. Anders and Larimer have tried to work out which of these superheavies is the most likely progenitor of xenon-136.

. . . .

PRESOLAR GRAINS: ISOTOPIC CLUES TO SOLAR SYSTEM ORIGIN

Hammond, Allen L.; *Science,* 192:772-773, 1976. (Copyright 1976 by the American Association for the Advancement of Science)

In retrospect, it should not be surprising to those who study the chemistry of the solar system that its material is inhomogeneous in composition and formed from more than one source. The galaxy is after all a turbulent place, the birth of a star not an event in isolation. But until a few years ago the converse had been widely assumed. The ground rules for interpreting terrestrial, lunar, and meteoritic chemistry included the dictum that processes within the solar system or the solar nebula must explain all observed phenomena.

Recently, however, a growing group of investigators has discovered significant variations in the isotopic composition of meteoritic bodies. As a result of this new evidence, a near-revolution in ideas about the origin of the solar system is taking place. At a recent meeting, specialists in the subject seemed intrigued and excited by (if not quite ready to adopt) a proposal that up to 2 percent of the atoms heavier than carbon in the solar nebula might have been created in a supernova explosion shortly before the nebula condensed to form planetary bodies. They appeared to accept almost without comment less ambitious models that assume the existence of "presolar" dust grains, of origin unspecified but different from that of the nebular gas cloud, incorporated into accreting planetesimals.

The anomalies discovered so far include variations of up to 5 percent in the isotopic mixture of common elements such as oxygen. Still larger anomalies have been found in rarer elements such as neon, xenon, krypton, and mercury. Smaller anomalies have been reported in magnesium, and the list is still far from complete. As the pattern of isotopic variations becomes clearer, theorists hope it will provide them not only with tracers to identify where in the solar nebula particular bodies were formed, but also with clues to the nucleosynthetic processes by which the particular mix of elements in our solar system was created. Already these newly observed isotopic patterns have narrowed the range of acceptable models for how the moon was made by establishing that it and the earth are geochemically more alike than was once believed.

The earth-moon similarity shows up in data obtained by Robert Clayton and his colleagues at the University of Chicago concerning the relative abundance of oxygen isotopes from those bodies and from meteorites. Meteorites in fact provide the main source of information on average compositions of materials from all over the solar system. Most investigators believe that the many types of meteorites found on the earth provide a representative sample of the solar system's raw materials, and much of the thinking about how planets are formed is based on what is inferred from the meteorites and their properties.

The most primitive type, known as carbonaceous chondrites, show no evidence of heating or melting and are believed to have condensed just as they are from the solar nebula, since their composition resembles that of the sun. The proportion of ^{16}O compared to the other oxygen isotopes in carbonaceous chondrites, however, is anomalously high compared to the mixture found on earth. Clayton explains the difference as due to the mixing of presolar grains of dust containing ^{16}O but almost no ^{17}O or ^{18}O with the solar nebula. Some fragments of these inhomogeneous meteorites contain as much as 5 percent of the ^{16}O-rich materials, while other planetary bodies, such as the earth, contain much less---as little as 0.2 percent. Exactly which minerals or compounds carry the ^{16}O-rich grains has not yet been determined, however.

A complication in interpreting the data is that chemical reactions and physical processes such as evaporation can alter the relative proportions of the isotopes found in samples from a common source---^{16}O evaporates faster than ^{17}O and ^{18}O, for example. These mass fractionation effects are enough to obscure the presence of anomalous material if only two isotopes are measured. Clayton's contribution was to measure all three and observe that values for samples related to each other through mass fractionation would vary in isotopic composition in a systematic way. Thus, data for samples from a single body in the solar system, such as the earth, plotted in terms of these ratios, fall on a single straight line (Fig. 1). Data from a variety of carbonaceous chondrites fall on a completely different line, which Clayton interprets as a mixing line between "ordinary" oxygen and essentially pure ^{16}O, presumably of a different nucleosynthetic origin.

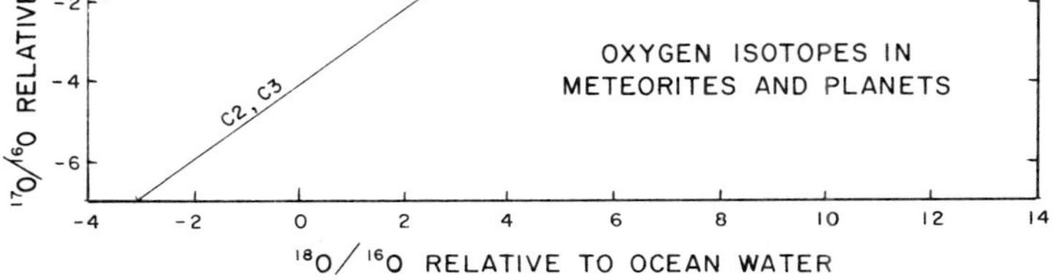

Oxygen-isotope ratios in different samples of cosmic materials. (Fig. 1)

Other classes of meteorites show still different proportions of anomalous ^{16}O. The largest class, known as ordinary chondrites, show evidence of some heating and metamorphic changes after their formation and are thus considered more highly evolved than the carbonaceous chondrites: they also show the lowest proportion of ^{16}O. Still more highly evolved are meteorites, including a group known as achondrites, which have undergone melting and show evidence of igneous processes similar to those on the earth, and which have an oxygen isotopic composition similar to that of the earth.

The isotopic patterns indicate that the earth was made of matter with a different chemical composition than the chondritic meteorites and could not have been formed from them. The highly evolved meteorites, according to this picture, would be more likely precursor objects. Since the isotopic anomalies presumably reflect inhomogeneities in the solar nebula, these different groups are thought to have originated in different regions of the solar system. Clayton's data indicate that the moon is also a part of the terrestrial group in its isotopic composition. Thus, the isotopic evidence supports models that propose a common origin for the earth-moon system and argues against models that propose capture of a moon originally formed in a different part of the solar system.

A second isotopic anomaly with important geophysical implications occurs in some mineral fragments found in the Allende meteorite---a body noted for the diversity of materials that it includes. According to Typhoon Lee of the California Institute of Technology, some of the Allende minerals show more ^{26}Mg than would be expected in a normal mixture of magnesium isotopes. The effect is a small one, not larger than 1.3 percent in the samples examined so far, but the amount of the excess appears to be correlated with the amount of aluminum in the mineral. Because of this correlation, many investigators believe that the ^{26}Mg anomaly arises from the decay of the unstable aluminum isotope ^{26}Al early in the history of the solar system.

The half-life of ^{26}Al is about 700,000 years and its decay has long been postulated as the heat source that caused the parent bodies of the achondrites to melt and give rise to the characteristic features of these objects. The melting process is believed, on the basis of radio-chemical dates, to have been completed within the first few million years after the solar system formed, 4.6 billion years ago. Other candidates for the heat source, such as bursts of radiation emitted by the sun, suffer the difficulty that the chondritic meteorites, formed at the same time, did not melt. Thus, the Allende minerals provide the first evidence that ^{26}Al may in fact have been present during the early years of the solar system.

What is still controversial is whether the ^{26}Al, with its geologically short half-life, was produced by a burst of high-energy protons after the solar system was formed, or was an extrasolar ingredient like Clayton's ^{16}O dust grains. An extrasolar source might be a supernova, since ^{26}Al is known to be produced in one shell of exploding stars. For extrasolar ^{26}Al to be incorporated in planetesimals, however, would require that the formation of the solar system take only about a million years, rather than the previously accepted figure of 1000 million years. Revising the time scale would also require revision of some radiochemical chronologies, such as those based on iodine and plutonium, and would raise questions about the rubidium-strontium and lead radiochemical dating methods. An extrasolar origin for ^{26}Al is thus disputed by Lee and other investigators, although the proton-burst origin he postulates is an admittedly imperfect model.

Irradiation processes within the solar system have also been proposed to explain high levels of ^{22}Ne and other anomalous patterns of neon isotopes in meteorites. The anomalies, discovered earlier by David Black of the Ames Research Center in Mountain View, California, are complicated because there are many plausible sources of neon, but there is now general agreement that as much as 10 percent of the ^{22}Ne cannot be explained and might be extrasolar. Robert Walker of Washington University, St. Louis, believes that production of the anomalous neon by a burst of protons is possible, but would require radiation of very special characteristics. In one model, for example, the protons would have had to be more energetic than 6 million electron volts to cause the observed effect and no others. He points out that there is no independent evidence of such an irradiative process, and that such explanations do not seem "overwhelmingly convincing." The alternative, again, seems to be an extrasolar source, such as a nova or a supernova close enough in time and space to the formation of the solar system that the anomalous neon was not thoroughly mixed throughout the nebular cloud.

A Supernova Trigger? That a supernova may in fact have been the source of all of the anomalies so far observed has been proposed by A. G. W. Cameron of Harvard University. His preliminary model---described as "very bold" by many of those at the AGU session---suggests that a supernova explosion triggered the collapse of a dense interstellar cloud to form the solar system. Isotopic anomalies are thus the consequences of nucleosynthesis in this one supernova, which may have contributed as much as 2 percent of the mass of heavy elements in the solar system. The triggering mechanism, as Cameron sees it, was the increase of heat and pressure that caused the interstellar cloud to compress and collapse; because the heating was uneven, density variations led to fragmentation of the cloud. Material ejected from the supernova mixed unevenly with the collapsing cloud, giving rise to the inhomogeneities observed by Clayton and others.

The nucleosynthetic processes within a supernova are not entirely understood, and some details are model-dependent, but the major nuclear reactions and their products within each of several different shells or regions of the expanding stellar envelope are known. What is called a type-II supernova, for example, makes nearly pure ^{16}O in two shells, and much smaller quantities of ^{17}O and ^{18}O in another shell. Neon is also manufactured in two different shells, according to Cameron, one producing mostly ^{22}Ne and the other ^{20}Ne. The radioactive ^{26}Al originates, as far as is known, only in a supernova, and in quantities that appear to correspond with what the isotopic evidence suggests. Isotopes of both krypton and xenon, for which anomalies have been observed, are also manufactured in supernovae---in fact, Cameron speculates, the xenon anomaly data may ultimately provide a basis for estimating the mass of the supernova. Although many of the details of his model remain to be worked out, Cameron believes that it can account for the observed phenomena and will, moreover, be easy to test as more isotopic anomalies are discovered and compared to supernova models.

The supernova model, and even the concept of presolar dust grains, have not yet eliminated other possible explanations for the isotopic anomalies. Other contenders include the proposal of now-extinct superheavy elements that underwent fission, which was advanced to explain part of the xenon anomaly. And there is some negative evidence---isotopes that do not show anomalies---that puts constraints on the theoretical flights of imagination. Lawrence Grossman of the University of Chicago, for example, looked for differences in the pattern of osmium isotopes in the Allende meteorite and terrestrial rocks, and found no large effect. Osmium is a very refractory element, evaporating at 1950°C, and its isotopes have different nucleosynthetic origins. If anomalies are due to presolar grains of dust that failed to evaporate in the solar nebula, Grossman argues, then osmium should be a good element to see them in. Thus, his result can be interpreted as placing a constraint on models of heavy element nucleosynthesis.

Only a few elements have been carefully examined for anomalous isotope patterns so far, so more definite clues to the origin of the solar system can be expected. One of the most obvious abundant elements to examine is silicon; like oxygen, it has three stable isotopes, the lightest of which is made by a different nucleosynthetic pathway than the others. As it happens, Clayton's technique for extracting oxygen from meteoritic rocks extracts silicon too. So he is in the process of reanalyzing hundreds of bottles of previously collected material, in the expectation of finding anomalously high or low amounts of ^{28}Si. Other investigators are also busy, because there now seems to be a consensus that the isotope data are no longer an awkward curiosity. Rather, now that the different pieces of evidence seem increasingly to point in a common direction, this area of research appears to be one of the most promising means of clarifying the prehistory of the solar system.

MYSTERIOUS METEORITES

Edmunds, M. G.; *Nature*, 263:95-96, 1976.

The early history of the Solar System has always provided ample scope for speculation, yet recent experimental investigations of the isotopic composition of certain meteorites have provided even more food for thought. A class of meteorites, the carbonaceous chondrites, has traditionally been regarded as the best available sample (except for some volatile elements) of the material out of which the Solar System formed. It came as rather a surprise, therefore, when work in the past few years showed that the relative isotopic abundances of particular elements in these meteorites are anomalous in that they can differ significantly from the so-called "cosmic" abundance ratios which are based on an amalgam of meteoritic, solar spectra and solar wind data. The workshop at Gregynog evolved into discussion of two major topics. The first problem was to sort out the experimental evidence to see exactly which elements really do show isotopic anomalies, and the second was to sift through possible theoretical mechanisms which could account for the origin of these anomalies. The three main possibilities examined were: chemical fractionation of isotopes occurring within some region of the early solar nebula out of which solid planets condensed; the production of particular isotopes in nuclear reactions caused by irradiation of the nebula by energetic particles; and the inclusion into meteorites of solid grains from outside the Solar System which had condensed in the stellar explosions which form elements. The grains could hold radioactive atoms and their decay products, carrying the isotopic anomalies from the nucleosynthesis sites.

It seems that definite anomalies have been demonstrated experimentally for several volatile elements. Anomalies for Ne and Xe are already well established. R. N. Clayton (University of Chicago) reported compelling evidence that enrichments of up to 5% in ^{16}O occur, with variations within a given meteorite implying a fine-scale, mineralogically-controlled heterogeneity. R. N. Clayton and J. Kerridge (University of California, Los Angeles) both presented arguments for ^{15}N anomalies, Kerridge concentrating on the fascinating possibility that the ^{15}N content of lunar soil decreases with decreasing age of the soil. Among the welter of possible mechanisms which were discussed for this correlation, the two most promising were time variations in either accretion of interstellar material rich in ^{15}N (for example from grains formed in nova explosions) or in the ^{15}N content of the solar wind incident on the moon and implanted into the soil.

An analysis of the meteorite Arapahoe was reported by F. Podosek (Washington University, St. Louis), showing a $^{129}Xe/^{132}Xe$ ratio which is only about half that found in younger chondrites, the lunar soil or present-day solar wind. He argued that the solar nebula must have possessed some inhomogeneity, perhaps by locking up the radioactive precursor of ^{129}Xe in dust and forming Arapahoe out of an ^{129}Xe-poor gas region. D. D. Clayton (Rice University, Texas and University College, Cardiff) was quick to suggest that the ^{129}Xe-rich dust could have been formed in supernova explosions. This idea met considerable opposition, on the basis of the argument that supernova-condensed grains would be unlikely to give the perfectly normal (compared with other chondrites) pattern of release of iodine gas isotopes which is observed when Arapahoe is heated.

Although anomalies in volatiles seem established, the existence of anomalies in refractory elements is much more doubtful. D. N. Schramm (Enrico Fermi Institute, Chicago) reviewed the ^{26}Mg anomaly measurements, and he concluded that there was definite enrichment in addition to chemical fractionation effects. But this was the only refractory element for which an anomaly seemed to be unambiguously demonstrated. Essentially negative results were reported by F. Begemann (Max-Planck-Institut, Mainz) in searches for anomalies in Hg and K, and by L. Grossman (University of Chicago) for Os.

Everyone agreed that meteorites are rather inhomogeneous bodies. R. Lewis and B. Srinivasan (University of Chicago) gave accounts of the separation treatments they have used (with E. Anders) to find within meteorites those minerals which actually carry the anomalous isotopes. One mineral known as "Q", which has been isolated but not identified, contains up to 80% of the trapped noble gases and 50% of the anomalous Xe, although it comprises less than 0.1% of the meteoritic mass. Anders' Chicago group prefers the interpretation that a volatile superheavy element underwent fission to produce the observed Xe anomalies, rather than believing that the minerals carrying the anomalies represent solid material from outside the solar system. O. K. Manuel (University of Missouri) repeatedly pointed out, however, that Xe anomalies are correlated with the trapped helium and neon content, implying that some kind of differentiation rather than fission has taken place. Evidence that fission of ^{244}Pu has produced some of the Xe is fairly strong. K. Marti (University of California, La Jolla), P. Pellas (Orsay) and Podosek, presented observations of fission tracks, but the number of tracks indicates only a third of the fission required to give all the trapped Xe. It is not clear whether the missing tracks have been destroyed, or if an alternative source of Xe is required.

....

MORE PUZZLES ABOUT THE EARLY SOLAR SYSTEM

Edmunds, M.; *Nature*, 273:337-338, 1978.

The determination of the relative abundances of isotopes in meteorites continues to produce unexpected results. A particularly copious and fruitful source of this early Solar System material has been the meteorite from Allende in Mexico. The relatively high abundance of ^{26}Mg, the decay product of ^{26}Al, in some aluminum-rich minerals in this meteorite has already prompted suggestions that heavy elements freshly made in a supernova explosion were incorporated into solid material during the formation of the Solar System. Anomalies in the isotopic composition of volatile elements like O and Xe have been known for some time, and now work by G. J. Wasserburg and his colleagues (Astrophys. J. Lett. 220, L15, L21; 1978) points to the first definite anomalies in refractory elements other than Mg.

The abundance anomalies exist in small distinct regions (or 'inclusions') in the meteorite, and inclusions showing two completely different types of anomaly have been found. The first type shows excess ^{26}Mg which, as already mentioned, implies contribution of freshly made supernova products. These inclusions show perfectly normal (that is terrestrial) abundance ratios of the three refractory elements analysed---Ca, Ba and Nd. The second type of inclusion appears to be highly chemically fractionated and shows anomalies in Mg isotope ratios which can be interpreted as either a deficiency in ^{26}Mg and ^{24}Mg, or an enhancement of ^{25}Mg. Only two of this second type of inclusion have been analysed in detail, but one of them shows excesses of ^{42}Ca, ^{135}Ba, ^{137}Ba, ^{145}Nd and a large excess of ^{48}Ca, together with deficiencies of four Nd isotopes. The other inclusion of this type shows a deficiency of ^{48}Ca and ^{135}Ba, although Nd and the remaining isotopes of Ca, Ba are normal.

The real problem in interpreting these anomalies arises because the isotopes involved are believed to be made in different nuclear processes. Although all these processes could occur during the extreme physical conditions accompanying the supernova explosion of a massive star, the processes take place in different regions of the exploding star. In the canonical model, shells (or 'zones') of material at different radii undergo different reaction chains. For example, ^{42}Ca is thought to result from a series of reactions starting with oxygen, while ^{48}Ca is believed to result from reactions in another zone. The Ba and Nd anomalies could be explained by addition of material from yet another zone where intense neutron fluxes give rise to the so-called 'rapid' neutron capture chain of reactions. The occurrence of peculiar ratios of isotopes from different nuclear sources within a single inclusion therefore implies a contribution from several sources to each inclusion. The variation of these anomalies between different inclusions implies considerable variation in the relative contribution of each source to the different parts of the early Solar System nebula of gas and dust out of which the solid material formed. If the excess ^{26}Mg anomalies are accepted as good evidence for a supernova explosion just before Solar System formation, then the implication is that material thrown out of each supernova zone was very inhomogeneously distributed into the nebula. The inhomogeneities must also have been incorporated into solid material before mixing processes could even out the isotopic (and presumably chemical) composition of the nebula. It is interesting that the isotopic anomalies are the same for the different minerals contained in each inclusion. This suggests that the inclusions as now seen must have been formed (or passed through a stage) where the contributions from each source were melted or vaporised and then recondensed or solidified to distribute the isotopes uniformly between each mineral.

The identification of the new refractory isotope anomalies is not yet completely unambiguous, since it depends on which isotope of an element is used as a standard against which to normalise the abundances of its other isotopes. Exactly how much fractionation has occurred remains uncertain, although the authors speculate that some of the apparent fractionation may in fact be just due to unequal contributions from the different nuclear sources. Results on the rare-earth elements are eagerly awaited, since the similarities of their chemical properties will reduce fractionation effects.

Isotopic anomalies undoubtedly exist, the range of elements showing them now includes O, Ne, Mg, Ca, Kr, Xe, Ba, Nd, and others may well follow. The chemical and nuclear inhomogeneity implied by these latest results may considerably complicate the building of models of the formation of the Solar System.

• Meteorite Biology

THE DISCOVERY OF ORGANIC REMAINS IN METEORITIC STONES

Birgham, Francis; *Popular Science Monthly,* 20:83-87, 1881.

The great problem, whether or not other celestial bodies besides our own planet are or in past ages have been inhabited by animate beings, must be a subject of the deepest interest to every thinking being. This question has for some time past been answered in the affirmative with great probability. The complete analogy of physical conditions which has been proved to exist in some other planets of our solar system, and which without doubt must also occur in innumerable planets of other solar systems, allowed the very probable deduction that not only on our own earth a higher organic process of evolution has taken place. Still, this conclusion by analogy had hitherto remained a simple, unproved hypothesis, although supported by good evidence.

But now at least it seems that we have obtained a direct answer to this question, and that we are able to see with our own eyes the veritable remains of animate beings from another celestial body.

It has been conclusively demonstrated that the meteoric stones which from time to time drop down on the earth have at no time formed a part of this planet, and it is now generally conceded that they are the remains of other celestial bodies---probably those of a destroyed planet.

In such meteoric stones, and especially in the class called chondrites, on account of the peculiar spherical inclosures found in them, the eminent German geologist, Dr. Hahn, has recently discovered an entire series of organic remains. By a laborious process of grinding down and polishing these fragments he succeeded in producing a large number of thin laminae or delicate stone shavings, which he subjected to a careful series of investigations under the most powerful microscopes. He has recently published a book on this subject, containing on thirty-two plates more than one hundred representations of these laminae of meteorites, every one of which contains different forms and figures, which Dr. Hahn positively identifies not as mineralogical but as organic, and, in fact, as zoological formations belonging to the different classes of sponges, corals, and crinoids. These pictures, which have been reproduced from the original laminae by photography, without any alterations or additions by a draughtsman, must cause great surprise to every geologist and paleontologist, who will at once recognize the structure of well-known coral types on several of the plates. The majority of the meteorites containing these forms are part of the celebrated great meteoric fall of Knyahinya in Hungary, which took place on the 9th of June, 1866.

Dr. Hahn, having sent the whole of his original shavings, comprising over three hundred specimens, to the noted zoologist Dr. Weinland for examination and determination, the latter has also published a report on this subject in a German scientific journal, and the present article is based on the statements of these two scientists.

The result of a thorough examination of these specimens, with a complete comparison of his own great collection of corals, fully convinced Dr. Weinland that a large number of the formations in question are without doubt remains of coral belonging to the class of the favositines, which on the earth are now to be found only in a fossil state, and then only in the oldest, or palaeolithic, stratum.

The terrestrial polypous branches of these favositines are composed of polypary tubes running parallel to each other. At the top, where the cups (calyx) open and the then living polyps are sitting, the coral branches of the favosite present a more or less regular network, composed of the walls of the different polyps. Besides, this class is especially characterized by the cross-partitions in the polypous tubes, as well as by the regular rows of minute holes in these walls, which facilitate the connection between each tube with its neighbors.

Now these polyparies (i.e., bundles of tubes entirely similar to those of the earth's favositines) can be found in a large number of Dr. Hahn's meteoric laminae, which originated not from a single but from many separate falls of meteorites. Many of them show with perfect clearness the very same cross-partitions and rows of holes at fixed intervals from each other with so much regularity that it is impossible to admit a coincidence. At the same time, no geologist would attempt to offer a mineralogical determination of these minute structures, partitions, and holes, which are already visible under a microscope magnifying two hundred times, but which may even still be traced up to four hundred and eighty times. It is, therefore, perfectly evident that the objects in question are organic remains, and, in fact, those of a class nearly related to our favosite corals.

Unfortunately, most of the laminae are cut parallel to the length of the polypary tubes, because the huge meteoric stones had to be broken up in order to obtain these shavings, whereby they generally split with the length of the coral-branches. Only one remarkably fine lamina, a perfect unicorn, in the whole collection, and also a part of the fall at Knyahinya, affords a full view from the top of the branch into the calyx of the polyparium and into the arrangement of the cups themselves. This specimen alone must bring conviction to every connoisseur of corals. It is represented on Plate X, Figs. 3 and 4, of Dr. Hahn's book, but the yellowish tint of the lamina prevented the object from coming out as clearly on the photograph as it can be seen on the original under a good microscope.

This object is evidently a complete small coral-branch of roundish shape, which stands with a broad base on another coral formation. The whole network of the calyx can here be seen with the greatest clearness. The cups are quite dark in the center, being filled with a black substance; then follows a whitish filling around this dark center, and then, plainly visible, the wall of each tube always as a sharply defined line, which is already visible with a low magnifying power. This network of lines, separating the single cups from each other, presents a variety of sizes and forms of the calyx, which, just like those of a great number of our corals, and especially of those of the Devonian favosites polymorphus, are very irregular, often larger or smaller than the average, with rounded or straight side-walls or smaller cups formed by the partition of one larger one. This ex-terrestrial coral has been named by Dr. Weinland, in honor of its discoverer, Hahnia meteoritica. All these coral structures in the meteorites are petrified, having taken the form of silicates of magnesia.

Another very notable peculiarity of these ex-terrestrial corals is their extreme smallness, for, in comparison with those of the earth's fauna, they represent a veritable pygmy animal kingdom. The just-described coral-branch of Hahnia meteoritica is but a white dot in the meteoric shaving, barely visible to the naked eye. Its greatest diameter measures but 0.90 millimetre, and the single cups average but 0.05 millimetre. These are dimensions quite unknown in any terrestrial corals, where a calyx of one millimetre diameter may already be called small. But we must still be prepared for yet quite different things in these ex-terrestrial organisms, for it is very possible that there may yet be found formations for which we can absolutely find no place in our systems of zoology. In fact, it is rather startling that the above-mentioned structures present forms of coral so nearly related to those of the earth, and we must, therefore, accept this fact as important evidence that an organic evolution of great similarity to that on our own earth has taken place on whatever planet from which these meteorites originated.

Besides corals, Dr. Weinland has also succeeded in finding a number of other organic remains in Dr. Hahn's specimens. The material for this investigation was very large, for the greater part of the specimens has evidently been fused together from organic fragments. This is especially the case with the great meteoric stone of Knyahinya, which weighed over two hundred pounds. Well-preserved forms, however, are scarce, for the meteoric material was principally composed of fragments and detritus, which greatly resembled, for instance, the youngest marine chalk in the Gulf of Mexico. But, after comparing a great number of laminae and attaining some practice in this work, Dr. Weinland succeeded in restoring certain often-repeated formations. The sponges especially were highly developed, and of these Dr. Weinland succeeded in actually determining three different genera. Of one characteristic bluish sponge, which occurred in several favorable shavings both as young and old specimens, he was able to make a drawing of its interior construction as easily as from a living specimen. He also thinks that he has discovered vegetable traces; at least, a remarkable, arched form, divided into two halves by a cross-partition, and measuring 0.8 millimetre in diameter, greatly resembles the shield-algae (cocconeis). But he is yet unable to decide whether the formations, claimed by Dr. Hahn as crinoids, really belong to this class, for some of them are certainly spongiae.

So far, not a single trace has been found of higher animal forms (mollusca, arthrozoa), but all the discovered organisms evidently indicate the primary formation of the celestial body from which they came. Then, again, this entire ex-terrestrial fauna hitherto discovered, which already comprises about fifty different species, and which originates from different meteoric falls, even from some during the last century, conveys the impression that it doubtlessly once formed part of a single ex-terrestrial-celestial body with a unique creation, which in by-gone ages seems to have been overtaken by a grand catastrophe, during which it was broken up into fragments.

Perhaps some readers of this article will ask why this remarkable discovery has not been made before, considering the great number of meteorites already collected in our mineralogical collections and the considerable number of scientists who have investigated them. Different circumstances will explain this question. Firstly, meteoric stones are always rare and costly objects, which are not easily sacrificed by their possessors, so that hitherto but very few of them have been broken up in order to obtain laminae suitable for microscopic investigation. Then, again, the latter have only been manufactured in limited numbers, so that the probability of discovering a favorable object in them could consequently be but very small. Dr. Hahn, however, has made extraordinary sacrifices, both in time and money, to obtain his specimens, of which he now possesses no less than six hundred. It must also be mentioned that hitherto the investigations of a few specimens have been made with a magnifying-glass only, and but seldom with powerful microscopes, such as Dr. Hahn employs.

Besides thus affording positive proof of the fact that other worlds are or have been inhabited by organic beings, this important discovery has also solved another interesting problem, which has long perplexed the scientific world. By the newest theory of the celebrated astronomer Schiaparelli regarding meteorites, the latter were supposed to emanate from incandescent comets and their tails. Now all the petrified organisms discovered by Dr. Hahn have been proved to belong to the subaqueous classes of animals, and indeed have lived in water which never froze entirely, and for which we would certainly have to look in vain in comets, which are now generally admitted to be in a state of active combustion.

This affords but another proof of the transcendent importance of this new and great discovery, the general results of which we have now placed before the readers of this journal.

ADDITIONAL NOTES ON THE QUESTION OF LIVING BACTERIA IN STONY METEORITES

Roy, Sharat K.; *Popular Astronomy,* 45:499-504, 1937.

In the October, 1936, issue of the C. S. R. M. in P. A., I have read the criticisms made by Professor Charles B. Lipman (1936) of my publication, "The Question of Living Bacteria in Stony Meteorites" (Roy, 1935). Ordinarily, I should prefer to disregard Professor Lipman's comments, but the subject---the alleged presence of living bacteria of extra-terrestrial origin in stony meteorites---has considerable popular appeal, and I have received a number of inquiries concerning my rejoinder. It is for these reasons that I feel that I should answer Professor Lipman and thus make available to the reading public the facts in the case. I realize that it would be difficult for those interested in the present subject to follow the arguments without having in their minds the contents of the original papers, but, since it is not practicable to republish them here, I have, as an alternative, listed them at the end of this paper for contextual reading. Many of Professor Lipman's criticisms are petty quibbles over points that can be quibbled about *ad infinitum*. I have, therefore, chosen the major points and those that can be discussed profitably. It is hoped that the discussion will clarify the situation once and for all.

Professor Lipman objects to my referring to his papers dealing with his "discovery" of living bacteria in ancient rocks and in anthracite coal, on the ground that his work on these subjects and that on stony meteorites are distinct. Where and what is the line of demarcation? Are not his papers, whether relating to rocks, coal, or meteorites, based on his alleged findings of living bacteria in them? In citing his previous papers, however, my object was not to discuss whether the experiments reported therein were similar to or distinct from his work on meteorites but to point out what he has done in the way of searching for bacteria in ancient objects and what corroborations he has received from those who have verified his findings and his ambitous interpretations. His report on the finding of living bacteria in ancient rocks---two from the Pre-Cambrian and one from the Pliocene---appeared in 1928. This finding, which he calls a "startling fact," was not checked by anyone and needed no checking, for after declaring that some of the organisms in the samples were indigenous to the rock, he concludes: "The question of whether they have relatively recently gained access to the interior of the rock or have always been there remains to be determined by further investigations." No reports of these promised investigations, if any, however, have since been published. Three years after the preceding paper, Professor Lipman (1931) reported the finding of living bacteria in anthracite coal and contended that these bacteria had existed in the coal ever since it was formed, some 250 million years ago. Following this announcement, Farrell and Turner (1932) attempted to verify these findings, but failed to do so except in coal that was fractured and thus was exposed to the ingress of bacteria through surface seepage. Professor Lipman charges me with having failed to mention that Farrell and Turner got results similar to his own, but that they chose "deliberately to interpret them [the results] differently." I quote here Farrell and Turner's conclusions as published by Farrell (1933): "Farrell and Turner (Jour. Bact., 1931 [1932?]) attempted to verify Lipman's findings on coal, but were unable to do so except in coal that was fractured and had become infiltrated with bacteria as a result of the seepage of surface and mine water. The microorganisms found in this cracked coal, and in the mine water and mine soil, were apparently identical with those described by Lipman, and were such as are commonly found in air, soil, and water. No bacteria were found in coal that was not fractured or cracked." A glance at the previously quoted passage will prove that I have not misinterpreted the results obtained by Farrell and Turner. True, they got similar results to Lipman's, but under what conditions? Only on coals that were cracked and had become contaminated by seepage of surface and mine water! No bacteria were found in coal that was not fractured or cracked!

Now with regard to living bacteria of extra-terrestrial origin in stony meteorites, Professor Lipman vehemently objects to my use of small specimens. He states (1936): "I refer to his [Roy's] choice of very small specimens of meteorites, the largest weighing only 8.65 grams. I have pointed out emphatically and on several occasions that the bacterial populations of rocks and similar materials are never more than very sparse." Bacteria, whether spore-formers or otherwise, are ubiquitous. Their average size is about 2 microns (1 micron or micromillimeter = 1/1,000 mm. = about 1/25,000 inch). The total weight of the meteorite specimens I have used is 23.61 grams, having a volume of approximately 15,640 cubic mm.---sufficient to house many hundreds of millions of bacteria. I do not know the bacterial population of rocks, and I cannot cite any authority, but since rocks vary widely in origin, texture, structure, and composition, which are obviously the chief factors that determine the scarcity, abundance, or absence of bacteria, it is safe to state that no general conclusion could be reached, even if extensive studies on the subject were made. Professor Lipman states that the bacterial population of rocks is "never more than very sparse" and attempts to substantiate his statement with the remark, "I have shown that in single specimens of rock like anthracite coal, ten to twenty times the size of the largest meteorite specimen which Roy studied, it often occurs that there are no bacteria." I believe that even Professor Lipman will be willing to admit that coal is not meteorite!

Apart from the untenability of Professor Lipman's analogy between coal and meteorite, he seems to contradict himself when he gives a list of the variety of bacteria he has found in a fragment of the Johnstown, Colorado, meteorite, weighing only 49 grams. The list is as follows (Lipman, 1932, p. 14): "In peptone soil extract, large variable rods were obtained in abundance (B. megatherium). Occasional coccus forms visible. In Na_2S peptone soil extract, very short rods or egg-shaped coccus forms, and also large variable rods in small numbers (B. megatherium). In addition, the Bristol's Algal Medium was examined and found in this case to contain large rods and large ovoid cells, and some fairly large coccus forms. Some cells there also appeared to be yeast-like or Torula-like." To say that the bacterial populations in meteorites "are never more than very sparse" and then to produce such an array of living celestial visitors in 49 grams must appear to everyone a strange form of logic! My reason for choosing small specimens, as I have mentioned in my previous paper, was that they could be handled more easily---an advantage which materially diminishes the chances of their being contaminated during the process of crushing. Furthermore, in spite of the diligent search of Chief Curator Henry W. Nichols and myself, we could not find any larger specimens that were completely crusted. Professor Lipman apparently, does not believe that the specimens I used were completely crusted. This is evidenced from his remark, "When one notes in the photograph which Roy gives of the specimens he used that at least two and perhaps three of the four were not completely crusted." He does not say that "one notes," but I believe he refers to the lighter spots in the photographs. Professor Lipman's suspicion is unwarranted and reveals only his unfamiliarity with meteorites. The lighter spots are thin films of burned material where the outer or heavier black crusts have been either pealed off or never formed. The crust of meteorites is not of equal thickness at every point. This condition is due to the fact that meteorites are the result of oft-repeated fracture, and hence their constantly changing surface is heated to different temperatures at different parts or points.

Not content with his criticism of incrustations of meteorites, Professor Lipman delves into still deeper phases of the subject---the origin, structure, and composition of meteorites. In his attempt to refute my statement that no bacteria were found in pebbles of peridotites and basalt, he states (1936), after omitting the word composition which I used, "My experience with densities of peridotites and basalts does not bear out Roy's claim for close resemblance between their structure and that of stony meteorites." It should have occurred to him that the terms structure and composition are not synonymous and that it is on composition, not structure, that the possibility of bacterial survival is dependent. What close resemblance there is between the composition of stony meteorites and terrestrial peridotites may be seen in the following passage which I cite from the work of the late Dr. George P. Merrill (1929, pp. 68-9): "These figures [average composition of stony meteorites, peridotites, and rocks of the earth's crust] are of interest when we consider that peridotites, which of all terrestrial rocks most nearly compare with meteorites, are of an igneous nature and of comparative insignificance as components of [the] earth's crust. They occur only as intrusives, that is, they have been forced up in a molten condition from unknown depths and intruded into and between the beds of overlying rocks. Their mineral nature is essentially the same as that of the meteorites, excepting that they frequently have undergone more or less alteration in which water and oxygen have taken a leading part, and contain no unoxidized metal. A striking and almost sensational similarity lies in the fact that they are sometimes diamond-bearing, as are also meteorites, though, so far as now known, only on an almost microscopic scale. The world's supply of diamonds both in South Africa and in the United States comes from terrestrial peridotites."

Professor Lipman then takes up the question of the origin of meteorites and states, in a positive vein: "Moreover, we know that basalts are igneous in origin, but no one knows the immediate origin of stony meteorites, even if we should grant that the matter from which they were formed originally was igneous in origin." Just whom Lipman includes in his we I do not know, but those who have studied meteorites and have become recognized authorities seem to know pretty definitely what the origin of meteorites is. As an example, I quote the late Dr. O. C. Farrington (1915, p. 205): "So far as the structure and composition of meteorites are concerned there can be no doubt that meteorites are of igneous origin. All terrestrial analogies indicate that in cosmic furnaces of some sort, fires glowed which gave meteorites the structures which they present to us. As between iron and stone meteorites, some differences of conditions existed which gave them somewhat different structures, since the iron meteorites show well-formed crystallization as if they had been maintained in a uniform condition of temperature, and that sufficient to keep them liquid or viscous for a long time, while the stone meteorites exhibit glass, chondritic structures, and other evidences of hasty crystallization which indicate rapid cooling." I may quote here also a passage from Dr. L. J. Spencer (1933, p. 297) which appeared as a review of Professor Lipman's previous paper (1932): "Organisms could not have originated inside an igneous mass; and if they had been introduced afterwards this could have happened when meteorites were lying in the earth's soil." Incidentally, Dr. Spencer is the leading student of meteorites today.

Professor Lipman questions even the nature and the number of bacteria I found in my culture media and control plates. He calls my findings a coincidence that "is just too good to be true!" Insinuating remarks such as this, without an explanation as to why it is "just too good to be true," are merely meaningless battles of words, not argument or evidence. I may, however, mention here that my experiments were carried out at the Bacteriological Laboratory of the University of Chicago, supported by Dr. N. Paul Hudson, Dr. Floyd Markham, and, to a lesser extent, Dr. James A. Harrison. Dr. Markham, whose interest and curiousity as to the results were no less than mine, observed every phase of the experiment to its minutest detail and actually cooperated during the process of crushing the meteorite specimens and inoculating the powder into various culture media. When growth appeared (in three of the twelve tubes inoculated, and two control plates), stained smears were made by me, Dr. Markham, and Dr. Harrison, all of which showed identical results---a coccus and a rod. The organisms were then subjected to an appropriate series of tests and were found to be Staphylococcus albus and Bacillus subtilis, both of which are met most frequently as contaminants in laboratories. Furthermore, Staphylococcus does not form spores, and hence cannot live indefinitely without metabolizing. My interpretation, therefore, that these organisms were contaminants rather than survivors of extra-terrestrial origin, is a safe, conservative, and, above all, a logical one.

The fundamental issue is whether or not extra-terrestrial organisms exist in meteorites. Professor Lipman syas that they do. He emphatically announces that "stony meteorites (aerolites) bring down with them from somewhere in space a few surviving bacteria, probably in spore form but not necessarily so, which can in many cases be made to grow on bacteriological media in the laboratory." Now the crucial question is how does he know that the bacteria he speaks of are extra-terrestrial? Why could not they be terrestrial? Each and every meteorite specimen that he has used had been in contact with the earth during an unknown interval of time and exposed to multiple agencies of bacterial ingress for periods ranging from eight to sixty-five years, at least. If extra-terrestrial bacteria could gain entrance into meteorites, why could not terrestrial ones? Both are bacteria and both had equal chances. It is obvious, therefore, that his conclusion that he has discovered living visitors from the skies is merely his own personal opinion and is not based on data that anyone searching for truth can or will ever accept.

Lastly, Professor Lipman declares himself to be the pioneer excursionist in the field of life beyond the earth. He asserts (1936): "When I undertook the study of the stony meteorites, there was no evidence, and there is not any now, so far as I am aware, that anybody had preceded me in a similar undertaking. In fact, even today, so far as I am aware, only one other person has worked on the problem and that is Mr. S. K. Roy, who decided to do so only after the paper telling of my findings appeared. In other words, I was and am the pioneer on this subject..." What this talk about pioneership has to do with the subject in question is not understandable. To be sure, I should make no attempt to deprive him of the honor of being a pioneer, if he were one. I am afraid, however, that he is not. The pioneer work of which he is a claimant and about which he speaks with such enthusiasm was done well over four decades ago by no less a savant than Louis Pasteur. Pasteur searched for microorganisms in fragments of a stony meteorite, Orgueil (a carbonaceous chondrite), which fell 1864 May 14, 8:00 p.m., at Orgueil, Tarn et Garonne, France, with negative results! Reference to Pasteur's investigations will be found in a paper, "Enseignements donnes a la Geologie par les volcans de la Lune," by the distinguished student of meteoritics, Stanislas Meunier (Scientia, March, 1925, p. 165). For the convenience of the reader I quote here from Meunier's paper the passage referring to Pasteur's work: "Ces masses, qu'elles soient de pierre, ou de fer, ou meme composees de ces deux substances, ou bien charbonneuses, n'ont jamais presente la moindre trace de corps organises, et le grand Pasteur qui vint au Museum me demander des echantillons de la meteorite charbonneuse d'Orgueil, que j'eus l'honneur de lui remettre, n'y decouvrit aucun germe organique."

METEORITES AND PLANETARY ORGANIC MATTER

Briggs, Michael H.; *Observatory,* 82:216-218, 1962.

It is widely agreed[1] that the source material for terrestrial petroleum deposits is biological debris of marine sediments. However, the recent discovery[2,3] of petroleum-like hydrocarbons within carbonaceous meteorites has raised an interesting problem. Either the meteorite hydrocarbons are part of the remains of an extra-terrestrial life-form, or they are abiotic compounds formed in space. Both alternatives have been strongly criticized, yet one is certainly true. Consequently the presence of these hydrocarbons in meteorites immediately raises the question of whether any of the organic constituents of terrestrial petroleum are compounds brought to Earth by meteorites. I wish to suggest that an extra-terrestrial origin for petroleum must now be given most serious consideration, and that the accumulation of meteorite-borne organic matter on planetary surfaces may be an important determinant of surface features.

Organic matter has now been demonstrated in two classes of extra-terrestrial debris, the carbonaceous chondrites[4] and metallic flakes derived from meteor showers[5]. The nature of the organic matter in the carbonaceous chondrites has been partly determined. It is a mixture of hydrocarbons, aromatic and heterocyclic derivatives, together with various low-molecular weight compounds[6,7]. However, the nature of the major fraction of the meteorite organic matter is still undetermined, though it has been suggested[8] to be high-molecular weight hydrocarbon polymers. As yet there is no evidence about the chemical nature of the amorphous organic attachments of the meteor fragments[5], but their survival, together with physical properties, suggests that they too may be a similar type of hydrocarbon polymer.

It is interesting that this type of material is known to be synthesized by certain discharges in methane[9], while waxy materials form in the electron-beam pathway of electron microscopes[10]. Consequently, the synthesis of hydrocarbon polymers by solar radiation in space is most likely.

It is difficult to estimate the total amount of organic substances added to the Earth by meteoric matter. Thus if only the carbonaceous chondrites are concerned, the twenty known meteorites of this class in collections contain some 6 x 10^3 grams of organic matter. However, most meteorites fall unobserved. Taking the estimate of Harrison

Brown[11] that the average rate of meteorite falls is of the order of 1.1 falls per year per 10^6 km^2, and assuming that carbonaceous chondrites constitute 3 per cent of all meteorite falls[4], it is clear that the average influx of carbonaceous meteorites over the whole surface of the Earth is about 17 per year. These carry in with them several kilograms of organic matter. When considered over the whole Earth, this amounts to about 10 micrograms of organic matter added to each km^2 per year. If meteorite falls have been reasonably constant throughout geological time, a total of the order of 10^{13} grams has been added to Earth's surface since the origin of the planet. This is several kilograms of organic matter per km^2.

Thus, considering the carbonaceous chondrites alone, it is clear that a very large amount of organic matter could be added to a planet's surface. On Earth it is probable that this meteoric organic matter has been largely decomposed by micro-organisms, but on a presumably lifeless planet, such as Venus, this material could accumulate on the surface. Hence it seems that Hoyle's suggestion[12] that the clouds of Venus are hydrocarbons may require reconsideration, for the above calculations show that carbonaceous meteorites could easily have added sufficient volatile hydrocarbons to the planet's surface to form a dense cloud layer. Similarly, on the surface of the Moon the dark maria may contain the remains of meteorite organic matter charred by the solar protonic wind. The cold hemisphere of Mercury may have accumulated similar materials, and the occasionally reported hazes of the planet[13] could be volatile organic matter of meteoric origin released from the surface by local radio-active heating.

So far I have considered only the carbonaceous chondrites on which reasonably reliable data are available. Yet these are comparatively rare objects. Much more abundant is meteoric dust. As mentioned above, organic attachments have now been detected on a high percentage of fragments collected from meteor showers. It is not possible as yet to estimate accurately the amount of organic material added to the Earth's surface by these fragments. However, if the presence of this material proves to be a general phenomenon for all types of meteor fragments, it is possible to make certain assumptions and derive a very approximate first estimate.

The total amount of meteoric materials accumulated by the Earth annually[14] is of the order of 2×10^6 metric tons (2×10^9 kg). If this material distributes equally over the planet, the average annual accumulation is about 4 kg/km^2. If this material has the average chemical composition of meteoric matter, it contains about 0.05 per cent by weight of carbon. Hence the annual meteoric deposition rate for this element is 2 g/km^2. It is difficult to estimate how much of this material is present as organic matter, but taking an arbitrary figure of 10 per cent, this gives 0.2 g/km^2 as the annual meteoric deposition rate of organic matter. Over the whole Earth this amounts to some 10^8 g of organic matter annually, or about 10^{17} g of organic matter since the origin of the planet. This is about ten thousand times greater than the amount brought in by carbonaceous meteorites.

Similar figures may apply for the other planets and augment the previous arguments concerning the possible role of this organic matter in altering planetary surfaces.

As there is a similarity between the hydrocarbon constituents of meteorites and those of terrestrial petroleum deposits, it is clear that meteorites could well be at least a partial source material for petroleum.

A possible objection[15] to meteorite hydrocarbons being a source material for petroleum is the fact that the C^{14} age of hydrocarbons from marine sediments approximates to the age of the sediment estimated by other methods. However, it has recently been shown[16] that meteorites contain C^{14}, and the isotopic abundance is similar to the amount found in modern terrestrial organisms. Hence, C^{14} determinations on sediment hydrocarbons cannot be used as a criterion of terrestrial biogenic origin.

Two other phenomena, also possibly connected with meteoric hydrocarbons, will be mentioned. Micropaleontologists have described many unusual small organic fragments found in sediments as microfossils of unknown origin and affinities[17]. At least some of these could be polymeric hydrocarbons derived from meteoric fragments. The initiation of rainfall by meteor particles has been discussed by several workers[18, 19], but there is no accepted explanation of the mechanism whereby the particles could act. However, the amorphous organic matter surrounding some of the fragments would seem to afford a suitable material to initiate droplet formation.

Most of the estimates given above are based on rather inadequate data, and some may be in error by several orders of magnitude. Nevertheless, the realization that organic material is present in meteoric matter would seem to be of possibly great importance to an understanding of many planetary and terrestrial phenomena.

References.

(1) W. G. Meinschein, Bull. Amer. Assoc. Petrol. Geol., 43, 925, 1959.
(2) W. G. Meinschein, B. S. Nagy and D. J. Hennessy, Science, 1962 (in press).
(3) B. Nagy, W. G. Meinschein and D. J. Hennessy, Ann. N.Y. Acad. Sci. 93, 25, 1961.
(4) M. H. Briggs, N. Z. Sci. Rev., 20, 36, 1962.
(5) D. W. Parkin, W. Hunter and A. E. Brownlow, Nature, 193, 193, 639, 1962.
(6) M. H. Briggs, Nature, 191, 1137, 1961.
(7) M. H. Briggs, Science and Culture, 1962 (in press).
(8) M. H. Briggs and G. B. Kitto, Nature, 193, 1126, 1962.
(9) A. T. Wilson, Nature, 188, 1007, 1960.
(10) E. Hall, "Introduction to Electron Microscopy", p. 299 (New York), 1953.
(11) H. Brown, J. Geophys. Res., 66, 1316, 1961.
(12) F. Hoyle, "Frontiers of Astronomy", p. 70 (London), 1955.
(13) E. M. Antoniadi, "La planete Mercure" (Paris), 1932.
(14) W. J. Thomsen, Sky & Telescope, 12, 147, 1953.
(15) W. G. Meinschein, personal communication, June, 1962.
(16) H. E. Suess and H. Wanke, Geochim, Cosmochim, Acta, 26, 475, 1962.
(17) D. J. Jones, "Introduction to Microfossils", chap. 3 (New York). 1956.
(18) E. G. Bowen, Aust. J. Phys., 6, 490, 1953; 10, 412, 1957.
(19) P. B. Storeb, Nature, 194, 524, 1962.

BIOLOGICAL MATERIALS IN METEORITES: A REVIEW

Urey, Harold C.; *Science*, 151:157-166, 1966.

Summary. If the materials discussed here were of terrestrial origin, it would be firmly suggested that the materials were of biological origin and indigenous to the samples. This by all odds would be the most simple and direct interpretation of the results. A contrary explanation would be that some organisms invaded these meteorites, which have fallen all over the earth during more than a century, and which have been stored on dry museum shelves, and that they grew vigorously for a short period and produced a record that in many ways duplicates what we find in very old terrestrial sediments, particularly in that they appear to lack important compounds such as those found in more recent soils and sediments of the earth---that is, chlorins---but contain compounds present in very old sediments---that is, porphyrins, fatty acids, hydrocarbons, and so on.

Many lines of evidence, as briefly reviewed above, strongly suggest that biogenic material exists in these meteorites and that it may be indigenous. Since we know that life has evolved on earth, the primary question, and by all odds the most important question, is decided for all samples of terrestrial origin. Certainly sufficient evidence has been found to justify further work on the subject of biological material in meteorites, and I am sure that such work will be carried out.

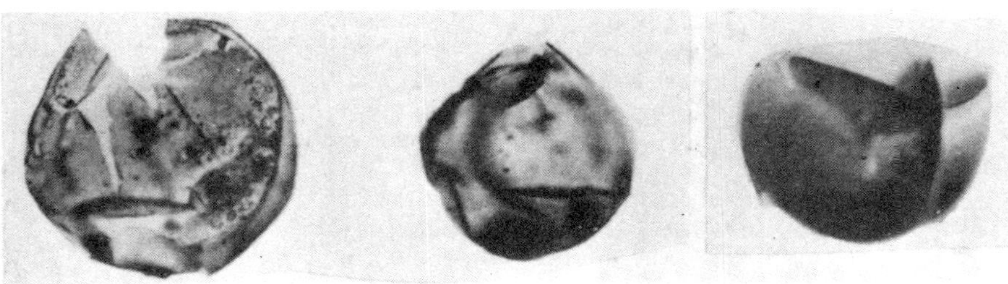

Acid-insoluble residues of organized elements from the Mighei meteorite

Origin of Meteorites with Fossils. Nagy, Meinschein, and Hennessy's paper in 1961 aroused great opposition because of the chemical composition of the meteorites. Those of us who had been working on meteorites for some years were certain that there could not be the residue of living things in them. Had the meteorites had the composition of sedimentary rocks on the earth, no great surprise would have been expressed. It would have been assumed that they came from some planet, probably in the asteroidal belt, which had been broken up some time in the past. The fact that the chemical composition of carbonaceous chondrites with respect to the nonvolatile fraction of primary matter, namely, the silicates and related compounds, is approximately that of the sun made it exceedingly difficult to understand how biological material could have originated in these objects.

We are asking whether the carbonaceous chondrites contain biological material closely related to terrestrial life; if so, such terrestrial life is of a type which, as we know it, exists only in water. Also, the evolution of the complicated and intricate processes of living things requires a source of free energy. Hence the surface of a planet which is large enough and has the proper temperature to retain liquid water, and which receives solar radiation, provides the only possible conditions for the evolution of life. If such a planet was partially covered with water, then oceans, rivers, and all the erosional processes found on the earth would be present. Sedimentary rocks should have been formed, and such sedimentary material should be arriving at the earth as meteorites. This is definitely not the case. No meteorite has ever been shown to be sedimentary. It should be noted that the falls of all the carbonaceous chondrites have been observed. If these fragile objects can be observed to fall and be recovered, then any sedimentary meteorites could also have been observed.

The other possibility is that the planet was completely covered with water. In this case we would expect no sedimentary rocks, but also the biological remains would be deposited only in thin layers at the bottoms of the oceans. The probability of securing a sample of this material would be very small, whereas, as noted above, the carbonaceous chondrites make up approximately 3 percent of the total observed falls. Hence this assumption is improbable or impossible.

There is left only the possibility that life evolved on one planetary object and was transferred to another planetary object of primitive composition. Of course the example of this in the solar system that immediately comes to mind is the earth, where we know life has evolved, and the nearby moon, which may have a composition consistent with that of the carbonaceous chondrites. It is an old suggestion that the meteorites have been coming from the moon, but the recent evidence, though not conclusive, is suggestive at least.

If the moon escaped from the earth it is not at all impossible that it could have been temporarily contaminated with terrestrial water. If the moon was captured by the earth, the process may have been very complicated and violent; such a capture hypothesis almost surely implies that many moon-like objects were about and that they and fragments of them were accumulated into the earth, a hypothesis that I put forward some time ago. If indeed the surface of the moon carries a residue of the ancient oceans of the earth at about the time that life was evolving, the Apollo Program should bring back fascinating samples which will teach us much in regard to the early history of the solar system, and in particular with regard to the origin of life. The possibility that water has been present on the moon has been pointed out recently by a number of students of the subject.

Some people are exceedingly skeptical with regard to the interpretation of life in meteorites, and probably the great importance of the subject justifies skepticism. On the other hand, one of the most fascinating results of all the planned exploration of the solar system would be the proof that life may exist somewhere else than on earth. It seems that we should be willing to consider the evidence for a residue of life in meteorites objectively, and that we should not draw final negative conclusions before all the evidence has been secured. It is not surprising that many investigators, including myself, are reluctant to accept these conclusions, for very considerable modifications of conventional hypotheses regarding the origin of meteorites and the composition of the lunar surface are involved. However, if the residue of indigenous biological activity in meteorites is indeed real, and if the meteorites do not come from the moon, other more complicated and possibly more interesting histories for these objects must be devised. In the meantime, until definite conclusions can be drawn, evidence should not be suppressed, and in fact people who obviously hope that life has existed in these objects should be encouraged to secure definite, positive evidence if they are able to do so.

• • ◦ • • • •

ORGANIC ANALYSIS OF THE PUEBLITO DE ALLENDE METEORITE

Levy, Ram L., et al; *Nature,* 227:148-150, 1970.

A blue-white bolide fell at Pueblito de Allende, Chihuahua, Mexico, on February 8, 1969. Several fragments of this meteorite were recovered shortly after the fall. Specific details of the recovery as well as a description of the physical appearance, mineralogical characteristics, and chemical analyses of these fragments have already been reported and the meteorite has been classified as a Type III carbonaceous chondrite.

The determination of organic compounds in meteorites, particularly the carbonaceous chondrites, has been of interest for more than a hundred years. Most of the meteorites investigated, however, were stored and handled in museums for many years before their analysis, and consequently the positive determination of organic material within the meteorite is always tainted with the suspicion of terrestrial contamination. The Pueblito de Allende meteorite fell at an opportune time, when various laboratories were preparing for an organic analysis of lunar material using highly refined analytical procedures.

Han et al. reported on the organic analysis of the meteorite. They crushed the sample and extracted the organics with benzene-methanol and analysed the extracts (and residue) with gas chromatography and mass spectrometry (GC/MS). Some samples were also dissolved in HF-HCl, extracted, and then analysed for organic content with GC/MS. Their measurements indicated that the inside of the meteorite contained, at most, a few p.p.b. ($1:10^9$) of extractable organics while the exterior surface (contaminated during landing, handling, and so on) contained approximately 0.5 p.p.m. organics.

We repeated the benzene-methanol extraction experiments of Han et al. on samples from the interior of the meteorite crushed to 150 μm grain size in a Carver press test cylinder. The complete analysis procedure including column chromatographic separation, solvent evaporation, experimental procedures, and GC/MS separation and analysis are described elsewhere. The chromatogram resulting from the extraction of 10 g of meteorite from the interior showed less than 1 p.p.m. of extractable organic compounds, a result which is in agreement with the report of Han et al.

. . . .

Organic compounds released when the Pueblito de Allende meteorite is heated can be the result of either terrestrial contaminants or organics native to the meteorite. Three important experimental observations suggest that much of the organic material recovered in the vaporization-pyrolysis experiments is indigenous to the meteorite and is not due to the pyrolysis of terrestrial contamination. They are (a) more organic material is on the inside of the meteorite than on the outside, (b) 75 per cent of the organic compounds are released at temperatures too low ($< 200^{\circ}$ C) to be thermal decomposition products from larger molecules, and (c) the primary pyrolysis products are alkylbenzenes and alkanes. It is difficult to think of a contamination mechanism which is capable of placing more organic compounds on the inside of the meteorite than on the outside. The organic material originally on the outside of the meteorite is probably lost during entry in the Earth's atmosphere. Normally, pyrolysis products consist of unsaturated molecules such as alkanes and styrene derivatives and not saturated compounds. When the meteorite is heated to 300° C, products characteristic of thermal fragmentation are observed. The saturated products, however, observed from the low temperature heating suggest that these products are formed by a non-pyrolytical mechanism. We therefore believe that a large fraction of the organic material released during heating of the Pueblito de Allende meteorite is present within the inorganic matrix in molecular form.

WHERE LIFE BEGINS

Wickramasinghe, Chandra; *New Scientist*, 74:119-121, 1977.

How did life originate on the Earth? If an unambiguous answer to this question is available, we may then extrapolate to more distant parts of the Universe. Unfortunately, there is yet no generally acceptable answer. However, I shall argue in this article that we may well be converging towards one.

A first step towards an answer to this fundamental question must involve an explanation of how the basic building blocks of all living organisms on Earth---the proteins and the nucleic acids---are formed. These latter compounds are in turn built up by linking together of simpler units---namely amino acids and nucleotides. So far scientists have, in the main, thought that molecular compounds such as amino acids must have been synthesised from a mixture of inorganic molecules: molecular hydrogen (H_2), water (H_2O), methane (CH_4) and ammonia (NH_3). The reason is that under the conditions of temperature, density and pressure that prevailed in a primitive Earth, these saturated molecules are the most highly preferred combinations in which H, C, N, O elements would occur.

Some violent energy source capable of breaking chemical bonds is required to cause the necessary re-arrangement of atoms in these simple molecules into amino acids and nucleotides. Such situations could arise, for instance, in intermittent flashes of lightning in a primitive terrestrial atmosphere. Indeed, laboratory experiments involving electric discharges and ultraviolet irradiation in mixtures with approprate combinations of inorganic molecules (H_2, H_2O, CH_4, NH_3) produce high yields of amino acids. There have also been laboratory studies indicating the feasibility of linking amino acids into protein-like structures, and of linking nucleotides into molecules similar to nucleic acids.

The necessity of starting from a primordial mixture of chemically saturated inorganic molecules involving H, C, N, O, uncontaminated with organic species, is not clear. It is probable that a large fraction of the Earth's crust resulted from the accretion of cometary dust, micrometeoroids and meteoritic particles, long after the initial condensation of an iron core. At the present time some 100 tons of meteoritic material enter the Earth's atmosphere every day. In earlier geological epochs the accumulation rate of such material could have been much higher. The main composition of these particles is iron and silicates of various kinds, with an admixture of trapped volatiles. Studies of cometary dust, meteoroids and meteorites suggest that these volatiles could have included many complex organic compounds and noble gases as well as water, methane and ammonia.

Hydrated silicates in meteorites could also have brought large quantities of volatiles including water, as well as traces of organic compounds. A class of meteorite known as the carbonaceous chondrites (which accounts for about half of all types of meteorite) is of special significance in the present context. These meteorites are believed to be the most primitive accumulation of solid material in the solar system. They contain a high carbon content, mainly in the form of coal-like aromatic polymers, but also including amino acids, polyformaldehyde and other intractable organic compounds in trace quantities. Table 1 shows the amino acids found in three carbonaceous chondrites.

Table 1. Identified meteoritic amino-acid compositions*

Amino acid	Murchison	Murray	Nagoya
	(per cent of total amino acids)		
Aspartic acid	3.4	5.5	10.1
Glutamic acid	6.6	4.8	20.3
Glycine	33.6	17.7	27.6
Alanine	14.0	6.6	7.8
a-Amino Isobutyric acid	19.4	50.7	0
B-Alanine	6.0	5.7	11.9

* Average content of amino acids in carbonaceous chondrites amounts to about 15 ppm. Average formaldehyde content is about 3 ppm.

It is of interest that a few of the 20 biologically important amino acids have actually been identified in these meteorites. Furthermore, there have recently been strong arguments and evidence to suggest that one-micrometre-sized clumps of heavily irradiated grains are present in these objects. That is inferred from discoveries of isotopic abundance anomalies in mineral separates from meteorites, and from crystallographic studies which reveal the presence of radiation tracks. The indications are that the individual grains in these clumps condensed in material expelled from nova and supernova explosions. In any event these grains are genuinely primordial to the solar nebula. In a series of recent articles in Nature, Fred Hoyle and I have proposed that the assembly of these grain clumps occurs in dense massive prestellar molecular clouds such as are present in the Orion nebula. Such clouds have been intensively studied by astronomers in recent years. They are believed to be sites of active star formation in the Galaxy. Prestellar molecular clouds and young stars, some shrouded by cocoons of dust and observable only by the infrared thermal radiation from the hot dust, co-exist with vast clouds of ionised gas associated with luminous stars. The total mass of material in a typical molecular cloud complex ranges from 10,000 to a million times the mass of the Sun, and such clouds are distributed more or less uniformly along the spiral arms in the central plane of the Galaxy.

Over the past decade, radio astronomical observations of prestellar cloud complexes have shown the existence of a rich variety of molecules, including quite complex organic molecules. Table II shows a list of molecules detected in dense molecular clouds. We note that the possible precursors of the simple amino acid glycine (NH_2CH_2COOH)---methanimine (H_2CHN) and formic acid ($HCOOH$)---have already been observed in dense molecular clouds. Even more complex molecules are likely to be present. The large departures from thermodynamic equilibrium in interstellar clouds are likely to be highly conductive to the assembly of exotic molecular species.

Tar-like Coatings. We envisage a situation in which inorganic dust grains, present in a collapsing prestellar cloud, accrete tar-like polymeric mantles including highly complex organic species. As collapse proceeds, collisions between these polymer-coated grains lead to the formation of 1 to 10-micrometre-sized clumps of grains. A large fraction of these grain clumps will survive the "switching on" of stars within the cloud and be dispersed, along with systematic gas flows, throughout the interstellar medium in the Galaxy. We have tentatively identified the grain clumps found in carbonaceous chondrites with interstellar grain clumps which might have served as condensation nuclei for solar system meteorites. As we shall see later, this identification appears to have been confirmed by recent data on the ultraviolet spectra of organic matter extracted from carbonaceous chondrites, and comparison of this data with the known absorption properties of interstellar dust.

The chemical composition of interstellar dust grains has been the subject of vigorous controversy among astronomers for several decades. Grains composed of frozen inorganic ices, graphite, iron, silicates, and organic polymers have been proposed by various authors at various times. Sub-micrometre-sized particles comprised of suitable combinations of such materials may be shown to account for all available data on the absorption and scattering properties of interstellar dust. But their precise composition is yet unknown. Interstellar dust clouds show up most strikingly as conspicuous dark patches and striations against the background of more or less uniform star fields, or extended regions of bright nebulosity. By measuring their extinction properties at a single wavelength we could show that the smeared-out density of dust grains is a few per cent of the density of all matter in interstellar space.

Table 2. Observed Interstellar molecules in dense molecular clouds

Inorganic		Organic
H_2 hydrogen	Diatomic	CH methylidyne
OH hydroxyl		CH^+ methylidyne ion
SiO silicon monoxide		CN cyanogen
SiS silicon sulphide		CO carbon monoxide
NS nitrogen sulphide		CS carbon monosulphide
SO sulphur monoxide		
H_2O water	Triatomic	CCH ethynal
N_2H		HCN hydrogen cyanide
H_2S hydrogen sulphide		HNC hydrogen Isocyanide
SO_2 sulphur dioxide		HCO^+ formyl ion
		HCO formyl
		OCS carbonyl sulphide
NH_3 ammonia	4-atomic	H_2CO formaldehyde
		HNCO isocyanic acid
		H_2CS thioformaldehyde
	5-atomic	H_2CHN methanimine
		H_2NCN cyanamide
		HCOOH formic acid
		HC_3N cyanoacetylene
	6-atomic	CH_3OH methanol
		CH_3CN cyanomethane
		$HCONH_2$ formamide
	7-atomic	CH_3NH_2 methylamine
		CH_3C_2H methylacetylene
		$HCOCH_3$ acetaldehyde
		H_2CCHCN vinyl cyanide
		HC_5N cyanodiacetylene
	8-atomic	$HCOOCH_3$ methyl formate
	9-atomic	$(CH_3)_2O$ dimethyl ether
		C_2H_5OH ethanol
	9-atomic	$(CH_3)_2O$ dimethyl ether
		C_2H_5OH ethanol

Furthermore, since the overall relative abundances of chemical elements in interstellar space must be closely similar to those in the Sun's atmosphere, we require the major contribution to the total dust mass to arise from the elements carbon, oxygen and nitrogen. This is because the C, N, O elements make up several per cent of the total hydrogen mass, whereas the proportion of other elements, eg Si, Mg, Fe, is significantly lower. Recent satellite observations of metallic ions in various stages of ionisation have led to the conclusion that most metals which can readily form refractory solids---eg Ca, Mg, Fe, Si---are strongly depleted from the gas-phase interstellar medium. There are also less C, N, O elements in the form of gas-phase ions, atoms and actually observed molecules, than we expect on the basis of normal cosmic abundances. A natural explanation is that these "missing" atoms are locked away in solid grains.

There is strong evidence to support the view that refractory grains composed to graphite (solid carbon), silicates and iron are formed in the atmosphere of cool red giant stars, and are expelled into the interstellar medium. In dense molecular clouds these grains could acquire more volatile mantles comprised of C, N, O elements in various combinations with hydrogen. The most stable and likely form of these molecular mantles is typified by suitably stabilised polyoxymethylene (formaldehyde) polymer $(H_2CO)_n$ and its co-polymers with appropriate conjugate species. An infrared spectral feature in dust clouds at wavelength close to 10 micrometres has been tentatively identified as arising from $(H_2CO)_n$ chains by A. Cooke and myself. Although formaldehyde-based polymers may occur most frequently, H_2, CO and H_2CO molecules being the most abundant and ubiquitous species, other organic polymer coatings could also occur. The average break-up temperature of organic polymers is 450°K, but at temperatures above 300° K they are already quite sticky---like pitch on a hot summer's day. This property is of crucial importance for the process of grain clumping in prestellar clouds, discussed earlier and, indirectly, for the commencement of biological activity in the Galaxy. Within grain clumps exceedingly complex molecules could be preserved from the dissociative action of interstellar radiation fields.

An Observational Clue. At first sight, the detection of very complex molecules in grain clumps might seem hopeless---at any rate by the usual radio astronomical techniques. The radio-wave transitions which characterise these molecules will be very difficult, if not impossible to predict. However, study of ultraviolet spectral data of organic compounds has revealed the existence of a striking signature in the form of a 200-angstrom-wide band centred on wavelength 2200 angstroms occurring in a wide class of highly complex molecules. The requirement for the presence of this band is the occurrence of conjugated double bonds of the types: C = C -C = C, C = C -C = O, C = C -C ≡ N, C C -C N etc. An example of this band for the case of mesityl oxide is shown in the uppermost curve of the Figure.

This absorption band at 2200 angstroms is strikingly similar to a feature at the same wavelength which occurs in the extinction curve of starlight due to interstellar matter. The interstellar extinction feature, originally discovered by T. P. Stecher in 1965, has been widely attributed to a resonance in small graphite particles. However, as D. P. Gilra has stressed, this explanation is not free of difficulty---stringent constraints must be imposed and these are not easy to justify. Grains have to be spherical with radii close to 150 angstroms, and they should not possess any mantles. If these constraints are violated the graphite particle resonance is no longer capable of accounting for the astronomical observations. The central wavelength of the resonance will be shifted away from 2200 angstroms.

A more viable explanation is that the 2200-angstrom interstellar absorption band is due to electronic transitions in exceedingly complex prebiotic molecules which are trapped within interstellar grain clumps. The mass density of such molecules required in order to reproduce the

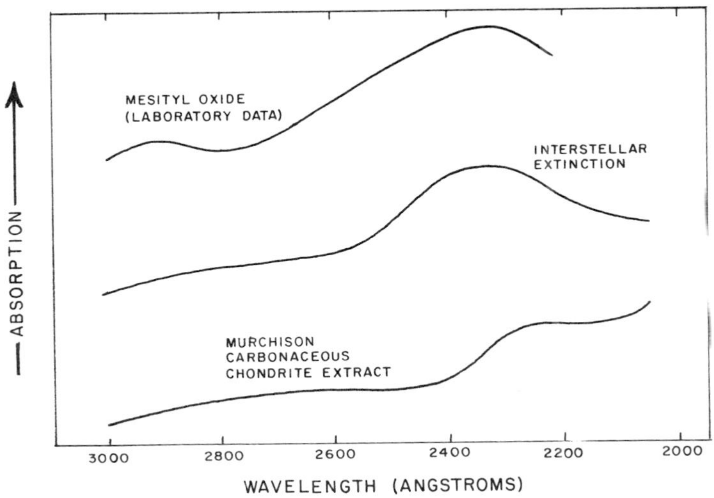

Ultraviolet spectral data for some organic materials compared with the extinction curve due to interstellar matter

astronomically observed band strength, corresponds to slightly less than 10 per cent of the available interstellar CNO elements in the form of complex molecules.

A strongly suggestive link between interstellar grain clumps, exobiology, meteorites and the transport of life to the Earth has come from recent spectroscopic studies of the extract of organic materials from the Murchison carbonaceous chondrite by A. Sakata, N. Nakagawa, S. Isobe, T. Iguchi, M. Morimoto, F. Hoyle and myself. The transmission spectrum of this meteoritic extract has revealed the occurrence of a 2200-angstrom band which is generally similar to that in the interstellar extinction curve, and to the laboratory band in "pre-biotic" molecules (bottom curve in Figure 9). This similarity of spectra has led Hoyle and me to discuss the interstellar origin of life in a series of papers recently published in Nature.

We regard pre-stellar molecular clouds, such as are present in the Orion nebula, as the most natural "cradles" of life. Processes occurring in such clouds lead to the commencement and dispersal of biological activity in the Galaxy. We start with inorganic dust grains---silicate, graphite, iron particles---mixed with mainly inorganic gases. Organic molecules of rapidly increasing complexity form as cloud collapse proceeds, and these are readily polymerised on the surfaces of inorganic grains. When the gas density gets sufficiently high, these polymer-coated grains frequently collide with one another---the collisions leading to the growth of grain clumps with imbedded prebiotic compounds. We argue that such clumps, which already have dimensions of a few micrometres, similar to the sizes of cells in simple microorganisms, are precursors of the most primitive biological system. A Darwinian-style evolution proceeds, basically involving a competition between the growth of grain clumps and their division and destruction in the interstellar environment. The simplest self-replicable system involving clumps of inorganic grains glued together with organic polymer coatings becomes most widespread in the Galaxy. Finally we could imagine that given the right conditions, the organic polymer films, which separate individual grains as well as surround entire clumps, evolve into biological cell walls.

In this context, the alleged identification of microfossils in carbonaceous chondrites is of interest. G. Claus and B. Nagy in 1961 claimed to have found organised cell-like structures microscopically similar to geological microfossils. Although this identification was subsequently disputed, J. Brooks and G. Shaw in 1969 found that the infrared spectrum of insoluble organic material in the Orgueil and Murray meteorites is identical with the spectra of pollen-like spores in Pre-Cambrian terrestrial sediments. It is possible that such spores represent primitive interstellar "protocells" in a state of suspended animation. Their release into the Earth's atmosphere may have led to the start of all life on our planet.

The details described here could be regarded as being yet somewhat speculative. But the recent experimental data on the connection between meteoritic and interstellar spectra, and on the spectra of complex prebiotic molecules, would appear to lend scientific respectability---even a degree of credibility---to such speculation. It would now seem most likely that the transformation of inorganic matter into primitive biological systems is occurring more or less continually in the space between stars.

Chapter 8

THE ZODIACAL LIGHT

INTRODUCTION

The conventional description of the zodiacal light tells us that it is a cone of diffuse light centered on the ecliptic, and seen rising above the western horizon after evening twilight and in the east before dawn. Often spectacular in the tropics, it is usually hard to detect in northern climes. Astronomers believe that the zodiacal light is simply sunlight reflected from a cloud of small particles residing in the plane of the ecliptic. The standard explanation of the zodiacal light is quite convincing, but it is contradicted by many anomalies.

Usually, the zodiacal light is confined to the neighborhood of the sun, which is of course below the horizon when the zodiacal light is seen, but sometimes the cone becomes a band extending across the whole night sky. The cone often seems asymmetric, too. Moreover, some observers claim that the zodiacal light is not centered on the ecliptic at all. And then, what causes the rarely seen rift or "dark patch" in the cone?

Long-term fluctuations in the zodiacal light are confirmed by many observers but hotly denied by others. Some claim that the brightness of the light is enhanced by the position of the moon and the level of solar activity. Even more remarkable are the rapid fluctuations (on the order of a few minutes) in the intensity of the zodiacal light. Possibly these short-term changes are related to variations in the solar plasma engulfing the earth. (See the companion volume, HANDBOOK OF UNUSUAL NATURAL PHENOMENA, for descriptions of transient sky-brightenings and anomalous auroral phenomena.)

• Unusual Distribution of the Zodiacal Light

OBSERVATIONS ON THE ZODIACAL LIGHT AT QUITO, ECUADOR, WITH DEDUCTIONS

Jones, George; *American Journal of Science*, 2:24:374-387, 1857.

I found there [Quito, Ecuador], that every thing distinctive in any celestial phenomenon, was brought out with a distinctiveness I had never before witnessed. Soon after reaching these altitudes, on going out, about midnight, to see whether I could find the Zodiacal Light over both horizons, as I had done when under peculiarly favorable circumstances in other parts of the world, I saw this light not only over the horizons but also quite across the sky. It formed a complete arch, passing near my zenith from the east to west; and was most distinctly and decidedly marked---looking as if a belt of thin white gauze were drawn across the heavens---being clearly visible, but still not perceptibly dimming the stars beyond. My eye soon became accustomed to tracing it: and, from that time on, I never failed to see this luminous arch, at every hour of the night, when the moon or clouds did not prevent. I soon perceived that a cross-section of it would not present the same intensity of light quite across: but that it was brightest at its central part or central longitudinal line, dimming thence towards the edges or boundaries, which latter I sometimes found it difficult to make out. At the central part (running lengthwise) the line of strongest light was so well marked that I soon began to draw it on my charts, where I consider it fully reliable, showing both the inclination of this arch or ring to the ecliptic and the places of the nodes. As a proof of its clearness of marking among the stars, I may mention, that on the three different occasions, when other persons who had never before noticed the Zodiacal Light came to observe with me, and I said to them merely, "You see that luminous arch up in the sky." "you see that, at its central part, the light is brighter than in any other portion: now I want you to tell me by what stars that brightest part passes," they never hesitated at once to give me the place of that strongest light, and to give it almost or quite exactly where it was in the habit of drawing it myself. So they did also with the boundaries of this luminous arch. Sometimes the light did not pass, by a regular dimming from this central line to the boundaries; but there was on each side a sudden offset or decided dimming about half way from it to the outer limits, whence it dimmed gradually again to the boundaries. I have repeated markings of this offset on my charts. It will be readily seen that I was not only, once every night, on the ecliptic; but was, six hours from that, at a distance of 23 or 24 degrees from that line: when this occurred towards the south, there was, on that side of the arch, beyond its usual boundaries, a faint and very dim light, as if I was then looking laterally at the nebulous ring. I did not ever notice this on the northern side. The light of this luminous arch was, at midnight, of a uniform character from one horizon to the other; a cold, white light, exactly like that of the Milky Way: and the arch had at that hour an uniform width the whole way across except at the horizons, where it widened out slightly; the nebulous substance at those lower extremes doubtless giving me reflected light, by which its outer edges of very diffuse matter were visible to the eye. This last sentence will convey to the reader's mind the idea that I did not consider the whole of this luminous arch as being reflected light: and I certainly, at an early stage of my observations on the Cordilleras, did begin to query whether this nebulous ring was not giving me light independently of the sun: that is---whether it was not self-luminous! My reasons for suspecting this arose from the great difference in color between the light near the horizon adjoining the sun, whether in the evening or morning, and in the other portions of the arch: and also from the most decided manner in which the former, which was evidently reflected light, usually turned towards me, whatever my position might be, but especially when I was farthest from the ecliptic: while the other portions of the Luminous Arch, though siding over towards me at my farthest removes, were influenced in a far inferior degree by my change of place. The light at the horizon next the sun, except in the middle hours of the night, was always a warm yellow light, such as it ordinarily has in our own latitudes: while in all other parts of the arch it was a cold, white light, exactly like that of the Milky Way. Indeed not only was this arch in color like the galaxy, but its brightness was often so great that the notes to my observations in the deep hours of the night several times say that it looked much like another Milky Way stretched across the sky.

One evening, while I was standing on the hill Yehinbia near Quito, at the summit of which my observations were made, I watched the sun's last rays on the snow-covered top of the distant Cayambi, a mountain scarcely inferior in elevation or grandeur to Chimborazo itself. All around me had, for a long while, been in shadow: the sun had set to the lofty peaks of Pichincha opposite to me, but still lingered on Cayambi, lighting it up with a yellowish hue: suddenly, the yellow light ceased, and the snow of Cayambi took a white color that chilled me to look at. I was struck with the great resemblance of this yellowish hue to that of the Zodiacal Light as seen over the horizons just after nightfall or before daybreak: while the subsequent white color exactly resembled that of the other portions of the Luminous Arch; or, in the dark hours of the night, of the whole of it, stretched across the sky.

The question agitating my mind in all these observations was an interesting one: for, if this white light was not reflected light but was an independent light, then we have ascertained a most important quality of nebulous matter: it must be self-luminous.

I could not, at any time, discover an interruption or break in the light of this luminous arch, such as might be produced by the shadow of our globe. I tried, first, without having ascertained exactly where such a shadow ought to fall, so as to avoid any self-deception on the subject; but I could see no shadow: then I noted, carefully, on the charts where the sun's opposition would fall, but still there was no shadow to be made out. A German observer---I believe it is Theodore J. C. A. Brorsen, ---who has seen this luminous arch repeatedly in Germany, says that the place opposite the sun was always brighter than any other part: I could never see any thing like that; the arch, to my eye, having a pretty uniform character quite across, except the yellow part at the horizons as already noticed. This yellow light, in a good sky, just after nightfall or just before day-break, would extend as high as my zenith, where it would imperceptibly change into the cold white light. In the former cases, it had the distinctions which I have made in my printed volume, of a great effulgence at the centre and the lowest part, and of a diffuse light without this: I sometimes thought that these parts were brighter than I had ever seen them at the level of the sea: but at Panama, on my way home, I witnessed such a splendid exhibition of Zodiacal Light as rather to make me change my opinion. I ought to add that, on this homeward journey, I saw this Luminous Arch, while at the level of the sea, at Payta, at Panama, off Jamaica, and off the northeastern end of Cuba. My eye by that time had become so experienced as easily to make it out in each of those places; and when, at Payta and at Panama, I pointed out this arch to others, they had no difficulty in, at once, making it out, and in giving me the places among the stars, of the central line and of the boundary on each side. I have little doubt that a person experienced in tracing it on the sky will be able to make it out in our own latitudes---say as far north as New York, in the evening in March. Brorsen informs us in the Astronomische Nachrichtung, No. 998, that he has seen the Zodiacal Light quite across the sky repeatedly at Septenberg in Germany: Baron Humboldt saw also a portion, if not all, of this Luminous Arch in 1803 in lat. about 13° N. (see Astron. Nachrichtung, No. 989): and I must have seen it repeatedly in my cruise in the Mississippi---for which see Report of Japan Expedition, vol. iii, pp. 55, 65 and 85, in all of which instances I was at a great distance from the ecliptic.

I had also at Quito one good exhibition of Zodiacal Light from the moon, the moon then being $1^d\ 1^h$ past full: also, on one occasion, at 9:30, had simultaneously the Zodiacal Light in the west from the sun and from the moon in the east, just before the rising of the latter: this light over at the east and west horizons being connected above by the Luminous Arch. I hope soon to be able to publish all the observations, 123 in number, made in this visit to South America: and also those of Mr. Herrick at New Haven and Prof. Moesta at Santiago in Chile made at the same time. But few of Prof. Moesta's have yet been received; but, both these and Mr. Herrick's sustain the conclusion in my published book, that, when the spectator is north of the ecliptic, the main body of the Zodiacal Light is north of that line: and south, when he is on its southern side.

THE MIDNIGHT ILLUMINATION ABOVE THE NORTHERN HORIZON NEAR THE TIME OF THE SUMMER SOLSTICE

Barnard, E. E.; *Astrophysical Journal*, 24:128-129, 1906.

This summer I have made observations of the illumination under the pole at midnight, near the summer solstice. This phenomenon was seen by Professor Newcomb from high altitudes in the mountains of Switzerland in the summer of 1905. There were but few nights on which it could be well seen here this summer. It was satisfactorily observed, however, in particular, on the night of June 22. Careful observations show that the light is far more extensive than my casual recollections of it would imply; and, unless the generally accepted ideas of the extent of our atmosphere, or the effect of the Sun in illuminating it at great altitudes, are entirely wrong, the illumination must be a manifestation of the zodiacal light, as suggested by Professor Newcomb.

1906, June 12. At midnight there was a distinct glow under the pole which reached 8° above the horizon and beyond. This glow was very diffused and extended for many degrees to the west, and to the Milky Way in the east. Near the horizon elsewhere the sky was dark.

Following are the observations made on the night of June 22 of this year.

$11^{h}\ 0^{m}$. The twilight glow is noticeable to the left of the north point.

$11^{h}\ 30^{m}$. It is easily seen as high as one-third the distance to Polaris, and is as conspicuous as the Milky Way between Cassiopeia and a Cygni. The limit of its visibility to the west would be on a vertical through B Ursae Majoris. The eastern limit is lost in the Milky Way. The light is conspicuous. It is a soft pale glow. In the past half-hour its motion to the right has been noticeable. It can easily be traced half-way to Polaris, and by averted vision diffuses even farther than that. The light is very soft and diffused with no definite limits. The sky elsewhere, which is unusually clear after several days of cloud and rain, is dark.

$12^{h}0^{m}$. The glow is so decidedly visible that one would not fail to notice it if looking toward the northern sky.

$13^{h}0^{m}$. The brightness has now passed to the right of the north point. The glow diffuses nearly to Polaris. It is easily seen half-way to Polaris, and gradually fades farther than this. To one-third the distance to Polaris it is very noticeable.

$13^{h}\ 40^{m}$. It is now away to the right of the north point; Capella is in a strong dawn, which extends 10° or 20° above and to the left of that star. The western limit would have the same azimuth as a Ursae Majoris. I have watched this illumination at frequent intervals tonight, and have seen it pass under the pole and move to the east. The sky is fine except near the horizon.

$14^{h}0^{m}$. The sky is now getting bright everywhere in the region of Capella, and the glow is widening and rising---true dawn is approaching. On this night both Mr. Barrett and Mr. Sullivan, who were working in the 40-inch dome, confirmed the observations.

June 28. The illumination passing under the pole, was visible all night.

July 17, $11^{h}\ 30^{m}$. The illumination under the pole well seen to one third the distance to the pole. At 12, haze all over the north prevented observations of the illumination.

It would seem, therefore, if this is the zodiacal light, that it extends at least 65° north and south of the Sun (assuming the southern extent to be the same as the northern), which is a considerably greater extent than that given it by Professor Newcomb.

ZODIACAL BAND

Mumford, George S.; *Sky and Telescope*, 33:94, 1967.

For about a century, astronomers have been aware of a very faint luminous band, situated along the ecliptic, which forms a feeble extension of the zodiacal light and passes through the gegenschein or counterglow. In Circular 1985 of the International Astronomical Union, a well-known Polish astronomer, K. Kordylewski, describes some novel observations of this feature:

"An expedition of the Polish Astronautical Society, under the direction of Dr. Kazimierz Kordylewski, Cracow, made visual observations of the zodiacal band between the gegenschein and the apex of the zodiacal light on 15 nights between 9 October and 4 December 1966 in Africa.

"The zodiacal band was not homogeneous, but exhibited a mottled structure, and in the constellations Capricornus-Aquarius and Leo, 90° from the nodes of the lunar orbit, it split into two parallel bands 5° apart.

"Ten members of the expedition independently made at least 1,000 determinations of the positions of individual mottles. From this it follows unquestionably that light-reflecting material exists, not only in the cloud satellites at the Lagrangian triangular points L_4 and L_5, but also along the entire lunar orbit. Thus the earth is, like Saturn, surrounded by a material ring.

....

ZODIACAL LIGHT OBSERVED BY SEA CAPTAIN

Hoffleit, Dorrit; *Sky and Telescope,* 12:156, 1953.

The zodiacal light, best seen in the tropics, has long been one of the controversial topics in solar-system astronomy. Recently a retired Dutch sea captain, Jan Drent, enrolled in a geophysics seminar at the University of California at Los Angeles. There he presented observations made during years of navigation to and from the Dutch East Indies. His data, reduced with the aid of geophysicists Robert E. Holzer and Louis Schlichter, indicate that the major axis of the zodiacal light lies in the general plane of the orbits of the major planets, instead of in that of the earth's orbit, where many astronomers had thought it to be.

• The False Zodiacal Light

CONFIRMATION OF OUTER ATMOSPHERE ASYMMETRY POSTULATED TO EXPLAIN THE FALSE ZODIACAL LIGHT

Hope, E. R.; *Nature*, 192:742, 1961.

In 1952 and afterwards[1-3], I suggested that the false zodiacal light ---an early morning pyramid of light above the western horizon, resembling the zodiacal light which at that time would be on the opposite side of the sky---might be explained if Earth's gaseous tail were driven off mainly from the western twilight rim of the globe. I can now point to some recent findings that support this idea[4,5].

A terrestrial gaseous tail has been repeatedly postulated (by Lomonosov in 1753, Houzeau in 1888, Evershed in 1902, Astapovich[6], Fesenkov[6], Divari[6], and many others) to explain the counterglow or gegenschein an antisolar luminosity with a variable brightness and a gaseous emission spectrum (Karimov[6], Tikhov), which Soviet parallax measurements place at an apparent distance of ~ 20 Earth radii. The gases forming such a tail could only be of atmospheric origin; presumably solar radiation pressure carries off atoms or molecules thermally expelled in high trajectories from the underlying denser atmosphere. I pictured "the solar heating steadily unwinding a ribbon of gas from the upper atmosphere into the ecliptic plane" (ref. 2, p. xviii).

The false zodiacal light, observed and named by Divari and Fesenkov in 1946 (ref. 6, p. 22) but previously described by Sykes[7] and by Schmid[7], is presumably a luminosity in the lower part of the tail, below the counterglow. The counterglow luminescence is thought to be excited by solar corpuscular emission in the region beyond the terrestrial magnetosphere (Astapovich[6]). To-day one may suspect that the nearer luminosity of the false zodiacal light is connected with the van Allen radiation, though other theories are possible[1,2,6].

Most remarkable is the asymmetry of the false zodiacal light: it is seen in the western sky as a post-midnight opening-out of the counterglow ellipse, but it is never seen in the eastern sky. If the gaseous tail itself is asymmetric, the reasons could be that (a) the rotation of Earth assists the solar radiation pressure at the evening twilight rim of Earth, but opposes it at the morning twilight rim; (b) at the evening twilight rim the atmosphere has been heated by the Sun, so that the supply of thermally accelerated exospheric particles, that is, the density of the exterior atmosphere, would here be significantly greater than on the morning rim (ref. 2, p. xvi).

Point (b) is now confirmed by the exospheric densities calculated from the orbital behaviour of artificial satellites. Beginning at ~ 200 km. height above the terrestrial ellipsoid, there is a minimum of density at 6 hr. true local time and a maximum at 14 hr. The difference between maximum and minimum steadily increases with altitude, and at 1,700 km., according to the model used by Romer[5], it exceeds two orders of magnitude. At 18 hr. the pressure remains high; at 700 km. height the supply of gas particles is ~ 7.5 times greater on the evening rim than on the morning rim of Earth (ref. 4, Fig. 4); at 1,700 km. it is ~ 12 times greater (ref. 5, Fig. 2).

The exterior atmosphere is chiefly maintained by solar energy absorbed and stored at lower (F-layer) levels. This explains the lag of the density maximum toward the afternoon and the continuing higher density in the evening.

The false zodiacal light, seen only in the west, must not be confused with the zodiacal twilight[8, 9]---the atmospheric component of the ordinary zodiacal light---which according to Divari's specific finding appears with equal intensity on the western and eastern horizons. Moreover the zodiacal twilight is concentrated toward the horizon, while the false zodiacal light, opening out from the counterglow (Divari[6], Fig. 10), is strong at higher angles of elevation.

It is also obvious that the false zodiacal light, extending from the counterglow, is unrelated to the cosmic dust-cloud component of the ordinary zodiacal light. Aside from the nearness, the variability and the emission spectrum of the counterglow as demonstrated by the Soviet authors, there is Roach and Rees's demonstration[10] that the counterglow and the cosmic zodiacal light lie in slightly different planes. The former is, as expected, in the ecliptic, while the cosmic dust cloud is symmetrical with a plane through the Sun determined mainly by Jupiter's gravitation[11]. (Long-lasting confusion has been caused by the reports of those who profess themselves unable to find the slightest asymmetry or variability of the zodiacal light with respect to the ecliptic. These reports may be dismissed as pointless, because there is no imaginable reason why the zodiacal dust cloud around the Sun should lie precisely in the ecliptic, which is merely the orbital plane of the Earth.)

1 Hope, E. R., in The Earth's Exterior Atmosphere and the Counterglow (Defence Res. Board, Canada, Pub. T65R, 1952).

2 Revised third edition of ref. 1 (1957).

3 Hope, E. R., Nature, 171, 555 (1953).

4 Priester, W., Martin, H. A., and Kramp, K., Nature, 188, 200 (1960).

5 Romer, Max, Nature, 191, 238 (1961).

6 Translated Russian papers (Defence Res. Board, Canada, Pub. T65R, 1957).

7 Sykes, W. M., J. Brit. Astro. Assoc., 15, 376 (1905). Schmid. F., Orion, 3, 147 (1951).

8 Brunner, W., Publikationen der Eidg. Sternwarte in Zurich, 6 (1935). Dauvillier, A., Le Magnetisme des Corps Celestes, 2, 141 (1954).

9 Divari, N. B., translated papers in Def. Res. Board, Canada, Pub. T302R (1958). Divari, N. B., Fesenkov, V. G., et al., translated papers in The Zodiacal Twilight (Def. Res. Board, Canada, Pub. T245R, 1958).

10 Roach, F. E., and Rees, M. H., in The Airglow and the Aurora (Pergamon Press, 1956).

11 Hoffmeister, C., Naturwiss., 38, 227 (1951).

• Anomalies in Shape and Brightness

SOME THOUGHTS ON THE ZODIACAL LIGHT

Bridges, Guy J.; *British Astronomical Association, Journal,* 13:242-245, 1903.

Humboldt, speaking of the Zodiacal Light in "Cosmos," says, "It is difficult to understand how so striking a natural phenomenon should have failed to attract the attention." Considering the regularity of its appearance, its brilliancy, and the number of observers now having their attention drawn to the evening sky, the writer of these brief notes finds the above quotation apt at this period. This fine ray, or rays, or corona, of luminous and transparent glow is now to be seen as darkness draws on, commencing at a point close to that of sunset and rising in an almost direct line to the Pleiades, attaining a much greater height as that cluster gets lower in the west every evening, when clearness of sky and absence of moon permit. In February the evening ray is supposed to be at its brightest, but any observer, who was fortunate enough to see it this month, will agree with the writer that it has been far brighter in March than in February this year. On March 22nd, at 8.20, the ray gave light enough to clearly distinguish red from green paint out of doors. Venus hung like a large golden lamp in a silver net. The following notes are intended only as suggestions from a non-scientific observer, and doubtless science can easily refute or endorse them.

Irregularities of the Zodiacal Light. It can usually be observed, when the light is fairly bright, that the ray is far better defined at its northern edge than at its southern. A clear-cut line is shown on the north compared to the south, when the light gradually fades into night sky.

The accompanying rough sketch indicates the prominent features of the appearance of March 22nd, time 8.20 p.m. It shows a rising of a lesser ray to the N. of the usual one, not often seen. Pulsations could be traced clearly in this ray, which did not appear long. Slight pulsations were traceable in the long ray at the upper part of the N. side. The south, as in the sketch, "melted" gradually and indefinitely into sky. Such an appearance as this is usual with a westerly wind veering to N.W. after rain, with a colder atmosphere and rising barometer, and when there is more electricity than normal in the atmosphere. These seem the atmospheric conditions which accompany the best displays of the Zodiacal Light, as a rule. It will be seen that the two columns have one feature in common, i.e., the definite N. side and the indefinite S. side. It would be interesting to hear if any other Members of the British Astronomical Association have observed this, and if there is any explanation. Not having seen the Zodiacal Light in tropical latitudes, perhaps the following suggestion is too "far-fetched":

The Dun Sky. One of the most remarkable features of the sunlit sky since Krakatoa has been that when the sun is high or low, when obscured by light clouds, instead of blue sky appearing between the clouds, as can be seen anywhere but in the sun's vicinity, or a white haze of light caught in vapour, the sky has become a dull brown-yellow. Sometimes

Sketch of the zodiacal light made on March 22, 1903, showing a "lesser ray" north of the common display

this is almost like bronze. White clouds crossing the zone of tan-coloured sky have a most peculiar effect, which must have been generally noticed as a new feature in skies, but of which the writer has never seen any explanation. Is it not probable that this dun mass is composed of a film of volcanic dust, forming a parallel belt round, or partially round, the globe? If this is not considered improbable, might not the Zodiacal Light be really a wider band of volcanic particles, perhaps attracted from our upper atmosphere, where they have been suspended between counter-attractions of magnetism, and gravitation? If this could be proved, or disproved, it might throw some light on the puzzle of the Zodiacal Light. But apart from the possible connexion with the Zodiacal Light, attention might well be called to the dun sky which appears in all weathers and with all winds, and seems much more an astronomical than a meteorological phenomenon. Is our sky turning brown?

The Five-rayed Sunsets. Correspondence in the papers has taken place about the rays observed after sunset. These rays are usually five in number. The third one, careful watching has apparently shown, has never quite died out, but has become, in twilight, what we understand by the Zodiacal Light. These rays stretch from west to east, and, on frosty evenings, are frequently brighter at first in the east than in the west. They are at no great height, not nearly so high, we may suppose, as the Martinique dust, because they appear almost immediately after the sun has disappeared, and have nearly vanished by the time the "glow" is at its height. They have, of course, nothing to do with clouds passing over the sun when below the horizon, for they can, when visible at all, be found always at the same relative distance apart, and they do not revolve, as they would were they caused by any moving vapour. Are they a sign of atmospheric collapse? Taken in regard to the dun sky and large amount of volcanic dust, which has been sent into and beyond the atmosphere, they might be accepted as "long distance" signals or messengers of a gradual change taking place in the atmosphere.

....

THE RIFT OR DARK PATCH IN THE ZODIACAL LIGHT

Dafter, Rosina; *British Astronomical Association, Journal,* 64:136-137, 1954.

I have twice seen the rift, or dark patch, that rarely appears in the centre of the zodiacal light.

The first occasion was in 1935 August. Going out one morning, I was greatly surprised to see, very clearly, right in the centre of the zodiacal light, an inky-black lozenge-shaped patch; it was of uniform intensity throughout. I watched it for some time, for it was a beautiful naked-eye apparition. As dawn approached, it slowly faded away. It was one of those remarkable celestial sights which, once seen, can never be forgotten. To provide a homely simile, it reminded me of a white tablecloth with ink spilt in the centre.

To my knowledge, only three observers have seen the phenomenon: Mr. Housman on Grisdale Pike, Mr. Coleman in India, and myself in Queensland. Mr. Housman gave exactly the same size to the patch as myself: 5 degrees across the zodiacal light, and 2 degrees from base to apex.

The cause of the patch remains unknown, but I would connect it with the Gegenschein at certain rare times. It is worthy of further investigation.

THE RIFT IN THE ZODIACAL LIGHT

Botley, Cicely M.; *British Astronomical Association, Journal,* 64:322, 1954.

Mrs. Dafter's communication in Journal, 64 (3), 156, calls to mind the interesting book Im Reich des Lichtes by H. Gruson (Braunschweig, 1895), to be seen in the library of the Royal Astronomical Society. Plate V of the book shows the zodiacal light as it was seen at Assuan in 1892, with a dark shaft in the centre, which, I think, was of short duration.

DISTRIBUTION OF LIGHT INTENSITY IN THE ZODIACAL LIGHT

Ross, A. D., and Ross, Mrs. A. D.; *British Astronomical Association, Journal*, 24:259-260, 1914.

The Zodiacal Light, while it may under favourable conditions be viewed from most countries, remains one of the least understood of astronomical phenomena. Although the spectroscope and telescope have clearly demonstrated its independence from the solar corona, the question as to its actual place in the solar system is still far from satisfactorily answered. Polariscope observations have shown that the Zodiacal Light is reflected solar light, and, while the particles causing the reflection are probably of meteoric nature, there is still considerable doubt as to the zone of distribution of these particles. The rate of fading off in the intensity of the light with increasing distance from the Sun is such as to indicate that the light is a solar appendage. But as the light has frequently been observed to extend more than 90° from the Sun, it would follow that the reflecting bodies must extend out from the Sun to a distance greater than that of the Earth from the Sun. At the same time, the Zodiacal Light has not been observed in all directions, as would be expected if the Earth were completely enveloped in the system of reflecting particles. The most probable explanation is that there is a circular zone surrounding the Sun at the distance of the Earth's orbit which is void of such particles, or in which, at least, the particles are comparatively scarce. There seems, in fact, to be a zonal gap in the meteoric swarm analogous to what occurs at the Cassini division in Saturn's ring system. The origin of such a gap may be due to one or other of two causes, viz., a removal of particles from that zone by the perturbing action of certain planets---just as Jupiter causes gaps in the system of orbits of the minor planets---or, more probably, by the Earth sweeping up the cosmic matter in its neighbourhood in accordance with the planetesimal hypothesis of cosmic evolution.

Whether such an explanation is correct or not, there is the further problem as to whether the plane of the somewhat discoid distribution of reflecting particles approximates to the plane of the Sun's equator or to that of the ecliptic. The problem may even be more complex if, as has been suggested, the distribution is bifid, the two bifurcations lying approximately over the two sunspot zones. Such a hypothesis was advanced by Dr. M. A. Veeder, and some recent observations lend colour to the theory.

The authors observed the Zodiacal Light on many evenings in August 1913, from Yallingup in Western Australia, a place lying some 8 miles south of Cape Naturaliste, and affording an exquisitely clear atmosphere. On each occasion the apex of the Zodiacal Light lay almost precisely on the ecliptic, but the distribution of the light intensity was decidedly unsymmetrical with respect to the ecliptic. The southern part was brighter and the southern edge was much more sharply defined. This is what one would expect if the mean plane of the cosmic particles within the Earth's orbit were slightly inclined to the ecliptic. It would also approximate to the appearance to be anticipated on Veeder's hypothesis. The diagram shows the appearance presented on the evening

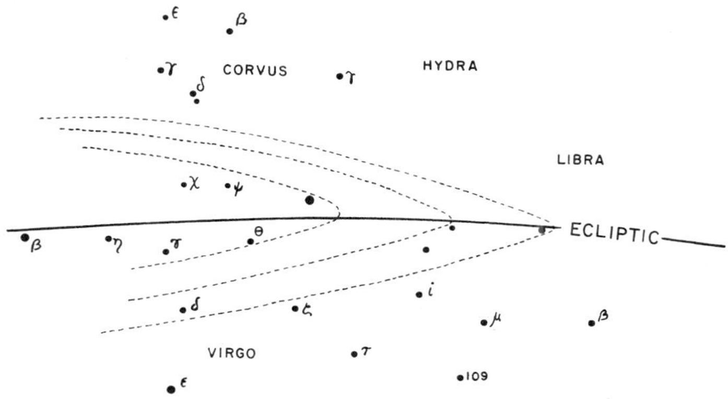

of 1913, August 20, at 7^h 50^m (Standard Time of 120° E. Long.). The outermost curve indicates the limit of distinct visibility of the light, while the inner curves are lines of equal intensity. The light was of an opalescent greenish colour, quite different from the bluish tinge characterising the Milky Way. It was as conspicuous as the brighter portions of the Milky Way in the constellation Argo, and, allowing for the greater atmospheric absorption due to lesser altitude, was probably intrinsically much brighter.

In the diagram the apex of the distinctly visible portion is given as about longitude 222°, or at an elongation of about 75° east of the Sun. The light could, however, be traced at least 5° further east, and later in the same evening the apex of the apparent beam would be about 84° or 85° east.

ZODIACAL LIGHT

E., R.; *British Astronomical Association, Journal,* 8:347-348, 1898.

1898, March 14. 7.0 L.M.T. Zodiacal light extending in a fine cone, passing S. of the Pleiades and reaching the Galaxy near ξ Tauri, roughly, midway between the top of Orion and β Tauri. The brightest portion is just S. of Aries, the portion near horizon embracing α and ο Piscium, η Piscium being just on the N. edge. 7.15. The fine delicate ray extension to the Galaxy is no longer visible. (This was probably due to the sensitiveness of my eye being diminished by the light of the lamp on the paper I was charting the light on.) The glow near Aries is very bright. There appears a faint bulging extension (A) embracing ξ and ο Tauri, and another similar but fainter one (B) covering δ and ε Arietis. The angle of the axis of cone to the horizon roughly measured with a protractor was found to be 81°. The brightest part is now distinctly higher than the Galaxy, and the light far softer and more regular. At times I fancy there is a dark bay under the triangle of

Aries, and S. of it, but I cannot be certain, as there is a low-lying bank of haze near the horizon. 7.30. γ Arietis is now just involved in the edge of the nebulosity. I can still (after resting my eyes in the dark) trace the ray extension to 14 Tauri S. of the Pleiades. There seems to be a thick faint extension (C) towards 41 Arietis, the edge of this extension roughly following the line joining α and 41 Arietis.

. . . .

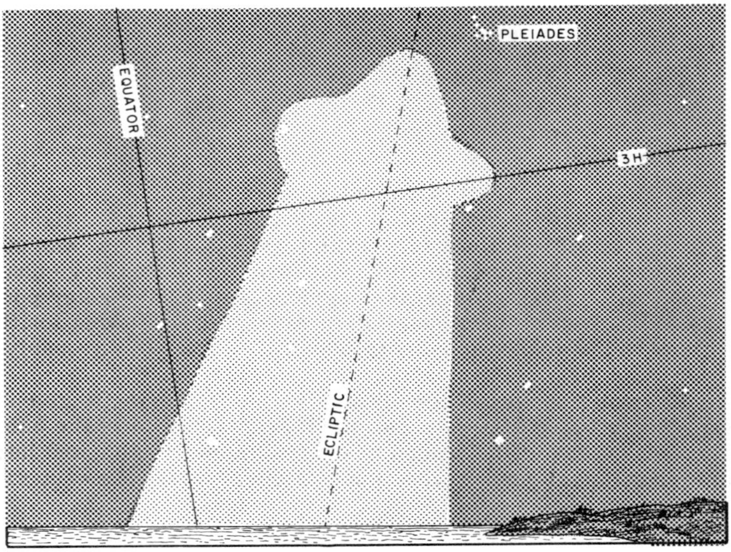

Peculiar bulges in the cone of the zodiacal light

ZODIACAL LIGHT

Anonymous; *British Astronomical Association, Journal,* 33:142, 1923.

On March 22, the Light was noted as being "extremely brilliant at Waziristan." On the same evening, at West Moors, its brilliancy, on the whole, far exceeded that of the Milky Way in Cassiopeia. The fact that the brightness was exceptionally great at two points on the surface of the Earth so far apart indicates a real increase of luminosity on that date.

On the other hand, the observations made on March 20 should be noted. On that evening the Light was seen at Rawalpindi, Kodaikanal, and West Moors, and appears to have differed little from its usual intensity. On the same evening at Waziristan the sky was clear, the stars brilliant, and the Galaxy bright. The climatic conditions were perfect, but the Zodiacal Light was totally invisible. S. A. C. remarks that this is "a very extraordinary occurrence." It must be noted that the distance from Waziristan to Rawalpindi is only about 200 miles.

• Time Behavior of the Zodiacal Light

THE ZODIACAL LIGHT

Muirhead, Henry; *Nature*, 38:618, 1888.

Mr. O. T. Sherman gives an interesting communication on the zodiacal light in Nature of October 18 (p. 594), and asks for reference to any observations. He alludes to Cassini. The following extract from a letter by Cassini may not have come under his notice: "It is a remarkable circumstance that since the end of the year 1688, when this light began to grow fainter, spots should have no longer appeared on the sun, while in the preceding years they were very frequent, which seems to support, in a manner, the conjecture that the light may arise from the same emanations as the spots and faculae of the sun". This does not quite tally with Mr. Sherman's notion that the maxima of the zodiacal light coincide with the minima of sunspots. May it not rather be that, supposing sun-spots to be largely occasioned by increased influx of meteoric matter falling into the sun, which matter gets sublimed and repulsed to augment the materials forming the zodiacal light, therefore the maxima of the latter may then lag behind the maxima of the sun-spots.

THE ZODIACAL LIGHT AND THE GEGENSCHEIN

B., G. J.; *British Astronomical Association, Journal,* 41:250-251, 1931.

In the Physical Review, 35, 1098, for May, 1930, E. O. Hulbert gives an account of an atmospheric theory of the Zodiacal Light and the Gegenschein of which an abstract is given in the following note:--- Attention is called to the fact that fluctuations occur in the intensity of the Zodiacal Light and that these fluctuations are connected with magnetic disturbances. This shows that the planet dust theory is open to objections. The purpose of the paper is to describe a theory of the A. L.

based on the action of molecules and ions in the outer fringe of the terrestrial atmosphere. The most extensive series of observations of the Z. L. are those made by the Rev. G. J. Jones while on a voyage in the Pacific Ocean from April, 1853, to April, 1855. The fluctuations in the Z. L., as recorded by Jones were plotted against the dates of magnetic storms observed at Greenwich. It was then seen that the variations in the Z. L. usually occurred during magnetic disturbances. There were 27 magnetic storms or groups of storms in the period covered by Jones's observations, 17 of which were followed by cases of unusual brightness of the Z. L. There are several cases of fluctuations in the Z. L. coinciding with magnetic storms between 1911 and 1929. It is estimated that the region of the Z. L. is ordinarily about 10 times as bright as the sky would be if there were no light from the stars.

The theory described in this paper assumes that fast flying atoms or molecules are sprayed out in all directions from the sunlit hemisphere of the earth which, after a number of hours, are ionised by the ultra-violet light of the sun. Under the action or solar radiation pressure and the earth's magnetic and gravitational fields ions at levels beyond 30,000 km. form a sort of oblong ring round the earth in the plane of the ecliptic. The ions absorb the sunlight in the far ultra-violet region of the solar spectrum and re-emit a portion of the absorbed energy as visible light. This is the Zodiacal Light. The difficulty with the spectrum is serious. If future observations show that the zodiacal spectrum is that of sunlight the above theory in the form now given will fall to the ground.

The ions which stream away from the sunlit side of the earth form a sheet mainly in the ecliptic plane. This is seen edge-on and forms the Zodiacal Band. The particles when they reach a distance of 1,000,000 km. from the earth are exposed to the full pressure of sunlight, and stream away from the sun like a comet's tail. When the terrestrial observer looks end-on at the stream he sees it as a patch of greater luminosity on the Zodiacal Band. This is the Gegenschein. An examination of Barnard's observations of the Gegenschein in 1882-99 shows roughly that the Gegenschein was larger during an epoch of magnetic storms, and smaller during magnetic calm. It is estimated that on this theory about one-millionth part of the earth's atmosphere escapes from the earth in a million years.

ZODIACAL LIGHT SECTION. INTERIM REPORT

Molesworth, P. B.; *British Astronomical Association, Journal*, 10:253-254, 1900.

There are one or two points I should like to call attention to in addition to those mentioned in my previous notice (British Astronomical Association Journal, Vol. IX., No. 6, p. 285):---

1. Pulsations in the Light.---Cassini (1683), Humboldt, Birt, and Lowe (1850), observers at Kew (1854), and Jones (1854), all speak of definite pulsations and variations in the Zodiacal Light at short intervals.

These may be illusions caused by the strain of long-continued gazing at a faint object, or by alterations in the brightness of the sky. At the same time the evidence in their favour is too strong to be contemptuously passed over.

The greatest care must, however, be exercised to guard against all possible causes of illusion. If the Galaxy is available, the Zodiacal Light and Galaxy should be compared, and areas of equal brightness selected in each. Should variation in the Zodiacal Light be suspected, these areas should be frequently compared to eliminate the effects of twilight.

2. Shifting of the Axis in Latitude.---If possible the boundaries of the Zodiacal Light should be determined more than once on the same date, and any apparent displacement of the axis in latitude should be recorded.

3. Influence of the Moon and Brighter Planets.---The influence of the moon on the Zodiacal Light is another point to be decided. Observations made as to how long before and after new moon the Zodiacal Light remains visible would give a fair gauge of the intensity of the light.

Jones gives several instances of the rising moon being preceded by a zodiacal cone, smaller than, but very much resembling, the morning Zodiacal Light, and quite distinct from atmospheric moonlight. Mr. Bayldon's observations also indicate something peculiar. Any appearance of this sort should be carefully noted.

. . . .

ZODIACAL LIGHT OBSERVATIONS

Glanville, W. E.; *Popular Astronomy*, 37:493-494, 1929.

1. The Fact of the Lunar Zodiacal Light: "March 22: At 7:30 o'clock this evening the light could be distinctly seen in the east. The moon was below the horizon. It (the light) was bright and was located where the Gegenschein ought to be seen but was much brighter than the Gegenschein. It appeared to be broader at the base than the Zodiacal Light."

"March 23: Moonrise, 8:30. At 8:00 o'clock the light was a well-defined cone pretty regular in outline. It gradually brightened until about 8:15 when it was brightest. Then as the moon rose it paled and spread."

Other Lunar Zodiacal Light observations were made on April 21, 22, 23, and 24. In his report for April 23 Mr. Hanahan says: "While the seeing was very fine tonight the light was not as bright as on April 20, 21, and 22, doubtless due to the moon not giving so much light."

2. The Gegenschein: "April 1: Gegenschein well seen but fainter than usual when conditions are as favorable as tonight. It appeared also smaller than usual"---due perhaps to its faintness.

"April 6: Zodiacal band extended across the heavens and joined the Gegenschein in the east. Seeing fine."

3. Fluctuations: "Evening Zodiacal Light, Feb. 4. The whole mass of the light tonight at several times would grow dim and after a

few minutes would become brighter. There was no sparkle or pulsation. As a whole it would wax and wane in brilliancy. Sometimes it would grow so faint as to be seen with difficulty."

"Feb. 13, evening. At times the Zodiacal Light paled so much as to be difficult to see and then the whole light would grow bright again."

"March 5, evening. The whole light dimmed and then brightened at times, but not regularly. There seems to be no rule. A maximum dimming is not followed by a maximum brightening."

"March 7, evening. No brightening or dimming tonight. Light steady. The earlier the observation, the wider the base of the cone especially when the ecliptic lies more nearly along the horizon."

More than once in the course of his reports Mr. Hanahan notes that the band of light could be seen clear across the sky and became fainter as night progressed. Also, Venus was in the western sky during this series of observations and Mr. Hanahan is moved to exclaim: "Venus is very troublesome." "I wish Venus were out of the way." Other observers have been similarly moved!

ZODIACAL LIGHT NOTES

Glanville, W. E.; *Popular Astronomy*, 38:124, 1930.

Fluctuations.---Under date of December 13, 1929, Professor I. Yamamoto, Imperial University, Kyoto, Japan, writes: "A singular experience was obtained by two of our observers recently. It was in the early morning of November 3 when Mr. Kamei, at Kyusyu in the west of the Empire, and Mr. Araki, near Okayama City by the Inland Sea, saw simultaneously a remarkable change of the Zodiacal Light within a few minutes. They are experienced observers and both living in the country, some 200 miles apart, so that they are certainly not influenced by any artificial or terrestrial disturbances. At present we are investigating the nature of these rapid changes in the Zodiacal Light." From the record kept at Cheltenham, Maryland, it appears that on November 3 a magnetic storm of strong intensity was noted. The coincidence of this disturbance with the fluctuations observed in Japan supports the theory of a possible connection between magnetic storms and Zodiacal Light fluctuations. This theory was discussed by Dr. E. O. Hulburt, Naval Research Laboratory, at the recent meeting of the American Physical Society.

Evening Zodiacal Light.---On January 3 from 7:00 to 7:15 the writer observed the Light passing south of the square of Pegasus and reaching an apex south of B Arietis, about 107° from the sun. The moon was three days old, just past apogee and not strong enough to blanket the Light, in fact it added diffused light in the zodiacal zone towards the horizon. During the time of observation the Light was steady. Clear sky. Main body of Light north of the ecliptic.

On January 4 at 7:30 the moon being four days old the Zodiacal Light was still visible though more diffused than on the previous evening. Its boundaries were not well defined but the contrasting relative darkness between the western side of the Square of Pegasus and the Milky Way in Cygnus left no doubt that the diffused light was zodiacal. Had the moon been in the perigee part of its orbit it is likely that no trace of the Zodiacal Light could have been detected. At 8:50, the moon then being about 5° from the horizon and the sky much darker than at 7:30, the Zodiacal Light was traced almost to the Pleiades.

Moon Zodiacal Light.---On December 21, moon twenty days old, moonrise 10:52, the writer scanned the eastern sky in Leo. As he watched at 10:30 he saw a faint illumination shaped like blunt-nosed cone extending almost to Regulus. At 10:40 the cone became elongated and reached almost to Praesepe where it remained until moonrise. Though extremely faint, the apparition was unmistakable. The best time to observe the moon Zodiacal Light is two or three days after opposition. When the moon is nearing the third quarter the illumination is proportionately faint and difficult to see. On this evening seeing conditions were excellent.

ZODIACAL LIGHT AND MAGNETIC STORMS

Hulbert, E. O.; *Popular Astronomy*, 38:215-216, 1930.

From April, 1853, to April, 1855, Rev. George Jones, U.S.N., observed the zodiacal light every night, weather and other things permitting. A comparison of his record with the Greenwich magnetic storm list gave the following results: In this period there were 26 magnetic storms, or storm groups, and 23 periods of abnormal zodiacal light, 16 of which followed within 3 days after a magnetic storm. Ten storms occurred on dates for which there were no zodiacal light observations, and there were 7 cases of abnormal zodiacal light with no accompanying magnetic storms listed. The abnormalities in the light were mainly fluctuations, and sometimes an unusual brilliance in the heavens. In the years 1911 to 1929 mention was found of five cases of abnormal zodiacal light. Four of these followed closely after magnetic storms, the fifth occurring at an epoch of magnetic calm. All in all the data give a rough average interval of less than two days for the lapse of time between the storm and the zodiacal light manifestation.

Of course abnormalities in zodiacal light may at times be due to local causes, such as variations in haze, etc., rather than to changes in the source of light itself, and in such cases there would be no connection between the zodiacal light behavior and magnetic disturbances. The foregoing facts which have been brought out, however, support the conclusion that the connection is genuine, and this suggests that zodiacal light is subject to the same influence as that which gives rise to the magnetic storm. The recent ultra-violet light theory of aurorae and magnetic storms attributed these phenomena to the action of molecules and ions of the earth's atmosphere under the influence of pulses of ultraviolet light from the sun. The extension of the theory to the zodiacal light may be done in two ways leading to two theories of zodiacal light behavior, each containing, however, very interesting difficulties.

The first theory assumes the view taken in certain textbooks that the zodiacal light is sunlight scattered or reflected from a great ring of particles around the sun in the plane of the ecliptic extending outside of the orbit of the earth. The ultra-violet solar flares which cause the magnetic storm would be supposed to render the ring of particles unusually luminous. This view requires that the zodiacal light be fluorescent light stimulated in the particles by the solar ultra-violet light and not reflected or scattered sunlight. The spectrum of the zodiacal light should therefore be different from that of sunlight, and yet the observations indicate that the two spectra are much the same. However, the zodiacal spectra were obtained with low dispersion, and the character of the spectrum is probably not known completely at the present time.

The second theory assumes that the zodiacal light comes from particles originating in our own atmosphere, an idea advanced long ago by Barnard and others. Fast flying atoms or molecules are sprayed out in all directions from the sunlit hemisphere of the earth, which after a number of hours are ionized by the sunlight. Once ionized they are guided by the magnetic field of the earth. Some of the ions may reach, or be formed at, levels above 50 thousand miles. Due to the action of solar radiation pressure and the earth's magnetic and gravitational fields there will be a concentration of ions, probably very rare, at high levels roughly in the plane of the ecliptic. The zodiacal light is the sunlight absorbed and re-emitted from the ions. The ultra-violet flare will cause changes in the number of the ions and in the light which they send down to the earth. The ions form a sort of an oblong ring around the earth which on the day side is perhaps 50,000 miles away and which on the night side stretches out to great distances. At the far end the ions continually stream away in the direction of the sun's rays, so that the ring merges into a sort of comet's tail which may be the Gegenschein.

THE ZODIACAL BAND

Bousfield, R. B.; *Royal Astronomical Society, Monthly Notices,* 94: 824-838, 1934.

Another curious phenomenon that I have witnessed frequently in mid-winter lies in the orderly alteration of relative intensities of the eastern and western cones. This was noted especially in 1931 July, when I made the following entry in my observation book:

"A curious feature observed repeatedly. The Western horizon seems to begin to rebrighten between 2.0 and 3.0 a.m. It reaches a high intensity about 3-1/2 hours before sunrise, being then bright while the East is still much less so. About two hours before sunrise the Eastern Cone begins to increase its intensity, while the West appears to go duller; and the East now outmatches the West till dawn. The moment of dawn can be adjudged to within a minute or two. Its sudden effect is unvarying and resembles the sudden hop of light in a room when the power-house has been sending a diminished current or starts up the engine. The change of maximum from West to East occurs long before this 'hop,' and it has no connection with the terrestrial dawn. That the West does brighten first and the East overtake it is a reality, which measurements with the lumeter bear out as a fact. But why so, is to me a mystery."

Equally curious is the building of the arch in February. It is an exact reversal of the long autumn process that extends from East to West, but a process that is completed under the eyes in a few hours. The evening apparition tails ever farther and farther eastwards till the zenith is passed near midnight. Finally, between 2 a.m. and 2.30 a.m. the East leaps up to meet it and the great bridge is complete. Gradual change in the conditions of observation (e.g. slope of the ecliptic, night-sky illumination, atmospheric transparency) must contribute to this result, as observations taken throughout a night and morning at a fixed altitude show, steady changes which would suggest as natural the gradual shortening of the Western Cone as the evening progressed.

. . . .

6. Lunar Zodiacal Light. In a monumental volume published by the Naval Department in Washington the Rev. George Jones discusses a number of observations that he has made suggesting illumination of zodiacal light matter by the Moon. I give in Appendix I extracts from my observation books on observations of the Western Cone in the early morning. They do not bear out, by special connection with lunar phases, the idea that the anomalies described above are to be explained in terms of a lunar contribution to the zodiacal light. For while the period of the New Moon frequently displays a dull cone and the period of the Full Moon frequently shows a bright light, no definite rule establishes itself. In particular, one observation on 1934 March 16, at 3.30, one night after New Moon, when a Western Cone was clearly visible with well-defined boundaries rising to an apex at η Virginis at an altitude of 45°, shows clearly that the Moon is not a necessary factor in rendering the Western Cone visible in the morning. Other observations made on the Eastern Cone in the evenings proved equally inconclusive.

The question of lunar zodiacal light is tied up with the position relative to the Earth of the matter forming the Zodiacal Band. If the lunar zodiacal light is finally confirmed, it means that part at least of the matter is near the Earth, presumably inside the lunar orbit. It occurred to me that such a distribution of the matter might be tested by the lifting of a shadow or the distribution of the matter might be tested by the lifting of a shadow or the darkening of the Band after a solar eclipse. For the annular eclipse of 1933 August 23, which ended close to sunset in Queensland, I endeavoured to organize a chain of observers from Northern Queensland down to Broken Hill in the south, arranging for charted observations on the night preceding the eclipse and on the eclipse night. Unfortunately owing to cloudy conditions over the whole of the Eastern Continent, with rain and thunderstorms, only two observations could be made. Mr. W. F. Gale observed shortly after dusk in Sydney across the harbour. He is a very experienced zodiacal light observer, but on the night of the eclipse he could see no sign whatever of the Western Cone. His observation might have been affected, as he notes, by the harbour lights and by a slight haze to westward.

I happened to be the other fortunate observer. At 8.40 p.m., presumably about 90 minutes after Mr. Gale's observation, the sky cleared completely for ten minutes, giving a very clear atmosphere; after that interval it clouded again and rained. But I had been waiting ready all evening and was able to take a double observation. I found that up to about 30° the base of the Western Cone was bright, at 0.0045 foot-candles, but at this altitude and beyond it up to the apex at 65° a sudden drop occurred to 0.0037 foot-candles; there was a clear line of demarcation which was noted each time the holophane lumeter swept across it. This was an observation which I have entirely failed to repeat on other evenings; on the following night, for instance, the light for the whole cone from close to its base up to its apex gave 0.0045 foot-candles. For some reason, not yet explained, on the night of the eclipse and on that night only this sharp line of varying brightness was found in the cone. A single observation of this kind cannot be regarded as conclusive, but it is at any rate sufficiently suggestive to call for more refined observations by astronomers suitably placed at some later eclipse.

7. Undulations. My own conviction is that the light of the Zodiacal Band is not constant. I have no doubt that it was brighter in 1931 than in 1932 or 1933. The morning arch in 1931 July was extraordinarily brilliant, the 1932 February arch was extremely bright, the 1933 February arch was less bright, though clearly seen; the 1934 January arch was seen, but it disappeared entirely in February and was not seen in March. In addition to these changes, which in the absence of independent support by other observers may have to be explained by varying conditions of observation, I have in common with Jones and other observers noted undulations in the brightness. These I have noticed more pronounced in the morning in the east: the waves are long and wide apart, while at night to the west they present more of a ripple effect, are closer set and harder to define. It is not a common effect. When I first noticed these undulations I had no knowledge that others had observed the phenomenon, but I found later that Jones described pulsations with a period of a minute or two in 1854, ascribing it to something intrinsically connected with the light itself, external to the atmosphere. Denning quotes a description of pulsations with a period of 30 seconds observed by Lowe in 1850, and attributes the effect to eye-strain. I have endeavoured to test this physiological explanation. After watching the undulations for some time without any attempt at strain, I turned to a portion of the Galaxy of large extent and similar luminosity. I gazed at it for a long time and even attempted to produce eye-strain and to tire the optic nerve. But the Galactic light remained absolutely steady under identical conditions of atmosphere and altitude. I could not perceive the faintest trace of undulation. Returning my gaze to the cone, it continued, majestically sweeping its widely spaced aerial rollers one after the other from base towards apex, one billow of light silently succeeding another.

The most striking observation that I have noted was on 1931 May 27, at 2.15. I quote from my notes:

"I was forcibly struck by undulations which stretched from north to south boundaries in the base and rolled up the narrowing shaft like ocean rollers being pushed up an estuary from the open sea. As they progressed in horizontal crests they narrowed in width, vanishing at the apex, as a wave might vanish at the end of a funnel-shaped inlet. There appeared to be several wave-crests simultaneously visible."

From observations on subsequent nights the height of the cone (not recorded for May 27 at 2.15) may be taken as 70°, and my memory is that at least three crests could be seen at one time and that it took about 2-1/2 seconds for a crest to run from base to apex.

It may not be without significance that on the following morning, 1931 May 28, between 3.0 and 3.30 there was a very strong horizon glow, so brilliantly illuminating the whole east and south-east that it threw all the trees and distant hills into bold relief.

....

OBSERVATIONS OF THE ZODIACAL LIGHT FROM A VERY HIGH ALTITUDE STATION

Blackwell, D. E., and Ingham, M. F.; *Royal Astronomical Society, Monthly Notices*, 122:143-155, 1961.

Introduction. Whilst at Chacaltaya the 1958 Cambridge Observatories expedition made a special study of the variation with time of the brightness of the zodiacal light and the surrounding night sky. Previous studies of this problem by other observers had given conflicting data and we hoped that the exceptional atmospheric conditions at Chacaltaya would enable us to decide whether or not the zodiacal light itself varies in brightness or position. Whilst this was not the primary problem, we also wished to examine the possibility of detecting corpuscular radiation directly, during its passage between the Sun and Earth, through the scattering of sunlight by free electrons in the corpuscular stream. The time of the expedition was particularly favourable for studies of this kind, for the observations were made at the peak of the solar cycle during a period when the solar activity was exceptionally great.

Previous observations of changes in the brightness of the zodiacal light. There have been many reports of variations in the brightness of the zodiacal light. We list some of these below, but at the same time we emphasize most strongly that studies of possible variations in the zodiacal light can be made only with great difficulty and any results should be treated with extreme caution. Worthwhile data can scarcely be obtained except from good photographic or photoelectric photometry carried out under excellent meteorological conditions in the tropics, and for this reason we reject almost all of the available visual observations, even though they have been made with the greatest care. The principal changes that have been reported are as follows.

(i) Short period fluctations. Reports of changes occurring during one evening within a time interval of a few minutes are common. The best known study of this kind is that of Jones (I) made between 1853 April 2, and 1855 April 21, during a voyage on the Pacific Ocean chiefly between latitudes $+20^{\circ}$ and $+40^{\circ}$. Jones's observations were made visually whilst he was on board ship, and he recorded the brightness and position of the zodiacal light together with changes in its brightness which he called "pulsations". Hulbert (2) has drawn attention to the correlation that exists between these times of zodiacal light variability and the times of magnetic activity as observed at Greenwich. Jones himself was unaware of any such relationship, and it is difficult to escape the conclusion that the phenomena described by him did actually occur.

There are other reports in the literature---see, for example, Hulburt (2)---of changes in the zodiacal light during periods of auroral activity, but we postpone until later the question of whether the observer is deceived by the presence of aurorae into thinking that the zodiacal light has changed.

(ii) Occasional enhancements of the zodiacal light during one whole night. Visual observations of this kind are not uncommon (2), and we refer here only to the unusually bright zodiacal light of 1896 March 4(3), which occurred during a period of auroral activity. Apparent changes of brightness from night to night occur in most detailed lists of photometric observations---see, for example, Roach (4) and Huruhata (5)---although an opinion is rarely expressed about the reality of these apparent changes.

(iii) An annual variation. An annual variation of brightness over a range of two to one has been observed photoelectrically by Elvey and Roach (6). Thom (7) has assembled many visual observations from a variety of sources and has found a similar variation.

(iv) Long period variations. Huruhata (5) has made photoelectric observations over a period of four years, finding an irregular variation from year to year amounting at most to a factor of seven. Thom (7), using the previously mentioned visual data, also finds a variation and some correlation with solar activity. Table IV of Paper I, which lists absolute brightness values obtained by various observers during the period 1952-8, using non-visual techniques, apparently supports this idea that there is a variation from year to year.

All of this evidence is contradicted by Regener (8) who cannot find any significant variation during fourteen months continuous observation over the period 1953-4. It is perhaps significant, however, that his observations were made during a period of very low solar activity. A daily record of the appearance of the zodiacal light at the Harvard College Observatory (latitude 42° N) ever the period 1876-1890, has been analysed by Searle (9) who states that the variations which seem to have occurred cannot be established with certainty, and any support for a correlation between the variation and the occurrence of aurorae is certainly very feeble. Barbier (10) also, working photoelectrically at Haute-Provence (latitude -44°), finds no evidence for a variation of zodiacal light brightness. He worked during the periods 1951-2 and 1952-3 when solar activity was low.

We conclude that it is not possible to deduce from the existing data whether or not the zodiacal light is of constant brightness.

The Chacaltaya Observations. The requirements for a site suitable for an investigation of possible variations of the zodiacal light are very stringent. Evidently the site must be near the geographical equator so that the varying inclination of the ecliptic to the horizon does not lead to systematic errors, and it must also be at a high altitude so that observations may be made close to the horizon, and have an excellent climate in which the extinction is low and varies little from night to night. In addition, it must be near the geomagnetic equator so that there is as little disturbance as possible from aurorae. Chacaltaya satisfies all of these requirements. In particular, it is at a low geomagnetic latitude (-3°) and so is especially suitable for the attempted observation of corpuscular radiation mentioned at the beginning of this paper.

The method of observing the zodiacal light and the result have already been described in Paper I. In this section we attempt to ascertain whether or not the observed variation in these data from day to day can be regarded as real.

Correlation between brightness measurements at λ 4470A and λ 6200A. The measurements of brightness in these two spectral regions have been made almost completely independently, and we may therefore test the reliability of these measurements by examining the correlation between them. To do this we have, for each photograph, averaged over all elongations the deviation from the mean brightness. In Fig. 1 we plot one set of measurements against the other and obtain a reasonably good correlation with a slope of unity, showing that the same relative change of brightness occurs in the two spectral regions.

Further, supposing that a perfect correlation may be expected to exist between the two quantities, the diagram shows that the probable

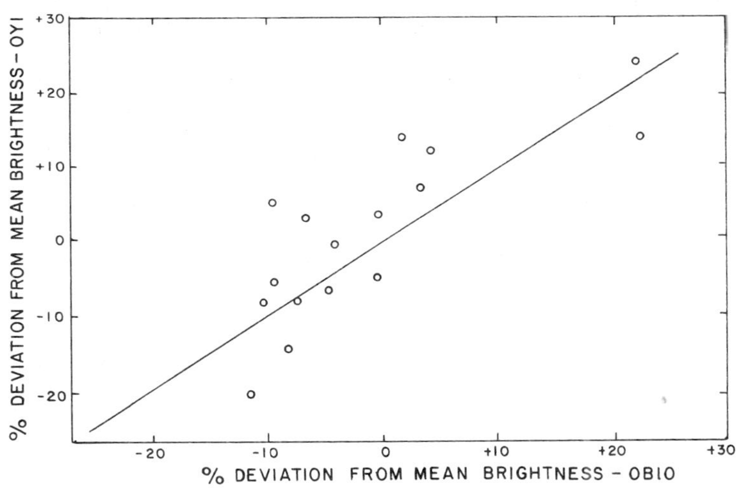

Correlation plot of the measurements of zodiacal light brightness changes in two spectral regions

error of such brightness measurements, averaged over one plate, is about 8 per cent.

Correlation between zodiacal light brightness and solar activity. As it now seems likely that the observed variations of brightness are real, we attempt to correlate them with solar activity, as indicated by the magnetic index K_p. We have been particularly fortunate during the expedition to have observed the zodiacal light before, during and after the magnetic storm following an important solar flare. The flare occurred on July 7 at 00^h 39^m U. T. and was recorded at Mitaka and Hawaii Observatories (II) as of importance 3^+. The intense magnetic storm which followed it on July 8-9 was one of the most important during the three year period 1957-9, the value of the planetary magnetic index K_p (II) remaining at 9 for 15 hours.

In Figs. 2 and 3 [omitted] we reproduce the plot of surface brightness against elongation for the two colours, which has already been given in Paper I, and show the points appropriate to days when the magnetic index K_p was greater than 4 (i.e. exceptionally high) at the time of observation; evidently there is a fair correlation between magnetic index and brightness in the two colours.

To investigate this relation further we have formed for each day the deviation of brightness from the mean averaged over elongation and over the two colours. In Fig. 4 these deviations are plotted as a function of magnetic index for the time of observation, showing a remarkably good correlation, the total change in brightness when the magnetic index changes between 0 and 9 being about 40 per cent. The correlation is not substantially altered when a phase difference of up to one day is introduced. These results apparently justify the suspicion that has existed for more than a century, that the zodiacal light itself does change during a time of auroral activity. We emphasize that this observed change in zodiacal light brightness is quite distinct from a change in general sky brightness which may or may not be due to a localized aurora. However, no rapid changes of the kind described by Jones were observed, either visually or photographically, and we are inclined to doubt whether these changes are associated with the zodiacal light itself, remembering that Jones often observed at rather high geomagnetic latitudes.

We have examined other published photometric data for the effect, but have found that the data are too scanty and the range of magnetic index too small show a good correlation.

Examination of the data for an annual variation of brightness. We are now able to use the graph of Fig. 4 to allow for the effect of magnetic activity on the zodiacal light brightness and to reduce the data to zero magnetic activity; the data may then be examined for an annual variation. Our observations are restricted to the period June 4-August 2, but this interval should be sufficient to show part of an annual variation, if it exists with an amplitude equal to that already reported. In Fig. 5 [omitted] we plot against date the relative brightness of the zodiacal light at elongation 40° reduced to zero magnetic activity---this elongation being chosen because the brightness values are most accurate here. Each point on the graph is a weighted mean of observations made in the blue and red regions. All of the photographs were taken at approximately the same time in the evening and reduced in the same way; a correction was made for extinction, using for this purpose observations of stars made with a photoelectric telescope. The graph shows no evidence for an annual variation of brightness such as has been observed by Elvey and Roach (6).

....

Conclusions. We find that following a solar flare there is an increase in zodiacal light brightness and also an increase in the brightness of part of the sky. The increase in sky brightness is attributed to the temporary emission from the Sun of a beam of corpuscular radiation of density 300 electrons cm^{-3}, and this view is supported by the colour of the increment in sky brightness and the place in the sky where the increase occurs. The increase can scarcely be due to an aurora because ot its colour and because of the rarity of aurorae at the low geomagnetic latitude of 3°S.

The increase in brightness of the zodiacal light is not accompanied by a shift of its position, which suggests that the extra emission is from the interplanetary dust cloud. This extra emission cannot be due to interaction between the dust cloud and the protons of the corpuscular stream because these carry insufficient energy. While it is possible that the increase is due to fluorescence of the interplanetary dust under the action of short wavelength radiation, the scarcity of the data precludes adequate discussion of this hypothesis.

References

1. G. Jones, Japan Expedition 1853/5, Vol. III, Washington, 1856.
2. E. O. Hulburt, Phys. Rev., 35, 1098, 1930.
3. E. G. Brenner, Observatory, 19, 206, 1896.
4. F. E. Roach, et al., Ap. J., 119, 253, 1954.
5. H. Huruhata, Publ. Astr. Soc., Japan, 2, 156, 1951.
6. C. T. Elvey and F. E. Roach, Ap. J., 85, 231, 1937.
7. A. Thom, Journ. B.A.A., 49, 103, 1939.
8. V. H. Regener, Ap. J., 122, 520, 1955.
9. A. Searle, Annals Harvard College Obs., 19, 165, 1893.
10. D. Barbier, Les Particules Solides dans les Astres. Sixth Liege Colloquium, p. 55, 1954.
11. Compilations of Solar-geophysical data. Nat. Bur. Stands. Colorado, U.S.A.
12. L. G. Jacchia, Nature, Lond., 183, 1662, 1959.
13. L. Biermann, Physics of comets. Fifth Liege Colloquium, p. 251, 1953.
14. J. W. Chamberlain, Ap. J., 120, 360, 566, 1954.
15. V. C. A. Ferraro, J. Geophys. Res., 57, 15, 1952.
16. V. C. A. Ferraro, Ann. de Geophys., 11, 284, 1955.
17. A. B. Meinel, Ap. J., 113, 50, 1951.
18. F. D. Kahn, M.N., 110, 483, 1950.

DOES ZODIACAL LIGHT INTENSITY VARY?

H., D. W.; *Nature,* 246:7-8, 1973.

Zodiacal light is the term used to describe the faint cones of light that extend away from the Sun in the ecliptic plane, or zodiac, of the sky. It is visible to the naked eye and can be best seen from a high mountain in the tropics where it appears in the western sky for an hour or so after the end of evening twilight, or in the east for a similar period before the beginning of dawn. The narrow cone of light stands up above the horizon and reaches about halfway to the zenith. It shares the apparent diurnal motion of the sky and during the night sinks below the western horizon until, two hours or so after its first appearance at the end of twilight, only the faint upper portions remain visible.

The cone is widest at the base, where it is of comparable intensity to the Milky Way, and narrows and grows fainter towards its apex. Studies of the spectrum of this light indicate that it is caused by sunlight scattered from small, solid interplanetary dust particles. These dust particles pervade the entire Solar System and by scattering sunlight they produce about one-third of the total light shining on the Earth's surface on a moonless night (the remainder is due to airglow and background starlight). Again using spectroscopic observations the Doppler shift of several of the Fraunhofer lines in the spectrum of the light can be measured. Radial velocities determined from these observations indicate that essentially all the dust particles are in direct orbits around the Sun.

It has been deduced from polarisation observations that the zodiacal light could be produced either by many very small particles (submicron) or by fewer larger particles having rough surfaces. The similarity between the spectra of zodiacal light and Sun light supports the latter hypothesis and it is now generally believed that most of the light comes from particles smaller than 10 μm (masses less than 10^{-9} g). Over a period of about 10,000 years these particles will be dragged into the Sun by the Poynting-Robertson effect, whereby the particles lose angular momentum due to the uniform re-radiation of adsorbed solar radiation. So if the dust cloud is in a steady-state condition there must be a continual source of small particles. The exact nature of this source is still in some doubt because the dust introduced must not only be of the right size but it must also be in the right place in the Solar System. A considerable fraction will come from the breakup of cometary nuclei as they pass close to the Sun and this process leads to the formation of streams of dust particles around the orbit of the comet. These dust streams then decay into the Solar System dust cloud. Asteroidal collisions could produce sufficient dust outside the Earth's orbit, which would slowly spiral into the Sun due to the Poynting-Robertson effect. Lunar ejecta produced by meteorite impact have also been suggested.

One way of deciding between these different possibilities is to investigate the temporal constancy of zodiacal light. Continual changes in the brightness as a function of, for example, the Earth's position around its orbit could indicate that the dust cloud was not equidense but was made up of a collection of individual toroidal particle streams. Constant brightness could, on the other hand, indicate that asteroidal debris was the main component of the cloud and/or that the meteor streams only contain large particles (mass greater than 10^{-6} g) which only over a period of about 10,000 years break up into smaller zodiacal dust particles and diffuse into the cloud. A third possibility is that new streams do contain many small particles but that these are perturbed from the stream orbit very quickly. It has also been suggested that the dust does not originate in either comets or asteroids but is just an extension of the outer solar corona and has a common physical origin. These last two suggestions would both give constant brightness.

Results of a satellite-borne experiment to measure the brightness of zodiacal light (6530 A) as a function of ecliptic latitude and the position of the Earth around its orbit are set out on page 26 of this issue of Nature. The satellite used, D2A 'Tournesol', was in a low altitude orbit and reaching a maximum height of 700 km above the Earth's surface. Levasseur and Blamont of the Service d'Aeronomie du Centre

National de la Recherche Scientifique obtain measurements of the zodiacal light intensity at 90° elongation, that is in the plane perpendicular to the Earth-Sun line, which agree very well on normal days (that is, for most solar longitudes) with previous measurements by other researchers. On certain days, however, the zodiacal light intensity increased by up to 100% above the normal value, and this increase lasted a few days. By making measurements in consecutive years Levasseur and Blamont found the variations in intensity to occur at the same positions around the Earth's orbit. They conclude that these increases were caused when the Earth passed through a meteor stream, the 100% increase on November 1971, which remained 50% above the normal level for six days around the maximum, being the result of an encounter with the Andromedid stream, a toroid of particles in space produced by the decay of comet Biela.

The April Lyrids have also been detected, and one very interesting fact emerging from the observations is that width of the stream of small dust particles detected by the zodiacal light observations appears to be twice as wide as the stream of much larger (mass $\sim 10^{-1}$ g) particles observed as visual meteors. Levasseur and Blamont conclude that the cloud of zodiacal dust particles is not uniform but consists of a collection of streams of particles, created by the decay of recent comets: these streams have a relatively short lifetime and defuse into a more general background on account of the Poynting-Robertson effect. Interparticle collisions and planetary perturbations can also cause this stream decay.

Interestingly, Alexander, Arthur, Bohn and Smith (Space Res., 13, 1973) have also detected increases of up to 100% in the density of picogram particles, in this case in selenocentric space, by using microphone sensors on board the satellites Lunar Explorer 35 and OGO III. These increases were found to coincide with the major meteor showers.

A slightly cautionary note, however, must be introduced. Sparrow and Ney (Science, N.Y., 181, 438; 1973), who have been measuring the intensity of the zodiacal light for four years from the satellite OSO-5, have failed to detect any temporal variations in the brightness or polarisation and have concluded that this variation if present at all is less than ± 10% of the mean value. Sparrow and Ney discount all the many previous observations which purport to show correlations between zodiacal light intensity and solar surface brightness, magnetic storms, lunar phases, the position of Comet Encke, the Earth's position around its orbit and the eleven-year solar activity cycle.

At present these contrasting observations make the choice of a hypothesis to explain the origin of the zodiacal dust cloud very difficult.

Chapter 9

MARS

INTRODUCTION

Of all the other planets that circle the sun, we know Mars the best. Though not as conspicuous in the night sky as Venus, there is no opaque atmosphere to hide its surface details from terrestrial and spacecraft telescopes. Thanks to the Mariner and Viking probes, we now have high-quality, close-up photos of the Martian surface. One would think that all the mysteries of Mars would have melted away under such scrutiny. But new mysteries have replaced some of the old ones, and many of the old enigmas persist down to the present day.

Martian mysteries fall into three broad categories: (1) permanent visible surface features; (2) transient Martian phenomena resembling the transient lunar phenomena (TLPs) of Chapter 6; and (3) the rather odd Martian moons. Whether permanent or transient the features and events seen on Mars emphasize that this is a dynamic planet with still-active atmospheric and geological processes. There have been, and perhaps still are, glaciers, floods, quakes---almost the full geophysical repertoire. Then there are spacecraft photos showing Martian phenomena with no terrestrial analogs. There may not be, as Percival Lowell once believed, a race of Martians maintaining a life-sustaining planet-wide canal system, but Mars today is no less intriguing to the planetary scientist than it was to Lowell.

FEATURES OF THE MARTIAN SURFACE

The question of the Martian canals, like that of UFOs, has been settled and resettled but is still with us. What did Schiaparelli, Lowell, and dozens of other astronomers really see? They say they saw images ranging from wispy streaks to a magnificent planet-wide network of clear-cut precise lines---some of them double. Photographs from earth proved that the so-called canals were not all subjective. However, Martian spacecraft have not confirmed a system of Martian canals. What then was seen and sometimes still is seen through the telescope? Interestingly enough, telescopic observations of Venus and Mercury sometimes give the impression of linear features. (See Chapters 3 and 4.)

Spacecraft photos have shown channels on Mars, but they all seem natural. Evidence of fluid erosion is widespread: channels, arroyos, and extensive drainage patterns. Water is the assumed eroding agent, but the present low atmospheric pressures on Mars suggest the possibility of other fluids. Equally puzzling is the origin and eventual disposition of this fluid.

Mars also displays a wide range of topographical curiosities---hillocks, ice formations, ridges, chaotic terrains, and so on. There is no room here to document the Martian volcanoes, ice caps, escarpments, and great canyons that emulate terrestrial features. Emphasis is on the more unique and anomalous features.

• Those Famous Canals

MARS AND HIS CANALS

Wilson, H. C.; *Sidereal Messenger*, 8:13-25, 1889.

Up to 1877 the observers appear to have confined themselves chiefly to delineating the outlines and variation of tint of the dark areas which immediately strike the eye when one examines the disc of Mars with a telescope of sufficient power, and to have neglected to examine carefully and persistently the bright reddish areas. During the opposition of 1877, which was exceptionally favorable, Mars being at its nearest approach to the sun and to the earth at the same time, M. Schiaparelli, the director of the Observatory at Milan, had the happy inspiration to concentrate his attention upon the great brilliant areas of the continents, in order to study their minutest details. He was rewarded by a brilliant discovery, of details so numerous and surprising in character that many astronomers are still, after the lapse of twelve years, incredulous of their reality. It was no less than a perfect network of very narrow dark lines, mostly straight, running across the continents in all directions, connecting the seas. To these Professor Schiaparelli gave the name of "canals," because, I suppose, of their straightness and from the fact of their connecting the seas. In 1879 he again saw the canals of 1877 and several new ones. No one else was able to see them, although many observers were provided with more powerful telescopes. Two or three saw some narrow markings on the planet, but it was not certain that these were the same that Schiaparelli had seen. During the opposition of 1881-82 Schiaparelli, following up his wonderful discoveries, struck a last blow to the confidence of the astronomers who hesitated to follow him; this time the canals appeared almost all to be double. A new canal appeared beside the old one, rigorously parallel in most cases, and starting, not from the same point of origin as if the old canal were simply divided into two component canals, but from a different point, as if a new canal had really been formed parallel to the first.

I have prepared a copy of a chart of Mars constructed by Professor Schiaparelli, from his observations made in 1882 and 1886, upon which you can see the network of fine lines, the canals, crossing the continents from sea to sea in all directions. They are somewhat exaggerated in distinctness in the drawing, so that you may see them with ease. This chart is upon what is called Mercator's projection, the regions near the poles being distorted out of all proportion, but the equatorial regions represented accurately. [See p. 385 for similar map.]

Will you compare the lines of the chart with Schiaparelli's own words of description:

> "These lines [the so-called canals] run from one to another of the dark spots of Mars, usually called <u>seas</u>, and form a very well-marked network over the bright part of the surface. Their arrangement seems constant and permanent (at least so far as can be judged by the observations of four and one-half years); but their appearance and the degree of their visibility is not always the same, depending on circumstances which we cannot at present discuss with full certainty. In 1879 many appeared which had not been seen in 1877 and in 1881-1882 all those which had been seen the first time were re-discovered, and other new lines as well. Their number could not be estimated as less than 60. Sometimes these lines or canals show themselves under the form of diffused and indistinct shading; at other times they appear as very definite and precise markings of uniform tone, as if they had been drawn with a pen. In most instances their curvature differs very little from a great circle, if indeed it does differ; some others however are much curved. The breadth of the finest

can hardly be estimated at less than 2° [70 miles] of a great circle but in some cases it reaches to about 4°. As to the length, that of the shortest is certainly less than 10°, others extend to 70° and 80°. The color is sometimes as dark as in the seas of Mars, but often it is brighter. Each canal terminates at its two extremities either in a sea or in another canal. I know of no instance where one end remains isolated in the midst of one of the bright areas of the surface without resting on lines and dark spaces.

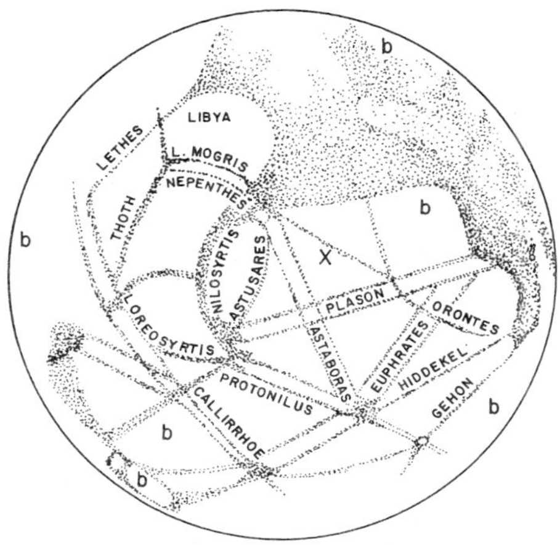

Schiaparelli's drawing of Mars, June 1888, using the 18-inch refractor at Milan

"Now in many of these lines it has fallen to me to observe the curious and unexpected circumstance of a doubling or reduplication; this happens in the following manner: To the right or left of a pre-existing line, which suffers no change from its previous direction or position, another line appears, nearly equal to the first and parallel to it; in some instances a slight difference of appearance being visible and sometimes also a slight divergence of direction. The distances between the pairs of lines formed in this manner varies from 6° to 12° of a great circle; there were also other lines which I suspected to be doubled, but the distance being less than 5° or 6° the telescope did not succeed in resolving them, and showed in that place a large, broad, and somewhat confused stripe. Sometimes a line is divided by another which intersects it into two districts or sections of unequal darkness and extent; in this case the companion line is divided into two sections in the very same way, with one exception no sensible irregularity of direction or of shape could be ascertained with the power used in these observations, which was always one of

417. Some of the pairs show so great a regularity that one would say that they were systems of parallel lines drawn by rule and compass. Perhaps however this regularity will not resist the use of a high magnifying power. In various instances, pairs are so connected the one with the other as to form a polygon of double lines with very pronounced angles, and such a series then occupies a great space. Two pairs sometimes cut each other without being interrupted; meeting then three by three they form at the points of triple intersection a network of which our telescope could only give an exact and complete resolution in one or two cases.

"Excluding those cases not well ascertained through the inability of the instrument to resolve objects so minute, the number of reduplications I observed in the last opposition is 20, of which 17 had been established in the course of one month, from January 19 to February 19, the mean date corresponding very nearly to the end of the second month after the vernal equinox of the planet. The phenomenon seems to be confined to a definite epoch, and it appears as if it took place simultaneously over the whole planet's surface occupied by the bright areas. No trace could be ascertained in 1877 during the weeks which preceeded or immediately followed the southern solstice of the planet. Only one isolated case presented itself in 1879, between December 24 and 26, and this was exactly reproduced under similar circumstances between January 11 and 12, 1882; it took place in the two lines named Nile I and Nile II on my chart of 1879. Both of these two epochs being close to the vernal equinox of Mars there is ground for believing that the phenomena of reduplication may be periodical, and perhaps connected with the position of the sun with respect to the axis of rotation of the planet."

When Schiaparelli's chart for 1882 was published it was received with incredulity and almost ridicule, by many. This was due in part no doubt to a fault in the engraving of the chart, which made the lines of the canals and the outlines of all the markings of the planet hard and unnatural, quite unlike the actual appearance in the telescope. But if you will compare the two charts you will see how difficult it is to reconcile the features of the one with those of the other. The different method of projection and the addition of so many new features, all exaggerated in distinctness, renders the whole aspect unfamiliar. Yet if one will start with a prominent marking as, for instance, Kaiser Sea (Syrtis Major), and follow the coast lines, he will find nearly every feature of the Proctor chart upon that of Schiaparelli and in its proper place. Perhaps the greatest differences apply to the long narrow inlets of Nasmyth, Bessel and Huggins. These are replaced on Schiaparelli's chart by narrow canals or pairs of canals (Protonibus, Ceraunus, Iris and Gigas). In fact if we examine the original drawings of Dawes we find that the appearance of these features corresponds very much more closely with the chart of Schiaparelli than with that of Proctor.

Up to 1886 no certain confirmation of the duplication of the canals or of their existence was obtained by any other astronomer, but in that year Professor Perrotin of the Observatory of Nice, succeeded with a 15-inch telescope in detecting fifteen of them, and witnessed also the duplication of several. It will give one an idea of the difficulty of seeing these objects, to know that Professor Perrotin, armed with so powerful an instrument, at first gave up the attempt after several days' fruitless effort; having renewed his trials still without success, he was about to give them up finally, when he succeeded in seeing the canal Phison which crosses Dawes continent.

During the present year both Perrotin and Schiaparelli were provided with still more powerful instruments, the latter with an 18-inch and the former with a 30-inch refractor, and the results obtained from the study of Mars' surface go to completely confirm Schiaparelli's discoveries. Several drawings of the planet by these two gentlemen have been published in recent periodicals and I have copied two of them for your benefit. The lower on the left hand (Fig. 2) is by Schiaparelli, from sketches made on June 2, 4 and 6. The most prominent marking near the center is easily recognized as Kaiser Sea (Syrtis Major and Nilosyrtis) of the chart. On the right is the curious forked bay of Dawes. The canals of the equatorial regions are drawn very much as they appear in the chart. Some old ones are omitted while several new appear. This (Astaboras) is straight and double instead of curved and single as in the chart. It is not really a separation of the old canal into two, but a new canal is seen starting from a different point of Kaiser Sea, and running parallel to the old one to a different point of the lake (Ismenius). The greatest difference is in the regions around the north pole. The system of canals is extended right up to the polar cap, and we notice across the ice cap itself two dusky streaks. This opposition was very favorable for the examination of the north polar regions, that pole being inclined considerably toward the earth, and the opposition occurring after the midsummer of the northern hemisphere of Mars, so that the polar snows were largely melted and the ice cap reduced to a minimum. It seems quite reasonable to suppose that these dark streaks may be open channels in the frozen polar sea. [Fig. omitted]

. . . .

What then are these strange markings, and why have they not been seen before? are questions which naturally arise. The discoveries of Schiaparelli were made with a telescope of only eight inches aperture. Why is it that so many observers armed with more powerful instruments have utterly failed to see them? Partly, perhaps, because their skies were not so transparent as that of Italy; partly because their eyes were not so keen as those of the Italian observer; partly also because the phenomena of the canals are periodic; but more because of lack of persistent, long-continued scrutiny of the planet's disc. Again Mars can be seen well only when near opposition, that is, about two months in the year and his distance at opposition varies from 34,000,000 to 64,000,000 miles because of the ellipticity of his orbit, so that a really favorable opportunity to observe minute details occurs only once in about fifteen years. Also, it is necessary for such observations that the sky of Mars be free from clouds over immense areas, whole continents in fact, and we can judge from terrestrial analogy that such a condition would be rare.

Several ingenious theories have been suggested to account for the canals. We can hardly admit that they are artificial canals constructed by the inhabitants. Their great width of seventy miles or more would prohibit that hypothesis, although we may suppose, with Proctor, those inhabitants to be of gigantic stature. Mr. Proctor suggested (M. N. xlviii, p. 307), that they might be rivers, the duplication being a diffraction effect, when mists hang over the river bed, but in doing so seemed to ignore their most characteristic features, their straightness, their frequent intersections, and the fact that they connect the seas, running from one to the other, and that the duplication is not the separation of the old canal, but the formation of a new one on one or the other side of the first.

M. Fizeau, the eminent physicist of Paris, explains them by the analogy of rifts in glaciers and considers the whole planet involved in a glacial epoch. He is led to this hypothesis by the presence of the dark canal in the north polar cap. Dr. Terby, of the observatory of Louvain, thinks this would necessitate a greater movement in the system of canals than has been observed. He finds no evidence of any change since they were discovered. Dr. Penard, a physicist of Geneva, suggests that the canals are immense fissures in the crust of Mars through which the waters flow from ocean to ocean. The mass of Mars, being about one-eighth that of the earth, has cooled off more rapidly, producing great fissures in the crust extending to a very great depth. Another writer suggests for the very same reason that the canals are ridges or wrinkles in the crust, or, in other words, mountain ranges. There are serious objections to each of these and we are still far from a satisfactory explanation. I am rather inclined to the view that these markings are in part, at least, rivers and chains of lakes, that the continents are low and flat and subject to extensive inundations at certain seasons of the Martial year. It seems possible in this way to account for the variations noticed in many of the markings and there appears to be no inconsistency in supposing evaporations and precipitation to be more rapid and abundant on Mars than upon the earth, because of the smaller force of gravity.

....

THE LINES ON MARS

Hussey, William J.; *Science,* 20:235, 1892.

In Science, Sept. 23, Mr. C. B. Warring communicates a theory to account for the gemination of the so-called canals of Mars. He suggests that the phenomenon may be due to a defect in the eye of the observer by reason of its possessing the power of double refraction in some or in all directions. That some eyes do possess the power of double refraction is a well-known fact. It is a defect which, I imagine, is much more common than is generally supposed. It may be suggested that data representing a large number of cases might show astigmatic eyes to possess the power of double refraction more frequently than others. I do not know that any data have been collected upon this point.

Concerning the existence of the canals of Mars and that they are sometimes really double, I have no doubt. My own recent work at the Lock Observatory has convinced me that they are not illusions due to imperfect eyesight. During the present opposition, I spent about thirty nights in the work on Mars, working with Professors Schaeberle and Campbell. On about half the nights I saw the so-called canals with more or less distinctness, but on only one occasion did I clearly see a canal double. This was August 17, when the canal called Ganges on Schiaparelli's map was clearly seen to be double, and was so drawn in my notebook. That the doubling was real and not apparent is evident from the fact that Professors Schaeberle and Campbell both saw the same canal double on the same night, and drew it so. Other canals, some of them nearly parallel to Ganges, were seen that night, but none of them appeared double.

Our work was done independently. In turn each went to the telescope, and made a drawing of what he saw. We did not see each others' drawings, nor did we talk of what we had seen. It was not until the next morning that we learned that each had seen Ganges double.

THE CANALS OF MARS

Maunder, E. Walter; *Knowledge*, 17:249-252, 1894.

Seventeen years ago our knowledge of Mars appeared to be in a very satisfactory state. The principal markings had been often and long observed, and had been found to be permanent. The Kaiser Sea, the Oculus, the Maraldi Sea, had been observed by Hooke and Cassini, Herschel and Schroeter, Beer and Madler, Dawes, Lockyer, Knobel and Green. The inference, from the annual waxing and waning of the white polar caps, that cloud, snow, and rain were features of the meteorology of the planet, had been confirmed by the testimony of the spectroscope, and that the white caps themselves were composed of ice or snow was a natural conclusion. It followed necessarily that there must be water on the surface of the planet, and the dark spots were considered as seas, leaving the brighter districts to be regarded as land. Mars was, in short, a smaller copy of our own world. "The analogy between Mars and the earth" was pronounced to be "by far the greatest in the whole solar system."

Schiaparelli's discovery of the "canals"---to use a term which, however misleading it has been, has now been too strongly sanctioned by custom to be easily changed---was the beginning of a new epoch in Martian observation, and its chief and most patent result has been to disturb our old conceptions of the analogy between Mars and the earth, and consequently to unsettle our notions of the physical condition of the planet. The reaction has gone so far that Prof. Schaeberle has reversed the old identification of the dark spots as seas, and claimed that the bright districts are to be thus regarded. The grounds for this change of view have not been generally accepted, and more is to be said for an objection raised by Prof. W. H. Pickering, that just as snow is not the only substance to give a bright white reflection---hoar-frost or cloud would serve as well---so water is not the only surface which would appear dark; forests and prairies would appear as at least sombre districts.

The "canals," however, have been the great element of disturbance. The report of their discovery was received in 1877 with very considerable mistrust. But time has been on the side of the patient and keen-eyed Italian astronomer. Each succeeding opposition has seen more of his "canals" identified by other astronomers; each construction of a new and more powerful telescope, each enlistment of a fresh earnest and skilled observer, has meant a further confirmation of his work. So that now there is a bulk of evidence in favour, not merely of some of Schiaparelli's "canals," but of his canal system as a whole, by no means lightly to be set aside.

The positive evidence has not, however, destroyed or overcome the negative. Both still hold the field, and both must be considered. The principal points against the actual existence of the "canals," as represented by Schiaparelli, may be summarized as follows:---

1. They are extremely narrow objects, approaching the theoretical limits of visibility, even when Mars is at its nearest approach. Thus a fine drawing reproduced in Flammarion's great monograph on "Mars," p. 568, shows "canals" of breadths not exceeding 0.04" or 0.05" of arc. The two branches of a doubled "canal" in one part of the sketch are distant from each other only 0.25". Nor is this all. In his account of his 1888 observations (maximum diameter of planet, 15.4"). he states that in a set of canals on Madler Continent he could make out, not merely the two banks of the canals, but also "very small undulations in their two banks, which could be distinguished from each other."

2. The distinctness of some of these objects, though so narrow, does not seem impaired by distance. Thus, in the opposition of 1877, the Indus was better seen by its discoverer when the planet had receded a great distance from opposition, and was very well seen when the planet was only 5.7" in diameter, and the canal 0.2" in breadth.

3. The great divergency between the descriptions of different observers. To take the one (and all important) feature of breadth. In the opposition of 1890, when Schiaparelli was observing "canals" down to a breadth of 0.05", Holden and Keeler, at the Lick Observatory, always saw the canals as dark, broad, somewhat diffused bands. We may judge what is meant by "broad bands," for the record goes on:---"In bad vision they were drawn in that way by Schaeberle also. Under good conditions, however, the latter observer described them as narrow lines, a second of arc or so in width." A second of arc corresponded at this time to a minimum of 6° of a Martian great circle; that is to say, to about the breadth of Herschel Strait, and very nearly to that of the Mare Sirenum, the narrower end of the Maraldi Sea, or to nearly double the breadth of the Nasmyth Inlet, or the point of the Kaiser Sea, markings seen even by the earliest observers. If Schaeberle could describe markings of such dimensions---markings twenty times as broad as Schiaparelli's narrowest canals---as "narrow lines," what must have been the breadth of the "broad bands" of his two colleagues? It is clear that the phenomena observed by the Lock observers were quite of a different order to those recorded by Schiaparelli.

4. The greatness and suddenness of the changes remarked in the "canal" system. The "gemination" or doubling of the "canals" has been remarked "to take place in a relatively short space of time, and by a rapid metamorphosis....Sometimes the metamorphosis has been completed in the interval of twenty-four hours between two consecutive ob-

servations. So far as the observer could judge, the phenomenon took place simultaneously along the entire length of the canal doubled." Schiaparelli himself draws attention to the strange and rapid changes taking place on the planet, and remarks: "Evidently the planet has fixed geographical details similar to those of the earth, with gulfs, canals, &c., on an irregular plan. There comes a certain moment and all this disappears, to give place to these grotesque polygons and geminations, which clearly represent approximately the former state; but it is a coarse, and, I might say, almost a ridiculous mask."

5. The "canals," when near the edge of the disc, are apt to be represented as much straighter than they could possibly be.

6. To these difficulties may be added that at the very time when to some observers the canal system was most developed, others have sometimes only been able to perceive the usual markings of the planet in their customary configuration.

How are we to explain these curious discrepancies?

First of all, many of these differences are to be explained by differences in "seeing" power, including in that term not merely atmospheric conditions, but instrumental and personal differences, such as the aperture of the telescope, its defining power, the magnification employed, and the keenness of sight and artistic skill of the observer. Two illustrations may be given of these. In 1892 a number of drawings by different observers were made of the De la Rue Ocean. The observers best equipped with instruments, and most favoured by observing conditions, recorded this ocean to be marked by the presence of one, two, three, or more islands. The better the conditions the more the number of islands represented; the smaller the aperture, and the less the experience of the observer, the fewer; so that the drawings showed every variety of representation, from an undivided greyish spot to an archipelago of five distinct and clearly separated markings.

Again, in Flammarion's "Mars," a series of representations are given of Herschel II. Strait. First in order of date come Beer and Madler's drawings of 1830, in which we see the "strait, not as a strait, but as a snake-like inlet, ending in a dark round spot. In 1862 Lockyer (Flammarion's "Mars," p. 155) represents the terminal spot as a nearly rectangular marking. In the same year Kaiser gives the northern edge of this rectangular spot a shaded appearance, as if he suspected the presence of the two "estuaries" (<u>Ibid.</u>, p. 174). In 1864 Dawes resolves this rectangular spot into the now well-known form of "Dawes' Forked Bay." In 1879 Schiaparelli (Flammarion's "Mars," p. 336) traces two canals flowing into the two arms of the Forked Bay.

These different aspects correspond precisely to the effect of improved "seeing," using that term in the larger sense adopted above. For if a careful drawing be made of the district under the aspect which it presented to Schiaparelli in 1879, and the drawing be approached from a distance, it will assume in succession precisely the apparent phases delineated by the above-named astronomers; allowance being made for the effect of foreshortening due to the difference of presentation of the planet, and to the inevitable inaccuracies of the drawings of even the most skilful artists.

I ventured to lay stress, in my paper on "The Tenuity of the Sun's Surroundings" in the March number of <u>Knowledge</u>, on the fact, which we so easily overlook, that "the smallest portion of the sun's surface visible by us as a separate entity, even as a mathematical point, is yet really a widely extended area." The same truth applies in its degree to the planet Mars. We have no right to assume, and yet we do habitually assume, that our telescopes reveal to us the ultimate structure of the surface of the planet.

An illustration of this point was afforded me some time ago, when a question arose as to the limit of visibility to the naked eye of sunspots. I was astonished to find that a group of spots had been recorded as seen directly, when their total area was much less than that of many well-defined circular spots that had entirely escaped scrutiny. A few experiments convinced me, however, that the observation was perfectly correct, and that it was often possible to see a straggling group of small unimportant spots, when a single spot of considerably greater total area would be invisible.

I then tried how small an object could be detected without optical assistance, the objects being always black marks (Indian ink) on white glazed paper, illuminated by dull diffused daylight.

The limit of my vision for a circular dot ranged from a diameter of 80" to 36" of arc. One of 20" was quite invisible; of 40", distinctly seen. This was decidedly smaller than I had anticipated. But the limit for a straight line, to my surprise, was as low as 7" or 8'; 12" was easy and conspicuous. More than this, a pair of lines, each only 4" in breadth, and the pair separated by say 20", was visible as a faint single line; two lines, even of only 3", meeting at a very acute angle, were visible after their separation had diminished below about 25". In each case the object was unmistakably discerned, and appeared as a line or dot; it was not, of course, defined so as to be seen in its true form.

Further, a chain of dots, each of 20", irregularly disposed along a straight line, the average interval between any two dots being three times the diameter of a dot, was easily seen as a continuous straight line, whilst a double chain of yet smaller dots, each 4" in diameter, and the two chains some 40" apart, was visible as a very faint continuous line.

The theoretical limit of visibility has been given as 40" or a little greater, a limit with which the above observations are really in tolerable accord; for when the angular diameter of the object fell much below 40" it was seen, not as a minute defined black dot, but as a grey diffused spot of about 40" or 45" in diameter. It would seem, then, that the smallest perceptible area is about 40", but that if there be within any such given area a sufficiency of dark markings, however individually minute, to turn the white to a decided grey, then that area will be visible as a grey spot. Two lines or a number of dots, easily visible as one object when close together, can readily be made invisible by a greater angular separation.

It seems to me that these rough experiments have a decided bearing upon the "canal system" and the supposed changes of Mars. It seems a violent hypothesis to call in inundations extending over many thousands of square miles to account for merely temporary changes, for sooner or later the old districts take on the old configuration, more especially since, as I have already shown, the meteorology of Mars must necessarily be a languid one. Indeed, it may happen that whilst several independent observers have recorded a change, others equally skilled have seen the planet as before.

. . . .

THE CANALS OF MARS PHOTOGRAPHED

Butler, Charles P.; *Knowledge*, 2:204-205, 1905.

A short time back the important news was telegraphed from Lowell Observatory that the much-criticised markings on the Martian surface had been successfully photographed by Mr. Lampland, and the last circular from the observatory not only confirms this, but contains an actual print from the negative showing the markings.

Many attempts have been made at the Lowell Observatory at Flagstaff, Arizona, to photograph the canals of Mars, commencing with the success of Mr. Douglass in 1901, when, by using a Wallace screen, he secured a good picture of the Mare Acidalium. Encouraging as the result was, there were no signs of any canal markings. The two chief difficulties were the variation of the atmospheric tremors, and the insufficient sensitiveness of the photographic plates. The endeavours to get rid of these errors resulted in the ordering of a film camera wherewith a succession of pictures could be rapidly taken behind a Wallace screen; and with this arrangement Mr. Lampland has got his interesting results. A most important item, however, has been the suitable cutting down of the aperture of the photographic telescope to suit the particular state of atmosphere at the time of observation.

From the many plates secured, the one taken on May 11 was selected for the reproduction sent with the circular. Side by side with the print is placed a photograph of a drawing by Professor Lowell, made before the camera was placed in position, and this serves the double purpose of showing the confirmation by the photograph of the objectivity of the visual observation, and at the same time of serving as a chart to it.

The print is enlarged 1.8 times from the original negative; and not only are the canals easily discernible, but it is evident that they are continuous lines, and not syntheses of other markings, as has been suggested by various writers.

An additional note by Mr. Lampland states that the photographs were obtained with the 24-inch Clark refractor of 386 inches focal length. The camera carries a negative enlarging lens, the equivalent focal length of the combination being 148 ft. The camera carries a plate holder for 3-1/4 x 4-1/4 plates, movable perpendicularly to the optical axis, thus permitting a dozen or more exposures on the planet, for the focal length given, to be made on the same plate.

A colour screen is placed immediately in front of the plate, separated by a small space to minimise the effect of small particles of dust or other extraneous matter. Cramer's isochromatic plates were used, this make being chosen on account of the fact that one of the maxima of the curve of sensitiveness of the plate coincides almost exactly with the vertex of the colour curve of the large objective. The best results have been obtained with the 24-inch stopped down to 9 or 12 inches, and the exposures were usually about eight seconds each with the 12-inch aperture.

ON SOME OBJECTIONS TO THE REALITY OF PROF. LOWELL'S CANAL SYSTEM OF MARS

Antoniadi, E. M.; *British Astronomical Association, Journal,* 20:194-197, 1910.

In 1888, Prof. Schiaparelli published a chart of Mars, based on his observations, on which we find a network of lines. But this map was only a schematical one, intended to convey an idea of the length and direction of the streaks, without giving their varying breadth and intensity.

Prof. Schiaparelli's "canals" showed the following widely differing appearances:---(1) Slight irregular shadings; (2) diffuse streaks; (3) grey bands with fainter edges; (4) streaks shaded on one side only; (5) streaks which were narrow near the "seas," and which expanded on the "continents" like "the tail of a comet"; (6) dark lines; and (7) black lines, with occasional irregularities. The majority of these markings seemed variable in width and intensity; while many of them were likely to appear sometimes double.

The objective basis underlying Prof. Schiaparelli's streaks was established in 1909 by the writer, who, while engaged in scrutinizing the appearance of the "continental regions" of the planet, in the most independent spirit possible, repeatedly witnessed the steady visibility of groups of complex irregular shadings in the exact positions of the "canals" detected at Milan.

In 1896, Prof. Lowell published a chart of Mars containing 183 canals crossing the "seas" as well as the "continents," all uniform and regular in appearance, and, in fact, quite unlike Prof. Schiaparelli's streaks. For, while the breadth of the "canals" seen at Milan varied between 1° and 5° or more, that of the Flagstaff lines ranged only from $1/2^{\circ}$ to $1\text{-}1/2^{\circ}$. Nor was their width shown to increase in high latitudes on Mercator's projection, which, by casting the poles to infinity, vastly exaggerates the size of real markings in the vicinity of the polar regions. Finally, the number of canals was greatly increased by the subsequent work of the Flagstaff observers, so that the late Prof. Newcomb, upon working out the area of the 398 canals seen in Arizona, found that it exceeded one-half of the total area of the planet.

Now, there are many objections to the reality of Prof. Lowell's geometrical network. But, as the space at our disposal is limited, we shall choose three objections only, leaving hardly any room for a reply.

Objection I.---The straight lines have an unnatural appearance.---Sharp, narrow, dark lines, obviously identical with those of Prof. Lowell, have frequently been noticed by the writer under tremulous seeing. But as their wonderful sharpness, which transcended the defining powers of even the great Meudon refractor, always contrasted with the

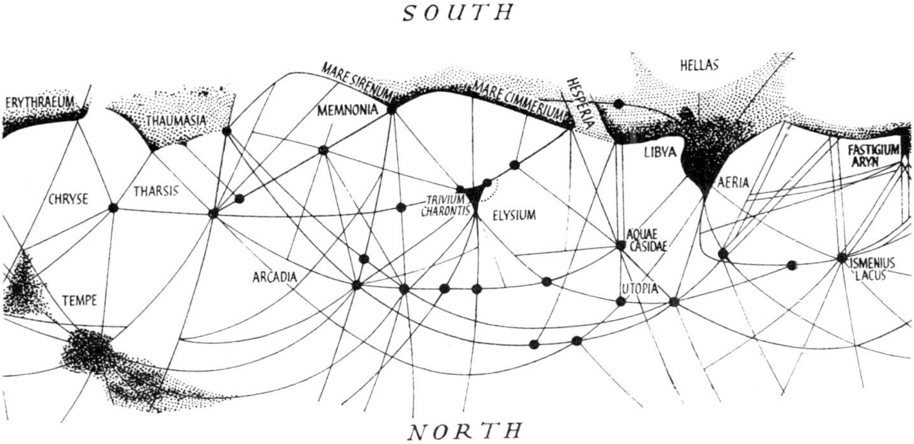

A Mercator Projection of Mars by Lowell, showing the major canals as he saw them. The black dots are oases.

very gentle blur enveloping all real markings of the planet, we safely conclude that such lines can by no means whatever belong to the Martian surface.

Objection II.---Very often the Flagstaff canals do not obey the laws of perspective.---A drawing of Mars, made in 1909 by Mr. Slipher, of the Lowell Observatory, shows many thread-like canals, of which four are straight near the limb. The appearances of lines on a sphere are familiar to all. Lines which are straight when transiting the central meridian must assume a curved form near the edges of the globe. Nor will the doctrine of chances at all justify the phenomenon. Geometrical reasoning and common sense oblige us, therefore, to pronounce illusive all canals defying the eternal laws of perspective.

Objection III.---Prof. Lowell's canal network disappears under the worst circumstances for it, that is, when much more delicate detail is quite plain.---Were large apertures to show less planetary detail than small ones, then the alleged existence of narrow lines on Mars might have been speciously argued with some glimmer of possibility. But, as shown by Prof. Barnard, great refractors reveal detail far beyond the reach of small telescopes, and this is a truth which no one, having used both kinds of instruments, could ever dream of questioning. Whenever definition was good at Meudon, the Martian "seas" and "continents" were variegated with innumerable complex markings, of which none have ever been drawn at Flagstaff. And yet nothing was seen to countenance the existence of a network of fine lines on Mars, then all but at its nearest approach to the Earth.

Such evidence is decisive. The visibility of delicate planetary detail is chiefly a question of separating power; and the separating power of the Henry 32.7-in., used invariably without stops, beats the Flagstaff 24-in. telescope, stopped down to 12 or 15 inches, in the ratio of 5 to 2. We readily recognise, even without proofs, the superiority of definition in the Arizona tablelands. No one, indeed, will question the importance of steady air in high stations, as such steadiness will lengthen the spells of perfect seeing, and permit a good scrutiny of all parts of the planet. But, even in Northern France, one does get sometimes perfect seeing, and, for some seven hours in 1909, we had at Meudon very good definition indeed; and then Mars appeared as if he were twice as near the Earth as at Flagstaff. And he appeared without a network.

The evidence of Prof. Barnard in 1894, and of Prof. Hale in 1909, is thoroughly confirmed by the writer's experience with the Henry glass. And evidence of this kind goes far to settle the question.

We have all applauded not only the great enthusiasm and magnificent example given to the world by Prof. Lowell, but also the numerous and valuable discoveries with which he has enriched our knowledge of Mars; his detection of quite a host of new objective "lakes," or "oases"; his new facts concerning the melting of the polar snows; his accurate delineations of the outlines of the Martian "seas"; his discoveries of small bays formed by the latter; and, last, but not least, his truly wonderful photographic success, which has set areography on a basis far more solid than was possible to attain by the various eye-estimates of the past. And if in our stern regard for the truth---our only concern---we are forced to dissent from him on the would-be existence of straight lines on Mars, we feel sure that he cannot take exception to criticism expressed in a perfectly fair and impartial spirit.

CURIOUS GEOMETRICAL FIGURES APPEARING UPON MARS

Pickering, William H.; *Scientific American*, 134:57-58, January 1926.

It is a rather curious coincidence that at each of the recent very near approaches of the earth to Mars, strikingly regular, although only temporary, geometrical figures should have appeared upon its surface. The well-known cross, centered in the approximate circle of Hellas, appeared to Schiaparelli in 1879. (Figure 1.) It may have appeared in the unusually close apparition of 1877, but in that year he saw only the circle and the single vertical canal. The diameter of the circle is 900 miles. The cross has of late years been replaced by an irregular curved structure.

The next very close approach of our planet occurred in 1892. In that year a regular pentagon with central radiating canals was seen in Arequipa (Figure 2.) One of the canals was missing, but it may have been too delicate for our 13-inch telescope, or it may have been covered by a temporary Martian cloud when we made our drawing. The center of the pentagon was located at Ascracus Lake, as it is now called, although then unnamed, and the diameter of the figure was 800 miles.

In the year 1909 no very complicated figure seems to have been seen

The Lowell Observatory published no special report on that apparition, as they had done for some of the earlier years. M. Jarry-Deloges on September 26 drew Figure 3, which is perhaps more interesting than striking, because large four-sided figures are most unusual upon Mars. The central meridianal canal is Laestrigon, the two lower ones Tartarus and Cerberus. The horizontal one, Aesculapius, has become more prominent of late years than formerly, while the two upper ones are rarely seen, but both have been confirmed by later observers. The length of Laestrigon is 1,200 miles. (Observations des Surfaces Planetaires, 2, Plate 9.)

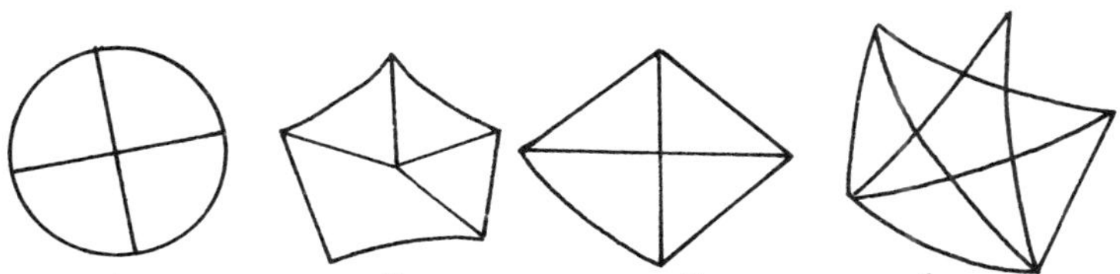

1 2 3 4

At the apparition of 1924 the earth passed closer to Mars than had been the case for over a century, and closer than will be the case again for a century more. This year an unusually large and complicated figure appeared. Again it was pentagonal, apparently a favorite figure with the supposed Martians, but in a still different place. The pentagon of Elysium, which later changed into a circle, is of course well known to all students of the planet. It measures 800 miles in diameter, but this one is much larger, measuring approximately double the size. Moreover it is not a true pentagon, although of that nature, but a five-pointed star with symmetrical appendages attached. (See Figure 4.)

. . . .

It is indeed curious that these complicated figures should occur on Mars and still more curious that they should all be only temporary, and should appear only at the very close apparitions of the planet, while at other apparitions the numerous other canals should present no symmetrical structures. There is of course a definite system to their customary arrangement that we are now gradually beginning to understand, and as above noted, the avoidance of large four-sided figures among them is of itself curious. Some people will doubtless believe that these designs are not due to mere accident, but are artificial, and constructed for our special edification, and as an announcement of the existence of intelligent life on their planet. If so, we wish the Martians would plant them out, or otherwise construct them, more frequently than once in every fifteen and a half years. If not due in the past to chance, we wonder very much what figure will appear at the next close opposition in 1939. However we must not expect too much of the Martians, and if they have been doing this sort of thing for the last 10,000 years or more, we must consider them to be far more persevering in their endeavors to communicate, than the inhabitants of our own self-satisfied, and very unresponsive planet.

THE SURFACE OF MARS

Anonymous; *Nature*, 147-30, 1941.

The close approach of Mars to the earth a year ago provided a favourable opportunity for close observation of the planet's disk from southern latitudes. Dr. E. C. Slipher of the Lowell Observatory, who obtained some 8,000 photographs from South Africa during the summer of 1939, describes his work in the Telescope of September-October 1940. The article is illustrated with nearly forty half-tone reproductions of selected plates; from these excellent copies some idea can be obtained of the fine detail reached by the original photographs. The irregular changes noted during the opposition were striking and are described in some detail; the seasonal changes followed their expected course, and are dealt with briefly. Dr. Slipher claims that his photographs record so many of the 'canals' and 'oases' in the positions and of the forms shown on Lowell's maps of the planet that the reality of these markings must now be regarded as beyond doubt. A recent suggestion that the canals are divisions between areas of unequal shading is discounted; no example has been discovered where the surface brightness on the two sides of a canal shows any visible difference. The colour-filter photographs suggest that the hazy north polar hood, which showed very rapid day-to-day variations of form during the 1939 opposition, is atmospheric in origin; it precedes and accompanies the deposition of the more brilliant white polar cap, and may consist of very fine ice spicules.

ON THE NATURE OF THE CANALS OF MARS

Sagan, Carl, and Pollack, James B.; *Nature*, 212:117-121, 1966.

Since their discovery in 1877 by Schiaparelli the canals of Mars have had a rather chequered astronomical career. According to those who systematically saw rectilinear markings on Mars (for example, Lowell and Pickering), the canals of Mars have the following properties:

(1) Although some canals are reported in dark areas of Mars, they are usually a feature of the bright areas.

(2) Many different observers---some working independently---draw the major canals in the same positions on Mars.

(3) The length of the canals varies from hundreds to thousands of kilometres.

(4) Some dispute existed (even among those who regularly saw canals) on their widths, Lowell in particular claiming widths down to and even below the Rayleigh resolution limit for visual observations, that is, several tens of kilometres.

(5) The short canals are described as linear, the long ones as following great or small circles on the globe of Mars.

(6) There are no reports of canals terminating in the bright areas.

(7) Canals terminate either at major dark areas, or at small circular dark areas known as oases; in some cases Lowell announced

more than a dozen canals emanating from a single oasis.

(8) The canals are described as undergoing seasonal changes very similar to those described by many observers for the dark areas; in particular the visibility and contrast of the canals against the surrounding bright background is enhanced in Martian spring and summer.

(9) Some canals are reported to geminate, that is, to appear double, particularly in the spring and summer; the gemination of the canals is described as occurring in a period as short as a few days.

A few of the larger classical canals are undoubted Martian features and have been photographed. Deuteronilus and Ceraunius---to be discussed later---fall into this category. Although they seem vaguely linear, their photographic appearance is much more diffuse than their early visual descriptions. In the next few paragraphs we shall be concerned chiefly with the less apparent fine canal network of Lowell et al.

The astronomical community rapidly polarized into a minority who could see these canals and a majority who could not. In the view of the second group the existence of the canals was highly suspect, and for them such dubious observational issues as the width or gemination of the canals did not arise. E. M. Antoniadi was a careful observer of Mars and an excellent draughtsman. The quality of his work can be extimated by comparing his drawings with the best of recent photographs of Mars. Some other drawings of Mars, for example those of Lowell, are not very impressive when subjected to such a test. Antoniadi found that under the very best seeing conditions the rectilinear markings were resolved into an array of disconnected, fine, mottled detail (see Fig. 1). Evidently under inferior seeing conditions the eye has a tendency to view the disconnected detail as continuous linear features. Other canals were found by Antoniadi to be fairly abrupt discontinuities between bright and dark areas. These observations, confirmed by later observers (see, for example, Dollfus) and consistent with contemporary knowledge of human vision, have convinced many that the canals of Mars are a psychophysiological rather than an astronomical problem.

There are, however, two facts which suggest that the problem is not so easily put to rest. We must first ask why canals are not seen when we are looking at disconnected fine detail near the limit of resolution for other objects---for example, Mercury or the Moon or Jupiter. Inspection of Fig. 1 shows that the disconnected fine detail is not randomly distributed over the field of view but rather is ordered in linear arrays. Second, the Mariner 4 photography of Mars, viewing the planet in a season of minimum canal visibility according to the early observers, nevertheless uncovered a variety of apparently rectilinear features. The two most prominent of these, commented on by many (see, for example, Burgess). The features are directed neither along the direction of video scan nor at right angles to this direction. The upper (northern) feature has a width of several resolution elements---that is, between 3 and 10 km. The lower (southern) has a width of about a resolution element, and is observed to traverse a large, heavily eroded crater. Such features bear little resemblance to the aspect of the canals presented in the righthand side of Fig. 1. There would thus appear to be two categories of "canals": one continuous, rectilinear, and probably of tectonic origin; the other linear but discontinuous.

Apart from Lowell's interpretation of the canals as the engineering

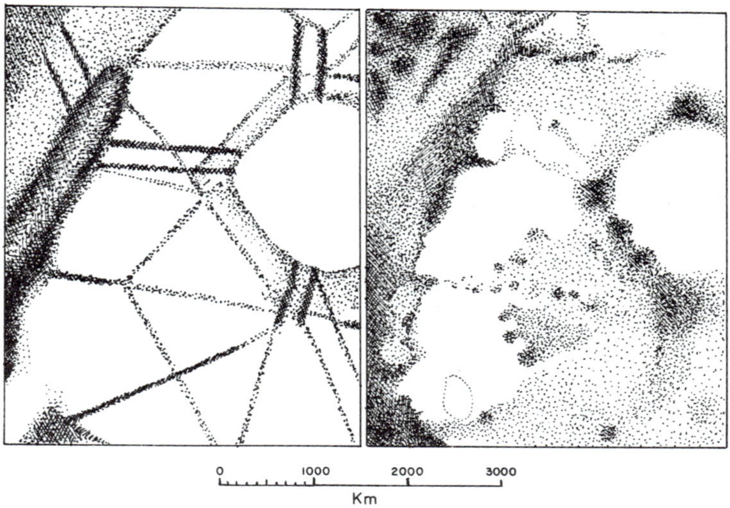

Two drawings of the Elysium region on Mars. Left: by Schiaparelli, from his 1877-1890 observations. Right: by Antoniadi, from his 1911, 1926, 1926 observations. (Fig. 1)

constructs of intelligent Martians, a variety of attempts has been made to understand the canals. All have encountered difficulties. Pickering and Wallace suggested that the canals were cracks in the crust of Mars produced during the contraction of the surface from an initially molten state. This interpretation is based on a now obsolete view of the evolution of planetary surfaces, and does not explain the absence of similarly prominent features on the Moon. Plassmann and more recently Oncley and Fulmer suggested that the canals of Mars were of the same nature as the rays of lunar craters. On a planet with as much drifting and wind-blown dust as Mars this is not a tractable suggestion. The lunar rays are known, however, to contain secondary impact craters which are ordered in more or less linear arrays. Sagan has suggested that the aligned disconnected detail on Mars may be due to crater sequences. The alignment might then be either accidental or the result of impact of secondary debris. Much of the resolved detail of the Martian canals (see Fig. 1), however, is of the order of 100 km across, and alignment of craters of this size is not to be found on the Moon except in rare cases. In a related suggestion Tombaugh has proposed that the craters of Mars are cracks resulting from impacts that produced the oases. Such cracks do not generally form in terrestrial high-velocity impact experiments, and in any event we would expect to see such cracks more prominently on the Moon than on Mars, which has winds and fine mobile dust. The fact that canals do not terminate in the bright areas would also be difficult to understand.

Gifford has suggested that the Martian canals are roughly rectilinear seif dunes, which should, in fact, be expected in the Martian bright areas. However, inspection of ground based descriptions of seif dunes (see Bagnold) and of photographs of seif dunes in North Africa and the Middle East obtained during the Gemini series of manned space flights in the United States shows that seif dunes almost always occur in collections of dozens of parallel sand ridges, each separated from the next by only a few dune widths. This is not the general character of the Martian canals. Furthermore, resolution of such dunes on Mars cannot be accomplished from the Earth, and the dunes are bright compared with the surroundings, rather than dark as are the Martian canals. For these reasons we must reject this interesting suggestion. Joly proposed that the canals are ridges drawn up gravitationally by asteroids on nearly grazing collisions with the Martian surface. This production mechanism seems exceedingly unlikely and the idea has not been further. More recently Katterfel'd has proposed that the arrangement of Martian canals and oases, particularly in the observations of Trumpler, strongly suggests that the canals are faults or escarpments. Similar ideas had earlier been expressed by Schiaparelli and Arrhenius. Hope has directed attention to terrestrial systems of rifts bounded by paired faults, which he feels may be related to the appearance of the canals during gemination. The suggestion that the major canals crossing the bright areas of Mars are rifts runs into several difficulties: chiefly, that drifting dust should long ago have filled such crevasses. Finally, mention may be made of an interesting suggestion by Wasiutynski that the overall pattern of major canals on Mars closely resembles the distribution of orogenic belts during previous epochs of mountain building on the Earth. Wasiutynski's view is not dissimilar to the views of Katterfel'd and of Hope except that Wasiutynski implies that some of the canals may be elevations rather than depressions. We note that this at least avoids the problem of filling by drifting dust.

. . . .

The southern of the two rectilinear features of Mariner 4, frame 11, cannot be used to test the radar results, because it is only one element resolution wide. The northern feature is wider, and is clearly bright on its northern face and dark on its southern face. From the crater illumination we see that the Sun is to the north. The sunward sides of ridges, but the opposite sides of clefts, will appear illuminated in such a situation. The Mariner 4 photography is thus consistent with the idea of rectilinear ridges on Mars. Other examples of ridges, similarly illuminated, can be found in other frames of the Mariner 4 photography.

If we accept Deuteronilus, Ceraunius and the Mariner 4 features as typical Martian canals, it would then appear that the Martian canals are ridges rising out of the prevailing dust and sand of the bright areas. Such ridges must be of tectonic origin, and it is of interest to note the opinions of Katterfel'd and Wasiutynski on the similarity of the Martian canal patterns to the pattern of terrestrial tectonic faults and orogenic zones. If the analogy of the Martian bright areas with the Earth's ocean basins is to be preserved, it would be useful to find ridges analogous to the canals of Mars in terrestrial ocean basins. Such ridges do exist. Best known is the mid-Atlantic ridge, a site of enhanced seismic activity, but many other oceanic ridge systems are known today despite the incompleteness of the exploration of the deep-sea floor (see, for example, Heezen and Menard). There are aseismic ridges that tend to emerge from the continental blocks, and related zones of linear faults with associated ridges. These zones are characteristically several thousand kilometres long, about 100 km wide, have a relief of 103 km, and are remarkably linear, several following great circles for more than 1,000 km.

. . . .

Several are narrow and approximately linear; they terminate near the continents, or from a non-continental region of common intersection with other ridges (compare canal property 7). The rift valley through part of the mid-Atlantic ridge may be seen.

If the oceans of the Earth were removed and the ocean basins filled with fine wind-blown dust, such ridges might well sometimes give the appearance of the right-hand side of Fig. 1, and sometimes of more continuous rectilinear features. Associated aspects of submarine geology are seamounts and islands, which also tend to be aligned along straight lines for thousands of kilometres, possibly indicating major faults in the Earth's crust. If analogous features exist in the Martian bright areas, they could conceivably be described as linear arrays of disconnected fine detail (see right-hand side, Fig. 1).

If the canals of Mars are ridges, then we would expect their visibility to be seasonally variable (property 8). We have suggested elsewhere that the seasonal changes of the Martian dark areas can be accounted for quantitatively by the scouring of fine particles off the highlands in the spring (thus darkening these regions) and by the redeposition of fine particles on the highlands by the higher-velocity wind systems of autumn and winter. If the Martian canals are ridges, they should experience the same types of seasonal changes that the dark areas do. Some terrestrial oceanic ridge systems continue geologically into the continents in possible analogy to the occasional extension of Martian canals into dark areas (property 1). It is even possible that the gemination of canals (property 9) may be connected with the presence of rift valleys in terrestrial submarine ridges. If drifting and wind-blown dust is present in differing amounts at the crests of Martian canals at different seasons of the year, it is conceivable that the rift valleys are sometimes visible (as when parallel ridges are uncovered by sand but the rift valley is filled) and sometimes not (as when the rift valley is itself largely free of fine particulate matter). The chief psychophysiological effect is probably the tendency to draw more rectilinear features than actually exist. With this exception, all canal properties seem understandable in the tectonic-ridge model presented here.

In summary, radar evidence and *Mariner* 4 photography suggest that the Martian canals are ridge systems or associated mountain chains in analogy to similar features in the terrestrial ocean basins. In such an interpretation much of the observed geometry and variability of the canals can be understood. Future radar observations as well as an improved understanding of terrestrial submarine relief would be useful in further testing these views.

MARTIAN CANAL PATTERNS

Anonymous; *Nature*, 222:818, 1969.

Mars is the only planet with surface features that can be mapped from the Earth without much difficulty, yet there are still only vague ideas about the nature of these features. The large dark areas ("maria") and the even larger white regions ("deserts") still lack satisfactory explanation. The so-called "canals", which frequently intersect in dark circular areas ("oases"), have provoked much speculation, often fanciful. It is surprising that little attention has been paid to the geographic distribution of the canals, particularly because an analysis of this distribution by Wells (Geophysical J., 17, 209; 1969) suggests that at least the meteoritic fracture theory for the origin of the canals can be ruled out.

Wells demonstrates that the canals can be mapped into two distinct but overlapping grid systems---one diagonal (trending NW-SE, NE-SW) and the other meriodional-latitudinal (trending N-S, E-W). Furthermore, because Mars, unlike the Moon, presents more than one face to the Earth, the grid pattern can be observed throughout the whole 360° of longitude. One feature of the diagonal grid system is that the elements are not generally precisely "perpendicular". Wells has thus been able to plot contour maps of the Martian surface bounding the three areas within which the northern apex angles of the diamond patterns were, respectively, breater than 90°, 80°-90° and less than 80° The isogonal maps produced were symmetric about the almost antipodal meridians 290° and 125°.

A comparison of this grid system with that of the Moon is interesting. During the past few years, Fielder (Quart. J. Geol. Soc., 119, 65; 1963, and other papers) has constructed a lunar diagonal grid system, comprising faults and crater chains, which is symmetrical about the central meridian perpendicular to the Earth axis. The distribution of the angles of intersection of the two elements of this system is similar to that for Mars. The fact that the Moon and Mars have similar grids thus suggests that, first, their origins are similar, and second, that the Martian canals are tectonic zones of weakness associated with such features as faults, ridges and crater chains. Furthermore, the existence of the Martian grid system throughout the full 360° of longitude suggests that a similar diagonal grid exists on the far side of the Moon. Implicit in all this is a global origin for the Martian canals, for it is unlikely that such consistent features would be produced by meteoritic impact and fracturing. This conclusion is also indirectly supported by the agreement between the tectonic surface relief of the Moon and that predicted from the (global) sub-crustal lunar convection cells postulated by Runcorn (Nature, 195, 1150; 1962, and other papers).

MARTIAN CANALS AFTER MARINER 9

Hughes, David W.; *Nature*, 258:288, 1975.

Ever since 1877 when Schiaparelli drew attention to the existence of canals on Mars they have been the subject of heated controversy and at times, distinct embarrassment to planetary astronomers. Mariner 9, the spacecraft that went into orbit around Mars in 1971, achieved a complete mapping of the surface down to a resolution of 1 km, so now the question, "where are the canals?" should be answerable. Sagan and Fox of the Laboratory for Planetary Studies, Cornell apply themselves to this question in a recent issue of Icarus (25, 602; 1975).

Many suggestions have been put forward to explain the observations, observations which cannot be discounted out of hand, because, even though the features are of intrinsically low contrast and small angular size usually observed under poor seeing conditions, many maps of tolerable mutual consistency have been produced by different astronomers.

Canals have been thought to be waterways or the vegetation along the banks of waterways constructed by an advanced civilisation. Natural valleys, surface cracks or ridges produced by a range of geological processes were another suggestion. Accidental alignments of surface features (for example chains of large craters) was a third. They may also be rectilinear sand dunes or dark rays emanating from craters, or simply albedo features, regions of differing reflectivity that are periodically covered and uncovered by windblown dust.

Sagan and Fox compare the Mariner 9 results with the canal cartography of Slipher (Mars, Sky Publishing Co., Cambridge, Mass; 1962). A striking correlation was found between the canal Agathodaemon and the great rift valley Valles Marineris but as the latter is a low albedo feature 5,000 km long and 100 km wide it is only to be expected that this is easily resolvable by ground-based observers. Thoth-Nepenthes is also easily discernible on both maps as is the ridge Ceraunius.

Returning to the hypotheses concerning canal origin, Sagan and Fox conclude that the resolution and surface cover of Mariner 9 would have detected a civilisation like Earth's if it had been present on Mars. Correspondences do occur between classical canals and valleys or ridges. Agathodaemon-Valles Marineris has been mentioned and Styx-Phlegra Montes and Ceraunius are other examples. Psychophysiological alignments of disconnected fine detail only seem probable in the case of the canal Oxia which seems to be the chain of craters Trouvelot, Rutherford, Becquerel, Curie and Sklodowska and an unnamed canal which goes through craters Galle, Wirtz, Helmholtz and Lohse. Very few crater rays were seen by Mariner 9 and sand dune fields do not seem to be sufficiently extensive to be observed from Earth. A few canals notably Cerberus and Thoth-Nepenthes do correspond to dark features on the Martian surface, these features being unconnected to topography. They might, however, be regions where fine bright dust does not settle or regions where the ground is dark due to surface roughness or some geochemical reason.

While a few canals do correspond to topographic and albedo features on Mars, Sagan and Fox conclude that the large majority of canals do not. Conversely there are many real surface features, for example large shield volcanoes like Olympus Mons, Arsia Mons and Pavonis Mons, where no canals were marked on the maps.

Sagan and Fox discount the canal observations and relegate them to monuments to the imprecision of the human eye labouring under difficult seeing conditions. But does the story really end here. Probably so, but the consistency of the diverse maps leaves a lingering unease. Maybe the facts that Mariner 9 could not distinguish between features which had less than 5% albedo contrast, that Mars was only observed for half a Martian year, or the violent dust storms that preceded Mariner 9's visit caused the real truth to be hidden.

• Possible Fluid-Eroded Channels

CHANNELS ON MARS

Sharp, Robert P., and Malin, Michael C.; *Geological Society of America, Bulletin,* 86:593-609, 1975.

Summary Discussion on the Origin and Implications of Martian Channels. Martian channels and channel-like features come in many sizes, shapes, and patterns. They are surely not of a single origin. Some may be created by influences acting from within the crust, others may result from external erosive processes, and some channels are probably composite, the result of exogenic modification of an endogenic feature.

Most Martian channels or channel-like features imply removal of material. An exception would be a channel of accretion made by greater accumulation along the margins than in the channel itself. Some lava channels on Mars are attributed to accretion, and certain terrestrial sedimentary channels are of accretional origin. The material missing from a channel may have moved down, up, or parallel with the surface, either laterally away from the channel or longitudinally along it. Relative downward motion can occur by tectonic faulting or warping, or by subsidence and collapse resulting from removal of underlying support through solution, deterioration of ground ice, or magma withdrawal. Upward movement occurs in the instance of explosions, impacts, or eolian deflation, with only the last seemingly worth serious consideration with respect to elongate depressions. Lateral movement can result from extension within the crust, and longitudinal removal usually occurs through transportation by some exogenic process.

Since some Martian channel-like features are closed at both ends, their creation through erosion would have to be by solution or deflation. Solution is suspect because of the need for abundant solvent and soluble rocks. Arguments are advanced for eolian erosion on Mars, but deflation as the sole cause of closed-end channels is contested by the highly localized erosional requirement, the need to reduce all material to extremely fine size, the likely inhibition of deflation by development of coarse surface armors, and the lack of evidence of associated eolian deposits. The most feasible means of creating closed-end channel-like features on Mars seems to be by downward movements of endogenic origin. As to the exogenic erosive processes which may have helped from other Martian channels, mass movements, wind, ground-ice sapping, ground-water sapping, and flowing liquids deserve consideration.

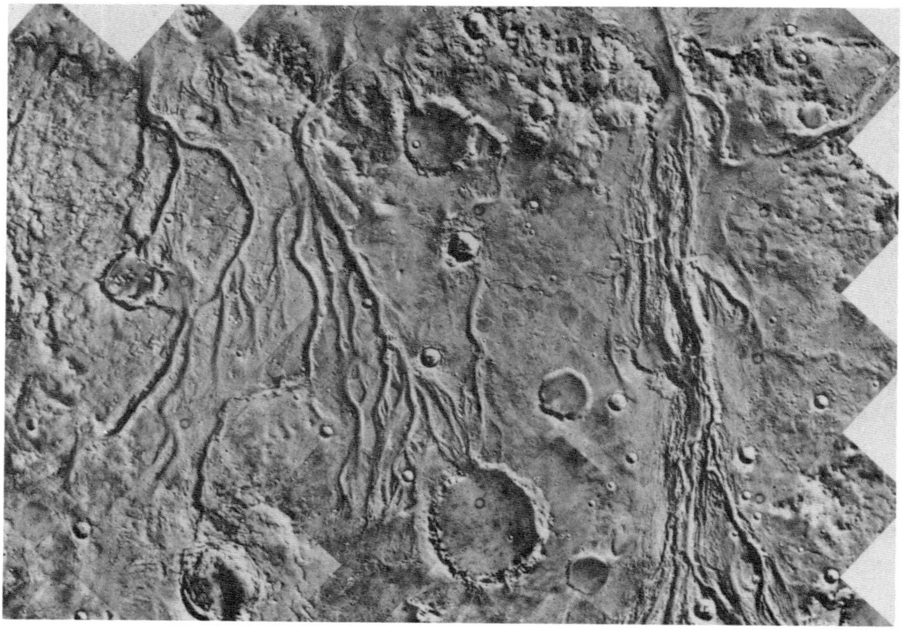

Martian channelled area just west of Chryse Planitia photographed from Viking Orbiter 1. (Courtesy NASA)

Erosional forms on steep faces may have been created by mass movements---for example, the parallel U-shaped features interpreted as avalanche chutes after Blackwelder and Matthes. However, as shown by Milton, there is little reason to attribute the large scoured channels of Mars to any mass-movement process, including density currents or volcanic glowing clouds.

The capacity of wind to create channels, other than the polar grooves and possibly the aureole of grooved terrain around Olympus Mons, remains an open question, but wind may have modified channels created by other processes. A possible example is Kasei channel in which the lineation and streamlining, attributed by Baker and Milton to fluvial flood action, may be the product of eolian activity.

Longitudinal erosion and transportation in any channel requires disposal of both the transported material and the transporting medium. In the instance of avalanche chutes, the material resides in detrital slopes below, and disposal of gravity is no problem. Material carried in eolian suspension can be widely dispersed, and air is also easily disposed of. However, eolian suspension is likely to be accompanied by saltation and traction, and this bed-load material should form accumulations not far removed from the erosional source. They are not yet distinguished on Mars in association with channels, although their recognition is admittedly difficult.

For channels supposedly cut by lava, disposal of the eroding medium is an acute problem. In addition to the eroded material, a much larger volume of lava must be accommodated in the debouchment area. Although unequivocable identification of lava fields on Mariner 9 photos is admittedly difficult, a great many eroded Martian channels have nothing suggestive of lava accumulations at their mouths.

The problems regarding water are not only its disposal but also its source and the environmental conditions that would permit it to act on the Martian surface. Disposal is not too difficult, as the water could have percolated into the ground or evaporated into the atmosphere. Under current conditions, the percolated water would freeze to ground ice, and any vaporized water, after perhaps having first been precipitated temporarily as local snow, would eventually be captured in the cold traps of polar ice. The problem then becomes the amount of ice that resides in the polar blankets, including layered deposits, a subject of considerable debate. Over a long time, however, significant amounts of water may have disassociated and escaped to outer space.

The suggestion that Martian water sources may have been largely lithospheric, rather than atmospheric, finds support in the following relationships. The large erosional channels are isolated features largely unaccompanied by other channels and unrelated to each other beyond the fact that they occur principally in heavily cratered older terrain and in equatorial or subequatorial regions. Even the smaller dendritic tributaries and slope gullies are localized. This suggests that water has been made available only locally. If the water were atmospheric, this would require significant precipitation only in spots. By contrast, the above channels all have possible local sources of lithospheric water. For Mangala and Ma'adim, the sources are large headwater seepage areas, and for Ares, it is chaotic terrain. For Nirgal channel, seepage into the headwater tributaries and into the trunk channel could have occurred. The larger channels on Ceraunius Dome have an obvious source in the central caldera.

The concept of water being supplied at localized spots of the Martian surface from underground sources thus has appeal. Parts of the Martian crust and surface became temporarily "juicy", at which time water was supplied to the surface in the form of seeps. In places, seepage could have nourished a system of gullies which dispersed the water widely over the terrain; in other situations, the seepage was integrated by branches and tributaries to feed a major channel; and in some places, the water may have been ponded and eventually released as a flood. Evidence for huge catastrophic floods is good in Mangala and Ares channels, and according to Baker and Milton also in Kasei channel, although the means of accumulating and storing water is a problem. Storage could have occurred in the headwaters seepage area of Mangala and within the extensive areas of depressed chaotic terrain at the head of Ares.

Martian lithospheric water might represent juvenile waters coming from a crystallizing magmatic body, or it might be melted primordial ground ice. Why the "juicy" spots should be randomly distributed is, perhaps, just a vagrancy of nature. Channels on Mars are concentrated into areas equator-ward of 40°N. and 60°S., with greatest abundance in equatorial areas, but this may reflect blanketing by younger deposits in sub-polar regions more than some environmental influence. Even in equatorial regions, catastrophic floods might experience difficulties under the currently rigorous environment, and a slower, continuing discharge would probably be ineffective. The fretting process that possibly plays a role in channel development would also suffer if it involved ground-water sapping and runoff.

Derivation of water from the lithosphere rather than the atmosphere only partly solves this environmental problem, but it may significantly reduce the degree of environmental amelioration required. Producing widespread precipitation from the atmosphere is one thing, and keeping water fluid on the surface is another, especially if the water is initially warm and contains some freezing point depressant. Rather than an atmospheric pressure of 1,000 mb, perhaps a fraction of that would suffice.

Irrespective of the degree of amelioration, the erosional channels do seem to require that at some time, possibly in the far distant past, the surface environment of Mars has been less rigorous than it is at present. This is what the channels appear to say in spite of various models, theories, and arguments to the contrary.

GEOLOGICAL OBSERVATIONS OF MARTIAN ARROYOS

Hartmann, William K.; *Journal of Geophysical Research*, 79:3951-3957, 1974.

Historic Variable Features Near Arroyo Systems. Early maps of Mars show a pattern of well-confirmed streaky markings different from that existing now. Examples are broad curved tails on Syrtis Major and Meridiani Sinus.

It is important to distinguish these from so-called 'canals.' The streaks were shown by many observers before canals were popularized. What appears to have happened is that many broad streaky markings were incorrectly charted by Schiaparelli and Lowell as narrow lines; other such lines were added where none exist today.

A point of interest here is that some, but not all, of the broad, streaky markings can be identified with canyons, arroyo systems, or areas containing arroyos. The best-known example is Coprates Chasma, shown as a stubby band or canal on virtually all maps of Mars. It is not a simple arroyo but serves to show that valleys can be the sites of dark deposits of long duration. In drawings by W. R. Dawes in 1864-1865, R. Proctor's map based on these drawings (1867), Schiaparelli's maps (circa 1890), Flammarion's map (1892), and Antoniadi's map (1930), extensive curved tails are shown extending north from Meridiani Sinus and Syrtis Major. The first is called Indus, and the second is called Nilosyrtis. Neither is prominent today, though patchy markings exist in these areas.

Indus has about the same curvature and approximate location as the Area Vallis. Today a substantial dark patch (not streak) lies outside the northeast border of this major arroyo. The region west of Area Vallis, about 30° in longitude and -15° in latitude, contains a number of north-south-trending major arroyos, including Tiu, Simud, and Shalbatana valles. This region is one of north-south-trending streaks shown prominently in the mid-1800's by Dawes and others before canals were popularized.

The streak or tail Nilosyrtis is less clearly associated with an arroyo system. Rather, it appears to be defined primarily by the contact between the plain and the cratered uplands. Nonetheless, as was remarked above, this contact is the site of arroyos such as Huo Hsing and Auqakuh valles, which parallel the contact in the upland.

In summary, a number of classical streaky variable features, less prominent today than they were 100 years ago, can be related to major arroyo systems. The preceding sections suffice to exclude any suggestion that fluvial episodes caused the variations in the last 10^2 years.

Therefore it is concluded that some streaky markings emphasized by early observers may be related to subaerial deposition of dark deposits in curvilinear patterns controlled by the topography of old curvilinear patterns controlled by the topography of old curvilinear arroyo and/or fracture systems. Because these structures are long in horizontal length but minor in vertical relief relative to global relief, they are associated with changeable deposition patterns, whose waxing and waning may be seen from earth. More permanent albedo markings are probably associated with interaction of prevailing winds with more substantial relief.

These conclusions are consistent with Milton's [1974b] conclusion that weathering has acted long enough to remove the mineralogic evidence (kaolinite) of the epoch of abundant surface water but not the morphologic evidence of it. Today's blowing Martian dust has been chemically modified since the days of substantial Martian rivers.

Conclusions. Arroyo systems on Mars are incompatible with the present climatological state of the planet and imply more widespread sources of liquids at some time in the past. Tributaries are important evidence in this regard.

The presence of craters in arroyos, combined with any of the current cratering models, suggests median ages of some 10^8 years (not less than 10^7 years) for fluvial activity in typical visible channels. Therefore acceptance of Ward-type obliquity periodicities to account for the main fluvial episodes would require many cycles of flow to have occurred in channel systems without removing the observed craters. More consistent with the evidence would be episodes with longer time scales of the order of 10^8 years.

Morphology of Martian arroyos and associated structures indicates that arroyos carried water that eroded banks, flowed in episodes considerably separated in time by many, many years, created overlapping and truncated channels, found crater walls resistant to erosion, and occasionally broke through crater walls to disturb crater floors. Breached craters may be sites of deep permafrost.

In the present Martian climatic regime, arroyo systems are sometimes associated with streaky variable dark deposits, visible from earth and produced by topographic disturbance of subaerial dust deposition processes.

PRIMITIVE ATMOSPHERE AND IMPLICATIONS FOR THE FORMATION OF CHANNELS ON MARS

Yung, Y. L., and Pinto, J. P.; *Nature*, 273:730-732, 1978.

The channels on Mars suggest that a flowing fluid has been present on the surface of the planet. It seems natural to assume that this fluid was water. The major difficulty, however, is that water freezes in climatic conditions like those now on Mars. It has been suggested that primitive Mars had a reducing atmosphere, composed mainly of methane. Such an atmosphere, as we show here, could be polymerised by solar ultraviolet radiation to produce higher hydrocarbons. These compounds are low viscosity liquids at today's temperature on Mars, and could contribute to the formation of channels. [Chemistry details are omitted.]

The channels on Mars are a complex phenomenon and probably no single explanation would suffice. Our photochemical model provides a modest amount of fluid (1-10m) for a brief interval of time (10-100 Myr). The existence of liquid alkanes complements rather than competes with other mechanisms for the formation of channels. A preliminary investigation suggests that the alkanes do not contribute significantly to greenhouse effects but their associated products such as C_4H_6O and C_3H_7N may. Also compounds like CH_3OH that can depress the freezing point of water may be present. We shall include these compounds in an extended study.

....

• Other Unusual Topographic Features

THE ORIGIN OF POLYGONAL TROUGH SYSTEMS ON MARS

Pechmann, James; *EOS*, 58:1182, 1977.

Abstract. Mariner and Viking photographs have revealed an extraordinary system of crudely polygonal troughs on the northern plains of Mars. Fifteen rose diagrams of trough azimuths show that locally they exhibit preferred orientations but overall the pattern is difficult to interpret. Average trough spacing is 5 to 10 kilometers and average width is 200 to 800 meters.

The widest troughs have flat floors and steep sides which suggests that they are either graben or tension cracks which have been partially filled.

The lower limit on crack depths for terrestrial tension crack systems appears to be about one tenth of the average crack spacing (Lachenbruch, personal communication). Thus, if the Martian troughs are tension cracks we would expect them to have depths of at least 500 to 1,000 meters. Mechanical analysis shows that it is difficult to generate such deep cracks by surficial tension due to thermal cooling and contraction in permafrost, dessication, or cooling in lava flows. It therefore appears likely that the troughs were formed in response to some deep-seated tension such as would result from tectonic uplift, horizontal movements of the crust, or expansion of the planet.

Troughs of very similar scale and morphology are found in the Caloris Basin on Mercury. Most are roughly concentric to the basin although some weak radial trends are present. These were interpreted by Strom et al. (1975, J. G. R. 80, 2482) as extensional graben formed by updoming of the basin floor. It is proposed that the polygonal troughs on Mars are similar tectonic features, although not necessarily related to uplift.

PYRAMIDAL STRUCTURES OF MARS

Gipson, Mack, Jr., and Ablordeppey, Victor K.; *Icarus*, 22:197-204, 1974.

Triangular and polygonal pyramid like structures have been observed on the martian surface. Located in the east central portion of Elysium Quadrangle (MC-15), these features are visible on the Mariner 9 photographs, B frames MTVS 4205-3 DAS 07794853 and MTVS 4296-24 DAS 12985882. The structures cast triangular and polygonal shadows. Steep-sided volcanic cones and impact craters occur only a few kilometers away. The mean diameter of the triangular pyramidal structures at the base is approximately 3.0 km, and the mean diameter of the polygonal structures is approximately 6.0 km.

The observed Martian structures tend to line up suggesting joint or fault control. However, they do not appear to be controlled by the visible faults. The structures appear to be either wind-faceted volcanic cones and blocks or solidified blocks which have been rotated in semiconsolidated lava.

. . . .

NORTH POLAR REGION OF MARS: IMAGING RESULTS FROM VIKING 2

Cutts, James A., et al; *Science*, 194:1329-1337, 1976.

Abstract. During October 1976, the Viking 2 orbiter acquired approximately 700 high-resolution images of the north polar region of Mars. These images confirm the existence at the north pole of extensive layered deposits largely covered over with deposits of perennial ice. An unconformity within the layered deposits suggests a complex history of climate change during their time of deposition. A pole-girdling accumulation of dunes composed of very dark materials is revealed for the first time by the Viking cameras. The entire region is devoid of fresh impact craters. Rapid rates of erosion or deposition are implied. A scenario for polar geological evolution, involving two types of climate change, is proposed.

Layered deposits in the north polar region of Mars. The regular banding may indicate cyclic climate variations. Photo area is 65 by 100 km. (Courtesy NASA)

VIKING: POLAR DUNES AND CAPTURED MOONS

Anonymous; *Science News*, 110:276-277, 1976.

Then there are the strange features typified by one that has been dubbed "the searchlight," an area formed by two diverging straight lines between which everything on the surface---dark markings and light ones---looks, well, different. "There appears to be some translucent cover over this region and its difficult furrows," says orbiter imaging team leader Michael H. Carr of the U. S. Geological Survey, "maybe a discontinuous ice deposit, but it must be extremely thin. And we can see the albedo features through it." Accumulated dust creates a somewhat similar impression in some parts of the equatorial regions, but there the cause seems readily apparent. The polar-cap versions, says Carr, "appear to be a somewhat different animal." [Excerpt]

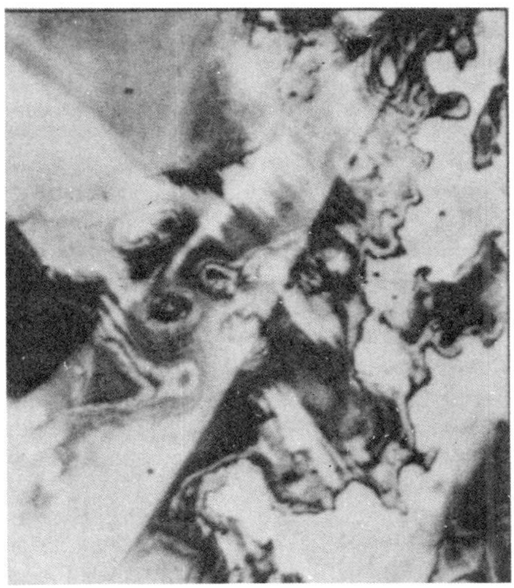

The so-called "searchlight" area in the Martian polar regions. (Courtesy NASA)

MARTIAN PERMAFROST FEATURES

Carr, Michael H., and Schaber, Gerald G.; *Journal of Geophysical Research,* 82:4039, 1977.

Abstract. The outgassing history of Mars and the prevailing temperature conditions suggest that ground ice may occur to depths of kilometers over large areas of the planet. The presence of permafrost is also indicated by several topographic features that resemble those found in periglacial regions of the earth. East of Hellas and in the Protonilus and Nilosyrtis regions there are features that resemble those formed on earth by gelifluction, the slow creep of near-surface materials aided by freeze-thaw of ground ice. In the south part of Chryse Planitia there are irregular depressions that resemble thermokarst features, and the pattern of tributaries to the equatorial canyons is suggestive of a sapping process that would result from the melting of ground ice. The morphology of ejecta around fresh Martian impact craters is distinctively different from that around lunar and Mercurian craters. Such differences could be ascribed to the presence of ground ice in the target materials. The convergence of these different observations supports permafrost conditions not only at present but also for much of the planet's history.

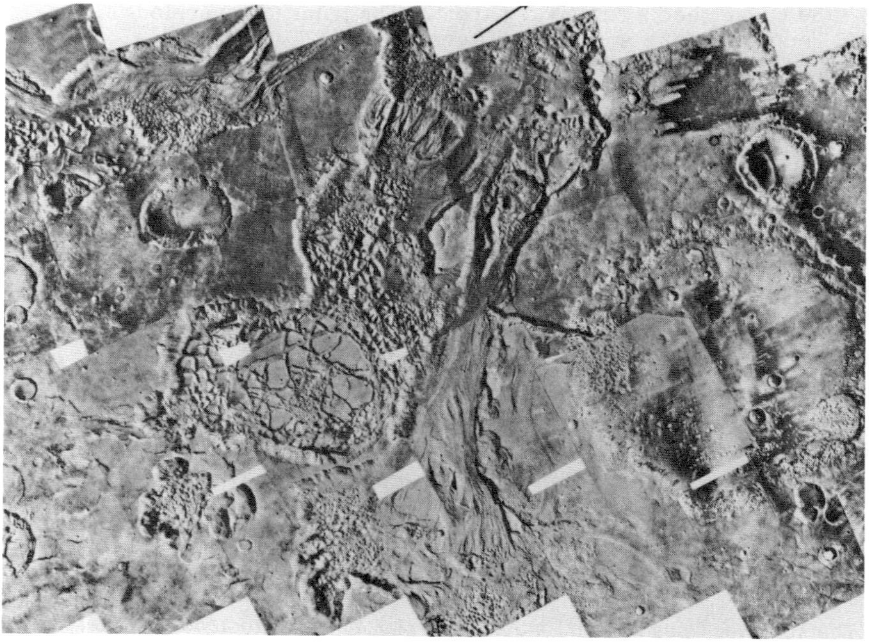

Apparent collapse feature on Mars, presumably caused by the melting of subterranean ice. The origin of such large quantities of fluid is an unsolved problem. (Courtesy Jet Propulsion Laboratory)

STRANGE HILLOCKS AND RIDGES ON MARS

Anonymous; *Science News*, 113:43, 1978.

No one knows what chain of events led to the strange appearance of this northwestward extension of the flanks of the huge volcano Arsia Mons in the Tharsis uplands. Countless hillocks, mostly 100 to 500 meters across, cover the flank's edge, which is surrounded by parallel ridges that run for hundreds of kilometers, undisturbed by craters, flow features or even variations in surface brightness. One hypothesis is that the hillocks were formed by a huge landslide, perhaps assisted by gravity since the surrounding plains slope downward about 0.5° to the northwest. (Ashflow deposits are deemed unlikely, due to the lack of signs that any of the material was blasted into place from vents in the volcano.) The ridges may be folds or "reverse faults" caused by the drag of the landslide over the underlying terrain, which would transmit an outward pressure perhaps capable of passing beneath surface features.

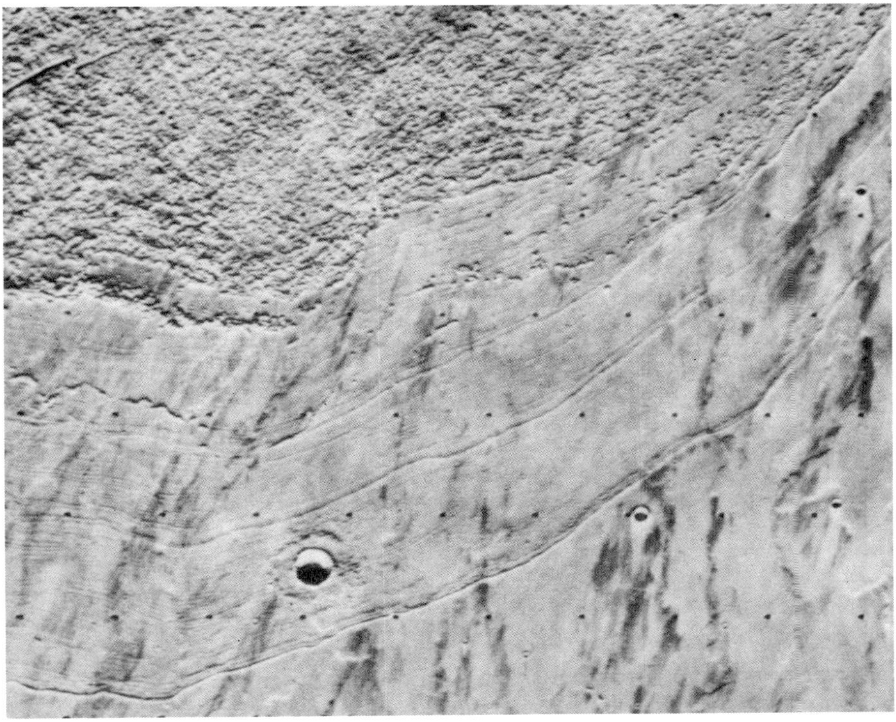

Puzzling hillocks and parallel ridges near the Martian volcano Arsia Mons. (Courtesy NASA)

TRANSIENT MARTIAN PHENOMENA

Ephemeral lights and patterns can be seen on Mars through the telescope despite the minimum distance of more than 35 million miles. To be seen over such distances, transient Martian events must be intense and/or widespread.

Temporary light and dark spots are the most common type of phenomena. Since Mars is an active planet with a thin, but still substantial atmosphere, clouds and dust storms are thought to be adequate explanations of most of these sightings. The brilliant flashes of light sometimes observed are more difficult to unravel. Are there still active volcanoes on Mars; or could there be large-scale electrical phenomena? Whatever the cause(s), the flashes must be very energetic to be so prominent over 35 million miles.

Blue-haze clearing is a well-established Martian puzzle. Dust, aerosols, or other substances in the Martian atmosphere could render surface details indistinct in the blue portion of the spectrum. But what would cause this obscuring substance to disappear sometimes, leaving the surface just as sharp in blue light as the rest of the spectrum? Some spacecraft photos of Mars show no blue-haze effects, which only deepens the mystery.

• Bright Spots and Flashes

THE BRIGHT PROJECTIONS ON MARS

Anonymous; *Observatory*, 17:295-296, 1894.

A telegram from M. Perrotin, of Nice, on Aug. 6, announcing the observation of a bright prominence beyond the terminator of Mars, and a subsequent announcement by Mr. Stanley Williams of a similar observation, have renewed scientific interest in this class of phenomena, and have also created some sensational interest in the public press. It may not be out of place to give categorically the recorded observations of this kind.

The first hint of such bright markings came from Prof. Schiaparelli, who, in 1888, in publishing some observations of white spots stated that the whiteness was always more pronounced when the spots were near the edge of the disk, but he did not observe the brightness <u>beyond</u> the terminator. In the same year M. Terby observed similar white spots, which were invisible until they approached the western edge of the disk, when they appeared very bright, and were apparently seen beyond the edge of the disk by irradiation; but it was not until 1890 that the phenomena of obvious projections were seen. On July 5 of that year from 10^h 0^m to 10^h 30^m Pacific Standard Time the observers at Lick saw a narrow elliptical white spot from 1".5 to 2".0 long projecting from and making a small angle with the line of the terminator. At 10^h 30^m the spot was certainly within the illuminated disk, still visible as a white patch on a dark background. On the next night July 6, from 8^h 3^m to 10^h 25^m Pacific Standard Time, two projections of the same kind were seen, one of which was no doubt the object observed on the preceding evening. The latitude of this principal projection was about 40°, and the apparent Martian longitude of the terminator during the times of observation ranged from 45° to 52° on the first night, and from 6° to 41° on the second. It is to be remarked that the additional spot seen on July 6 was situated at the end of a long bright stripe on the surface of the planet.

During the opposition of 1892 these phenomena were again seen. M. Perrotin, at Nice, saw bright projections beyond the terminator on June 10, from 15^h 12^m to 16^h 17^m Nice Mean Time in latitude about -30°, the longitude of the terminator during this hour ranging from 212° to 228°, on July 2 from 14^h 10^m to 14^h 40^m, and on July 3 from 14^h 11^m to 15^h 6^m, in latitude -50° on each evening, from which it may reasonably be inferred that these objects, whatever they were, were in the same region of the planet (about longitude 335°). M. Perrotin says that these were seen with great distinctness, and that it is scarcely possible to consider them the results of illusion.

At this opposition the observers at Lick were Professors Hussey and Campbell; the former observed from July 2 to 17, and states that on no night when he looked did he fail to see a prominence. Prof. Campbell observed them on five nights between July 10 and 17, but after this date no one at Mount Hamilton saw these projections, although they were looked for up to August 17. All the prominences observed at Lick were between latitudes -30° to -50°, except one or two at about latitude -25°. The most prominent were observed on July 11 and 13, the same object being doubtless observed on the two evenings (latitude -47°, longitude 357°) in a region known as Noachis.

From Arequipa, where Mr. W. H. Pickering was observing Mars in 1892, there are no detailed reports of similar observations. In 'Astronomy and Astro-Physics' for December 1892 he says: "Clouds have on several occasions been observed to project beyond the terminator and also beyond the limb. The height of some of these clouds has been measured, and it appears that some of them attained an altitude of at least twenty miles."

Since 1892 no observations of this kind have been made until August 6 of this year, when, as mentioned above, M. Perrotin announced an observation of a brilliant projection beyond the terminator, and again on August 19 Mr. Stanley Williams observed a similar phenomenon; the American observations of the planet will be waited for with interest. As to the cause of these, there appear to be two rival explanations. One, as suggested in Mr. W. H. Pickering's report of his observations, that these luminosities are bright clouds; and in favour of this theory he says that the small mass of the planet is not inconsistent with clouds at such great altitudes as these observations would require. On the other hand, the observations show in many cases that these bright objects are permanent on the planet, a quality which we find difficult to attach to clouds.

The other theory advanced by Prof. Campbell in an article in Number 35, of the 'Publications of the Astronomical Society of the Pacific,' and to which we are indebted for much of this, is that these projections are due to mountain-chains lying across the terminator of the planet, in some cases covered with snow, in others not necessarily so. A chain of mountains parallel to the planet's equator, by reason of their length, would give the permanence of the phenomenon which has been observed, as on July 11 and 13, 1892, when the luminosity was observed for two hours or more, during which time 800 or 900 miles of the planet passed over the terminator. Also it will be seen, on considering the geometry of the question, that the observed distance from the terminator is not a direct measure of the altitude of the mountain, if the object is such, but it is a quantity into which the length of the chain enters with much effect.

Prof. Campbell suggests that the Martian latitude and longitude of every such observed prominence should be computed with accuracy, and when this determined position is on the central meridian, the planet may be scrutinized to find whether it falls on a dark or bright marking, which will supply some evidence as to the character of the region.

MARKING ON MARS

Anonymous; *British Astronomical Association, Journal,* 11:125-126, 1900.

A Kiel telegram announces that Mr. Douglass, at Flagstaff, Arizona, observed a projection on the northern edge of the Icarium Mare, which remained visible for 70 minutes. According to the "Standard" of December 27, M. de Fonvielle relates in the "Matin" that "a series of bright lights suddenly appeared in a straight line extending for several hundred kilometres. These gigantic fires burned without interruption for one hour and ten minutes, and then disappeared as suddenly as they had come." M. de Fonvielle evidently believes in a message from Mars, and says that if the Martians really lighted these fires, it is indispensable for the astronomers of this world to let them know that they have been understood, and that we count on their intelligence to succeed in understanding us, and in creating an alphabet. But his views are not in accord with those of M. Flammarion. A telegram from New York appearing in the "Standard" of January 4, says that M. Tesla is prepared to construct a machine which will carry sufficient energy to Mars to operate delicate telephone and telegraph instruments.

The daily press has made much capital out of these telegrams, as was the case eight years ago when a similar projection was noticed both at the Lick Observatory and at Nice. Nor was that the first time that projections had been observed. Schroeter saw some on September 21, 1798, and on November 12, 1800. In the "Observatory," XI., 298, will be found a letter from Terby, describing three bright projections seen several times up to the date of his letter, June 4, 1888. In "Comptes

Rendus," CXV., 379, Perrotin states that he saw bright projections on June 10, July 2 and 3, 1892, and that the same had been seen by the Lick observers. Several projections were seen also in 1894, 1896, and 1897, by Members of the Mars Section, and are recorded in the "Memoirs" of the Association, IV., iv., 119, 120; VI., iii., 67, 68. The appearance is most probably caused by brilliant clouds projecting beyond the terminator or illuminated tops of ranges of mountains on the planet.

EXPLANATION OF THE SUPPOSED SIGNALS FROM MARS OF DECEMBER 7 AND 8, 1900

Lowell, Percival; *Popular Astronomy*, 10:185-194, 1902.

1. On a certain morning in December, 1900, paragraphs appeared in the papers throughout the United States with the startling announcement that Mars had been signaling the Earth the night before. Lights, it was reported, had suddenly shone out upon the surface of the planet, lasted for a time and then vanished. What the signals meant was not so forthcoming. Vividness of headline made up for meagreness of news.

Interest was not confined to the United States. Reportorial inquisitiveness was as rife in the Old World as in the New, and Europe was behind America in the receipt of the message only the time necessary for its transmittal.

2. To broaden one's horizon is a good thing; and to broaden it beyond the bounds where horizon itself disappears, a still better one. But the broadening is apt to come not in a way we expect, and to prove the more broadening for that reason. I hope, therefore, not seriously to lessen interest in the phenomena by saying that they were certainly not what they were popularly taken to be, and were with equal certainty much which was not supposed and is quite as interesting.

The innocent cause of the misrepresentation was a dispatch sent from Flagstaff to the writer and communicated by him through the usual channels to the astronomical world. The signaling part of it was a tale added by journalistic ingenuity at the time that profession became possessed of the subject. The original dispatch read:

> Projection observed last night over Icarium Mare, lasting seventy minutes.
>
> (Signed) Douglass.

3. Projections in the case of one heavenly body, the Moon, are not unfamiliar objects. On almost any night when that body shows a terminator, that is a sunset or sunrise edge, a keen eye can detect one or more of them along it without telescopic aid. With Mars the phenomenon is much less common and, though many such projections have in the last few years been seen upon the planet, the sight is one of some rarity.

4. In the case of the Moon it is possible to find out the cause of the projections. By magnification through a telescope the little knob that breaks the otherwise uniform boundary of light and shade is seen to resolve itself into the tip of a mountain peak or the summit of a crater

wall, which catches the light while the lower ground at its foot is plunged in shadow, and so seems to project beyond the rest of the disk. With Mars no such forthright determination of the problem is possible. For no magnification we can apply is potent enough to disclose of itself the character of the country. We are, therefore, obliged to reason upon what we see.

5. Taking lunar analogies for guide, it was generally inferred that the martian projections too were due to mountain peaks. From which of course it followed, or as one may say preceded, that there were mountains on Mars. But the Flagstaff observations of 1894 showed that, on general principles, this was very improbable. The study of the surface markings led the writer to a general theory about the character of the planet, in which mountains not only found no place but to which they were decidedly opposed. At the same time that the theory suggested itself, but independent of it, Mr. Douglass observed several projections, and conceived and published another explanation for them, and this one proved consonant with what the theory demanded, to-wit: that, instead of being due to mountains upon the planet's surface, they were due to clouds floating in the planet's air. He showed that the observations were thus much better explained; in fact, that his observations could hardly be accounted for with probability on the mountain hypothesis at all.

6. The opposition of 1894 was very prolific of projections, over four hundred being seen at Flagstaff in the course of nine months. The next opposition was not so good; while in that which has just passed, that of 1900-1, only two were detected. It was these two which gave rise to the notion of signals from the planet.

Now the variability in the number seen at different oppositions should have materially shaken faith in the mountain explanation. Mountains are permanent affairs, and if they he high enough to catch the light and show as protuberances at one time they should do the like at another. The change in the inclination of the disk would not materially alter their visibility. But it is one of the humorous anomalies about human nature that general reasoning affects minds so little when applied to unfamiliar matters, while in familiar ones it is the guiding principle of life.

7. Argument from the two projections of the last opposition is, on the other hand, particular. Although they were but two in number, testimony in the case is very much to the point. Indeed, their isolated character helps to make their cogency the clearer.

On December 7, at $16^h\ 15^m$ S. M. T., Mr. Douglass suddenly noticed a projection on the terminator of the planet, a little to north of the Sabaeus Sinus. The phase loss at the time was $36^o.4$. As he continued to watch it the projection increased. The distance of its tip from the edge of the terminator passed successively through the values 2/3, 1, 1-1/3 and 1-1/2 of a thread; the thread used being the stationary spider's thread of the micrometer. Meanwhile he was busy taking the position angle of the tangent to the terminator, at the point directly under it, at intervals of a few minutes.

. . . .

18. The season of the martian year at which these clouds occurred is of interest. On December 7, 1901, it was April 26 in the northern hemisphere of Mars. The Sun had gone north of the equator and was

then overhead on the fourteenth parallel of latitude. The heat equator was a little behind it. Apparently then a current bearing the clouds was setting toward the heat equator from within the tropics to the south, where the season corresponded to the end of October. This current was deflected some eighty degrees to the east, and became an east-by-north wind.

19. Its origin may have been local. A little to the south of where the cloud first appeared lies the long east-and-west stretch of the Sabaeus Sinus or Icarium Mare. Now the form of the cloud was of the same general shape---a cloud stretching east and west five times as far as it did north and south. The Icarium Mare is undoubtedly a great tract of vegetation, where moisture would be held and whence it could accordingly be given off. Arising there, either from seasonal or temporal cause, the vapor would gather into a cloud and proceed to float away over the desert regions to the north. If this, then, is what happened in the case before us, we may conceive the cloud as having been generated on the 6th of December over the Icarium Mare, rising to a height of thirteen miles, and then traveling east by north at about twenty-seven miles an hour off into the desert of Aeria, there to dissipate after an existence of three or four days. That it was a phenomenon of capricious not of regular production is shown by its not having been repeated ---that is, it partook of the subtle unpredictability of cloud

BRILLIANT WHITE SPOTS ON MARS

Anonymous; *Nature*, 89:17, 1912.

In No. 10, vol. xix., of Popular Astronomy Mr. L. J. Wilson, who observes Mars with an 11-inch reflector, cites several occasions during October and November, 1911, when his observations at Nashville, Tenn., revealed the presence of very conspicuous and brilliant white spots on the planet's disc; such spots were seen, on October 14, in the region following Hesperia. Comparing his recent observations with those made during 1909, Mr. Wilson concludes that the frequent formation of such spots is an unusual feature of the present apparition.

A LUMINOUS EXTENSION AT THE TERMINATOR OF MARS

Van Biesbroeck, G.; *Popular Astronomy*, 32:589-591, 1924.

In the evening of Oct. 27, under five atmospheric conditions, the 40-inch telescope was turned on Mars near the meridian, and at once a curious appearance was noted on the terminator. The accompanying sketch, Plate XXVIII, shows the general aspect of the planet. [Not reproduced] The drawing was made around 14^h 15^m G. M. T., at which time the central meridian was at 171°.4 longitude. The dominant dark band is Mare Sirenum, branching to the left towards the Solis Lacus which has just passed out of sight, while toward the right the dark belt is continued by Mare Cimmerium. Farther south the irregular contour of Mare Chronium forms roughly a second dark belt, above which Thyle I and II appear as brighter areas extending to high latitudes not far from the polar cap. The diameter of the latter was measured as 1".14, which corresponds to 520 km (327 miles). The northern hemisphere has no prominent dark markings in this longitude. Trivium Charontis begins to show on the morning side; a couple of hours later it stood near the central meridian, followed by Cerberus, the two being almost equally dark. Just below and a little preceding Titanum Sinus there was a brilliant spot about the same size as the polar cap, but not quite as luminous. A little north of this comes a faint belt of mottled gray patches corresponding to the location of Nodus Gordii, Lucus Maricae, Ammonium, etc., but at no time was there the slightest impression of anything like the sharp circular notches (lakes), and very much less of the thread-like geometric canals with which this and other regions of the planet have been abundantly covered by certain observers. The same conclusion was repeatedly arrived at during this whole opposition, at each opportunity that I got of examining the planet under good conditions. On the present night the seeing was very nearly perfect for several hours and faint close double stars of less than 0".3 separation were readily measured at the same altitude as Mars, which could be advantageously examined with a power of 750.

However, the object of this note is to describe more especially the curious appendage to the terminator which I had already noticed at the first glance at the planet (13^h 55^m G. M. T.). Starting from a point about 5° south of the Martian equator the terminator loses its normal elliptic shape and bulges out into the shadow-side of the planet; this extension stretches southward over 3".5 or 1600 km and at about half of its length it gradually separates from the normal terminator; the southern end is a cusp separated by 0".5 from the terminator and the space between the latter and the cusp appears as dark as the shadow-side of the planet. The luminous extension itself was not very brilliant, although projected on black background; it seemed to have the same dull steel gray color as that of the adjacent Mare. The whole appearance was quite striking and several colleague observers, who were called in, checked its presence independently. The observing conditions staying highly satisfactory, a careful watch was kept on the object. From its location it was found that it roughly extended along the meridian of 250°, from latitude -5° to -32°, beginning therefore at the northern part of Ethiopis, stretching across Hesperia and Mare Tyrrhenum as far south as Ausonia. As time went on the extension even increased in length and reached a maximum about 15^h; at 15^h 05^m the elevation above the terminator was measured as 0".55, and the whole length 4".06, or 1870 km. At that time the luminous appendage, slightly curved like the terminator, was entirely detached from the latter except for a tiny isthmus at the north end; its shape reminded me of the appearance of some low flat prominences. At the same time there was a suspicion that the object had shifted bodily to the south. This was confirmed by micrometric measures on the estimated middle point:

Greenwich Mean Time		Mean Latitude	---Length---		
			in arc	in km	in miles
h	m	°	"		
14	15	-18	3.50	1610	1010
15	5	-19	4.06	1870	1170
15	50	-22	3.07	1410	880
16	10	-25	2.30	1060	660
16	30	-27	1.5	690	430
16	50		last impression.		
17	0		gone.		

The further development is illustrated by the four small drawings representing successive appearances of the same luminous appendage. It became narrower and shorter after 16^h. At $16^h\ 30^m$ the separation from the terminator was no longer clearly visible; the planet was then two and one-half hours west of the meridian and the low altitude began to affect the seeing. However, the bulging out of the terminator was still quite plain but it was evidently on the wane. A last impression of a small elevation over the terminator was obtained at $16^h\ 50^m$. At $17^h\ 00^m$ the terminator appeared again normal, but the low altitude of the planet made the observation difficult. The next night, Oct. 28, conditions were not so favorable anymore, yet good enough for making sure that the same appearance did not recur after one rotation.

There are on record a number of similar luminous appearances at the terminator at previous oppositions: Lowell in "Mars and its Canals," p. 101, gives a vivid description of the one of May 25, 1903, which he considered as one of the most remarkable. The present one is a still more striking one, both by its size and especially by its extraordinary persistence during three hours from the beginning of the observations, at which moment it was already fully developed.

Lowell interprets the phenomenon as a cloud of sand-dust, led into this explanation by the ochre color of the cloud as seen beyond the terminator. I did not have the impression of such a color on this occasion, although under the fine conditions of that night the rich variety of colors of the disc came out very beautifully. I interpret the phenomenon as an extended high cloud or fog, illuminated in the morning sky before the subjacent territory is fully brightened up and dissipating under the heat of the sun as it moves into daylight. This might account for the separation between cloud and terminator. It would, however, require a fog of very great extent over the night-side of the planet, namely from about 250° to 290° in longitude; in latitude I assume the cloud to extend in an oblique line from -18° to -27°, so that, as the rotation brings in successive parts of the fog, the front of which melts away in sunlight, the luminous part at the terminator gradually shifts to the south.

Whatever the interpretation of the phenomenon, I thought it worth while recording it at once, in the hope that further information will have been obtained by observers in other localities.

APPARENT FLASHES SEEN ON MARS

Wilson, Latimer J.; *Popular Astronomy*, 45:430-432, 1937.

On May 30, 1937, during a part of the interval between 4^h 35^m and 6^h 0^m G. C. T., while Mars was being observed through a 12-inch aluminized reflector, 250 magnification, with unusually good seeing, what appeared to be a series of bright flashes was seen extending across the south polar cap about 1" north of the southern rim of the disk. They were irregularly intermittent and were estimated to be about one magnitude brighter than the rest of the cap. They were entirely unexpected, but when once seen they became the chief concern of the observation period.

Measured from the drawings and photographs the thickness of the south polar cap on the central meridian was about 1".87. The flashes extended almost midway between the northern edge of the cap and the southern rim of the disk. A line of tiny white spots seemed to extend across the cap, some of the spots coalescing to swell into a brilliant white spot which quickly became yellow, then red-yellow, the phenomenon passing from left to right across the polar cap. Because of the tranquil state of the atmosphere and the general steadiness of the well-defined disk of the planet, the phenomenon was interpreted as genuine. A Barlow eyepiece was used in which there is considerable chromatic aberration when the image is not in the center of the field and this, of course, should be considered, but, after watching a number of the flashes, I decided that the color changes were real.

There were two centers, or sources from which the flashes came, one in about longitude 190 degrees, and the other in 212 degrees. Plotting the positions and allowing for the northern planetocentric position of the earth, it appears that the string of white spots and the flashes were along south latitude 55-65 degrees.

Drawings were made at 4^h 35^m, 4^h 55^m, 5^h 30^m (when flashes occurred) and at 6^h 0^m G. C. T. No flashes were seen after 5^h 40^m. Observations secured under good conditions when this region was presented favorably on the dates preceding and following May 30, although indicating unusual brightness in this area of the south polar region, disclosed no similar phenomena. Unfortunately, the photographs secured through the writer's telescope are hardly adequate to settle the question as to the precise nature of the phenomena.

Observation of the planet since 1909 has not disclosed to the writer a similar polar aspect, though at other times small, twinkling white spots have been seen in or at the edge of both caps. Professor William H. Pickering writes that many years ago he saw effects similar to those which occurred on May 30, and therefore confirms them as having occurred in the past.

Reference to Flammarion's map of Mars (Popular Astronomy, April, 1916) indicates that the apparent flashes in south latitude 55-65 degrees were in the vicinity of Thyle I and Thyle II. As a tentative explanation it is suggested that possibly this region is the northern escarpment of a plateau, Mare Chronium being a deep depression. Slopes facing the sun and earth covered with ice might reflect sunlight in a series of flashes, the rotation of Mars bringing new reflecting surfaces into view, under a cloudless sky.

It would be very interesting to know if any other observers employing comparable magnification and aperture witnessed the phenomena described here. The white spots and the effect of flashing brightness were not seen in June, July, or August, though carefully looked for but under less favorable conditions. The angle between earth, sun, and Mars was different at the later dates, and a repetition would hardly be expected.

Mr. Walter H. Haas, Alliance, Ohio, employed a 6-inch aperture under conditions of hazy atmosphere, did not see the polar phenomena of May 30, though his description of Mare Cimmerion is confirmed by the writer. Mr. John H. DeWitt, Jr., and his brother, Mr. Ward DeWitt, some distance from the writer's location here in Nashville, noticed unusual whiteness of the southern cap, but report that they did not notice any flashes.

MARS, 1954—UNUSUAL OBSERVATIONS

Avigliano, D. P.; *Strolling Astronomer,* 10:26-30, 1956.

Unusual Bright Spots. What is without a doubt the most unusual observation report we received in 1954 is one made by Tsuneo Saheki. It records a truly unique brightening of a small area on Mars, the Edom Promontorium. We print here the observation as recorded in Saheki's own words:

"On the night of July 1, 1954 at about $13^h\ 15^m$ plus 1^m U. T. when I had been carefully looking at the whole disc of Mars getting ready to begin a drawing of the Martian markings, the lighter yellow patch of Edom Promontorium was seen in its usual form and aspect. Suddenly it began to increase in brightness, changing in color to white. In about 3 seconds it reached its maximum brightness when it showed as very bright (about half the brightness of the south polar cap). At this time it showed as a white, larger round form isolating the Sinus Meridiani from the dark band of the Sinus Sabaeus.

"Soon the glare spot began quickly to decrease in brightness and in about 5 seconds it returned to its usual yellow appearance. The brightness of the spot seemed to be somewhat fainter than the bright spot that I observed in 1951 and the duration of this brightening was much shorter than that in 1951. I can state that this bright spot was not illusive but true, because the seeing conditions were rather good at the time of the observation; I recorded them at 6-7 on Pickering's Standard Scale." See Figure 6. [Fig. omitted]

In June, 1937, the Japanese observer S. Mayeda saw a tiny bright spot at the location of the dark spot, Sithonius Lacus and in December, 1951 Saheki saw a white glare spot at the approximate center of Tithonius Lacus (to which he refers in the above report). These two previously recorded glare spots were seen near the limb of the Martian disc. The 1954 glare spot, however, was well on the disc at the time of the observation (see Figure 7 by Saheki for a full disc view of Mars near the time that the glare spot was seen---also note on this drawing the

Bright flare on Mars, June 4, 1937. (Adapted from Sky and Telescope, *14:144-146, 1955)*

two tiny oases, the Antigones Fons following the tip of the Syrtis Major and the Lex Fons at the tip of the following fork of the Meridiani Sinus, plus other delicate details). Saheki notes that his eye was not fatigued from observing at the time of his observation of the glare spot and that what was seen could not have been caused by effects in the Earth's atmosphere or effects in the eye itself such as Muscae Volantes (flying flies). For more details on these glare spots see the article, "Martian Phenomena Suggesting Volcanic Activity" by Saheki in the February, 1955 issue of Sky and Telescope, pp. 144-146. This observation by Saheki shows excellently what is possible for an experienced observer of Mars to achieve through persistent effort in the study of the planet. The area of Edom Promontorium has been seen as very white on many occasions in the past, but to this Recorder's knowledge such a sudden increase and decrease in brightness in this area as seen by Saheki is indeed unique.

Another somewhat unusual bright area that was seen in 1954 (not of such an inexplicable nature, however, as Saheki's glare spot recorded above) was an area that persisted for several days seen in the latter part of July, 1954. It was observed as a roundish, brighter spot between the two canals, the Nilokeras I and the Nilokeras II. The following observers reported this area on the dates noted (U. T. dates):

July 17, 1954. Seen as a bright spot near C. M. (Bohannon).
July 19, 1954. Seen as a lighter yellow spot, near C. M. (Avigliano).
July 20, 1954. Seen as a brighter yellow spot well on disc. (Avigliano). See drawing number 3 on the full page of Avigliano's drawings which appeared on page 41 of the March-April, 1955 issue of

The Strolling Astronomer.
July 24, 1954. Seen as a lighter yellow spot near limb. (Saheki). See drawing number 4 on the full page of Saheki's drawings which appeared on page 81 of the July-August, 1955 issue of The Strolling Astronomer.

As most of the observers recorded this spot as yellowish in tint it most probably was either an unusually prominent yellow cloud or a much brighter desert region. This spot was not seen at other presentations of this general area both before and after the dates noted.

....

BRIGHT FLARES ON MARS

Anonymous; *Sky and Telescope*, 39:83, 1970.

Recent articles by German and Russian astronomers have advocated conflicting interpretations of Martian flares. The classic instance of this phenomenon was seen on June 4, 1937, by Sizuo Mayeda in Japan. Observing the Alcyonius region of Mars with an 8-inch reflector, he saw an intense point of light suddenly appear near Sithonius Lacus. Far brighter than the planet's polar cap, it scintillated like a star and disappeared after about five minutes.

Writing in the February, 1955, issue of this magazine, the Japanese planetary expert Tsuneo Saheki reported two additional cases and suggested that the phenomenon might be due to volcanic activity on Mars.

This idea has been elaborated by Holger Heuseler of the Wilhelm Foerster Observatory in Berlin. His article in a recent issue of the German magazine Die Sterne lists 13 observations of bright Martian flares between 1937 and 1967, all but one by Japanese observers. The recorded durations range from five seconds to more than 40 minutes.

Mr. Heuseler has plotted the locations of these events on a Mars map, finding a markedly nonrandom distribution. For example, three of the 13 flares were in Edom, a bright region just north of Sabaeus Sinus. He attributes the phenomena to active volcanism on Mars, which he believes to be less intense than that on Earth, however.

The Soviet astronomer V. C. Davydov argues against a volcanistic interpretation in two articles in Astronomicheskij Zhurnal, 46, 623 and and 1074, 1969. From a critical reanalysis of the Mayeda observation, he shows that the bright patch could have been a directed reflection of solar rays from a cloud of ice crystals in the Martian atmosphere, giving a parhelic halo. Alternatively, the directed reflection could have been in principle produced by reflection from a nearly vertical surface situated upon the planet itself. The short duration of the visible flare would be explained by the planet's rotation. Other flares than Mayeda's can be similarly interpreted.

Davydov cites six more Martian flare observations not noted by Heuseler. These sightings were made in 1896 by J. M. Offord (England), 1900 and 1903 by Percival Lowell (United States), 1911 by V. and G.

Fournier (Algeria), 1924 by N. P. Barabaschev (U. S. S. R.), and in 1956 by V. P. Bedenko and A. P. Kutireva (U. S. S. R.).

THE SIGNIFICANCE OF BRIGHT SPOTS OBSERVED DURING THE 1971 MARTIAN DUST STORM

D'Alli, Richard E.; *Icarus,* 31:146-156, 1977.

Abstract. Mariner 9 photographs of the southern hemisphere of Mars taken during the 1971 planet-wide dust storm display circular bright spots at a time when all near-surface features were totally obscured. Correlating the positions and diameters of these spots with topography shows that they correspond to craters. About half of all the large craters in the study area were brightened. The associated craters are large and flat-floored, have significant rim uplift, and contain dark splotches on their floors. The depth/diameter relationship of the bright spot craters is comparable to that of a planet-wide sample. Depth may not be important in selectively brightening certain craters. The visibility of bright spots in A-camera photographs is strongly dependent on the wavelength of the filter used during exposure. It is proposed that bright spots result from the multiple scattering of incident light in dust clouds entrained within craters during dust storms. The appearance of the dust clouds is a function of the availability of a dust supply and, perhaps, air turbulence generated by winds flowing over upraised rims and rough crater floors. Bright spots persist during the final stage of the planet-wide dust storm. If bright spots are dust clouds, this persistence demonstrates that crater interiors are the last regions of clearing.

• Transient Dark Areas

A DARK ENTERING WEDGE AT THE TERMINATOR OF MARS

Hamilton, G. H.; *Popular Astronomy,* 33:287-289, 1925.

In the December issue of Popular Astronomy Dr. Van Biesbroeck records a luminous extension at the terminator of Mars; and at the end of his notes concerning this phenomenon hopes that other observers in other localities may also have perceived it.

This extension, though interesting in itself, was not the most singular object at that time, but was simply an addition to the main occurrence.

I had already made a drawing on that night at $7^h\ 45^m$ E. S. T., longitude of the center of the disk 216°, but saw nothing strange. A little after 8^h, while I was studying the disk, I suddenly saw---at least it impressed me as being sudden---a very definite break in the terminator, a jagged saw-toothed cut, in the vicinity of Hellas, just appearing from the unilluminated portion of the disk. At that time it may well have been over that region preceding Hellas, since the terminator side of the planet in that latitude was fairly bright, lacking in markings of note and desert-like in colour. This dark wedge was so definite and black---as dark as the night sky in the field---that my first thought was that it must be some optical trouble in the telescope.

Professor Pickering was in the house at that time and I asked my wife, who was with me, to have a look. She described the marking exactly as I saw it. I then called Professor Pickering and he proceeded to make a drawing. It was then that the bright protuberance was definitely noted by him. I had suspected it, but thought it a matter of contrast until he mentioned it.

The marking lasted from about $8^h\ 18^m$ until $9^h\ 58^m$, when my notes state, "slight cloud on Hellas near where dark wedge appeared earlier in the evening. This marking has now completely gone, though while seen it was as dark as the phase portion of Mars." My notes also say that it was still there at $9^h\ 25^m$.

During the period between my first general drawing and my last, this phenomenon remained and while it lasted I was able to get three sketches of it on the terminator with rough drawings of the region near by.

It was found not to be a surface marking, since it remained on the terminator during the period of observation. As the three sketches show, the surface detail gradually appeared from under the markings as the planet revolved. For that day, at least, the marking travelled from approximately Hellas preceding to Hellas center, and at its disappearance left a small bright area---perhaps cloud---in its stead.

At first thought I took the dark marking to be the shadow of the cloud-like protuberance but, as can be seen from the drawings, any shadow of this cloud could not be on the sunward side, and would in fact not be seen at all, since the shadow would miss the planet's surface altogether.

On October 23, with much better seeing, I was able to draw this same phenomenon, or a similar one, in about the same position in latitude as that of Dr. Van Biesbroeck's protuberances of the 27th. In this drawing there is no sign of the projection, but it may easily have been overlooked in my attempt to get the dark marking properly placed on a drawing with all the other surface features of the planet.

In all my drawings and sketches of it, I have shown this dark wedge and luminous extension farther south in latitude than that of the Yerkes. On October 20, my three drawings were made under poor conditions of seeing---ranging from 8 to 6 on the Pickering standard scale of 12, where anything lower than 6 in the scale of seeing is too poor to attempt observation. The fact of my first observations being made under these conditions shows how definite the marking was, to be able to stand out from the surrounding detail in the manner that it did. I was able to suspect a slight motion southward of the marking, but the detail that

I was able to compare it with was too faint at the time to be thoroughly certain. At all events when seen again on October 23, under good conditions (Seeing 11 on same scale), the marking was much closer to the equator, and my next observation of it on October 27, the same day as that made at Yerkes, it was again much farther south.

This opens up the question as to whether it moved southward over a period of a few days, to again reappear and follow the same course along the terminator, or, since Dr. Van Biesbroeck seems to have seen another object, in a lower latitude and an hour later, it may be that there was a succession of these objects following each other southward. Mine being faint and far south at $13^h\ 15^m$ G. M. T. may have been at the end of its career, while that observed by Dr. Van Biesbroeck at $14^h\ 15^m$ an hour later and much farther north may have been a first appearance and, if it had been possible to follow it in the western sky for a sufficient time, might have reached the position in which I saw the other object.

I am curious as to the reason that the entering wedge was not seen at Yerkes as it was seen here. The dark marking as seen there seems only to have been the division between the terminator proper and the luminous extension. In the case of the observations at Mandeville the dark marking extended appreciably into the illuminated disk, and surface detail was seen not only preceding it but north and south of it. The shape of this marking was also very distinct, appearing as a very accentuated "Beak of the Sirens," though farther south of course, and pointing in the same direction or nearly so as the "Beak" is often seen.

....

MARS, 1954—UNUSUAL OBSERVATIONS

Avigliano, D. P.; *Strolling Astronomer,* 10:26-30, 1956.

Unusual Dark Spots. Notable among the reports received from Dr. James C. Bartlette, Jr. were two showing very small intensely dark spots on the Martian disc. As Bartlett also reported two similar observations at the 1952 apparition of Mars, we will list all four of these observations. The quotes are directly from Bartlett's original reports.

Spot number one. "May 1, 1952---Tyrrhenum dark spot.

> Near to the southern terminus of Mare Tyrrhenum I was surprised to find a very small, intensely dark spot looking much like a satellite shadow, though perhaps not quite as black. I feel certain of the existence of this spot."

The reference Bartlett makes regarding the spot looking like a satellite shadow is that it resembled the shadows thrown on Jupiter by the Jovian satellites. The usual tests of objectivity were employed by Bartlett (moving the image through the field, revolving the eyepiece, etc.) with the result that Bartlett was convinced that the spot was truly on the planet.

Spot number two. "July 3, 1952---Hellespontus dark spot.

> A minute black spot---like a satellite shadow---was suspected

on the N. edge of the Hellespontus at about latitude -35°. This was so minute that I cannot be absolutely sure of its existence, though it recalls the similar anomalous spot observed near Mare Tyrrhenum." See Figure 2.

Spot number three. "July 7, 1954---Castorius black spot. At about the position of Castorius Lacus, fixed from its relation to Propontis, and therefore far north on the disc, I distinctly saw a very small, round, intensely black spot, perhaps not more than 2" diameter---a mere dot. This appears to have been of the same nature as the two black spots seen in July and April of 1952 associated with the Hellespontus and Mare Tyrrhenum respectively." See Figure 3.

Spot number four. "September 23, 1954---Ascuris Lacus black spot. With 300X a much darker, blackish spot came out near or at the site of Ascuris Lacus." See Figure 4.

Regarding this latter observation Bartlett writes, "It is to be noted that the apparent diameter of the (Martian) disc was then only 12".6 though with 300X an apparent diameter in excess of 53' was obtained. This was the same apparent diameter as on July 3, 1952, when with only 3.5" aperture and 100X, yielding an apparent diameter of 20', the Hellespontus spot was discovered."

Confirmation of Bartlett's dark spot of July 7, 1954 was obtained independently by Avigliano on July 9, 1954 with an 8" Cave reflector (see Figure 5). Concerning this observation Avigliano's notes read in part: "The oasis near the position of Euxinus Lacus appeared abnormally dark, much darker than the other oases seen." The C.M. of Mars at the time of Avigliano's report was almost identical with that of Bartlett's observation; and as the positions of the dark spot as seen by each observer check so closely (compare Figures 3 and 5), the abnormal darkening of the oasis (most probably either Euxinus Lacus or Castorius Lacus) must certainly have taken place over a period lasting for at least the intervening time between the two reports. It is interesting to note that Avigliano drew the dark oasis as a somewhat larger spot than Bartlett shows it. Also noteworthy is the fact that the 1952 darkenings appeared in the S. Hemisphere of Mars while the 1954 darkenings were seen in its N. hemisphere. What may account for the darkening of these and other similar areas seen on Mars by past observers we will leave for those with more speculative minds to ponder.

....

• The Martian Blue Haze

BLUE HAZE AND THE VERTICAL STRUCTURE OF THE MARTIAN ATMOSPHERE

Hess, Seymour L.; *Astrophysical Journal*, 127:743-750, 1958.

1. Introduction. It has been known for many years that photographs of Mars taken in blue light do not usually show those surface features of the planet which can be photographed in yellow or red light. The accepted interpretation of this observation is that a cloud or haze layer exists in the Martian atmosphere and that this haze scatters short wave lengths of light more strongly than long wave lengths. Thus the surface detail is more obscured the shorter the wave length. This explanation received strong substantiation when Slipher (1937) found that there were occasional periods during which the surface markings could be photographed in blue light. Only a cloud layer of some sort could exhibit such variability in transparency. As a result, the phenomenon has come to be called the "blue haze."

Over the years since this first observation of "blue clearing," a number of other cases of temporary dissipation have been recorded. The characteristics of these clearings have been (1) the clearing reveals large portions of the surface; (2) the clearing usually occurs for several days near the opposition date and can be detected at most oppositions; (3) it takes only a few nights for the haze to dissipate; and (4) it takes only a few nights for the haze to re-establish itself.

. . . .

Abstract. The atmospheric haze which usually obscures the surface of Mars in blue light is probably a condensate rather than a dust because it can clear away rapidly. Carbon dioxide and water vapor are both present in the atmosphere, and condensation of each into the solid phase has been suggested to explain the blue haze. Calculations are carried out by assuming a convective atmosphere with a surface temperature of 10° C for the subsolar point and a lapse rate of 3.7° C/km. When the optical properties of clouds produced in such an atmosphere are determined with the aid of the Mie theory of scattering, it is found that (1) carbon dioxide ice clouds are far too opaque to be the cause of the blue haze; (2) water ice clouds can cause the blue haze if the surface frost point lies near -90° C. The second result is consistent with the failure of all modern attempts to detect water vapor spectroscopically in the Martian atmosphere, since the total percipitable water content corresponding to such a frost point is only $2\ \mu$. These results indicate that the author's earlier suggestion of carbon dioxide ice clouds should be rejected.

THE SURFACE OF MARS

Leighton, Robert B.; *Scientific American*, 222:27-41, May 1970.

A curious characteristic of earth-based photography of Mars is that most photographs taken through a blue filter show greatly reduced contrast between light and dark surface features. On rare occasions, however, the contrast is much enhanced---a phenomenon known as blue-clearing. It has been postulated that a blue scattering layer in the atmosphere is responsible. The new Mariner pictures taken through a blue filter show no evidence of being obscured by a blue haze when compared with overlapping views taken through red and green filters; the surface of the planet is clearly visible in all colors. Nevertheless, pictures of Mars taken at the same time from the earth show a typical amount of obscuration in the blue. Thus we know that Mariners 6 and 7 did not happen to fly by Mars at the time of a rare blue-clearing. The unusual effects seen when Mars is photographed from the earth through a blue filter remain unexplained. (p. 37)

THE MOONS OF MARS

An enduring Martian enigma has been the literary references to the planet's two moons well before their official discovery in 1877 by Asaph Hall. The literature here is extensive and full of guesses (often wild) about how Swift, Kepler, and others might have predicted what Hall saw through his telescope in 1877. We will probably never know the truth here.

A more modern Martian mystery concerns the strange grooves spacecraft cameras discovered on the surface of the moon Phobos. The most likely hypothesis here is meteor impact but loose ends do remain.

• Possible Pre-1877 Observations

THE MOONS OF MARS

Lamont, Roscoe; *Popular Astronomy*, 33:496-499, 1925.

During the summer of 1924 when Mars was nearest the earth, numerous references were made in the papers to the statement in Jonathan Swift's "Gulliver's Travels," published in 1726, that the astronomers of Laputa had discovered that Mars has two moons, the inner one revolving about the planet in 10 hours and the outer one in 21-1/2 hours; no such bodies being known to the astronomers of this earth until 1877, when Professor Hall at the Washington Observatory discovered the two moons; the inner one being found to revolve in 7 hours 39 minutes and the outer one in 30 hours 18 minutes. Exclamations such as "Marvellous!" "Was Swift a Wizard?" "A Second Mother Shipton!" led me to make a little inquiry into the matter with the following result.

After Galileo discovered four satellites of Jupiter in 1610, Kepler stated that probably Venus, Mars and Saturn, as well as the earth and Jupiter, had moons revolving about them. As the earth has one moon and Jupiter was found to have four, two would be about the right number for Mars, and it is not strange if the astronomers of Laputa, about 1726 discovered the two moons. And now it will be well to quote just what Swift says in "Gulliver's Travels." This is never done. In the "Voyage to Laputa," Chapter III, it is said:

> They have likewise discovered two lesser stars, or satellites, which revolve about Mars; whereof the innermost is distant from the center of the primary planet exactly three of his diameters, and the outermost, five; the former revolves in the space of ten hours, and the latter in twenty-one and a half; so that the squares of their periodical times are very near in the same proportion with the cubes of their distance from the center of Mars; which evidently shows them to be governed by the same law of gravitation that influences the other heavenly bodies.

This shows that Swift was something of an astronomer and understood Kepler's third law, which he states, and how to make use of it. The inner moon was said to be three diameters of the planet from the center of Mars. The diameter of Mars according to modern astronomers is 4230 miles, but in Swift's time this length was not very accurately known, and in Roger Long's astronomy, Vol. 1, page 339, published in 1742, the diameter of Mars is given as 4800 miles. Three times this is 14,400 miles. Suppose another moon to revolve about the earth at a distance from the center of the earth of 14,400 miles. Our present moon revolves in 27-1/3 days at a distance of 238,840 miles. How long would it take another moon to go around the earth at a distance of 14,400 miles from the earth's center? By applying Kepler's third law which the Laputans used in such cases, that the squares of the periodic times are proportional to the cubes of the distances from the center of the earth, we have,

$$x^2/27\text{-}1/3^2 = 14400^3/238840^3$$

and from this equation $\underline{x}$ is found to equal .4046 days or 9 hours 43 minutes, nearly 10 hours. Why, then, should not a moon revolve around Mars in 10 hours at a distance from its center of 14,400 miles? It wouldn't do it, but the Laputans may have thought it would. Gulliver says of the Laputans, in Chapter II, "They are very bad reasoners," and so they may have made the inner moon revolve around Mars in 10 hours because it would do the same thing around the earth at the same distance. The outer moon was distant from the center of Mars five diameters of the planet. How long will it take that moon to revolve around Mars? Kepler's third law will give it $x^2/10^2 = 5^3/3^3$. Gulliver said that the period was 21-1/2 hours, and that the squares of the times of revolution of the two moons were nearly in the same proportion with the cubes of their distances from the center of Mars. This equation gives 21 hours, 30 minutes, 59.6 seconds. The Laputans were bad reasoners, and they always went out accompanied by a flapper, so Gulliver says, but they knew how to use Kepler's third law and that was all that was needed.

The reason why Swift took the distances of the moons from the center of Mars as 3 and 5 times the planet's diameter was probably this: He had read in David Gregory's astronomy, published in 1713 (Vol. 1, page 25, of the second edition of 1726), the following about Jupiter's satellites: "Jupiter has four; the innermost of which revolves about in 1-1/4 of a day, at the distance of 5-2/3 semi-diameters of Jupiter from his center; the second revolves in 3-1/5 days at the distance of 9 semi-diameters." 5-2/3 and 9 semi-diameters are 2-5/6 and 4-1/2 diameters, so Swift took those of Mars as 3 and 5.

Or, having fixed the distance of the inner moon from the center of Mars at three diameters of the planet, Swift may have reasoned like this: A moon at a distance from Jupiter's center of about three of his diameters revolves in 1-1/4 of a day or 30 hours. A moon at a distance from the earth's center of three of the earth's diameters would revolve in about 20 hours (as he would find). So let's have a moon at a distance from the center of Mars of three of his diameters go around in 10 hours.

In Sir Robert Ball's "Reminiscences and Letters," edited by his son, there is the following, page 17.1:

> Dr. Bernard, Bishop of Ossory (who was then Dean of St. Patrick's), wrote to him in April, 1908:
>
> 'A friend tells me that he has been informed on good authority that when Swift made his wonderful statement in Gulliver's Travels (Pt. III, A Voyage to Laputa, Ch. 3) about the number of satellites of Mars and their periodic time, he was not merely romancing, but writing down what some man of science had told him. I feel sure this is not the case, and would be very glad to have your view.'
>
> The reply was as follows:
>
> 'No doubt Kepler interpreted Galileo's anagram about Saturn as a triple object to mean that Mars had two satellites. We are told that Micromegas (1752) saw them, but Voltaire obviously borrowed this idea from Swift. Cyrano de Bergerac seems also to have predicted them in the same way as Swift.
>
> 'But it is, I believe, quite certain that these are all random guesses. It is impossible that they could be seen with instruments of a date earlier than that of the famous telescope at Washington by which they were discovered in 1877.
>
> 'No doubt the notion that Mars had satellites was suggested by the consideration that
>
> Venus had no satellites.
> Earth " 1.
> Mars " x.
> Jupiter" 4.

'It did not require any great genius to say that x was probably 2. It is absolutely certain that no one predicted the periodic times, and the most astonishing 'shot' in the whole of science is Gulliver's statement that the period of one of these satellites was so little as ten hours. Had Swift consulted an average astronomer he would have been told, 'Oh, yes, certainly, let Mars have two moons. That is quite reasonable, but the ten-hour period is preposterously short. Make it ten days and it will also look quite reasonable.' The fortunate circumstance was that Swift drew on his own genius and not on the scientific conventions.

'My own theory is that Swift wrote (under Arbuthnot's advice) ten days, but by a clerical error ten hours was printed!'

If Swift wrote ten days and the printer made a mistake, how did Swift draw on his own genius? And what kind of a being is "an average astronomer" who would tell Swift that a moon, at a distance from the center of Mars of three of his diameters, would require ten days to revolve about the planet? One of Swift's Brobdingnags would walk it in less time than that. "An undevout astronomer is mad," so Young tells us, but what would he have said of that "average astronomer"? Swift stated to Pope in November, 1726, that an Irish bishop had said that "Gulliver's Travels" were so full of improbable lies that he hardly believed a word in the book. Swift would have said something very similar to that "average astronomer," as he had already seen in Gregory's astronomy that the innermost satellite of Jupiter, at a distance from the center of Jupiter of 5-2/3 semi-diameters of the planet (250,000 miles), revolved in 1-1/4 of a day. According to Ball's theory, Swift wrote ten days and the printer by mistake made it ten hours. We then have the inner moon revolving in ten days and the outer one in 21-1/2 hours. Or maybe Ball thought the printer made two mistakes.

It is quite likely, as Sir Robert Ball says, that Swift wrote under Arbuthnot's advice (but not ten days). In Boswell's life of Johnson it is said: "Talking of the eminent writers in Queen Anne's reign, he observed, 'I think Dr. Arbuthnot the first among them. He was the most universal genius, being an excellent physician, a man of deep learning, and a man of much humor.'" Swift and Arbuthnot were great friends, and what Swift lacked Arbuthnot could supply. His "Essay on the Usefulness of Mathematical Learning" is often quoted.

But when it comes to discovering moons, Anna Kingsford has got Swift, Voltaire and Cyrano de Bergerac beaten before they start. In her book "Clothed with the Sun," published in 1889, page 282, she says:

> Here is Jupiter! It has nine moons! Yes, nine. Some are exceedingly small. And, oh, how red it is! It has so much iron. And what enormous men and women! There is evil there too. For evil is wherever are matter and limitation. But the people of Jupiter are far better than we on earth. They know much more; they are much wiser.

No doubt about it. In 1889 astronomers on this earth only knew of four moons of Jupiter, those discovered by Galileo in 1610, but the ninth one was discovered in 1914 by Professor Seth B. Nicholson at the Lick Observatory. It is "exceedingly small"; supposed to be about fifteen miles in diameter; just about a good football for Swift's Brobdingnags, but no "average astronomer" would have found it.

OBSERVATIONS TENDING TO PROVE THE EXISTENCE OF A RING OF ASTEROIDS AROUND THE PLANET MARS

Lamey, Ch.; *Comptes Rendus*, 85:538-539, 1877.

Compiler's Translation. I find in my record of astronomical observations, under the dates of October 24, 1864 and January 3, 1865, the remark that red gleams were situated on each side of the disc of the planet Mars corresponding nearly to the equatorial plane. These appearances and considerations of another order reminded me of the gleam due to the existence of a ring of large asteroids, which encircles the planet and which produce a certain analogy with the crepe ring of Saturn.

• The Grooves of Phobos

CHAIN CRATERS ON PHOBOS

Anonymous; *Astronomy*, 5:55, January 1977.

Viking has discovered another mystery in the most unexpected place---on one of the two small Martian moons. Mariner 9's mapping of both Phobos (12 x 14 x 17 miles, or 20 x 23 x 28 kilometers) and Deimos (6 x 7 x 10 miles or 10 x 12 x 16 kilometers) showed many craters and left most investigators with the impression that they were merely rocky chunks that bore the scars of meteorite impacts. There was a puzzling feature on Phobos that a few analysts noticed but, without better data, could say little about.

At the limit of resolution were a few small crater pits that seemed to align in one or two chains. This was unusual, because crater chains on the moon are traditionally explained as volcanic pits---small eruption sites strung along fracture lines. Yet Phobos apparently is too small to generate heat and conventional volcanic activity.

Viking's high resolution photos have revealed that the crater chains are real and part of an extensive system of parallel grooves, a few hundred yards wide. There may be a tendency for the grooves to lie parallel to the direction of the satellite's orbital motion, although there appear to be several swarms with somewhat different orientations. Scientists are at a loss to explain them. Theories being discussed include: grooves left by much smaller satellite debris also orbiting Mars (though the grooves seem to follow contours of Phobos' surface too closely for this to be tenable); fractures radiating from an impact crater not yet recognized (perhaps on the side of Phobos still poorly photographed), or fractures created in the body of the Martian satellite when it was part of a hypothetical larger body that spawned both Martian moons, perhaps during a catastrophic impact.

Photograph of Phobos taken by Viking Orbiter 1 on May 27, 1977. The right-hand photo has been computer-enhanced. (Courtesy NASA)

TIDAL STRESSES MADE PHOBOS GROOVY

Anonymous; *New Scientist*, 74:394, 1977.

The recent series of encounters between the Viking 1 orbiter and the Martian moon Phobos seems to have yielded a remarkably complete portrait of the little moon. The Viking scientists are now converging on a view of Phobos as a 27-km-long lump of the most primitive stuff of the solar system, probably captured from the asteroid belt soon after the planets were formed. The new interpretation was presented to the Royal Astronomical Society last Friday by Viking team members Joseph Veverka and William Michael.

An important new factor is the mass of Phobos, measured for the first time as the Viking orbiter swung within 97 km of it in the February series of flybys. It comes out at 1.1×10^{19} g, making the density about 2.1 g cm^{-3}. This surprisingly low value is identical to that of the most primitive meteorites known, the carbonaceous chondrites. And there is other evidence that Phobos is made of carbonaceous chondrite material, which is probably the nearest thing we have to the original dust from which the planets and moons condensed.

. . . .

The photos also show Phobos' mysterious parallel grooves in greater detail. They are now seen to be concentrically arranged around the long axis of the moon, which points towards Mars. This highly regular pattern suggests a single explanation---they must represent cracks in the moon induced by tidal stresses. Some of the grooves have little craters strung along them, and the Viking analysis tentatively interpret these as due to outgassing of vapours from the cracks. Suitable volatiles are readily available in carbonaceous chondrites, which contain up to 20 per cent of water in their material.

This implies though that Phobos can't have originated where it is now, for unless current ideas about the origin of the solar system are drastically wrong, anything with that amount of volatile material belongs much further out from the Sun---in the outer asteroid belt. So Phobos is probably an asteroid, perturbed in towards the Sun in the early stages of the solar system, and somehow captured by Mars.

. . . .

However the capture happened, Phobos' history did not end there, as the Viking pictures show traces of subsequent dramatic events. The surface is saturated with impact craters, and there are long bands of smaller craters parallel to the equator, which the Viking analysts now interpret as the scars of Phobos' scooping up debris in its orbit around Mars---debris which could have been ejected from Phobos itself by earlier impacts. There is also a second set of grooves, running underneath the main set at a different angle---possibly evidence for a super-impact some time in the past which knocked the moon away from its original axis.

. . . .

ORIGIN OF THE GROOVES ON PHOBOS

Thomas, P., et al; *Nature,* 273:282-284, 1978.

The characteristic surface features of the inner martian satellite Phobos are impact craters and long linear depressions, here called grooves. The grooves have been explained as surface manifestations of deep-seated fractures, but there has been a divergence of opinion as to the cause of the fracturing. Soter and Harris have attributed the fracturing to tidal stresses induced by Mars; Pollack and Burns have suggested fracturing by drag forces during the hypothetical capture of the satellite by Mars; Veverka et al. have attributed it to a large, nearly catastrophic cratering event. We present here new data on the grooves which tend to support the cratering hypothesis and seem to rule out the tidal mechanism.

. . . .

Figure 1 is a map of the distribution of grooves on Phobos. For reference we have sketched in the outline of the largest topographical feature on Phobos---the 10-km crater Stickney. The region west of Stickney, between longitudes 70° and 160°W, has been imaged at resolutions of only ~150 m, whereas imagery with a resolution of ~50 m exists for the remainder of the satellite. Thus, no significance should be attached to the apparent low density of grooves west of Stickney evident in Fig. 1.

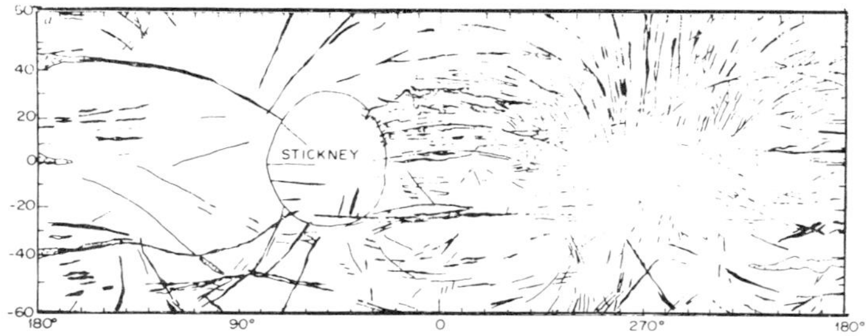

The striking pattern in Fig. 1 is that the grooves emanate from all sides of Stickney and run continuously towards a region near 270° W where they die out. This area, although not exactly 180° away from the centre of Stickney, is on line with a surface normal from its centre when one considers the actual three-dimensional shape of Phobos. (Phobos is approximately a triaxial ellipsoid with principal axes 27 x 21 x 19 km across; the longest axis coincides with the 0° - 180° longitude direction.)

The remarkable pattern of grooves seen in Fig. 1 strongly suggests that the formation of Stickney and of the grooves were intimately related events. This idea is supported by the typical morphology of the grooves. Individual grooves are best developed in the neighbourhood of Stickney and taper out near longitude 270°. Most grooves are 100-200 m wide and 10-20 m deep, with concave-up cross-profiles. The largest grooves, 700 m wide and 90 m deep, are near Stickney. Only near the antipodal point near longitude 270° are the grooves consistently less than 100 m in width.

The grooves have a more complex morphology than that expected of simple fractures. They commonly have a beaded or pitted appearance due to subtle depressions along their lengths. Some grooves, especially between longitudes 170° and 220°W have wide, irregularly bounded segments of hummocky topography, some of which seem to rise above the surrounding surface. Some straight-walled groove segments seem to have subtle, slightly raised rims. Thus, the morphology of the grooves is not that of simple fractures or of fault blocks. The grooves almost certainly represent some response of the Phobos regolith to deeper fractures, and it is probable that the formation of the grooves involved a variety of processes of which deep-seated fracturing was dominant. Attempts to elucidate some of these processes are under way. Whatever the modifying processes, the grooves seem to be closely related to the formation of Stickney.

Chapter 10

JUPITER

INTRODUCTION

The giant planet of the solar system is Jupiter. The naked eye sees Jupiter as a bright jewel in the night sky. Some observers believe they can even see the four Galilean satellites without a telescope. With a little optical help these four large satellites become bright planets circling a miniature sun---a solar system within the solar system.

The mysteries of Jupiter are many. Is it a huge sphere of ices or an aspiring star with an internal energy source? What is the Great Red Spot and does its character vary with solar activity? How do Jupiter's satellites modulate radio emissions from the planet? The foregoing are modern-day problems attacked by radio telescopes and probe-borne instruments. There are also many perplexities left over from the days when the telescope was king. The frequent observation of double satellite shadows on Jupiter's disk does not seem to have a good explanation. A fascinating assortment of curious spots, disk irregularities, and satellite distortion remain unattended. Jupiter is alien, unterrestrial, and full of puzzles.

SURFACE FEATURES

The terrestrial planets live up to their name---all have solid cratered surfaces and seem made from earth-stuff. Low-density Jupiter, however, presents us with masses of swirling chemicals that assume bands and eddies of various hues. The surface is always changing and transient Jovian phenomena are to be expected. Thus, only a few exceptional transients are recorded here. Of special interest is the possible variation of Jovian features with the solar cycle, a synchronism also detected in terrestrial planet phenomena, such as surface luminescence.

In keeping with its turbulent surface Jupiter appears to radiate more energy than it receives from the sun. Just what generates this energy is unknown. The planet actually may be a tiny star that is expiring or trying to come to life. Jupiter seems to have a strong magnetic field, and particles trapped in its radiation belts emit considerable radio energy. The strength of these emissions appears to be modulated by the Jovian satellites in some way. It is curious that Jupiter, the solar system in miniature, should have its emissions affected by attending bodies, just as some claim the solar activity cycle is controlled by the planets!

• Disc Irregularities

IS JUPITER "HUMPY"?

Downing, A. M. W.; *British Astronomical Association, Journal*, 21: 150-151, 1910.

In his "Discussion of the Eclipses of Jupiter's Satellites, 1878-1903," published in Vol. LII. of the Annals of Harvard College Observatory, Prof. Sampson finds certain abnormal residuals occurring in his comparison of the results deduced from the revised theory and elements of the orbits of the four great satellites of Jupiter with the observations. Prof. Sampson has discussed these residuals with great care in Chapter X. of the above-mentioned volume of the Harvard Annals. As the data on which these researches are founded are the refined photometric observations of the eclipses of the satellites made at the Harvard College Observatory, and as Prof. Sampson has now completed and published his epoch-making work containing the tables designed for calculating the positions and phenomena of the satellites between the dates 1850 and 2000, it was hoped that the time had arrived when accurate predictions of the eclipses would be possible. But a consideration of the residuals referred to shows that this is a vain hope, and that errors amounting occasionally to as much as 30^s (or even more) in the computed times of the eclipses of Satellite I. must be expected, with, of course, still larger errors in the cases of the more distant satellites.

Prof. Sampson's discussion of the residuals in the case of Satellite I. is very complete, and it is to the final one of his suggested explanations of the discordances in question that I wish to draw the attention of the Members, and especially of those who belong to the Jupiter Section.

Discrepancies might, of course, be due to omitted terms in the theory with which the observations were compared, but an examination shows that no periodic term that can be entertained could improve some of the most glaring cases.

With respect to possible numerical errors in the calculations, Prof. Sampson expresses his confidence that none of any importance could have escaped detection, owing to the various checks employed in the progress of the work.

After due consideration Prof. Sampson is led to attribute the great part of these residuals, even when not noticeably large (i.e., those amounting to, or greater than, $\sim 2^s$ for Satellite I.), to a physical cause.

As to the nature of this physical cause, a simple supposition would be that the surface of Jupiter is not a spheroid of revolution with an accuracy sufficient for these observations. It is, however, by no means easy to bring this supposition to a crucial test. An examination of the residuals, on the view that the figure of Jupiter is an ellipsoid of three unequal axes, leads to a negative result. Each place where regularity in the residuals was suspected proved delusive, and it must be left to the future to decide whether there is anything of the sort that can be reduced to rule.

There remains for consideration the supposition that the surface that casts the shadow might be bounded by irregular and non-permanent cloud-like protuberances of considerable elevation. With respect to this idea, there is some ground for admitting that such irregularities exist. Observers have recorded several occasions on which they have notices the shadow of a satellite elongated temporarily as it crossed the disc. Mr. Stanley Williams has recorded the observation of a "large protuberant mass." If such a protuberance were to anticipate the time of disappearance of Satellite I. at eclipse by 30^s, its height must be about 1/140 of the radius, or 300 miles. The equivalent in arc would be 0''. 13. If the protuberance did not extend more than about 5,000 miles on either side of its summit, it would be carried completely out of sight in 12 minutes, and there are, unfortunately, no cases in which two eclipses have been observed under such conditions as to supply a test. There are, however, a number of cases in which the decrease of light during the progress of an eclipse, as measured by the photometer, was found to be abnormally quick or slow, and these are exactly the features which a protuberant mass would produce. Even in an eclipse of Satellite I. (during the progress of which, on the average, 27 measures of the fading or increasing light were made) there is ample time for such a phenomenon to declare itself. And the oscillations of light that have occasionally been noticed in the eclipse observations might be ascribed to minor irregularities on the surface of Jupiter with, perhaps, more probability than to unequal brightness on the satellite's disc.

The conclusions to which Prof. Sampson is led may be thus summarised:---

(1) The photometric method of observation of the eclipses gives the instant of half brightness with an accuracy distinctly greater than the residuals from gravitational theory would indicate.

(2) The anomalies noticed in the observations are real features due to the departure of Jupiter's figure from a perfect spheroid.

(3) These departures are probably of an irregular and transient character; on the average they may amount to 1/500 of the radius, but exceptionally to about three times as much.

We see, therefore, that there is a certain amount of indirect evidence in favour of the "humpiness" of Jupiter.

A CURIOUS FEATURE ON JUPITER

Anonymous; *Nature*, 102:432, 1919.

On the night of January 16, at about 9 p.m., Mr. Frank Sargent, of Bristol, observed a luminous protuberance on the eastern edge of Jupiter. It was situated on the equatorial side of the north equatorial belt. He watched it for some time, and it was visible as a white spot well within the limb of Jupiter, but grew fainter as it advanced further on the disc. Clouds interfered and prevented a transit being taken, but on the following night Mr. Sargent re-detected the object, and it was on the central meridian at about 6.46, though so faint as to be scarcely perceptible. He saw it projecting from the western limb at about 9.5 p.m., when it was quite bright and very easily distinguishable. Luminous projections of this kind are often visible on Mars, and are effects of irradiation, but, in the case of Jupiter, where the atmosphere is considerably denser, the conditions are very different, and it seems probable that the feature observed on Jupiter may have been a real prominence, or it would have been obliterated amid the dense vapours on the limb of the planet.

• Anomalous Spots

THE PERIODICITY OF JUPITER'S MARKINGS

Noble, William; *Observatory*, 18:337, 1895.

I may perhaps be permitted to supplement Mr. Waugh's letter in the August number of the 'Observatory,' by saying that my attention was first called to the possibly cyclical character of Jupiter's markings by the discovery of the curious old engraving of the planet in Seller's 'Atlas Coelestis,' concerning which I read a paper before the Royal Astronomical Society in June 1887 ('R. A. S. Monthly Notices,' vol. xlvii. p. 515)---a comparison of this engraving with a drawing of my own made in 1858, with another by the late Mr. Lassell of the same date, which appears on p. 52 of vol. xix. of the 'Monthly Notices,' and with many others of the now famous "red spot," which subsequently appeared, inducing a strong conviction in my mind that this particular region of the planet was probably the seat of the periodical appearance of a notable disturbance in the shape of a spot, sometimes dark and sometimes vividly coloured. My 25 drawings of the planet were selected as typical ones, for comparison with others taken at more or less distant intervals.

. . . .

PERIODIC DISTURBANCES IN THE NORTHERN HEMISPHERE OF JUPITER

Williams, A. Stanley; *Observatory*, 23:176-177, 1900.

In the Monthly Notices, vol. lix. p. 76, there is an important paper by Mr. W. F. Denning dealing with the subject of periodically recurrent disturbance in the region of the north temperate belt of Jupiter. In 1880, and again in the years 1891-92, there occurred great outbreaks of spots on this belt, perhaps the most remarkable fact connected with them being the enormous velocity of the spots relative to other markings, this velocity being very considerably greater than that of the great equatorial current. In the paper above cited Mr. Denning discusses some previous occasions upon which spots had been observed in former years in this region, and comes to the conclusion that outbreaks of the kind recur at intervals of a little more than 10 years, and calls the attention of observers to the importance of examining Jupiter in February 1901, in order to ascertain if there is any trace visible of a similar outbreak of spots.

It seems to me, however, that the evidence in favour of this period is not very strong, and that, if there is a periodical disturbance here, the period is more likely to be about 12 years than about 10 years. The north temperate belt is seldom or never entirely free from spots, so that the mere fact of spots having been shown by certain early observations and drawings really means very little in itself. Then the circumstances connected with the remarkable oblique streak observed in 1860, upon which Mr. Denning appears to lay much stress, seem strongly to indicate that the phenomena in this case originated at the northern boundary of the great equatorial current, and not in the region of the north temperate belt at all. On the assumption of some eruption giving rise to the formation of a dark spot just to the north of the north edge of the swift equatorial current, and of this spot becoming partly entangled in the latter, the natural result would be for the material of the spot to be drawn rapidly out into an oblique streak, exactly as was observed in 1860.

It would appear, therefore, that we have only two real, indisputable facts to go upon with regard to periodicity, namely, that an outbreak of swiftly moving spots occurred in 1880, and another one in 1891-92. So that if there should be any regular recurrence of the phenomenon in question, which I believe will turn out to be the case, the true period is more likely to be nearer 12 years than 10 years. It is therefore highly important that observers should keep a close watch on the region of the north temperate belt during the next few years. According to my views, the next eruption of these swiftly moving spots is not likely to occur before 1903, whilst, according to Mr. Denning's period of about 10 years, some traces of the outbreak should be apparent in February of next year, or even, perhaps, during the present year.

ANOTHER RED SPOT ON JUPITER

Green, Nathaniel E.; *Astronomical Register*, 23:116, 1885.

There is now no doubt about the character of the mark following the great red spot to which reference was made in last number of the Register, it is evidently a small one of the same kind. It was observed this evening under favourable circumstances. The old red on the meridian about 7.30 and the new red following about 9.0, so there is 1h 30m. between them; and the new spot is best seen when the old one is disappearing at the limb. The position of the new spot is on the polar edge of the southern belt, and its size about half the length by one third the breadth of its forerunner.

RAPIDLY MOVING SPOTS ON THE PLANET JUPITER

Anonymous; *Nature*, 144:748, 1939.

Mr. B. M. Peek, president of the British Astromical Association and director of the Jupiter Section has observed an outbreak of small dark spots of projections at the south edge of the North Temperate Belt of Jupiter. They appear to be rotating at such a speed that a complete rotation would take place at about 9 hours 50-1/2 minutes. There is a remarkable similarity between these spots and those which occurred in 1880, 1891 and 1929. Astronomers in possession of telescopes with apertures of 8 inches or more should be able to see these spots.

DISTURBANCES ON JUPITER

Anonymous; *Nature*, 122:743, 1928.

Rev. T. E. R. Phillips spoke on this subject at the October meeting of the British Astronomical Association. Mr. B. M. Peek observed a curious marking south of the south tropical belt early in August. This expelled a number of small dark spots that travelled at a great speed in the direction of increased longitude. They gave a rotation period of 9 h. 59 m., which is the greatest ever recorded. They were carefully watched as they approached the Great Red Spot. The majority of them were deflected into a curved path which went round the Spot on the north side. As they passed the narrow passage between the Spot and the equatorial belt, they were drawn out into elongated ovals, suggesting a strong current through the narrows. They were followed for a short distance after this and they melted away. A few of the spots hazarded the direct path across the Red Spot, but they suffered for their temerity, as they were lost to sight and never reappeared. Mr. Peek's original marking also produced a region of irregular disturbance, which travelled, though much more slowly, in the opposite direction, that of diminishing longitude.

. . . .

ROW OF DARK SPOTS ON JUPITER

Anonymous; *Science,* 69:sup xiv, January 18, 1929.

On the planet Jupiter, now shining brightly in the southern evening sky, within the last few months there has appeared a peculiar row of dark spots, strung like beads on a dark thin line. They have been observed by E. C. Slipher, of the Lowell Observatory at Flagstaff, Arizona. These marks on Jupiter, he said, are indications of some immense turbulence on the planet. The most amazing thing about them, he said, was that they did not remain in the same place on the planet, but lagged behind the rest of the visible surface, so that their longitude was continually changing. The photographs made by Dr. Slipher were taken through various colored filters, so that some were made in red and some in violet light. Unlike Mars, the same details are shown in the blue photographs as in those made with red light, so that he expressed the belief that these markings are all high above Jupiter and near the outside of the atmosphere.

A STRANGE PHENOMENON

Terry, F.; *Observatory,* 5:23, 1882.

Allow me to draw the attention of the readers of the 'Observatory' to the following very singular observation, which I made yesterday evening the 18th instant.

At $6^h\ 33^m$, Brussels mean time, I brought Jupiter into the field of my telescope, which then carried a power of 120, in order to make a second observation of the planet, when I was much surprised to see the field cut in two by a bright red line. It traversed the field of view in a vertical direction, passing between Jupiter and the two satellites, which were at the time to the west of him, and it remained visible for some three or four seconds after I first remarked it. This red line was exactly similar to the red rays which we see during auroras; however, up to $10^h\ 15^m$, I was not able to detect any trace of any such phenomenon. Was this strange appearance, perhaps, due to the passage of a shooting-star?

• Red-Spot Variations

SUDDEN CHANGES IN THE MOTION OF JUPITER'S GREAT RED SPOT DURING 1962

Peek, B. M.; *Royal Astronomical Society, Monthly Notices,* 130:423-427, 1966.

Summary. Fifty-seven visual determinations of the System II longitude of Jupiter's Great Red Spot, made during the last four months of 1962, have been plotted and analysed.

The totally independent testimony of a number of observers provides what appears to be overwhelming evidence that during about three weeks centred near October 20, the longitude of the Red Spot, which previously and subsequently was sensibly stationary in System II, increased by approximately 4°.

On the evidence available it seems impossible to extend the duration of the transition beyond a maximum of five weeks.

These conclusions augment rather than conflict with those recently published by B. A. Smith and C. W. Tombaugh, which were derived from the measurement of a number of photographs of Jupiter, obtained by them during the same period.

CORRELATION BETWEEN SOLAR ACTIVITY AND THE BRIGHTNESS OF JUPITER'S GREAT RED SPOT

Graf, E. R., et al; *Nature*, 218:857, 1968.

Interest in Jupiter has been stimulated by recent investigations. A correlation between solar activity likely to affect the ultraviolet radiation in the solar system and the activities of the Jovian great red spot would be of some significance for those interested in the planet. In fact, examination of the sunspot cycle and the relative brightness of the Jovian red spot reveals just such a correlation.

Fig. 1 shows the curves for both Zurich sunspot numbers and the relative brightness of the Jovian red spot between 1892 and 1947. The data for the relative brightness of the red spot have been taken from Peek and his technique for indicating the intensity of the spot, based on observations, has been adopted.

Fig. 1 shows there is a pronounced correlation between the cyclic maxima and minima of the two curves during the period for which data were compared. In conclusion, it may be worth noting that a maximum in the current sunspot cycle is now anticipated, and also that recent observations have revealed a high intensity of the Jovian red spot.

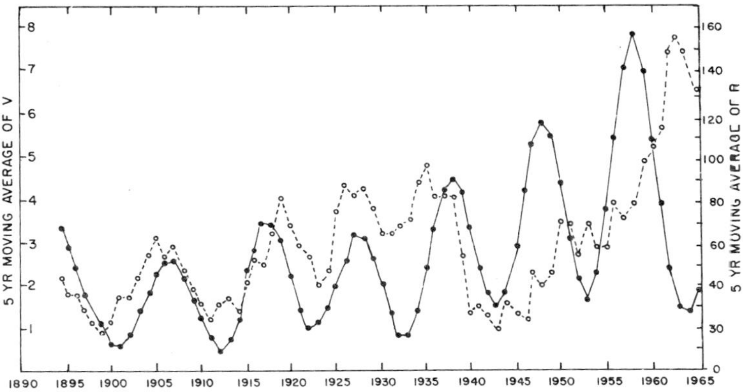

Sunspot numbers (solid line) versus Red Spot brightness (dashed line).

RELATION BETWEEN THE VISIBILITY OF JUPITER'S RED SPOT AND SOLAR ACTIVITY

Basu, D.; *Nature*, 222:69-70, 1969.

The visibility of Jupiter's red spot in relation to solar activity has recently been widely discussed in a series of letters. The original claim of dependence of the red spot on ultraviolet radiation from the Sun noted by Graf, Smith and McDevitt for the period 1892-1947 was discarded by Argyle on the basis of very low value of correlation coefficient ($r = 0.27$) between the visibility (V) and Zurich relative sunspot number (R_Z). The inclusion of recent observations of Jupiter by Reese in the analysis of Solberg and Chapman extends the data to 1967 and shows even a lower value of the coefficient of correlation ($r = 0.16$). The use of some other aspect of solar activity instead of sunspot numbers---such as solar particles, suggested by Scott Smith---is not suitable, for necessary data are not available as far back as 1892. Because the sunspot is known as the index of both wave as well as particle radiations from the Sun, however, a re-investigation of the relationship with the only available sunspot data was considered worthwhile.

Examination of the curves for the visibility of Jupiter's red spot and the solar cycle given by Graf et al. shows no apparent correlation because of the presence of some irregular fluctuations in the visibility function. In the analysis here the irregular fluctuations have been removed by evaluating 5 yr moving averages of the values of V. To achieve uniformity in data the values of R_Z were also subjected to the 5 yr moving average. The modified curves for the entire period 1894-1965 are shown in Fig. 1. [See above]

A striking feature of the curves is the close correspondence between the two phenomena from 1894 to 1945. For this period the correlation coefficient (with moving average values) is r 0.5714 and the standard error for the 52 values compared is $\pm$ 0.14, which shows that r is more than four times the value of σ_r. This correlation is clearly significant and indicates that the variation of the visibility of Jupiter's red spot does have a component dependent on solar activity.

It is interesting, however, that subsequent values of $\underline{V}$ exhibit a completely different correspondence which does not fit into the previous period and consequently the correlation coefficient for the whole period 1894-1965 is reduced to r = 0.2433. Nevertheless, considering the behaviour of the red spot from 1894 to 1945, as revealed by the present analysis, it is difficult to conclude that there is no correlation between Jovian and solar activities.

The apparent discontinuity around 1947 in the cyclic variation of the 5 yr moving averages of $\underline{V}$ raises a second problem. Because the values of $\underline{V}$ provided by Peek (1892-1947) and Reese (1948-67) have been scaled by the same criteria, it must be admitted that Jupiter's red spot begins to show a clear inconsistency around 1947. It is interesting to note in this connexion that this is also the time when the Sun shows an unexpected large enhancement in sunspot numbers. It is perhaps worth investigating whether the two might have a common cause external to both, such as local variation in density of interstellar medium.

• Jovian Radio Emissions

IO-RELATED RADIO EMISSION FROM JUPITER

Dulk, George A.; *Science*, 148:1585-1589, 1965.

Bigg discovered that Jupiter's satellite Io exerts a strong influence on Jupiter's decametric radio emission. In making his discovery, Bigg analyzed data obtained on the Boulder (Colorado) spectrograph in the years 1961 through 1963. This report is an extension of Bigg's work to the years 1960 and 1964. It shows that Io affects both the probability of emission and the character of the emission spectrum and that the emission probability approaches 1.0 when Io's position and the Jupiter longitude are simultaneously favorable, and deduce that "early-source" and Io-related "main-source" emission are simultaneously emitted into two cones once every 12.9 hours when the plane of Jupiter's magnetic axis swings past Io.

....

The most probable explanation for the Io effect on Jupiter's radio emission is that Io disturbs the orbits of particles trapped on nearby field lines of Jupiter's magnetosphere, causing them to precipitate into Jupiter's atmosphere or ionosphere where they generate the radio emission. The emission is highly directive, and is received on the earth only if the earth is within the emission pattern. The narrow-band features of the emission are consistent with the hypothesis that a small group of field lines in Io's vicinity are disturbed; however, the broad-band emission often present implies that field lines are disturbed over a fairly wide longitude range, a fairly thick L-shell, or both. The implications of Fig. 4 [Not reproduced], together with similar implications from third-source and fourth-source emissions, are that Io can induce emission into the ecliptic plane when Io is over the hemisphere $110^o < \lambda_{III} < 290^o$. No emission is induced into the ecliptic when Io is over the other hemisphere, and implication that any emission induced by Io when it is over that hemisphere ($290^o < \lambda_{III} < 110^o$) goes into a sheet which does not intersect the ecliptic. Together with the observation that Jupiter's dipole is tilted toward the earth when the LCM $\sim 200^o$, this implies that Jupiter's dipole is also displaced; for otherwise we should observe emission with characteristics similar to the early source (for example) when Io is symmetrically situated over the other hemisphere. A displaced dipole has previously been postulated by Warwick, and observed by Berge and Morris.

The relation between Io-related and Io-independent emissions is not clear. Since both types of emission have many characteristics in common, it is unlikely that they are generated by different emission processes. Perhaps Io-independent emissions are due to particles disturbed by a different mechanism, are generated on a different L-shell, or are generated on field lines leading to a different longitude region on Jupiter. Warwick has suggested that so-called Io-independent events may be generated by Io, but that there is much broader beaming of the resultant emission. Alternatively, the longitude range affected by the Io disturbance may occasionally be large, perhaps because the disturbance at a given longitude decays with a time constant of several hours.

It is difficult to understand how Io's influence can extend to a large distance from its surface. Its gravitational force is small compared to Jupiter's only 0.1 of Jupiter's radius (R_J) from Io. Its motion through Jupiter's field lines is sub-Alfvenic, so that an ordinary Mach-type shock wave will not exist. A possible agent to extend Io's influence is a magnetic field. If Io has a magnetic moment equal to the earth's magnetic moment, Io's field will be confined by Jupiter's field to about 0.2 R_J of Io's, although Jupiter's field lines will be distorted at a distance several times greater than 0.2 R_J, on the assumption that Jupiter's magnetic moment is about 4×10^{30} gauss cm^3. Io need not have an internal magnetic field source to have a magnetic moment; it may contain a quantity of highly permeable material to interact with Jupiter's field. Or a magnetospheric tail may be formed near Io in a manner similar to that suggested by Gold for the moon, where the motion of the body through a plasma and magnetic field induces the magnetospheric tail.

Even if Io influences only a small region near its position, it will still affect particles in a fairly thick L-shell if Jupiter's magnetic dipole is displaced horizontally (perpendicular to the rotational axis). For instance, if Jupiter's dipole is displaced 0.3 R_J away from the rotational axis, in the course of one Jupiter rotation Io will affect particles in the shell $5.6 < L < 6.2$. Berge and Morris' observations indicate the dipole is displaced 0.5 R_J along the rotational axis and about 0.4 R_J horizontally. Warwick, by a different line of reasoning, derived a vertical displacement of 0.7 R_J and a horizontal displacement of about 0.2 R_J. Therefore, a horizontal displacement of 0.3 R_J is not unreasonable, and the affected L-shell would be 0.6 R_J thick.

Though such a displaced dipole may help explain some of the narrow-band, drifting features of the spectra, it does not explain the Io-related broad-band emission or the simultaneous emission of early- and main-source spectra with largely different character. I suggest that the broad-band emission provides evidence that Io's influence is not confined to its near vicinity, but influences a region a few tenths of a Jupiter radius in thickness or perhaps 40° wide in longitude. A magnetic field or magnetospheric tail at Io is the probable agent to extend the influence.

The Io effect enables us to predict when most of the major radio events will occur, the duration and highest frequency of emission, and even the character of the emission spectrum. Quiet times can also be predicted. There is some evidence for year to year changes in the character of the emission and Io effect. Emission in 1964 was more patchy, weaker, and less frequent than in 1963 or 1962, and the Io effect was stronger in 1964. The significance of these changes has yet to be determined.

• Jupiter's Internal Energy

ON A PHOTOGRAPH OF JUPITER'S SPECTRUM, SHOWING EVIDENCE OF INTRINSIC LIGHT FROM THAT PLANET

Draper, Henry; *American Journal of Science,* 3:20:118-120, 1880.

There has been for some years a discussion as to whether the planet Jupiter shone to any perceptible extent by his own intrinsic light, or whether the illumination was altogether derived from the sun. Some facts seem to point to the conclusion that it is not improbable that Jupiter is still hot enough to give out light, though perhaps only in a periodic or eruptive manner.

It is obvious that spectroscopic investigation may be usefully employed in the examination of this question and I have incidentally, in the progress of an allied inquiry, made a photograph which has sufficient interest to be submitted to the inspection of the Astronomical Society.

If the light of Jupiter be in large part the result of his own incandescence, it is certain that the spectrum must differ from that of the sun, unless the improbable hypothesis be advanced that the same elements, in the same proportions and under the same physical conditions, are present in both bodies. Most of the photographs I have made of the spectrum of Jupiter, answer this question decidedly, and from their close resemblance to the spectrum of the sun indicate that, under the average circumstances of observation, almost all the light coming to the earth from Jupiter must be merely reflected light originating in the sun. For this reason I have used the spectrum of Jupiter as a reference spectrum on many of my stellar spectrum photographs.

But on one occasion, viz: on September 27, 1879, a spectrum of Jupiter with a comparison spectrum of the moon was obtained which shows a different state of things.

. . . .

A casual inspection will satisfy any one that such modifications in the intensity of the background are readily perceptible in the original negative. They seem to me to point out two things that are occurring: first, an absorption of solar light in the equatorial regions of the planet; and second, a production of intrinsic light at the same place. We can reconcile these apparently opposing statements by the hypothesis that the temperature of the incandescent substances producing light at the equatorial regions of Jupiter did not suffice for the emission of the more refrangible rays, and that there were present materials which absorbed those rays from the sunlight falling on the planet.

. . . .

THE INTERNAL POWERS AND EFFECTIVE TEMPERATURES OF JUPITER AND SATURN

Aumann, H. H., et al; *Astrophysical Journal*, 157:L69-L72, 1969.

Abstract. The total power emitted by Jupiter and Saturn has been measured by observing the planets from a jet aircraft at 15-km altitude with a telescope system open from 1.5 to 350 μ. The two planets were found to radiate 2.7 and 2.4 times the amount of power they receive from the Sun, respectively. These new results put observational restraints on models for the internal structures and atmospheres of the two planets.

JUPITER: HIS LIMB DARKENING AND THE MAGNITUDE OF HIS INTERNAL ENERGY SOURCE

Trafton, Laurence M., and Wildey, Robert L.; *Science*, 168:1214-1215, 1970.

Abstract. The most accurate infrared photometric observations (8 to 14 microns) to date of the average limb darkening of Jupiter have been combined with the most refined deduction of jovian model atmospheres in which flux constancy has been closely maintained in the upper regime of radiative equilibrium and a much more accurate approximation of the 10- and 16-micron vibration-rotation bands of ammonia has been incorporated. The theoretically predicted emergent specific intensity has been multiplied by the spectral response function and folded (mathematically convolved---intersmeared) with the spatial response function of the atmosphere-telescope-photometer combination. The resulting comparison indicates that Jupiter is radiating from three to four times as much power as the planet is receiving from the sun.

IDIOSYNCRACIES OF THE JOVIAN SATELLITES

The Jovian Galilean satellites resemble the terrestrial planets in size. Indeed, many a science fiction story has peopled them with alien cultures. Such speculations are presumptuous because very little is known about these moons. Through the telescope they present vague markings that come and go. Even their shapes seem poorly defined; some competent astronomers have seen them as ellipses and other shapes.

The transits of the satellites across the huge disk of Jupiter engender two types of enigmatic phenomena: (1) occasional dark transits, where the orbs are jet black rather than sunlit; and (2) double shadows on the face of Jupiter. Subjective factors may be at work in both cases, as may be physical interactions between the satellites and Jupiter's extended atmosphere.

Like most of the other solar system planets, Jupiter has suggested (rarely) that it has a faint ring. Once more we have the strange effect of objects (satellites) hanging on the rim of the occulting body (Jupiter). This is analogous with stars seen within the rim of the moon. (Chapter 6). It is instructive that, at the limits of seeing, the eye sees all manner of unusual lines, vague shapes, and unexpected interactions of images.

• Odd Shapes and Surface Features

W. H. PICKERING AND THE SATELLITES OF JUPITER

Ashbrook, Joseph; *Sky and Telescope*, 26:335-336, 1963.

Compiler's Digest. William H. Pickering (1858-1938) was the brother of E. C. Pickering (1846-1919), the leading astronomer of his time. Instead of following his brother's footsteps into the realms of stellar photometry and spectral classification, William H. Pickering turned to visual studies of the moon and planets. Although a skillful and persistent observer, he is remembered for his rather bizarre theories and speculations based upon his lengthy telescopic studies; viz., the existence of fauna and flora on the moon. Less well known are his curious findings regarding the shapes of the four Galilean satellites of Jupiter. While observing from Arequipa, Peru, in 1892, with a 13-inch refractor, Pickering's micrometer measurements of Satellite I suggested that the satellite's disc was not circular. Here is what he wrote:

> At first glance I noticed to my surprise that its disc was not circular, but very elliptical. A brief computation the next morning showed that if my measurements were correct, that the polar flattening would correspond to a rotation period of about forty minutes, assuming a uniform density. Observations on the next evening confirmed my first measurements. Some of the other satellites were also measured, and I then returned to the first one, when to my astonishment, instead of showing an elliptical disc, it showed one that was perfectly circular, precisely like the other satellites. I could scarcely believe my eyes, but as I continued to watch and measure, I saw the disc gradually lengthen again and assume the elliptical form, and I then understood what had really been found. The 1st satellite has the form of a prolate spheroid or ellipsoid, or in popular parlance is 'egg-shaped.' The two minor axes are approximately equal, and the satellite revolves about one of them, or as we may say, it revolves 'end over end.'

Pickering also discovered that the discs of the other three Galilean satellites also varied between circles and ellipses in regular ways. But other astronomers, such as E. E. Barnard, could not duplicate Pickering's results. Nevertheless, Pickering and his associates continued to observe variable shapes. Pickering finally concluded that the Galilean satellites were not solid, "but appeared to be concentrated meteor swarms, or heaps of dust, like Saturn's rings."

During 1925 and 1926, J. Stebbins and T. S. Jacobsen of Lick Observatory applied photoelectric photometry to the problem and convinced themselves that Pickering and his associates were not seeing true changes in disc geometry. The Galilean satellites, according to Stebbins and Jacobsen, are certainly solid objects with periods of axial rotation equal to their periods of orbital revolution about Jupiter. Their brightnesses do vary markedly, and many astronomers now believe that

Pickering's apparently anomalous observations are really due to subjective factors associated with these changes in brightness.

DIPLOPIA AND EGG-SHAPED JOVIAN SATELLITES

Antoniadi, E. M.; *British Astronomical Association, Journal*, 9:86, 1898.

Prof. W. H. Pickering's very extraordinary 1893 observations of rapid changes in the outline of Jupiter's third satellite, which was seen alternately round and elliptical, are present in the mind of every student of the planet Jupiter. On 1896, April 17, I accidentally had the opportunity of confirming the accuracy of the American astronomer's observations. The third satellite appeared, in spite of myself, and at the first glance, egg-shaped, with a tendency to duplicity (Fig. 1). The longest dimensions made, when I started observing, an angle of some 45° with the planet's orbital plane; but the value of this angle was slowly and steadily decreasing. The appearance lasted for about an hour, when the elongation gradually subsided into the circular form. At first I thought I was witnessing an occultation of a satellite by another, but reference to the "Connaisance des Temps," showed that this was not the case; so I thought myself victim of some illusion, though of what kind, of course, I could not say.

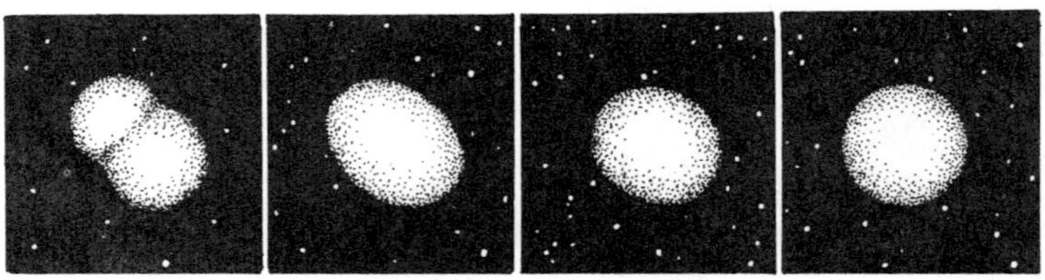

$9^h\ 11^m$ $9^h\ 51^m$ $10^h\ 6^m$ $10^h\ 11^m$

In my recent papers on gemination, published in the last volume of the "Journal," I had suggested the idea that in the case of the presence of a direction of geminatory preponderance, the disk of Mars ought to appear slightly elongated. But between the optical elongation of the Martian disk and the elliptical contours of the Jovian satellites there was only a step. I made this step, by which I will now venture to offer optical gemination as a possible explanation of the remarkable phenomena observed by Prof. Pickering. I must add that the validity of this interpretation does not rest entirely on reasoning, inasmuch as, on June 12 last, I had simply to act on the focussing tube in order to elongate and double the Jovian satellites at random.

FOUR INDEPENDENT DRAWINGS OF GANYMEDE

Haas, Walter H.; *Sky and Telescope,* 9:59, 1950.

During the recent convention of western amateur astronomers in Los Angeles, Calif., four members of the Association of Lunar and Planetary Observers had an opportunity to use the Griffith Observatory 12-inch Zeiss refractor. The instrument was made available through the courtesy of Dr. Dinsmore Alter, director of the observatory. On the evening of August 23, 1949 (local date) we began observations by directing the telescope at Jupiter. The image found was a truly excellent one, and eight dark belts were easily visible (north equatorial, south equatorial north and south components, south temperate, north temperate, north north temperate, south south temperate, and equatorial band). In the south equatorial belt we saw a number of tiny dark spots considerably smaller than the disks of the satellites. Their diameters were probably only 0".3 or 0".4, and they were undoubtedly beyond the grasp of most telescopes of ordinary sizes. Since the view was so good, we turned our attention to Jupiter III or Ganymede and were pleasantly surprised to find markings on it that were almost easy. The writer had never before seen detail on the satellite so well, even with an 18-inch refractor.

We then each made a drawing of the satellite, and these are reproduced in the accompanying picture. They are completely independent; no observer had any idea what the others had seen when he was at the telescope. The magnification employed was 555 times. The views shown are all simply inverted ones with south at the top and west at the left. The disk of Jupiter III was oriented by noting the direction of other satellites from it. The sky was clear. The seeing, or atmospheric steadiness, was noted on a scale of 0 to 10 with 10 best. Data on the four drawings follow:

No.	Observer	Time (p.m. PST)	Seeing
1	Haas	11:25-11:37	4
2	Brinckman	11:50	4
3	Cave	11:52-11:58	4-6*
4	Cragg	12:00	6-7

*Seldom 4

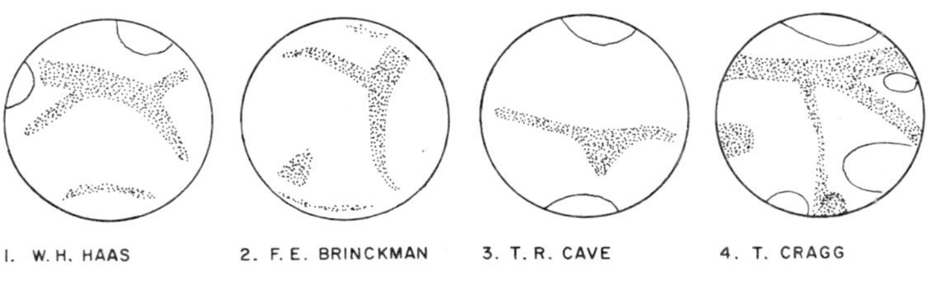

1. W.H. HAAS 2. F.E. BRINCKMAN 3. T.R. CAVE 4. T. CRAGG

Four independent drawings of Ganymede markings

General impressions of detail on the satellite are perhaps best summarized in Cragg's joking remark the next day: "We now call it Mars, Junior." Haas more formally noted while observing: "The resemblance to Mars when the central meridian of longitude is about 250° (Syrtis Major, Hellas, polar caps, etc.) is indeed striking at times."

Every observer but Brinckman saw white caps near both the top and the bottom of the disk. These may be regarded as north and south polar caps; for variations in the brightness of Ganymede have a period equal to the period of its revolution around Jupiter, and this result has been interpreted to mean that the satellite always keeps the same face toward its primary. Its axis of rotation must hence be very nearly perpendicular to the plane of its orbit, and the white areas must be near the geographic poles. Haas thought the south cap much brighter than the north cap, but the others do not confirm this difference. It will be seen that Cave and Haas drew the polar caps not diametrically opposite each other; one may wonder whether there is any analogy to the similar aspect sometimes found on Mars when one of its polar caps is atmospheric.

Three of the observers agree surprisingly well that the most conspicuous dark markings took the form of a large, near-horizontal Y and that the two arms of this Y met its base in the upper right (south-east) quadrant of the disk. In fact, the Y is evidently imperfectly shown on Cave's drawing too, though the junction point is there in the lower half of the disk. It appears probable that a dark feature drawn by Brinckman near the north limb and suggestive of the Mare Acidalium on Mars is also present on Cragg's drawing. The greater amount of detail on Cragg's drawing than on the others may be due to the better seeing when he was at the telescope.

MOONS OF JUPITER: IO SEEMS TO PLAY AN IMPORTANT ROLE

Metz, William D.; *Science*, 183:293, 1974. (Copyright 1974 by the American Association for the Advancement of Science)

Jupiter and its Galilean moons were observed from the ground long before the space program began, and few people expected startling discoveries from terrestrial telescopes. But a recent observation of the innermost Galilean moon has motivated a theory that seems to correlate many older observations from the earth as well as some of the new discoveries by Pioneer 10. Last year Bob Brown, at Harvard University, Cambridge, Massachusetts, discovered evidence of sodium on the moon Io. The initial observation was so straightforward that it could have been done 50 years ago. Furthermore, the existence of atomic sodium on a planetesimal seemed so unlikely that few people believed the initial observation until it was confirmed by high-resolution studies at the Smithsonian Astrophysical Observatory, Mount Hopkins, Arizona, by Brown and Fred Chaffee. A new model for sodium emission that was prepared just before Pioneer 10 reached Jupiter predicted several of Pioneer's discoveries, most notably the existence of an ionosphere on Io.

The innermost of the four large Galilean moons of Jupiter, Io has a 42-hour orbit around Jupiter and a mass comparable to that of the earth's moon, and for some time ammonia ice has been suspected on Io. To explain the observations of sodium emission, Michael McElroy, Yuk Ling Yung, and Bob Brown at Harvard propose that the surface of Io is covered with a layer of ammonia ice containing traces of sodium, and perhaps potassium and calcium. Even at the low surface temperature of Io, some ammonia would evaporate. Ultraviolet sunlight would then dissociate the ammonia to produce molecular nitrogen, which could excite sodium atoms to produce the emissions that have been observed. (Nitrogen would do this most efficiently if it were excited to a metastable state, perhaps by a mechanism similar to the auroral process on the earth.) The dissociation of ammonia would also produce atomic hydrogen, which is so light it would escape from the gravitational attraction of Io.

McElroy and his associates predict that there should be an extensive cloud of hydrogen, nitrogen, and sodium around Jupiter in the vicinity of Io's orbit. Pioneer 10 was not designed to find nitrogen or sodium, but a torus of hydrogen with a diameter of about the diameter of Io's orbit has been observed. The sodium emission model also requires hydrogen in the atmosphere of Io, as possibly detected by Pioneer 10, and the existence of an ionosphere formed by photoionization of sodium. The discovery of an ionosphere for Io was, of course, one of the most definitive early results from Pioneer.

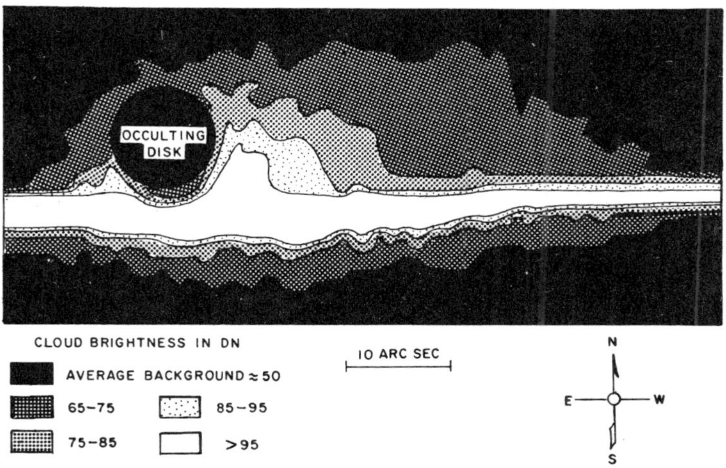

Map of sodium emission from Io

Two other phenomena associated with Io can be incorporated within the sodium emission model. After Io emerges from Jupiter's shadow, it is unusually bright for about 15 minutes and then gets dimmer.

Several people have suggested that ammonia from the atmosphere falls as snow onto Io while it is on the night side of Jupiter, then the brilliant snow evaporates as Io emerges into sunlight.

Another striking observation known for several years is that the bursts of radio noise from Jupiter are strongly correlated with the position of Io in its orbit. To explain this, Peter Goldreich of California Institute of Technology, Pasadena, and D. Lynden-Bell University of Cambridge, England, proposed that if Io had a conducting ionosphere, it could generate a large voltage as it cut through the magnetic field lines of Jupiter. The voltage could drive a plasma process, perhaps like the aurora, which would produce the observed radio noise. The sodium emission model provides the needed ionosphere, of course. In return, the generator model provides a method of exciting nitrogen and also heating the atmosphere to high temperatures needed for sodium to escape. No evidence for a torus of sodium around Jupiter has been found, but Lawrence Trafton at the University of Texas, Austin, and associates have made Earth-based observations that suggest a cloud of sodium emission around Io extending to about 20 times the diameter of the moon.

Other possible implications of the new model, to be published in the Astrophysical Journal (Letters), are that the gases from Io could be the source of trapped radiation belts, and that high-temperature neutral gases could move out far beyond Io's orbit, become ionized, and perhaps cause the stretching of the magnetic field that is observed.

The new observations suggest that Io affects the environment of Jupiter in many ways. Jupiter is unique, but the moons of Jupiter, and perhaps Saturn and Neptune, may be much more significant than their sizes alone suggest.

ODD POLARISATION HINTS AT ICY CRATERS ON JOVIAN SATELLITES

Anonymous; *New Scientist*, 79:340, 1978.

Modelling their observations in Icarus (vol 34, p 268), Ostro and Pettengill attempt to ferret out the "mystery process" responsible. Single-bounce backscatter reverses the sense of polarisation (which is not observed) while multiple scattering would rapidly dilute the echo signals (which are in fact "surprisingly strong"). Double-bounce backscatter is a strong contender for the mystery process but only if a special geometry is postulated.

The best model appears to be a surface pocked with nearly hemispherical craters, thickly covered with ice or any icy clathrate. Incoming radar waves would bounce off one crater wall at about 45° from the bottom, bounce off the other side at the equivalent point, and return (albeit diminished in power) to the Puerto Rican receiver. Craters 1 km in diameter and 0.3 km deep would fit the bill, and their existence would come as no surprise after all the evidence of cratering in the inner Solar System. (Excerpt.)

• Dark Transits and Double Shadows

DARK TRANSIT OF ONE OF JUPITER'S SATELLITES

Anonymous; *Sidereal Messenger*, 3:126-127, 1884.

"Last night I observed the transit of Jupiter's fourth satellite; the first contact was about two minutes later than the Almanac time, and the satellite was about eight minutes getting completely on the planet's disc. The air was very steady and definition good. The satellite entered the white portion just south of the great, dark-red belt. I observed the satellite for about five minutes after the internal contact, and saw nothing unusual, but on the contrary, could only see the bright satellite on the white belt with difficulty. At the Almanac time of II occultation disappearance (8^h 52^m L. M. T.), Mr. W. H. Lowden took the instrument to observe the phenomenon, and he at once announced that there was a shadow of a satellite on the planet. Thinking he must be mistaken, I again looked and found, as he had said, a black spot 'as black as a drop of ink,' and I then noticed that this spot occupied about the position of the satellite then in transit. I then thought I must have made an error, but upon referring to the Ephemeris, I found that no shadow should be on at that time, and that consequently it must be the fourth satellite projected on the disk as a black spot instead of the usual bead of light. This phenomenon so occupied our attention that the occultation disappearance was allowed to pass unnoticed. Thereafter we watched the spot, at intervals for nearly an hour, during which time it remained absolutely black."

It will be noticed that Mr. Burckhalter observed the satellite enter as a white disc on the body of the planet, and that it was subsequently seen black. He observed with a ten and a half-inch reflector, but does not report the magnifying power employed, and has furnished no drawings.

Professor Davidson then referred to the earlier notices of these or similar phenomena. It was first reported by Cassini in 1665, and afterwards more fully described by Maraldi, in 1707.

It was not again mentioned until 1796, perhaps from inattention to the subject; but more recently the great telescopes have revealed these peculiarities. South saw two satellites on the planet, but of a chocolate color. Bond, on the 28th of January, 1848, observed the transits of I and II, whilst III itself was seen as a black spot between the two shadows, and not to be distinguished from them except by the place it occupied; it was smaller than its shadow in the proportion of 3 to 5, not duskish, simply, but quite black, like the shadow."

Again, on the 18th of May, 1848, Bond saw the III satellite enter the disc very bright; twenty minutes later it was hardly perceptible; a little after it was a dark spot; then for two and a half hours it was perfectly black and nearly round.

The transit of IV on March 25th, 1873, was remarkable for its absolute blackness, and was seen by many observers in England.

Several dark transits of I were recorded in England in 1880, and it is quite likely that the literature upon the subject of Jupiter may reveal other notices.

Several solutions of this curious problem have been offered, and were mentioned by the speaker, but none have been satisfactory.

BLACK TRANSIT OF JUPITER'S THIRD SATELLITE

Swift, Lewis; *Sidereal Messenger*, 9:474, 1890.

On the evening of July 21, 1890, I observed the most intensely dark transit of a Jovian satellite that it has ever been my fortune to witness. Both satellite III and its shadow were on the face of the planet, the former showing a round disc of dense blackness, its limb sharp and well defined, while the shadow was less black in hue, being brownish in tint, and apparently not exactly circular in form, with definition of the limb somewhat less distinct. As other work demanded my attention, I did not renew the observation to see if it underwent any change of color as it neared the edge of the disc. This general phenomenon I have witnessed before, but never saw the satellite so strikingly black as on this occasion.

OBSERVATIONS OF THE PLANET JUPITER AND HIS SATELLITES DURING 1890

Barnard, E. E.; *Royal Astronomical Society, Monthly Notices,* 51:543-549, 1891.

I send a series of sketches of the appearances of Satellite IV. when emerging from transit 1890, Aug. 13. [Not reproduced]

On this occasion as the satellite emerged it seemed to leave behind it a small dark spot which apparently remained stationary and diminished in size, as the satellite receded, until it became lost through exceeding smallness.

In connection with the black transits of III. and IV. I have often seen I. transiting as a dark or dusky spot. On Sept. 8, 1890, with the 12-inch this satellite presented a remarkable aspect while in dark transit. I noticed that it appeared elongated in a direction nearly perpendicular to the belts of Jupiter.

With high powers (500 and 700) and perfect definition the satellite appeared distinctly double, the components clearly separated. At my request Mr. Burnham kindly examined the satellite with me, and we both distinctly saw the phenomenon of apparent duplicity (see Ast. Nach. 2995). In reference to the appearance of the satellite---whatever may be the explanation---Mr. Burnham has no hesitation in stating that it was as distinctly double as any double star that he has seen. The distance between the centres of the two images was about 1", and the position angle at transit 173° ± . The south component was very slightly the smaller (Plate 14, fig. 4). It was not possible to follow the satellite closely, because of interruption by a number of visitors, and the great telescope was not available for observing the phenomenon.

I have offered two explanations of this phenomenon (A. N. 2995).

1st. That the satellite had a white belt on it parallel to those of Jupiter, or

2nd. That the satellite is actually double. I do not now think that the first of these theories is satisfactory, as it would require a far whiter belt than we have reason to believe will exist on the satellite, reasoning from our knowledge of the phenomena of Jupiter itself. I am strongly inclined to favour the theory of actual duplicity.

It may be asked, if the satellite is double why has it not so been seen when projected on the sky, if the distance is as much as one second? In the case of two stars this distance would be very easy with any considerable telescope. We must remember, however, that in the case of a double satellite we should have two sensible discs, which would perhaps never be well enough defined to show a maximum separation of only one second, while the greater part of the time the distance would be much less than this. If, however, the satellite were projected on a background brighter than itself it would appear reduced to its minimum size, so that if duplicity existed the chances for detecting it would be far more favourable. To better illustrate my meaning: Mercury is a very brilliant object to the naked eye about the times of its greatest elongation. Yet when this planet crosses the Sun it is shorn of its light and cannot be seen with the unaided eye, though it is then a fifth nearer to us. On this same principle if the satellite consists of two moons the largeness of their ill-defined discs would make them appear as one in a telescope. Transferred, however, to a bright surface, the confusing light is got rid of and the size of the object is reduced to a minimum, and if double then will be the most favourable time to detect it.

If this satellite is really double, the components at the time of my observation were probably at their greatest elongation. If so this would indicate a distance from centre to centre of some 2,000 miles. The individual components would be about 1,000 miles in diameter, and their orbits nearly perpendicular to that of Jupiter, with a period of revolution of a few hours.

I trust that this satellite at its transits this year will receive the closest attention with powerful telescopes. It should be examined with the highest powers, and the direction of elongation, if any, carefully noted. If double, I have not much hope of the components ever being seen when projected on the sky.

In reference to the abnormal or black transits of the third and fourth satellites, from an experience of many years in actually observing the phenomenon I should feel much hesitation in endorsing the theory that they are due alone to contrast. I have seen these satellites cross the same regions of Jupiter's disc in bright transit and again like drops of ink. Assuming the albedo of the satellite to be approximately constant, that of Jupiter would need to vary so vastly that it must rank as a variable star.

In speaking upon this matter Mr. John Tebbutt, in Pub. A. S. P., makes a statement that is certainly not upheld by observation. He says,

"Further, a dark phase does not take place when the satellite crosses the polar regions of the planet." In Pub. A. S. P., No. 5, pp. 108, 109, I have given a series of observations of the phenomena of Jupiter and his satellites, made with a 5-inch refractor at Nashville, Tenn., from which I quote---

"1880, December 30, $8^h\ 30^m$, Satellite III. seen in a high south latitude as a small black spot; continued visible as a black spot until near p. limb, and only lost its blackness at $9^h\ 4^m$."

"1886, May 8, $9^h\ 20^m$. Satellite IV. near north pole; very black."

Perhaps there would be no trouble in finding other instances where the satellites have been seen black in the polar regions.

I have never been able, in twelve years' time, to witness the phenomenon of a satellite shining through the limb of Jupiter---i.e. visible through the planet after the satellite had disappeared behind it in occultation. I have carefully watched a number of occultations with this subject specially in view, but, though the seeing on many of these occasions was perfect, no trace of any of the satellites could ever be seen after the last contact at occultation. I have, therefore, come to the conclusion that this phenomenon, which has been reported by several observers, is exceedingly rare, or it is wholly beyond the reach of a first-class 5- or 12-inch telescope.

. . . .

DOUBLE SHADOW OF JUPITER'S SATELLITE I

Hoffman, H. O.; *Sidereal Messenger*, 10:517, 1891.

On the night of Nov. 14, after the egress of Jupiter's Satellite I, I was watching the shadow as it hung on the edge of the great red spot, and while intently observing, the round shadow seemed to become oblong, in which form it stayed for at least thirty minutes and then the shadow seemed to open and presented a double shadow, both round and black, the one much smaller than the other.

Was it really a double shadow or was the second shadow a smaller interior body?

It might have been the result of the larger shadow's contact with a dark spot in the belt, but the little black dot remained after the egress of the shadow of I.

DOUBLE SHADOW OF JUPITER'S SATELLITE I

Hulbert, H. S.; *Astronomy and Astrophysics*, 11:87, 1892.

In regard to Mr. Hoffman's observation of the double shadow of Jupiter's Satellite I, reported in the last Messenger, permit me to say.

that I made a similar observation on the evening of Sept. 29. It was in transit nearing egress and it appeared as a white disk against the dark southern equatorial belt; following it was the usual shadow and at an equal distance from this was a second shadow, smaller and not so dark as the true one, and surrounded by a faint penumbra. It would seem to me that there is some intimate connection between this doubling of the shadow of Satellite I, as seen by Mr. Hoffman and myself, and the elongation or doubling of the satellite itself, as seen by Mr. Barnard on Sept. 8, 1890, and Aug. 3, 1891, and which he explains on the supposition of a white belt encircling the equator of the satellite.

UNUSUAL PHENOMENON OBSERVED ON JUPITER, SEPT. 29, 1891

Hulbert, Henry S.; *Astronomical Society of the Pacific, Publications*, 3:380, 1891.

On the evening of September 29, 1891 at 7:02 P. M. Central Standard Time, I noticed the following interesting phenomenon on Jupiter. Satellite I was in transit, showing a well-defined white disk against the darker background of the southern equatorial belt. Following this was the customary shadow, and following this at an equal distance was a second shadow. This secondary shadow was a trifle smaller and considerably fainter than the true shadow and seemed to be surrounded by a faint penumbra. This secondary shadow disappeared simultaneously with the egress of the true shadow. The seeing was exceptionally good and the air very steady. My instrument is a three-inch [refractor] by Brashear mounted equatorially with clock, etc., and powers of 126, 160 and 212 diameters.

ON THE PHENOMENON OF THE TRANSIT OF THE FIRST SATELLITE OF JUPITER .....

Barnard, E. E.; *Royal Astronomical Society, Monthly Notices*, 52:156, 1892.

In a letter which I have received from Mr. A. Stanley Williams he informs me that he has sent a communication to the Royal Astronomical Society in which he endeavours to explain the apparent duplicity of the first satellite of Jupiter at its transit 1890 September 8, by supposing the phenomenon to have been a close conjunction of the satellite with a small spot which he had seen three days earlier on the planet.

Leaving aside the fact that it is wholly improbable that two experienced observers should have been so mistaken in a matter of this kind, I would say that the phenomenon of apparent duplicity was watched for upwards of half an hour before the visitors interrupted (from before sidereal $18^h\ 30^m$ to about $19^h\ 7^m$), part of which time Mr. Burnham observed with me. During that interval no relative motion was detected. At the transit of I, the relative motion of the satellite and a spot on Jupiter would have amounted to 0".15 each minute of time---a displacement which would have been only too apparent in a few minutes with the 12-inch and the high magnifying power employed. It is, therefore, apparent that Mr. Williams's explanation can have no bearing on the apparent duplicity of Satellite I at its transit 1890 September 8.

I consider the observation of the double transit an important one, the explanation of which will perhaps be still more important.

It should be accepted as unquestionable that the phenomenon of 1890 September 8 was wholly connected with the satellite. One or the other of the two probable explanations which I have given in Monthly Notices, No. 9, vol. li., will doubtless be found in the end to be the true solution of the matter. It is unfortunate that the transits of this object still occur over a dark portion of Jupiter. As soon as these are transferred to a bright region we may expect to know something more definite.

THE DOUBLING OF JOVIAN SATELLITE SHADOWS

Wend, Richard E.; *Strolling Astronomer,* 19:79-80, 1965.

This discussion is based upon three photographs of Jupiter taken by Mr. Phillip R. Glaser on October 26, 1963, 18 days past opposition. The film was Ektachrome X, ASA 64. Exposure was two seconds, by eyepiece projection (approximately f/70) with an 8-inch f/7 reflector, mirror made by Frank Vaughn.

On the original photographs the polar diameter of Jupiter was 3.8 mms., and the equatorial diameter was 4.1 mms. The diameter of the satellite shadow, Jupiter I (IO), was 0.1 mms. on the first and third photographs made.

The instrument was well cooled down, and seeing was suitable for photography.

The first photograph (not published here) showed the black shadow of Io on the limb of Jupiter. The shadow was round. To an observer positioned so that the Jovian terminator was near his central meridian, the low angle of the sun would produce an oval shadow. From Earth Jupiter is nearly "full", and our line of sight is only a few degrees away from the Sun's so that the shadow appears round. The belts, the Red Spot, and good limb definition were also present on this photograph.

The second photograph was taken 16 minutes later (3^h20^m U.T.), and the shadow of Io was now doubled! It measures 0.1 X 0.2 mms. on the original. The original shows this doubling more clearly than the copy published here as Figure 7. [Not reproduced]

Again, note the belts, the Red Spot, and limb areas. With an 8-inch at f/70, seeing must be good to get image quality like this; and if the odd satellite shadow were due to poor seeing, other detail in the photograph would suffer, and a poor photograph would result. Similarly, if the mechanical system were accidentally bumped during the exposure, the entire planetary image would be degraded; the limb region particularly would be noticeably affected. Critical examination of the original transparency under high magnification does not indicate an emulsion defect at this point; such defects are usually more contrasty and sharper edged than a true image.

The ephemeris for this date indicates that Io itself crossed Jupiter's limb 27 minutes ahead of its shadow, and was well ahead of the shadow at 3^h20^m, U.T. so that this is not a case of a satellite and its shadow being close together, a phenomenon that can happen only at opposition.

The third photograph was taken three minutes after the second (and 19 minutes after the first); the shadow of Io again presented its normal aspect.

This phenomenon of doubling has been observed visually in the past (although I haven't heard of its being photographed). In the book "The Planet Jupiter" by Peek, the author describes on page 268 a visual observation of his own: "For minutes together, except for a few seconds now and again of normal vision, the shadow of Io was traversing the SEB_n accompanied by a companion shadow quite as sharply defined and only a little smaller, on the SEB_s." Note that this object is the same satellite shadow we are discussing here.

If the effect is spurious, it wouldn't be surprising to find one of the images less intense than the other since a random effect would be less likely to duplicate exactly the conditions creating the undistorted image. The photograph, however, shows both components with the same intensity.

It should be pointed out here that color film does not have unlimited latitude. Correctly exposed for the bright surface detail, the film's threshold of sensitivity is above the inky blackness of a satellite shadow. Thus there is room for some variation of intensity, as long as the brightest variation is still below the film's threshold at the particular exposure.

Since the immediately preceeding and following photographs don't show the effect, it doesn't seem likely that the angle of vision is responsible; the effect would be observed more often, too, if such were the case.

One suggestion that has been made from time to time is that the two shadows are both real, and one is located on a lower Jovian cloud level. If this is so, the upper layer must be dense enough to record a dark shadow on the film, but thin enough to transmit sufficient light to illuminate the lower layer to approximately the same brightness as the upper layer. These requirements appear incompatible. Also, if the upper layer is dense enough to provide a dark image, scattering of light by this layer should prevent the shadow on the lower layer from being as sharp-edged as the one above. The lower shadow would be much less contrasty, and more spread out. This also does not appear to be true.

Thus we are left, for the time being, with the reluctant conclusion that the doubling of satellite shadows must be an effect of seeing. Since Jupiter's disk is only about 50" in diameter, and the area of shadow doubling only about 2" in diameter, it is difficult to account for such precise localization of a seeing effect.

Atmospheric seeing is not likely to be the cause because it is too gross, moving across larger areas of the field. The haphazard doubling would not hold still long enough to register on a photograph, particularly with an exposure longer than a second. Variable air density in the telescope itself is Mr. Peek's nomination for the culprit. In a reflector, perhaps a small cell of air, precisely located and of the correct density and affecting a small area of the field for approximately half of the exposure time, could by refraction produce a satellite shadow of significant contrast. It really doesn't seem likely, but the alternatives are even less attractive.

If this can happen, then one would think that people who observe large numbers of double stars, or variable stars (and their comparison stars), would notice an occasional spurious doubling of stellar images. I have personally questioned four of the top AAVSO observers, and they have not noticed any doubling.

Photographs should also turn up some false stellar images, particularly during a hunt for faint novae, comets, or other objects not on star charts. It is assuredly uncomfortable to have to accept the possibility that some detail in astronomical photographs is spurious, and it will not always be possible to identify such false images.

ANOTHER DOUBLED JOVIAN SATELLITE SHADOW

Anonymous; *Strolling Astronomer,* 19:143, 1965.

An observation by Reverend Kenneth J. Delano becomes extremely interesting in connection with Mr. Richard E. Wend's article "The Doubling of Jovian Satellite Shadows", Str. A., Vol. 19, Nos. 5-6, pp. 79-80. On the night before this issue reached him Mr. Delano witnessed the same phenomenon for the first time! It was again the shadow of J. I which was involved. The telescope was a 12.5-inch reflector at 300X, seeing 2 and transparency 5. The observer notes: "On three occasions between $0^h\ 40^m$ and $0^h\ 50^m$, U.T., March 22, 1966, in a second or two of steadier seeing when Io's shadow was most distinctly seen, Io's shadow appeared doubled. As I recall, Io's shadow was on the SEB, either in the zone separating the SEB_s from the SEB_n or on the SEB_s itself. Its companion shadow, which was no momentarily glimpsed, was at least one-half as dark or large. It lay to the southeast and was contingent or nearly so. Really everything about the observation was brief and uncertain other than the simple fact that at three moments last night Io's shadow gave the appearance of being double."

Mr. Delano further remarks. "...I'm convinced that satellite shadow transits, and in particular those of Io, merit more attentive observation. It has been suggested that what is seen is the satellite's shadow on two different layers of the Jovian atmosphere, but would it be any more difficult to assume that Io has its own huge cloud-like satellite, going around it or has a magnetospheric tail (as suggested in Science magazine, Vol. 148, June 18, 1965, pg. 1588)? Perhaps such a cloud satellite or magnetospheric tail is so constituted as to absorb (thus able to cast a shadow) but not reflect (thus not visually detectable) the sun's light. The sun's absorbed light might be released at radio wavelengths instead, bearing in mind that a relationship has already been found between Io and radio noise from Jupiter."

HOT SHADOWS OF JUPITER

Wildey, Robert L.; *Science,* 147:1035-1036, 1965.

Abstract. On the evenings of 26 October and 15 December 1962, while the disk of Jupiter was being scanned for thermal emission in the 8- to 14-micron wavelength region, a large enhancement was discovered in the emission from shadows cast on Jupiter by the Jovian satellites Ganymede and Europa. However, on the evening of 14 December 1964, the shadow of satellite Io was observed and no enhancement was detected. The effect is thus variable with time.

• Occultation Phenomena

AN OCCULTATION PHENOMENON

Holmes, Edwin; *British Astronomical Association, Journal,* 14:25-27, 1903.

A phenomenon in connection with the occultations of the satellites of Jupiter mentioned in "Celestial Objects," and by Mr. Proctor, and also seen and casually noted by some of our own Members, is, I think, of a sufficiently interesting character to deserve more attentive observation. I allude to the apparent visibility of a satellite through the limb of the planet.

On Friday, October 3, the third satellite was due to be occulted at $10^h\ 25^m$. I watched it carefully, but without thought of this question. After occultation began, it appeared to pass entirely within the limb, and took exactly the appearance of a transit before it finally vanished. It was visible with 12-1/40in. and power of 226 for at least one full minute after it was clean within the apparent outline of the planet. It was considerably brighter than the edge, upon which it appeared as if projected, and looked merely as if slightly obscured.

The Rev. T. W. Webb, in "Celestial Objects," details a number of instances, of which I quote the more definite ones. On 1863, April 26, Wray, with an 8-in. object glass, saw II. projected distinctly within the limb for nearly 20 seconds. On 1857, January 3, Hodgson saw I. projected on the disc for nearly one minute, with clear space round it. Carlisle, at Stonyhurst, saw II. within the limb for about 45 seconds after last contact in occultation, 1878, September 20, and in the same year Todd and Ringwood, at Adelaide, observed, with an 8-in. achromatic and splendid definition, three certain and two doubtful cases of projection of I. and II. as though seen through the edge of the disc. Mr. Webb adds: "We are at once reminded of the projection of stars on the Moon, but all such anomalies are worth recording. Todd says that the satellites are visible through the dusky, hidden by the bright, parts of the limb, but this might result from projection on a different background. Neison is of opinion that they are seen through a spurious limb produced by irradiation, and extending perhaps 0''.5 beyond the true."

When Mr. Webb suggests that the visibility through the bright part of the limb, and invisibility through the dark "results from projection on a different background," it seems to me he, for the moment, forgot that the satellite is not "projected on the background" at all, but is seen through the limb. Of course there would be less contrast through the parts more nearly equal in brilliancy to the disc of the satellite. The estimate that the spurious limb extends 0''.5 beyond the true is correct in principle, but the brevity of the expression did not allow of a full demonstration as to amount under various conditions.

In our own Jovian reports, I find that Mr. Freeman, February 25, 1893, found Satellite III. "hang on the limb for about two minutes. Boiling limbs." This, perhaps, is hardly definite as to being seen through the limb. But on March 22, 1895, Mr. Goodacre found Satellite I. "appeared fully two minutes on limb as a faint white projection." I am not quite clear whether this is a case of visibility within the limb or outside it. On February 23, 1896, Mr. Henry Ellis noted "internal contact" of I. at $9^h\ 37^m$, and satellite gone at $9^h\ 41^m$, from which it appears he saw it through the limb for four minutes, unless "internal contact" is a printer's error. Finally, Captain Molesworth, April 12, 1898, saw II. "as a bright point while just inside." This is the only certain observation of the phenomenon in our reports, which is curious when so many able observers are engaged, and appears to indicate that it seldom occurs.

Mr. Proctor, in his "Flowers of the Sky," relates that Mr. Todd and his assistant, Mr. Ringwood, with a fine 8-in. Cooke, on two occasions

> observed that a satellite, when passing behind the planet's edge, did not disappear at once, but remained visible through the edge for about two minutes. The same satellite behaved thus on each occasion, viz., the satellite nearest the planet. As this satellite travels at the rate of about 645 miles per minute, it follows that the satellite was seen through a depth of nearly 1,300 miles, or, after making all possible allowance for optical illusions, some 900 or 1,000 miles.

He goes on to argue that refraction cannot be great in the air of Jupiter, or the satellite would have been altogether distorted, and he draws other conclusions. These two observations are identical with two of Mr. Webb's. It is always dangerous to differ from Mr. Proctor, and every word he wrote is worthy of serious consideration, but I think he let his pen run away with him in the above statement. He has treated the matter as if we were looking down vertically on a satellite which was plunging also vertically into the Jovian atmosphere. If the satellite is seen by us through Jupiter's atmosphere, when it is apparently 1,300 miles within the limb, it is evident that we see it through the chord of an arc of which 1,300 miles is the versed sine, and, therefore, roughly, through 20,000 miles. I have not calculated it exactly, because one limb would be seen through much more than the other, and the amount would vary rapidly from second to second; but if anyone alleges error, I am not particular to a few thousand miles more or less.

But I think it is scarcely conceivable that a satellite could be visible through 20,000 miles of a cloud-laden atmosphere such as that of Jupiter, so I fall back on Neison's irradiation theory with a little addition, as furnishing the explanation. The amount of irradiation will evidently depend upon the aperture of the telescope in conjunction with the brilliancy of the object. It is dependent on the same cause that produces the spurious discs of stars, and, in the absence of evidence as to any difference of amount, I treat it as the same. This will give us at each limb under good seeing conditions, 4''.56 divided by the aperture in inches, again divided by two. This, in my case, is 0''.19. Now this, on October 3, represented at Jupiter's then distance about 360 miles, and, as the satellite travels about 430 miles per minute, it is evident such irradiation alone would not explain the observation.

But for many months now definition has been very bad, and on the night in question double stars, 1'' apart, were with difficulty seen double. They were not in violent motion---not boiling---but were nebulous balls rather than stars, and I submit that under these conditions the limb of the planet also would partake of the same enlargement, scarcely measurable, perhaps, because with sufficiently high powers the definition would not be good enough for measurement. Judging from the stars, the spreading amounted to nearly half a second, or 900 miles, amply sufficient to allow the satellite to be visible for nearly two minutes.

If irradiation only were concerned, the phenomenon ought to be observed at every occultation, and I think, in spite of the two cases where splendid definition is mentioned, it will, by future observations, be found it occurs under conditions when the limb of Jupiter is softer than usual in appearance.

PECULIAR DISAPPEARANCE OF JUPITER'S SATELLITE, III

Truman, O. H.; *Popular Astronomy*, 24:263-264, 1916.

Some time since I saw a rather curious and rare occurrence, which, while it may have little or no significance, I think is worth giving an account of.

On December 1, 1915, at 8 hrs. 37 min. C.S.T., the aspect of Jupiter and his satellites was about as on the inclosed sketch. The distance from III to Jupiter was estimated by eye to be about 3-1/2 diameters of the planet, which would make it about 140'', and from II to III was about 15'' to 20''.

Within a very few seconds of 8 hrs. 37 min. 15 sec. C.S.T., as the satellites were being watched with 300 power on our five-inch refractor, with Jupiter himself in the field, III entirely or almost entirely winked out. II and Jupiter were not affected. The disappearance and reappearance seemed gradual, and extended over a space of at least half a

Unexplained eclipse of one of Jupiter's satellites. Note: text mentions satellite III in eclipse, but original drawing shows IV.

second: more probably 3/5 or 3/4 second---much longer than it takes to wink the eye. The seeing was very good---9 on my scale.

The question as to whether the disappearance and reappearance were actually gradual is very important. I have since experimented with the occultation of an artificial star, and I found that the disappearance had a tendency to seem gradual; and it was the disappearance which most certainly seemed gradual in the phenomenon observed; yet I know that the instant the thing happened, the apparent leisureliness of it was the most striking feature---the air of deliberation and matter-of-fact-ness. I can still, twenty days after, picture the occurrence to my mind.

When one comes to inquire into the cause, a bird at once suggests itself. But with a bird, the occultation would have lasted only so long as it would have taken him to traverse a cylinder five inches in diameter, and at the speed that birds fly, the time would hardly have been long enough to make itself felt over persistence of vision, unless indeed the bird happened to be flying right along the line of sight, or nearly so, which is improbable. More than that, I do not know whether birds are accustomed to flying at that time of night and year, that is, except owls. But this bird must have been at a considerable distance from the telescope in order to avoid occulting II, for an angle of 20'' makes an arc of less than 1.2 inches, even at 1000 ft. distance, and I believe the owl flies low.

But if one, to avoid the improbabilities noted above, turns to occultation by some kind of a body in interplanetary space, the question arises, how could a body large enough to make an occultation lasting as long as that seen, escape being itself seen? For anything that intercepts light usually reflects light.

Altogether the incident seems very curious, and it would be interesting to know whether any others have seen anything similar. I have heard mention of mysterious occultations of stars, but I have never seen an account of one, and I think they are usually ascribed to birds.

The above account is written from full notes made within a few minutes of the occultation, while the facts were still very vivid in my mind.

ABSENCE OF POST-ECLIPSE BRIGHTENING OF IO AND EUROPA IN 1970

Fallon, Frederick W., and Murphy, Robert E.; *Icarus*, 15:492-496, 1971.

Abstract. Observations of four eclipse reappearances of Io, three of Europa, and one of Ganymede were made from the Mauna Kea Observatory during the 1970 apparition. A single beam photometer was used on a 24-in. reflector for all of the observations. Wavelengths from λ 4100 to λ 6500 Å were used with aperture stops of 7 and 15 arc-sec.

No post-eclipse brightening was observed on any of the satellites. Upper limits to such effects are estimated at $0^m.01$ for Io and Europa. No estimate is made for Ganymede since it was only observed once. These upper limits may be contrasted to the post-eclipse brightening of Io of about one-tenth of a magnitude reported by Binder and Cruikshank (1964). One of the observations was simultaneous with that of Franz and Millis (1971) who also found no brightening. It is suggested that the phenomenon may be sporadic in view of the recent positive observations by O'Leary and Veverka (1971) and by Johnson (1971).

THE POST-ECLIPSE BRIGHTENING OF IO

Cruikshank, D. P., and Murphy, Robert E.; *Icarus*, 20:7-17, 1973.

Abstract. We review the photometric work on eclipse reappearances of Io. New observations of eclipse reappearances of Io confirm the post-eclipse brightness anomaly reported by Binder and Cruikshank (1964)

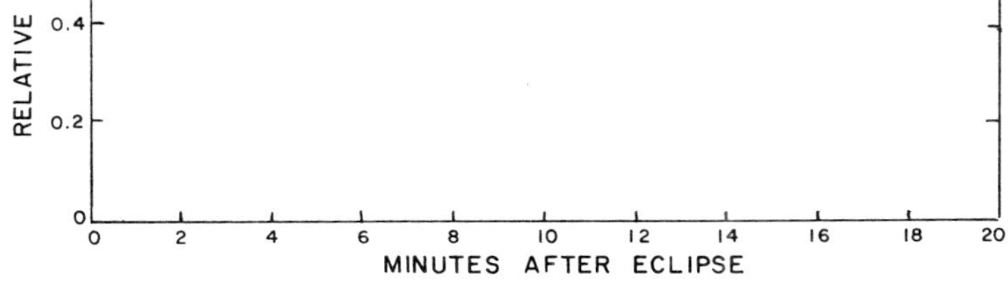

Typical brightening curve of a Jovian satellite after eclipse. (Adapted from New Scientist, *60:239)*

but testify to its intermittent nature. A post-eclipse anomaly of approximately 0.07 mag was observed on two occasions in 1972, while observations of Europa and Ganymede showed no brightness anomaly greater than 0.01 mag. The atmospheric condensation model for the anomaly on Io is reviewed in terms of the quantity of frost required to produce the effect and the corresponding amount of gas liberated to the atmosphere upon sublimation. The observational data and the results from a stellar occultation are in general accord with the theoretical predictions of the stability of heavy gases on Io, while both observational and theoretical criteria are satisfied by a tenuous atmosphere of a heavy gas such as methane or ammonia having a surface pressure $\sim 10^{-7}$ bar.

• Jupiter's Ring (?)

A RING AROUND JUPITER?

Anonymous; *New Scientist*, 21:363, 1964.

Astronomers have always thought that the famous ring-system of Saturn is unique in the solar system; a faint ring round Neptune, reported during the past century by Lassell, Challis and Bond, seems to be definitely non-existent. Now, however, the Russian astronomer S. Vsekhsvyatsky, of the Ukraine, has made the unexpected suggestion that Jupiter too may be associated with a very thin ring, composed of material of low density.

Vsekhsvyatsky has made a close study of drawings of Jupiter done by experienced observers in many countries, and believes that a very thin, broken streak in the region of the planet's equator indicates the existence of a ring. He considers that it has been formed as a result of gigantic outbursts in Jupiter itself, and that its composition is identical with that of the ring-system of Saturn.

These conclusions are likely to be hotly challenged. There is no theoretical bar to a Jovian ring, but if it exists it is extremely elusive, and confirmation is likely to prove a very difficult matter.

Chapter 11

SATURN

INTRODUCTION

In a good telescope, Saturn is a spectacular sight. The rings, of course, make all the difference. Although other planets may have ring systems, Saturn's are bright and vivid. The three major rings are thousands of miles in width but only a few miles thick. Such obvious features are registered in all the textbooks. The idiosyncracies of the rings are rarely mentioned; for example, the spots or "knots" that sometimes appear, the elusive extra inner and outer rings, and the peculiar shadows cast by the planet on the rings that seem to defy geometry. Explanations of these exotic phenomena are hard to come by. Actually, the origin and constitution of the rings are not known with certainty. It is not beyond the bounds of possibility that planetary rings are more common than supposed and that those circling Saturn are just the most obvious ones at the current stage of solar system development.

Saturn's strongly banded sphere seems to have attracted less attention than the rings; and our list of anomalies is limited to one---the white spots that come and go. As with Jupiter, astronomers contemplate a seething, circulating atmosphere for Saturn. Amidst such turbulence, transient phenomena are likely. The precise nature of Saturn from core to cloud tops remains a mystery. Its surface temperature (92°K) seems higher than solar radiation can maintain, suggesting that Saturn, like Jupiter, has an internal heat source.

• Elusive Additional Rings

ON THE OBSERVATIONS OF THE REPORTED DUSKY RING OUTSIDE THE BRIGHT RINGS OF THE PLANET SATURN

Baum, R. M.; *British Astronomical Association, Journal,* 64:192-196, 1954.

The first intimation of a dark annular appendage exterior to the bright ring of the Saturnian system was obtained in 1907[1] by the French astronomer, M. G. Fournier. Using the 28-cm refractor of the Jarry-Desloges Observatory on Mount Revard (altitude 1550 m) in Savory, he made his first observation of the 'new' ring on 1907 September 5^d 10^h 15^m. Subsequently it was again recorded by him on September 7^d 10^h 25^m when under very fine seeing conditions the aspect of the 'new' dark ring was described thus: 'Tout autour de l'anneau dans les meilleurs moments de calme on apercoit une zone lumineuse tres pale, surtout bien visible a gauch et au-dessous de l'anneau, ainsi qu'a ses extremites. Cette remarque avait deja ete faite le 5 septembre et prise pour une erreur de mise au point; mais le presence reelle de cette zone n'est pas douteuse.'[2]

This discovery, though given prominence in several publications of the time, would appear to have passed the better part of a year without any notice being paid to it, a fact borne out by the events of the 1908 apparition of the planet.

In September, 1908, M. E. Schaer of the Geneva Observatory, completed a Cassegrain reflector with which, during the autumn and winter months of that year, he obtained a series of observations on the globe and rings of Saturn.[3] Thus it was that, whilst engaged in this pursuit, the Swiss astronomer detected what appeared, to him at least, to be a new feature---namely a dark ring surrounding the known bright rings. This was during the early part of October and, judging by the telegram sent to the Astronomische Nachrichten,[4] it would seem that the earlier discovery by Fournier had either been overlooked or was not known to the Swiss observer.

Following the receipt of an announcement from the Geneva Observatory, on 1908 October 10 the rings were examined at Greenwich with the 28-inch refractor by three observers, namely Bowyer, Lewis, and Eddington.[5] In view of their partial confirmation of the exterior dark ring these observations are here considered in detail.

The first Greenwich observation was obtained by Bowyer on October 10 between 10^h 30^m and 11^h 50^m. U.T. in bright moonlight. A dusky edge to the outer boundary of the rings was noticed---a feature that was perceptible on the following limb of the globe where the ring crossed the planet. Lewis observing from 11^h 45^m to 11^h 55^m on the same night became aware of two features that were different from what he remembered of the Saturnian aspect of 1895; these were (a) the north-preceding edge of the outer ring had a hazy border to it, and (b) the north-following edge appeared fairly sharp and dark whilst the south edge, preceding and following, showed as quite normal. On October 11 at 11^h 16^m Lewis observed a portion of the ring-edge to be bordered by a faint fringe, while at $11^h\ 20^m$ this feature was seen at both the preceding and following edges. Bowyer between $11^h\ 10^m$ and $11^h\ 35^m$ on the same evening suspected a faint fuzzy aspect at the north-preceding border. October 12, between $10^h\ 20^m$ and $10^h\ 30^m$, yielded another observation of the new ring by Lewis who detailed its appearance as a faint fringe to the northern outer reaches of the ring system. On the same evening between $10^h\ 38^m$ and $10^h\ 50^m$ Bowyer saw the north edge of the bright rings bordered by a dusky ring. Observing on the 15th between $10^h\ 10^m$ and $11^h\ 15^m$ he considered the rings to be much sharper in definition than on the 12th, and during fine periods of seeing noticed the outer dark ring quite distinctly. At $11^h\ 10^m$ on the same evening, Ring A had a dull edge to it on the north side but was bright in the south---only in steady moments was the 'new' ring observed. Eddington, also on the 15th, over the period $10^h\ 45^m$ to $11^h\ 10^m$ found traces of the new ring at the northern boundary of the bright rings; the southern reaches appeared normal, nor could any indication of the ring be found projected on the globe, as was observed on the 10th by Bowyer. Some further observations on October 22, 27 and 30 gave indications of the new appendage.[6]

Probably the last observations of the exterior ring were made in 1909 January by Schaer, who found the feature decidedly more easily detected than previously.[7]

On the other hand the renowned American observer E. E. Barnard at the Yerkes Observatory, employing the 40-inch refractor, was unable to find any trace whatsoever of the exterior ring.[8] This negative testimony by an experienced observer of planetary detail would appear to have some considerable bearing upon the subject, especially as in 1908 Barnard was making observations on the ring system in the hope of ascertaining the cause and nature of the bright 'beads' or condensations seen on the dark part of the rings during the previous year. Through July, August, September and October, 1908, Barnard obtained observations, but reported nothing that would suggest the new feature. During November and December of the same year cometary photography took up the greater part of this observer's time---rather unfortunately as at the time Saturn was in good position for observation. On January 12 and 19, with seeing at 3 on a scale of 5 for both dates, the rings were examined for the new object but without success. A reduction of aperture on the 10th to 35-inches gave much better definition---from 3 to 4.[9]

Following this period of active scrutiny of the ring system, with respect to the dark exterior appendage, nothing further was reported of this feature, and it was generally assumed that some form of illusion had given rise to a subjective phenomenon resulting in the illusory aspect of the dark ring.

In 1951 the study of the planet Saturn was taken up by the writer at Chester. During the course of that year several observations were obtained and acquaintance made with the subject of Saturnigraphic study. Prominent amongst the features observed in this period were the luminous condensations on the rings, set equidistantly on either side of the globe in the ring-plane. The instrument used during the year was a 3-inch refractor.

The following year the observations were continued, but this time with a 6.5-inch reflecting telescope. Seeing conditions over this period were generally poor and opportunities to study the planet and its rings few and far between. In April, however, some unusually good moments ensued giving rise to an excellent opportunity to study the planet. The first observation was made on 1952 April 1^d 20^h 00^m U. T., when with an eyepiece of x216, with seeing at 4 on a scale of 10, considerable detail was noticed. At 20^h 40^m on this evening seeing rose to 6 and, with this happy conspiration of atmospheric circumstances, traces of some dusky nebulous matter in the form of an additional ring were recorded beyond Ring A, which the writer could not recall having observed in 1951. The appearance was so unexpected that strong doubts as to its objective character were entertained, resulting in its being dismissed as some form of illusion though consideration of the steel engraving aspect of the global details suggested the improbability of such views. However all doubts were finally dispelled later the same evening at 21^h 15^m when the dusky ring was again strongly noticed as a pale nebulous feature extending beyond the bounds of the bright ring system. Further traces of a dusky fringe were found on April 11 between 20^h 26^m and 21^h 22^m. A narrow penumbral feature showed beyond the bright rings on April 13 at 21^h 05^m U. T. with seeing at 4-5 on a scale of 10. The last view of the new ring in 1952 was obtained on April 16^d 22^h 07^m when the aspect was described in these words: 'a diffuse fringe extends the ring system beyond its normal limits, in good moments this feature shows as a definite dusky ring---quite unmistakeable as such; there is no doubt as to the objectivity of this "new" feature.'

In 1953 the dark ring was vaguely seen on June 10^d 21^h 30^m U. T., with seeing at 5 on a scale of 10, with the 3-inch Cooke objective of the Grosvenor Museum, Chester, charged with an eyepiece of x120. The Cassini division was well seen at this time, whilst the global details were sharply defined, the North Equatorial Belt showing a wealth of delicate tracings, spots and wisp-like shades. On June 13^d 22^h 00^m U. T., seeing 3-4 on a scale of 10, using the 6.5-inch reflector charged with an eyepiece of x216, a beautiful view of the exterior ring was obtained, its appearance being reminiscent of that of 1952 April 16. A further observation on June 27 at 21^h 40^m to 21^h 45^m, with seeing at 7, again revealed the dark ring, its aspect being quite obvious. A last study in 1953 June 29^d 21^h 45^m U. T. with a 9-inch reflector, eyepiece x120, in uncertain seeing, gave some vague but decidedly poor views of the outer ring. One interesting fact of the 1953 observations was that the exterior ring appeared rather fainter than in 1952 and, in the opinion of the writer, would have been missed altogether if knowledge of its existence had not been obtained in 1952.

Confirmation of the 1952-1953 observations of the writer are apparently not found wanting, for it would appear that two well-known American observers, T. R. Cave, Jnr, and T. Cragg of Mount Wilson,[10] also obtained evidence pointing to the existence of the outer dark ring in the period 1952 to 1953. No precise details of these studies have as yet been released, apart from the first barren announcement, so it is quite possible that these observers noted the ring long before the writer did. No other reports to support these observations have been received, so far as the writer is aware though it is possible that the ring has been observed by others and dismissed by them as a subjective phenomenon.

As to the nature of the observed appearances, there seems to be little doubt as to their objectivity and as a consequence close examination of the ring system is called for. An important consideration that may indicate the cause of the very pale aspect of the ring lies in two facts: (a) the reported observations of the ring have only been forthcoming when the ring system has been observed edgewise, and (b) the progressive decline in brightness of the ring in 1953 as compared to 1952. As regards (a) the answer may be that the particles composing the ring are widely separated---that there are fewer particles area for area, than in the bright rings of the system. Thus when the ring plane is passing through the Earth and the system is decreasing in breadth the rare particles in the outer ring are compressed by perspective into a much smaller area, resulting in a decided increase of brightness and therefore visibility at this ring-phase. As a consequence, it would account for the progressive decline in brightness noted in (b), due to the opening of the rings and, therefore, the apparent increasing distance of the outer ring particles from each other. This secular change contrasts with the short-period changes of brightness indicated in the observations. These may be the result of the varying distribution of the dust particles composing the ring, brought about by rotation---a not improbable factor, as Fournier in 1907 actually observed a steady displacement of the particles in his observation of September 7.[11] Another and very important factor to consider is our own atmosphere, the unsteadiness of which can successfully mask any trace of the diaphanous outer ring.

For the observation of the exterior dark ring excellent seeing conditions are essential requisites, together with fine transparency. Close and detailed scrutiny of the ring system is called for, especially with large instruments, and it is hoped that attention will be devoted to this end during the present apparition.

References

(1) Jarry-Desloges, Observations des Surfaces Planetaires, Fas. 1, 1907, pp. 106, 1908.
(2) Ibid, pp. 106-107.
(3) A. N. No. 4331.
(4) A. N. No. 4276.
(5) M. N. R. A. S., 69 (1), 39-41, 1908.
(6) Ibid, pp. 40-41.
(7) A. N. No. 4331.
(8) M. N. R. A. S., 69 pp. 621-624.
(9) Ibid, pp. 622.
(10) Mitteilungen fur Planetenbeobachter, 6 (3), 16-17, and The Strolling Astronomer, 6, (11), 160, 1952.
(11) Jarry-Desloges, Observations des Surfaces Planetaires, Fas. 1, 1907, pp. 108, 1908.

AN = Astronomische Nachrichten

CONCERNING THE "D" RING OF SATURN

Feibelman, W. A.; *Nature*, 214:793-794, 1967.

In his excellent chronological review book of observations, The Planet Saturn, Alexander[1] compares the outer, "D" ring of Saturn to the

Loch Ness Monster: some see it, and some do not. During the second half of the nineteenth century a number of visual observations were reported by experienced observers but the issue seems to have been settled by Barnard[2] in 1909, using the 40 in. Yerkes refractor visually and getting negative results. Apparently few, if any, attempts have been made photographically to detect a ring outside the well known A, B and C rings.

The recent edge-on configuration of the ring system was an appropriate time to investigate the problem of the hypothetical "D" ring photographically. It is known that when seen nearly edge-on, the A ring, normally fainter than the B ring, can sometimes appear brighter than B. Similarly, an outer "D" ring might appear relatively bright at the time, while when in the open position it may be completely unobservable.

Although there are theoretical arguments against the existence of an outer "D" ring, for example, the sweeping effect of the inner moons (compare (Alfven[3]), there is at least one modern observation of a belated and gradual emergence of the satellite Iapetus from the shadow of the A ring[4], suggesting an extension or at least a gradual tapering of the outer edge of the A ring.

On six nights between October 27, 1966, and January 16, 1967, about fifty photographs of Saturn of considerable length were taken with the 30 in. refractor at the Allegheny Observatory of the University of Pittsburgh. On all moderately long (5-10 min) and very long (30 min) exposures on backed Kodak '103a-0' without filter between October 27 and December 12, 1966, a very thin extension of the nearly edge-on ring system can be seen. The thin line extends to more than twice the known ring diameter and has an approximate equivalent brightness of $m_{pg} =$ 15/sq. sec of arc or fainter. When two or three exposures are viewed in superposition the thin line is clearly seen when in registration. After December 12 the thin line became more difficult to detect and by January 16 no trace of it could be found. Printed reproduction of the thin line would be very difficult, but two of many microdensitometer traces are presented in Figs. 1a and 2a. These two tracings were for November 14 and December 12, 1966, respectively, and Figs. 1b and 2B show the locations of the scans X-X. The positions of the satellites are indicated and were obtained from short exposures. (For the December 12 sketch, Titan to the east of the planet and Hyperion to the west have been omitted.) The inner cross-hatched circle indicates the extent of the greatly over-exposed image of Saturn, while the outer cross-hatched area represents the photographic halo. The over-exposed image causes a general density gradient on the densitometer traces which accounts for the sloping background level, but a noticeable increase in density at the position of the thin line is consistently evident on some thirty scans at different positions.

A tentative interpretation of the thin line is that it represents a tenuous, outer ring seen only briefly when the ring plane is nearly edge-on to our line of sight. Clearly more observations at the next edge-on orientation are needed, but in the meantime re-examination of existing plate material taken with other telescopes appears desirable. One of the reasons why the thin line may have escaped previous observations is the fact that most large reflectors have their secondary mirror supports in an east-west, north-south direction so that the diffraction cross resulting from long exposures would hide the thin line. (Notable exceptions:

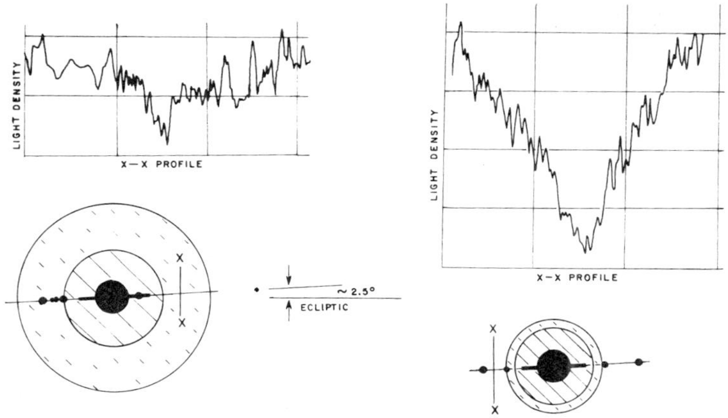

Microdensitometer traces of Saturn's rings

the Lick 120-in. and McDonald 82-in. reflectors the secondary supports of which are at 45° angles to the meridian.)

While it should be re-emphasized that the interpretation of the present data is tentative only, it is unlikely that the observations can be attributed to instrumental effects or guiding errors for the following reasons.

(1) No diffraction spikes are seen on any of the exposures. (2) The thin line lies exactly in the plane of the rings, about 2.5° to the horizontal. Guiding errors are usually in right ascension. (3) Background stars do not show any guiding errors or extensions. (4) The thin line is not likely to be caused by adjacency effects in the emulsion between closely spaced images of satellites. On at least one night when four bright satellites were west of the planet and two were widely spaced in the east, the thin line showed up very well in the east. (5) To avoid internal reflexions, no filters were used. (6) The thin line observations were made when the Sun was above and the Earth below the plane of the rings.

. . . .

1 Alexander, A. F. O'D., The Planet Saturn (Macmillan, New York, 1962).
2 Ibid. 317.
3 Alfven, H., On the Origin of the Solar System (Clarendon Press, Oxford, 1954).
4 Ellison, W. F. A., quoted in ref. 1, 344.

• Spots and Projections on the Rings

WHITE SPOT ON SATURN'S RINGS

Anonymous; *Observatory*, 12:195, 1889.

Dr. Terby noticed on March $6^d\ 9^h\ 12^m$ a white spot on Saturn's rings. It remained for more than an hour in the same place, and then clouds interrupted the observations. On March $12^d\ 9^h\ 20^m$ the spot was still in the same place and remained till further observation became impossible. He used an 8-inch Grubb refractor, powers from 150 to 280.

Mr. Taylor, who is now working Mr. Common's 5-foot reflector, kindly sends us the following note:-

"On Thursday, March 14, a telegram was received from Prof. Kreuger, via Aberdeen, giving the information that Terby announces 'region blanche sur anneau Saturne, contre ombre globe.' Friday the 15th being fairly clear, the 5-foot was used on Saturn, the air being unsteady and a thin haze covering the sky. No markings on the rings other than those usually seen could be made out, although the definition was considered very satisfactory. Bad weather prevented anything further being done until Monday, March 25, when the sky was very clear and the definition excellent. Saturn bore the highest powers, details of the ring and ball being very prominent; but no 'region blanche' on the ring could be made out. The shadow of the planet on the ring was not perfectly regular, but had a curious notched outline, extending further from the planet on the middle ring than on the outer one. The gauze ring was very brilliant; the satellites were all visible; but nothing corresponding to Terby's announcement was seen."

BRIGHT KNOTS ON SATURN'S RINGS

Butler, Charles P.; *Knowledge,* 5:13, 1908.

The observations of Saturn during November resulted in the detection of several interesting novelties, which, if found to be true, may have far-reaching consequences on the constitution of the planet's system. The original message from Professor Lowell, of the Flagstaff Observatory, reads as follows:-

"Condensations in Saturn's rings confirmed here and measured repeatedly. They are visible, symmetric, and permanent. Outer situated near the outer edge, ansa b, and inner at middle of ansa c. A conspicuous relative gap also detected and measured at 1.56 radius from the centre of the planet. Ring easily seen. Placed further south from shadow at west than at east." A later telegram read: "Ring shadow on Saturn bisected, black medial line; phenomenon explicable by extra-plane particles only."

If these observations indicate real aggregations of the ring material, it appears that we have here a unique opportunity of studying the results of interactions or, possibly, collisions between celestial bodies, both physical and constitutional. If the spectrum of these condensations be photographed by anyone possessing the necessary instruments and favoured with a fine sky, the result of collisions, and, therefore, presumably, vaporisation of part of the materials, would be indicated by the presence of bright radiation lines superposed on the Fraunhofer spectrum due to reflected sunlight, or if the amount of vapour be sufficient, we might find absorption lines of increased intensity at the corresponding positions.

A PROJECTION ON SATURN'S OUTER RING

Anonymous; *Nature*, 85:248, 1910.

During the total eclipse of the moon on November 16 M. Jonckheere directed the 35-cm. equatorial of the Hem Observatory to Saturn, and found a bright projection extending outwards from the eastern extremity of the exterior ring. The projection was best seen with low powers (100 and 200), and its intensity decreased gradually, going from the outer edge of ring A on to the background of the sky. On November 20 and 24 the same projection was seen with difficulty (Astronomische Nachrichten, No. 4461).

POSSIBLE CHANGES IN SATURN'S RINGS

Anonymous; *Nature*, 88:388-389, 1912.

At the meeting of the Royal Astronomical Society, held on Friday last, it was announced that a telegram had been received, by Sir David Gill, from Prof. Todd, in which he said:---"Near the extremities of the major axes of the bright outer ring of Saturn, with the aid of a powerful telescope, I have observed a certain sparkling flocculence which I interpreted to be a dissipation of the ring."

Commenting on this message, the Rev. T. E. R. Phillips stated that he had observed Saturn the previous night, but had failed to note any extraordinary feature such as was described in Prof. Todd's message; it was, however, possible that the affected section of the ring was not then in view. He also added that, according to the accepted view of the constitution of the rings, disturbances of some kind were likely to occur from time to time, and that these might be revealed by irregularities in the shape of the shadow of the ball on the rings. At the previous apparition he had seen such irregularities, but recently he had thought the shadow perfectly uniform. Other observers, who had been able to see the Encke's division easily, had also noted nothing irregular or unusual.

A Daily Mail inquiry at Greenwich elicited the suggestion that the phenomenon may have been produced by the collision of two of the particles froming the ring, the heat generated by the impact possibly raising the particles to incandescence. Owing to the comparatively large separation of the particles and their uniform motion, such collisions would not be of frequent occurrence.

THE WHITE SPOT ON SATURN'S RINGS

Johnson, Hugh M.; *British Astronomical Association, Journal*, 51: 309-312, 1941.

The purpose of this brief paper is to renew the attention of those who still occasionally observe Saturn to a phenomenon of that planet's rings, doubtless unimportant, but intriguing though apparently overlooked for a half-century. The object concerned is a rather bright and whitish spot or area bordering the shadow of the planet on the rings, to be seen in a fixed position on the rings both before and after opposition. I first casually noticed it early in the 1939 apparition of Saturn using my 8-inch reflector and 213 x, but interest in it remained dormant until latter 1940 when some old issues of The Sidereal Messenger came on hand disclosing several observational accounts of the spot excited by its discovery in March 1889 by the planetary observer Terby of Belgium. Below are quoted extracts typical of these accounts, with some interjected opinions concerning the spot's origin.

(1) "The White Spot on Saturn's Ring. . . . was observed at this Observatory on the evening of March 14 both by Professor Brooks. . . and myself. In consequence, however, of its faintness, and of the bright moonlight in which it was viewed, it was a difficult object; but as we both saw it in the same position and of the same size and shape, there could be no doubt in the mind of either of us that we had seen the 'spot,' which appeared as a narrow band extending across both outer rings, its western boundary being in contact with the black notch termed the shadow of the ball on the ring. . . . As we found the spot in the same place as at discovery, it cannot belong to the ring itself. . . . We are led, therefore, to believe that the phenomenon must be produced by reflected sunlight from the globe of the planet, though in just what manner produced we are not able to determine."---Lewis Swift, Warner Observatory, Rochester, N.Y., March 16, 1889.

(2) "My first view of it was through Dr. Swift's 16-inch. . . . Since my return home I have given it very careful study and it has, at intervals, been a comparatively conspicuous object. . . . The brightness appears to me to be variable. Pulsations of the light of this 'white region' have been noticed at irregular intervals, ranging from two to seven minutes. Its appearance at my last observation was that of two small nearly semicircular white 'tufts,' where the ring is cut by the shadow of the globe."---W. R. Brooks, Smith Observatory, Geneva,

N.Y., March 19, 1889.

(3) "This phenomenon is still visible, and has been distinctly seen on every occasion that the planet has been observed. It is visible with all powers from 80 to 450 on the 10-1/8-inch equatorial. With a magnifying power of 150 it is most conspicuous. To me 'the white region' has appeared somewhat smaller of late, as if the shadow of the ball was encroaching upon the white spot....I do not think the contrast theory will explain this white spot, for an observation made a few nights ago through large thin clouds proved that the white region was intrinsically brighter than the rest of the ring.... These large drifting clouds acted like a graduated photometric wedge of coloured glass, gradually obscuring the planet and as gradually allowing it to reappear. In every case the white spot was the last to disappear at the planet's obscuration, and the first to reappear as the planet became visible."---W. R. Brooks, Smith Observatory, Geneva, N.Y., April 19, 1889.

(4) "The Astronomical Journal, No. 191, contains a report by Professor Holden of the observations made at Lick Observatory with both the 36-inch and 12-inch refractors. On the best nights the observers were able to detect nothing abnormal, but when the definition was poor an ill-defined, yellowish 'lump' could be seen 'extending across rings A and B about half as wide as the shadow, and close to it.' 'A similar appearance was seen across both A and B where they met the ball of the planet on the s.p. side.' On March 21, at the suggestion of Mr. Schaeberle, the experiment was tried of placing an occulting bar in the eyepiece and bringing it over difficult parts of the planet. Wherever the bar was placed a brighter confused lump seemed to rise and border the bar. If the bar was over the ball, this brighter border extended across the disk. If it was placed on the ring the appearance was just that of the lump bordering the shadow of the ball on the rings. The observers, therefore, concluded that there was nothing abnormal upon the rings, but that the appearances seen were due to bad atmospheric conditions alone."* Editorial Comment, May 1889.

Reinvigorated by these early observations, I began to observe Saturn more often, making special notes on the spot 20 times between 1940 October 13 and 1941 March 29, U.T. At the same time several additional observations were forwarded to me by Messrs. W. H. Haas of Ohio, F. R. Vaughn, jun., of Iowa, and D. P. Barcroft of California. We attempted to gain impressions of the spot's size, shape, brightness and whiteness, etc., and variations in these qualities; some conflict resulted. To me, using the 8-inch reflector and 213x, the spot was always present in all grades of seeing; the ratio of the area of the spot to that of the shadow averaged near 1/2 during the period of observation, varying unprogressively between 1/3 and 1, being largest perhaps but not most conspicuous in poorest seeing though even then never exceeding unity; the spot was somewhat diffuse though occasionally was "sharply bounded," the contour being "irregularly spindle-shaped"; the spot was usually considered to be the whitest part of the whole Saturnian system, "being whiter than the outer half of ring B or the equatorial zone on the planet itself." On February 7, using an unsilvered 6-1/2-inch reflector and 177x, the spot was still visible. On March 5, using again the 8-inch reflector but only 71x: "The shadow and spot are rather difficult due to smallness. The white spot might not be now noticed did I not know previously of its existence. It appears as an almost star-like whitening on the ring, and is made quite bright with this low power."

On several occasions a suggestion of whitening was given at the unshadowed junction of the rings and dusky limb also. Unfortunately no critical observations of the spot (or spots) have been made at or near opposition when a minimum of shadow would be present. It was not possible to determine just how Cassini's division was affected where it passed through the white spot because of its delicacy there: I thought perhaps it was blurred.

The observations of Messrs. Haas and Vaughn support mine in general as given above, with the important exception that they make the area of the spot 2 to 3 times that of the shadow, or about 5 times greater than my estimates! Four previous drawings of Saturn by Mr. Haas made since 1939 uphold his recent estimates of the size of the spot. This sharp disagreement is difficult to explain, as the telescopes used were not dissimilar: Mr. Haas employed a 6-inch reflector and a 12-inch refractor, and Mr. Vaughn an 8-inch reflector. With the 12-inch refractor and 150 x Mr. Haas saw the spot undiminished in size in full daylight on February 19. Mr. Barcroft, using his 6-inch reflector, apparently was not so successful in holding the "spot." Beyond the shadow he saw instead a "wisp-like" formation, or formations, across the rings (the outside boundary of the "spot"?).

Perhaps the most important point about this phenomenon is that it seems to be a very good case of another rather obvious feature commonly overlooked by the best observers in a field of science (assuming that the 1889, 1939-41 spot exists every apparition). For it must have been quite unknown before 1889, and, in personal investigation, I have failed to find the white spot on any of those reproductions of drawings made between 1889 and 1939 which I have seen, even by such excellent workers as E. M. Antoniadi, E. E. Barnard, W. F. Denning, P. Lowell and T. E. R. Phillips. Even those who have made special studies of the shape of the shadow of the planet on the rings do not seem to have drawn attention to the spot. The presence of a white area on the rings thus situated might do much to produce the "concavity" of this shadow, often commented upon as being contrary to the form geometrically required.

As to theory explaining the white spot, the idea of contrast with the shadow lends itself most readily, and it was quickly called a "white contrast spot" when first seen here in 1939. Our 1940-41 apparition results do not strongly suggest any other explanation. However, the general apparent great whiteness of the region, its having been seen in daylight sky and again with a telescope of low light-grasp, and Brooks's cloud observation in quotation (3), all would make the spot a somewhat curious example of that class.

Note by Director of Saturn Section. The appearance of a spot at the point where the shadow of the globe of Saturn crosses the rings has been frequently noticed by observers of the planet, who have, practically unanimously, considered it as an effect of contrast or irradiation causing an apparent extension of the ring into the shadow. It is difficult to imagine any other explanation; if it were a natural, and real, feature of the ring, it would share in its rotation, and could not retain permanently the position in which it is always seen. This bright spot is by no means hard to see; I have seen it on many occasions, and always under much the same aspect; as far as my experience goes, it is visible with any power or aperture that will show Cassini's division well, and is more prominent some months before or after opposition than near to that time. As the planet is now admirably placed for observation, and approaching opposition, observers of Saturn will find it interesting, and possibly important, to watch any variations in the brightness of this spot that may take place, and especially the date of its last visibility before opposition, as the shadow becomes narrower and finally disappears. It will also be of interest to look out for variations in the colour of the spot; it has, during the present apparition, seemed to me to be distinctly whiter than ring B, which to me is always somewhat yellow; probably an effect of its greater brightness.

The observations mentioned by Mr. Johnson represent very much what observers have from time to time seen; the "pulsations" recorded in its intensity by Mr. W. R. Brooks in 1889 seem easily explained by slight variations in the quality of the image, due to variations in the seeing; to my eyes the visibility of the spot is singularly independent of atmospheric conditions.

The fact that this bright spot is not shown on drawings made of the planet seems to me to suggest that observers generally have regarded it as of a subjective nature, and that they treated it as such when recording what they saw; as Mr. Johnson remarks, the spot may quite well be the cause of the apparent concavity of the outline of the shadow, which has been frequently remarked.

I shall be very glad to receive any further observations of this interesting appearance, which should be easily seen under present conditions for some time to come.

Mr. Johnson deserves the thanks of those interested in Saturn for calling attention to this curious and interesting appearance. (M. A. Ainslie)

---

* The effects of (a) "contrast" with the shadow, (b) superposition of an occulting bar, possibly optical, and (c) poor atmospheric conditions, would seem to be distinct.

• Changes in Ring Brightness

BRIGHTNESS VARIATIONS OF SATURN'S RINGS

Hughes, David W.; *Nature,* 261:191, 1976.

Astronomers, since 1921, have occasionally reported that the brightness of Saturn's rings varies with azimuthal angle (the angle θ in Fig. 1). The degree of variation is a matter of considerable debate, however, and papers have been published claiming that the eastern ansa of the inner ring B is brighter than the western one, fainter than the western one, and that each ansa is fainter than the ring position directly in front of Saturn (that is $\theta = 90^{\circ}$). The outer ring A is also found to vary in brightness, but in a more complicated way.

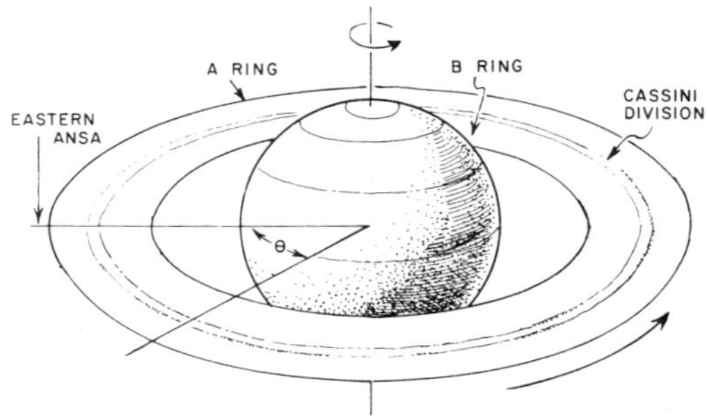

The brightness of Saturn's A ring varies with the azimuthal angle θ, but the brightness of the B ring seems constant.

Hoping to solve this ambiguity Karl Lumme and William Irvine of the Department of Physics and Astronomy, University of Massachusetts, have made careful photometric studies of some 100 good-quality Saturn plates from the New Mexico State University and Lowell Observatory collections. The preliminary results are reported in a recent edition of the Astrophysical Journal (204, L55; 1976). The brightness of each ring as a function of θ has been examined for each of six wavelengths in the band 3,600 to 8,000 A and also for phase angles, a, (the angle Earth-Saturn-Sun) larger or smaller than 1° and for ring tilt angles, B, larger or smaller than 20°. Results show that at 5,500 A, $a > 1^{\circ}$ and $B > 20^{\circ}$, the outer ring A has a maximum of brightness at $\theta = 45^{\circ}$ and 225° and a minimum (about 10% lower) at $\theta = 135^{\circ}$ and 315°. The effects seem to be less pronounced at shorter wavelengths. For the inner B ring Lumme and Irvine find small maxima in brightness for $\theta = 0^{\circ}$ and 180°. This is caused by the image spreading on the photographic plate due to the finite resolving power of the telescope, atmospheric effects and photographic irradiation on the plate. When corrections are made for this spreading, the resulting brightness is found to be independent of θ.

The authors conclude from their results that the particles in ring A may have rotation periods locked to their period of revolution about Saturn. So just like the Moon they always point the same face to the central planet. The above data also indicate for the first time that the average properties of the particles in ring A may differ from those in ring B.

Morrison (Icarus, 22, 57; 1974), who studied the rings by infrared radiometry, found that the particles at the western ansa had a higher brightness temperature than those at the eastern ansa. He proposed that the particles have uneven albedoes, the brighter sides being orientated towards their direction of orbital motion.

The nature and size of the particles in the rings are still poorly understood. Pollack (Space Sci. Rev. 18, 3; 1975) reviewed recent visual, infrared, radio and radar data and concluded tentatively that the mean particle size is around 10 cm, the bulk composition being, in the main, water ice. These particles occupy between 0.5 and 1% of the total volume of the rings. The physical thickness of the rings is about 100 m but the rings do bulge slightly at some of the period resonances with satellites Mimas and Titan. The total mass of material in the rings is small. Pollack's figures leading to a value in the region of 10^{19} g, which is about the mass of a large comet and is over a million times less massive than the Moon.

It is hoped that this picture will be greatly clarified in the next decade when the data from the Pioneer and Marioner Jupiter/Saturn space probes have been received and analysed.

• Curious Shadows on Saturn's Rings

ON THE REVERSED CURVATURE OF THE SHADOW ON SATURN'S RINGS

Jenks, Aldro; *Sidereal Messenger*, 9:255-260, 1890.

On the evening of April 25th, 1889, at about 8:30 p.m., I was examining Saturn with a power of about 180 on a 4-1/8-inch achromatic by Brashear, when, much to my surprise, I found the shadow of the globe on the rings curved the wrong way, i.e. from the globe, as shown in the following drawing, "Fig. I."

Thinking my eyes might be deceiving me I called my wife, and without telling her what I had seen, requested her to describe the shape of the shadow. She described the shadow as having its right hand edge curved away from the planet.

I wrote to Professor Comstock of the Washburn Observatory about it, and was informed by him that while my observation of Saturn was unusual, it was far from being unprecendented; that the same appearance was observed in 1875 with the 26-inch achromatic at Washington, and that Webb, in "Celestial Objects for Common Telescopes," says: "The outline of this shadow has often been found curved the wrong way for its perspective." Professor Comstock also adds, "I do not know that any satisfactory explanation for this anomaly has ever been given."

"Wrong curvature" of planetary shadow on Saturn's rings

In seeking the cause of this phenomenon the first explanation which presents itself is, that it is due to some personal idiosyncrasy, or peculiarity of vision of the observer; but when several observers see the same appearance, and this without previous intimation of what they are expected to see, it would seem to exclude that source of error.

As several eye-pieces were used it could not arise from distortion produced by the eye-piece. The objective is an excellent one, and has never shown any distortion of an object; besides, the same appearance has been seen through telescopes of the highest excellence.

It might be thought to be due to atmospheric causes, but it is not to be supposed that on every occasion when this appearance has been observed, the shadow only would be distorted by atmospheric causes.

Excluding these sources of error, we are forced to the conclusion that the cause of this phenomenon must be sought in some physical peculiarity of the ring system itself.

Long ago, Secchi pointed out that a reversed curvature of the shadow would result from a slight convexity in the ring. This is easily verified. With a lamp, globe and paper rings a little experiment will satisfy anyone that when the portion of the ring on which the shadow falls is made convex, a reversed curvature of the shadow always ensues, placing the eye and lamp in the same position relatively to the globe and ring that the earth and sun occupy to Saturn; and we also find, there is no other shape which can be given to the ring that will produce this effect.

We seem justified, then, in assuming that this appearance is produced by such a shape of the rings. This, however, is not satisfactory. The real question arises, what is the explanation of that explanation? It is evident the surface of the ring cannot be permanently convex, because, in that event, the shadow would always be found curved the wrong way; nor can any portion of the ring be permanently convex, or we should have this appearance at every revolution of the ring, and as the ring revolves in a little over ten and one half hours, it could be seen at some time on every clear night.

. . . .

PHENOMENA OBSERVED UPON SATURN .....

Trouvelot, E. L.; *Sidereal Messenger,* 10:74-82, 119-125, 171-179, 1891.

II. The Gradual Invasion of the Illuminated Part of the Ring by a Shadow. From Oct. 6, 1877, when the sun was 1° 49' to the north of the plane of the ring, until Feb. 6, 1878, the day of its crossing the same plane, the illuminated surface of the ring kept gradually decreasing in size, in proportion as the height of the sun decreased, so that Feb. 5, the evening before the passage, the bright surface was no longer shown except by a narrow luminous thread, very difficult to recognise because of its extreme thinness.

The phenomenon consisted in a gradual invasion of the anterior surface of the ring by something which resembled a shadow cast by an opaque body, and which, little by little, advanced upon it and obscured it.

The 6th of October the phenomenon was already rendered apparent by a pronounced eclipse of the part of the ring which crossed the planet. December 18, when the height of the sun was reduced to + 0° 44', the phenomenon was much more accentuated, and all the part of the ring which crossed the ball, as well as the parts near the ansae were obscured. Jan. 25, 1878, when the elevation of the sun was not more than 0° 12', the encroaching shadow stretched so far upon the ring that it reached even to the posterior part, and seemed to mingle with that which the ball then cast upon it to the east in such a way that the eastern ansa seemed entirely separated from the planet by a dark gap, which in size and position corresponded with the shadow cast by Saturn upon its rings. Besides, the angular shape of this shadow, very easy to recognise, left no doubt about its identity. February 4, when the sun was not more than + 0° 5' high, the lighted surface formed only a luminous streak sharply defined and of extreme thinness, which the 5th of February, the evening before the passage of the sun through the plane of the rings, was very difficult to distinguish, and seemed, at moments, discontinuous and formed of luminous globules hardly perceptible.

Evidently the phenomena in question could not be explained by the increasing obliquity of the surface of the ring, caused by the progressive lowering of the earth towards its plane; in the first place, because the gradual decrease of the illuminated part of the ring was not at all in proportion to the lowering of our globe, but took place more rapidly; and also, because this illuminated surface, instead of being a perfect ellipse, had an irregular form. And then, the elevation of the earth, which the 6th of October was + 3° 31' and kept increasing until Nov. 16 when it was + 3° 55', was still + 1° 20' the 5th of February, 1878, when the illuminated portion of the ring was reduced to an interrupted luminous streak, and so narrow that it was difficult to distinguish it. Now when the earth occupies an identical position, or even one inferior to that which it held at the last date, the ring is still a very remarkable object, subtending an angle of 0''.86, and upon which one still distinguishes easily the opening of the ansae, and which cannot pass unnoticed even with instruments of small aperture, while at the same time the height of the sun is scarcely less than 2°, which is the projection of

Saturn and its rings for Feb. 4, 1878, shows the ring as it must have appeared if its surface had been flat and consequently exposed to the rays of the sun.

Neither can the phenomenon be attributed to an error of observation, when the shadow cast by the ring upon the ball would have been taken for and confused with it. In fact, the 6th of October this shadow cast toward the north already began to disengage itself from the nebulous ring C, from which it was separated by a narrow thread of light belonging to the globe. This thread of light separating the shadow from the ring C increased from day to day in the same measure that the shadow thus cast decreased; so that, if by an error, the thread of light belonging to the globe had been confounded with the ring, the illuminated surface of the latter ought to have grown larger instead of smaller, as the observation showed it. Besides the darkening of the ring stretched out beyond the globe, which amply suffices to show that it was distinct from the shadow cast by the ring since that could not reach beyond the globe.

The close connection observed between the increase of the shadow on the illuminated surface of the ring and the lowering of the sun on its horizon shows clearly that that body was the principal cause of the phenomenon.

In reflecting upon the causes which would be capable of producing such a darkening on the ring, we hardly find more than one which can explain it and even that one offers some serious difficulties. We attribute the observed phenomenon to the elevation above the general level of a zone slightly inclined toward the planet. Supposing that the rings had a flat surface, it is evident that this surface would receive the solar rays and consequently would remain visible as long as the sun remained above it.

We have shown elsewhere (Bulletin Astronomique, t. II, p. 16 and following, 1885) that the form of the shadow of the ball cast on the rings could not be reconciled with a surface perfectly flat, and we have shown that the level of this surface is changeable and varies often in height. From the form of the edges of this shadow we have decided that the maximum thickness of these rings is found on the ring B, at some distance from the division of Cassini. Now if, inside of this division, the ring B possesses a variable zone of a considerable height, it will result that, when the sun approaches the plane of the ring, this more elevated zone, intercepting the solar rays, will cast a shadow behind, which will extend farther and farther in proportion as the sun descends, and which will end by covering almost all the part within this zone nearly the half of the surface of the ring A, the half most distant from the sun. (pp. 79-82)

SECONDARY SHADOW ON SATURN'S RINGS

Anonymous; *Nature*, 71:401, 1905.

During a series of observations of Saturn made at Aosta (Italy) in October, November, and December, 1904, Signors M. Amann and Cl.

Rozet observed a secondary shadow, other than that of the planet, projected on to the illuminated surface of the rings. First seen on October 20, this shadow was thinner and much less accentuated than that of the planet, whilst its curvature was in the opposite sense to that of the latter body. From October 20 to November 15, despite the fact that numerous opportunities of observing it occurred, the shadow was not seen, but from the latter date until the end of December it was shown on twenty-six drawings of the system. On seven drawings made between December 22 and 27, the shadow appeared bifurcated where it traversed the inner ring, and on November 28 and 29 a third line of shadow, narrower and feebler than the preceding and much further from the planet, was seen (Comptes rendus, No. 5, 1905).

A NOTE ON THE APPEARANCE OF SATURN AT OPPOSITION

Bartrum, C. O.; *British Astronomical Association, Journal,* 24:359-362, 1914.

The following paper contains an attempt to explain the appearance, at and about the time of opposition, of a shadow of the planet on the ring on both the following and preceding sides when, from geometrical considerations, there should be no such shadow, or a shadow on only one side. A further attempt is made to explain the dark spots or peaks at the points where the Cassini Division of the ring meets the above shadows.

I. The Double Shadow. The so-called false shadow is believed by the writer to be no shadow, but to be the limb of the planet, sufficiently dark at opposition phase when projected on the bright ring to cause the appearance seen.

According to a well-known law, the illumination of a surface varies with the cosine of the angle of incidence of the light. This law applies to an ideally smooth surface, but may generally be taken as approximately true. Again, the angular intensity of luminosity of an element of surface is independent of its inclination to the line of sight. In Fig. 1, if abp is a section through the planet in the plane of its orbit, OS being the

1.

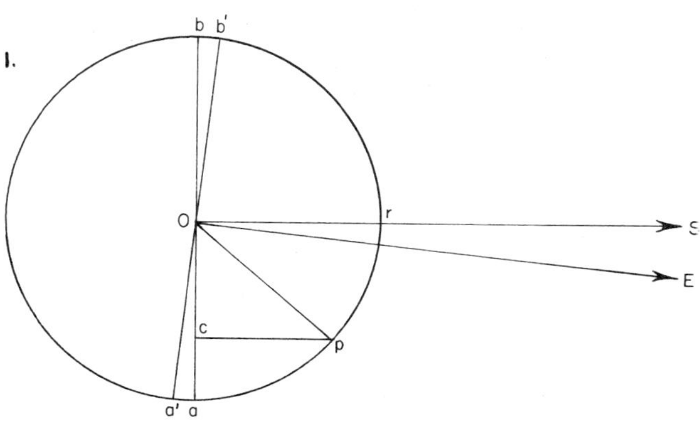

radius vector, and ab a diameter perpendicular to the latter, ∠ cpo is the angle of incidence of the Sun's light, and the illumination at any point p bears to that at r the ratio cp: Or. Hence, at any point near to a and b the illumination is small. Now at opposition the points a and b are on the limb of Saturn as seen from the Earth. It is to be expected then that the extreme edge will be very dark if this law of illumination holds good for this planet. Confirmation of the suggestion that the limb of Saturn is very dusky when tangentially illuminated is found in the fact that in the neighbourhood of the poles, where the limb continues to be so illuminated, the duskiness also continues. On the other hand, this duskiness near the poles may be due to a lower albedo in these latitudes. In the equatorial regions where projected against the sky a narrow dark edge would not be easily noticeable, but further south, where the limb crosses the brilliant ring, such a dark edge would seem to be a sufficient explanation of the apparent double shadow.

Before and after opposition the line of sight from the Earth makes an angle with the direction of the Sun's light. In Fig. 1, if OE is the direction of the Earth, and a' b' is perpendicular to OE, the points a' and b' will be on the limb, a' being a complete darkness, and b' having an illumination nearly proportional to bb'. Now, in the plane of the orbit at the late opposition the angle SOE changed by about 0.12° per day. As then, on this supposition, the dark edge is very narrow, a few days only would be expected to make a considerable difference in the illumination of the limb. For only a few days before opposition would the so-called false shadow be apparent on the following side, and for only a few days after this phase would it continue on the preceding side; the former period would also be expected to be rather longer than the latter, owing to the apparent inclination of the major axis of the ring to the plane of the orbit.

The above argument is based on the supposition that the illumination is proportional to the cosine of the angle of incidence of the Sun's light, but the intensity of apparent luminosity may fall off at the limb in a greater proportion than this law would account for if the light has to traverse an absorbent atmosphere.

II. The Dark Spots or Notches. If the conclusion arrived at above be accepted, this phenomenon can be accounted for by the fact that the image of the system of Saturn is somewhat blurred, even under the best conditions of observation. An examination of the beautiful photographs taken by Barnard and Lowell within a few days of the opposition of 1911 shows in several ways that the bright parts of the image are diffuse at their edges, spreading into the adjacent darkness. When, for instance, the outer margin of the ring passes behind the ball (Fig. 2aa) it shows

2

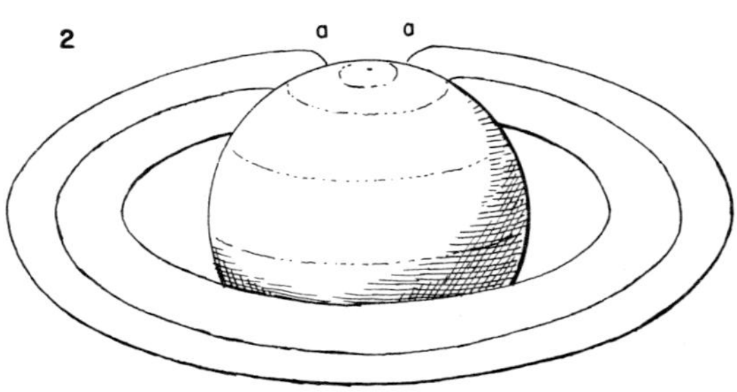

a rapid bend inwards. This blurring, or want of exact definition, may have several causes, viz., want of atmospheric homogeneity, irradiation on the retina in the case of visual observation, and halation on the plate in that of a photograph. Possibly diffraction may add its share. Each of these effects will expand every point of light into a small circle. The effect this will have upon an image can be fairly well imitated by viewing an object through a lens or a telescope slightly out of focus, or by throwing a picture of the object on to a lantern screen similarly out of focus. An experiment of this kind shows at once that a want of definition accounts for the phenomenon under consideration. If the curves in Fig. 3 be projected upon a screen and the lantern lens be racked slightly out of focus, a dark spot like that seen on Saturn appears at a. Or, again, if a lantern plate be covered with black paper and a figure like that on Fig. 4 be cut out to represent the Cassini Division meeting the limb of Saturn, a projection is formed at the point of meeting when the lantern is out of focus.

3

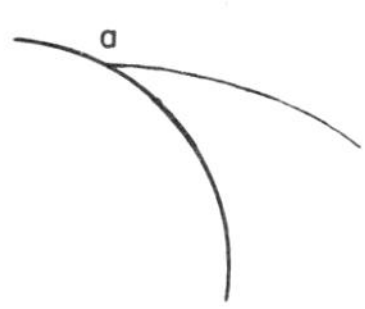

4

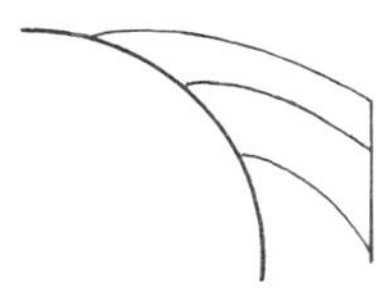

Further to test the soundness of this explanation the writer made a model of Saturn and his rings of a tennis ball, knitting needles, and a circle of cardboard, on which the Cassini Division was drawn in black ink. He placed this model against a piece of black velvet in the rays of a lantern, having removed the condenser and lens so that a narrow beam of light passed direct from the incandescent lime through diaphragms upon the model. He then photographed it with the camera as near to the beam of light as it could be placed, to represent the conditions at opposition, making several exposures with different deviations from accurate focussing. The resulting photographs show the dark spots as seen in a telescopic view of Saturn and in a photograph of the planet.

It has been observed that the shadow of the ball on the rings before and after opposition does not show a continuous curve within and without the Cassini Division. It has also been remarked that the sections of the two rings by the shadow are seen to be concave towards the ring. Both these phenomena receive their explanation on the above hypothesis.

• White Spots on Saturn

WHITE SPOT ON SATURN

Denning, W. F.; *Scientific American*, 89:79, 1903.

On July 1, after observing Jupiter for some time, I directed my 10-inch reflector to Saturn, and found the details sharply defined. The dusky north polar cap was very distinct, and so was the dark belt on the north side of the equator. The belt was darkest and more strongly outlined on its southern side, probably by contrast with the bright equatorial zone. I soon noticed a large bright spot on the north side of the belt, and in a position nearing the western limb of the planet. It was followed by a diffused dusky marking. The luminous spot must have been on the planet's central meridian at about 14h. 1m., but this is only a rough estimate, as the marking was far past transit when I first saw it. It is to be hoped that this feature will prove fairly durable, in which case it may be expected to furnish an excellent means of redetermining the rotation period of Saturn.

A telegram from Kiel which has been widely published, states that Barnard, of the Yerkes Observatory, saw a white spot in Saturn's N. hemisphere central on June 23, 15h. 47.8m. Williams Bay time. Allowing for the difference of longitude, this would be 21h. 42m. G.M.T. Adding eighteen rotations of Saturn of about 10h. 14m. will bring us to the time when the spot was approximately in transit as observed at Bristol, and there seems no doubt as to the identity of the objects.

This disturbance on Saturn will recall Prof. Asaph Hall's white spot seen in the winter of 1876-7, and followed from December 7 to January 2. A number of transits of this object were observed by Hall, Eastman, Newcomb, Edgecomb, and A. G. Clark, and from the data obtained the former found the rotation period of Saturn to be 10h. 14m. 23.8s. ± 2.30s. mean time.

The spot lengthened out into a bright belt, and soon lost its distinctive character.

THE WHITE SPOT ON SATURN

Anonymous; *Science*, 78:6, 1933.

A large white spot, so immense that it could engulf an object over twice the diameter of the earth, has appeared suddenly on the equator of Saturn, the ringed planet and second largest of the solar system.

It was discovered by John E. Willis, of the U.S. Naval Observatory, at 12:18 a.m. Saturday, August 5, while he was observing a transit of Saturn as a routine operation, using the fixed six-inch telescope. Although he observed the planet for only about a minute before it left the field of view, Mr. Willis recognized the spot and called upon other astronomers at the observatory to check his discovery. Principal Astronomer H. E. Burton turned the 26-inch telescope and a smaller 12-inch telescope upon the planet and confirmed the discovery. B. P. Sharpless also made confirming observations.

Captain J. F. Hellweg, superintendent of the observatory, reported the discovery to Harvard College Observatory, whence it was bulletined to observatories throughout the world in order that they might join in the study of this unusual phenomenon.

Just what causes the white spot is unknown. Saturn is made of the lightest stuff of any planet, with a density one eighth that of the earth or seven tenths that of water. The equator of the planet, which is in line with the plane of its rings, is known to move faster in rotation than the other parts of the planet. Perhaps the white spot is a gigantic whirl in Saturn's equatorial belt.

On Saturday morning it was estimated that the spot was about a tenth the diameter of the planet, but on Saturday night Principal Astronomer Burton, with the 26-inch telescope, found that the spot was much larger. While difficult to measure because not well defined, the spot seemed to be about 20,000 miles long and 12,000 miles wide, being formed by a sort of extension in the brighter equatorial belt of the planet. It is expected that it will continue to be observable for several weeks.

Only twice before do astronomical records show that such spots have been observed on Saturn. The first was seen through the same 26-inch Naval Observatory telescope in 1876 by the late Professor Asaph Hall, Sr., just three years after the telescope was built. By using the spot as a mark of reference, Professor Hall was able to make the determination of the period of rotation of Saturn that is now quoted in astronomical tables, 10 hours, 14 minutes, 24 seconds. The 1876 Hall spot was not so large as the Willis spot now visible. Mr. Willis used the Hall determination of Saturn's period of rotation in predicting when the spot would return to visibility, and Saturn night's observations showed that the Willis spot is revolving around on the planet in about the expected time. The second spot on Saturn was discovered by the late Professor E. E. Barnard, of the Yerkes Observatory, in 1903. Unlike the Hall and Willis spots, the Barnard spot was considerably north of the planet's equator.

The spots are useful in making measurements of Saturn's rotation, but Sir William Herschel made a close determination of the planet's period of rotation in 1794, without the aid of a spot. He came within 2 minutes of the value later determined by Professor Hall.

• Extended Glow about Saturn

EXTENDED GLOW PRECEDING REAPPEARANCE OF SATURN DURING A LUNAR OCCULTATION

Reed, G., et al; *Nature,* 247:447-448, 1974.

The recent occultation of Saturn by the Moon, visible across Canada in the early morning of Wednesday, October 17, 1973 provided the opportunity to observe and photograph the appearance of the planet from behind the dark lunar limb. Teams of observers from the Calgary Centre of the Royal Astronomical Society of Canada and the University of Calgary Physics Department established telescopes at various sites in Alberta but poor weather restricted photography of this event to times rather late in the reappearance and to a "close approach" sequence at the University of Calgary Rothney Astrophysical Observatory south of the shadow path.

The present observation was made from the northernmost site at Ardrossan, near Edmonton, Alberta (53° 33' 45" N, 113° 07' 50" W) where a team of four observers were spaced along a line at right angles to the shadow path several hundred metres apart in an effort to observe photoelectrically as well as visually the grazing occultation of the 8 mag Titan by the south lunar limb. This attempt was thwarted by instrument problems and by intermittent clouds, but the sky cleared considerably by the predicted time of Saturn reappearance with only light haze surrounding the 66% sunlit Moon. Cameras were mounted on three of the four telescopes at this site and observers were linked by intercom to each other and to a tape recorder on which WWV time signals as well as observer comments were recorded. The fourth observer was to provide careful visual observations and detailed timing of the reappearance and was equipped with a 6 inch f/9 Newtonian telescope equipped with a 12 mm eyepiece and a x2 Barlow lens (overall magnification x230) to ensure that light from the bright limb did not obscure the observation in the 12 arc min field of view at the moment of egress.

The observation upon which this report is based was made by the visual observer at a time between 1 s and 2 s before the appearance at the dark lunar limb of the first pin-point of light from Saturn's rings. At this time, the observer's eye was attracted to a faint glow which he probably first saw by averted vision but which delineated the dark limb and which appeared like seeing "a campfire on the other side of a treeless hill on a dark night". The extent of the glow was estimated afterwards to have been a little greater than the east-west extent of the rings. The pin-point of light from the rings which appeared in the centre of the faint glow rapidly brightened and became bigger as the rings became visible following the third contact. The glow increased slightly in width and showed easily detectable Moon limb curvature on its sharp edge but at about 4 s beyond third contact the glow was indistinguishable against the increasing ring brilliance.

Superficial explanations for this glow can be quickly dismissed. If internal reflection within the telescope of light from the bright lunar surface had caused the glow, it was certainly a coincidence that it not only appeared in the exact position where the rings showed themselves but also was aligned with the direction and position of the dark Moon edge. Mist in the Earth's atmosphere could have produced a hazy glow just as the first light of the rings appeared but this would almost certainly have been circular from the initially observed edge of the rings. This must remain a distinct possibility but it is at least as tempting to associate this glow with the planet and a tenuous outer atmosphere surrounding Saturn and its rings which is normally invisible against the planet's brightness. Observation of a faint outer ring) has been referred to several times this century but not confirmed by modern photography. These observations may be related to the present result. Furthermore a very similar observation to that seen for Saturn was reported by Brock for the planet Jupiter when it was occulted by the south-limb mountains of the Moon. At that time a glow was noted to be moving along above the mountain range during the grazing occultation. This glow associated with Jupiter was reported to vanish immediately the planet appeared and it was not seen by observers situated one quarter of a mile north and south of the observer.

Thus, a tentative conclusion for these unique visual observations must be that these giant planets have extended atmospheres associated with them, an idea suggested for Jupiter by Peek who postulated an upper atmosphere consisting of light elements and extending to great heights above the cloud layers. Estimates of total pressure of gas above the cloudtops from spectroscopic data vary from 2 to about 13 atm. Such gaseous atmospheres would not be expected to scatter sunlight but could presumably emit light equivalent to airglow or even auroral luminosity under the action of the solar wind. The existence of a magnetosphere and Van Allen belts is inferred from the intense decimetric radio bursts detected from Jupiter. The requisite particle flux for excitation of such luminosity is thus present, at least for Jupiter. Radio bursts from Saturn are much weaker but the similarity of this planet to Jupiter would lead one to believe that such luminosity is possible from Saturn also.

In the case of the present observation of Saturn, if we assume 1.5 s for the time between appearance of the glow and that of the rings, the extension beyond the rings is approximately 5,000 km or 4% of the planet's diameter. The observer's report of the glow accompanying Jupiter would indicate a glow extending about 1,000 km above the cloud tops of Jupiter, or less than 1% of the planet's diameter. The geometry of these events was such that the glow was observed close to the polar regions of both planets.

It is obvious that repeated measurements of this phenomenon, particularly careful photoelectric observations, are necessary in order to verify the existence of the glows but these can be made only very infrequently. Of much greater importance, bearing in mind the approach of the Pioneer spacecraft to Jupiter, is that extended glows should be searched for in data from these deep space probes.

• Brightness Changes of Saturn's Satellites

UBV PHOTOMETRY OF IAPETUS

Millis, R. L.; *Icarus*, 18:247-252, 1973.

Introduction. The remarkable brightness variation of Saturn's satellite Iapetus was discovered by J. D. Cassini over three centuries ago. More recent observations by visual observers have revealed that the satellite's brightness varies smoothly over a range of about two magnitudes, reaching maximum brightness near greatest western elongation and minimum brightness near greatest eastern elongation. Comparison of the more accurate visual light curves suggests that the amplitude of brightness variation may be variable from one apparition to the next (Widorn, 1952).

Even though the range of Iapetus' brightness variation greatly exceeds that of any other satellite, photoelectric observers have given Iapetus scant attention. To the writer's knowledge, only two sets of photoelectric data have been published. UBV measurements made on 17 nights between 1951 and 1953 have been discussed by Harris (1961). In addition to generally confirming the results of earlier observers, Harris' data showed that the color variation of Iapetus amounted to only a few hundredths of a magnitude, the object being redder when fainter. Unfortunately, the phase coverage of these observations was limited, and the satellite was not observed near minimum light. More recently McCord and his collaborators (1971) have observed Iapetus through a series of narrow-band interference filters between 0.4 and 1.1 μm. Their brightness measurements in the visual region of the spectrum agree reasonably well with Harris', but do not significantly extend the orbital phase coverage.

In this paper UBV measurements of Iapetus on 34 nights well distributed in orbital phase and spanning almost two successive revolutions are reported.

....

Conclusions. The observed differences between two successive cycles of Iapetus' light curve can be explained in terms of a two-hemisphere model with the two hemispheres having different photometric phase functions. This interpretation is consistent with existing polarization and albedo measurements, and when taken together with these implies that the leading hemisphere of Iapetus is similar in many of its optical properties to the moon and certain asteroids, while the trailing hemisphere is more comparable in its optical properties to the inner Galilean satellites.

An alternative explanation of the non-repeatability of the light curve is that an intrinsic change in the reflectivity of the dark side of Iapetus occurred within a time period of 80 days. The writer considers this to be unlikely, but additional observations over a period of years will be required to settle the matter conclusively.

NEW EVIDENCE FOR THE VARIABILITY OF TITAN

Noland, M., et al; *Astrophysical Journal*, 194:L157-L158, 1974.

I. Introduction. Several authors have suggested recently that the brightness of Titan, Saturn's largest satellite, may be secularly variable. The question has been reviewed by Andersson, who attempts to reduce available photometry to a common scale, using the well-established assumptions that the brightness of Titan is independent of its orbital position, and that the phase coefficient of Titan is about 0.005 mag per degree in the V. In particular his analysis indicates that between 1967.9 and 1973.9 Titan has brightened fairly uniformly by 0.10 mag. from V(1, 0)= -1.20 mag to V(1, 0) = -1.30 mag, where V(1, 0) is the magnitude reduced to unit distance from the Sun and Earth and to zero phase angle.

Andersson's results are based on a compilation of heterogeneous data, obtained by different observers, using a variety of instruments. Ideally, one would like to compare observations made by the same observers using the same equipment. In addition, since the spectrum of Titan has steep gradients, a narrow-band filter, such as the Stromgren y, is preferable to a wide-band filter, such as the V of the UBV system. In this Letter we report new y-filter photometry of Titan, which was made in 1974 with the Mauna Kea 61-cm telescope, and compare it with the 1972 y-filter observations of Noland et al., which were made with the same instrument.

II. New Observations. The instrumentation and techniques used are as described by Noland et al. Our primary standard for the 1974 observations, 114 Tau, was also observed during 1972, enabling us to convert the recent measurements to our previous system of magnitudes relative to 37 Tau.

....

The resultant average magnitudes are given in table 1 [not reproduced] and show a brightening of ~ 0.04 mag between 1972.7 and 1974.2. This is consistent with Andersson's analysis, which indicates a brightening of $dV/dt \simeq -0.02$ mag per year during the period 1967.9-1973.9.

III. Discussion. At least five causes for the variability of Titan can be suggested:

a) Fluctuations in the solar constant. This is unlikely since other well-observed objects, such as the Galilean satellites, have not shown comparable brightenings in recent years.

b) Random fluctuations in cloud cover. There is much indirect evidence and some visual evidence that Titan has a cloudy atmosphere. Fluctuations in cloud patterns could easily account for brightness changes of 10-20 percent. Such changes are known to be responsible for changes in the brightness of Jupiter.

c) Seasonal (29.5 year) fluctuations in cloud cover. At perihelion Saturn is 6 percent closer to the Sun than at aphelion. The difference in insolation might affect the extent of cloud cover.

d) Fluctuations in cloud cover related to the 11-year solar cycle. The intensity of short ultraviolet radiation and the solar wind could affect the structure and composition of the Titanian clouds.

e) Changes in aspect. As seen from Earth, the polar axis of Titan oscillates through 28° with a period of 29.5 years. If the polar regions of Titan are brighter than the equatorial regions, Titan should appear brightest when one of its poles is seen at maximum tilt (1974) and faintest when the tilt is zero (1965/1966).

At present there is no way to distinguish among these possibilities. Although Andersson discusses observations dating back to 1896, the earlier coverage is spotty and inconclusive. The only clear trend is the brightening from 1968 to the present. Since Saturn not only attains perihelion in 1974, but also is at maximum tilt, a decrease in brightness during the next few years would lend support to a periodic explanation related to aspect or season. Continued observations are clearly in order.

....

Chapter 12

THE PLANETS BEYOND SATURN

INTRODUCTION

Beyond Saturn a few bright wanderers ply the night sky. They are so distant that anomalies are difficult to discern. Except for the common suspicion of rings (recent discoveries involving Pluto are not included here), there are no connecting links among Uranus, Neptune, and Pluto. Each has its own personality.

• Uranus

THE STRANGE RINGS OF URANUS

Anonymous; *Science News*, 111:245, 1977.

The recent discovery of apparent rings around the planet Uranus has grown more interesting still, with the possibility that there may be more rings than originally inferred and that the largest ring may be strangely irregular in shape.

The March 10 observations were in the form of a series of "dips" in the measured light from a star (SAO 158687) shortly before and after it was blocked out by Uranus itself. The blockage, called an occultation, was recorded by James Elliot of Cornell University, flying over the Indian Ocean aboard a NASA airborne observatory, as well as Robert Millis of Lowell Observatory, working at Perth Observatory in Australia, and other astronomers in India, Japan and South Africa.

Elliot's original speculation was that the inbound and outbound "mini-occultations" represented a number of individual objects, or moons, with diameters as large as 100 kilometers. Study and comparison of the observations, however, showed that the spacing and duration of the inbound mini-occultations was almost identical to those recorded after the planet had passed from in front of the star. It is extremely unlikely that individual moons would be spaced so regularly, so it is now felt that the starlight was blocked by the two sides of a series of rings. The light was not completely shut off---only reduced by about 50 to 90 percent---and Elliot concludes that the individual ring particles are probably smaller than 2 kilometers across.

There seem to be five principal rings, which Elliot has labeled (beginning with the innermost one) alpha, beta, gamma, delta and epsilon, together occupying a band about 7,000 kilometers wide. Each of

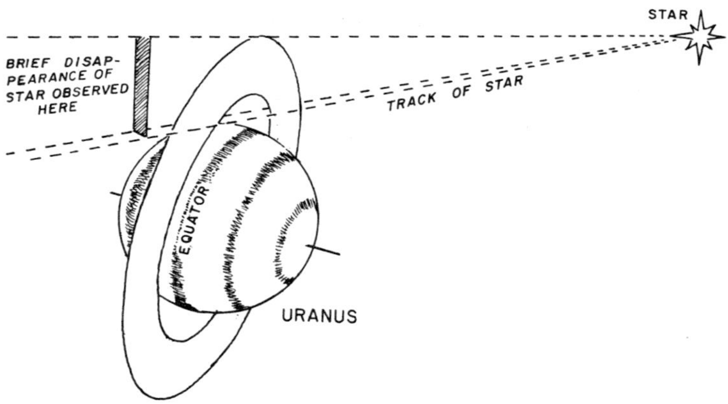

Rings of Uranus were detected in 1977 during the occultation of the star SAO 158687.

the four inner rings appears to be perhaps 10 kilometers wide, while the outer ring may be as wide as 100 kilometers. The outer ring may also be either thicker or more dense, since it occulted about 90 percent of the light from the star, compared with about 50 percent for each of the others.

Further study has now made the strange structures (far narrower than the rings of Saturn, the largest of which is more than 25,000 kilometers wide) seem even stranger.

Millis's data from Perth, although they missed the alpha and delta rings while the telescope was being recentered (only the inbound occultations were visible from Perth), show two additional partial occultations inside the alpha ring. This could be evidence of two more rings, with the "shallowness" of the occultations possibly indicating that these rings are narrower or less dense than the others.

Still more perplexing is the matter of the wide outermost, or epsilon, ring. The inbound occultation, says Elliot, lasted about 7 to 8 seconds. The outbound occultation of presumably the same ring, however, lasted only 3 seconds. This translates to mean that the ring's inner edge is about 600 kilometers closer to the planet on the outbound side.

• Neptune

RING OF NEPTUNE

Challis, J.; *Astronomische Nachrichten*, 25:231, 1847.

I have been able with the Northumberland Telescope to verify Mr. Lassell's suspicion that the planet has a Ring. I first received the impression of a Ring on Jan. 12. Two independent drawings made by myself and my assistant Mr. Morgan gave the following representation of its appearance and position:---

The angle made by the diameter of the Ring with a parallel of declination was 66° by a not very satisfactory measurement taken on Jan. 15. The ratio of the diameter of the Ring to that of the Planet is

Ring of 1847

by estimation that of 3 to 2. I am unable to account for my not having noticed the Ring earlier.

NEPTUNE, ITS SUPPOSED RING AND SATELLITE

Anonymous; *American Journal of Science*, 2:4:287, 1847.

Several European astronomers have pronounced in favor of the existence of a ring around the planet Neptune. Mr. Lassell, of Liverpool, observing with his Newtonian reflector, of two-feet aperture, first announced its existence, in October, 1846; and in January last, Prof. Challis, of Cambridge, using the large Northumberland reflector, was disposed to believe Mr. L.'s assertion. The ratio of the diameter of the ring to that of the planet is about that of 3 to 2. The angle made by the axis of the ring with a parallel of declination, in S. preceding or N. following quarter, is about 65°. Other observers, however, with equal means, cannot detect any such ring.

Mr. Lassell has announced to the London Times, his verification of the existence of a satellite of Neptune, suspected last November.

REFLECTIONS CONCERNING NEPTUNE'S RING

Hoyt, William G.; *Sky and Telescope*, 35:284-285, 1978.

In an intriguing but little-known episode more than a century ago, Neptune was once found to have a ring, and an eminent astronomer confirmed it. The discoverer of Neptune's "ring" was the English amateur William Lassell (1799-1880), the same man who detected Neptune's satellite Triton and the third and fourth satellites of Uranus. The confirmation was provided by Rev. James Challis (1803-1882), director of Cambridge University Observatory and a key figure in the controversy over priorities following the discovery of Neptune in September, 1846.

Lassell first observed the Neptunian ring early the next month, with the 24-inch Newtonian reflector of his Starfield Observatory at Liverpool. On October 3rd, while examining Neptune, he was "struck with its shape, which was evidently not merely that of a round ball." On the 10th, with various magnifications up to 567x, he "received many distinct impressions that the planet was surrounded by an obliquely situated ring....having its major axis nearly at right angles to a parallel of declination, or in other words to the course of the planet through the field."

....

During the latter part of the 19th century and the early 20th century, extensive measurements of the diameter of Neptune and of the position of Triton were made by several astronomers with large refractors of optical quality much superior to Lassell's reflector. The fact that this repeated scrutiny did not reveal a ring demonstrates clearly that Neptune does not have one fitting the description given by Lassell and Challis.

....

NEPTUNE BRIGHTENS UP HIS IMAGE

Anonymous; *New Scientist*, 73:393, 1977.

Neptune, 30 times farther away from the Sun than ourselves and invisible to the naked eye, is very much an unknown world. But new observations at Kitt Peak National Observatory, reported at a recent meeting in Honolulu, have indicated for the first time that the planet has changeable weather. Dr. Richard Joyce of KPNO and his colleagues discovered last year that Neptune, which has an exceptionally high reflectance, had increased its infrared brightness by a factor of over four in the year between April 1975 and March 1976.

....

• Pluto

PLUTO: PUZZLE PLANET OF THE SOLAR SYSTEM

Anonymous; *New Scientist*, 55:74, 1972.

Ever since its discovery in 1930 Pluto has presented a series of puzzles to astronomers. To start with, there is now serious doubt about whether Pluto has the correct mass and orbit to explain the very perturbations which led to its discovery. With an orbital period of roughly 250 years it will be a long time yet before exact details of the influence of Pluto on the outer planets can be calculated. But, even though the planet has been studied for only one sixth of an orbit, there is no doubt that the evidence building up points to discrepancies between prediction and observation. The naive conclusion is that there is yet another planet in the outer reaches of the solar system.

Second, Pluto's mass is disturbingly small and its density---according to the simplest calculations---is disturbingly large. All the other outer planets are gas giants, but Pluto has a mass rather less than that of the Earth and a diameter of only about 3600 miles, according to direct observation. These figures give such a ridiculously large density for the planet that many astronomers prefer to believe that Pluto has a true diameter much greater than apparent.

What observers measure as the diameter of the planet, they argue, is really the size of a bright highlight near the centre of a highly reflecting sphere. But why should such a planet, probably cratered in a way similar to the Moon and Mars, have albedo large enough to produce this optical illusion? At present, the most widely held view is that the diameter measurements are right and that it is the mass of Pluto which has been wrongly estimated. If that is correct, Pluto cannot be the trans-Neptunian planet sought to explain the discrepancies in the orbits of the gas giants, and its discovery was thus entirely fortuitous.

....

IS PLUTO NO BIGGER THAN THE MOON?

Anonymous; *New Scientist*, 72:372, 1976.

New spectroscopic observations, published this week in Science (vol 194, p 835) by three astronomers at the University of Hawaii, reveal that Pluto may be a much smaller and lighter body than hitherto imagined. No one knows for sure how big Pluto is---estimates have placed its diameter somewhere between 5000 and 7000 km---nor its mass, tentatively put at 0.18 Earth masses. On that basis Pluto's mean density, which provides us with some indication of what its composition might be, comes out to between 2 and 4 g per cu. cm. For comparison, the rocky Earth's mean density is some 5.5 g per cu. cm.

The Hawaiian team, D. P. Cruikshank, C. B. Pilcher and D. Morrison, have looked at Pluto in the near-infrared part of the spectrum with a 4-metre telescope at Kitt Peak National Observatory. Absorption spectra strongly indicate that Pluto's surface is at least partially, and may be completely, covered in frozen methane. This frost, the three workers claim, is quite distinct from either water or ammonia ice. The cover is apparently not complete as the brightness of the planet varies by some 20 per cent as Pluto rotates. However, the ice makes it much shinier than the dark rocky surface assumed in previous calculations of the planet's diameter made from its stellar magnitude (approximately 14). In other words, a much smaller body will reflect the same amount of sunlight.

Comparing Pluto's reflectance with other bodies known to be frost covered would give the most distant planet a maximum diameter of only 3300 km, rather less than that of the Moon, or a minimum diameter of a mere 2800 km. If, as seems likely, Pluto represents a low-temperature "condensate" of the primordial solar nebula, its density may lie between only 1 and 2 g per cu. cm. If that is so, the planet's mass would be far too small to have produced the perturbations of neighbouring Neptune or Uranus, which were supposed to have provided the prime clue leading to the discovery of Pluto in 1930.

Chapter 13

ENIGMATIC OBJECTS

INTRODUCTION

Astronomers usually have little difficulty in classifying objects seen in the heavens as stars, planets, asteroids, comets, or meteors. The speeds with which these objects move across the sky and the paths they take are generally diagnostic. However, every once in a while some apparition passes into a telescope's field of view that does not meet the criteria for any of these "acceptable" denizens of the solar system--- in short, it is an "enigmatic object."

Enigmatic objects may be bright or dark; if the latter, they are seen passing in front of the sun or moon. Often their proper motions are large and trajectories unusual. They are here today and gone tomorrow. Other enigmatic objects hang motionless in the sky for a brief moment and disappear. The occasionally bright objects seen near the sun, such as the widely observed 1921 object, are typical of this genre.

What are the enigmatic objects? Some observations reported here are likely errors of observation and others can probably be written off as "abnormal" comets, meteors, and the like. In other words, identification criteria can be stretched enough to accommodate them. The remainder, though, must be natural or artificial objects that are just "passing through," from where and to where we just do not know.

Because enigmatic objects by definition are difficult to classify, they follow in order of their appearance.

HUTH'S "MOVING STAR" OF 1801-2

Anonymous; *Nature*, 14:291-292, 1876.

At the beginning of the present century, when, although Bode and some few others had been looking forward to such a discovery, astronomers generally were startled by Piazzi's accidental detection of the small planet Ceres, we read of observations of more than one so-called "moving star," which, after progressing slowly for a short interval, finally disappeared. The most singular narrative refers to an object said to have been remarked by Hofrath Huth, at Frankfort-on-the-Oder, on the night from December 2 to 3, 1801, particulars of which were communicated to Bode in several letters during the ensuing five weeks. If the observations are bona fide, there is yet a mystery attaching to the object to which they relate. Huth was one of the three independent discoverers of the periodical comet now known as Encke's, on October 20, 1805, Pons and Bouvard sharing with him an almost simultaneous discovery, and he did other astronomical work. Writing to Bode on December 5, he says: "In the night from the 2nd to the 3rd of this month, I saw with my 2-1/2-feet Dollond, in a triangle with θ and δ Leonis to the south-west, a star with faint reddish light, round, and admitting of being magnified. I could not discern any trace of it with the naked eye; it had three small stars in its neighbourhood." He writes again on the 15th, that unfavourable weather had allowed of his observing the object only on three occasions, which appear to be on the early mornings of the 3rd, 13th and 14th, and he concludes from his observations that it had a slow retrograde motion to the south-west. From the 13th to the 14th, by eye-estimate, it had retrograded 4' of arc, and from the 3rd to the 13th at most 30'. He forwarded to Bode at this time a diagram of the neighbouring telescopic stars. On December 21 he writes again that he had only succeeded in observing his moving star on one additional night, that of December 19-20, when he found it "near four stars apparently situate to the westward, about half a diameter of the full moon below a smaller one." Its path appeared directed toward i Leonis and towards the ecliptic. He adds: "Of the motion of this planet-like star I can now no longer doubt, since I have observed a difference of 5/6° nearly, between its positions on the 3rd and 20th." In a fourth letter, dated 1802, January 12, he informs Bode that he had seen the star on two later nights, those of the 1st and 2nd of the same month from 11h. to 14h., with many telescopic stars in its vicinity, of which he enclosed a diagram, by eye-estimate only, with the path of the object.

He mentions that on January 1 the star was even smaller than one of the satellites of Jupiter, and on the following night he had difficulty in perceiving it in close proximity to a star towards which it was moving. On the 5th he could discern only now and then, to the right of the star, on the left of which it was situated on the 1st and 2nd of January, and at a very small distance from it, a glimmer, but the star's former place on the left was vacant. He concludes that the object must have been receding from the earth, and might perhaps have been more distinct and larger before December 3. On the night of January 6 there was no trace of it. He closes this final letter by saying that he would have gladly learned that some other astronomer had observed this star and confirmed its motion, and expressing his regret that Bode had not succeeded in finding it. On the latter point Bode remarks that the weather during December had been but very rarely favourable for observation, and in the few moments that the sky was clear he had occupied himself with his "Seeker" and Dollond, partly in giving attention to the neighbourhood of Huth's star, and partly to the region in which Ceres was expected to be recovered on her second appearance. He also remarks on the imperfect manner in which the star's positions had been communicated to him, but concludes that "without doubt it was a distant comet," and its great distance caused it to appear without nebulosity. He supposes it on December 3 to have been in longitude 156° 20', with latitude 10° 40' north, and on January 2 in 154° 20', with latitude 8° 50'. Huth's rough diagrams are reproduced in the Berliner Fahrbuch, 1805, but they are on a very small scale, and no two persons are likely, perhaps, to agree as to the inferences to be drawn from them. We may remark, however, that the arc of great circle between Bode's extreme positions exceeds the length of the path, as described in Huth's letters. The following places result from an examination of the figures with the particular view to identify several of the telescopic stars entered in the larger diagram:---

1801, Dec. 3.	Longitude	157°.0	Latitude	10°.5
" " 14.	"	156°.7	"	- 9°.9
1802, Jan. 1.	"	156°.2	"	9°.1

Calculations founded upon the deductions from Huth's diagram lead to no satisfactory, indeed no probable, results. The ordinary formulae fail, but the distance of such an object could hardly have been great.

With regard to the bona fides of Huth's observations, it is worthy of remark that he wrote several letters to Bode, while according to his own showing, observations would have been very practicable, but for the unusual prevalence of clouded skies; while there is no doubt of the looseness with which he gave its positions.

CACCIATORE'S SUPPOSED PLANET OF 1835

Anonymous; *Nature*, 18:261, 1878.

It might have been expected that long ere this, if the object twice observed at Palermo in May, 1835, were really a planet, it would have been recovered by one or other of the astronomers who have occupied themselves with the examination of the ecliptical region of the sky.

The particulars of the Palermo observations were communicated by Cacciatore to Valz in a letter dated September 19, 1836, and at an earlier period to the late Admiral Smyth, as will be known to readers of the "Cycle of Celestial Objects." Valz sent a copy of Cacciatore's letter to Schumnacher, who published it in No. 600 of the Astronomische Nachrichten. When observing the star 503 of Mayer's catalogue with the Ramsden circle, on May 11, 1835, it was noted down that a smaller star of the eighth magnitude followed Mayer's star two seconds of time, and was about 2-1/2' to the south. Such entries were frequently made by Piazzi, when observing with the same instrument, as may be seen from his catalogue, but although No. 503 occurs there, no mention is made of a star near it. On the next fine night, May 14, observing Mayer's star again, the assistant, according to custom, read out the note made on May 11: "Seguita da una altra di 8 per 2" circa di A. R. circa 2-1/2' al sud." No star was then visible in this position even in a dark field, but one of the eighth magnitude preceded Mayer's star nine seconds of time, only 1-1/3' to the south. Cacciatore says he intended to repeat the observation on the following evening, the weather promising to continue fine. Returning to the library he found that no one of the four small planets known at that time was in the observed position, and he appears to have considered the object either a planet beyond Uranus or a comet, remarking: "Onde con impazienza attendeva il dimani." But the night of May 15 proved unfavourable, rain setting in, followed by clouded skies for upwards of a fortnight, and not until June 2 could an observation be attempted, "Ma la stella era involta nel crepusculo feci varj tentativi fuori del meridiano, non transcurai ogni mezzo per riconoscere la mia osservazione." Cacciatore says his assistants were unsuccessful on other evenings to the end of June. The search was repeated in the first five months of 1836, but to no purpose.

Valz first showed that a body with the observed positions on May 11 and 14, could not be a distant planet, as Cacciatore had conjectured, but rather a pretty near member of the minor-planet group, which, on the hypothesis of a circular orbit, might have a period of revolution of about three years, with the ascending node of the orbit in longitude 339° 36' and an inclination of 3° 22' to the plane of the ecliptic. In 1849 Dr. Luther repeated the calculation with the following results:---Radius of orbit, 2.1055; ascending node, 343° 20'; inclination, 3° 37'; period, 1,116 days; and from these elements Oeltzen computed a zodiac for the planet, or a table indicating with right ascension as argument, the northern and southern limits of declination (Astron. Nach. No. 662). It is certain that any determination of the position of the orbit from Cacciatore's data must be open to considerable uncertainty, and hence a search for his supposed planet amongst the one hundred and eighty-eight planets now discovered would not be decisive one way or the other if confined to similarity in the position of the nodes and the inclination; places must be calculated for the epoch of Cacciatore's observation for such planets as could by possibility pass near Mayer's star. An attempt in this direction has failed to identify the object. That a minor planet which so far from opposition attains the brightness of stars of the eighth magnitude can still remain unknown to us is, to say the least, very improbable. Must we leave Cacciatore's star in the same category as those reported to have been observed by Huth in 1801 and Reissig in 1803, to which reference has been made in this column?

SINGULAR PHENOMENON

Anonymous; *Scientific American*, 8:333, 1853.

We have received a letter from Professor A. C. Carnes, of Burritt College, Tenn., with the following account of a singular phenomenon, that was seen by a number of the students, on June 1st, at 4-1/2 A. M., just as the sun was rising:---

"Two luminous spots were seen, one about 2° north of the sun, and the other about 30 minutes further in the same direction. When seen, the first had the appearance of a small new moon; the other that of a large star.---The small one soon diminished, and became invisible; the other assumed a globular shape, and then elongated parallel with the horizon. The first then became visible again, and increased rapidly in size, while the other diminished, and the two spots kept changing thus for about half an hour. There was considerable wind at the time, and light fleecy clouds passed by, showing the lights to be confined to one place."

The students have asked for an explanation, but neither the President nor Professors are satisfied as to the character of the lights, but think that electricity has something to do with it. The phenomenon was certainly not an electrical one, so far we can judge, and possible was produced by distant clouds of moisture.

MYSTERIOUS OBJECT NEAR THE SUN

Silverman, Sam; *Sky and Telescope*, 42:3, 1971.

In compiling information relating to sky brightness during total solar eclipses, we have come across an observation that is similar (but not necessarily related) to the sighting of the mysterious object near the sun on August 7, 1921 (June issue, page 352). The occasion was the total eclipse of August 7, 1869.

"A small but exceedingly bright point, like a star, was witnessed by Messrs. Farrell, Phelps, and Locklin, and Mrs. Farrell, during the period of totality. It appeared near the limits of the corona, below the moon's disk---direct vision---and in the region of the anvil-shaped protuberance. Mr. Farrell judged it to be about one-sixth---some say one-tenth---the size of Mercury; which latter star was almost directly to the right of it, on a line parallel to the horizon.

"With the exception of Mrs. Farrell, all located the star a little to the right of the red prominence, or at about 230° from the north point, reckoning by the east. Each of the observers mentioned feels very positive that what he saw was truly a star."

The quotation is from the Observations of the U. S. Naval Observatory for 1867, Appendix II, page 180. In the American Journal of Science, 99, 139, 1870, it is pointed out that the observed position matches that of Pi Cancri. That star, however, is magnitude 5.4---too faint to have been seen with the naked eye during a total eclipse.

A CURIOUS ASTRONOMICAL PHENOMENON

Anonymous; *Scientific American*, 40:294, 1879.

Under the date of April 13, Mr. Henry Harrison, of Jersey City, sends to the New York Tribune the following communication:

"At about 8:30 o'clock last evening, as I was searching for Brorsen's comet, I suddenly hit upon an object which I supposed to be a planetary nebula, very much resembling that near Beta Ursa Majoris, nearly on a line north, between the Pleiades and the variable star Algol. Being somewhat in doubt as to the existence of such a nebula in that region, I started the driving clock, noted the right ascension and declination, which were 2h. 34m. and 37°N., searched the catalogues, but found no such object recorded. By this time I found the object gone out of the field, but soon found it again, when it had gained four min. in R. A., its declination being unchanged. A half hour or so later, watching it constantly with amazement, I found it had gained the same amount. I no longer trusted to my own vision, but called a friend to confirm what undoubtedly was there. He saw it, and we both began to speculate as to its physical composition. A comet it could not be, because of its rapid motion from N. W. to S. E., nor could it have been a cloud, because it maintained not only its shape, diameter and density, but also its luminosity, and in the absence of both sun and moon a batch of cloud viewed with a telescope would have no definition, form, or illumination. Still following it as it slowly swept toward Alpha Auriga, I found that a calculation of R. A. at 9:35 was 3h. 4m. N. D. 37°.

"In order to obtain more knowledge about this wonderful phenomenon, for such I must call it, I concluded to telegraph at once to the Naval Observatory at Washington to set the circles, as I calculated, about 7h. 8m. west of the meridian and declination north 37°, which position it would occupy by the time the message would reach the observatory. Returning from the telegraph office at 10:45, its altitude must have reached a height of 40°. Still it maintained its form and brilliancy. I must say here, that before sending a dispatch to the Naval Observatory, I thought that the object might be a reflection, but this thought was rapidly removed by placing a pasteboard tube of eighteen inches in length over the objective, but onward it moved with independent motion. Even the two-inch finder showed it faintly. At 10:45 I noticed its declination to be 37° 6'; at 11:30 it neared the double star Alpha Gemini, and rapidly passing before a star of the sixth or seventh magnitude, seemed not to obscure the latter, but showed the star almost as brilliant as immediately after its passage. The declination now was 37° 28'.

At 12 o'clock I retired for a short time, after a wearisome chase of nearly three and a half hours. At 2:10 o'clock this morning I found it, after a search of about twenty minutes, in the zenith. It now seemed to be more brilliant than at any previous observation, its declination being now 37° 40', and I fancied I could see it with the unaided eye, but cannot be positive of this. I must confess, although absurd, the thought entered my mind that one of the planetary nebulae, tired of its position, was seeking another and a better home."

THE CURIOUS ASTRONOMICAL PHENOMENON

Harrison, Henry; *Scientific American Supplement,* 7:2884-2885, 1879.

Seeing that you have honored me with a reprint of my letter to the New York Tribune, dated April 13, in which I gave a superficial description of a comet-like object moving through space from northwest to northeast, and being daily questioned by astronomers as to correct and more details of data, I take pleasure, now that the matter seems to become one of general interest, in giving a more minute statement, and at the same time vindicate myself---being accused by some of the "learned professional astronomers" for not having published my discovery through the proper scientific channel and in the proper manner. I am as yet in the infancy of astronomical researches and totally inexperienced in the discoveries of "wonders," such as "volcanoes in action on the moon's surface," "second companion to Polaris," "satellite to Venus," or "a third satellite to Mars," and a thousand other wonderful discoveries that are yearly reported to the Smithsonian Institute or the Naval Observatory, and thrown into the waste basket. I have never made any discoveries, never intend to. I have an utter detestation for people who speculate in scientific matters, who stir up the world with dispatches of their discoveries on one day, and the next become the laughing stock of an intelligent community.

I did not think that the above phenomenon was anything but of a meteoric nature, and coupled with this belief, which I at present adhere to, it would have been ridiculous for me or any one else to have made a great outcry. The publication of my letter to the Tribune, and the message sent by me to Prof. Hall, were both urged by a personal friend, whom I called into the observatory to see the object, otherwise it would this day only be known to my immediate friends; the coolness with which my dispatch was received at the Washington Observatory (as it was no little trouble for me to send it so late at night, living over a mile from the telegraph office) has compelled me to regret my publicity on my part. (The presumption of astronomers that I had discovered Brorsen's comet ought to have been abandoned immediately, from the fact that the motion of Brorsen's comet was a little over a degree per day, whereas this object moved with a rapidity of over 2 minutes in R. A. to one minute of time, and on its path passed the above named comet by about 4° at 9 h. 10 m. local time, having then already moved 78 m. of R. A. since I first saw it.) There is one fact, however, which reconciles me to it, and that is the fact that my object was also seen by Mr. J. Spencer Devoe, of Manhattanville, N. Y., who published a letter to that effect a few days afterward.

ON AN OBJECT SEEN NEAR COMET B, 1881, ON JUNE 10, 1881.

Bone, W.; *Royal Astronomical Society, Monthly Notices,* 42:105, 1882.

On June 10, 1881, whilst measuring the position of the Comet, then visible here at 5^h 52^m mean time of place, I noticed a peculiar discordance in each succeeding measure, and at length found that the star (?) from which I was measuring was a rapidly-moving body. At first I was inclined to believe it the result of refraction, but this should have affected both Comet and star nearly equally. On more careful inspection I found it was somewhat discoid, but its light, although bright, was diffused and hazy. It moved through 6' of arc in 34^m 34^s of time, in a northerly direction. I immediately telegraphed down full particulars to the Melbourne Observatory, and asked for instructions. Bad weather prevented me from searching for it next morning, and in the evening I could not succeed in again picking it up, neither could I find it where seen on the preceeding evening. I never received any answer from the Melbourne Observatory; but when in Melbourne a few days since I called there, and on reminding Mr. E. J. White of the circumstance, he said that Dr. Gould had stated he saw the nucleus split into two about that time; but I have since ascertained that it was so observed at Cincinnati on July 6.

This struck me as so remarkable that, acting on the principle of your society, "Quicquid nitet notandum," I determined to send you my record of the observation.

BARNARD'S "UNEXPLAINED OBSERVATION"

Ashbrook, Joseph; *Sky and Telescope*, 15:356, 1956.

Let us watch Barnard at work in the 36-inch dome on the night of August 12-13, 1892. There was the curious triple nebula NGC 6302, which Barnard had discovered 12 years earlier with his 5-inch refractor while comet hunting, and now wished to draw. Then he turned the telescope on Mars, to make micrometer measurements of its satellites. Next he observed Jupiter, timing an occultation of its second satellite, and examining the great red spot. Finally, in the morning twilight, Barnard turned to the planet Venus. In his own words:

"I saw a star in the field with the planet. This star was estimated to be of at least the 7th magnitude. The position was so low that it was necessary to stand upon the high railing of a tall observing chair. It was not possible to make any measures, as I had to hold onto the telescope with both hands to keep from falling. The star was estimated to be 1' south of Venus and $14^s \pm$ preceding, at 4^h 50^m a.m. Standard Pacific Time.... [From this, Barnard deduces the position 6^h 50^m 21^s, $+17^o$ 13'.6 for 1855.0.]

"There seems to be no considerable star near this place, and the object does not agree with any BD star.

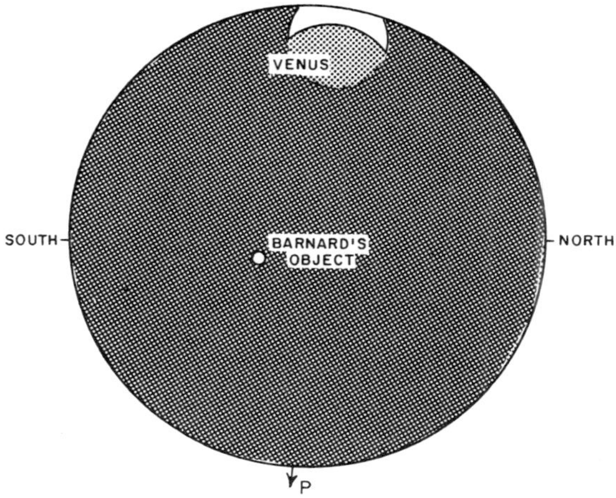

Barnard's Object near Venus through the Lick 36-inch reflector

"The observation was made in broad daylight, a half hour before sunrise."

Barnard could not identify the object. It was not an intramercurial planet, for the field was 38 degrees from the sun. The bright asteroids Ceres, Pallas, Juno, and Vesta, were all elsewhere in the sky, he pointed out. The starlike appearance of the image ruled out the possibility that it could have been a ghost of Venus.

Could there be a mistake in the date of his record, so the position of the object, referred to the moving planet, could now be altered to correspond to a known star? No this doubt is definitely removed by Barnard's records of other phenomena timed on the same night.

The account appears watertight, and the only reasonable assumption is that the 7th-magnitude star seen by Barnard was a nova, as F. J. M. Stratton has suggested. Appearing in the daylight sky, even a bright nova could have escaped ordinary detection, fading to inconspicuousness before this part of the sky again became accessible by night. The position observed by Barnard lies in the Milky Way in Gemini, a reasonable location for a new star.

. . . .

HAS THE MOON A SATELLITE?

Anonymous; New York *Times*, 4, July 23, 1896.

Geneva, July 22.---Prof. William R. Brooks, director of Smith

Observatory, while observing the moon last night with the large telescope, made a most interesting and unique discovery. A dark round object was seen to pass rather slowly across the moon in a horizontal direction. Prof. Brooks believes that it was the passage of a dark meteor between the earth and the moon far beyond the earth's atmosphere, so that it remained non-luminous. The observation is now in astronomical records.

NOTE FROM DR. L. SWIFT

Swift, L.; *Astronomical Journal,* 17:8, 1896.

On the evening of Sept. 20, I turned to look out of the window, some 20 feet distant, to see the sun descend behind a spur of the Sierra Madre range, about one-quarter of its diameter having disappeared. Instantly I noticed a luminous object about one degree above the sun. Thinking it was caused by refraction from a knot in the glass, I moved my head, but found the object remained stationary. Going out on the veranda the object was seen more distinctly. At first it occurred to me that it might be a small fire on the mountain, but this idea was quickly dispelled, as one-half of the sun's disk was still above the mountain, and the object still higher. Seizing an opera-glass I saw that it had a very much fainter companion some 30' north, but it could not be seen without the glass. In about four minutes after the sun had set the two objects also disappeared behind the mountain.

Both objects were seen by some fifteen people. Yesterday morning I attempted to refind the bright one with the 4-1/2-inch comet-seeker, but failed, as the sun is not visible from the Observatory until 15° above the horizon. Last evening I began a search with the comet-seeker with the sun 10° in altitude, but nothing was seen until fully one-half of the sun had disappeared when I caught sight of it, but its faintness surprised me, for it was less bright through the telescope than with the naked eye the previous evening, though it is possible, and, perhaps probable, that it was the companion that was seen.

I only saw it for 5 seconds, and rushing into the Observatory to get a view of it with the 14-inch which I had previously set to near the place, I found the eye-piece was beyond my reach. Repairing to the comet-seeker I found both sun and object had set. I estimate that the object disappeared simultaneously with the sun's upper limb.

The only thing that perplexes me about this strange affair is, that the sun, object, and companion, were not in line, but must have deviated from it by 30° or more.

A REMARKABLE OBJECT IN PERSEUS

Espin, T. E.; *Royal Astronomical Society, Monthly Notices,* 58:334-335, 1898.

On the night of January 16, while sweeping for red stars and stars with remarkable spectra, I passed suddenly from the starry background into what appeared to be a cloud. Although the night was very clear, yet I felt convinced it was a cloud, and continued my sweep. At the end of the sweep the telescope, as is usual, was moved 40' south and the return sweep made. I again came upon the peculiar obscuration. My suspicions were aroused, as it seemed strange that on a clear sky there should be a small cloud, and that stationary. I waited twenty minutes, and then re-examined the object, and found it still there. I marked it on the B. D. charts, and next morning turned to the New General Catalogue of Nebulae and the Addenda, but could not find it. On January 24 I re-examined it, and by rapid sweeping laid down its limits. They are as follows:

In R. A. from $4^h\ 23^m\ 30^s$ to $4^h\ 28^m\ 30^s$
Decl. " $+ 50^o\ 15'$ " $+ 51^o\ 14'$ 1885.

This gives as its centre:

P. A. $4^h\ 26^m\ 0^s$. Decl. $+ 50^o\ 44'.5$ (1855).

It was observed again on January 25. The blotting out of the stars was very marked, and the object seemed more remarkable than ever. It is elliptical, major axis $P = 336^o$. It was also observed on February 16 and February 17. An attempt was made to photograph the region on January 24, but the night was very unsteady, and though the exposure was carried on for two hours, the plate only gives stars to the 12th magnitude. N. G. C. 1624, which had been picked up independently, had left a trace on the plate, but there was no trace of the new object. A photo was again attempted on February 17, but at the end of twenty-five minutes it clouded up. Meantime I had written to Mr. Heath at Edinburgh, and he informs me that Dr. Halm found it on February 17, with the 6 in. refractor, without much difficulty. Mr. Heath also saw it, and says: "To both of us it conveyed the impression of an attenuated cloud-like object or haze, producing a difference of colour from that of the neighbouring sky, and, as you remark, dimming the tiny stars which appear to shine through it."

SIGNS OF THE SKY

Markwick, E. E.; *English Mechanic*, 86:100, 1907.

The "Cambrian Natural Observer" for 1905---an excellent little publication which is specially devoted to the consideration of phenomena such as are dealt with in these articles---(at p. 33) describes the curious phenomenon of a dark object seen traversing the sky above the town of Llangollen. It was intensely black, about two miles above the earth's surface, and travelled at the rate of about twenty miles an hour. Could it have been a balloon? It was seen on September 2, 1905.

SOME ASTRONOMICAL CURIOSITIES

Gore, J. E.; *Scientific American Supplement*, 67:362-363, 1909.

On November 29th, 1905, Sir David Gill observed a fireball with an apparent diameter equal to that of the moon, which remained visible for five minutes and then disappeared in a hazy sky. Observed from another place, Mr. Fuller found that the meteor was visible two hours later! Sir David Gill stated that he does not know of any similar phenomenon.

During the night of July 21st-22nd, 1896, Mr. William Brooks, the well-known astronomer, and director of the Smith Observatory at Geneva (New York), saw a round dark body pass slowly across the moon's bright disk, the moon being nearly full at the time. The apparent diameter of the object was about one minute of arc, and the duration of the transit 3 or 4 seconds, the direction of motion being from east to west. On August 22nd of the same year, Mr. Guthman (an American observer) saw a meteor crossing the sun's disk, the transit lasting about 8 seconds.

A meteor which appeared in Italy on July 7th, 1872 was shown by Prof. von Niessl to have an ascending path toward the latter end of its course! The length of its path was computed to be 683 miles. When first seen its height above the earth was about 42 miles and when it disappeared its height had increased to about 98 miles, showing that its motion was divided upward!

BRIGHT OBJECTS OBSERVED NEAR THE SUN

Steavenson, W. H.; *British Astronomical Association, Journal*, 25:36-37, 1914.

Though the phenomenon described in this paper was probably not astronomical but purely terrestrial in origin, the excuses pleaded by the writer in bringing the matter forward are that the appearances were encountered in the ordinary course of observation; that similar observations have from time to time been reported by other observers, especially within recent months; and that a satisfactory explanation of them is still lacking.

At about midday on 1914, September 9, the writer's 3-in. equatorial had just been pointed at Mercury, then about 8° from the Sun, when a round bright object, about the apparent size of Mercury, but much brighter, suddenly appeared at the S.f. margin of the field, passed centrally across the latter, and left it at the opposite, or N p, side. The object took approximately three seconds to cross the field, which was 25' in diameter, the power in use being one of 120. A few seconds later another appeared, following the same position angle as regards the direction of its motion, though it did not pass quite centrally across the field. After this more appeared, and the display was kept up, at the rate of about one every five seconds, till at least three o'clock, when the sky became overcast. In the three hours during which they were seen several hundreds were observed, and, as their motion was not excessively rapid, the writer was enabled fairly easily to note their chief characteristics, which were as follows:---

The average time taken to cross the field centrally was about three seconds; some took five or six, others no more than one, or even half a second. Their apparent diameter averaged about 5''. The smaller ones were 2'' or 3'' in diameter, while the largest were 8'' or 10''. Most of the latter required a slight racking out of the eye-piece (not more than one-tenth of an inch), in order to bring them into focus; but the smaller ones all focussed in infinity, as was of course the case with Mercury. In shape about half of them were perfectly round, but the rest were elongated with thickened ends, being somewhat dumb-bell shaped. It was particularly noted that the direction of their elongation was always at right angles to a line drawn to the Sun, which suggests that the elongation observed was due to phase. All the objects were well defined, and the prevailing colour was yellowish white. The vast majority moved in one position-angle, namely, 135°-315° ±. None of them varied more than 45° from this direction. The telescope was pointed for several minutes three or four degrees east and west of Mercury, but no bright objects were seen. All the objects observed were intrinsically a good deal brighter than Venus, and, of course, far outshone Mercury.

After observing the display for about an hour the writer sent a wire to the Rev. T. E. R. Phillips at Ashtead. Unfortunately, however, clouds arrived a minute or two before the wire, so that he was unable to attempt a confirmation of the observation, except at a point some distance from the Sun, where, like the writer, he encountered nothing unusual. On the following day the writer observed Mercury for several hours, but saw nothing of the kind seen the day before.

. . . .

[See also pp. 271-272 for "telescopic meteors."]

A LUMINOUS OBJECT SEEN ON MAY 4, 1916

Perrine, C. D.; *Astronomical Society of the Pacific, Publications,* 28: 176-179, 1916.

About 9^{h} 10^{m} in the evening of May 4th (Cordoba mean time) Dr. Glancy called my attention to a peculiar object in the southeastern sky.

It was then a bright streak just below a Pavonis, sensibly straight, about 8° or 10° long and one-half to one degree in width. It was more sharply defined toward the west, that extremity resembling the head of a large bright comet, but without any well defined condensation or nucleus. In general appearance to the naked eye and with the exception of a changed position it was an exact counterpart on a smaller scale of Halley's comet when this object was near to the Earth and its head near enough to the horizon to suffer greatly by absorption from the atmosphere.

After a few minutes' examination with field glasses a series of pointings was made as long as it could be seen with the finder of the 12-inch equatorial. These pointings were upon the western end of the object with respect to terrestrial objects, or the eastern end with respect to the stars.

These observations are as follows:

Cordoba Mean Time	R. A.	Decl.	Remarks
$8^h\ 52^m$	$20^h\ 18^m.5$	-57°	Casually observed and set down from memory.
9 34	21 02 .3	-55 .8	3 pointings.
9 46	21 21 .2	-55 .5	2 pointings.
9 58	21 27	-55-1/4	1 pointing, very uncertain.

The first observation was made by Dr. Glancy; the last three by Dr. Perrine.

When first seen (estimated to be at $8^h\ 52^m \pm 2^m$) by Dr. Glancy, the western extremity of the object was in apparent coincidence with the 2d magnitude star a Pavonis. It was soon seen that it was approaching the horizon and would shortly be lost to view---the diurnal motion not sufficiently counteracting its own motion.

In the intervals between points it was studied with field-glasses and with the naked eye. It grew fainter as it approached the horizon. After it had disappeared a low-lying bank of haze or cloud was observed at that point. During the earlier stages of its apparition there seemed few or no signs of streamers which are often observable in the tails of comets. At about the middle of its period of visibility when it had grown perceptibly fainter the portion midway looked, with field-glasses, like the main streamer of the middle portion of a comet's tail. It retained its rectilinear form as long as it could be seen.

There were no clouds, with the possible exception mentioned above, in any part of the sky at any time during the apparition of this object nor at any time during the night in question so far as I know. The sky was of its usual transparency and work was in progress with the astrographic telescope (on chart plates) and with the meridian circle as well as earlier with the 12-inch equatorial with which Dr. Glancy observed Neujmin's comet.

The appearance of the object was not that of an ordinary aqueous cloud. I think there can be no doubt that it was a self-luminous object.

There were no means of photographing it or of examing its spectrum in the short time available.

Upon plotting the observed pointings including the estimated position by Dr. Glancy at $8^h\ 52^m$ it is found that they all appear to fall upon a great circle passing thru the Sun within their probable errors. The observed motion was 10° on a great circle toward the Sun in the interval of one hour.

Explanations of the nature of this object seem to be limited by the circumstances to two:---a, that it was a comet traveling very close to the Earth and toward the Sun or b, that it was the cloud left by a meteor.

The appearance is better explained by the assumption that it was a comet. There were no changes of form such as usually occur in meteor clouds---the object was sensibly straight and streamer-like thruout its entire apparition. The extremity nearest the Sun's place was as sharply defined as, and similar in shape to, the head of a large comet close to the Earth, while the opposite portion faded out gradually like the tail of a comet. There was no pronounced condensation and so far as I could determine no nucleus altho one stellar object of apprximately sixth magnitude was suspected to have motion. It was not possible under the circumstances, however, to verify the fact. The axis of the object was at first nearly parallel to the horizon, but later became more inclined, the west (terrestrial) end being nearest the horizon.

The tremendous geocentric motion demanded, while not impossible, would be, nevertheless, very unusual.

Some time after the apparition of this object several meteors were noticed, one at least of which was visible in the southeastern sky, coming from a radiant to the north and perhaps 30° or 40° above the horizon.

The observed motion extrapolated would have carried the object almost to the Sun's place at sunrise. There seemed a possibility, however, that it might slow up and that it would be visible before sunrise. It was therefore, carefully looked for, in a good sky, but nothing seen of it.

Considering all the circumstances, particularly that its motion would carry it rapidly into the northern sky if it were a comet, it was deemed best to warn northern observers. Hence the following telegram was sent to Harvard College Observatory:

> May 5, 1916.
>
> Bright object visible here last night nine to ten. Moved from alpha Pavonis ten degrees toward Sun's place. Tail. Possibly a comet.

It is not considered, however, that its nature can be established without observations from other points. Requests for observations were sent to the newspapers but no reports have been received.

Dr. Glancy has investigated the question of an orbit for the positions, which is also interesting from a theoretical standpoint on account of the possibility of such a close approach, and her results follow.

NOTE ON A SOLAR OBSERVATION

Markwick, E. E.; *British Astronomical Association, Journal,* 27:188-189, 1917.

I am rather interested in a spot observed on the Sun on April 5. On that date at 6^h 10^m p.m., G.M.T., using a hand draw telescope with object-glass 1.5 inches in diameter, and low power microscope inverting eye-piece magnifying about nine times, or perhaps a trifle more, the telescope being mounted on a small pillar and claw stand, a small spot was seen, not quite half way from the centre to the vertex, as shown on the accompanying diagram. The orientation of this is only approximately correct. As far as I could make out this spot was absolutely solitary on the disc; I could not trace any outliers or dots in its vicinity. It was very small, greyish, not quite black, and, so far as could be judged, circular. With 43 years' experience of Sun-spots I have always found that even a small solitary dot, such as this, generally gives some evidence, such as a slight irregularity in shape, or the presence of faculae, or other dots in the vicinity, that it is part of the photosphere and not a foreign body seen projected on the disc. But in this case, possibly owing to my present very limited optical resources, I gained the impression that this spot might have been something other than a Sun-spot. It was watched for some time, and regarded as a min. vis. for the telescope used, in the way of a dark spot on a bright background. I may say that the definition on the Sun with this small glass is wonderfully sharp and clear.

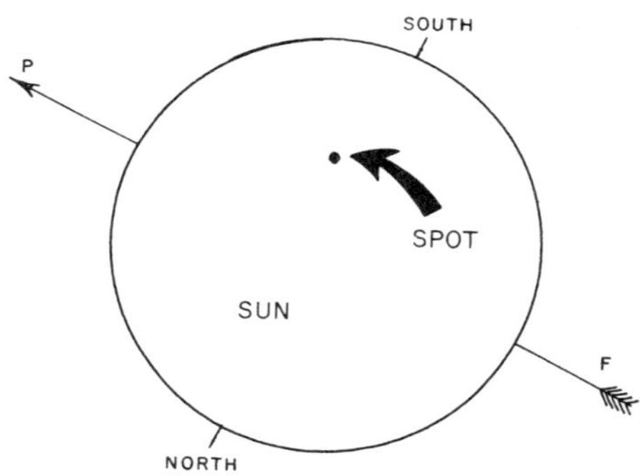

Peculiar, perfectly round solar spot of Arpil 5, 1917

On the previous day, April 4, at $5^h\ 30^m$ p.m. with the same glass, after a careful survey, there was not the slightest sign of a spot on the solar disc; and, again, on April 6, at $9^h\ 34^m$ a.m., there was the same state of things as on the 4th, not a sign of a spot. The spot described above must therefore (if it belonged to the solar disc) have burst out and died away between the dates named.

By the courtesy of Mr. E. W. Maunder, I am able to state that there was no sign of a spot on a photograph of the Sun taken at Greenwich on April 5, at $11^h\ 17^m$, or on another taken at $11^h\ 20^m$. This still further limits the period of visibility of the spot.

My chief object in reporting this observation is to ascertain whether any of our members observed the spot in question, and so definitely settle the matter as to its nature. I can quite imagine that in the early days of the telescope, and much later even, a spot of this nature might well have been taken for an intra-mercurial planet, or other body, between us and the Sun. The time available for its observation in my case was too brief to enable any change in position to be noticed. Finally, I may add that the possibility of its being a speck of dust or dirt in the eyepiece is ruled out by the fact that as light clouds passed over the Sun's disc, the spot faded away, to reappear again when he shone clear once more. Also the spot kept its place on the disc, however much the latter shifted, or was shifted in the field of view of the telescope. At the time of observation there were horizontal strata of cirrus cloud about, with clear gaps between.

OBSERVATIONS OF AN UNIDENTIFIED OBJECT SEEN NEAR THE SUN ON SUNDAY, AUGUST 7, 1921

Campbell, W. W., et al; *Astronomical Society of the Pacific, Publications,* 33:258-262, 1921.

The sunset phenomena on Sunday evening, August 7th, were especially interesting, and Captain Eddie V. Rickenbacher, Major Reed Chambers and Mrs. Chambers, Professor Henry Norris Russell, director of the Princeton Observatory, New Jersey, Mrs. Campbell and Director Campbell, were observing them from the porch of Mr. Campbell's residence. There were highly colored clouds at considerable altitudes in the western sky, and a shallow stratum of thin clouds adjoining the horizon, with clear sky between the two cloud areas described.

As soon as the Sun's lower limb touched the apparent horizon, which consisted of haze, smoke and light clouds, the Sun's image began to pass thru the interesting series of geometrical figures so frequently observed here, especially in the summer months. Starting with the ellipse, the succeeding figures resembled the oldfashioned Rochester lamp shade, the straw hat now in vogue with men, an extremely elongated ellipse, the cigar form, and finally, at disappearance, the knitting needle form. The last figure endured for at least a minute of time. Just as this figure was disappearing Major Chambers said, "What star is that to the left of the Sun?" Captain Rickenbacher said that he had been watching that star for several minutes, but had not mentioned it because he supposed it was well known. Major Chambers said that he had seen it at least a minute or two and had not spoken of it for the same reason. Mrs. Chambers, Professor Russell and Director Campbell saw the starlike body immediately. Mrs. Campbell went into the house to secure a pair of binoculars. Within a minute she had handed them to Director Campbell, who saw the object in the binoculars not more than two seconds before it disappeared behind the horizon cloud stratum.

The five observers agreed that the object was starlike. Director Campbell remarked that it was probably Mercury, but Professor Russell said at once that it was entirely too bright for Mercury. Shortly following its disappearance the observers compared notes as to the position of the object and as to its color. Captain Rickenbacher said it was six solar diameters from the Sun, and that the line joining Sun and object made an angle of 45 degrees with the vertical line thru the Sun's center. Major Chambers indicated graphically the angle, and his estimate agreed with that made by Director Campbell as 50 degrees southerly from the vertical. Russell and Campbell estimated the distance from the Sun's center at three degrees. Campbell thought that the color of the object was yellower than a star in that position would be, but it must be remembered that the Sun at disappearance had a true zenith distance approximating 93 degrees and that the true zenith distance of the object must have been between 91 and 92 degrees. Venus has frequently been observed to set below the western mountains at approximately that zenith distance, as a bright object and with color not distinctly orange or red. Russell and Campbell agreed that the object observed was brighter than Venus would have been if seen in the same position and circumstances as the strange body.

The observers had no reason to doubt that the object was genuinely celestial, as in the binoculars its diameter still seemed stellar. It seemed to partake of the diurnal motion of the stars in that it moved down toward the lower cloud stratum and disappeared behind it. To imagine an object in our own atmosphere which would have supplied all of the phenomena described seemed entirely too difficult.

There was no further opportunity to study the object. After dark Professor Russell, from the Nautical Almanac, determined the position of Mercury to have been below and to the right of the Sun. All of the other bright planets were likewise accounted for. The question of notifying other observatories by telegraph was carefully considered. What could the object be? Its brightness, as seen on the sunset sky, its position close to the Sun and its sudden appearance as an unobserved object, suggested strongly the nucleus of a comet which had approached the Sun in such a direction as to escape notice. There seemed no alternative except that of a new star, such as, but much brighter than, the one which suddenly appeared in June, 1918, in the constellation Aquila. Clearly it was out duty not to withhold information from astronomers in other parts of the world, awaiting confirmation on Monday evening, but the facts should be communicated, in order that no opportunities for observing the strange body should be sacrificed. A message, composed by Professor Russell and Director Campbell, read as follows:

> Star-like object certainly brighter than Venus three degrees east, one degree south of Sun seen several minutes before and at sunset by naked eye. Five observers. Set behind low clouds. Unquestionably celestial object. Chances favor nucleus bright comet, less probably nova.

Careful search for the object at sunrise Monday morning, sunset Monday evening and at sunrise Tuesday morning was unsuccessful. If the object is a comet, its position could not be predicted because the orbit is totally unknown. It might be too near the Sun, either between Sun and Earth or on the far side of the Sun. If the object was a new star, it should have been observable Monday evening unless the brightness was greatly reduced. In the latter case there will be little chance of observing the object in the bright sky surrounding the Sun, for several weeks to come. The position of the object, R. A. $9^h 22^m$, Dec. 15-1/2°, about 40° from the galactic plane, militates against the nova hypothesis. (W. W. Campbell)

Letters Relating to the Preceding Note

(The following letter is reprinted from The English Mechanic, issue for August 19, 1921, page 49, copy having been supplied thru

Miss Calvert and Professor Barnard.)

NEW STAR, OR COMET, OR WHAT

A few days ago the statement appeared in our newspapers that an observer at the Lick Observatory had seen, on the evening of August 7th, a bright star near the Sun, and that it was visible without optical aid. My experiences on that evening may be worth recording. About 8:30 on that Sunday evening I noticed a bit of very clean sky over the place of the recently set Sun, and thinking I might pick up Jupiter and Saturn, I took my binoculars (power about 3) and commenced to search. I soon alighted upon a bright object, which I at the moment thought was Jupiter, but the next moment I saw was not a planet at all; neither did I think it was a star. It was elongated in the direction of the Sun, and was of a distinct reddish tinge. I judged it to be about 6° from the Sun and a very little south of him. It was unfortunate that I was not able to get the telescope on the object, as it was too low. I only held it for about three minutes, as clouds came and hid it.

If the object which I saw was the same as that seen by the American observer, and if our estimates of its position were fairly accurate, then the object must have passed inwards towards the Sun some 3° in five or six hours. It must also have greatly increased in brightness in that time, as the object which I saw could not be compared to Venus. (S. Fellows)

(The following letter has been received thru Professor S. I. Baily, Director of Harvard College Observatory.)

Dear Sir:

Reading an article by Dr. Russell in Scientific American for September 6, 1921, about the new object observed at Lick Observatory, I wish to give you my observation of August 6, 1921, 5:50 p.m. Eastern standard time. Having been deeply interested in astronomy for the last twenty-five years I casually glanced at western sky at that time and saw a very bright (silvery-white) object. Not being familiar with the position of Venus on that date I thought it was Venus, and yet it shone altogether too bright for Venus. *****The Sun at this time was obscured by buildings and the object was five to seven degrees east of Sun, and would say as many degrees south of ecliptic, the Moon being in sky at the time. At this hour in broad daylight this was a striking object. Did not scintillate. This must have been the same object seen at Lick Observatory three or four hours later, although they report it much nearer to Sun. This may be due to an enormous speed. A relative, Miss Crow, of Jackson, Michigan, asked me the name of the very bright star setting in the west about 7:30 p.m. the same date; this fixes the observation and date clearly in my mind. (H. C. Emmert)

A cablegram purported to be dispatched from Heidelberg, Germany, and printed in the American newspapers about the 10th of August, stated that at Konigstuhl the Earth had been observed to pass thru the tail of a comet on the night of August 8th. Inasmuch as a message to this effect has apparently not been sent out by Dr. Max Wolf, Director of the Konigstuhl Observatory, it seems that the report is entitled to small weight. The sender may have been the Wolff News Agency rather than Professor Wolf.

A postcard received from Dr. Max Wolf at Mount Hamilton on September 22, 1921, reads as follows: "Your bright object at sunset

August 7th has been observed too at Plauen (lat. = + 50° 30', long. 12°7'E. of Gr.), Germany: August 7th, $7^h35^m.0$ Gr. M. T., in R. A. $11^h6^m.7$ and Decl. +7°9'." At the time and place stated the Sun would still be above the horizon and the object referred to must have been very bright to be visible.

In the English Mechanic for September 2, 1921, the astronomical editor, Colonel E. E. Markwick, writes:

> I am sure all readers will congratulate Mr. S. Fellows on his wonderfully interesting and lucky observation of the mysterious object seen near the Sun on August 7th. Taking his observation in conjunction with that made at the Lick Observatory, there is, to my mind, no reasonable doubt that it was a brilliant comet that was seen.
>
> Professor E. E. Barnard writes me as follows:
>
> The case of this comet is not exceptional, as was shown by the Tewfik comet at the total eclipse of May 17, 1882, which was seen and photographed in Egypt, near the Sun (during total eclipse). I, among others, hunted faithfully for that comet morning and night for a long time. Of course if the present comet was a sungrazer it would move in a very narrow orbit, the whole of which might lie in daylight. It may be discovered yet in the southern hemisphere. (W. W. C.)

The astronomical column in Nature, September 8, 1921, reports an additional observation made by Lieutenant Day and others at Ferndown, Dorset, England, on Sunday, August 7th, about 7 p.m. Greenwich Mean Time. The object was estimated to have magnitude minus 2 and to be 4 degrees distant from the Sun.

The editor of Nature comments that a comet of retrograde motion, moving nearly in the plane of the ecliptic, having small perihelion distance, and approaching the Sun from behind, might remain in close proximity to the Sun during the whole time that it was bright.

A further note in Nature makes it probable that the "comet's tail" reported from Jonigstuhl as observed on the night of August 8th was an auroral streamer.

No other observations worth quoting have come to my notice. There appears to be no reason to doubt that the object was truly celestial. (W. W. C.)

THE BRIGHT OBJECT NEAR THE SUN

Anonymous; *Nature*, 107:793, 1921.

Three of the five observers of this object, referred to in last week's issue, p. 759, were Prof. Campbell and his wife, and Prof. H. N. Russell, who is staying at the Lick Observatory. The object was seen shortly before sunset; the fact that it partook of the diurnal motion indicated that it was a celestial body. Prof. Campbell observed it with binoculars, and noted that it still appeared stellar, which favoured its being a nova. If so, it is probably the most brilliant since that of Tycho Brahe. The approximate position is R. A. 9h. 22m., N. decl. 16°. The galactic latitude is about 40°. The object does not appear to have been seen since August 7. It may be recalled that the great 1882 comet and that of January, 1910, were seen close to the sun.

A report from Konigstuhl Observatory, near Heidelberg, states that on the night of August 8-9 a number of luminous bands lay across a clear sky from W. N. W. to E. S. E.; they moved slowly towards N. N. E., growing paler as the dawn came. It was conjectured that it might be the tail of the light object seen at Lick Observatory on August 7, passing very near the earth. It will be recalled that a somewhat similar phenomenon was reported when the earth passed through the tail of the great comet of 1861 on June 30 of that year.

It seems possible, however, that the present streamers may have been auroral, as the cometary nature of the Lick Observatory object is still in doubt.

THE BRIGHT OBJECT NEAR THE SUN

Anonymous; *Nature*, 108:69, 1921.

Two observations of this object were made in England on August 7, some hours before that at Mount Hamilton. The first (communicated by Col. Markwick) was made by Lieut. F. C. Nelson Day and others at Ferndown, Dorset, about 7^{h}. G. M. T. Its magnitude was estimated as *minus 2* and its distance from the sun as 4°.

Mr. S. Fellows observed it at Wolverhampton with binoculars shortly after sunset (*Eng. Mech*. August 19). He noted it as reddish, elongated towards the sun, from which it was distant 6°. The estimates of distance are probably too rough to use for the deduction of motion. It may be noted that a comet with retrograde motion near the plane of the ecliptic and small perihelion distance approaching the sun from behind might remain in close proximity to it the whole time that it was bright.

Astr. Nach. No. 5116 contains full particulars of the luminous bands seen at the Konigstuhl and Sonneberg Observatories on August 8d. 12h. G. M. T. It appears probable that they were auroral, especially as Prof. M. Wolf noted a similar appearance on August 5, 11h. 15m. to 11h. 36m. G. M. T. It was "a long, very bright cloud west of the Pleiades, brightest near Arietis....it faded rapidly, only a trace visible at 11h. 36m.

THE BRIGHT OBJECT OF AUGUST 7

Anonymous; *Nature*, 108:98, 1921.

Astr. Nach. No. 5118 contains a further observation of this object made at Plauen, Vogtland, by the daughter of Prof. E. Kaiser and several others. It appeared like Venus at its greatest brilliancy, low in the evening sky shortly after sunset. Its position with regard to distant terrestrial objects was accurately noted. At G. M. T. 7h. 35m. its azimuth was 98° 22.6' from south to west, and its apparent altitude 2° 35.9', or 2° 21.9' corrected for refraction. Plauen is in N. lat. 50° 29' 45", long. E. Greenwich 12° 7' 11"; Prof. M. Wolf deduces that the R. A. of the body was 11h. 6.7m., N. decl. 7° 9'. This gives longitude 165.0°, N. lat 1° 20'. It may be mentioned that the place of Jupiter was R. A. 11h. 23.9 m., N. decl. 5° 6', but it does not seem possible that this could have been the body observed.

The Lick Observatory position at G. M. T. 15h. was about R. A. 9h. 22m., N. decl. 15.8°, or longitude 138°, N. lat. 0.4°.

It is to be noted that the object observed in England was much closer to the sun (estimates of distance 6° and 4° respectively) than the Plauen object, so that an element of doubt remains. But the latter observation was made in a much more exact manner, so it deserves greater weight. If we assume the identity of the Plauen and Lick objects, and that the motion was parabolic, then the maximum distance from the earth on August 7 was 0.005 in astronomical units, or about twice the moon's distance. It appears unlikely that a comet at this small distance would have such a well-defined stellar appearance.

THE BRIGHT OBJECT NEAR THE SUN

Anonymous; *Nature*, 108:193, 1921.

Dr. W. Bell Dawson writes from Ottawa stating that he saw a bright object low in the west in unusually clear air just after sunset on September 4, which he assumes to be the same as that observed at Mount Hamilton on August 7. The identity seems unlikely; Dr. Dawson's object may have been Mercury (R. A. 11h. 36m.: N. Decl. 3° 38'), or Jupiter (R. A. 11h. 45m.: N. Decl. 2° 48'; the Sun was in R. A. 10h. 53m.: N. Decl. 7° 6'). However, it is well to put the observation on record, the mystery of the Mount Hamilton object being still unsolved.

WILK'S COMETARY OBJECT

Anonymous; *Popular Astronomy*, 34:538-539, 1926.

A cablegram from Professor E. Stromgren announced the discovery by Wilk, at Cracow, of a comet in the following position: September 1.9069 U.T., R.A. $15^h53^m12^s$, Dec. + 3° 55'. Magnitude 6. "Direct motion one degree in four minutes." A radiogram was received at Harvard College Observatory, direct from Cracow, confirming the announcement.

We had no success here in locating this unusual cometary object. The magnitude 6 given by the discoverer placed it at the limit of naked-eye visibility near e Serpentis. But the unusual remark "direct hourly motion 15°" showed at once that this was a very uncommon appearance. In fact on receiving the message in the evening of September 2, I thought there was some erroneous interpretation and that 15° per day instead of per hour was already exceptional enough. Through haze and clouds I therefore exposed a pair of plates centered about 20° east of e Serpentis with a wide-angle Ross lens. Only stars down to 8^M were recorded but no unexpected object was noticed. Cloudy weather interfered with further search the next few days but in the meantime the rapid angular motion was confirmed by a card from the Harvard College Observatory. I do not think there is any record of a celestial object, other than a meteor, showing such a rapid apparent motion across the sky. At that rate it was next to impossible to tell in what direction the object would have to be expected on the following days. If we consider it as a comet passing close to the earth its speed must have been about 42 km/sec. Taking into account the component of the earth's velocity I made a rough estimate showing that the object must have been at a distance from the earth less than twice the distance of the moon" I had supposed the object to move at right angles to the line of sight; otherwise the distance had to be still smaller. Under that assumption it was found that after an interval as short as twenty-four hours the object would have slowed down almost to a standstill somewhere in Aquila, but that at the same time it would have decreased in brightness by at least five magnitudes. On September 7 a plate was exposed for two hours by Messrs. Bobrovnikoff and Morgan on the region thus indicated but nothing suspicious was found.

Possibly more detailed information will enable conputers to predict more closely where the object was to be expected; unless this can be done no further search of the plates can be attempted with any chance of success.

The object reminds us of the strange cometary appearance that was noticed May 4, 1916, at Dordoba by Perrine and Miss Glancy (Pub. A. S. P., 28, 176, 1916). Its displacement amounted to 10° per hours. The object might of course have been an unusual form of meteor; some of these have sometimes been followed for several hours. I cannot refrain from thinking also of a terrestrial source as a conceivable explanation. It is curious to note in this connection that from the time given for the observation the altitude of the object is found to have been only 4° above the horizon of Cracow, and that any stationary source of light (pilot light of a captive balloon, automobile headlight on a mountain, mirage of a terrestrial light, or what not) would have shown relatively to the stars an hourly motion of 15° in the direction indicated by the discoverer.

But it may be better not to anticipate anything about Wilk's unusual discovery until further information about the circumstances of the observation are available. Such information is awaited with great interest.

A PECULIAR CELESTIAL APPARITION

D., P.; *British Astronomical Association, Journal,* 37:285-286, 1927.

In H. C. O. Bulletin, No. 845, Hertzsprung gives particulars of the examination of a photograph taken in 1900 with a 1-inch Cooke camera showing what he describes as the "most serious riddle I have met with during the examination of thousands of plates." The object in question shows as two round nebulous images, there having been two exposures of 13^m and $29^m\ 20^s$, taken with an interval between of 40^s, during which the driving clock was stopped. The first image is more concentrated towards the centre than the second one, which is larger in size. The position is R. A. $4^h\ 1^m$, Dec. + 59° 34' in the constellation Camelopardus.

Hertzsprung states: "It is hard to compare the intensity of the hazy objects with stars, but the images are easily visible with the naked eye on the plate and may correspond to a star of about the sixth magnitude.

"There can hardly be any doubt that with only one exposure on the plate the object would have been considered as an accidental spot on the plate. This again emphasises the desirability of having two identical instruments next to each other so that the two plates can be taken simultaneously of the same field. In the ordinary way of taking only one plate it is just the most unusual incidents that are likely to escape recognition.

"It seems hard to escape the conclusion that we have in the present case to deal with a celestial phenomenon. The fact that the character of the images, taken within less than an hour, is distinctly different, tends to show that the event cannot well have taken place at stellar distances, but is more likely to have occurred inside our own planetary system. The explanation that the phenomenon is due to a collision between two bodies would account for the short duration, but even not regarding the improbability, it seems unsatisfactory as the images are perfectly round.

"The images are about 38° from the ecliptic, and galactic latitude is about 6°. As no trace of a similar object has been found on plates taken either the night before or the night after December 15, 1900, the object cannot well be a comet."

As Hamlet says, there are more things in heaven and earth than are dreamt of in our philosophy!

[See also Sky and Telescope, 34:382-383, 1967.]

COMET OR ASTEROID?

Anonymous; *Science*, 95:sup 10, April 3, 1942.

The fast-moving object in the constellation of Leo, the lion, discovered on March 12 by Dr. Y. Vaeisaelae, of Turku, Finland, has been confirmed and photographed by astronomers at the Lowell Observatory at Flagstaff, Arizona. However, it is not yet known whether the new object is a comet or an asteroid.

A week after its discovery the Lowell observations, made by H. L. Giclas, show that the object is moving about one minute of arc every fifteen minutes, which is rather fast for an asteroid, especially since its apparent path is at right angles to the ecliptic and directly across the sky from the sun. If it is a comet it may have already come nearest the earth, for its motion is slowing down, which may indicate the object is going away. On the other hand, an asteroid (flying mountain) might appear to move this way, if it had an orbit of high inclination to the earth's orbit.

In a letter to Harvard Observatory, clearing-house for astronomical news in the western hemisphere, Dr. V. M. Slipher, director of Lowell Observatory, writes: "Herewith are two positions of the fast moving object, obtained by H. L. Giclas, of our staff. He photographed it both with the 13-inch search telescope and with our 9-inch Schmidt of 22-inch focus. On both negatives the images are not stellar but are somewhat diffused and have the appearance of comet trails."

The positions are: March 18 at 1:27 A.M. EWT, 11 hours 12 minutes 45 seconds; plus 11 degrees 41 minutes; March 19 at 2:08 A.M. EWT, 11 hours 15 minutes 3 seconds; plus 13 degrees 24.5 minutes.

This indicates that in the week after its discovery, the object has moved fifteen degrees or one twenty-fourth of the way around the sky. Its motion is just west of north. However, only large telescopes can see it, as it is of the thirteenth magnitude.

AN UNUSUAL SKY PHENOMENON

Anonymous; *Strolling Astronomer*, 9:48, 1955.

Under this title the leading article in Vol. 2, No. 25 of Vega, Mr. R. M. Baum's excellent publication, describes a really amazing observation by Mr. Harold Hill, Dean Brook House, Abbeylakes, Near Wigan, Lancashire, England. Whatever one may think of "flying saucers"---and it is perhaps difficult by this time to say much new about them---we here have a careful observation by an outstanding lunar observer of more than twenty years' experience in astronomical work. Mr. Hill will give his own opinion as to what was seen upon request. We here summarize from Vega the observation itself.

On July 8, 1954 near 20^h 30^m, U.T., at his observatory in Abbeylakes Mr. Hill suddenly noticed at an altitude of about 55° in the southern sky a bright, apparently stationary, star-like object. With 33X and 50X on a 12-inch reflector the "star" was resolved into a bright silvery object of indeterminate shape, accompanied to the left by a loose cluster, with perhaps 15 to 20 members, of what appeared to be minute stars of varying magnitudes. These were constantly moving about, a "hive of activity". After a short break caused by clouds, Mr. Hill saw two bright objects close together. The cluster had vanished, though perhaps merely outside the field of view of the telescope. "My attention was directed to the two bright objects which showed sensible motion. They were separating as I watched and not only appeared to be gyrating and flashing but also exhibited a slow pendulum movement about each other in a manner similar to dancing partners! The increased movements of the objects made following with the 12-inch a difficult affair, and they passed out of the field as an adjustment had to be made to the telescope." They were also lost to Mrs. Hill, who had been simultaneously watching with binoculars. Clouds ended the observation at 20^h 55^m, U. T. Mr. Hill's impression was that the objects were at a very great height. "The two principal members defied description, they flashed as they gyrated as though metallic surfaces were reflecting the rays of the Sun (now below the horizon). Probably the closest analogy to what was seen is to be found in the flashing of the facets of a diamond as it is turned in strong light."

AN OBSERVATION OF AN UNIDENTIFIED CELESTIAL OBJECT

Clark, Frank C.; *Strolling Astronomer,* 10:67-68, 1956.

The following observation has been contributed by Mr. Frank C. Clark, 210 E. Fleming, Las Cruces, N. Mex. As with other unconfirmed observations of unconventional objects, the true nature of what was seen is likely to remain in doubt. However, we think that this account should interest our readers; and it would be interesting to have their opinions on the interpretation of what Mr. Clark saw. We quote his report:

"It was about 9:30 P. M., M. S. T. on the night of September 8, 1956 (4^h 30^m, U. T. on September 9). I was observing the planet Mars through a 12-1/2-inch reflector at full aperture using a magnification of about 300 diameters without a filter. The driving clock was disengaged. The sky was partly cloudy but the planet could usually be observed in full brilliance for several minutes at a time between clouds.

"At this time a faint star-like object of about 11th magnitude was observed to pass through the field of vision in a direction opposite to the apparent drift of Mars and about 1 minute of arc below the lower limb of the planet. This corresponds to a true position of about 1 minute of arc above Mars and traveling east. The field of vision at the chord of transit was about 5 minutes of arc and the object was observed for about 10 seconds of time before passing out of the field. The telescope was moved so as to follow the object which once again was observed to transit the field in about 10 seconds. Mars of course was not in the field during this second observation. Once again the telescope was moved in an attempt to track the object. A glimpse was caught but only for an instant, possibly because of cloud. As far as could be ascertained the object was of a yellowish color.

"During 10 seconds of time the Earth rotates about 2-1/2' of arc. Since the field was 5' of arc, the object moved 7-1/2' of arc in 10 seconds of time or 1 degree in 1 minute and 20 seconds. Assuming a satellite, this corresponds to a period of revolution of about 7 hours and 48 minutes, a distance of about 8,500 miles from the Earth's surface and a mean orbital velocity of about 9,800 miles per hour.

"These orbital data were read from a chart on page 63 of the Bell Aircraft Co. booklet Pocket Data for Rocket Engines (1954). The distance from the Earth is altitude rather than geocentric distance, and therefore a foreshortening correction must be applied. However it is likely that this discrepancy is less than the probable error resulting from inaccuracies in the estimated time of field transit."

FAST-MOVING ASTEROID?

Anonymous; *Sky and Telescope*, 41:153, 1971.

On the morning of January 22nd, C. Roger Lynds of Kitt Peak National Observatory in Arizona had an unusual experience while working at the 84-inch reflector. The instrument was pointing toward the constellation Sextans when he became aware of a moving object that resembled a star of magnitude 10-1/2 or 11.

Dr. Lynds kept the object in view for several minutes as it traveled northeastward among the stars, passing within a second of arc of 9th-magnitude BD + 2°2309 at 10:47 Universal time. Its motion was roughly estimated as two seconds of arc per second of time. "The object was stellar in appearance and less than 1".5 in diameter," according to the announcement in Circular No. 2303 of the International Astronomical Union, where the body is tentatively thought to be a fast-moving minor planet passing near the earth.

An upper limit to the object's distance from Earth is easily derived of its orbital velocity is assumed to be 42 kilometers per second (the velocity in a parabolic orbit at one astronomical unit from the sun). The distance at the time of Dr. Lynds's observation would have been 4.3 million kilometers (about 10 times as far as the moon) if the asteroid were moving at right angles to the line of sight. If the motion were in a direction 30 degrees out of the line of sight, this upper limit would be halved.

CHRISTOPHER SCHEINER'S OBSERVATIONS OF AN OBJECT NEAR JUPITER

Ashbrook, Joseph; *Sky and Telescope*, 42:344-345, 1971.

On April 14, 1612, Scheiner wrote a letter to his patron Mark Welser, prefect of Augsburg, describing his telescopic observations of Jupiter and its satellites on nine nights between March 29th and April 8th. For each night he gave a drawing of the field of view of his telescope, two of which are copied here.

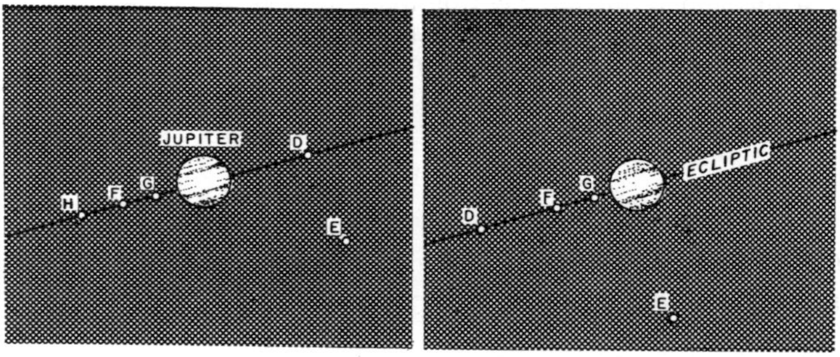

Scheiner's drawing of Jupiter and his satellites as seen at Inglostadt on March 30, 1612, at 9 P.M., D, F, G, and H are the Galilean satellites. E is the enigmatic object considered a fifth Galilean satellite by Scheiner.

....

The drawing that Scheiner made on March 29th shows only Jupiter and two satellites. But the next night all four were plotted and, in addition, a fifth object was seen, estimated to be about eight minutes of arc west of Jupiter and six minutes south. It was "very bright and large, like any of Jupiter's satellites."

This object, labeled E on the sketches, was again recorded on March 31, April 1, 3, 5, 6, 7, and 8, growing steadily fainter until on the last date it could be seen only with difficulty in a very clear sky. Scheiner looked for E again on later occasions but could not find it, although conditions were favorable.

From his rough estimates of the motion of E relative to the planet, Scheiner concluded that what he had seen was not a fixed star but a new satellite. Actually, it is difficult to concur in his reasoning, and it can be conjectured that the desire to make an interesting discovery guided him to that conclusion. At any rate, Scheiner's E cannot be any of the satellites of Jupiter that we now know. What was it?

....

Chapter 14

COMETS

INTRODUCTION

Comets are currently conceived as icy conglomerations that flare up as they approach the sun under the influence of the solar heat and solar wind. This conventional picture is contradicted by many maverick comets. Some develop brilliant tails at great distances from the sun; others have no discernable nucleus at all. Not infrequently, comets split; more rarely, they explode. They are definitely not well-behaved denizens of space.

Where do comets come from? Most astronomers suppose that a reservoir of comets exists outside Pluto's orbit, but a few nonconformists opt for the eruption of comets from the giant planets, particularly Jupiter.

The most famous comets are periodic. But few are as punctual as they should be if their journeys are unhampered in space. There must be some massive body "out there" beyond Pluto that perturbs cometary orbits---or so goes conventional wisdom.

The ancients believed comets to be important omens. Strangely enough, a few studies do relate the frequency of comets to the solar cycle and terrestrial events! The bits of celestial fluff that are comets could scarcely exert forces strong enough to affect the sun and earth---at least none of the forces we know.

ENCOUNTERS OF THE EARTH WITH COMETS

Lyttleton, R. A.; *Nature,* 245:144-145, 1973.

The proposal by Urey that tektites may originate through Earth-comet collisions has recently been extended to suggest that the energy released may be sufficient to produce periods of intense geological activity. But the hypothesis is crucially dependent on the picture adopted for the structure of comets, and Urey assumes that the fundamental permanent heart-and-soul of a comet is a solid nucleus of kilometric size, with activity of the comet resulting from solar radiation and possibly interaction with the solar wind. If that were so, such an object impacting the Earth would not produce essentially different effects from those by a meteorite or asteroid of comparable mass.

In spite of the publicity devoted to the icy nucleus model, observations of comets make it clear that the hypothesis is untenable. These considerations have recently been reviewed in detail, but it may be worth mentioning again some of the more inescapable instances. Comet 1927 IV remained visible out to 11.5 a.u., where solar heating would be negligible, and an object only kilometres in diameter would not be detectable at such distance. Comets do not appear as point-sources but expand up and grow fainter, merging with the background sky and vanishing through excessive diffuseness of their light. Again Comet 1729 had perihelion distance of more than 4 a.u. yet was visible to the unaided eye, and Comet Humason (1962 VIII) remained far outside the orbit of Mars yet developed a splendid tail. On the other hand, Ikeya-Seki (1965 VIII) developed a negligible tail when inside the orbit of Mercury. This difficulty was dealt with _ad hoc_ at the time by claiming that the comet was "worn out", without explaining how this could be known for a freshly discovered comet. But immediately after perihelion it produced a tail 20° long visible to the unaided eye, as was predicted on the dust swarm hypothesis.

Another telling consideration is that only about 60% of comets have shown signs of any nucleus, and even for these not at all times nor at every return. The angular sizes of some nuclei when measurable are far too large to correspond to solid bodies, and though occasionally one may seem starlike to the eye there is no possibility of resolving an object of only kilometric size. The nucleus of 1882 II shortly after perihelion suddenly became a brilliant streak 200,000 km long, which then divided into five or six separate comets, resembling "a string of pearls". Yet another more recent observation unfavourable to the nucleus picture was Comet Bennett (1969i) which in Ha was found to have an overall extent in excess of 10^{11} cm. Radiometric measures, interpreted as if relating to a nucleus, indicated a temperature of 700 to 800 K at about 0.55 a.u. from the Sun: what sort of ice could this be? Yet again, during intense meteor showers (the debris of comets), there has been no instance of any meteorite reaching ground level in the shower orbit, decisive evidence that no large particles exist in comets. Finally, no feasible mechanism whereby such icy snowballs could form in the Solar System has been forthcoming.

On the other hand, from more than a century of observations before 1950, the structure inferred was that of vast swarms of widely separated meteoric particles, and this receives theoretical support from the accretion process for the formation of comets from interstellar dust. All the numerous accounts and interpretations by former skilled observers are still available, and it is only by ignoring this mass of evidence, not to mention modern observations, that the icy nucleus model can survive at all. Even the short period Comet d'Arrest has a coma 10^{10} cm or larger at times. But it is not likely that the actual limit to the particle distribution should be the mere optical limit, as the vast extent of Comet Bennett seen in Ha would suggest.

....

COMET OUTBURSTS

Whitney, Charles; *Astrophysical Journal*, 122:190-195, 1955.

Abstract. Several comet outbursts accompanied by the ejection of spherical halos are investigated. Special attention is paid to the mass and energy ejected in a typical halo. On the basis of the results, it is suggested that the energy of an outburst may be derived from normal insolation. The activity can possibly be explained in terms of the icy-conglomerate comet nucleus.

1. The Observations. The purpose of the present investigation is to study the phenomenon of the ejection of spherical halos from cometary nuclei. A typical cometary outburst to be considered here consists in a brightening of the comet by 2-5 mag. and the expansion from the nucleus of a spherical halo with a velocity between 0.1 and 1.0 km/sec. The ejection of halos by comets is not an extremely rare occurrence, and certain comets appear particularly prone to it. The comet Schwassmann-Wachmann (1925 II), notable also for its orbit ($e = 0.14$, $q = 5.51$), is an extreme example. It shows an estimated two or three such outbursts per year.

Data concerning the adequately observed outbursts of three comets are compiled in Table 1. The first and second columns are self-explanatory. The third column shows the number of magnitudes by which the comet increased in brightness during the burst. The fourth column gives the magnitude of the comet at the height of the burst, reduced to the standard distances of 1 A from the sun and the earth and corrected for the phase angle at which the comet was observed. Table 2 contains the measured expansion velocities of the halos for these four outbursts.

The tabulated velocities are close to thermal velocities at the temperature of cometary nuclei (0.4 km/sec for molecular weight 20 at 6 A from the sun). N. Richter has recently derived velocities of 0.1-0.5 km/sec for the halos associated with bursts of comet Schwassmann-Wachmann. Bobrovnikoff (1954) reached a similar conclusion from data on 57 halos.

....

"SPONTANEOUSLY" SPLIT COMETS

Harwit, Martin; *Astrophysical Journal*, 151:789-790, 1968.

Comets---particularly long-period comets---split surprisingly often. Stefanik (1966) has listed thirteen well-documented cases of comet fracture. He points out that only a few of these comets split through the action of todal forces, in a close approach to the Sun or to Jupiter. Others split for no apparent reason, sometimes before perihelion, sometimes after, sometimes close to the Sun, sometimes at great distance. As Stefanik remarks, the most notable property of these comets is that all but two have parabolic orbits and are "new" in Oort's (1950) sense of the word.

The purpose of the present Note is to point out that fracture in these comets always takes place close to the ecliptic plane, and to offer an explanation. For most of the comets that split through a non-tidal process Table 1 lists the orbital inclination i, perihelion distance q, the estimated date of the split (which Mr. Stefanik very kindly supplied), the date at which the comet crossed the ecliptic plane, the distance h from the ecliptic plane, and the distance from the Sun at which the comet split. It is very difficult to estimate the exact time of splitting, particularly for comets that split near the Sun, because there the solar gravitational force produces (Opik 1966) an acceleration that depends on the particular configuration of the fragments, relative to the solar radius vector, at the instant of fracture.

Table 1. Data on Split Comets

Comet	i	q	T_{split}	$T_{ecl.\ cross}$	h_{split} (a.u.)	r_{split} (a.u.)
1860 I	79°68	1.199	Before 2/17/1860	1/18/1860	0.2	1.2
1899 I	146.26	0.327	5/ 6/1899	4/12/1899	0.4	0.71
1905 IV	4.28	3.340	Before 3/ 1/1906		Small	3.34
1915 II	54.79	1.005	Before 5/ 1/1915	5/12/1915	0.14	1.56
1947 XII	138.51	0.110	12/ 6/1947	12/ 2/1947	0.2	0.20
1951 II	87.89	0.72	(1/30-2/12) 1951	12/21/1950	0.7	0.72
1955 V	107.50	0.885	9/ 3/1955	8/17/1955	0.6	1.04
1957 VI	33.20	4.46	1/ 1/1957	6/ 5/1957	2	4.90
1846 II	12.58	0.865	9/13/1844		Small	4.68
1916 I	15.53	1.558	1/31/1916		Small	1.56

Figure 1 shows the comets' approximate location---relative to the ecliptic plane---at the time they split. The high concentration near the ecliptic plane is clear.

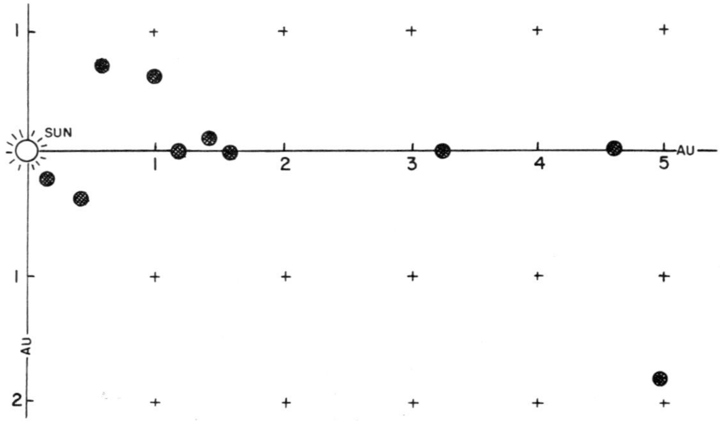

Locations where comets have split relative to the plane of the ecliptic (Fig. 1)

A possible explanation for this concentration might be found in collisions of these comets with interplanetary "boulders"---roughly 10-meter-sized meteorites that might be expected in interplanetary space. Harwit (1967) has discussed the possibility of finding such objects in large numbers. Although the kinetic energy of impact of a boulder colliding with a parabolic comet could be high enough to produce the observed velocities of separation, it is more likely that the impact itself only triggers the action. The heating produced in the impact (we deal here with impact crater sizes well over 100 meters in diameter) could lead to explosive exothermic reactions among cometary materials; it is also possible that strains built up after many eons by radioactive heating in the comets' interiors could be released by a triggering action and result in the split.

The collision hypothesis can explain why new comets are particularly likely to split. The nuclei of these comets may have diameters of the order of 50 km, perhaps five times as great as typical diameters of periodic comets. This increases the collision probability by a factor of 25. Moreover, since short-period comets---and presumably boulders ---predominantly orbit the Sun in low-inclination and low-eccentricity orbits, the impact frequency per orbit about the Sun and impact velocity in collisions with short-period comets will be low. The kinetic energy available in impacts of boulders on new comets would be about one hundred times as great as on short-period comets.

On the collision hypothesis one expects that splitting would be somewhat more probable for comets having retrograde rather than direct orbits. The available statistics are too few to show such an effect, but this prediction of the theory constitutes a crucial test of its validity. It is possible that sudden flaring up of previously inactive comets can also be explained by means of collisions. It might be interesting to see whether these rapid changes in comet activity also occur when comets are close to the ecliptic plane. Such an investigation is likely to be complicated because available descriptions will not always give a clear-cut differentiation between activity caused by release of new comet material through collisions, and activity produced through ionization and/or dispersal of previously liberated material by an increase in the solar wind.

SHOULD A COMET WAG ITS TAIL?

Anonymous; *New Scientist*, 20:501, 1963.

Daniel Malaise of the Astrophysical Institute of the University of Liege, Belgium, reports in the Astronomical Journal (Vol. 68. p. 561) his studies of the comet Burnham 1960 II, made at the Observatore de Haute Provence, France. From his observations, and with the aid of a computer, Malaise has deduced that the tail of the comet, which was about 1.8 million miles long, swung from side to side in the plane of the comet's orbit, through an angle of 15 degrees. It did so quite regularly, with a period of four days. Although fluctuations of comets' tails have been observed before, this is the first time that a definite wagging motion has been observed. Malaise also found that the comet's tail faded out in a time of three hours and regained its brightness the next day.

Astrophysicists are abandoning the old idea that the tail of a comet, which always points away from the Sun, was produced by the pressure of sunlight on the diffuse matter of the comet. The existence of a solar wind---a continuous stream of ionised particles originated in the Sun---was predicted by L. Biermann about ten years ago on the basis of fluctuations in comets' tails. Observations by spacecraft have since amply confirmed the presence of the solar wind in interplanetary space, and earlier this year D. B. Beard and M. P. Nakada (see New Scientist, 15 August, p. 323) gave an up-to-date explanation of how the combined action of the solar wind and the interplanetary magnetic field produced the streaming tail.

Malaise is not sure how to explain the tail-wagging by the comet Burnham. He remarks that it is difficult to imagine why the solar wind velocity would sweep back and forth with a periodicity of four days which cannot be correlated with anything in the solar system. Another possibility is that the nucleus in the comet's head is rotating with a period of four days and that features in the tail are related to the nucleus rather than directly to the solar wind: even so, it is not at all clear why the tail should wag.

It is Malaise's opinion that the surface condition and mechanical structure of the nucleus will play a decisive role in explaining the detailed structure of the tail---and suggests that the launching of a space probe to bypass an active comet would be a decisive step towards increasing our knowledge of the physics of comets. Meanwhile, he is looking for other comets that wag their tails in a regular way.

AURORA IN A COMET'S TAIL?

Anonymous; *Sky and Telescope*, 3:11, June 1944.

Julie M. Vinter-Hansen, writing in the Publications of the Astronomical Society of the Pacific, reports on some interesting comet observations published originally in the International Astronomical Union Circulars, received recently from Europe. Arend, at Uccle, Belgium, noticed a nebulous object of 13th magnitude with a nucleus, close to Comet Whipple (1942g) on March 29, 1943. Measurements indicated that the position of the object must have coincided with that of the comet on March 28.5. Dr. Brunner-Hagger, at Zurich, Switzerland, had photographs of the comet on nearby dates which seemed to confirm the hypothesis that the object had split off from the comet on March 28.5.

But Brunner-Hagger noted that a fine aurora had been seen on March 27.8. Terrestrial aurorae are caused by corpuscular radiation from the sun. Suppose the diffuse object in the comet's tail had the same origin? From the time elapsed between the aurora and the formation of the cometary object, he deduced that the solar corpuscles travel at the rate of about one astronomical unit in 30 hours. Other Comet Whipple photographs showing similar spots were available for the period February 28th to March 4th. Solar eruptions had occurred February 26th-28th. Again correlating the two phenomena, Dr. Brunner-Hagger found similar velocities for the corpuscles.

COMETS FROM THE ERUPTIONS OF JUPITER

Anonymous; *Science*, 76:sup 8, July 1, 1932.

Many of the comets that appear in the night sky to the telescopes of astronomers may really be the product of eruptions from the surface of the planet Jupiter within the last few centuries, if the theory just proposed by S. Vsessviatsky is correct. In a communication to The Observatory, British astronomical journal, Dr. Vsessviatsky, who is connected with the Astronomical Institute of Moscow, renews this suggestion, which was originally made a number of years ago by Richard Proctor, a famous English astronomer.

The "capture theory," held by many astronomers, supposes that these comets originally came into the solar system in parabolic orbits from vast distances. When one happened to pass close to Jupiter, that planet, with its great mass, pulled it out of its former path by gravitational attraction. After that the comet moved in an elliptical path, between the region of the sun and the orbit of Jupiter. Dr. Vsessviatsky points out that if this were the case it would be very rarely that a comet entering the solar system would happen to pass close enough to Jupiter to be pulled into the elliptical orbit. He estimates that it would only happen to something like one in 100,000 comets, but actually, he declares, there are about sixteen of these short-period comets to a hundred parabolic ones. Also, he says, the diminution in brightness of the short-period comets indicates that their age is a matter of only a few centuries.

"The conclusion follows," writes Dr. Vsessviatsky, "that the date of birth of a short-period comet does not precede by very long the date of its discovery, whereas captures by the giant planets would occur only at very long intervals." He also points out that all of these comets are moving around the sun in the same direction as Jupiter, and the other planets, whereas if they were captured some would probably be moving in the other direction, or "retrograde." The connection between these comets and Jupiter can be fully explained by the hypothesis that they are the product of eruptions taking place on Jupiter's surface.

Besides the family of comets related to Jupiter, there are others apparently connected with Saturn, and possibly with Uranus and Neptune. The theory suggests that very active processes are in progress on the surface of the large planets, resulting in the frequent expulsion of matter.

NEW PLANET MAY BE ADDED TO THE SUN'S FAMILY OF NINE

Anonymous; *Science Newsletter*, 41:361, 1942.

A new planet about the size of Pluto but a little less distant may be added to the sun's family, which already contains nine members.

No such tenth member has yet been found, but the attraction of such a body would account for the three days' delay in the return of Halley's comet in 1910, according to the calculations of Dr. R. S. Richardson of Mt. Wilson Observatory reported in the Publications of the Astronomical Society of the Pacific (February). And he has told astronomers just where to look for the new body. It should now be found, he says, at about right ascension 16 hours, declination minus 20 degrees. These figures enable astronomers to point their instruments directly at the suspected spot.

At first, Dr. Richardson thought Pluto might be the culprit that held back Halley's comet. But it turned out that Pluto was miles away at the time---in fact, more than three billion miles away. This was the very closest the planet ever got to the comet, and it happened in October, 1901. Also Pluto is so tiny, only about as big as the earth! Finally, Dr. Richardson's very careful calculations showed that if Pluto had any effect at all, it was in the wrong direction; it would have hastened rather than delayed the comet.

. . . .

What is needed to explain the comet's dilatory behavior, Dr. Richardson found, is a planet whose orbit just grazes the furthermost tip of Halley's orbit, grazes it by about 9,000,000 miles. The planet would be about the size of Pluto or the earth. A larger planet at a greater distance would also do the trick, but the planet must be small because otherwise it would have been discovered.

If this planet is found, it will be the third to have been predicted mathematically and afterwards discovered. The other two are Neptune, discovered in 1846 from the calculations of Leverrier and Adams, and Pluto, discovered in 1930 from the calculations of Lowell and Pickering.

A FORMER MAJOR PLANET OF THE SOLAR SYSTEM

Van Flandern, T. C.; *EOS*, 57:280, 1976.

Abstract. Recent dynamical calculations by M. W. Ovenden have demonstrated the former existence of a 90-Earth-mass planet in the asteroid belt until $16x10^6$ years ago. These calculations have now been strikingly confirmed by the discovery that very-long-period comets apparently originated in the explosion of such a planet at that epoch. The evidence from comets is present in the distributions of each orbital element; but the most compelling evidence comes from a backwards integration of 60 well-observed very-long-period comets to their previous perihelion passage, which shows that most of these intersect at nearly the same point on the heliocentric celestial sphere. Taken in conjunction with the already existing evidence, these new results leave little room to doubt that a Saturn-sized planet did exist between Mars and Jupiter 16 million years ago, and then violently exploded.

COMETS AND CLIMATIC CHANGES

Anonymous; *Nature,* 176:1152-1153, 1955.

In a paper entitled "Evidences Cometaires des Variations des Climats" (Bull. Astro. Inst. Czechoslovakia, 6, No. 1; 1955), F. Link and Z. Linkova deal with statistical research relating to the discovery of comets, correlated with solar activity. They have found that the annual number of comets discovered varies in the course of the solar cycle of eleven years. This variation is inversely proportional to rainfall, with two maxima during the eleven years, according to Helman's investigations, and more recently it has been found that the number of comets discovered during a cycle is greater in the even than in the odd cycles. The numbers of comets discovered between 1755 and 1953 are arranged in cycles 1-18, and then the mean number of discoveries per year in the successive cycles is given separately for even and odd cycles. The influence of the meteorological conditions on the observations of comets is shown by means of a graph. Solar activity being greater during the odd cycles, the circulation in the terrestrial atmosphere is increased, and in consequence observations become more difficult; as a result of this, there is a decrease in the number of comets observed.

This close connexion between the number of comets discovered and the meteorological conditions (the latter depending on solar activity) led the authors to carry out investigations on climatic variations since 1611, and the principle underlying this research is based on the assumption that the number of comets discovered is a function of three factors: the cosmic factor, that is, the actual number of comets that approached the Earth; the climatic factor, or the atmospheric conditions in civilized countries, which only need be included, as observations of comets in uncivilized countries would be useless for statistical purposes; the social factor or human activity (in particular, astronomical activity) in civilized countires. While the first factor can be regarded as constant, the second would increase in a more-or-less continuous manner in time, and the third would be revealed in the slope of the curve N = f(A), N being the number of comets discovered from zero year up to the year A. Both these quantities have been taken from a catalogue by Baldet ("Ann. Bur. Long", Supp. 1950), and a searching analysis of the results leads to certain conclusions on the meteorological conditions between the years 600 B. C. -A. D. 1740. A comparison is made between these and the data given by C. E. P. Brooks in "Climate through the Ages", covering about a thousand years up to comparatively recent times, and there is a close accordance between them. The Czechoslovak workers hope to carry out further investigations on the subject.

A REMARKABLE COINCIDENCE

Campbell, W. W.; *Science*, 46:36-37, 1917.

The most remarkable coincidence known to me relates to the discovery of Perrine's second comet. I published the facts in the case in The Observatory, Vol. 26, pp. 293-94, 1903, where they were made familiar to many astronomers. On describing the coincidence recently to a group of my colleagues in other sciences they urged strongly that I republish the facts in a journal of more general character, and thus make known the occurrence to students in other subjects.

Professor Charles D. Perrine, of the Lick Observatory staff, discovered the first of his many comets on November 17, 1895. This was Comet c 1895. He observed it night after night until December 20, 1895, when it was lost to sight in the glare of the sun's rays. The orbit of the comet was accurately determined, and its path for the early months of 1896 was computed and published in advance. I had the pleasure of assisting Mr. Perrine when he first looked for its reappearance from behind the sun, on the morning (just before dawn) of January 30, 1896. He found it at once, in the predicted position, and as an object easily visible in medium-sized telescopes. Because the comet was following its predicted path so closely we decided not to squander money in cabling the fact of its reobservation to European observers. Perrine observed his comet morning after morning as weather permitted, for fifteen days, until on February 14 a cablegram was received from Kiel, Germany, announcing that Lamp had reobserved Perrine's Comet c 1895 that morning. The cablegram in cipher code was received at the Lick Observatory by one of the astronomers, in perfect order as shown by the control word; but in converting the cabled right ascension of the comet from degrees and minutes of arc into hours and minutes of time the translator made an error of 24 minutes of time, equivalent to 6° of arc. The erroneous translation was handed to Perrine. He compared this with what he knew to be the real position of Comet c 1895, by virtue of his observations in the preceding half month, and saw that there was a discrepancy of about 24 minutes of time. Inasmuch as the check word in the cablegram was correct he judged that the object observed by Lamp in Kiel must be a different comet from his own. The following morning was clear and he pointed the 12-inch telescope to the position that was handed to him. In looking through the finder of the telescope he saw an eighth magnitude comet in the field of view. This did not surprise him. He observed the position of the new comet, and we transmitted the observation by telegraph and cable, as usual, as belonging to a new comet discovered by Lamp in Liel. This new object was at once known as Comet a 1896. Naturally considerable mystery existed (see Astronomical Journal, Vol. 16, p. 56, 1896, and Astronomische Nachrichten, Vol. 139, pp. 365-66, 1896). Several weeks elapsed before the tangled situation was unravelled at Mount Hamilton by our looking up the original cipher cablegram and detecting the error of 24 minutes in the conversion of arc into time, made after the cipher message had been translated and checked.

It is a surprising fact that the error should have directed the telescope upon an unknown comet, but the surprise increases when we consider another attendant fact. The new comet was moving amongst the stars very rapidly; more than 2° east in right ascension and more than 3° north in declination, daily. When the cablegram was written in Kiel on the morning of the fourteenth the new comet was six or seven degrees from the cabled position. When the erroneous position was handed to Perrine on the morning of the fourteenth the new comet was three degrees from that position. When the first opportunity came, the following morning, to examine the erroneous position, the rapidly-traveling comet had moved into that position. Had the tesescope been pointed to that position on any other morning whatsoever, the celestial visitor would have been far outside the finder field, and the chances are fair that it would have come and gone unseen. The cabled Kiel position of reobservation of Comet c 1895 and Perrine's position of Comet a 1896 were:

Comet c 1895, Feb. 14, R. A. = 19 h. 45 m.,
Dec. = -2° 23' (correct translation).
Comet c 1895, Feb. 14, R. A. = 19 h. 21 m.,
Dec. = -2° 23' (erroneous translation).
Comet a 1896, Feb. 15, R. A. = 19 h. 22 m.,
Dec. = -2° 49'.

The angular radius of the finder field was about 1° .3.

I doubt whether another case of coincidence as remarkable as this one is on record in the literature of astronomy.

Chapter 15

BODE'S LAW AND OTHER SOLAR-SYSTEM REGULARITIES

INTRODUCTION

Astronomy has always occupied an exalted place in the hierarchy of science. The supposed regularity of the motions of the stars and planets and the grand mathematical laws describing these motions are basic to astronomy's favored position. Beyond the orderly realm of celestial mechanics, however, are rules-of-thumb and inspired laws that owe little to basic physics. Bode's Law is a classic example. Here are presented some of the competitors to Bode's Law and the curious resonances of the solar system. If these relationships are true expressions of nature, there should be some unrecognized physical force behind the rules-of-thumb. This is worth searching for.

The long-term stability of the solar system has been proclaimed loudly and longly. Like the Law of Uniformity in geology, solar system stability has been considered essential for the scientific description of solar system evolution. Instability implies chaos, and chaos does not yield easily to the scientific method, but it may exist nevertheless.

• Solar-System Laws and Resonances

KEPLER AND THE RESONANT STRUCTURE OF THE SOLAR SYSTEM

Gingerich, Owen; *Icarus*, 11:111-113, 1969.

A. M. Molchanov is by no means the first to seek for the "resonant structure of the solar system"; his research bears an interesting parallel to a thread that runs through the entire work of Johannes Kepler. Both authors have addressed themselves to a deeply fundamental question: Why are the planets arranged in their present pattern? Even in the 17th century this was an unorthodox query, and Kepler's vision of the "intelligible but inaudible" harmonies sung by the planets has generally been dismissed as sheer mysticism or numerology. Nevertheless when a serious consideration of this question is reopened on the pages of Icarus, it is worthwhile to review briefly the verve and imagination that characterized Kepler's inquiry.

As a young teacher in Graz (Austria) fresh from the seminary in Tubingen, Kepler posed for himself three unusual questions: Why are there only six planets? Why are they arranged this way? Why do they move with these periods? The first part of the answer, he tells us, came as a flash during a class lecture. There are six planets because there are five, and only five, regular solids. God the Great Geometer has provided an archetype for the solar system by arranging these figures in a concentric fashion separated and surrounded by six spheres, suitably spaced for the planetary circuits.

Astonishingly, the scheme works with fair accuracy when space is allowed for the eccentricities of the planetary paths. The figures are given in Table 1 as tabulated in Kepler's Mysterium Cosmographicum of 1596 (Chap. 14). Kepler was obliged to compromise the elegance of his system by adopting the second value for Mercury, which is the radius of a sphere inscribed in the square formed by the edges of the octahedron, rather than in the octahedron itself. With this concession, everything fits within 5°, except Jupiter, at which "no one will wonder, considering such a great distance." This marvelous remark reflects the state of mind of a convinced theoretician, and the fact that better values of the orbital parameters improve the agreement is perhaps irrelevant.

The Mysterium Cosmographicum arrived from the printer early in 1597, and within a few months Kepler sent a copy to the great Tycho Brahe, specifically hoping to obtain in return better observational data.

Table 1. Ratios of the Planetary Orbits (Assuming the Innermost Part of the Outer Orbit to be 1000)

Planets	Regular solid	Computed by Kepler	According to Copernicus
Saturn			
	Cube	577	635
Jupiter			
	Tetrahedron	333	333
Mars			
	Dodecahedron	795	757
Earth			
	Icosahedron	795	794
Venus			
	Octahedron	577 or 707	723
Mercury			

In Kepler's view, it was divine fate that brought them together in Prague and that set him onto an analysis of Mars. This endeavor, documented in the Astronomia Nova (1609), led to the discovery of Mars' elliptical orbit and the law of areas.

Kepler's successful attack on Mars, clearly founded on the finest observational data of his day, might have turned his interest from the Platonic pursuit of nested solids. The Astronomia Nova is his most enduring contribution to astronomy, but to the end of his life he ranked the nest of spheres and polyhedra among his highest achievements. The research on the Martian orbit was in many respects merely an arduous digression from his more profound search for the archetypal harmonics of the universe.

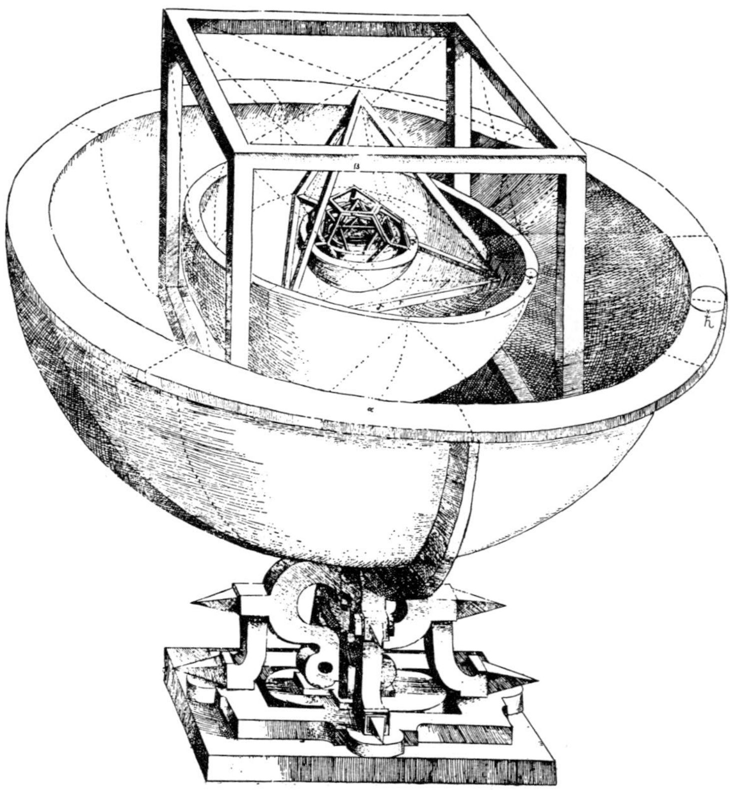

Kepler's geometrical vision of the solar system

Armed with more accurate parameters, Kepler again plunged into his quest for the ultimate order of the planetary systems, this time following up some initial ideas on the third question. Why do the planets move with these particular periods?

Kepler first sought in vain for harmonic ratios between the planetary periods. Commenting on this failure, he wrote in his Harmonices Mundi (1619, Bk. 5, Chap. 4) that apparently God the Creator did not wish to introduce harmonic proportions between the periodic times. But, he continued, "since God has established nothing without geometric beauty which was not bound beforehand by some law of necessity, we easily conclude that the periodic times got their appropriate lengths because of a prior archetype."

Success finally crowned Kepler's search when he dismissed the times of the total journeys around the Sun and concentrated on arcs near the aphelia and perihelia. The movements, when halved a suitable number of times, reduced to the ratios of the musician's major scale! Sing out, Urania," cried Kepler, "while I ascend through the harmonic scale of celestial motions to the heights where the true archetype of the fabric of the universe is hidden" (1619, Bk. 5, Chap. 7).

Of course, Kepler noted, a planet could not be at aphelion and perihelion at the same time; hence the silent harmonics did not sound simultaneously but only from time to time as the planets wheeled in their generally dissonant courses around the Sun. Swept on by the grandeur of his vision, he exclaimed: "It should no longer seem strange that man, the ape of his Creator, has finally discovered how to sing polyphonically, an art unknown to the ancients. With this symphony of voices man can play through the eternity of time in less than an hour and can taste in small measure through human arts the delight of God the supreme Artist by calling forth that very sweet sensation of pleasure in the music which imitates God" (1619, Bk. 5, Chap. 7).

Few modern scientists expose their motivations so candidly, but the sweetness of celestial harmonies can still be savored.

LAW CONNECTING MOTIONS IN PLANETARY SYSTEM

Anonymous; *Nature*, 60:597, 1899.

M. Ch. V. Zenger, of Prague, has recently put forward the results of work he has been engaged on for some years past, and a part dealing with the relations existing between the "time of a planet's revolution" and its position in the solar system appears in the Bulletin de la Soc. Ast. de France, October 1899, pp. 431-434. He finds that the orbital movements of the planets and also of some periodical comets have a simple law connecting them with the time of the sun's rotation. If "r" is the time of rotation of the central controlling body, then "R," the time of orbital revolution of the planet, is given by the relation

$R = n\frac{r}{2}$; where "n" is a whole integer, different for each body.

Taking Faye's value for the solar rotation = 25.2 days, $\frac{r}{2} = 12.6$

days, and the author gives the following data:---

	Mercury	Venus	Earth	Eros	Mars	Jupiter	Saturn	Uranus	Neptune
N	7	18	29	51	54	344	854	2436	4776
R	88.2d.	226.8d.	365.4	642.6	680.4	4344.4	10765.4	30693.6	60177.6

Between the earth and Eros, the author mentions the possible existence of a hitherto unknown planet for which n = 40, and the period of revolution of which would therefore be about 500.4 days.

Several tables are also given showing the conformation of the satellites of the various planets to a similar relation, and the author considers the whole as helping to confirm his electrical theory of the solar system.

DOES AN UNKNOWN FORCE PERPETUATE DIURNAL ROTATION?

Barr, John W.; *Popular Astronomy,* 27:694, 1919.

The spheres which, with the sun, constitute the solar system, may be divided into two classes: those having diurnal rotation, like our earth; and those which lack such rotation, like Venus, Mercury, our moon, and presumably all the other satellites. Now, a remarkable circumstance is, that diurnal rotation is always coincident with the presence of one or more satellites which revolve around the rotating body; whereas all bodies that lack satellites also lack rotation. This condition is the very opposite to that which we should be led to expect, especially in a comparison of the dead state of Venus and Mercury with that of living planets, whereon notwithstanding the greater proximity of the former to the sun, the brake of the tides should have been far more potent, due to the attraction of their moons plus that of the sun.

Do not these facts suggest the propriety of an inquiry whether, sustained diurnal rotation being always coincident and concurrent with the presence of satellites, it may not be also contingent upon their presence; whether there may not exist some unknown (X) force, or some unfamiliar law directing known forces, which operates, in the case of planets having satellites, to counteract by positive acceleration the negative acceleration due to tidal friction; whether perhaps the earth, for example, as regards rotation, has not attained a state of stable equilibrium, insuring days of uniform length through incalculable periods of astronomical time. In such a case the force opposing the retarding effects of tidal friction would be, not unaided primordial momentum, but primordial momentum plus X.

If we assume the existence of such an X force, it is only necessary to define it as a force of minor magnitude, just a sufficient auxiliary to primordial momentum to postpone or perhaps forever prevent the death of the planets.

Proof of X would necessarily include proof of the proposition that when, under certain conditions, one body revolves around another, rotation of the latter is accelerated.

The moon and other satellites may be powerful magnets which point their same poles always toward their primaries; and that this, if the distance be not too great, would accelerate the primaries' rotation seems certain. An iron sphere so suspended as to be free to move will be made to rotate if the pole of a magnet is revolved around it at proper distance. This may be proved by a simple experiment.

Or perhaps the following proposition may some day be accepted· "When (while) their common center of gravity resides within one of two bodies having relative angular motion, the rotation of that body is accelerated." In the light of our present knowledge, such an hypothesis as this would appear visionary indeed; and yet, as the center of gravity of the system which each planet forms with its moon, or with each of its moons where there are more than one, is invariably situated within the body of the planet itself, this circumstance might well have some direct bearing on the question of the existence of the postulated X force.

A RELATION BETWEEN THE MEAN DISTANCES OF THE PLANETS FROM THE SUN

Caswell, A. E.; *Science*, 69:384, 1929.

In 1772 Bode drew the attention of the scientific world to an empirical law, previously discovered by Titius, relating the mean distances of the planets from the sun. If we write the following number series: 0, 3, 6, 12, 24, 48, 96, 192, 384, and add 4 to each member of the series we obtain the following numbers, which are very nearly proportional to the mean distances of the planets from the sun: 4, 7, 10, 16, 28, 52, 100, 196, 388.

The first term of this series, which corresponds to the planet Mercury, does not belong to the series but should have the value 5.5 instead of 4. Moreover, the actual distance of Neptune is less than four fifths the expected distance. Nevertheless, Bode's law has served a useful purpose inasmuch as it suggested the existence of an unknown planet in the fifth position and thereby led to the discovery of the host of asteroids.

The writer has discovered another simple relation between the planetary distances, and so far as he is aware this relation has not been reported hitherto. It suggests the possibility that the orbits of the planets may be "quantized" somewhat after the manner of the electronic orbits in the Bohr atom. For this reason it may prove to have some theoretical importance.

The mean distances of the planets from the sun are proportional to the squares of simple integral numbers. The four innermost planets are represented by four successive integers, viz., 3, 4, 5 and 6. The space between Mars and the average mean distance of the asteroids (taken as 2.7 astronomical units) corresponds to a difference of 2 between the corresponding integers, that between the asteroids and Jupiter to a difference of 3, that between Jupiter and Saturn to a difference of 4, that between Saturn and Uranus to a difference of 6, and that between Uranus and Neptune to a second difference of 6. The last two planets do not fit into the law quite as well as the others, but on the whole the agreement is good, and can scarcely be accidental.

The following table gives the data upon which the preceding statements are based. The numbers given in the third column of the table are obtained by dividing the mean distances in astronomical units by 0.0425, and extracting the square roots of the quotients.

Table

	Distance from sun in astronomical units	Square root of comparative distance	Nearest integer	Percentage deviation
Mercury	0.3871	3.018	3	0.60
Venus	0.7233	4.125	4	3.13
Earth	1.0000	4.851	5	-2.98
Mars	1.5237	5.988	6	-0.20
Planetoids	2.7(?)	7.97	8	-0.4(?)
Jupiter	5.2028	11.06	11	0.59
Saturn	9.5388	14.98	15	-0.13
Uranus	19.1910	21.25	21	1.19
Neptune	30.0707	26.60	27	-1.50

SOME NEW LAWS FOR THE SOLAR SYSTEM

Malisoff, William M.; *Science,* 70:328-329, 1929.

Apropos of A. E. Caswell's suggested law, namely, "the mean distances of the planets from the sun are proportional to the squares of simple integral numbers," the writer wishes to point out the following corrections, extensions and other new laws.

(1) The percentage deviation from proportionality to the squares of the integers is double that indicated by him.

(2) Since the earth's distance is taken as a standard in all measurements, one would expect a good reason for not assuming its distance to correspond to a perfect square of an integer (in this case 5). If this is done the deviations from the above law are as high as 12 per cent.

(3) One would expect similar relations to hold for the satellites of the planets. For the satellites of Mars the ratio $5^2:8^2$ holds quite closely. For the four satellites of Uranus the ratio $5^2:6^2:8^2:9^2$ holds poorly. But for the satellites of Saturn and Jupiter one must either omit several or resort to initial numbers greater than 5^2 for the nearest satellite. Of course, if large integers are to be admitted one may get as close a fit as one pleases for almost any distribution of distances. On the whole the evidence from this source is unfavorable to a deep-seated significance for the relation cited.

(4) The writer would point out a relation that depends strictly on

the square root of the distance of a planet from the sun or a satellite from its planet. It is the velocity, which varies inversely as the square root of the distance from the axis of revolution. For the planetary system one could then state as a law. The velocities of the planets are inversely in proportion to simple integral numbers. Thus,

Planet	Period	Mean velocity	30.3/mean velocity	Nearest integer
Mercury	0.2408	10.1006	3	3
Venus	0.6152	7.3872	4.1	4
Earth	1.000	6.2832	4.83	5
Mars	1.88	5.0924	5.95	6
Jupiter	11.86	2.7563	11.0	11
Saturn	29.46	2.0344	14.9	15
Uranus	84.01	1.4346	21.1	21
Neptune	164.6	1.1464	26.5	27

(5) Another law may be stated, as a consequence of Kepler's third law and the distance relation, namely, the periods of the planets are proportional to the cubes of simple integral numbers.

The same integers as above are involved.

(6) In this connection the writer would bring to the attention of American scientists an effort by Viktor Goldschmit to elucidate some of the numerical regularities in the distances of planets and satellites from their axes of revolution. The journal in which it occurs is not generally known. He observes the distances of the planets to be quite closely in the sequence 1/13, 1/7, 1/5, 1/3, 1, 2, 4, 6. The four larger planets are considered to have condensed together before the group of the four smaller ones. A mathematical treatment strictly analogous to the phenomena of standing waves in sound, the distribution of lines in spectra, the progress of crystallization and similar phenomena gives the same law of harmonic relations of distances not only for the planets but also for satellites. The harmonic sequences are as follows:

Condensations	0	1/3	1/2	2/3	1	3/2	2	3
Large planets	0		1/2		1		2	3
Small planets	0	1/3		2/3	1		2	
Jupiter's satellites	0		1/2	2/3	1		2	
Uranus' satellites	0		1/2	2/3	1	3/2		
Saturn inner satel.	0		1/2	2/3	1	3/2	2	
Saturn outer satel.	0				1	(6/5)		3
Earth moon	0				1			

The sequences are brought into line by transformations derived from considerations of the dominance of certain positions, as the 0, 1 and ∞ in condensation.

(7) The writer would further approach the question of the regularities of the spacing of the planets from another basic point of view. In brief, considering the accurate correspondence of the velocity of a planet or satellite inversely to the square root of its distance from the axis of revolution, we may conceive the propagation of a wave of velocity at the initiation of revolution to follow the law of a logarithmic spiral. We should then expect the distances of the planets as well as their velocities

to be represented as the radii vectors of a logarithmic spiral. The law we would propose is: A small integral number of geometric means will determine the positions and velocities of all the planets and satellites. This is suggested by the well-known properties of such spirals. Symbolically,

$$\log d = nk, \text{ or } e^{nk} = d$$

where d is the distance, k is a constant and n is a simple integer. A fairly accurate straight line plot is obtained by using values of n for the entire series between Mercury and Neptune, 1, 4, 5, 7, 10, 13, 16, 19 and 21. The interval of 3 predominates. The velocity, v, is related to the above by the relation

$$\log v = k' - \frac{nk}{2}$$

where k' is another constant and n the same simple integer as above. It follows also that a similar logarithmic relation exists for the periods, that is, a small number of geometric means will determine the periods. If p stands for the period, then

$$\log p = \frac{3}{2}nk + k''$$

where k'' is still another constant and n is the same simple integer as as above.

Similar relations may be derived for centripetal force and the like, but enough has been given here to indicate the underlying principle.

A NEW LAW OF SATELLITE DISTANCES

Penniston, J. B.; *Science,* 71:512-513, 1930.

Besides the celebrated Bode's (or Titius) law there have been a number of attempts to establish a law governing the distances of satellites from their central body, including two discussions of the subject in Science in 1929. My approach to this subject was made about four years ago in somewhat the same manner as that of Dr. A. E. Caswell who holds "the mean distances of the planets from the sun are proportional to the squares of simple integral numbers." I added, however, to the square of the integer the integer itself, thus assuming that the terms differ from the squares of integers by a progressively changing amount. For example, adding to each of the integers 1, 2, 3, 4, 5.... its square, we obtain the values, 2, 6, 12, 20, 30.... This is simplified by dividing throughout by 2, giving us the series 1, 3, 6, 10, 15.... Those familiar with Bernoulli's Tabula Combinatoria will recognize the series as the ternaries of his table.

The following table shows the results for all the satellite systems, including the planets as satellites of the sun, of the solar system where there are at least two known satellites on which to base comparisons. The base term for the relative distances in each satellite system is 10 except in the case of Mars where it is one. Distances under Bode's law

Table Illustrating Laws of Satellite Distances

System	Satellite	Relative Distance	New Law	Bode's Law	Caswell's Law
Sun	Mercury	3.87	3	4	3.82
"	Venus	7.23	6	7	6.80
"	Earth	10.0	10	10	10.6
"	Mars	15.2	15	16	15.3
"	Ceres	27.7	28	28	27.2
"	Jupiter	52.0	55	52	51.4
"	Saturn	95.3	91	100	95.6
"	Uranus	191.0	190	196	187.0
"	Neptune	300.0	300	--	310.0
"	"Planet X"	400.0	406	388	408.0
		430.0	435	388	435.0
Mars	Phobos	1.00	1	--	1.70
"	Deimos	3.22	3	4	3.82
Jupiter	V	2.71	3	4	3.82
"	I (Io)	6.27	6	7	6.80
"	II (Europa)	10.0	10	10	10.6
"	III (Ganymede)	15.8	15	16	15.3
"	IV (Callisto)	27.9	28	28	27.2
"	VI	169.4	171	--	170.0
"	VII	173.0	171	196	170.0
"	VIII	348.5	351	--	357.00
"	IX	371.0	378	388	382.0
Saturn	Mimas	10.0	10	10	10.6
"	Enceladus	12.8	--	--	--
"	Tethys	15.8	15	16	15.3
"	Dione	20.3	21	--	20.8
"	Rhea	28.0	28	28	27.2
"	Titan	66.0	66	52	61.2
"	Themis	78.1	78	--	83.1
"	Hyperion	79.0	78	100	83.1
"	Iapetus	190.0	190	196	187.0
"	Phoebe	698.0	703	772	712.0
Uranus	Ariel	10.0	10	10	10.6
"	Umbriel	14.1	15	16	15.3
"	Titania	22.8	21	--	20.8
"	Oberon	30.4	28	28	27.2

(represented by the series 4, 7, 10, 16, 28, 52, 100, 196, 388 and 772) are included not only for comparison of the results with those of the new law but also because that law gives very good values for some of the planetary satellites as well as for those of the sun. Likewise the distances under Dr. Caswell's law are included. These are obtained as explained in his article in Science.

The new "Planet X" has been officially reported in a circular from the Lowell Observatory as having a distance from the sun of 40 to 43 astronomical units. Therefore double values have been included in the table until a more accurate determination of the distance is made. Enceladus, the second satellite of Saturn, does not, apparently, belong to any of the three series, but it is here included for completeness. The orbit of Themis is not accurately determined, but it seems that there may be a faint satellite at about the distance given. If in the satellite system of Mars 10 be taken as the relative distance for Deimos, better values will be obtained for the Bode and Caswell series, but the value of one was adopted for Phobos instead because of the near correspondence of the distance of that satellite to that set by Roche's limit. Ceres, the largest of the asteroids, is included in the table although it may belong to a system of harmonics rather than to the fundamental series. Referring to the table, it is seen, of course, that not all of the successive terms of the new law are represented in any one satellite system nor are the same terms represented in different systems.

It is important to notice that Bode's law gives fairly good values for at least five of Jupiter's satellites, for four of those of Saturn and for three of those of Uranus. This fact seems to be overlooked in nearly all modern popular discussions although it was referred to in some articles fifty years ago. Sir James Jeans, like many others, has said "...it seems more than likely that it [Bode's law] is a mere coincidence with no underlying rational explanation." Yet Bode's law is an approximation for several terms both of Dr. Caswell's law and of my own, and Dr. Caswell believes that his series "suggests the possibility that the orbits of the planets may be 'quantized' somewhat after the manner of the electronic orbits in the Bohr atom." The quantum principle is important in wave-theory, and my own solution resulted from studies in that field. The new series can be derived by taking the fiction, employed in the mathematics of a vibrating sphere, of a double source of wave-action of suitable strength at the center of the sphere and by modifying this concept to that of two sources of wave-action whose distance apart is relatively small in comparison with the length of the waves set up in the surrounding medium. Then, by disregarding the distance apart of the two sources, it is possible to develop the series by superposition. This assumes that the solar system can be treated somewhat on the order of the Schrodinger atom rather than that of Bohr. A French writer, Lieutenant-Colonel Delaunay, also believes wave-action is important in this problem. Finally, Victor Goldschmidt, of Heidelberg, according to Dr. Malisoff, has shown that "a mathematical treatment strictly analogous to the phenomena of standing waves in sound, the distribution of lines in spectra, the progress of crystallization and similar phenomena gives the same law of harmonic relations of distances not only for the planets but also for satellites." Referring to Goldschmidt's original article, I find that he has represented the distances of the planets as follows: Mercury 3.90; Venus 7.10; Earth 10.0; Mars 16.7; Jupiter 50.0; Saturn 100; Uranus 200; Neptune 300, and (following his method) "Planet X" 400. Dr. Caswell has given it as his opinion that "on the whole the agreement is good, and can scarcely be accidental." After considering all this evidence independently arrived at by different investigators I am disposed to agree with him and offer as my opinion that Bode's law, as a first approximation, may have its origin in actual causal phenomena.

DISTANCES OF PLANETS FROM THE SUN.....

Richardson, D. E.; *Popular Astronomy,* 53:14-26, 1945.

Abstract. It is found that the mean distances of planets from the sun and of satellites from their primaries in the satellite systems of Jupiter, Saturn, and Uranus are in obeisance to a common law of variation with respect to planet (or satellite) ordinal number. The principal variation of distance with respect to ordinal number is exponential and found to be given by $(1.728)^n$. This principal variation is modified by a second variation, individual values of which, for a given system, equal the lengths of radii vectors of an ellipse corresponding to polar angles $(4\pi/13)n$. While each system has its own characteristic distribution ellipse, the various ellipses have certain properties in common. With the exception of a few orbital bodies that subtend unusually small solid angles at the centers of their respective systems, the probable error of computed distances is 0.20 of one per cent, the greatest single error being 0.764 of one per cent.

NONGRAVITATIONAL FORCES AND RESONANCES IN THE SOLAR SYSTEM

Jefferys, William H.; *Astronomical Journal*, 72:872-875, 1967.

Abstract. A discussion of the problem of resonances in the solar system is given, with particular attention to the role that may be played by nongravitational forces in the formation of resonant situations. A discussion of various ideas for the origin of the Kirkwood gaps, in particular, shows that it may be necessary to invoke such forces in order to obtain a complete theory. It is proposed that collisions between asteroids may be important in the formation of the gaps.

Introduction. The existence of resonance phenomena in the solar system is a long-established fact, with a history going back to the discovery of the "great inequality" between Jupiter and Saturn. The old list of such phenomena---the pointing of the moon's face towards the earth, the Trojan asteroids, the Kirkwood gaps, and the gaps in Saturn's rings, the "great inequality" involving the Galilean satellites of Jupiter---to name a few---has been augmented recently by a probable resonance between the periods of Mercury's rotation and revolution, by a fairly well confirmed resonance between Neptune and Pluto and by a possible resonance between the rotation period of Venus and its syndodic period of revolution with respect to the earth.

A striking feature about such resonant configurations is the apparent necessity to invoke other than purely gravitational forces in order to explain the existence of so many examples of them in the solar system. For example, in their exhaustive study Goldreich and Peale investigated the role of such forces in explaining the present rotational state of some of the bodies in the solar system. Goldreich has also invoked such a process (tidal interaction with Jupiter) in order to explain the previously mentioned "great inequality" among the Galilean satellites.

. . . .

ON THE ORIGIN OF COMMENSURABILITIES IN THE SOLAR SYSTEM—II

Dermott, S. F.; *Royal Astronomical Society, Monthly Notices,* 141: 373-376, 1968.

Summary. A systematic search for regularity in the major satellite systems has revealed that the orbital periods of the regular satellites are closely approximated by the relation, $T_n = T_o A^n$ where T_n is the orbital period of the nth satellite. It must be allowed though, that in any one system there are a small number of vacancies. For all systems $\underline{A}$ is the square root of a small integer and it is suggested that T_o is related to the rotational period of the primary. The relation can be applied to the planetary system but there are some anomalies. It is suggested that this regularity, which is related to the preference for near-commensurability among pairs of mean motions in the solar system, is a condition of formation rather than the result of evolution and thus could be of considerable cosmogonic importance.

SPIN-ORBIT RESONANCE OF THE INNER PLANETS

Campbell, Philip M.; *Science*, 165:930, 1969.

Radar measurements of the axial rotation of Mercury and Venus indicate that the spin and orbital motions of the inner planets are in resonance. Mercury's sidereal spin angular velocity is apparently 3/2 of its mean orbital motion about the sun. The spin angular velocity of Venus is retrograde and apparently synchronized with successive close approaches with the earth.

Shapiro observes that, "The length of the day on Venus is therefore about 117 days, apparently by coincidence almost exactly two-thirds of the length of the day on Mercury." I suggest that this commensurability may not be a coincidence, but may be associated with the mechanism responsible for locking Venus's spin period into resonance with the earth's orbit.

The necessary condition for capture into a spin resonance is the presence of an appropriate term in the tidal torque which will damp oscillations about the resonance state. This term is present in the case of Mercury, and the 3/2 resonance is fairly well understood. In the case of Venus, however, the appropriate damping term is absent. There is no obvious reason why the spin of Venus should be synchronized with the earth's orbit rather than with its own orbit about the sun, as in the case of Mercury.

Venus makes four axial rotations as seen from the earth in one synodic period of 583.9 days. The synodic spin period of Venus, 146.0 days, is something of a magic number among the inner planets. This period is almost exactly 2/5 of the earth's orbital period and is very close to 1-2/3 of Mercury's orbital period. The synodic period of Mercury as seen from Venus is 144.5 days. This suggests that Mercury's orbital motion is nearly commensurate with the earth-Venus spin resonance. If we assume that at some time in the past a triple conjunction of the earth, Venus, and Mercury occurred, one finds that Mercury is again very close to the earth-Venus-sun line after 583.9 days when the Venus spin resonance occurs.

The relation of Mercury's orbit to the earth-Venus spin resonance could provide the necessary mechanism for trapping Venus's spin into this commensurability. Although Mercury's mass is only 1/20 that of the earth, the small additional torques could have a cumulative effect when applied at the proper frequency.

The proof of this hypothesis would require a fairly difficult calculation of the capture probability along the lines set forth by Goldreich and Peale, including the additional tidal torques of Mercury. It might then be possible to show that this particular resonance occurred because of the combined tidal action of Mercury and the earth on Venus at a time when the three orbits were properly synchronized.

DENSITIES OF THE TERRESTRIAL PLANETS

McCrea, W. H.; *Nature*, 224:28-29, 1969.

Littleton has suggested that the Earth, Moon and Mars may originally have formed a single rotationally unstable planet, and that Mercury and Venus may have formed another such body. He has shown that this is possible in accordance with the theory of rotating fluid masses and with the dynamics of the solar system. This indicates that it might be instructive to consider some numberical properties of the two bodies that would be got by reassembling the material of these two sets of present planets. (For conciseness I here call the Moon a planet.) When we use the latest available determinations of the parameters of these planets, the outcome is indeed remarkable.

The most recent comprehensive revision of the required parameters, on the basis of modern radar and optical observations, seems to be that published in 1967 by Ash, Shapiro and Smith. They derive new values of all the required masses (with the solar mass as the unit) and of the radii of Mercury and Venus. Here I use the values they get for the "general relativity fit", but the differences between these and the "Newtonian fit" are very small.

As regards parameters not evaluated by these authors, for the volume of Mars, I use the value inferred by Lyttleton from the work of de Vaucouleurs; for the volumes of the Earth and the Moon, I use the values listed by Allen. Also I use the value of the solar mass given by Allen in order to express the results in familiar units. The results are shown in Table 1. The density p(MV) of Mercury and Venus combined is simply the total mass divided by the total volume of the two bodies; similarly, for the density of the Earth, Moon and Mars combined, p(EMM). The density p(T) of all combined is the total mass of all these bodies divided by the total volume. The values are rounded off to the number of decimals shown.

The remarkable result is the near equality of the densities of the combined systems. Were Mercury and Venus combined into one body of the same total volume, and the Earth, Moon and Mars into another, we see that their densities would differ by less than 2 parts in 1,000 from the mean. So far as densities serve to indicate composition, we should have two bodies of effectively identical composition.

The numerical result does seem to be significant for several reasons. (a) If the same calculation is applied to the values of the planetary parameters tabulated by Allen, which were the best available only a few years ago, then in place of the densities $\rho(MV) = 5.25$ and $\rho(EMM) = 5.27$ shown in Table 1, we find $\rho(MV) = 5.13$ and $\rho(EMM) = 5.28$. Thus the effect of the much improved precision, due mainly to the radar measurements, is to reduce the difference between the combined densities from 0.15 to 0.02. This is good evidence that the difference really is very small. (b) Although the difference is so small, the published standard errors indicate that it could be significant. It is in the direction $\rho(EMM) > \rho(MV)$, and this is what we should expect if the combined chemical compositions are indeed the same. For the compressibility of the material would have most effect in the case of the Earth, which is the most massive of all the individual bodies, and overall there must be a little more compression in the bodies EMM taken together than in the bodies MV taken together. Thus the fact that there is this very small difference in the densities is actually evidence in favour of identical combined chemical compositions. (c) As we shall now see, the result makes sense of the apparent vagaries of the densities of the individual bodies, and in doing so lends further support to Lyttleton's suggestion.

Table 1. Masses (Multiples of 10^{27} g) and Densities (g cm^{-3} of Terrestrial Planets and of Combinations of These

	Mass	Density			Mass	Density
Mercury	0.330	5.47		Earth	5.974	5.51
Venus	4.872	5.24		Moon	0.073	3.34
				Mars	0.639	3.90
MV combined	5.202	5.25	$\rho(MV)$	EMM combined	6.687	5.27 $\rho(EMM)$

All combined: mass = 11.889; density = 5.26 = $\rho(T)$

The values shown in Table 1 are plotted in Figure 1. This shows that, as these planets exist now, there is no simple pattern in their densities as associated with their masses. Neither is there any clear trend in the densities as associated with the distances from the Sun. Now the variety of densities demonstrates that there are wide differences in the chemical compositions of these bodies. This would be astonishing were the bodies formed separately out of the same raw material (in the absence of a correlation with mass, that might arise from different rates of loss of certain constituents, or of a correlation with distance from the Sun, that might result from a sorting out of constituents of the raw material). The difficulty is elegantly resolved if the planets as they

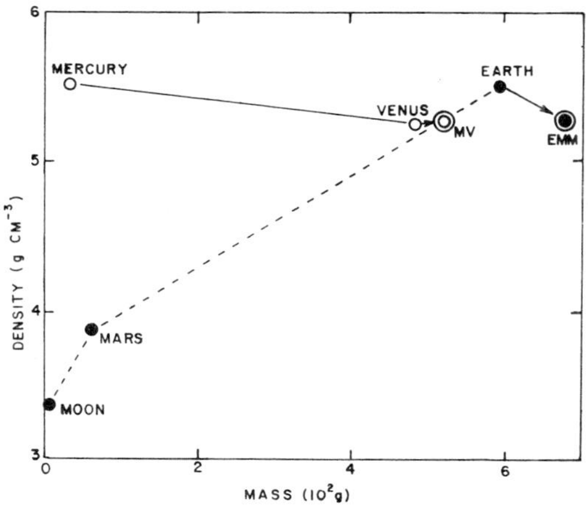

Masses and densities of the terrestrial planets. MV = Mercury and Venus combined. EMM = earth, moon, and Mars combined.

exist now were in fact produced by the break-up of two bodies having identical overall chemical composition. This is a compelling argument in support of this process of formation.

I recently pointed out that the densities of the Earth, Moon and Mars are apparently understandable if these bodies resulted from the break-up of a single rotationally unstable planet. In that case, the smaller portions have the lower densities, and this seemed to be a natural consequence of their coming from the outer part (mantle) of the unstable body after its material had undergone some segregation. Here the original planet was considered to become unstable as a whole and then to break up in accordance with Lyttleton's discussion. In the MV case, on the other hand, if the core of the original planet became rotationally unstable, before the body as a whole became unstable, then Lyttleton's ideas would have to be applied to the core and not to the whole body. If the core broke up in accordance with those ideas, then the smaller part would carry away with it proportionately less of the surrounding lighter (mantle) material and so this would ultimately produce the portion having the higher density. As regards the general notion that fission of a rotationally unstable core might be of astronomical significance, Roxburgh has already considered this as the fundamental process in a theory of the formation of a certain class of binary stars. As regards the possibility of some difference in rotational behaviour between the MV and the EMM systems, Lyttleton has in this context recalled that the Earth and Mars rotate with periods both about 24 h, whereas Mercury and Venus have very much longer periods of axial rotation.

The main point is that, in order to start with a common supply of raw material and to finish up with several planets possessed of a variety of masses and densities showing no simple correlations among themselves, we know no alternative but to suppose that one or more bodies were formed as an intermediate stage, that segregation of material occurred within them, and that these bodies then broke up in various ways to yield the existing planets. The only mechanism we know for such a break-up in rotational instability. From the work of Lyttleton and others, we know the possible outcomes; in particular, the break-up of one unstable body yields two main fragments of disparate mass. In order to produce the planets under discussion, two intermediate bodies are therefore needed. These, being formed directly out of the raw material, would have identical compositions. The foregoing discussion makes it highly likely that two such bodies did in fact exist.

The raw material, and the bodies formed directly from it as the intermediate bodies just mentioned, would have included much light gas, mainly hydrogen and helium. Such gas would have been lost in the subsequent evolution. This would not affect any of the arguments because the loss would affect both intermediate bodies in the same way. It is, however, a crucial feature in two respects. In the first place, it implies that the intermediate bodies were formed by condensation from relatively slowly rotating bodies of more diffuse material. This would explain how they became unstable, as a result of the increasing angular velocity produced by increasing condensation. In fact, it is in principle the only way we know in which a body could become rotationally unstable in nature. Were it built up "in the cold" directly from planetesimals, which is the other means of forming planets frequently discussed, it would never be unstable; contrariwise, this is a means of avoiding the "angular momentum problem" that leads to instability. In the second place, the process of condensation must result in segregation of materials in the manner discussed in previous work. Thus the unstable body would also be a body of nonuniform composition whose break-up would produce fragments possessing different densities. Conversely, if we regard the characteristics of the present terrestrial planets as evidence for the earlier existence of rotationally unstable planetary bodies, this would afford evidence that such bodies were originally in a much more diffuse state containing a large excess of hydrogen and helium.

CRITIQUE OF "THE RESONANT STRUCTURE OF THE SOLAR SYSTEM" . . . . .

Backus, G. E.; *Icarus,* 11:88-92, 1969.

Abstract. A. M. Molchanov's eight resonance relations among the orbital frequencies of the nine planets in the solar system are satisfied not exactly but with a relative error of 4.5×10^{-4}. It is shown that with such a large error, those relations can be the result of chance. Such relations could be expected among any nine numbers chosen at random.

I. Introduction. In a recent paper Molchanov (1968) lists eight "resonance relations" among the observed frequencies of revolution of the nine planets. By a resonance relation he means an equation of the form

$$\sum_{i=1}^{9} n_i x_i = 0, \tag{1}$$

where $x_1, \ldots, x_9$ are the planetary orbital frequencies and $n_1, \ldots, n_9$ are small integers. He points out that eight such relations together

Table I. Frequencies and Resonance Vectors of the Solar System

	Planet	x_i^{obs}	x_i^{calc}	$n_i^{(1)}$	$n_i^{(2)}$	$n_i^{(3)}$	$n_i^{(4)}$	$n_i^{(5)}$	$n_i^{(6)}$	$n_i^{(7)}$	$n_i^{(8)}$
1	Mercury	0.902345	0.902473	1	0	0	0	0	0	0	0
2	Venus	0.353262	0.353233	-1	1	0	0	0	0	0	0
3	Earth	0.217321	0.216971	-2	0	1	0	0	0	0	0
4	Mars	0.115546	0.115298	-1	-3	-2	1	0	0	0	0
5	Jupiter	0.018321	0.018343	0	0	1	-6	2	1	0	0
6	Saturn	0.007378	0.007337	0	-1	-1	0	-5	0	0	0
7	Uranus	0.002587	0.002620	0	0	1	-2	0	-7	1	1
8	Neptune	0.001319	0.001310	0	0	0	0	0	0	-2	0
9	Pluto	0.000875	0.000873	0	0	0	0	0	0	0	-3
	$E^{(r)} \times 10^4$			-11.1	-7.55	-2.41	4.46	-2.46	2.13	-0.508	-0.386

with knowledge of the scale of the system suffice to determine the frequencies. Of course if these eight resonances are real they are of great importance in understanding the dynamics of the solar system.

In fact, as Molchanov points out, his eight resonance relations (1) are not satisfied exactly by the observed planetary orbital frequencies. The sums on the left in (1) do not vanish, but are typically of the order of 10^{-4} times the orbital frequency of Jupiter. We will try to show in this paper that such an error is too large to rule out the possibility that the eight resonance relations are the result of chance. Any nine positive numbers chosen at random will very likely satisfy eight resonance relations of Molchanov's sort, with errors no larger than those generated by the solar system data.

....

RESONANCES IN COMPLEX SYSTEMS: A REPLY TO CRITIQUES

Molchanov, A. M.; *Icarus*, 11:95–103, 1969.

Abstract. The conclusions reached by Backus and Henon, that resonance relations in the solar system of the sort proposed by Molchanov are a result of chance, is based on a very crude statistical model. A more accurate model gives a value $P \sim 10^{-10}$ for the probability of chance formation of systems similar to the solar system.

CAPTURE RESONANCE OF THE ASTEROID 1685 TORO BY THE EARTH

Danielson, L., and Ip, W.-H.; *Science*, 176:906–907, 1972.

In the solar system there are many resonance phenomena due to different mechanisms. The motions of two planets or satellites can couple together so that the ratio of their orbital periods has the value of p/q where p and q are small integers. This kind of coupling also occurs between the spin of one body and the orbital motion of another. The most remarkable example is the orbital coupling between Neptune and Pluto. Due to its large eccentricity, the orbit of Neptune. In spite of this the two bodies can never collide because, as found by Cohen and Hubbard, these two planets are in a 3/2 resonance such that the distance between them is always larger than 18 astronomical units (A. U.). This effect implies that Neptune and Pluto may have coexisted since the beginning of the solar system.

This result leads to the question whether stray bodies could also be stored in the inner part of the interplanetary space by similar effects, namely, by resonances with the Earth or Venus, or both. In this report we discuss the asteroid 1685 Toro.

The present orbital elements of Toro show that its period is almost exactly 1.6 years. Further, the relative positions of Toro and the Earth exhibit a certain symmetry, including close approaches. This capture resonance, if we may call it so, has been investigated in more detail by integrating the perturbed orbit of Toro for 100 years backward and forward, that is, for a total of 200 years. The perturbations by Venus, Earth, Mars, Jupiter, and Saturn are included in the orbit integration according to Cowell's method.

. . . .

Naturally, a time span of 200 years is insufficient for determining whether Toro is caught in a permanent resonance. Since we know that Toro comes at least as close as 0.15 A.U. to the Earth, we consider an investigation based on the secular variation of the elements rather unreliable. Further calculations are being made in order to study the stability of the resonance. One of the objects of studying orbits like that of Toro in detail is that a resonance orbit such as this is one of the possible precapture orbits for the Moon. Another is that the existence of orbits like Toro's may be important for determining the lifetimes of the Apollo asteroids.

In summary, a calculation of the orbit of the asteroid 1685 Toro over 200 years shows that it is in at least a temporary 8/5 resonance with the Earth. The amplitude of the libration around the equilibrium position is about 8°, and the period of libration is about 144 years. Toro is also very close to 13/5 resonance with Venus.

BODE'S LAW AND THE MISSING PLANET

Ovenden, M. W.; *Nature*, 239:508-509, 1972.

Exactly two hundred years ago, Titius published a mnemonic for the mean distances of the planets from the Sun. His rule was

$$a_i \propto 2^{(i-2)} \cdot 3 + 2^2$$

where a_i is the major semi-axis of the orbit of the ith planet from the Sun. Titius's law represents the distances of the then known planets with an accuracy of a few per cent, provided that (i) for Mercury, we take $i = -\infty$ instead of $i = 1$ and (ii) the orbital $i = 5$ is left vacant. The law made three valid predictions. Uranus (discovered by William Herschel in 1781) fits the orbital $i = 8$. After the discovery of Uranus, Bode publicized the law, which became known as Bode's Law. The search for a planet for $i = 5$ culminated in the discovery of the first asteroid, Ceres, in 1801. Recognition of a similar law for the satellite system of Saturn led to the discovery of Hyperion in 1848.

While no doubt Bode's law as it stands is not of precise physical significance, it has long been asked whether the distribution of planetary (and satellite) orbits is due to chance, or to some physical principle that would operate in any planetary system. Possibilities for such a physical process are (i) properties of the mechanism of formation of the solar system, (ii) mutual planetary perturbations subsequent to the formation of the solar system, such perturbations being either the simple point-mass gravitational perturbations, or else perturbations involving dissipative forces (such as tidal action).

I have shown that the point-mass perturbations (without dissipative forces) are a sufficient explanation of the existing distribution of satellite and planetary orbits, affording a reason for the present planetary distances to an accuracy $\sim 1\%$. The following is a brief summary of my arguments.

The intuitively obvious conclusion that a set of planets will rapidly change their configuration when their mutual interactions are strong---and will change only slowly when their interactions are weak---is confirmed by numerical integrations of simulated satellite systems. I generalize this result into the "Principle of Least Interaction Action".

A satellite (or planetary) system of N point masses moving under their mutual gravitational attractions will spend most of its time close to a configuration for which the time-mean of the action associated with the mutual interactions of the satellites is an overall minimum. [Mathematical details are omitted here.]

The following results have been obtained, which confirm the essential validity of the techniques used. (i) The present distribution of the five satellites of Uranus is within 5% of the calculated minimum interaction action distribution, with the exception of the innermost body, UV, whose mass is given to only one significant figure. The discrepancy can be reduced by increasing the mass of UV. The time-scale is $\sim 2 \times 10^9$ yr. (ii) The five inner satellites of Jupiter form a quasi-isolated system. Their present distribution is also within 5% of the distribution of minimum interaction action, except again for the innermost satellite, JV. In this case no mass at all is tabulated for JV, and the assumed mass of 10^{-6}x the mass of Jupiter may well be wrong. The evolutionary time-scale is $\sim 6 \times 10^7$ yr. (iii) The triplets UV, UI, UII and JI, JII, JIII show the Laplace relationship between the mean motions

$$\text{Jupiter } n_1 - 3n_2 + 2n_3 = n_1 (2 \times 10^{-7}) \simeq 0$$
$$\text{Uranus } n_5 - 3n_1 + 2n_2 = n_5 (3 \times 10^{-4}) \simeq 0$$

The slower evolution of the Uranian system is the reason for the larger proportional discrepancy from zero of the resonant relationship. The time required to approach within the observed precision of resonance is 2×10^9 yr for the Jovian system and 6×10^9 yr for the Uranian system. It is remarkable that two systems, for which the relative discrepancies differ by four orders of magnitude, should not merely give the same evolution time to better than an order of magnitude but that this time should be the age of the solar system, determined from meteorites to be 4.5×10^9 yr.

I now apply the technique to the planetary system, taking first the system of Jupiter, Saturn, Uranus and Neptune (JSUN) . Line 4 of Table 1 shows the minimum interaction action configuration for this system. Line 3 shows the present major semi-axes. The discrepancies are large, despite the fact that the time-scale for evolution of this system to minimum $\sim 5 \times 10^8$ yr.

The discrepancies may be precisely removed if it is assumed that material (referred to as A) of mass 90 ± 5 $M_\oplus$ (where $M_\oplus$ is the mass of the Earth) had been in the asteroid belt from the beginning of the solar system until 1.6 ± 0.4 x 10^7 yr ago, and that it then dissipated. Since then, the JSUN system has been evolving towards its new minimum interaction action distribution. Line 1 of Table 1 shows the minimum interaction action distribution of the system AJSUN, which it would have reached on a time-scale ~ 5 x 10^8 yr. Line 2 of Table 1 shows the configuration that JSUN had 1.6 x 10^7 yr ago, as found from my evolutionary sequence.

Table 1. Major Semi-Axes (A.U.) of Outer Planets

A	J	S	U	N	
2.794	5.211	9.509	19.46	29.71	Minimum config.
--	5.206	9.506	19.40	29.74	-1.6 x 10^7 yr
--	5.203	9.539	19.18	30.06	Present t = 0
--	5.15	10.5	12.7	37.3	Minimum config. (+ 2.8 x 10^8 yr)

$M_A = 90\ M_\oplus$.

The time-scale for the evolution of the residue of A (taken to be 0.1 $M_\oplus$) presently in the asteroid belt is such that changes of only 0.05 A.U. are to be expected in 10^7 yr. The fact that the minimum configuration has A close to the present mean major semi-axes of the asteroid belt is an additional piece of evidence of the internal consistency of the interpretation. It should be noted that the sequence of minimum configurations as a function of M_A and the evolutionary sequence of JSUN are both one-parameter systems, so a match of five separate quantities to a precision ~ 0.3% is very unlikely by chance. Not only that, but the time found for the dissipation of A is a physically-meaningful time, being in reasonable agreement with the longest cosmic-ray shielding times for chondrite and achondrite meteorites. The picture is therefore consistent with the break-up of a single body into pieces at this epoch, some already below the critical cosmic-ray shielding size, others breaking to that size by subsequent collisions.

The confirmatory evidence of the former presence of A comes from considering now the system of the terrestrial planets Mercury, Venus, Earth and Mars from Jupiter, if the hypothesis is correct, the present positions of MeVEMsJ should correspond to the evolution, through 1.6 x 10^7 yr, of this system from the minimum interaction configuration of MeVEMsAJ since the evolution time of MeVEMsAJ ~10^9 yr. In fact, the characteristic time of evolution of the MeVEMsJ system ~ 8 x 10^{11} yr, so that effectively the present positions of MeVEMsJ should correspond to the minimum interaction action distribution of MeVEMsAJ. The agreement shown in Table 2 is good. The major discrepancies are for the planets Mercury and Mars, which have high eccentricities (0.205 and 0.093 respectively). The errors are thus no larger than would be expected from the approximations made in calculating R for an elliptical orbit as though it were a circular orbit.

Table 2. Axes of Inner Planets

	Me	V	E	Ms	A	J
Minimum config.	0.394	0.719	1.00	1.49	4.79	5.20
Present config.	0.387	0.723	1.00	1.52	--	5.20
Minimum config.	0.461	0.696	1.02	1.24	--	5.21

$M_A = 90\ M_\oplus$.

I claim that the approximate treatment described above is adequate to deal with the problem, and that the agreement of observation with prediction and the complete internal consistency of the evolutionary picture give compelling reasons for accepting the essential correctness of the hypotheses put forward.

Two major problems remain. First, what physical process caused the sudden dissipation of A? From the point of view of the dynamical arguments presented here, it is probably true that A was always in the form of a ring. But while it may be difficult to "explode" a planet, it would seem even more difficult to dissipate a ring suddenly after it has been quiescent for 4.5×10^9 yr. Second, only $\sim 0.1\ M_\oplus$ seems now to reside in the asteroid belt. What has happened to the other $89.9\ M_\oplus$?

THE TITIUS-BODE LAW AND THE EVOLUTION OF THE SOLAR SYSTEM

Nieto, M. M.; Pensee, 4:5-7, Summer 1974 (Also:*Icarus*, 25:171-174, 1974)

I have been asked to discuss the Titius-Bode Law of Planetary Distances: does it shed any light on whether or not there has been any recent large-scale evolution in the composition of the solar system? I will aim to stick directly to this question---even though it may not be a question. Let me explain that enigmatic statement.

Before we can know whether the subject of this talk means anything, we have to know whether the "Titius-Bode Law of Planetary Distance" is indeed a "law." If it is not a "law," then clearly the "law" says nothing about anything, we can skip the rest of this talk, drop the discussion, and concentrate on something useful, like Watergate.

In modern notation, the original law can be stated this way: the distance to the nth planet is

$$R_n = (4) + (3)\,(2)^n, \quad n = -\infty,\ 0,\ 1,\ 2, \ldots \qquad (1)$$

where $n = \infty$ for Mercury, 0 for Venus, 1 for Earth, etc. The law is normalized to 10 for the Earth. It "predicted" the discovery of the asteroid belts, Uranus, and was used as the basis for computations to discovery Neptune, even though it failed badly for Neptune and Pluto (see Table 1). [Table omitted]

The question thus arises, were the early successes of the law just coincidences, the true worth of the law being the "failures" for Neptune and Pluto?

As I emphasized in my recent book on the subject, before one can begin to answer that question one must realize that equation (1), with its ad hoc first term, should not be thought of as the only possible parametrization, but only as a first guess. In fact, if one looks at the actual distances of the planets from the Sun and searches for the best pure geometric progression fit, one finds that the geometric progression

$$R_n = 3.34\,(1.73)^n, \quad n = 0,\ 1,\ 2, \ldots \qquad (2)$$

gives the fit shown in Figure 1. The actual distances oscillate about

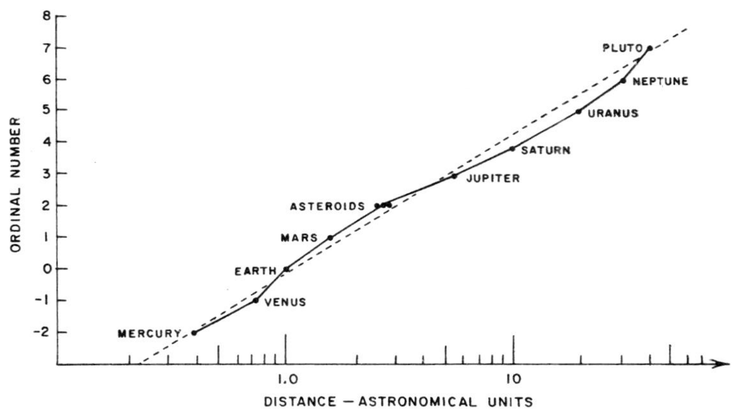

Planet ordinal number versus distance from the sun

the pure geometric progression, so that one can represent the "true" curve as

$$R_n = 3.34\ (1.73)^n\ f(n), \tag{3}$$

where f(n) is a function which oscillates about 1. (Similar distributions can be made for the satellite systems of the major planets.)

From all this, can we decide whether or not the "law" is a "law"? The answer, unfortunately, is "NO." The "predictions" of the law are tantalizingly close to the observations (especially given the early history of the law, when it predicted Uranus and the asteroids). But they are not right on top of the planets. Further, by the time you add in the periodic oscillation about the geometric progression one is left with "intuition" as to the over all validity, and "intuition" presently goes both ways. One needs a theory that predicts (in whatever mathematical format) both the progression and the oscillations about it.

My own feelings are that if the law is a "law," then the split into the geometric progression and the oscillation come from two separate processes. The geometric progression, I feel, would come from the early history of the solar system, when the material that would eventually form the planets was in the form of a rotating disk of gas and dust around the proto-Sun. Then, as in a theory like Hoyle's energy and angular momentum would be transferred to the disk by the "magnetic brake" mechanism. This mechanism is possible because the whole system would be a plasma, so that a magnetic coupling could be made between the rapidly rotating proto-Sun and the disk. This would slow down the Sun and speed up the disk and make it larger.

In such a system, turbulence would be set up. Further, it is known that such systems can set up regular turbulence patterns, which in turn would lead to preferred condensation at the edges of the patterns. This as explained elsewhere, could in principle set up the geometric progression.

The point I have made is that I do not see how point-gravitational and/or tidal interactions, either as or after the planets were formed, could set up the geometric progression. The problem there is that such interactions tend to set up resonances and/or stable patterns that have preferred or antipreferred distances which involve ratios of periods of orbits---not a geometric progression of 1.73. (Examples of this phenomenon are the rings of Saturn and Mimas or the Trojan asteroids and Jupiter.)

The oscillation about the geometric progression represents, I believe, a point-gravitational or tidal evolution starting from after the planets were formed, and slowly reaching equilibrium. Calculations that have been performed on the time scales of such evolutions all seem to indicate that this evolution would have been completed long ago.

Thus (always keeping in mind that we don't know whether the Titius-Bode law is a law) if one can believe in the law, then this would indicate that there has been no large-scale evolution in the composition of the solar system in the recent past.

Now let me come to two discussions of the Titius-Bode Law which would take exception to these conclusions:

The first was put forward by Bailey, just before the first Apollo landing on the Moon in 1969. It raised an old question which dates back centuries, the idea of a missing planet in the interior of the solar system. Bailey proposed that perhaps the Moon is a lost planet, "Luna," which perhaps once existed in an unstable orbit between Mercury and Venus, and which was captured by the Earth. Bailey did not date this process, except to state that capture should have occurred in the last half of the age of the solar system.

The connection with the Titius-Bode Law comes about by considering the original form of equation (1) and then figuring out what changes in Mercury's orbit would have occurred by the transfer and capture. Bailey fits the original form with the "original location" of Luna and the "original location" of Mercury. The problems are what I consider to be the ad hoc nature of the first term in the Titius-Bode Law, and the detailed problems of how and when capture occurred. If one were to choose, one could like the capture to have been recent, but one just cannot show it.

The other theory involving the Titius-Bode Law was put forward recently by Michael Ovenden. He placed in a new perspective the idea of Olbers that the asteroid belts are the remnants of an ancient planet which was disrupted, for some reason. Ovenden tried to explain the Titius-Bode Law by looking at the gravitational evolution of the solar system from a set of initial conditions to its present configuration. He found that he could do that if he assumed that there used to be a planet of 90 Earth masses where the asteroid belts now exist, and that 16 million years ago the planet, for some unexplained reason, disrupted, leaving the remaining system to evolve gravitationally to the present configuration.

The critical point, of course, is what could have caused the explosion, and left the asteroid belts with only 0.1 Earth masses, all essentially in the plane of the ecliptic? Napier and Dodd looked into this question and, given the energy and angular momentum constraints, decided that the event was almost impossible to reconstruct using gravitational, nuclear, or chemical interactions.

So there you have it. The Titius-Bode Law, since its validity as a physical law is by no means proven, cannot give you a definite answer as to implications for possible recent, large-scale evolution of the solar system. But I think you can make the statement that if you argue that the law is indeed a law, and that the solar system obeys this law, then it speaks against recent large-scale evolution.

Let me put it this way: If God were to come and tell us that the law is a law and that the planets obey it, then the present configuration of the planets is something which "had to come about." On the other hand, since recent large-scale changes in the configuration of the solar system would have to be a random or accidental event (something like just happening to be under a rock when it fell), how could this chance happening have finally brought the planets into agreement with a law which they already were supposed to obey independently?

To sum up, I think you can believe in the Titius-Bode Law, or recent large-scale evolution, or neither of them. But I don't think you can believe in both of them.

STABILITY OF THE SUN-EARTH-MOON SYSTEM

Szebehely, V., and McKenzie, R.; *Astronomical Journal,* 82:303-305, 1977.

Abstract. The models of the restricted and general problems of three bodies are used to determine the stability of the Sun-Earth-Moon system by means of surfaces of zero velocity. Hill's result is verified by the model of the restricted problem as long as the ratio m_E/m_S 2.52×10^{-6}. The model of the general problem, on the other hand, contradicts this result and we show that the eccentricity of the Earth's orbit renders the system unstable by opening the surface of zero velocity. It may be concluded, therefore, that the Moon may escape from the Earth and may become a planet or, in reverse, that the planetary origin and the capture of the Moon by the Earth becomes a strong dynamic possibility.

Chapter 16

STARS, GALAXIES, AND COSMOLOGY

INTRODUCTION

The universe beyond the solar system is immense, complex, and subtle---so different that it seems to defy terrestrially formulated laws. The comfortable world of Newtonian physics and the usual conservation laws are shattered by quasars, pulsars, and black holes. Even as astronomy struggles with these foundation-shaking ideas, new anomalies keep appearing: unidentified X-ray bursts, gravitational radiation, hierarchies of galaxies, etc. A sampling of these anomalies as well as a few old-fashioned curiosities are presented in this last chapter in the following order:

- Anomalous celestial objects
- Cosmological enigmas
- Unusual extraterrestrial radiations
- Tests of relativity
- Physical laws and constants

ANOMALOUS CELESTIAL OBJECTS

The simplest way to classify those points of light and radio energy fixed on the celestial sphere is to ask whether they seem to be single objects (stars) or those great island universes of stars called galaxies. Single, telescopically unresolved stars would not seem to offer much opportunity for the identification of anomalies, but even here there are flickering stars, jumping stars, disappearing stars, and so on. As the telescope is turned on objects that are apparently beyond our own galaxy, some strange entities come into our ken. There are all manner of exotic galaxies that possess jet structures, dumbbell shapes, and wildly varying emissions. All these mavericks are supposed to be well beyond our own galaxy by virtue of their red shifts. But even this foundation stone of astronomy must be questioned, too.

Returning to the more familiar confines of our own stellar system, the question of an unseen companion to the sun comes to the fore. Next there is the usually dismissed problem of electrical phenomena in stellar astronomy. The Drayson episode comes last. It is an all-but-forgotten annoyance afflicting the astronomers of a century ago, but its lesson should not be forgotten.

• Anomalous Stars

THE SEVENTH STAR OF THE PLEIADES

Anonymous; *Science*, 86:sup 6, July 23, 1938.

Corroboration of a world-wide legend, rooted in ancient mythology, that once the six resplendent star "sisters" of the Pleiades numbered seven, has been offered by Dr. William A. Calder, of Harvard Observatory. The star "Pleione," identified by astronomers as "Number Seven" of this group, has been suspected in the past as the mysteriously disappeared sister, and careful comparative measurements of stellar magnitudes in this region by Dr. Calder tend to confirm this suggestion.

Pleione was observed to diminish in light about a sixth of a magnitude in a slow continuing decline during the winters of 1935-36-37. Before this investigation, slight changes in the brightness of certain members of the group had been suspected. In this study, Harvard cameras utilized a potassium-hydride photoelectric cell, permitting very exact detection of slow or minute variations in the star light. In all, the relative brightnesses of twenty-five of the most conspicuous stars in the Pleiades region were observed during the three winters of the survey. The report includes a reminder, which is not elaborate, that the spectrum of Pleione formerly had emission lines and resembled that of P Cygni, a star that was at one time a nova. In recent years the bright lines of Pleione have disappeared.

"That some change has taken place in the Pleiades is borne out by tradition," Dr. Calder pointed out. "Almost all nations of the earth have legends about the 'seven who are now six.' The surprising universality of this impression is difficult to explain unless a now diminished seventh Pleiad formerly was conspicuous." Six Pleiades are normally visible to the unaided eye, but under exceptional conditions double this number have been noted. Telescopes show a population of several hundred stars which, for the most part, are members of a physically related aggregation, as is shown by a general unanimity of motion.

NATURE OF HOAG'S OBJECT

Gribbin, John; *Nature*, 250:623-624, 1974.

It is now nearly a quarter of a century since Hoag discovered a remarkable astronomical object which appeared to be a "perfect halo" surrounding a diffuse nucleus. But only now is Hoag's object, as it has come to be known, receiving the attention which its peculiarity merits. According to O'Connell, Scargle and Sargent (Astrophys. J., 191, 61-62; 1974) the object is certainly worth investigating and "one of the more exotic possible interpretations" is that the ring is the image of a background galaxy produced by a gravitational lens effect.

Hoag's object is clearly visible on Sky Survey prints (at a = 15 h 15.0 min; $\beta = +21^{\circ}$ 46') but it has not previously been studied in detail. The core of the object is a fuzzy circle 6" in diameter, and the surrounding blue ring or annulus has inner diameter 28" and outer diameter 45" with an axial ratio of 0.93 ± 0.05. The ring is structureless, and there is no sign of connecting material bridging the gap between core and annulus.

In the study now reported by O'Connell et al. new information has been obtained from spectrograms taken with the 200-inch telescope and the Lick 120-inch. The observed redshift of 12,740 km s^{-1} and an assumed Hubble constant of 75 km s^{-1} Mpc^{-1} imply a distance of 170 Mpc, corresponding to a linear diameter of 5 kpc for the core and inner and outer radii of 23 kpc and 37 kpc for the ring.

Almost all of the properties of the core seem to be in line with those expected for compact galaxies, but there seems little doubt that the ring is "an independent structural component ... and not simply the unobscured rim of a normal SO disk". Several other similar systems are known, and O'Connell et al.cite the examples of NGC6028, NGC2859 and IC5285. "There is no reason to suppose that radiation from these rings is anything other than direct starlight", and in the case of Hoag's object the radiation certainly cannot be reflected light from the core, since the ring is "at least" as luminous as the core.

The interpretation which seems to be favoured by O'Connell et al. involves gravitational encounters between galaxies which could initiate star formation in "trapped" rings; they point out that a pair of galaxies near Hoag's object on the sky could, if they are at the same distance as the object, have passed by it 10^8 to 10^8 yr ago, just right to account for the apparent age of the stars in the ring.

CURIOUS ASTRONOMICAL RECORD

Botley, Cicely M.; *Sky and Telescope*, 33:339, 1967.

I have come across a curious astronomical record that is just 750 years old. The original source is the medieval Latin chronicle of Burchard and Conrad, which is quoted in a German magazine, Wochenschrift fur Astronomie, Meteorologie, und Geographie, 1881, page 328.

In condensed form, this is their story:

"A. D. 1217. That autumn, in the early evening, a wonderful sign was seen in a certain star in the west. This star was located a little west of south, in what astrologers call Ariadne's Crown [Corona Borealis]. As we ourselves observed, it was originally a faint star that for a time shone with a greater light, and then returned to its original faintness. There was also a very bright ray reaching up the sky, like a large tall beam. This was seen for many days that autumn."

Could the star that became temporarily bright have been the recurrent nova T Coronae Borealis during an unrecognized maximum in A. C. 1217? The two recorded outbursts of this nova were in 1866 (magnitude 2) and 1946 (magnitude 3). I suspect that the narrative may deal with two distinct phenomena. The "very bright ray" suggests a comet tail, but no bright comet seems to have been recorded that year.

Instead, the beam may well have been the zodiacal light, somewhat enhanced by auroral activity. In Prior's Annals of Dunstable, there is a record of a canon of Dunstable Priory, Bedfordshire, England, seeing on October 27, 1217 (Old Style), "a very large cross passing with great magnificence from west to east." If we take the term "many days" to refer to the star, the beam could have been a record of the same auroral ray.

THE ANOMALOUS CEPHEID MYSTERY

Anonymous; *Science News*, 110:308, 1976.

Variable stars have fascinated astronomers at least since the medieval Arabs gave Algol its ghostly name. In recent years observation of variables has been very helpful in the formation of astrophysical and cosmological theories. Now, a study of an unusual class of variables leads two astronomers from the Hale Observatories, Robert Zinn and Leonard Searle, to suggest either that the two members of a binary star system may sometimes coalesce into a single star or that something strange has happened in the evolution of some small galaxies that are close to our own. Those at least appear to be the most tenable hypotheses of six brought forward by Zinn and Searle in the Nov. 1 Astrophysical Journal to explain the character of a group of variable stars they refer to as "anomalous Cepheids."

There are several classifications of what might be called well-behaved Cepheids, including classical, globular-cluster and RR Lyrae varieties. The classical and globular-cluster Cepheids have the useful characteristic that their intrinsic brightness varies according to the length of the period of their variation. Thus, by observing the periods of these kinds of Cepheids in another galaxy, astronomers can figure out their intrinsic brightnesses. Comparing intrinsic and apparent brightnesses will yield the distance to the galaxy by a simple law of optics. Thus these kinds of Cepheids are important for establishing cosmological distance scales. (The intrinsic brightness of RR Lyrae types is apparently the same---Zinn and Searle assume it so---no matter what their periods may be.)

What Zinn and Searle call anomalous Cepheids do not obey any of these laws. The distribution of the anomalous Cepheids is a bit curious. They are found in all the dwarf spheroidal galaxies near our own that have been searched: the Draco, Sculptor, Leo II and Ursa Minor systems. They are abundant in the Small Magellanic Cloud, but none have been found in the Large Magellanic Cloud. They are also virtually nonexistent in the globular clusters of stars that form a halo around our galaxy. It is this distribution that makes them a mystery so long as they are to be regarded as a single astrophysical class with a common sort of origin.

In an attempt to determine something about the origin of the anomalous Cepheids, Zinn and Searle studied seven stars in the Draco galaxy, five anomalous Cepheids and two RR Lyrae stars. The main purpose was to discover the mass of the anomalous Cepheids by comparison with the RR Lyrae stars. (There are relations between density and period and luminosity and radius that allow this to be done.) A hypothesis to be tested had suggested that the properties of the anomalous Cepheids could be explained by regarding them as having approximately twice the mass of the sun and being very poor in metals. That hypothesis might work for the anomalous Cepheids in the Small Megallanic Cloud, where stars are known to be still forming out of metal-poor gas, but it runs counter to what is believed about conditions in the dwarf galaxies.

It turns out that the anomalous Cepheids in the Draco galaxy are heavier than its RR Lyrae stars. This, Zinn and Searle state, is the main result of their investigation, and it tends to prove the hypothesis they started with. They give six possible explanations for this circumstance and the most telling objection to each. The two that are least objectionable are that the progenitors of the anomalous Cepheids were close binaries that somehow coalesced into a single, large-mass star or that the anomalous Cepheids are more massive because they were born so; that is, the anomalous Cepheids are younger than the RR Lyrae stars.

Either explanation involves a dilemma: It appears from previous studies that at least one kind of binary can coalesce into a single star, but why should these systems be abundant in the dwarf galaxies and virtually absent in the globular clusters and the Large Magellanic cloud?

On the other hand, if the anomalous Cepheids are younger than the RR Lyrae stars, how can they appear in a system like Draco, which is too loosely constructed and has too weak a gravitational field to have retained the gas from which young stars might form? This objection can be avoided if one assumes, as several astronomers have proposed, that all these dwarf galaxies were once part of a single system (called the Greater Magellanic Galaxy) that somehow broke apart. If so, the dwarfs might have retained gas from the parent system, or the Cepheids may have formed in the parent system before the breakup. But for them all to have been part of one progentitor, they should lie more or less in the same plane. Unfortunately, Sculptor does not lie in one of the two Greater Magellanic planes that have been proposed, and Leo II does not lie in the other.

Thus, although the investigation seems to have verified the hypothesis it began with, that verification has raised an even greater mystery, and the solution of that mystery should have an important bearing on the theory of binary stars, the theory of galactic morphology or both.

FLICKERING STARS SIGNAL NEW CHALLENGE TO ASTRONOMERS

Anonymous; *New Scientist*, 70:526, 1976.

Do large numbers of stars show small oscillations in size like those recently reported for the Sun? This is one interpretation being discussed for unpublished observations which reveal erratic brightness changes of a few per cent in stars hitherto considered to be constant in light output. When fully available, these results are certain to send astronomers scurrying to their telescopes in search of further additions to this entirely new category of variable stars.

The observations have been made over the past four years by Norman Walker of the Royal Greenwich Observatory. He used a sensitive photometer built by the RGO's electronic department, capable of measuring light fluctuations of a few thousandths of a magnitude. The photometer was attached to the 30-cm telescope at Granada University Observatory in the Sierra Nevada mountains of southern Spain, chosen because its 8600 ft (2600 m) altitude provides particularly clear observing conditions.

Walker was looking for possible light variations in a rare, chemically peculiar family known as the Lambda Bootis stars. Instead, he found to his amazement that his comparison stars were varying in brightness by a few hundredths of a magnitude, and in one case up to six hundredths of a magnitude.

....

A co-discoverer of the Sun oscillations, Dr. George Isaak of Birmingham University, predicted at the Herstmonceux Conference in March that other stars should show similar oscillations, thus causing slight brightness variations which he planned to look for. It is not at present clear whether the flickering-star observations, which Isaak did not know of at the time of his prediction, provide evidence of such an effect.

One problem is that brightness variations measured by Norman Walker have amplitudes 10 to 100 times that expected for the Sun. Isaak is willing to concede that some effect other than oscillation of the stars may account for the measured brightness variations, but notes: "Whatever property the Sun has, we find some stars have much more of it." Only detailed analysis will reveal whether the observed "random" fluctuations are the result of several periodicities beating together, which would support the idea that the stars are oscillating.

Norman Walker, by contrast refers to the brightness changes in most of his stars as a "flaring", which he believes may be associated with large-scale turbulence. "The surface of these stars is violently active." he says. Since these stars are hotter and younger than the Sun, and have stronger magnetic fields, they can have bigger flares than the Sun.

....

THE PECULIAR VARIABLE STAR V SAGITTAE

Anonymous; *Sky and Telescope*, 18:632, 1959.

One of the most intriguing of variable stars is V Sagittae. Though it varies in a cycle of some 530 days, during which it changes from magnitude about 10 to 13, it also has a superposed shorter variation of a magnitude or so in eight to 18 days. In addition, sudden changes often take place within a few minutes, while on other nights it may remain nearly constant for some time.

This star is located in the northeast corner of Sagitta, at right ascension $20^h\ 18^m.0$, declination $+20^{\circ}\ 57'$ (1950 coordinates). Lydia Petrovna Ceraski, working at Moscow, discovered its variability in 1902. Its spectrum is peculiar, and the prototype star of its variation is Z Andromedae. V Sagittae can become as bright as magnitude 9.5 at maximum, while it has been as faint as 13.9.

. . . .

THE TEMPORARY STARS

Packer, David E.; *British Astronomical Association, Journal.*4:364-368, 1894.

Perhaps no class of celestial objects possess greater interest than the so-called temporary stars. Their comparatively rare occurrence and the mysterious character of their spectrum, render them objects of special interest, and any new light that may be thrown upon their nature and constitution cannot fail to be useful. They are generally represented to us as expending their energy in one great outburst of radiance, thence subsiding, more or less irregularly, into lasting insignificance, never to appear again.

Latterly, I have had occasion to study the records of a number of these strange objects collected principally from the Chinese Annals, and I find evidence of a remarkable character, tending to show that in some of these bodies a recurrence of the phenomenon has been recorded after a lapse of several hundred years, and in one or two cases a decided inclination to a periodical character has been found. These Chinese observations are taken from the "Encyclopaedia" of Ma Tuan Lin and the "She Ke." They range from about B.C. 214 to A.D. 1640. Their authenticity is well sustained by such eminent investigators as Biot (Connaissance des Temps, 1846), Humboldt (Cosmos), and Williams (Chinese Observations of Comets, &c.), to whom I am indebted for the translated records (see "Monthly Notices," Vol. 33, p. 370). Many of these ancient observations are very vague, but others are sufficiently detailed to allow of an approximate position to be obtained. When two or three such appearances are recorded, whose approximate position in space fairly agree, the natural inference is that they either represent repeated outbursts of the same star, or simple outbursts from a group of temporary stars situated in that region. That temporary stars do show a tendency to group together has been noticed before, but it appears doubtful whether the association would be of such a close character as the observations would require, and the periodical tendency of these recurrent outbursts strengthen the inference that they represent one and the same star.

The most remarkable instances have been given below where the observations referring to each individual star are grouped together in chronological order with the authority appended. All vague or doubtful records, or any indicating a cometary character to the object have been carefully omitted. Most of these recorded appearances have already been recognised as probably referring to temporary stars rather than to comets. It will be noticed that all the included objects lie within the zodiacal region. Particular attention appears to have been paid to this region by the ancient observers, probably on account of its assumed importance in their astrological predictions.

I. in Virgo:

A. D. 222, November 4. A strange star seen to the left of space bounded by stars in Leo and Virgo within space between B and n Virginis. Williams.

A. D. 837, May 3. A strange star seen within Twan Mun (between B and n Virginis) near the stars Ping (v, E, π Virginis). On June 17 the strange star within Twan Mun disappeared (two separate records). She Ke. Biot. Williams.

A. D. 1871. X Virginis, R. A. $11^{h} 56.7^{m} + 9^{\circ}.38'$ (1900). A strange variable, 7.5 mag. which diminished to 12 mag., and has now vanished. Discovered by Prof. Peters. This object singularly agrees in position with the star of A. D. 837.

II. in Scorpio:

A. D. 303, March. A strange star appeared in the S. D. (sidereal division) of μ^2 Scorpii. She Ke. In the tail of the Scorpion. M. T. L. In October it disappeared.

A. D. 1203, July 28---August 6. In Scorpio's tail near μ^2 Scorpii, a new star of a bluish-white tint without any luminous nebulosity, resembling Saturn. Biot.

A. D. 1604, September 30. A star seen in the degrees of S. D. Wei (e, μ, v Scorpii). It resembled a round ball of a reddish yellow tint; seen in the S. W. till November, when no longer visible. In 1605, January 14, it again appeared in the S. E. in S. D. Wei; in the second moon (March?) it gradually faded; in the eighth moon, day Thing Mun (August 21) it disappeared. Biot has S. D. Fang for the second S. D. Wei, but Williams says it is S. D. Wei in the "She Ke."

This object has been supposed by some to be an erroneous observation of Kepler's Nova (A. D. 1604) in Ophiuchus. Humboldt questions their identity, and the doubt is confirmed by the general agreement of position with the earlier recorded apparitions. Intervals of 810 and 401 years are represented. I would suggest a mean period of 403 years, subject to an irregularity of several years. *

* There is another observation recorded by Biot, A. D. 1224, August 24. An extraordinary star appeared in the division of μ^2 Scorpii. This if not a meteor, may probably refer to the same object.

III. in Scorpio:

B. C. 134, June. A strange star was seen in S. D. Fang (B, δ, π Scorpii). She Ke.

A. D. 1584, July 1. A strange star appeared in S. D. Fang, not far from π Scorpii. Biot.

It is thought that the earlier apparition may have been the object that induced Hipparchus to compile the first known catalogue, the date of which was B. C. 125.

IV. in Sagittarius:

A. D. 386. Strange star seen between λ and ϕ Sagittarius, where the star remained without motion from April to July. M. T. L.

A. D. 1011, February 8. New star in Sagittarius between σ and ϕ. M. T. L.

As λ, ϕ, and σ Sagittarius form a triangle, it is probable that the two apparitions represent one and the same star with an interval of 625 years. Biot prefers σ, τ, ξ, ϕ Sagittarius instead of λ, ϕ, σ.

V. in Leo:

A. D. 70, December. A strange star appeared in the head of Leo, in the group Hien Youen (a, γ, e, n, ζ Leonis) it was visible for 48 days. Biot.

A. D. 101, December 30. A small yellowish-blue star seen in the head of Leo in the group a, γ, e, n, ζ Leonis. Biot.

Probably a primary and subsidiary outburst with a 30-year interval. In this region was situated Flamsteed's missing 25 Leonis, which Pigott searched for in 1783-5. Its position for 1790 was R. A. 9^h 46.8 + 12° 20' 46".

VI. Between Gemini and Orion:

A. D. 837, April 29. A strange star seen in the lower part of S. D. Tsing (γ, ε, λ, μ, &c. Geminorum) to the east. May 21. The strange star in the lower part of S. D. Tsing disappeared. (Two separate notices) "She Ke." Williams.

A. D. 1054, July 4. An extraordinary star S. E. of ζ Tauri, disappeared at end of year. Biot.

These two apparitions probably refer to the remarkable star between x^1, x^3 Orionis (referred to in the February number of the British Astronomical Association "Journal"). Allowing for precession it can be shown that on the date of the earlier observation the lower part of Gemini was over the N. W. horizon, and its preceding and following portions pointing in a N. E. and S. W. direction, the term "to the east" would probably refer to the preceding portion below μ where Hevel's star was located.

Admitting that the phenomena of A. D. 837 and 1054 were abnormal apparitions of Hevel's star, I find that a mean period of 76 years, Epoch 1667, April 29, agrees fairly with the recorded observations as tabulated below:

Authority	o.	c.	c-o.
	A. D.	A. D.	
Chinese Obs.	837	831	6 yr.
Chinese Obs.	1054	1059	-5 yr.

Hevelius	1667	1667	--
Dr. Bevis	1738-45	1743	--
Mr. Shackleton	1894, Feb. 24	1895	10 mo.

The columns above refer to the observed and calculated dates of appearance and their difference, according to the above period. The irregularities though large are not proportionally greater than is known to exist in many known long-period variables.

It would be interesting to know whether any other observer noticed a star in this position during the past spring. Owing to its proximity to U Orionis the region since 1885 must have been many times under observation, and any abnormal appearance could hardly have escaped notice were its apparitions more frequent. The region might repay watching for some time to come.

Other instances are not wanting, but from a vagueness in the records the identity is less certain. Thus the strange stars of A. D. 1138 June, in the S. D. of B Arietis (Biot), and of A. D. 1012, May---August, in the south of Aries (Hepidannus), may be apparitions of one star with 126 years interval. Also the Nova of A. D. 1230, December---1231, March between Ophiuchus and Serpens, may have been an earlier apparition of Kepler's Nova of A. D. 1604, in the same region. The astronomer, Ismail Bouillard (Bullialdus), refers in an old work to the "extreme brilliancy" of the "Star Nebulosa" (Andromeda Nebula) is A. D. 1664, contrasted with its dull and obscure appearance in A. D. 1666, "Astronomische Nachrichten," 2687; "Ast. Reg.," Vol. XXIII., 255.

This observation has been quoted by Charlier as a probable earlier outburst of Nova Andromeda, 1885, with an interval of 221 years.

These ancient records somewhat modify the view generally held that these mysterious objects show a decided tendency towards the galactic zone. Of 39 such recorded apparitions seven have been shown to be probable re-appearances, and of nine the accounts are too vague for any definite determination. Of the remaining 23 records, I find that 14 are distributed along the galactic zone, and 9 scattered about the non-galactic regions, with a tendency to group towards the nebular regions of Leo, Virgo, and Coma Berenices.

Another remarkable feature is the strange tendency to group together in point of time. These synchronous appearances have been discussed by Humboldt, who has puzzled himself about them. I hope, in a future paper, to discuss these matters, and to point out a probable explanation.

JUMPING STARS

Anonymous; *Observatory,* 11:385-386, 1888.

Some time ago a correspondent of 'Nature' (1888, May 31, p. 102) called attention to a "curious apparent motion of the Moon seen in Australia." The description is as follows:---"We saw such a curious phenomenon on Sunday night, about 10:30. Miss C---, Miss H---, and I were sitting in the balcony when we noticed the Moon apparently dancing up and down. It was on the wane, so looked so extraordinary. The motion was visible only when she was behind a narrow stratum of cloud and continued at intervals for thirty minutes. I felt quite sea-sick with watching it, and Miss H---was so frightened; she thought there might be an earthquake coming, so went to bed in her clothes to be ready for an emergency." The letter does not seem to have prompted the communication of similar experiences, and possibly the phenomenon is an extremely rare one; but it has been observed more than once in the case of stars, and was called by Humboldt, who seems first to have noticed it, "Sternschwanken." In Astr. Nach. No. 2841, Herr G. D. E. Weyer made a curious observation on the subject; he noticed Antares, on an evening when other stars were scintillating strongly, apparently moving in jerks to the left and then to the right, with intervals of rest. The star was at an altitude of about 4° or 5°, and the apparent motion 2° or 3°. But he found that on directing a telescope to the star, the apparent motion ceased; though on looking again with the naked eye there were still the same jumps, sufficient to have carried the star right out of the field of the telescope. Herr Weyer remarks that he was at the time tired with computation, and that the phenomenon appears to have been subjective in his case, though it agreed well with other accounts of Sternschwanken. In Astr. Nach. Bd. 38, p. 371, is another account by E. Vogel, who saw Venus jumping or dancing in this way on 1853, July 1, from the Tayhonu Mountains; while on August 4, Sirius, and in September, Regulus showed the same irregularities. In writing to Humboldt, Herr Vogel sums up his observations as follows:

1. The phenomenon is best seen in strong twilight, when 2.3 magnitude stars are barely visible, and at an altitude of 5° or 6°.
2. The motion is then parallel to the horizon.
3. The star oscillates rapidly 4 or 5 times in the same direction and then remains 5 or 6 seconds at the further extremity of the arc before returning to mean position.
4. In weak twilight the motion is apparently at an inclination of 45° with the horizon.
5. The dancing does not occur in the absence of twilight or at an altitude of more than 10°.

We should be glad to hear if any of our readers have noticed this phenomenon.

JUMPING STARS

Fillinger, Harriet H.; *Popular Astronomy*, 47:450-452, 1939.

During the past few months I have had the rare privilege of seeing "jumping stars" on two occasions. It has been suggested to me that you might be interested in a brief account of them.

On the night of February 20, 1939, or, rather, early on the morning of February 21 I had the pleasure of seeing two of these jumping stars.

The day had been an excessively hot one for the time of year but after dark the temperature had gone down rapidly so that by midnight the temperature was near normal for the season and it was quite cold between one and one-thirty a.m., during the time the stars were observed.

Vega, the star showing the fastest and most erratic motion, was only a degree or so above the steep side of a mountain behind which it was rising. In addition to showing rapid vertical and horizontal motions through an angular distance of about two degrees it had a circular motion which, most probably, was due to a combination of the other two motions which were taking place rapidly. All of the stars were unusually bright and scintillated more than usual. Vega, which also showed an unusual amount of scintillation, showed rapid changes in color. It would remain stationary for a few minutes at a time and then, suddenly, would begin to dart up and down in a vertical fashion or back and forth in a horizontal line. It might go through these motions several times and then begin to run around very rapidly in a circle having a diameter of about two degrees. These "jumpings" and circular motions continued until the star was perhaps ten degrees above the horizon. It was indeed one of the most interesting and uncanny phenomena I have ever witnessed.

At the time of the above observation Jupiter was almost opposite Vega but higher above the western horizon. About two a.m. we noted that Jupiter was showing considerable motion. Its motion was chiefly back and forth along a diurnal circle. Once or twice its motions were almost perpendicular to the horizon and perhaps twice it moved in an almost horizontal fashion. Once it must have "dropped" three degrees in one "jump." It did not show the most peculiar circular motion.

About six weeks after the above date, the night watchman of the college reported seeing the phenomenon a second time. This time he observed a star in the southern sky and another almost opposite it in the northern sky. On both of the occasions noted bright stars almost opposite each other were seen to jump. Since the night watchman does not know the stars by name and since he did not report this occurrence until some time after noting it, I cannot give a date or accurate information about the jumping stars, but I do not doubt this observation as he first called the matter to my attention in February and in May.

Two nights ago, May 9, about nine o'clock in the evening, I had the pleasure of watching for the second evening a jumping star. After a hot day there was a sudden drop in the temperature and by nine o'clock it was relatively quite cool. Sirius was low in the western sky and was showing appreciable motion. The sky was rather cloudy and not many stars were visible, but those which were visible were showing an unusual amount of scintillation. The cloud just below Sirius faded into the mountain top and made it difficult to estimate the distance of the star above the horizon, but it was, perhaps, about seven degrees above the mountain top. The star soon dropped behind a cloud so we were able to watch it only a very few minutes. Its motions were back and forth along an ellipse of very high eccentricity instead of back and forth along a line. The major axis of this ellipse was about a degree or slightly less in length. The star's motion in this ellipse was not quite continuous. Since the eastern sky was quite cloudy we were not able to determine whether a star opposite Sirius showed any motion.

Three accounts of jumping stars in the literature are known to me and are to be found in the volume of <u>The Observatory</u>, issued in 1888, number 141, page 385; 142, pages 404-405; and 143, pages 433-434. These are brief accounts of stars as noted in Australia, England, and India. There is also some reference to jumping stars in the 1938 issue of this same journal, I believe, but this issue of the journal is not available to me so I cannot give the reference.

It was indeed a privilege to observe such unbelievable phenomena. The circular motion, in particular, was so incredible that had there not been three of us looking at it and seeing the same things each of us would have questioned our eyes or our sanity, or both. Two of us observed Sirius on May 9.

I should be interested in knowing whether it is customary to observe jumping stars in opposite parts of the sky whenever the phenomenon is observable and also whether convection currents in the air are responsible for the apparent motions of the stars on these occasions.

A PECULIAR TELESCOPIC PHENOMENON

Doig, Peter; *British Astronomical Association, Journal,* 40:115, 1930.

Triangular Star Discs. In the year 1867 the celebrated observer, the Rev. W. R. Dawes, published a memorable paper on his double-star work (Monthly Notices, R. A. S., 27, 232) in which a very interesting paragraph occurs regarding the occasional appearance of star discs as triangles with bases roughly parallel to the horizon. He states that the famous W. Struve had informed him that he also had often seen this but could not account for it in any way. The phenomenon was noted with two object glasses on the same night and was also remarked by the great telescope maker Alvan Clark, who was on a visit to Dawes at the time (in 1859), Areturus appearing strongly triangular to both of them. The most peculiar feature of the occurrence was that the triangular tendency seemed to be strongest always when the wind was in the East or South-East. In the same volume of the Monthly Notices, p. 281, the Astronomer Royal, Airy, has a short paper in which he attributes the phenomenon to a physiological cause, fatigue or indisposition, to be expected, he remarks, with an East wind. An interesting discussion on the matter is reported in the Astronomical Register, 5, 123, 1867.--- Huggins and Capt. Noble taking part and stating that they had seen the phenomenon but that they did not agree with the physiological explanation suggested by Airy. Certainly its appearance to two observers and through two different object glasses, as mentioned above, does not suggest a physiological or instrumental origin; and that its occurrence is not confined to refractors is shown by Grover's account (in the same journal, 6, 118, 1868) of a triangular disc being once "strikingly conspicuous in the case of Aldebaran, on which occasion the wind was South East": instrument 6-1/2-inch, with reflector, power 150. Have any of our present-day telescopists observed similar deformation of stellar discs? I have not seen any recent reference to such observations nor, in a fair amount of miscellaneous reading, come across any satisfactory explanation.

THE RED RECTANGLE

Maran, Stephen, P.; *Natural History*, 84:93+, November 1975.

An unusual nebula called the "red rectangle" has been found around a double star known as ADS 4954. Nebulae are clouds of intersteallar gas or dust, and several thousand have been identified in our Milky Way galaxy. Only a few, however, such as the Orion Nebula, are bright enough to be visible to the naked eye, the rest are either telescopic objects or dark nebulae that do not shine but are recognizable on sky photographs as black spots in crowded star fields. Most nebulae are soft-edged and either round or very irregularly shaped. The red rectangle, by contrast, has relatively crisp, linear edges and, despite its name, is shaped like an hourglass. It is also an intense source of infrared radiation. (Its not-quite-accurate designation comes from a survey photograph taken in the 1950s on which it appears to be rectangular.) About 1,100 light-years from earth, near the southern border of the constellation Monoceros, "the Unicorn," and visibly dim even though it is more than two million times larger than the diameter of our sun, the red rectangle is of interest to astronomers because infrared sources are associated with the poorly understood interstellar and circumstellar dust clouds of our galaxy. The object, accordingly, had no less than fifteen telescopes trained on it last year at major observatories from Canada to Chile.

....

The salient findings leading to that possibility can be briefly summarized. Photographs made with the 4-meter (159-inch) Mayall telescope at Kitt Peak National Observatory near Sells, Arizona, showed that the "rectangle" actually has an hourglass shape, with the binary star ADS 4954 at its waist. The two lobes of the hourglass appear to be shining by reflected light from the binary. By contrast, the infrared radiation does not come from the whole object but from a small cloud of dust at the waist that is heated by the twin stars. Although ADS 4954 was discovered in 1915, one member of the pair has not been sighted since 1962. Nevertheless, the unseen star is believed to be present and quite luminous, although temporarily obscured from view on earth by a dense fragment of the dust cloud. We infer this because the one star that we do see now could not heat the dust cloud sufficiently for it to give off the measured amount of infrared light.

Based on its shape, the red rectangle appears to be a member of a small and not fully understood class of celestial objects known technically as "cometary nebulae." This class is so rare that it is omitted from three of the leading English-language books on nebulae. The name derives from the resemblance of some of these objects to a comet with a star at its head. Being shaped like an hourglass, the red rectangle is classified as a biconical cometary nebula, suggesting two wide, fan-tailed comets, head to head.

The nature of these rare nebulae has been debated by astronomers for several decades. The principal argument concerns the relationship of the stars to the nebulae; were these young stars formed by condensation from the material of the nebulae or, on the contrary, were the nebulae produced by the expulsion of matter from the stars? The nature of the nebular glow has also been questioned. Many astronomers believe that the light comes from the stars at the "cometary" head and is reflected by dust particles in the nebulae. Others, noting that the spectrum of the nebular light is sometimes different from that of the nebula's star, have suggested that the light is produced, at least in part, by some as yet unknown process in the nebular gas itself. Also of concern has been the wide range in the appearance of cometary nebulae: some are arc shaped; others are described as resembling commas; some have the classical cometary, or fan, shape; and still others are biconical or hourglass shaped, like the red rectangle. Could all of these shapes be variations of the same kind of object?

....

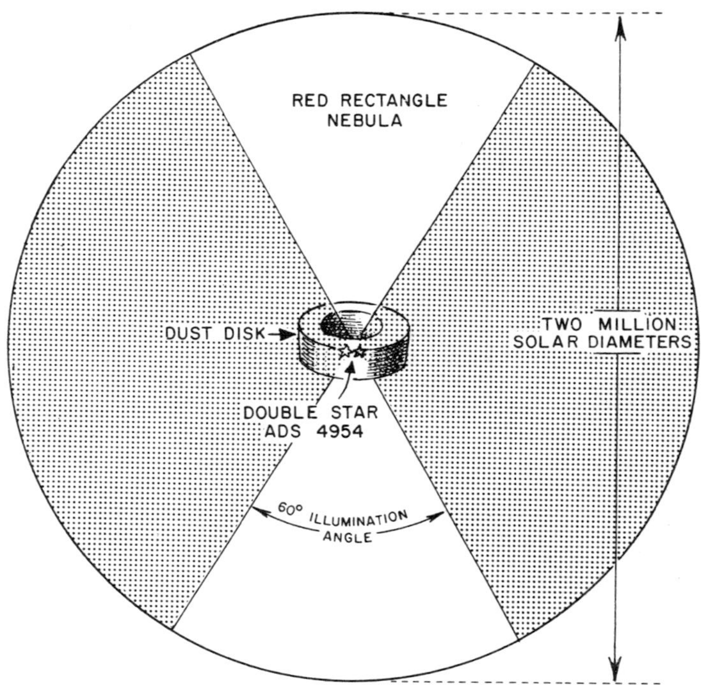

MASS-LUMINOSITY DISCORDANCES

Hoffleit, Dorrit; *Sky and Telescope*, 8:125, 1949.

In a paper on the masses and mass ratios of close binary systems published in the Annales d'Astrophysique, Dr. Otto Struve drawn attention to groups of stars that appear to violate to an enormous degree the ordinarily well-defined mass-luminosity relationship. For example, the close pairs known as W Ursae Majoris systems have nearly equal component stars, their light curves indicating a difference of only 7/10 of a magnitude, whereas the mass-luminosity relation and the mass ratios of the pairs would require a difference of three magnitudes. Can the discrepancy be explained by a common envelope through which an adjustment in the outer layers takes place, rendering the components spectroscopically and photometrically more nearly alike than would be consistent with the mass-luminosity relation?

Some Algol-type eclipsing systems are likewise puzzling. Their spectra and light curves indicate absolute magnitudes of +3 while the masses found would require absolute magnitudes of +13! It is important to note that when the spectra of both component stars are observed it is impossible to reconcile them with the masses of both components and with the mass-luminosity relation. Dr. Struve comments, "We should probably recognize that at least the fainter components of the Algol-type binaries are stars of a kind not previously recognized in astronomy. Since they apparently do not occur among visual double stars it is possible that they are abnormal only because they have not had time to recover from the catastrophe which led to the formation of the binary and to adjust themselves to conditions appropriate for their masses. It would carry us too far to speculate further on this subject, except to say that if the adjustment is not now complete there is no stability and hence not even a certainty that the present masses of the components are safe from dissipation." Although slow streams of gas moving around the binary system might account for the discrepancies noted, such interpretations are not yet conclusive.

A third group of discordant systems occurs near the upper end of the mass-luminosity relation---among the stars of high mass. Examples are Plaskett's star (HD 47129) and Pearce's star (HD 698). In the former, however, Dr. Struve found that the intensities of the spectral lines and the radial velocities of the fainter components undergo erratic changes so that the lines cannot be used for the determination of the mass ratio. On the other hand, HD 698 gives a serious discrepancy between the absolute magnitude obtained from the mass-luminosity relation and the magnitude inferred from the hydrogen lines of the spectrum. From the evidence obtained from Plaskett's star, Dr. Struve believes it is possible that a reinterpretation of the measurements of the spectrum of the secondary component is necessary and that the recorded discrepancy between mass and luminosity has been caused by distorted velocity curves.

ANTI-GLITCH SNAGS PULSAR THEORIES

Anonymous; *Science News*, 110:5, 1976.

Astronomers are unveiling a whole new can of worms in pulsar behavior which is guaranteed to give any theorist in the field gray hairs. In 1968, when pulsars were first observed, theorists instantaneously assigned the phenomenon to spinning neutron stars. They envisaged the pulsing to be a lighthouse effect caused by a cone of radiation emitted by the star. In 1969, the Crab and Vela pulsars (among the most studied pulsars) exhibited sudden increases in their pulsing frequency, later dubbed "glitches." After some momentary head-scratching, theorists decided that glitches are probably caused by "starquakes." Following a starquake, they reasoned, the star would settle down, under the influence of its own gravity, to a smaller diameter. The effect on the spin rate would be the same as for a spinning ice skater who suddenly pulls in his arms. In fact, this explanation has successfully described, in impressive detail, most of the pulsar glitches since observed.

Nature, however, seems to have reaffirmed its supremacy over the mortal deliberations of the theoretical scientist. Gordon E. Gullahorn and John M. Rankin of Cornell University reported at the meeting of the American Astronomical Society last week in Haverford, Pa., observations using the Arecibo radio telescope that show some anomalous pulsar behavior. They have seen several examples where a pulsar frequency suddenly decreases. This is totally contrary to the established quake model. The observation seems to suggest (if one maintains the starquake picture) that the star suffers a physical change that increases its effective diameter. Under the effect of a pure gravitational field, such a distortion is not energetically preferred.

The same two observers also report that the pulsar PSR 0823 + 26 not only has "anti-glitched," as described, but also has undergone a sudden "phase jump," so that suddenly, one day, its pulses began to arrive earlier than expected (see diagram). The anomalies don't even stop there, however, because whatever caused this pulsar's sudden phase jump has left its frequency unchanged. By the admission of several experts in the field, these observed peculiarities may portend the death of the standard starquake model.

. . . .

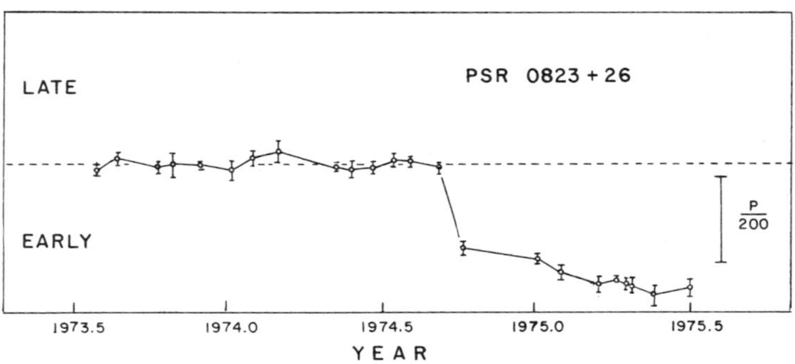

Pulsar antiglitch

A SLOW PULSAR AND COSMOLOGY

Anonymous; *Science News*, 110:280, 1976.

According to theory, pulsars are spinning neutron stars. It is said they spin fast when they are formed and slow down as they age, so the spin rate is often used to estimate the age of a pulsar.

But there's an unusual pulsar that spins extremely slowly JP 1953 + 29. According to Victor N. Mansfield and John M. Rankin of Cornell University, writing in National Astronomy and Ionosphere Center publication 64, this pulsar is spinning down so slowly that when the effects of its motion across the sky on its apparent (to us) spin rate are considered, it may actually be spinning up from the point of view of someone riding along with it.

That result is so shocking that the two investigators let it lie there and say they would rather consider the pulsar one with an extremely slow spin-down rate. They then point out then that circumstance could provide a check on some cosmological theories that propose that the force of gravity weakens as the universe ages. If gravity weakens, the pulsar's diameter should increase, and that should affect the spin-down rate. Mansfield and Rankin calculate that the cosmologies proposed by P. A. M. Dirac in 1938 and by Fred Hoyle and J. V. Narlikar in the 1970s, but not Dirac's 1974 proposal, are in conflict with the measured spin-down rate of JP 1953 + 29.

INFRARED OBJECT WHICH VARIES

Anonymous; *Nature*, 225:502, 1970.

There seems to be no easy explanation for the odd behaviour of the infrared object which E. E. Becklin et al. of California Institute of Technology have found to be fluctuating in brightness by two magnitudes, equivalent to a factor of more than six (Astrophys. J. Lett., 158, L133; 1969). The favoured model for several infrared objects in the galaxy---that they are surrounded by a dust cloud which absorbs the energy from the central star and re-radiates it at infrared wavelengths---gives no hint as to why there should be such striking variability. But the magnitude and period of the fluctuation are like that of the Mira variables, and one possibility is that the central star is a Mira variable. The dust cloud would then have to be opaque enough to obscure certain strong absorption bands which are prominent features in the infrared spectra of long period variables but which are absent in the object IRC + 10216 under study.

At 5 microns the object is the brightest thing outside the solar system, and clearly it is going to be an embarrassment to infrared astronomy if its nature cannot be cleared up fairly speedily, at least in outline. At visual wavelengths the object is insignificant---fainter than eighteenth magnitude---but on a plate taken by Arp at the 200-inch it has an elliptical non-stellar appearance. It has been observed at 2.2 microns since 1965 using a 62-inch infrared telescope and these are data reported by

Becklin _et al_. The time scale of the brightness variations seems to be 600 days or so.

The location of IRC + 10216 is still a mystery. Absence of proper motion and of parallax sets a lower limit on distance of 100 parsec and this is, of course, still within the galaxy. Assuming a distance of 200 parsec---a typical distance for a nearby member of the galaxy---the luminosity comes out to be similar to that of the infrared source in the Orion Nebula and to long-period Mira variables. But Becklin _et al_. cannot rule out that the object is extrangalactic, in which case its size as inferred from the 600 day fluctuation places it at 200 kiloparsec with a luminosity comparable with that of the galaxy.

Clearly Becklin _et al_. favour the view that IRC + 10216 is in the galaxy, even though it refuses to fit any particular classification. Its similarity to the infrared point source in the Orion Nebula suggests that it may be a protostar, yet it is in an unlikely area where there is no gas or dust and no other young stars and the brightness variations are awkward. One interesting speculation is that IRC + 10216 may be a Mira variable evolving into a planetary nebula, and the clue here is the elliptical nebulosity which surrounds the object.

A CURIOUS FEATURE OF THE RADIO SKY

Brown, R. Hanbury, et al; *Observatory,* 80:191-197, 1960.

Introduction. For several years it has been known that the radio-frequency isophotes of the sky show a spur of comparatively intense radiation which appears to emerge from the galactic plane at about $l^{II} = 30^{\circ}$ and to run upwards towards the north galactic pole. So far it has proved impossible to identify this curious feature with anything observed optically and, in our opinion, neither of the two existing theories of its origin is satisfactory.

We shall first review these two theories briefly and then put forward some alternative ideas which are worth consideration and which draw attention to the need for further observations.

The Spur. In Fig. 1 we have sketched the outline of the spur in celestial co-ordinates. These sketches have been made from surveys at 38 Mc/s and 158 Mc/s. Unfortunately our picture is incomplete because neither of these two surveys extends below declination -20°.

The outside edge of the spur, that is to say the edge remote from the galactic centre, is shown by a solid line. This edge is remarkably well defined and to illustrate this we have shown in Fig. 2 a few scans across the spur made with a 2° beam at 158 Mc/s. The inside edge is not clearly defined and is complicated by the presence of large irregular patches of comparatively intense radiation. We have shown the approximate area of the spur by hatching; over most of its path its width is roughly 10°.

Beyond the line marked AA' on the sketch the outline of the spur is more uncertain. The difficulty of tracing it beyond this line is increased by the presence of a large patch of intense radiation, centred rough-

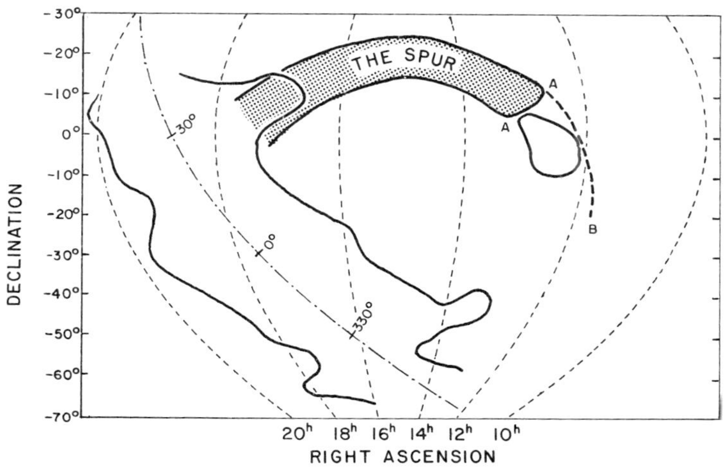

"The Spur", a region of intense radio emission

ly on the position $12^h\ 30^m$, -2°, which is shown in the figure. From an inspection of the isophotes we have formed the conclusion that this patch is associated with the spur and that the outside edge continues along the broken line AB.

An inspection of the few available surveys which show the spur indicates that its spectrum is non-thermal. Representative values of the brightness temperature of a point in the spur are $10,000^o$K. at 38 Mc/s decreasing to about 15^oK. at 400 Mc/s. Although the brightness is very patchy it is greatest near the galactic plane at $l^{II} = 30^o$ and decreases with increasing galactic latitude.

. . . .

Conclusions. The two existing theories of the spur do not seem to be promising; they are based on what we believe to be an incorrect interpretation of the radio observations. As an alternative we suggest that the radio results may be explained more reasonably either by an irregularity in the local spiral structure or by the remnant of a supernova. It is obvious that these suggestions do not exhaust the possibilities, for example it could be argued that the spur is some large-scale effect in the corona of the Galaxy; but we have confined our suggestion to phenomena for which there is some independent evidence and have not postulated anything radically new.

There is as yet no optical evidence to support either of the models we have put forward. In our view the second model, the remnant of supernova, is the more probable explanation; it is possible that an object of this type can have such a low optical brightness that it has escaped detection.

The origin of the spur is in itself a fascinating problem, but it is also important to the general study of radio emission from the Galaxy that this problem should be solved. Further progress must obviously depend upon more extensive radio surveys particularly in the southern hemisphere; in addition, we would urge strongly that the spur, and in particular the sharp edge, should be examined photographically with instruments capable of detecting the lowest possible surface brightness.

• Peculiar Galaxies

THE EXTRA-GALACTIC FERMENT

Anonymous; *Mosaic*, 9:18-27, May/June 1978.

Exotic Galaxies. In addition to the typical spiral and elliptical galaxies of the universe, astronomers in the last several decades have found a variety of strange, remote objects that radiate with enormous power.

They include the Seyfert and radio galaxies, BL Lacertae objects (see "The V(ery) L(arge) A(rray) Turns On," Mosaic, Volume 9, Number 2) and quasars. Measurements of the red shift of the emission lines of their optical spectra show that they are receding from the Milky Way galaxy at velocities of thousands and occasionally hundreds of thousands of kilometers per second.

The most distant object found to date is a quasar, Q1442 + 101, which has a velocity of recession of about 270,000 kilometers per second, or 91 percent of the velocity of light itself. This puts it at a distance of 10.6 billion light-years, if current estimates of the universe's expansion are correct.

If this and other quasars radiate isotropically---that is, equally in all directions---they must shine with power on the order of 100 trillion trillion trillion (10^{38}) kilowatts. This is about 10,000 times the entire radiated power of the Milky Way galaxy.

A startling feature of the quasars and other exotic sources is that their intensity often fluctuates in short intervals, sometimes days. This means their energy source is confined to small regions, at most a few light-days in diameter. (The solar system is about 11 light-hours across.)

What chiefly puzzles astronomers is how such prodigious amounts of power can be generated in such confined regions. For all the energy they release, such thermonuclear fusion processes as those in the sun and stars have an efficiency of only about one percent. More compact and energetic sources are required. Astronomers have suggested multiple star collisions, chain reactions of supernovas, giant pulsars, lagging cores of super-condensed material left over from the original

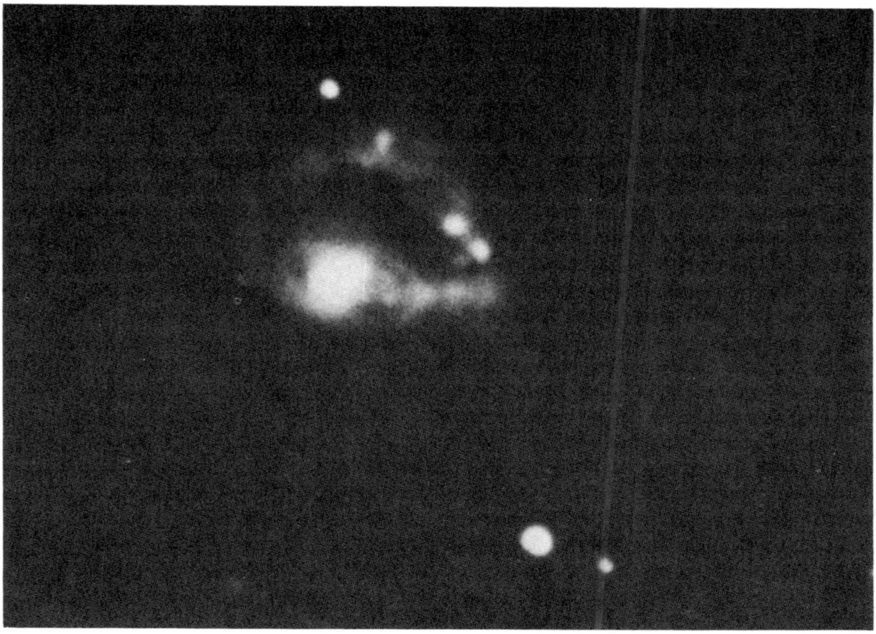

Exotic galaxy. This fuzzy ring galaxy has a violently active Seyfert nucleus.

big bang and even "white holes" where energy is pouring into this universe from some postulated parallel universe.

The explanation that is increasingly attractive to astrophysicists, however, involves the black hole, presumed to be a supermassive object which has collapsed to unimaginable density in the overwhelming force of its own gravitational field (see "Stellar Ontogeny:...to Ashes" and "Black hole in a galaxy" in this Mosaic). Gas and plasma spiraling inward toward a black hole would be compressed and heated to enormous temperatures and might conceivably radiate at the levels observed in the quasars and other exotic sources.

While the energy source of the exotic objects remains one of the biggest unknowns in astronomy today, astronomers have learned quite a bit about other aspects of their nature. There is growing evidence that they reflect features of the two familiar types of ordinary galaxies: The radio galaxies and BL Lacertids appear to be elliptical galaxies in an explosive phase, while the Seyferts, and possibly the quasars, are spirals---the latter so distant that only their bright cores can be seen. It is possible that all galaxies may experience these outbursts at some time in their history and that they may have occurred far more frequently in the early life of the universe than they do today.

Puzzles, Old and New. While the environmental revolution has been the most arresting development of the last several years in extragalactic astronomy, it is only the latest in a long list of problems and puzzles that have confounded the practitioners of this science. Many of these are long-standing: the metal abundances in the ancient stars of the globular clusters in the halo of the Milky Way galaxy; the mass discrepancies in galaxies and clusters of galaxies. Other problems are more recent: the source of energy for extraordinarily powerful radio galaxies and quasars discovered in the last several decades (see "Exotic galaxies" box accompanying this article).

The oldest objects in the Milky Way galaxy are the dense, globular star clusters of the galactic halo. Most ancient of these is a cluster designated M92, whose reddish light indicates an age of 10 billion to 20 billion years. Since the properties of these tight clusters vary in relation to their distance from the center of the galaxy, and because they share many of the properties of the general halo population, astronomers believe they are indigenous elements of the galaxy and not itinerants that have wandered indiscriminately into its gravitational embrace. Accordingly, this star cluster must have formed soon after the universe itself.

But there is a difficulty here. The small, dim, supposedly pristine, first-generation (Population II) stars composing M92 and the other globular clusters contain some heavy metals. True, the abundance is only about one percent of that of such second-generation (Population I) stars as the sun, which formed in part out of the heavy-element debris of earlier, more pristine stars.

But if the first-generation stars of these globular clusters are as old as the galaxies (presumed to have formed simultaneously and very early in the life of the universe), then how can they have any iron or other heavy elements at all? The well-established theory of nucleosynthesis requires that supernovas "process" lighter nuclei into iron and heavier nuclei (see "Stellar Ontogeny:...to Ashes" and "Stellar Ontogeny: from Dust..." in this Mosaic). If this is correct, and primeval stars in the globular clusters show heavy elements, then the oldest first-generation stellar populations in the universe must have been preceded by even earlier, short-lived stars sufficiently massive to end in supernovas.

Calling the problem a "flagrant scandal that we so rarely mention in public," King demanded at the Yale conference: "Where is the first generation? We believe firmly that the big bang made only hydrogen and helium; then where did we get the heavy elements that are in Population II stars? Was there a [pre-first-generation] Population III that has left no visible traces behind?"

Another hoary problem in extragalactic astronomy is the discrepancy between the observed masses of galaxies and clusters, on the one hand, and the actual behavior of those bodies, on theother hand. If the laws of Kepler and Newton apply to these aggregates---and we must believe they do---then there appears to be less mass present in the universe than is necessary for the galaxies to be stable; most of the mass of the universe must presently be beyond the reach of the most sensitive detectors we have been able to deploy (see "One Universe, Indivisible" in this Mosaic).

"The most serious problem in extragalactic astronomy today is the notorious 'missing mass,'" King told the Yale conference. "There are rich clusters of galaxies where it is quite clear that the total gravitating mass is about ten times what we can account for in the conventional masses of the individual galaxies....

"The other missing mass problem shows up in the outer parts of spiral galaxies, where the rotation curves have clearly never heard of Kepler.... The rotation curves say there is a large amount of mass out there, [but] it emits no light by which we can study its nature.

"The missing mass problem is extremely disquieting.... We are talking about 90 percent of the mass of the universe, present but not speaking. Can we really claim to know anything about the nature of the universe if we don't know the properties, or even the nature, of 90 percent of its material?"

Violent Galaxies. The problem of accounting for the enormous power radiated by several types of extragalactic objects (see "Exotic galaxies" box accompanying this article) is a relative newcomer to the list of seemingly intractable problems that arise in the study of galaxies. It has proved to be an exceptional challenge.

When the quasars were first discovered in the nineteen-sixties, they confronted astronomers and astrophysicists with an acute dilemma: If their enormous red shifts truly represented distance, nothing known in physics could explain their source of energy. Indeed, the very existence of such a compact but colossal source of energy seemed for a time to challenge the known body of physical principles, and a variety of fanciful notions like the "white hole" hypothesis (see "Exotic galaxies" box accompanying this article) were seriously considered in some quarters.

On the other hand, if the red shifts displayed by the objects were false indicators of recession velocity, then the sources could be nearby and the problem of the energy source would go away. But the implications of this explanation were even more horrifying to astronomers. If some entirely unknown physical mechanism could mimic the Doppler displacement of the emission lines of a receding object, then the whole concept of an expanding universe would be thrown into question; the Hubble scale of cosmic distances---an essential tool for both astronomers and cosmologists---would have to be discarded.

This latter alternative looked like a more direct road to hell than the former. Though they'd have taken the road if they had to---if the data pointed unequivocally to it---most astronomers have concluded that the quasar red shifts do indicate distance. (As it has turned out, the weight of the evidence accumulated since the local versus cosmological debate peaked in the late nineteen-sixties has tended to support the judgment that the quasar red shifts are valid distance indicators.) This left the astrophysicists with a fearful mess to clear up---the source-nature of such energy emission, without which there can be no coherent theory of the dynamics and evolution of the early universe---but it saved the larger body of extragalactic and cosmological knowledge so painstakingly assembled over the past half century. (pp. 24-26)

ODDBALLS AND GALAXY FORMATION

Binney, James; *Nature*, 255:275-276, 1975.

There seems to be no limit to the number of weird objects one can turn up in the sky. We now have catalogues compiled by Zwicky, Vorontsove-Velyaminov and Arp, containing over ten thousand peculiar extragalactic objects. These range from almost stellar, highly luminous emission line objects to groups of distorted, apparently interacting galaxies. Many of the objects identified as peculiar on optical plates turn up also as unusually strong emitters in other ranges of the electromagnetic spectrum. Undoubtedly these objects would well reward greater study. Indeed three recent papers suggest that they may be able to teach us something about one of the greatest enigmas of modern cosmology: the formation of galaxies.

Freeman and de Vaucouleurs (Astrophys. J., 194, 569; 1974) suggest that pairs of an irregular galaxy and a spheroidal galaxy could be produced by the collision of spiral galaxies with an hypothesised intergalactic cloud of neutral hydrogen (see Nature, 254, 181; 1975). Freeman and de Vaucouleurs suggest that the Universe may be filled, not necessarily uniformly, by a few hundred clouds of neutral hydrogen per massive galaxy. If the radii and masses of these clouds lie in a narrow region around 10^5 light-years and 10^9 solar masses they claim one can account for the observed density of spheroidal-irregular galaxy pairs in terms of the compression and fragmentation of the gas involved in the collision of the gaseous part of a spiral galaxy with one of these clouds. Stars formed from this gas would form the irregular or ring galaxy while the stellar component of the spiral galaxy would pass on as a spheroidal system.

....

All this tinkering with the population of dwarf galaxies does not of course directly help us to understand the more fundamental formation of giant galaxies like our own. Nothing daunted by the failure of cosmologists to come up with a satisfactory mechanism for triggering the formation of condensed objects in an expanding Universe, Meszaros (Astr. Astrophys., 38, 5; 1975) has explored the possibility that a primordial population of black holes of about solar mass might have precipitated the much later collapse of galactic masses. He finds that if the total mass of black holes in the present-day Universe were between one and ten times the mass contained in visible galaxies, the statistical fluctuations in the number density of black holes in the early Universe may have caused perturbations in the overall matter density of the Universe to grow from the end of the first 10^{-4}s of our history. These black holes would now be clustered around galaxies in enormous numbers (about 10^{12} around the Milky Way) forming a massive halo of the type whose existence has been suspected on other grounds (Ostriker and Peebles, Astrophys. J., 186, 467; 1973). Unfortunately the detection of these black holes is well nigh impossible and the physics underlying theories of their formation rather doubtful.

The real problems of galaxy formation remain very much insolved. The greatest difficulty is that we still have no idea what induced the formation of the first bound objects in an expanding Universe. One suspects that the origin of these local inhomogeneities in the structure of the Universe will be understood only when we know why the Universe is so remarkably smooth and homogeneous on a large scale and a major effort is now going into the possibility that this isotropy is the result of quantum effects back in the period when the Compton wavelength of an elementary particle was comparable with the radius of the then visible Universe. The other urgent task in cosmology is to understand how in the more recent past great masses of gas collapsed to form galaxies and clusters of galaxies. This task is particularly exciting because there is every reason to believe that we should be able to actually see this happening by looking out into distant space. What we require are detailed models of the collapse to enable us to identify a nascent galaxy when we see one.

THE NATURE OF HOAG'S OBJECT

O'Connell, Robert W., et al; *Astrophysical Journal,* 191:61-62, 1974.

Some time ago, Hoag (1950) discovered a remarkable object at $a = 15^h\ 15^m 0$, $\delta = +21^\circ\ 46'$ (1950) which he described as a "perfect halo" surrounding a diffuse nucleus. At first glance, it appeared to be a planetary nebula; however, its galactic coordinates ($l^{II} = 30$, $b^{II} = +56^\circ$) and color were not typical of that class of object. Therefore, Hoag suggested "as a most conservative alternative" that it was a "pathological" galaxy. Curiously, although plainly visible on the Palomar Sky Survey, Hoag's object is not included in the Morphological Catalogue of Galaxies (Vorontsov-Velyaminov and Arhipova 1963), the Catalogue of Galaxies and Clusters of Galaxies (Zwicky and Herzog 1963), or the Catalogue of Galactic Planetary Nebulae (Perek and Kohoutek 1967). It does not lie within the boundary of a recognized cluster of galaxies.

A print of Hoag's object taken with the 36-inch Crossley telescope of Lick Observatory is reproduced in figure 1 (plate 10). The object consists of a slightly diffuse circular core 6" in diameter surrounded by a symmetric annulus with inner and outer diameters of 28" and 45", respectively, and an axial ratio of 0.93 ± 0.05. No luminous material is visible between the core and the inner edge of the ring, and little structure is apparent in the ring.

We have obtained several spectrograms of the object with the Cassegrain image-tube spectrograph on the 200-inch (5-m) telescope of Hale Observatories and the prime-focus spectrograph on the Lick 120-inch (3-m) telescope. The core exhibits the Ca II H and K lines and the G band in absorption at a redshift of 12,740 ± 50 km s^{-1}. The [O II] λ 3727 line is not visible. Adopting a Hubble constant of 75 km s^{-1} Mpc^{-1}, we find the distance to Hoag's object to be 170 Mpc. Hence, the core of the object is 5 kpc in diameter, and the annulus has inner and outer diameters of 23 kpc and 37 kpc, respectively. Therefore, Hoag's object is indeed a pathological galaxy.

. . . .

One of the more exotic possible interpretations of Hoag's object---namely, that the ring is the image of a background galaxy produced by the gravitational focusing of the core---is also worth considering. This was originally proposed by Hoag (1950)and can be explored using Liebes's (1964) analysis of gravitational lenses. We find that the properties of the hypothetical background source inferred from our observations of the ring are consistent with those of the most luminous Sc galaxies. However, the inordinately high mass for the core of Hoag's object, $\mathfrak{M} < 10^{12} \mathfrak{M}_{\odot}$. Even in the absence of this observation, the model is rendered implausible by the extremely high mass-to-light ratio, $\mathfrak{M} / L_V \sim 1500 (\mathfrak{M} L_V)_{\odot}$, implied for the core coupled with the improbability of the necessary 8" alignment of centers of two relatively rare objects and the existence of a number of other systems with similar features.

NORMAL GALAXY EMITS JETS

Anonymous; *New Scientist*, 71:694, 1976.

For many years Halton Arp has been trying to convince his fellow astronomers that perfectly normal galaxies are in the habit of ejecting sizable objects---perhaps even quasars---into intergalactic space. On the whole he has not succeeded. But now he presents some prize specimens which, if not very sizable, have evidently been ejected from the nucleus of an archetypally normal spiral galaxy, called NGC 1097.

The ejecta are three narrow, straight jets, all pointing directly away from the galaxy's nucleus. The projected lengths of the longest two are 80 and 50 kiloparasecs respectively, and they end in what might fancifully be seen as a ricochet and a puff of smoke. The third is shorter, and lies diametrically opposite the second.

So far as we know, this galaxy is unique. A few other spirals have long jets, but in most cases they seem to be the debris of intergalactic collisions. The only galaxy that may be displaying a comparable phenomenon is NGC 4258, a nearby spiral with a pair of broad curved ridges of hot gas and radio-emitting plasma. They apparently represent some 10^8 solar masses of gas, ejected from the nucleus around 18 million years ago. Unfortunately we do not yet know what the NGC 1097 jets are made of; it is hard to believe that they consist of stars, but they don't shine in the H_{α} light of hot gas.

....

How could the nucleus of NGC 1097---which looks fairly innocuous now---have produced these jets? If they really do lie in the plane of the galaxy, then most of the current ejection models will not do, because they are designed for radio galaxies---which probably spurt out their symmetrical radio lobes each way along the rotational axis, not along the equator. But there is one process which does eject things in the equatorial plane. This is the gravitational slingshot, in which colossal superstars in the nucleus fling each other out as they pass in a near-collision. This might do very well for NGC 1097---but the conclusion depends on the jet composition. (Astrophysical Journal Letters, vol 207, p 147.)

NEW LIGHT ON QUASARS: UNRAVELING THE MYSTERY OF BL LACERTAE

Metz, William D.; *Science,* 200:1031-1033, 1978.

For nearly as long as quasars have been a puzzle, some equally curious objects named---or actually misnamed---BL Lacertae have been on the agenda of unsolved astrophysical problems. Quasars were discovered first, and until recently they have captured more of the limelight. But the situation is changing as a result of new observations, and the objects known as BL Lacs, for short, are moving to center stage. "Perhaps they would have attracted more attention if they had had a dramatic name like blazars," quipped Columbia's Ed Spiegel during a recent after-dinner speech on the subject. What is becoming clear is that both BL Lacs and quasars are astronomical powerhouses of uncertain origin, at once superluminous and extremely compact.

The first BL Lac object was originally studied in 1929, and at the time it was thought to be a variable star. As is the astronomer's custom for variable stars, it was named according to the constellation in which it is located, Lacerta (meaning "the lizard"), and designated by a two-letter code indicating that it was the 90th variable star discovered in that constellation. Not until 1969, when it was found that the presumed star was also a radio source, did the stellar classification come into question. The radio emission was a strong clue that BL Lacertae was not a star in our galaxy since most objects that have its type of radio spectrum either are in other galaxies or are quasars. More objects like BL Lacertae were quickly found, and by 1972 astronomers realized that the erstwhile variable star was one member of a class of objects which were almost certainly extragalactic. They looked starlike in an optical telescope, were extremely variable in brightness, and were also powerful and compact radio sources. But until quite recently the peculiar BL Lac objects could not be placed at any particular location in the universe because they seemed to have no trace of the spectral features by which cosmological (large astronomical) distances are measured.

The question of distance was an acute one. If BL Lacs were as remote as quasars, then the problem of explaining how they produce enormous power in a small volume might be even more severe than it is for the most violent quasars. (Several lines of evidence indicate that the energy equivalent of 10 billion suns must be produced inside a region the size of the earth's solar system.) In fact, when one of the brightest BL Lac objects flared up in 1975, it was for several months the most luminous object known in the universe. But that is getting ahead of the story.

The primary reason that BL Lacs were set aside as a separate class of objects 6 years ago was that their spectra were nearly devoid of sharp features. Like quasars, BL Lacs appeared pointlike on photographic maps of the sky, but, unlike quasars, they had no sharp lines in their optical spectra attributable to either emission or absorption of light from ionized gases. Without emission or absorption features, no red shifts could be determined for BL Lacs and the basic cosmological distance scale was not applicable. (The red shift is a measure of the Doppler effect on the spectral lines of various atoms and ions, shifting the lines from the ultraviolet into the red portion of the spectrum. The magnitude of the shift is proportional to the distance of objects receding from our galaxy, as was first shown by Hubble.)

The Case of the Missing Red Shift. Without any features by which to measure a red shift, the troublesome conclusions about the power of BL Lacs were open to question. If the objects were closer than assumed, then the energy needed to produce the apparent brightness would be markedly reduced. The power production problem would thus abate. Certain eminent astrophysicists, however, added objection to uncertainty. These skeptics, championed by Geoffrey Burbridge, a British theoretical astrophysicist who was recently named head of Kitt Peak National Observatory, questioned whether red shifts would be a valid measure of distance for objects as unusual as quasars and BL Lacs, since the red shift-distance relation was derived from normal, quiescent galaxies. There was no evidence that either quasars or BL Lacs were embedded in the middle of normal galaxies, although many astronomers believed that they were.

The developments of the past year and a half have erased these uncertainties, and caused astronomers to rethink the role of BL Lacs altogether. BL Lacs have been found to have faint but observable spectral lines and they have been shown to lie in galaxies. Even Burbridge now accepts the conclusion that BL Lacs are at cosmological distances. Thus energy production in BL Lacs must now be seen as a problem at least as challenging as that of energy production in quasars. Furthermore, there are physical reasons to think that the characteristics that minimize (but, it is now clear, do not entirely eliminate) the spectral lines in BL Lacs may make it easier to unravel the nature of the energy source.

As more BL Lacs are found with weak spectral lines, the differences between BL Lacs and quasars are diminishing. At a recent conference called to survey the progress made in the past year, the discussion indicated a consensus that there is no significant difference between BL Lacs and violent quasars. The feeling was strong enough that Martin Rees, the head of the Institute of Astronomy at Cambridge, England, and perhaps the most noted astrophysicist actively modeling such processes, suggested that it was time to invent a new classification scheme. Ken Kellerman, a prominent observationalist who has been looking at the energy source of such objects with the high magnification available from very long baseline radio telescopes, went a step further to say that the name BL Lac should "never be used again." While the astronomers could not agree on new definitions, it seemed that BL Lacs might constitute an extreme form of quasar.

The properties of BL Lacs are certainly more variable than those of many quasars. Over the course of a few months, the optical energy output of BL Lacertae itself---which is by far the most thoroughly studied object in the class---will vary from the equivalent of one very luminous giant galaxy to the equivalent of ten such galaxies. The light from BL Lacs in general is also more polarized than the light from quasars, and that polarization may change quite rapidly---often varying significantly in the course of a day. The radio characteristics of BL Lacs tend to be similar to those of quasars, however. Both tend to emit radiation that has the spectral characteristics of a nonthermal source (meaning that the powerhouse inside must be generating radio power by a rather exotic mechanism), and the source of the radio emissions tends to be extremely compact in both cases. This comparison does not, of course, apply to the large subset of quasars which do not emit radio waves.

Optical observations, as well as radio observations, can be used to put a limit on the size of the central powerhouse. The time scale for luminosity and polarization variations is a measure of size because the travel time of light is an indicator of distance. An object that changes its light output in a day cannot be much larger than the solar system, which requires half a day for light to cross.

Class distinctions based on spectral lines began crumbling in 1974 and collapsed last year. Using an image tube scanner capable of recording extremely faint optical spectra, two astronomers at Lick Observatory, near Santa Cruz, California, showed that BL Lacertae itself had very weak spectral features that could be identified with both emission and absorption lines. The work was published in March 1977. Working with the Lick 3-meter telescope, Joseph Miller and Steven Hawley obtained a spectrum of BL Lacertae from which they could determine a red shift. The lines were extremely faint but they were there. The possibility of a line feature in the BL Lacertae spectrum had been raised by J. B. Oke and James Gunn 3 years earlier. A report of weak emission lines has also been published for one other BL Lac object, designated 1400 + 142, and at least four others are now known. Speakers at the conference in Pittsburgh concluded, and the general consensus of participants did not dispute, that most BL Lac objects have weak emission and absorption features.

If the measurements of BL Lacertae had stopped with the determination of the red shift of the strong central source, the debate over the powerhouse of BL Lacs would have been stuck on the same arguing point as the debate over quasars---namely whether the red shifts were indicative of distance or not. But Miller and Hawley, along with Howard French at Lick Observatory, proceeded to examine the fuzz around the center of BL Lacertae and proved to the satisfaction of the astrophysical community that the bright nucleus is the center of an elliptical galaxy which appears normal in every respect.

Suppressing the intense light from the center of BL Lacertae sufficiently well to see the comparatively faint galaxy around it was no mean feat. An annular aperture, made by evaporating aluminum over a ring placed on a glass microscope slide, was used to block out the light from the center of BL Lacertae. Miller and his colleagues again used the Lick image tube scanner and observed the object on two successive nights when the distortion from the atmosphere was minimal. To further reduce atmospheric effects, they only observed BL Lacertae when it was almost directly overhead. They found that the galaxy around the nucleus of BL Lacertae has a normal stellar spectrum and---most importantly---has the same red shift as the bright nucleus, namely 0.07.

Finding the Galaxy Around BL Lacertae. The observation not only showed that the exceedingly bright source identified as BL Lacertae had somehow evolved in the center of an apparently normal galaxy, it also effectively quashed the argument that BL Lacertae might be less powerful than it appeared.

Geoffrey Burbridge, for one, accepts the cosmological interpretation of the BL Lacertae red shift, although he is still skeptical about quasars. ("No one has ever shown that quasars are in the centers of galaxies," he told the astronomers gathered at Pittsburgh. "The evidence isn't there.")

With the possibility of anomalous red shifts cleared away, the inquiry into the nature of BL Lacs is proceeding to the physical description of the central source. Some 30 or 40 BL Lacs have been identified so far, and they have generally been less remote and less luminous than quasars, but this may be an observational bias rather than a reflection of the true distribution. Nevertheless, BL Lacs as a class are more variable than quasars, and at their peak luminosity may outshine them. The nucleus of BL Lacertae itself fluctuates between one-fourth and ten times the strength of the surrounding galaxy. When a BL Lac object designated as AO 0235 + 164 flared up in 1975, it increased in brightness more than 100-fold in a few months. Because it has several sets of spectral lines with different red shifts, its distance and power are uncertain. But at its peak, it may have been emitting 10^{48} ergs per second, more than 10^4 times the luminosity of our galaxy and equal to the brightest known quasar, according to Arthur Wolfe at Pittsburgh.

BL Lac objects more distant than those known today could be even brighter, but two considerations affect their detection. The emission lines are so faint that it would be extremely difficult to determine red shifts for any BL Lac objects that were as far away as distant quasars. Furthermore, instrumental limitations make it difficult or impossible to see surrounding galaxies associated with the more distant BL Lacs known now.

The size of the central source producing all this energy can be delimited but has not yet been determined. Radio measurements made with a very long baseline between telescopes, on the order of 3000 miles, show that the source is smaller than a few milliseconds of arc in angular extent. This indicates that the powerhouse is less than 30 or 40 light-years across. The rapidity of the energy variations indicate a central core that is only a few light-months across. The polarization fluctuations indicate a size more on the order of light-days. Among the quasars, only the subset known as optically violent variables and typified by objects such as 3C 446, 3C 279, and 3C 345 approach the rapidity of BL Lac variations. These exhibit light variations occurring in a few months and polarization changes on the time scale of weeks. In either case, the problem is to explain how these objects produce more than 100 times the energy of a giant elliptical galaxy in a region less than one-millionth the diameter.

The Knottiest Problem. Astronomers attending the conference on BL Lacs at Pittsburgh did not burst forth with new insight into this problem, which is one of the knottiest in astronomy. (Black holes, supermassive stars, and dense concentrations of normal stars are the three explanations that have been proffered.) But they did agree that it might be easier to learn the nature of the power source from BL

Lacs than from quasars. Emission features are the key, again.

The faintness of BL Lac emission lines is generally interpreted to mean that there is relatively little gas associated with the energy source, while the prominence of emission features in quasar spectra indicates that there is a great deal of gas present (or that the central source is proportionately weaker.) This distinction can be made because the radio data indicate that the radiation from the ultimate source is nonthermal, probably produced by the process called synchrotron emission. The emission lines, however, are thermal in character.

In many respects, BL Lacs look like quasars in which the synchrotron light has been boosted to such a level that it swamps the emission features. Whether one thinks of a BL Lac as a quasar with very little gas or as a quasar with an anomalously large synchrotron component, the effect is that the synchrotron radiation is easier to study. For this reason, Geoffrey Burbridge, Martin Rees, and others refer to BL Lacs as the best available chance to see the "bare machine."

Building on a model they proposed several years ago, Rees and Roger Blandford, from Caltech, quickly developed an explanation for the relative strengths of the continuum and line radiation by postulating that BL Lacs and quasars are identical objects oriented in different ways to the line of sight from the earth. Their explanation is that the continuum radiation produced by the central object propagates out into space in a fairly tight beam, while the emission line radiation from the gas surrounding the central object propagates in all directions. If the beam is directed toward our galaxy, the continuum radiation dominates and the object is called a BL Lac. If the beam is directed away, we see mostly line emissions and call the object a quasar. This idea is new and was embraced with steadfast neutrality by the Pittsburgh attendees. Problems could arise with the model if there proved to be nearly as many BL Lacs as quasars, however. An analysis of the population question by B. Setti, of the University of Bologna, indicated that, even though observational biases make it difficult to determine the statistical prevalence of BL Lacs, the two types of objects may be equally abundant. Setti concluded, with certain limitations, that there appeared to be 100 quasars per cubic megaparsec of the universe and more than 30 BL Lacs.

Other models were also discussed, including the suggestion by Stirling Colgate that the core of a BL Lac consists of a high-density region of stars that energize the surrounding medium by frequent supernova explosions. A new finding about the polarization features of BL Lacs, however, would put severe constraints on this and every other model. J. E. Ledden, of the Virginia Polytechnic Institute, reported that during the great outburst of AO 0235 + 164 the plane of radio polarization had slowly rotated through 130° in the course of several months. This argues strongly that a single, coherent source of radiation was at work, not a disconnected group of stars. The synchrotron process will produce the required coherence if all the synchrotron radiation comes from a single object. (The question of optical coherence is still open, but it is nevertheless mind-boggling to think that for some months AO may have been a 10^{41} watt laser!)

No doubt more models for BL Lacs will be proposed, more observations of their special properties will be made, and radio and optical astronomers working together will try to advance the data base for these objects to the level of completeness that characterizes quasars. At the same time, the present categorizations may change. Martin Rees suggested at Pittsburgh the "need, rather than carrying around all this classical baggage, for a more sensible and logical classification scheme for active galaxies."

The story of BL Lacs is not only the story of a small group of astronomical oddities that have found their way into the mainstream of astrophysical inquiry. It is also an outstanding example of the progression of thought that appears to be occurring throughout astronomy, as many different high-powered objects that have been discovered by different observational techniques---radio, optical, and infrared---are being seen more and more as manifestations of the same phenomenon, powered on different scales by the same machine.

GALAXIES AS GRAVITATIONAL LENSES

Sadeh, Dror; *Science,* 158:1176-1178, 1967.

Abstract. The probability that a galaxy gathers light from another remote galaxy, and deflects and focuses it toward an observer on Earth, is calculated according to various cosmologic models. I pose the question of whether an object called a quasar is a single, intrinsically luminous entity or the result of accidental alignment, along the line of sight, of two normal galaxies, the more distant of which has its light amplified by the gravitational-lens effect of the nearer galaxy. If galaxies are distributed at random in the universe, the former alternative is true. But, if we assume that most galaxies exist in pairs, we can find about 30 galaxies occurring exactly one behind the other in such a way as to enable amplification of the order of 50. This model explains also the variations in intensity in quasars, but fails to explain others of their observed properties.

GALAXIES AS GRAVITATIONAL LENSES

Barnothy, J., and Barnothy, M. F.; *Science*, 162:348-352, 1968.

Abstract. Of all the galaxies in the visible part of the universe, 500 million are seen through intervening galaxies. In some instances the foreground galaxy will act as a gravitational lens and produce distorted and (in brightness) greatly amplified images of the galaxy behind it; such images may simulate starlike superluminous objects such as quasars (quasi-stellar objects). The number of gravitational lenses is several times greater than the number of quasars yet observed. In other instances the superposition of the image upon a visible foreground galaxy may simulate morphological configurations resembling N-type, dumb-bell, spiral, or barred-spiral galaxies.

• Does the Sun Have a Companion Star?

IS LUCIFER A PARTNER OF THE SUN?

Anonymous; *New Scientist*, 68:373, 1975.

Does the Sun have a distant small stellar companion? Over the past few years astronomers have realised that our Galaxy may contain large numbers of dwarf stars, with masses less than one thirtieth the mass of the Sun, in which nuclear reactions produce little energy. Such stars would be only faintly luminous at infrared wavelengths due to gravitational contraction and slow cooling. Now Kris Davidson of the University of Minnesota has raised the possibility of one or more such stars being loosely associated with the solar system.

In a paper in Icarus (vol. 26, p. 99), Davidson argues that such an object could have a maximum mass of about one hundredth the mass of the Sun without perturbing the outer planets perceptibly. A lower mass limit for the dwarf would be somewhere near to Jupiter's mass at around one thousandth of a solar mass.

The companion star would lie at least 700 times the Earth's distance from the Sun and more probably some thousands of times that distance. Binary star systems are known with such large separations between the component stars so this suggestion is quite feasible. Comets are known to travel to these outer reaches of the solar system and interactions between an infrared dwarf and comets might disperse the comets out of the solar system, or contrarily, perturb them to their observed eccentric orbits.

Davidson shows that, if we are lucky, we could observe such an infrared star as its radiation flux would be detectable with present technology. He even proposes the name "Lucifer" for the Sun's companion, should it turn out to be there.

HAS THE SUN A COMPANION STAR?

Harrison, E. R.; *Nature*, 270:324-326, 1977.

Pulsars are accurate timekeepers. They are believed to be rotating neutron stars, with strong magnetic fields, and the energy they radiate is at the expense of their rotational kinetic energy. As each pulsar ages, its period P (relative to the Solar System barycentre) slowly increases, and its period derivative $\dot{P}(= dP/dt)$ slowly decreases. Certain interesting pulsars (displayed in Table 1) have anomalously small period derivatives, and rather surprisingly, are found grouped together in the same region of the sky (shown in Fig. 1). I suggest here, as an explanation of the peculiar properties of these pulsars, that the barycentre of the Solar System is accelerated, possibly because the Sun is a member of a binary system and has a hitherto undetected companion star. [Table 1 omitted]

....

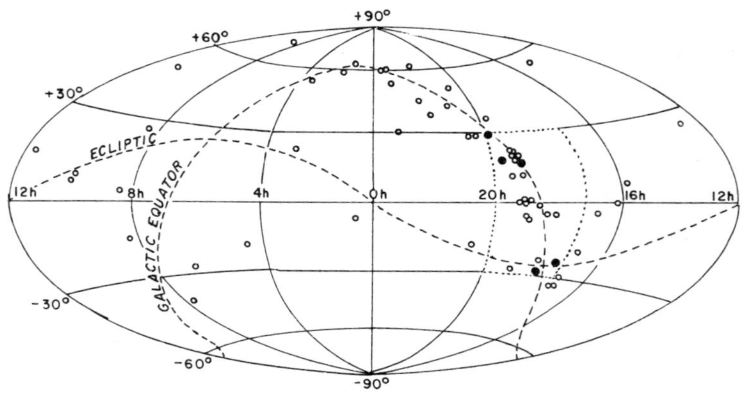

Nonuniform distribution of pulsars. Open circles: $10^{-14} > dP/dt > 10^{-16}$. Filled circles: $dP/dt < 10^{-16}$ (Fig. 1)

Most nearby stars are members of systems that contain two or more stars, and several stars have unseen companions detected astrometrically. The possibility of the Sun possessing a faint companion star has so far not been entirely eliminated. A hypothetical companion star of, say, an apparent magnitude $m_V = 10$ (much too faint to be seen by the naked eye), has a luminosity $\sim 10^{-9}$ times that of the Sun. It is conceivable that the companion star is a crystallised white dwarf that has cooled rather rapidly to a surface temperature of less than 10^3 K during the lifetime of the Solar System. A red (or even black) dwarf of mass $M \lesssim 0.1$, that has failed to ignite hydrogen (but may be burning deuterium), has an exceedingly low luminosity and is therefore also a plausible candidate. If, however, the companion is a neutron star or black hole, produced by a supernova in the early history of the Solar System, then it is difficult to understand how our weakly bound binary sistem has survived such a violent event. It is more likely that a companion neutron star or black hole is in hyperbolic orbit, and the present close encounter is a transitory phenomenon lasting only several thousands of year. The companion star is therefore presumably either a faint white or red dwarf in closed orbit about the Sun, or a gas-accreting nearby neutron star or black hole in open orbit. It should not be difficult to detect low-luminosity objects of this kind having large annual parallax and proper motion.

Obviously, a star in closed orbit about the Sun must lie close to the ecliptic plane so as not to disturb the planets excessively out of their common orbital plane. According to current theories on star formation, stars and their planets condense in interstellar gas clouds, and therefore planets probably revolve in the same orbital plane as the parent stars of a binary system. A companion star orbiting close to the ecliptic is not inconsistent with the distribution of pulsars of $\dot{P} < 10^{-16}$ shown in Fig. 1.

A companion star in closed orbit would produce various long-term effects within the Solar System. For example, the apsidal period U (the period of precession of perihelion) of a planetary orbit of period P_p is

$$U = \frac{4}{3}(1 + M^{-1})\ P^2_c P_v{}^1 \qquad (5)$$

and hence, with $M \sim 1$, $P_p \sim 1$ yr, $P_c \sim 10^4$ yr, this equation yields $U \sim 10^8$ yr, and the precession amounts 0.5 arc s per century (or 0.1 arc s per century in the case of Mercury). Short-term perturbations in the motions of Neptune and Pluto are significant, but as far as I can determine, are not unacceptably large.

Long-term changes in climate are conceivably attributable to periodicities in the Earth's orbit, that could be enhanced by the presence of a companion star, particularly if its orbit has appreciable ellipticity. Furthermore, cometary motions will undoubtedly be affected. According to Oort's comet-cloud theory, the comets extend out to a distance 10^5 AU, and the comet cloud therefore envelopes the proposed binary system of the Sun and its companion star. Oort has suggested that the nearby stars, at a few light years distance, perturb the comet cloud and as a result comets continually diffuse into orbits of comparatively small perihelia. The proposed binary system would obviously play an even more effective role in stirring the cometary orbits, and could also explain why the comet cloud, during the lifetime of the Solar System, has expanded to its present size.

Has the Sun a companion star? I find it hard to believe that a star so close can exist and yet remain undiscovered. On the other hand, pulsar observations of extraordinary precision imply that it might exist, and therefore a search for a companion star is perhaps worth undertaking.

. . . .

HAS THE SUN REALLY GOT A COMPANION STAR?

Henrichs, H. F., and Staller, R. F. A.; *Nature,* 273:132–134, 1978.

Harrison has pointed out that the six radio pulsars with the smallest observed spin-down rates ($\dot{P}$) are grouped together in a relatively small region of the sky, roughly in the direction of the galactic centre. To explain this particular effect he suggested that the intrinsic values of $\dot{P}$ are actually larger than the observed ones, because the barycentre of the Solar System would be accelerated ($v \simeq 10^{-6}$ cm s^{-2}) in that direction. Harrison then suggested that this acceleration would be due to an hitherto undetected companion star of the Sun. He proposed a faint white, red (or even black) dwarf in a closed orbit, rather than a gas-accreting neutron star or black hole in an open orbit. We show here that a companion producing such an acceleration should almost certainly have been discovered because of its brightness in the infrared or visible region of the spectrum, depending on the character of the object. We conclude that Harrison's explanation is probably incorrect, and therefore we presume that the observed spin-down rates of these six pulsars (including the binary pulsar) are intrinsic and that their extreme values are due to the particular way in which the pulsars were formed. We show that the apparent clustering of these radio pulsars is a consequence of a strong observational selection effect and conclude that the distribution of all radio pulsars seem to be symmetric around the galactic centre.

....

ON COMPANIONS AND COMETS

Kirk, J.; *Nature,* 274:667-669, 1978.

Harrison has recently hypothesised that the Sun possesses a companion star, in order to explain an anomaly in the distribution on the sky of pulsars which lose speed very slowly. This is a controversial suggestion mainly because proper motion sky surveys down to at least 14th mag should have detected such an object. However, one can imagine various objects which might fulfill the constraint of not being visible. Harrison suggested that crystallised white dwarfs, red dwarfs and black dwarfs could be in a bound orbit about the Sun, and neutron stars and black holes would more likely be in an unbound orbit, due to the supposedly explosive natures of their births. Whatever the luminosities of these objects they must produce an acceleration of the barycentre of the solar system to remove the pulsar anomaly. We discuss here the effect of this acceleration on the orbits of the 'new' comets, and how it rules out the possibility of a solar companion in a bound orbit.

According to the Oort theory of comets a cloud of cometary material exists around the Solar System at a distance $\sim 10^4$ AU. This is a considerable part of the distance to the nearest star, but the cloud moves with the velocity of the Solar System and is therefore identified with the Sun. A new comet occurs when a part of this cloud is perturbed by a passing star which changes the perihelion distance of its orbit, bringing it into the range where it may become visible (~ 6 AU). After experiencing the perturbation, the comet makes the long journey inwards to the Sun, taking a time approximately equal to the free-fall time $\sim 10^5$ yr to complete. Most of this time is spent in the outer part of the orbit at large distance and low velocity. We consider here the effect of a Sun's companion on the orbits of the comets in two cases: For a companion in a bound orbit, and for a companion in an unbound orbit. [Mathematical details are omitted.]

In conclusion, the orbits of the new comets are so close to being parabolic that the existence of a companion in bound orbit about the Sun can be excluded. Furthermore, this sets a lower limit on the velocity of a companion in hyperbolic orbit about the Sun. This lower limit is, however, insufficient to rule out the possibility that the Sun is undergoing an encounter with an object from the galactic halo.

• Electrical Charges on Stars

EXISTENCE OF NET ELECTRICAL CHARGES ON STARS

Bailey, V. A.; *Nature,* 186:508–510, 1960.

It has been found possible to account for the known orders of magnitude of five different astronomical phenomena and the directions relating to three of them by means of the single hypothesis that a star like the Sun of mass M_s gm. carries a net negative charge, $-Q_s$, which is given by the formula:

$$Q_s = B_8 G^{1/2} M_s \text{ e.s.u.} \quad (1)$$

where G is the constant of gravitation and B_s is a pure number of the order of 0.03.

Thus, for the Sun we have:

$$Q_s = 1.5 \times 10^{28} \text{ e.s.u.} \quad (2)$$

The same hypothesis leads to simple qualitative or semi-quantitative explanations of at least the thirteen other phenomena considered later.

The first five phenomena are as follow: (1) the magnetization of the Earth; (2) the outer Van Allen belt of radiation; (3) the maximum energy of about 5×10^{18} eV, found for a primary cosmic ray particle; (4) the Sun's north polar magnetic field vector H_p; (5) the approximate equality of the ratio P/U for the Sun to the mean value of this ratio for Blackett's group of five magnetic stars, where P is the magnetic moment and U is the angular momentum of a star.

The phenomena 1-4 arise as follows from the presence of the solar charge -1.5×10^{28}.

The velocity of translation of the Sun relative to the Earth is v_s -29.8 km./sec. and consequently its charge, $-Q_s$, sets up in a non-rotating frame of reference accompanying the Earth a magnetic field

H_r which is perpendicular to the ecliptic, directed from north to south and has near the Earth the intensity 6.9×10^{-3} gauss.

The magnitude and direction of H_r are approximately those required by J. S. Chatterjee for the initial field, of unknown source, which saturates the highly permeable magnetic material of a 20-km. shell near the top of the Earth's crust and thus imparts to the Earth its observed magnetization. For Chatterjee requires a "field of the order of 0.1 to 0.01 gauss".

Moreover, the field H_r not only magnetizes the Earth but also neutralizes the resulting external terrestrial field in a narrow and approximately toroidal region which lies nearly in the Earth's equatorial plane and has a radius equal to about 3.7 earth radii. The magnetic field in and near this region is such that it could act as a "magnetic bottle" which is partly accessible to energetic charged particles from much further out. These results account very well for the existence of the outer Van Allen belt which has a maximum density at about 3.6 Earth radii.

The Sun's charge also produces an electric field which has the potential $V = -3.0 \times 10^{17}$ volts at the Earth's orbit. This field causes free protons and other nuclei in distant space to approach the Sun and move in orbits round it. Those nuclei which lose little energy on the way to the Earth's orbit will then reach this orbit with an energy of 3.0×10^{17} Z eV., where Ze is the charge on the nuclei concerned. For an iron nucleus this energy amounts to 8×10^{18} eV. Also the orbits of all the nuclei will be oriented at random. These results account very well for the observed maximum energy 5×10^{18} eV, of a primary cosmic ray particle and for the isotropy of cosmic rays.

The rotation of the negatively charged Sun imparts to it a magnetic moment which is directed from north to south. This in turn produces a north polar magnetic field vector H_p which would have the magnitude -0.75 gauss if the Sun's charge were distributed in the same way as the mass and if the mass and angular velocity were distributed as in Chapman's model A. A uniform distribution of Q_8 throughout the Sun would yield the value -8.5 gauss. These results agree in order of magnitude and direction with the observations of Hale and Langer who, according to a report by H. W. Babcock, found a mean value of -4 gauss. They also agree with the most recent observations of H. D. Babcock. The presumably temporary reversal of the direction of H_p, during six years, which was observed by the Babcocks suggests that our estimated value of H_p corresponds to the constant component of the Sun's magnetic field, and that there exist also an oscillating component which varies with the solar cycle and a component due to local turbulence.

The fact (5) would arise from the hypothetical stellar charges specified by formula (1) if the Sun and the average of Blackett's five stars had axially symmetrical distributions of the angular velocity ω, the charge density σ and the mass density ρ such that at a point (r, θ) the quantities $\omega\sigma$ and $\omega\rho$ are each functions of θ and r/a alone, where a is the diameter of the star. The latter assumption is the reasonable one that magnetic stars are homologous rotating bodies.

It would be a remarkable coincidence if these five phenomena were all explained correctly in their orders of magnitude and their related directions by means of a single hypothesis which is entirely false, even if the last phenomenon calls for a supplementary assumption of homology in magnetic stars.

....

LIGHTNING, SOLAR FLARES AND RADIO GALAXIES

Bruce, C. E. R.; *Nature*, 209:286, 1966.

In 1925, when it was still believed that a potential of n x 10^9 V was necessary to cause a lightning flash, C. T. R. Wilson put forward the suggestion that the soft cosmic rays up to about 10^9 eV might be accounted for by runaway electrons in thundercloud fields. When it was pointed out that a lightning flash can occur with a total thundercloud potential of n x 10^7 V, the suggestion ceased to have much significance as applied to thunderclouds. However, in presenting a paper on these long leader strokes at the Belfast meeting of the British Association in 1952, I pointed out that the Wilson mechanism might apply in the much longer cosmic electrical discharges, in which relatively large electric fields must be built up by the short-circuiting of these much more extensive electric fields by the leader stroke of such discharges as solar flares.

In an earlier communication, it was emphasized that temperatures of the order of 10^8 °K must be reached in these last electrical discharges, and in a later communication it was pointed out that observational evidence had actually been obtained for the association of such temperatures with outbursts of solar flares.

The velocity of the jets of gas generated by these discharges is of the same order as that of the thermal velocities of the ions, namely, about 10^8 cm/sec, and thus accounted for the delay of about 1 day between the solar outburst and the main magnetic storm at the Earth. The associated electron velocities will be about 40 times this value or about 4 x 10^9 cm/sec, and it is now suggested that some of these electrons become the runaway electrons which account for the arrival at the Earth of a soft cosmic ray component about half an hour after the observed solar outburst. Their average speed is thus accelerated by a factor of about 2 in the solar electric field.

This consideration would confirm the view arrived at from an investigation of cometary phenomena that the polarity of the solar field is the same as that of the thunderstorm field; that is, the regions farthest out in the gravitational field become positive.

This view of the existence of two major velocities of propagation of two different types of activity, a high speed electron component and a plasma jet, has interesting possible applications in the investigation of radio galaxies in which the same two components may be expected to occur. Observations on the gas velocities made by Seyfert of about 4 x 10^8 cm/sec suggest the existence of plasma jets at temperatures of about 4 x 10^{30} K. Associated with these we should expect to find active haloes due to the runaway electrons from the main discharges and reaching dimensions of the order of ten times the latter. This is just what is observed, and in the case of Virgo A (NGC 4486) "the general direction of the extension is the same as that of the 'jet' which optical observations have shown issuing from the nucleus of NGC 4486". Shain writes in the same contribution that "when angular sizes of a number of radio galaxies are determined, any optical search for associated galaxies must extend to galaxies having one-tenth or less of the radio diameters."

LIGHTNING, NOVAE AND QUASARS

Bruce, C. E. R.; *Nature,* 209:798, 1966.

Quasars 'or something' have been a desideratum of the electrical discharge theory of cosmic atmospheric phenomena and universal evolution since its inception in 1941, and the absence of any reference to galaxies which are radiating an inordinate amount of energy was one of the main difficulties facing the theory. As the years went on, I had to conclude that the energy must be largely in the unobservable range, for example, X-rays or γ-rays. However, in the original outline of these ideas published in 1944 (ref. 1a), I emphasized that galaxies exist which possess an emission line spectrum, and that in all probability they are those which the theory required. In a recent review article[2] on quasars, this is still their only optical characteristic mentioned.

On the theory, they are analogous to the naked eye phase of novae, that is, they represent the phase around peak current in the discharge[1b], during the neutralization of the electric field built up slowly during lifetime of a star or galaxy, probably by asymmetrical impacts between grains[1c], though the exact nature of these field-building processes is not yet agreed even in thunderclouds[1d].

In a paper presented to the second U. S. Air Force Conference on Atmospheric Electricity in 1958 (ref. 1e) I pointed out that this phase of the discharge lasts for about 10 μsec in the lightning discharge, from 1 to 10 min in a solar flare[1f], about 10 days in a nova[1b], and might be expected to last for 10^6 years in the corresponding galactic discharge. However, if, as seems reasonable in view of the geometrical similarity between the two cases, we assume that this phase will occupy a similar fraction of the total life of the outburst on the stellar and galactic scales, we obtain an estimate of 10^4-10^5 years for the peak current phase of the galactic discharge, which agrees well with Greenstein's range of estimates of 10^3-10^7 years for the duration of quasars[3].

Another characteristic which quasars and novae have in common is the extreme rapidity with which their radiation can vary during the outburst, due very probably to the pinching out of the discharge. In Nova Hercules 1934, for example, the brightness of the star fell abruptly by about nine magnitudes soon after its peak. Thereafter its brightness increased somewhat more slowly until it approached its pre-extinction value.

From the photographs[4] of the well-known 'jet' or discharge channel in the radio source Virgo A, NGC4486, it would appear that these pinches also occur on the galactic scale, since the jet is divided into three 'sausages' by the pinching of the channel in two places.

A corresponding phenomenon was recently photographed in the lightning flash[5], so these varied observations would appear to afford another interesting link-up between phenomena on these three very different scales[1g].

The characteristics of the phenomena will be different from those already considered in relation to the pinching of laboratory discharges, as we have now to consider the breakdown of an electrostatic field[1j]. As indicated in the recent article just referred to [1g], huge electrostatic as well as electromagnetic forces are involved.

It may also be noted that the discharge theory of necessity introduced circular magnetic fields in and around the spiral arms of the galaxies from its initiation in 1941. In 1953 magnetic fields along the arms were introduced by Chandresekhar and Fermi[6], but as Cowling emphasized[7], without a proposed origin of these fields, that did not take us much farther. However, in 1961 it was proposed[8] to transform these last fields into circular fields around the arms in agreement with those originally introduced in 1941 by the electrical discharge-theory.

Furthermore, in 1961 (ref. 1k) I estimated the total energy liberated by a radio galaxy as 10^{60} ergs when the value available for comparison, derived from Virgo A. was 10^{58} ergs (ref. 9). From a survey of all the available data, Heeschen[10] later arrived at a value of 10^{60} ergs in agreement with the earlier value deduced from the electrical discharge theory.

1 Bruce, C. E. R., a, A New Approach in Astrophysics and Cosmogomy (London 1944); b, Observatory, 69, 193 (1949); c, Phil. Mag., 46, 1123 (1955); d, J. Franklin Inst., 268, 425 (1959); e, Rec. Adv. Atmospheric Electricity, edit. by Smith, L. G., 467 (Pergamon Press, 1959); f, Observatory, 69, 110 (1949); g, Nature (in the press); h, Nature, 147, 805 (1941); j. Proc. Roy. Soc., A., 183, 228 (1944); k, J. Inst. Elect. Eng., 7, 513 (1961).

2 Burbidge, G. R., Burbidge, E. M., and Sandage, A. R., Rev. Mod. Phys. 35, 947 (1963).

3 Greenstein, J. L., Scientific American, 209 (No. 6), 54 (1963).

4 Baade, W., and Minkowski, R., Astrophys. J., 119, 215 (1954).

5 Matthias, B. T., and Bochsbaum, S. J., Nature, 194, 327 (1962).

6 Chandrasekhar, S., and Fermi, E., Astrophys. J., 118, 113 (1953).

7 Cowling, T. G., Magnetohydrodynamics, 31 (Interscience Publishers, Inc., New York, 1957).

8 Hoyle, F., and Ireland, J. G., Mon. Not. Roy. Astro. Soc., 122, 35 (1961).

9 Burbidge, G. R., Nature, 190, 41 (1961).

10 Heeschen, D. S., Scientific American, 206, 41 (1962).

• The Drayson Flap

"THE CHANGE OF CLIMATE AND ITS CAUSE"

Davidson, M.; *Knowledge,* 12:220-222, 1915.

Not long ago a book entitled "The Change in the Climate and its Cause," by Major R. A. Marriott, D. S. O., made its appearance, and I have written this article in the hope that some light may be thrown upon the views held there---views which, I believe, will not find general acceptance in the astronomical world.

In criticising this book it is necessary to go back a good many years. In 1873 a work entitled "The Cause, Date, and Duration of the Last Glacial Epoch" appeared, the author being the late Major-General A. W. Drayson. The following year another volume came out as a sequel to this, the title being "The Cause of the Supposed Proper Motion of the Fixed Stars"; and in 1888 "Thirty Thousand Years of the Earth's Past History" was published. In 1911 Admiral Sir Algernon F. R. De Horsey brought out "Draysonia," which was intended to explain more fully Drayson's system. It contains many examples, fully worked out, to show the application of Drayson's theory to the determination of the obliquity of the ecliptic, stellar positions, precession, and so on. Major Marriott's work gives a further explanation of the theory, and its bearing upon climatal changes: these latter are cited as corroborative evidence of the truth of Drayson's theory.

As it is impossible within the limitations of the article to deal very fully with the subject, I shall endeavour briefly to indicate that this theory is, and to show that its so-called applications are more apparent than real. In doing so, we must assume a certain amount of astronomical knowledge on the part of the readers, as, otherwise, too much space would necessarily be devoted to ordinary explanations. Drayson believed that, in addition to the daily rotation of the Earth upon its axis, there was also a slow rotation taking place around another axis, the arc between the poles of the two axes being 29° 25' 47". As the pole of the ecliptic is distant from the pole of the heavens by about 23° 27' 3" at present, the pole of the second axis of rotation would be 6° from the pole of the ecliptic. A complete revolution would be accomplished, he believed, in thirty-one thousand six hundred and eighty-two years; and as the obliquity of the ecliptic is measured by the arc between the pole of the heavens and the pole of the ecliptic, this obliquity would vary considerably, its greatest and least values being the sum and difference respectively of 29° 25' 47" and 6°, that is, 35° 25' 47" and 23° 25' 47".

For an explanation of the causes of precession and nutation I must refer readers to any elementary text-book of astronomy. The conical movement of the Earth's axis, by means of which it is carried around the pole of the ecliptic in twenty-five thousand eight hundred and sixty-seven years, appears to have puzzled both Drayson and his followers. Thus, on page 92 of "Thirty Thousand Years of the Earth's Past History," he gravely says: "The mere statement that the Earth's axis traces a conical movement, when no mention is made of which pole remains fixed, is a vague statement, utterly deficient in detail. That such a movement, if it occurred, would throw the centre of gravity out of its orbit did not seem to have been even considered." We need do no more than point out that each semi-axis of the Earth describes a similar cone, the apex of each cone being at the centre of the Earth, which point we may regard as fixed. It is difficult to understand how such a question should have been raised.

We shall now inquire into the problems which are solved by the aid of this system---problems which will be found in all the works cited. First of all, the question of the obliquity of the ecliptic will occupy our attention: a calculation is shown on page 37 of Marriott's book, and on page 28 of De Horsey's "Draysonia." The formulae used are, I imagine, based upon the elementary formula for the solution of a spherical triangle, $\cos a = \cos b \cos c + \sin b \sin c \cos A$, where the symbols correspond with the following in the first work mentioned:

a OE, that is, the distance of the pole of the heavens from the pole of the ecliptic, or the obliquity of the ecliptic.

b OC, that is, the distance of the pole of the heavens from the pole of the second axis of rotation, which, according to Drayson, is always $29^{\circ}\ 25'\ 47''$.

c CE, that is, the distance of the pole of the second axis of rotation from the pole of the ecliptic, which is also constant, being always 6°.

A angle ECO.---This angle is found by assuming that it changes $40''.9$ annually, and that its value will be zero in A. D. 2295.

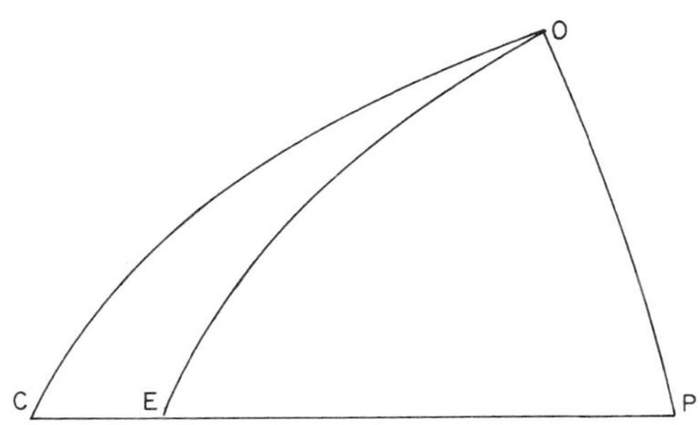

When the formulae are used by all three authors for computing the obliquity, formulae which are adapted for logarithmic computation, but which are based upon the above, we must admit that values are found which are nearly correct. For this reason some might be led to attribute a certain importance to the theory, if, as Major Marriott says, "Drayson can compute the obliquity of the ecliptic for any year, past or future, without observation." We shall now proceed to show that such a computation cannot be made with accuracy.

We should notice, in the first place, that the side c is only 6°, and that therefore its sine is small. In the next place, the angle A was less than 8° in A. D. 1600, and will be only $3^{\circ}\ 20'\ 29''$ in A. D. 2000, decreasing gradually until it becomes zero in A. D. 2295. Now it is clear that within certain limits we can change A without producing great changes in the value of cos a, for the value of sin b sin c is only .05136. In A.D. 1600 cos A = .9905675, and in A.D. 1900 cos A= .9969515; hence, even over this period, there will not be much alteration in the value of the expression sin b sin c cos A, and, as we have just pointed out, it always remains small, never exceeding .05136. For this reason there is nothing remarkable in the formula applying approximately over comparatively short periods of time. To test its accuracy we should try to apply it over long epochs, and compare the conclusions with facts obtained by observation. I have done this, and give the results below. The "zero year," that is, the year when the obliquity was a minimum, is taken to be A.D. 2296 by Drayson, 2295 by Marriott, and 2294.75 by De Horsey. The difference, however, is of no practical importance, not affecting the results of the calculations, except in fractions of a second. I have taken 2296 in my calculations (see Table).

Date.	Obliquity calculated by Drayson's Method.	Obliquity found by Observation.	Names of Observers.	Error by Drayson's Method.
B.C. 1100	24° 59' 46"	23° 54' 3"	Tcheon Kong	1° 5' 43"
B.C. 160	24 16 33	23 45 52	Chinese	0 30 41
A.D. 827	23 43 40	23 33 52	Arabians at Baghdad	0 9 48

Probably it will be argued that the determinations by the early observers are untrustworthy. It may be true that the very earliest are correct only to the nearest minute of arc; but this still leaves a very wide margin of error by Drayson's method of calculation. The figures require little comment. An error of more than one degree is too serious to ignore. Even the more modern determination of the obliquity by the Arabians shows that Drayson's method is only approximate to within about ten minutes of arc. It seems very remarkable that Drayson and his admirers should not have tried to apply the method over long periods.

As we come up to more recent times the error diminishes. This is to be expected, because the diminution of the angle A causes the formula we have given to approach more closely to the form cos a = cos b cos c sin b sin c, or a = b -c, that is, a = 29° 25' 47" -6° = 23° 25' 47", which is not far from the obliquity in recent times. By assuming that the formula held exactly for long periods as it holds approximately for short periods, an obliquity of 35° 25' 47" was found for 13544 B.C. As is well known, astronomers do not believe that the obliquity ever reached such a value; a variation of about 1° 18' 41" on either side of 23° 17' 17", the mean value, is the limit. The discussion of the reasons for assigning these limits is quite beyond the space allotted for this article. In spite of Major Marriott's criticism on page 64, where he points out that La Place could not get away from the assumption that the pole of the ecliptic was the centre of the curve described by the pole of the heavens, we are still willing to accept the results of his investigations on the changes of the obliquity being limited to a total variation of less than 3°.

It scarcely seems necessary to dwell on the other problems which are supposed to be capable of solution by this method. We have shown that it breaks down utterly when applied to the obliquity. Does the same thing hold with reference to precession and to the positions of the stars? De Horsey gives a formula on page 20 of "Draysonia" which may be expressed as follows: $\pi = 20.0529 \cos a$, where π is the annual precession in declination and a the right ascension of the star. This formula gives results which are fairly accurate, but this approximate accuracy can be explained by the fact that it is almost identical with the formula an astronomer would use for finding the precession in declination, assuming the precession of the equinoxes to take place in accordance with the usually accepted facts. Without going into the proof, I may merely state that the annual precession in declination can be represented by the formula $P = 50''.2 \cos a \sin w$, where w is the obliquity of the ecliptic, and a the right ascension of the star. If we take w to be $23^{\circ}\ 27'\ 4''$, this reduces to $P = 19.978 \cos a$, which is almost identical with De Horsey's formula. Here, again, if we are willing to ignore small errors, the formula based upon Drayson's system appears to give nearly correct results. If it were applied for a long period of time, when w had varied by about 30', the method could make no claim even to an approximation. We may point out that Admiral De Horsey appears to be under an utter misapprehension about the subject of aberration. On page 21 he says that he will possibly be told that the motion of a star in right ascension and declination is properly termed "aberration." I think no astronomers are likely to confuse precession with aberration, which is an entirely different phenomenon, depending upon the orbital velocity of the Earth and the velocity of light. "Some of us may suffer from aberration of the intellect; perhaps I do," he continues. Certainly those who fail to distinguish between aberration and precession lay themselves open to the charge.

The method professes to find the right ascension and the declination of stars for any date, past or future. This, inter alia, will be found under the comparative summary given on page 17 of Major Marriott's work. That it can do so over comparatively short periods of time with small errors none of us will deny. But when we remember that the method ignores the proper motion of the fixed stars, or, rather, professes to explain the "so-called proper motions," we must receive it with some hesitation. In Chapter X of his work, published in 1874, Drayson explains "The Cause of the so-called Proper Motion of the Fixed Stars." The second axis of rotation is the key to the problem, and, apparently, astronomers had fallen into serious errors by overlooking this important consideration. It is certainly difficult to discuss the whole question of stellar motions in an article like this; but if Major Marriott believes that assigning proper motions to the stars was only a scheme for overcoming a difficulty, and that "every star whose position did not agree with the theory was assigned a proper motion of its own" (page 42, "The Change in the Climate"), may I refer him to the spectroscopic method of determining stellar motions? A recent work, by Dr. Campbell, Director of the Lick Observatory, was published in 1913, and gives a very good account of this important branch. Adherents of the second axis of rotation theory would do well to study carefully Dr. Campbell's "Stellar Motions, with Special Reference to Motions determined by Means of the Spectrograph." They will not, perhaps, find the necessity for appealing to a second axis of rotation to explain the motions of the stars.

The final point which we propose to discuss is that of climatal changes. The second rotation of the Earth would have brought about the condition of maximum obliquity about the year 13544 B.C., an obliquity amounting to 35° 25' 47". Thus the Arctic Circle would have been brought down to about latitude 54-1/2° at this time, and in middle latitudes an arctic climate would prevail in winter, and tropical conditions in the summer. There would follow from these variations in climate annual migrations of the animals, and in this way we are able to account for the remains of tropical and arctic animals being found in the same drift. Major Marriott cites the views of American geologists on pages 69 and 70, the evidence apparently showing that the last Glacial epoch was of more modern date than had generally been supposed. Drayson's theory would give an obliquity twenty-three thousand years ago of a magnitude sufficient to inaugurate a Glacial period, and the obliquity had diminished sufficiently seven thousand years ago to cause the glacial conditions gradually to pass away. The author is struck by the apparent agreement between the astronomical and the geological evidence. "The empirical science of geology," he says, "is thus confirmed by the exact science of geometry with astonishing precision" (page 72).

If we understand this theory of climatic changes aright, the heat of the summers during the period of great obliquity liberated the icebergs formed in the winter, and these were scattered into localities of lower latitudes. If an arctic winter had continued uninterruptedly for thousands of years, no icebergs would be liberated from arctic regions or be seen in the Atlantic. The hot summers which would prevail during the time of great obliquity were as necessary to account for the dispersion of icebergs as the cold of winter to produce ice. When the obliquity has reached a minimum, in A.D. 2295, Major Marriott promises us the apex of the genial period when the contrast between summer and winter will be least. The fact that Grimsby fishermen encountered floating ice in great quantities round the Faroe Islands some years ago, and also that Iceland had no snow and scarcely any frost in mid-December, 1910, are given as corroborative evidence, amongst others, of an approach of genial conditions.

That the small diminution in the obliquity is responsible for these facts will not, I think, receive general acceptance. The fallacy of post hoc, ergo proper hoc, is one into which we are all liable to fall at times. Why the diminution of the obliquity should be held responsible for local variations of climate, or, indeed, for a universal change, is not quite clear. Atmospheric dust, sunspots, volcanic activity, variations in both oceanic and aerial currents, and many other agencies, have far more influence in producing changes of climate than a diminution or an increase of the obliquity, if, indeed, this has any practical effect at all. Again, is Major Marriott prepared to defend the theory of the annual migrations of the animals advocated on page 47 of "Thirty Thousand Years of the Earth's Past History"? We can scarcely describe animals like the elephant, rhinoceros, and hippopotamus as migratory. Even if we grant the possibility of these animals performing great annual migrations, how are we to explain the presence of two very different floras and molluscuous faunas in the Pleistocene deposits of Middle Europe on any theory of "annual migrations"? The case against the annual migration theory has been stated by various writers, amongst whom we may refer to the late Lord Avebury, in Chapter IX of the last edition of "Prehistoric Times"; to Professor J. Geikie, in his "Prehistoric Europe," page 65, as well as in "The Great Ice Age." The arguments will be found very clearly summed up in these works.

Enough has probably been said to show that the theory of a second axis of rotation is quite untenable, and its acceptance leads to contradictions, inconsistencies, and grotesque hypotheses. It is, indeed, clothed with a semblance of reality when it appears to give accurate results for the obliquity, precession, and so on. As we have shown, these results are approximate only over short periods, and are utterly untrustworthy for long epochs. Finally, the dynamical problems raised in the mind of the present writer by a consideration of a body like our Earth, possessing at the same time two axes of rotation, are of such a formidable nature that he must candidly admit them to be entirely beyond his powers of adequate solution.

COSMOLOGICAL ENIGMAS

Cosmology touches upon many physical laws and concepts of the universe long held dear by science. Of late these "touches" have been heavy handed---almost shattering. This section opens with a survey paper by a noted astronomer that immediately gets down to cases by chipping away at the favored Big Bang theory. After this heresy, selected articles focus on the following questions:

- Are the red shifts really a measure of distance?
- How can quasars generate the prodigious quantities of energy assigned them?
- Can quasar features really move faster than light?
- How did the isotopic abundance anomalies arise?
- Where is the supposed missing mass of the universe?
- Do black holes and white holes really exist?

These are all fundamental questions, and we do not have convincing answers to most of them. Even some proposed answers, such as the Big Bang hypothesis, create even grander questions; viz., What happened or existed before the Big Bang?

Deep space, the serene sea populated by Jeans' island universes, now seems a seething tempest full of entities possessing unbelievable properties.

• Cosmological Overview

WAS THERE REALLY A BIG BANG?

Burbidge, G.; *Nature,* 233:36-40, 1971.

In any given field of science, activity has its peaks and its valleys. Many view activity as synonymous with progress. If one judges by the number of papers currently appearing and by the interest of the younger astrophysicists, activity in cosmology is at a peak at present. To the outsider and even to many inside, it looks, at first sight, as though considerable progress has been made in recent years, so that the outline of the way in which the universe has evolved is understood and only the details need to be explained. Is there any justification for this? Those who have read widely in the earlier and more recent cosmological literature, as well as those who rely on North's account, are aware that views of cosmology at any epoch are largely determined by the ideas of a few strong individuals, rather than by an objective appraisal of the information available. Today a similar situation prevails, and it is further complicated because we are often trying to use classes of objects to test cosmological theories long before we understand the physics of them.

In my view, modern developments have been influenced by several factors. They include new observational discoveries combined with an extremely simple minded approach to them, a deep seated conviction by many that general relativity is correct in all details, a belief that astronomy has nothing to teach us about fundamental physics, and, last but not least, a deep seated hostility by both observers and theoreticians in the astronomical community toward the steady state theory. Usually this last point is not mentioned in polite society, but it is blatantly obvious in Dingle's remarks while president of the Royal Astronomical Society; it is often apparent in private discussions, and a further manifestation is a complete disregard of the theory when the state of cosmology is reviewed.

What are the major scientific events which make up the case for a big bang? They started, of course, with the work of Einstein in 1917. Incidentally, he, as did his contemporaries, took it for granted that the universe in the large is unchanging, that is, he assumed that it is in a steady state. But it was Lemaitre and, independently, Friedmann who showed that the Einstein universe is unstable so that, when perturbed in the direction of expansion, it will continue to expand, tending to a de Sitter universe in the limit. Lemaitre's analysis showed that there is an infinity of models satisfying Einstein's equations. Similar solutions are obtained in the scalar-tensor theory. The most interesting model is the one which starts at a finite time in a state of infinite density. It was therefore inferred that, according to the theory of general relativity, the universe must be expanding, and most likely began in a state of high density. Almost immediately following this major prediction of general relativity, Hubble and Humason conclusively established the redshift-apparent magnitude relation, which gave observational proof that the universe is expanding. Thus by 1930 Einstein's theory had made what many considered to be the greatest successful prediction in the history of science. For the next twenty years the drive in observational cosmology, almost completely in the hands of Hubble and his colleagues, was to attempt to decide from observation which of the myriad of general relativistic models best represent the universe. Soon, however, there developed an apparent paradox associated with the ages of different systems, because it appeared that the ages of various astronomical objects, notably the Earth, seemed to be greater than the age of the universe derived from the rate of expansion obtained by Hubble. Two possible paths of escape from this dilemma were the Lemaitre model with its arbitrarily long waiting period, and the steady state cosmology requiring the continual creation of matter.

The present situation with regard to this age difficulty is that we now realize that the uncertainties associated with the determination of H and those associated with obtaining the ages of the oldest stars are great enough that we no longer have any strong evidence that there is any problem. Most recently it has been stated that H = 50 km/s MPC^{-1}, or $H^{-1} = 20 \times 10^9$ years, while the age of an old galactic cluster, NGC 188, is given as 10×10^9 years, and the age of the elements obtained from the radioactive isotopes is $\sim 10 \times 10^9$ years, with uncertainties depending on assumptions made about the sequence of events leading to the isolation of U and Th in the solar system.

The prediction that the universe is expanding, followed by the discovery of the expansion, was the first major success of cosmology based on general relativity. The second came from the comparatively elementary discussion of the physics of the early stage in a dense expanding universe---the primaeval fireball---by Gamow and his colleagues, and the discovery in 1965, by Penzias and Wilson, of background radiation which may well be relict radiation from this fireball. These two predictions, followed by two major discoveries which, at first sight, appeared to confirm the theory, provide the basic scientific case for evolving models of the universe.

If we wish to account for the expansion without invoking an initially dense phase, we are left with either the steady state model in which matter is created or with an oscillating model. The steady state alternative seemed attractive to some, especially "when taken in conjunction with aesthetic objections to the creation of the universe in the remote past. For it is against the spirit of scientific enquiry to regard observable effects as arising from causes unknown (and unknowable) to science, and this in principle is what creation in the past implies. There is apparently a great gulf between those who find creation in the past acceptable, but continuous creation unacceptable, and those who have no such strong feelings. It is my belief that much of the conflict in cosmology in modern times has arisen because of this division which has little to do with science. More to the point, we must critically evaluate the detailed evidence which has accrued in the past twenty years concerning the question of whether or not we have definitely established that the universe is evolving.

Determination of q_0. In 1956 Humason, Mayall, and Sandage gave a detailed account of the status of the Hubble relation. At that time the largest redshifts which had been measured were z = 0.2. In the fifteen years that have alapsed since then, Sandage and his collaborators have worked intensively on the problem of refining the measurements of H and q_0. Many progress reports on this work have been given by Sandage, see also Peach. As far as the value of H is concerned there is some reason to believe that we are beginning to converge toward a realistic value some ten times smaller than that obtained by Hubble in 1936. As far as the determination of the deceleration parameter q_0 is concerned, the position is not so good. There are two basic reasons for this. First, very few clusters with redshifts greater than 0.2 have been discovered. The most recent plot given by Sandage contains forty-two first-ranked galaxies in clusters, and only three of these, with $z = 0.29$, 0.36 and 0.46, are in excess of $z = 0.2$. The three clusters were observed by Baum more than ten years ago, and a detailed discussion of his data has never been given. Many people believe that the radio astronomer's ability to detect more distant objects will lead, or has led, to an improvement of the situation. The only major achievement of radio astronomy in this connexion so far was the identification, in 1960, of 3C 295 with a galaxy with a redshift of $z = 0.46$. This source lies in a cluster, and Baum obtained photometry which, combined with the redshift, gave this last point on the Hubble diagram.

About 100 galaxies associated with radio sources and 200 quasi-stellar objects, most of which are radio sources, have been optically identified, and redshifts have been obtained. None of these, with the exception of the cluster around 3C 295, has given us any information about the classical Hubble diagram beyond $z = 0.2$. This is because the radio sources are associated with galaxies with redshifts > 0.2, or they are associated with galaxies which obviously cannot be treated as "standard candles", or they have been identified with QSOs which show a very large dispersion in the redshift-apparent magnitude diagram. Using the radio properties as a guide, we can single out a certain class of objects, but it is not clear at all that they can be used for cosmological investigations of the classical kind. If we make a table of optical types of radio galaxies as a function of increasing redshift, we see that, at small redshifts, almost all types of galaxy are being identified. At intermediate redshifts the galaxies are mostly giant ellipticals, while at the largest redshifts, between 0.2 and 0.3, the galaxies with published redshifts have nearly all been classified as N-type, meaning that a large part of the optical energy that they are emitting is due to non-thermal processes and hot gas rather than stars. Thus they cannot yet be used as standard candles, as the data available at present are inadequate.

....

Quasi-stellar Objects. The QSOs are the only extragalactic objects so far discovered which have very large redshifts. Therefore they are of the greatest importance for cosmology provided that their redshifts are largely due to the expansion of the universe and do not have any other explanation. If the redshifts are of cosmological origin then the luminosity-volume test previously mentioned for the QSOs in the 3CR catalogue, or a similar analysis also carried out by Schmidt for a sample of QSOs which are not powerful radio sources, leads to the conclusion that they are not distributed uniformly in space, but that the space density is very strongly dependent on z and hence on epoch.

This means that they provide powerful evidence against a simple steady state universe, and if one wishes to preserve the steady state model it must be argued that it is a fluctuating steady state universe in which evolution can occur within an epoch corresponding to redshifts up to 2. In any case the cut-off of redshifts beyond 2 is of great cosmological significance and some have attempted to relate this to the appropriate epoch in a universe of Lemaitre type.

None of these arguments has any validity unless it can be demonstrated conclusively that the redshifts of the QSOs are of cosmological origin, and this has simply not been achieved. There are a number of arguments that can be adduced for and against the cosmological hypothesis but there is no certain answer. Until this is reached, use of QSOs for cosmological investigation is premature.

Microwave Background Radiation. As mentioned above, the discovery of microwave background radiation, following the prediction that primaeval fireball radiation generated in a big bang would still be present, was one of the strongest pieces of evidence for such a beginning. Is the information that we now have on the microwave background still compatible with this interpretation? Observations made at wavelengths between a few millimetres and about 20 cm fit quite well on a black-body curve with a temperature of about 2.7 K. The energy density of the radiation is then about 4×10^{-13} erg cm^{-3}. If the black-body form is confirmed, this will be the strongest evidence in favour of an evolving universe. But it is not yet certain whether or not the spectrum is really of black-body form, because the wavelength range directly observed, and not in dispute, is the Rayleigh-Jeans part of the curve where $I(\nu) \propto \nu^2$.

The observations which have been made either directly or indirectly close to and just beyond the peak of the black-body curve are in conflict unless it is argued that some very strong line radiations are present. The indirect observation of the radiation field in the galactic plane which is exciting the CN, CH and CH interstellar molecules suggests that the temperature is less than ~ 8 K at 9.36 mm, less than 5 K at 0.56 mm, less than ~ 4 K at 1.3 mm, and it appears to be about 2.8 K at 2.64 mm. Direct observation from rockets and a balloon above the Earth's atmosphere at wavelengths in the range 0.4-1.3 mm gives a flux which corresponds to a black-body temperature of about 6 K. It is possible that the radiation detected by these observers is concentrated in a line which could be produced in a geocorona, in the Galaxy, or have a cosmological origin. The interpretation of the results is unclear at present.

If it turns out that the microwave radiation field departs strongly from the black-body form, then either extra components are present, or the radiation has a quite different origin. In both cases the answer must be that the excess radiation, or the total flux, ultimately arises in discrete sources. A number of investigations have been made in which these possibilities have been explored. The sources are most likely to be the nuclei of galaxies which radiate largely at infrared and microwave frequencies.

It was the discovery of the microwave background radiation that led to a large number of theoretical investigations concerned with the early history of the universe. What should be made clear, however, is that the existence of this radiation is very powerful evidence for an initially dense state, provided that the radiation has a black-body spectrum as is predicted in the simple theory. If it turns out that it departs strongly from the black-body shape, many attempts will undoubtedly be made to explain it in terms of a primordial radiation field plus large contributions from discrete sources, or even a more complex big bang. But this must not obscure the important point that the absence of a black-body form would mean that the prediction has not been fulfilled and the strong and direct evidence for an initial dense state will have disappeared.

This concludes my discussion of direct observational evidence bearing on whether or not the universe is evolving and began in a dense state. I believe that if one attempts to evaluate this evidence objectively there is still no really conclusive evidence in favour of such a universe.

Nucleosynthesis and Origin of Galaxies. Another approach can be taken in looking for evidence which bears on the question of whether or not the universe began in a dense state. The bulk of the mass that we see is condensed in the form of stars and galaxies with a wide range of masses and angular momenta. This matter is not distributed uniformly in space but is condensed into dense aggregates. Moreover, it has a complicated chemical composition. The major problems of cosmogony are therefore to understand the origin of the elements, and the formation of discrete objects---galaxies and other compact massive objects.

Does the existence of these features of the universe suggest that their origin is closely connected with a big bang?

I shall consider first the problem of nucleosynthesis. Early work on this problem made little progress because knowledge of nuclear physics was still comparatively primitive. It was Gamow, Alpher and Herman who first made a serious attempt to explain the synthesis of the heavier elements from an initial dense cloud of baryons, leptons and radiation in a hot big bang. It is well known that this attempt failed due to the difficulty associated with building beyond mass 4. In the 1950s a serious attempt was made to understand the relative abundances of all of the isotopes on the assumption that they have nearly all been synthesized in the interior of stars. This theory has been, in large part, highly successful and is generally accepted. It was originally thought that the fact that heavy elements are synthesized from lighter ones in stars meant that there is a slow build-up of heavy elements as a galaxy grows older, and that the relative abundance would be correlated with the ages of the stars. It is now clear that while this effect is operating, a large amount of nucleosynthesis takes place early in the life of a galaxy and the processes going on steadily throughout its life are in some ways of lesser importance.

From the point of view of cosmology the most important result is that very detailed calculations made by Wagoner, Fowler and Hoyle have demonstrated conclusively that the bulk of the elements cannot have been made in a big bang. The only important element, from the point of view of nuclear physics, that could have been synthesized in a big bang is helium. Thus soon after the discovery of the microwave background radiation there was a serious attempt to argue that the helium that is detected in stars and gaseous nebulae in our own galaxy and in other galaxies is primordial helium made in a big bang. The argument appeared plausible because, on the one hand, it was suggested that the helium/

hydrogen ratio is the same in different celestial objects (stars, gaseous nebulae in our own galaxy, and a few gaseous nebulae in one or two nearby galaxies) and is 25-30 per cent by mass, while, on the other hand, the fractional amount of helium relative to hydrogen that is made in a big bang is close to about 27 per cent. This simple argument broke down, however, as soon as detailed studies of different aspects of the problem were made. First of all, it was pointed out that there appear to be stars which contain very little helium so that perhaps there is no universal helium abundance. While this argument is still open, detailed studies of big bang models led to the conclusion that there are ways of making very little helium in such an initial process. Fowler has recently emphasized that the helium-hydrogen ratio is determined completely by the choice of baryon and lepton numbers in a big bang which has nothing to do with the theory of general relativity, but is tied only to initial conditions which lie outside the known laws of physics. Thus, while the helium abundance in the universe might be related to the occurrence of a big bang, there is no direct evidence that the two are related, and there are a number of alternative ways of understanding the large helium abundance found in some stars and galaxies. For example, it appears that helium could have been synthesized in massive objects evolving in the nuclear regions of galaxies.

Finally, we come to the vexed question of the origin of galaxies. As the big bang bandwagon has gained momentum, an increasing number of investigations have been carried out in which attempts are made to explain the condensation of dense objects from an initial cloud of matter and radiation which is expanding. It has been known for many years that this is very difficult to understand, and the investigations have now reached the point where it is generally accepted that the existence of dense objects cannot be understood unless very large density fluctuations in a highly turbulent medium, or otherwise, are invoked in the first place. There is again no physical understanding of the situation; it is a condition which is put in, in a hypothetical state, to explain a major property of the universe. Thus these "theories" amount to nothing more than the statement that protogalaxies have a cosmological origin, and their origin cannot be understood any better than can the original baryons and leptons in an evolving universe. At this level therefore these theories are in no better shape than the apparently much more radical views of Ambartsumian or Hoyle and Narlikar who believe that the origin of galaxies is tied closely to the properties of the dense active nuclei and that perhaps the initial states of galaxies were high density states. Hoyle and Narlikar have even dared to suggest that perhaps "new" physics is involved. What the debate largely reduces to is that on one side there is an apparent belief by the majority that creation in the distant past is acceptable, but that creation at recent epochs is unthinkable, while on the other it is thought that there is little to choose between the two points of view, and that only with much more original work can we hope to resolve the issue. It is clear that the evolving universe concept gains no support from attempts to understand either the origin of the elements or the origin of galaxies.

Was there really a big bang? I believe that the answer clearly must be that we do not know, and that if we are ever to find an answer much more effort must be devoted to cosmological tests, with a much more openminded approach, and that much more original thinking must be done to attempt to explain the large amount of observational material, and not only that material that can be used in a narrow sense to fit preconceived ideas. Probably the best argument in favour of a beginning is the general result that the ages of many stars in our galaxy are approximately equal to H^{-1}. Probably the strongest argument against a big bang is that when we come to the universe in total and the large number of complex condensed objects in it, the theory is able to explain so little.

• Red-Shift Problems

PECULIAR GALAXIES AND RADIO SOURCES

Arp, Halton; *Science,* 151:1214-1216, 1966.

Abstract. Pairs of radio sources which are separated by from 2° to 6° on the sky have been investigated. In a number of cases peculiar galaxies have been found approximately midway along a line joining the two radio sources. The central peculiar galaxies belong mainly to a certain class in the recently compiled Atlas of Peculiar Galaxies. Among the radio sources so far associated with the peculiar galaxies are at least five known quasars. These quasars are indicated to be not at cosmological distances (that is, red shifts not caused by expansion of the universe) because the central peculiar galaxies are only at distances of 10 to 100 megaparsecs. The absolute magnitudes of these quasars are indicated to be in the range of brightness of normal galaxies and downward. Some of the radio sources which have been found to be associated with peculiar galaxies are galaxies themselves. It is therefore implied that ejection of material took place within or near the parent peculiar galaxies with speeds between 10^2 and 10^4 kilometers per second. After traveling for times of the order of 10^7 to 10^9 years, the luminous matter (galaxies) and radio sources (plasma) have reached their observed separations from the central peculiar galaxy. The large red shifts measured for the quasars would seem to be either (i) gravitational, (ii) collapse velocities of clouds of material falling toward the center of these compact galaxies, or (iii) some as yet unknown cause.

OBSERVATIONAL PARADOXES IN EXTRAGALACTIC ASTRONOMY

Arp, Halton; *Science*, 174:1189-1200, 1971. (Copyright 1971 by the American Association for the Advancement of Science)

Quasi-Stellar Radio Sources. When starlike sources of radio noise were discovered, the feature that made these "quasars" so remarkable was their very high red shifts. Estimating their distances from the relation between red shift and distance for normal galaxies yielded luminosities that were of the order of hundreds to thousands of times brighter than those of normal galaxies. The brightest quasar in apparent magnitude, 3C 273, for example, was calculated on this basis to be radiating about 10^{47} ergs per second, an amount that would be generated by converting completely into energy 1 solar mass per year. At their red-shift distances some quasar would be even more luminous. Known energy-generation mechanisms, of course, are much less efficient than any hypothetical total conversion of mass to energy.

An additional difficulty is that some quasars are observed to vary in optical and radio brightness. This observation means that the amount of energy that undergoes a time fluctuation must be enclosed within a region that has a maximum diameter of, in some cases, no more than a few light-days. This limitation leads to exceedingly high energy densities and difficulties in getting the photons of light out of the object without excessive numbers of collisions with electrons. (p. 1189)

. . . .

Summarizing the Discordant Evidence. The first point that should be stressed is that, even though the observations contradict some of our basic assumptions about galaxies, the contradictions so far involve only special kinds of objects: quasars, compact galaxies, peculiar galaxies, and ejected material. There is no evidence to indicate that the brightest E galaxies in clusters or the largest Sb galaxies which dominate groups are not normal. There is no reason to believe that these two types of galaxies have not condensed at an early epoch and that they have red shifts which conform perfectly well to the relation between red shift and distance. If we come to understand the paradoxical observations about the younger, more compact classes of galaxies, this new understanding may modify to a great or small extent our estimates of the expansion rate of the universe. On the other hand, it is possible that a component of nonvelocity red shift may pertain to special objects without changing at all the presently accepted expansion rate.

With that introductory reservation I would summarize the discordant observational evidence on the objects I have discussed as follows:

1) Quasars with lower red shift are not found in any of a large sample of rich clusters of galaxies. Quasars are not distributed on the sky, or in red shift, in a way in which we would expect a very distant population of galaxies to be distributed.

2) Quasars and compact objects are associated with nearby galaxies and have moderately low luminosities.

3) A whole class of objects, including quasars, compact galaxies, blankfield radio sources, luminous material, lines of galaxies, and companion galaxies appear to be ejected in opposite directions from the nuclei of large active galaxies.

4) Lines of galaxies and compact galaxies give evidence of being stable only on time scales hundreds of times less than the normally accepted ages of galaxies and therefore are presumed to be quite young.

5) There appears to be a continuity of physical characteristics going from quasars through compact galaxies to companion galaxies. Quasars seem to have very large nonvelocity red shifts, but companion galaxies also have nonvelocity red shifts. The nonvelocity red shifts of the companion galaxies, although very small, are systematically positive and characteristic of a wide range of kinds of companion galaxies.

Fundamental Paradoxes. One paradox that these observations present is that of the ejection of galaxies from nuclei of other galaxies. Obviously a normal assemblage of stars cannot be hurled about like a snowball because the members of such an assemblage do not have the cohesion to stick together. Thus either the ejected galaxy must emerge as a compact coherent body from which stars later differentiate and separate, or some diffuse, gaseous material must emerge, at which point there must take place very rapid condensation of stars within the emerging gas. Both of these models would seem very strange in view of current assumptions about star formation in the present epoch by condensation in rather quiescent, outer regions of galaxies.

Another, perhaps more fundamental, paradox is: What is the cause of the nonvelocity red shift? It may be possible, by using known physical mechanisms, to explain this type of red shift. The models, however, would probably be very complicated. For example, Sistero showed that collimated photons overtaking and scattering off relativistic electrons will give red shifts where ($\Delta\lambda / \lambda$) is constant throughout the spectrum. Since relativistic electrons are very forward-scattering, this mechanism would also circumvent the usual second objection to variants of the "tired light" theory, namely, that the apparent angular diameter of the source is enlarged appreciably by numbers of scatterings. Some support for Compton scattering models might be forthcoming from the fact that, in order for photons to penetrate the densities of relativistic electrons that are derived on the assumption of cosmological distances, the electrons and photons would indeed have to be moving in closely similar directions. But a difficulty arises in that the nucleus or source of the spectrum must be shielded from direct view by the observer. This restriction suggests that the nucleus is shielded by nondiscrete blue scattering from electrons in other directions. Another possibility is that the nucleus is shielded by dust. This mechanism in turn introduces the possibility that a nucleus seen reflected from a dust cloud moving away from the observer would be shifted by the velocity of the moving reflector. Rees has mentioned the additional possible mechanism of redshifting within an optically thick, expanding shell around a spectral source. He has also mentioned the possibility that we see objects ejected with high velocities but that, for some reason, possibly dust obscuration in front of the objects, we do not "see" the approaching velocity.

In fact, it seems that there are a number of possible explanations that can be derived from conventional physics. Although these explanations could give only very complex and intricate solutions, they cannot at the present time be ruled out as solutions. It would be of utmost importance to see if these models could all be demonstrated to be very unlikely explanations (as the gravitational red-shift models were shown to be). The importance of this step would lie in the existence of new physics which would then be implied. It is possible that in the year 1971 we do not know all the physics there is to know. Perhaps in the nucleus of a galaxy or in the process of ejection physical conditions are encountered that are so extreme that local geometry is affected or clocks run slow. If there were, in fact, new physics to be discovered, it would most naturally be in the realms where observations most fundamentally contradict the current theoretical expectations. But, of course, before we consider seriously new physics, we must first exhaust the more conventional alternatives.

Whose Move Now? Most of the evidence that I have reviewed here is in the literature in various places. The reason for bringing it all together here and trying to evaluate it on more than a purely technical level is that there seems to have been some sort of impasse reached in astronomy with regard to this evidence, and it is of interest to examine why this is so and how the deadlock might be broken.

As I have stressed throughout this article, if the observations are correct, some of our fundamental assumptions are wrong. The only escape from this conclusion is to say that each association of discordant red shifts and observed ejection phenomena is accidental. For example, if the string of quasars coming out of the exploding galaxy is not accidental, then quasars are closer than cosmological distances and their red shifts are due to some other cause. I have tried to reduce these questions to crucial sets of associations---either "yes" or "no," with the decision hinging on the facts of the observation. Many astronomers either reject these cases as "selected" accidents or adopt a neutral attitude. But it seems to me that the accumulation of evidence is now very difficult to either reject or ignore.

It could be argued that each case is somewhat different---that sometimes quasars, sometimes compact radio galaxies, sometimes companion galaxies are involved in associations. The answer to this argument is that all these objects are varieties of galaxies and that there is a continuity of attributes involving all these various kinds of extragalactic objects. In these associations it is usually the most extreme and peculiar objects that have the most discordant red shifts. Actually this observation strengthens the conclusion that the associations are significant, because, if there were accidental associations, they should, for the most part, involve normal objects, which are the kinds of objects most frequently observed. Instead they tend to involve peculiar objects---objects that are the least likely to be involved accidentally. But physically we know the least about these peculiar objects, and they are the ones for which there is the greatest a priori chance that new and unknown physical mechanisms are at work.

In the end, however, we must all agree that the ultimate criterion for science is experiment and observation. If the observational paradoxes discussed in this article can be demonstrated to be false or accidental, then we can say that the paradoxes are solved on the basis of our present knowledge. If the observations stand, then we must conclude that something new of vast importance is happening and we should get on with the exciting job of finding out more about it. (1198-1200)

ON THE NATURE OF MASS

Hoyle, F., and Narlikar, J. V.; *Nature*, 233:41-44, 1971.

Evidence concerning "discrepant redshifts" has accumulated fairly rapidly in recent months. The case where it is hardest to deny the evidence was reported at the beginning of the present year by Arp. The galaxy NGC 7603 was found to be connected to an appendage by two arms which intersect at the appendage $z = 0.056$. In the customary Doppler interpretation of the redshift this difference exceeds 8,000 km s^{-1}. A redshift difference of the same order was found many years ago by Minkowski, Humason and Zwicky in "Zwicky's triple system". The galaxies in this system are also connected by a bridge.

If it is accepted that a discrepancy of $\sim$ 8,000 km s^{-1} exists for NGC 7603, then there seems no good reason for denying the two cases with discrepancies $\sim$ 20,000 km s^{-1} obtained by Sargent. Both the latter cases are compact groups of galaxies in which one member has the discrepant redshift. Less than ten such groups have been examined. The probability of a more distant galaxy being projected against a nearer cluster by chance is small. Arp (private communication) has obtained what appears to be a jet connecting the quasi-stellar object Makarian 205, $z = 0.07$, to the galaxy NGC 4319. The discrepancy is again of order 20,000 km s^{-1}.

These cases have been observed of a quasi-stellar object apparently associated with a small compact cluster of faint galaxies. In one case Gunn has reported the same redshift for the QSO as for one of the galaxies of the cluster. In the other two cases there appear to be large redshift differences (Hazard, Jauncey, Sargent and Gunn, private communication).

In a recent preprint Burbidge, Burbidge, Solomon and Strittmatter have drawn attention to a bridge which appears on the Palomar sky prints to connect the quasi-stellar object PHL 1226 to the bright galaxy IC 1746. The redshift of PGL 1226 is $z = 0.404$, whereas that of the galaxy has not yet been measured, but must be small because of its brightness.

Because all objects are projected on the sky there must be some apparently peculiar juxtapositions of objects that are really at very different distances. This has hitherto led to a situation in which all but a very few astronomers have dismissed apparent redshift discrepancies as simply unusual projection effects. Yet there has to be a point of balance in one's judgment as to the extent of the chance juxtapositions one is willing to accept. One of us was aware at an early stage of the case of Zwicky's system (Zwicky, private communication) but it seemed that, in spite of the bridge, there might be a very peculiar chance juxtaposition. But the case of VV 172 formed the point of balance and NGC 7603 has turned the balance.

If a highly convincing theory of discrepant redshifts were available then we think there is little doubt that the data would today be considered reasonably clear-cut. It has been the absence of such a theory that has caused most astronomers to prefer to believe in unusual projection effects. If, as seems very possible, the accumulation of data forces us over a watershed (not only in our thinking but in the history of astronomy) it will clearly become necessary to arrive at a theory of discrepant redshifts. We wish to emphasize the need for a thoroughly radical assessment of the problem, considering it unlikely that a satisfactory theory will be achieved by a small change in our concepts. Explicitly, we do not think discrepant redshifts will be explained adequately either as simple Doppler peculiar motions or as excess reddening due to gravitation.

THE GERITOL UNIVERSE: TIRED LIGHT

Anonymous; *Science News*, 108:277, 1975.

For several classes of objects, it seems that in different regions of sky different apparent redshift-distance relations exist. (If the class of objects is defined narrowly enough, they will all have about the same intrinsic brightness, and so distance can be estimated from the apparent brightness of each, independent of redshift.) At first this seemed to indicate a possible lopsided expansion, but now three astronomers at the Institut Henri Poincare in Paris, Hiroshi Karoji, Laurent Nottale and Jean Pierre Vigier, say the matter goes beyond that. Their latest results point toward a belief in the so-called "tired-light" hypothesis.

Their latest paper, which has been submitted to Astrophysical Journal Letters, deals with the apparent motion of faint radio galaxies (magnitudes 13.0 to 15.5). The observations show "an even more curious distribution" of redshifts than observations of other classes of objects. In summary they say, "Everything goes as if light emitted from distant sources is redshifted when it travels through clusters of galaxies.

"[This] evidently favors the existence of a 'tired-light' mechanism first discussed by Hubble and Tolman" in 1935 and recently taken up by some other observers. (It is called tired light because light loses energy when it is shifted toward the red.)

Tired light would be a totally new phenomenon in the behavior of light not foreseen in current theories of how a stream of photons behaves in a gravitational (or perhaps other kind of force) field. "If this is true," Vigier writes in a letter to Science News, "the idea of universal expansion itself is in deep trouble, and one would have to come back .... to the static cylindrical model proposed initially by Einstein himself."

QUASAR REDSHIFTS SEEM TO COME IN BUNCHES

Anonymous; *New Scientist*, 51:612, 1971.

Some astronomers have claimed that the redshifts of quasars are not uniformly spread out but, instead, tend to cluster about certain values. It now looks as if the workers who found those "periodicities" were right. The suggestion was at first refuted by detailed mathematical tests, but evidence accumulating over the last couple of years has favoured the interpretation that peaks in the redshift distribution are related to one another.

Dr. K. G. Karlsson, of Uppsala University, has found no less than five peaks at particular redshifts. These critical values at which quasar redshifts congregate form a geometric series, and it is particularly interesting that the most recently determined redshifts lie close to one or other of the peaks (Astronomy and Astrophysics, vol. 13, p. 333).

Of course, one would like to know just how much these new discoveries have been influenced by the knowledge that the peaks are there ---when cosmologists try to fit some peculiar set of spectral features to a redshift they tend to try out the most likely redshifts first. Thus a self-sustaining process could easily encourage the identification of all the redshifts corresponding to the well-known values before other less common redshifts are dug out from a batch of spectra.

However, if Karlsson is right, and it certainly looks that way, he has found a relation between the more common quasar redshifts, particularly around $z = 1.95$ and a peak in the distribution at the very small value of $z = 0.06$. This can only imply a link between quasars and galaxies, because objects with redshifts as low as 0.06 are galaxies, not quasars. His results suggest strongly that redshifts are an intrinsic property of quasars and do not necessarily indicate their distances from us. Moreover, in peculiar galaxies, at least, the same sort of effect occurs. As so often recently, the conclusion seems to be that quasars, peculiar galaxies, and probably even normal galaxies are related and form some sort of (possibly evolutionary) chain.

CONTROVERSY OVER THE EXTRAGALACTIC DISTANCE SCALE

Rowan-Robinson, M.; *Nature*, 264:603-604, 1976.

The redshift-distance relation for galaxies discovered by Hubble and Lundmark in the 1920s is the most tangible evidence we have for an expanding universe. The constant of proportionality, the Hubble constant, can be interpreted as the expansion time-scale of the universe. Yet in spite of the consumption of many thousands of hours of telescope time, the extragalactic distance scale remains a subject of controversy, highlighted by a disappointingly inconclusive International Astronomical Union symposium at Paris in September.

At the symposium Tammann summarised the massive programme of galaxy distance measurements that he and Sandage have been pursuing, culminating in a value for the Hubble constant, H_0 of around 50 km s^{-1} per Mpc (for a recent review see Nature, 262, 97; 1976). Yet almost all other workers reporting at the symposium disagreed with the Sandage and Tammann distance scale, preferring a value for H_0 closer to 80. For example, van den Bergh and de Vaucouleurs reported significantly lower distances for members of the Local Group of galaxies, to which our Milky Way galaxy belongs. Hanes' work on globular clusters in galaxies in the Virgo cluster, the nearest of the great clusters of galaxies that determine the Hubble relation on the large scale, led to a distance estimate for Virgo of 12.5 Mpc (1 Mpc = 3.26 million light years) instead of the Sandage and Tammann value of 22. And Tully and Fisher used their 21-cm neutral hydrogen line method to obtain a value for H_0 of 75.

It is therefore fascinating to see in the latest Astrophysical Journal (210, 7; 1976) Sandage and Tammann's version of this last method. They take Tully and Fisher's data, reanalyse and recalibrate it, and emerge with a Hubble constant of 50.3 ± 4.2. The 21-cm method depends on a correlation found by Tully and Fisher between the 21-cm hydrogen line width and the optical luminosity of a spiral galaxy (one has to have some galaxies with distances known more directly to calibrate this relation). The apparent optical brightness of a galaxy then gives the distance through the inverse square law.

Sandage and Tammann use the method to obtain the distance of the Virgo cluster, and compare the result with six other methods (including, ironically, an earlier estimate by Hanes which agreed with theirs). The adopted Virgo distance, 22 Mpc, is then combined in three different ways with the mean velocity of the cluster to yield $H_0 = 50.3 \pm 7.0$, 50.3 ± 4.2, 50.3 ± 4.2. The authors remark "That the values of H_0 from all three methods are the same is clearly fortuitous". If the agreement between these three means at so high a significance level is indeed simply good luck, then heaven protect these authors from bad luck.

Do these disagreements over H_0, which appear even when the same data is used, matter? Well, first, the actual value of H_0 matters to cosmologists, because by comparing it with the age of our own and other galaxies they can deduce how long galaxies took to form. The big bang picture would be in trouble if the expansion time was less than the age of our Galaxy. Second, discrepancies which are such a large multiple of the quoted errors do not enhance the status of observational astronomy. At the moment the onus seems to be on the opponents of the Sandage and Tammann scale to assemble a similar body of observational data. Theoreticians will have to maintain the scepticism about H_0 that the wiser among them have always shown in the past.

AN ODD COUPLE

Anonymous; *Science News*, 112:11, 1977.

Halton Arp of the Hale Observatories has been collecting instances of the pairing or grouping of astronomical bodies with very different redshifts in their light. He does this in furtherance of his contention that redshifts are not always due to the relative speed and therefore distance of the objects but may come from other causes. If that is true, then objects with different redshifts may be at the same distance and so can be physically linked.

At the meeting, Arp presented an instance where a high-redshift object appears to be in front of a low-redshift one. There appears to be a bright object with a redshift corresponding to a speed of 13,300 kilometers per second in front of the E galaxy NGC 1199, which has a redshift corresponding to 2,600 kilometers per second. Arp argues that the bright object is in front of the E galaxy because it is surrounded by a disk of dark matter that appears spectroscopically as if it is absorbing light from the E galaxy. If the bright object were behind the E galaxy, it would add to the brightness of the image, and the absorption by the dark disk would not be seen. Therefore, most of its redshift must come from some cause other than motion of recession, he says.

NGC-1199

Arp, Halton M.; *Astronomy,* 6:15, September 1978.

The E galaxy NGC-1199 is the brightest member of a small cluster of galaxies. Among its companions is a galaxy so compact that, at first glance, it looks like a star. Such a galaxy is not likely to appear by itself in space, nor is it part of a more distant cluster of galaxies. It would be expected---by its appearance---to be a member of the small group of bright galaxies of which NGC-1199 is the dominant member. But the redshift of the compact peculiar galaxy is 13,300 km per second, whereas the redshift of the central NGC-1199 galaxy is only 2,600 km per second.

In the beginning, an investigation of possible interaction between the compact galaxy and NGC-1199 revealed a circular shadow around the compact as if it were in front of NGC-1199 and absorbing the light of the E galaxy behind it. From 1965 to 1976, I photographed it with the 200 inch telescope at Palomar, the four meter telescope at Kitt Peak National Observatory and the four meter telescope at Cerro-Tololo Interamerican Observatory until I was certain that I had the best possible photographic representation of this silhouetted galaxy. Finally, I measured the absorption ring with the new linear response Silicon Image Tube Spectrograph on the 200 inch telescope.

With all these techniques, I now believe that I have demonstrated that the compact galaxy is in front of the E galaxy, despite its much higher redshift. Like other examples of galaxies and quasars with anomalously high redshifts, the physical mechanism for producing this intrinsic redshift is not presently known.

• Quasar Energetics and Faster-Than-Light Motion

QUASARS FLARE SHARPLY: EXPLAINING THE ENERGY GETS HARDER

Metz, William D.; *Science*, 189:129, 1975. (Copyright 1975 by the American Association for the Advancement of Science)

During the last two decades, astronomers have realized that extraordinarily powerful and violent events are commonplace occurrences in the universe. The enormous energies of quasars, large radio sources, and other peculiar objects are continuing puzzlements. Recent research at Harvard University has magnified the problem of the quasars by showing that one of the brighter ones underwent a previously unsuspected flare in 1937, and for a period of several months was the most luminous object known in the universe.

In April of 1937, quasar 3C279 intensified very quickly from its normal optical brightness of 15th or 16th magnitude until it reached a maximum brightness of magnitude + 11.3. The quasar is thought to be more than 5 billion light-years away, assuming that its red shift (0.536) is equally as good a measure of distance as are the red shifts of normal galaxies. At such a distance, the absolute magnitude of 3C279 at the peak of its flare must have been -31.4. (The magnitude scale of optical brightness is logarithmic, with negative values brighter than positive ones, and five magnitudes equivalent to a factor of 100.) At such a high optical luminosity, the quasar would have been emitting 10^{48} erg/sec.

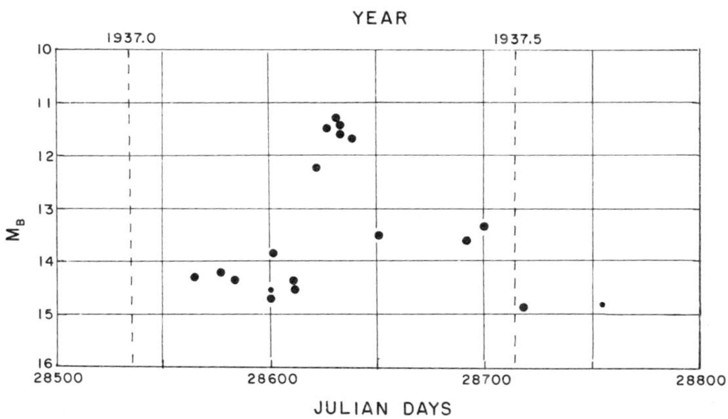

Changes in optical brightness of quasar 3C279 in 1937

For comparison with other objects in the universe, the absolute visual magnitude of the sun is + 4.7, that of the entire Milky Way is -20; the absolute magnitude of a giant elliptical galaxy is typically -22, and that of a bright quasar -26. The flare that occurred in 1937, shown in the figure below, was a phenomenon equivalent in effect to turning on thousands of giant galaxies in a matter of months and then turning them off again. But because the event happened so quickly, the short distance that light could have traveled in that time suggests that it occurred in a region far smaller than a single giant galaxy. "The flare was absolutely huge," according to Harry van der Laan at Leiden Observatory, Netherlands, and constitutes a whole new limit for the energy problem in astronomy.

The photographic plates that contained the record of the depression-era flare of 3C279 were recorded as part of a systematic observing program pursued by the Harvard College Observatory for almost 70 years, from 1885 to 1954, and recently renewed. From the observatory site in Cambridge, Massachusetts, and from the Agassiz station about 40 miles away, Harvard astronomers conducted a regular patrol of the Northern Hemisphere. Regular coverage of the Southern Hemisphere was carried out at the Harvard southern station, which was located in Arequipa, Peru, until 1924 and then moved to Bloemfontein, South Africa. About 500,000 plates were made during the time the sky patrol was active. Every portion of the sky was typically photographed with wide-angle cameras several times a year.

The plates have been carefully stored on three floors of the old observatory building, which now adjoins the building of the Smithsonian Astrophysical Observatory. Two years ago, with the encouragement of George Field, who sits as director of both institutions, Harvard astronomers began looking into the plate collection for data that would be relevant to some of the newer scientific frontiers---particularly high-energy astrophysics. William Liller and associates at Harvard found evidence for a regular optical period in the brightest x-ray source in the sky, Sco X-1, and also began a survey of the variability of the optical emissions of the brighter quasars. William Liller, Martha Liller, and Lola Eachus found evidence for flares not only in 3C279, which exhibits the most pronounced flare discovered so far, but also in quasar PKS 1510-089, and in the quasar-like object MA 0829 + 047. There is also some evidence from the Harvard plates that quasar 3C279 flares regularly every 7 years, since brightness peaks were recorded in 1944 and 1951.

The power produced by quasar 3C279 peaked at a value greater than that known for any other object. Over one period of 13 days in 1937, the quasar emitted an average of 10^{41} watts, assuming it is at the cosmological distance given by its red shift. This is an enormous power, comparable to that discussed in trying to explain the double radio sources (Science, 27 June 1975). The radio sources continually emit 10^{37} watts. But Geoffrey Burbidge, at the University of California, San Diego, suggests that it may be just as much of a problem to explain how radio sources can emit 10^{37} watts for millions of years as to explain how quasars can emit 10^{41} watts for a few days or months.

As one of the school of astronomers who question whether the red shifts of quasars are indicators of their cosmological distances, Burbidge also questions whether 3C279 really became as bright as magnitude -31.4. But he thinks that the discovery is nonetheless significant. Either 3C279 is much closer, according to Burbidge, or else "you cannot appeal to the evolutionary processes of normal stars" to explain it.

"SUPERRELATIVISTIC' QUASARS AGAIN

Anonymous; *Science News,* 110:40, 1976.

When examined with radio telescopes, several of the quasars seem to show two components flying apart at high velocity. In a few cases, the velocity appears to exceed that of light. Superluminal velocities would destroy one of the foundation stones of modern physics, so a number of ingenious attempts to explain them away have been put forth. One of the simplest and most appealing of these is the "Christmas-tree effect." It says that what is seen is not motion of two components, but precisely timed flashes of a number of components that give the illusion of such motion.

A blow against the Christmas-tree idea is now struck by 13 astronomers from the United States and Sweden (Astrophysical Journal 206: L78). For almost four years, they studied the quasar 3C 345, using very-long-baseline interferometry involving antennas in California, Massachusetts, West Virginia and Sweden. The high resolution of the VLBI technique allows them to maintain that the brightness distribution of the source "was clearly dominated by two components" [italics theirs]. They call their result "difficult to reconcile with the so-called Christmas-tree model" but do not drop the other shoe by suggesting an alternate explanation. They leave 3C 345 flying apart at apparently 2.5 times the speed of light.

SUPER-LUMINAL EXPANSION IN EXTRAGALACTIC RADIO SOURCES

Blandford, R. D., et al; *Nature*, 267:211-213, 1977.

Abstract. Very long baseline (VLB) investigations of compact extragalactic radio sources can often be interpreted in terms of two nonthermal components apparently separating with speeds in excess of the speed of light. Given the conventional physical laws and the cosmological interpretation of large redshifts, the available morphological evidence, while inconclusive, is most compatible with a general class of models in which a signal propagating with a speed close to that of light is scattered by an appropriately shaped screen. Possible observations are discussed which might discriminate between this and alternative explanations.

• Background Radiation

FREAK RESULT VERIFIED

Anonymous; *Nature*, 223:779-780, 1969.

A plea to take seriously last year's surprise measurement of an unexpectedly high value for the far infrared background radiation is published in the latest Astrophysical Journal Letters by Houck and Harwit of Cornell (157, L45; 1969). Last year's measurement, announced in November by Houck and Harwit in company with Shivanandan (Phys. Rev. Lett., 21, 1460; 1968), caused disbelief in some, and induced others to rush into print with analyses of the implications. The literature has been short on explanations, however. This ought to be remedied by the latest paper from Cornell which confirms the original rocket result with remarkable ease, although sceptics will still want to see the experiment repeated using independent equipment.

What needed to be verified was the observation that above the absorption of the atmosphere the infrared flux between 0.4 and 1.3 mm fails to correspond to the predicted level by nearly two orders of magnitude. Instead of falling on the spectrum for a black body at 3 K---the temperature of the background radiation which Penzias and Wilson discovered at microwave frequencies---the far infrared measurement corresponds to a temperature between four and six degrees greater. In the latest experiment Houck and Harwit used a slightly altered telescope, launched by the same type of rocket as before, to produce an identical result. No increase in signal is detected when the telescope scans across such features as the Crab and Orion nebulae, or across the ecliptic and galactic planes. And neither telescope scanned the goings-on at the centre of the Galaxy which are known to be a prolific source of infrared. The indications are that the radiation is in fact isotropic.

By underpinning the earlier measurement, Houck and Harwit give support to the views of Cowsik and Pal (Phys. Rev. Lett., 22, 550; 1969) and of Shen (ibid., 22, 568; 1969) who say that the anomalous infrared radiation may be the cause of a flux of high energy gamma rays which Clark, Garmire and Kraushaar claim to have detected (Astrophys. J. Lett., 153, 203; 1968). What happens is that electrons in interstellar space give some of their energy to infrared photons by collisions, lifting the photons into the gamma ray part of the spectrum.

Indeed, Houck and Harwit show signs of accepting the high energy gamma rays as indirect support for the existence of the anomalous infrared flux.

But explanations of the origin of the infrared are hard to come by. One possibility is that transitions in ions of abundant elements such as carbon, nitrogen and oxygen boost the background radiation at wavelengths shorter than 0.5 mm (Petrosian, Baheall and Salpeter, Astrophys. J. Lett., 155, L57; 1969). Another idea, of Houck and Harwit, is that infrared radiation from galactic nuclei may be responsible.

Yet doubt has been cast on the infrared measurements by Bortolet, Clauser and Thaddeus who argue that if the abnormal flux is real it will be absorbed by interstellar material, and excited states thus produced will lead to optical absorption lines (Phys. Rev. Lett., 22, 307; 1969). Measurements of the absorption lines allow upper limits which are compatible with the standard 3 K background to be placed on the infrared flux at two points in the band covered by the Cornell measurements. So far the only way to surmount this obstacle seems to be to say that the infrared flux is concentrated into spectral lines---certainly Bortolet et al. rule out a continuum.

Clearly the background radiation at wavelengths shorter than the 2 mm peak of the 3 K black body spectrum needs more observations. It would be good, for example, to see a verification of the infrared peak with different equipment, and to check the observation of high energy gamma rays. Otherwise the house of cards which is building up may collapse.

HOW HOT IS THE UNIVERSE?

Anonymous; *Nature*, 226:111-112, 1970.

As techniques improve, measurements of the background radiation are beginning to come in, which, taken collectively, are making cosmologists sit up in their armchairs. Until recently the temperature of the 3 K background radiation was inferred from measurements taken on the long wavelength side of 2 mm, where the peak of a 3 K blackbody background would occur. Beginning with the measurement by Penzias and Wilson at 7 cm the measurements fit the blackbody curve well down to 8.5 mm, which was the limit until 1968.

But with the development of better detectors for the infrared part of the spectrum, the situation is changing. To begin with, there was the rocket measurement from NRL and Cornell which gave a blackbody temperature of 8.3 K for the background signal in the band between 0.4 and 1.3 mm (Shivanandan, Houck and Harwit, Phys. Rev. Lett., 21, 1460; 1968 and Houck and Harwit, Astrophys. J. Lett., 157, L45; 1969). But there have also been measurements in favour of the status quo. The Princeton group were the first to push the measurements to shorter wavelengths than 8.5 mm, and their result at 3.3 mm is sufficiently close to the 3 K curve not to cause any excitement (Boynton, Stokes and Wilkinson, Phys. Rev. Lett., 21, 462; 1968). Soon after, a team from Columbia and the Goddard Institute clashed with the result of Shivanandan et al. in inferring upper limits consistent with a 3 K background at three isolated wavelengths between 0.35 and 1.32 mm (Bortolot, Clauser and Thaddeus, Phys. Rev. Lett., 22, 307; 1969).

The latest development comes from MIT and this time the chief result is to cast doubt on the blackbody nature of the background radiation (Muehlner and Weiss, Phys. Rev. Lett., 24, 742; 1970). To obtain readings in the vicinity of the expected peak at 2 mm means finding some way to combat the absorption by water vapour which is strong at these wavelengths, and this is why the Princeton group took their radiometer to the High Altitude Observatory in Colorado, and the MIT group flew theirs in a balloon to a height of 40 km. To keep the noise low the MIT indium antimonide detector was cooled by liquid helium, and two filters could be interposed in the field of view to give information about the wavelength dependence of the radiation down to about 0.5 mm.

The reason for the excitement is, of course, that the blackbody nature of the radiation and its equivalent temperature of 3 K have lately become the chief exhibits in favour of a big-bang cosmology. The odd thing about the MIT data is that surprisingly different fluxes are recorded when the two filters are interchanged, although the differences in the wavelengths which they pass are small. One filter cuts off wavelengths less than about 1 mm, and the second cuts off wavelengths about 0.2 mm shorter. With both filters out of the line of sight the spectral response drops smoothly from 1 mm to 0.5 mm. But the fluxes recorded for each spectral response at the disposal of the experimenters do not correspond to the ratios expected from radiation governed by a blackbody law, and this indication that the background is nonthermal is the odd feature of the experiment.

It is possible that what is being recorded is emission features superimposed on the 3 K curve. This is an explanation already put forward to reconcile the large flux in the 0.4 to 1.3 mm band recorded by the NRL-Cornell group with the apparently incompatible measurements of Bortolot et al., and is not ruled out by the MIT group.

Precisely what temperatures are to be associated with the new data ---assuming the radiation is blackbody, which it seems not to be--- depends on the pairing of the flux measurements to allow for sources of radiation in the equipment and the atmosphere. From raw data which indicate that temperatures in the three spectral responses are not higher than 7.4, 8 and 4.7 K, Muehlner and Weiss obtain 5.5, 7 and 3.6 K after corrections. Obviously there is a need for more experiments in this awkward region where the 3 K spectrum peaks---unfortunately a second MIT balloon failed. It remains to be seen whether the infrared measurements will force a revision of ideas about the background radiation or whether they will come to be regarded as a shoal of red herrings.

HOW SPECIAL IS THE UNIVERSE?

Davies, P. C. W.; *Nature*, 249:208-209, 1974.

An examination of the delicacy of advanced life forms and the requirement of highly special conditions for their existence would seem to support the long-standing tradition that the Universe has been created in a singularly convenient form for the presence of intelligent life.

It is truly remarkable that modern theoretical physics can make a contribution to this philosophical question. For example, it is sometimes pointed out how various naturally occurring numbers must be restricted in their values in order to be consistent with the existence of living matter. Dyson has remarked (Scient. Am., 225, 51-59, 1971) that an increase in the coupling constant of the strong interaction by only a few percent would have resulted in a catastrophic synthesis of most of the hydrogen in the Universe during the early stages of the big-bang, thereby depriving stable stars such as the Sun of their energy source, making life seemingly impossible. Another important naturally occurring number is the ratio between the electrical and gravitational forces between an electron and a proton (a staggering 10^{39}). Strangely, this number coincides with the age of the Universe measured in atomic units of time. Carter has argued (unpublished work) that this coincidence is necessary for the existence of intelligent life, on the basis that the lifetime of stable stars depends on the first number, and that this must be somewhere near the time required for intelligent life to evolve, which is of course just equal to the present age of the Universe. Thus the approximate equality of the two numbers is not a coincidence, but a direct result of our own existence.

The application of this philosophy to the initial state of the Universe rather than to the values of the constants of nature has been made by Collins and Hawking (Astrophys. J., 180, 317-334; 1973) who addressed themselves to one of the outstanding mysteries of modern cosmology. Why, out of all the possible distributions of matter, are we living in a Universe in which this distribution is so accurately isotropic? For it is a remarkable fact that whichever direction we look, the 3 K background radiation has the same intensity to one part in 1,000. This fact is all the more puzzling when it is realised that such widely separated regions of the Universe cannot have even been casually connected when this radiation was emitted, owing to the existence of a so-called particle horizon across which no information can pass. Collins and Hawking have pointed out, however, that although the Universe appears isotropic now, it need not always have been so. Restricting their discussion to homogeneous cosmological models, they show that there is a class of cosmologies which become isotropic as they expand from the initial big bang. The most relevant discovery of their work is that the set of initial states of the Universe for this to happen is of measure zero in the set of all possible initial conditions. In layman's language, there can be no deviation at all from these special initial states if the Universe is to become isotropic. Moreover, it is just this same class of cosmological models which permits the growth of galaxies and stars by ordinary gravitational condensation, and hence permits the existence of life. So Collins and Hawking conjecture that man's own existence is a consequence of the isotropy of the Universe (see Nature, 242, 304, 1973).

....

• Isotope Abundance Anomalies

THE MYSTERY OF THE COSMIC HELIUM ABUNDANCE

Hoyle, F., and Tayler, R. J.; *Nature*, 203, 1108-1110, 1964.

It is usually supposed that the original material of the Galaxy was pristine material. Even solar material is usually regarded as 'uncooked', apart from the small concentrations of heavy elements amounting to about 2 per cent by mass which are believed on good grounds to have been produced by nuclear reactions in stars. However, the presence of helium, in a ratio by mass to hydrogen of about 1:2, shows that this is not strictly the case. Granted this, it is still often assumed in astrophysics that the 'cooking' has been of a mild degree, involving temperatures of less than 10^8 K, such as occurs inside main-sequence stars. However, if present observations of a uniformly high helium content in our Galaxy and its neighbours are correct, it is difficult to suppose that all the helium has been produced in ordinary stars.

It is the purpose of this article to suggest that mild 'cooking' is not enough and that most, if not all, of the material of our everyday world, of the Sun, of the stars in our Galaxy and probably of the whole local group of galaxies, if not the whole Universe, has been 'cooked' to a temperature in excess of 10^{10} K. The conclusion is reached that: (i) the Universe had a singular origin or is oscillatory, or (ii) the occurrence of massive objects has been more frequent than has hitherto been supposed.

. . . .

THE PUZZLE OF GALACTIC DEUTERIUM

Mitton, Simon; *New Scientist*, 57:537-538, 1973.

One of astronomy's basic aims is to discover the densities and distributions of the chemical elements inside the Milky Way. From a study of the chemical composition of stars and gas clouds they hope to deduce the original chemical composition of our Galaxy, and to describe how some 16 billion years of subsequent evolution has modified the chemical elements. Until recently this problem could only be tackled by the classical tools of optical spectroscopy, often involving a detailed analysis of the atmospheres of individual stars. To this information the chemical analyses of Earth and Moon rocks, as well as of meteorites, add an additional constraint on theorising. However, the detection, over the past three years, of line radiation from many complex molecules deep in space has now added a further dimension to the intriguing task of charting the chemicals in the Milky Way. Several research groups in the US have just announced the discovery of radiation from deuterium---the second lightest atom---in cosmic gas clouds. Previously this substance had only been observed on Earth and in the cloudy atmosphere of Jupiter.

Deuterium is the heavy isotope of hydrogen composed of one proton, one neutron, and an outer electron. Modern theories of the synthesis of the chemical elements hold that all the deuterium in the universe was created within the first few seconds of the Big Bang. Thus the detection and investigation of deuterium in celestial sources can, in principle, provide clues to the conditions which prevailed in the earliest history of the universe. The snag is that deuterium gets cooked inside the stars: basic nuclear reactions fuelling stars like the Sun rapidly destroy deuterium during the conversion of hydrogen to helium and heavier elements. Given a model for galactic and stellar evolution, however, it is possible to calculate how much deuterium should still be left in the interstellar medium at the present time. Opinions on the deuterium abundance have ranged from zero up to about 0.001 per cent of the abundance of atomic hydrogen. The actual amount of deuterium discovered in the last two months has surprised astronomers, and stimulated renewed interest in how this substance may be formed in the cosmic environment.

....

Cosmological Implications. The large abundance of deuterium that definitely exists in Orion and at the galactic centre has bearings on cosmology. This is because the amount of deuterium that can be produced in the earliest epoch of the universe is critically dependent on the mean density of matter within the universe. The observed amount of material in the form of stars, dust, and galaxies, and the existence of the background radiation at a temperature of 2.7 K constrain the deuterium/hydrogen ratio to less than 10^{-5}. The values turned in by the radio astronomers are quite definitely at variance with this figure---by three orders of magnitude!

What then is the answer to the deuterium problem? As with most problems in astronomy there are several ways out of the maze, but no one knows which will lead to blind alleys and which will reveal interesting physical properties. An obvious point, put forward by the discoverers, is that they may have stumbled on unusually rich concentrations of deuterium in Orion and the galactic centre. However, the cloud in Orion has dimensions of the order of a light year and a mass one thousand times that of the Sun, so this explanation is not exactly easy to sustain. Evidently it will require more observations to resolve this particular point.

If deuterium does turn out to be fairly abundant in the Galaxy, then the "special case" argument will collapse, and theorists will have to seek new ways of making deuterium. This point is discussed in a recent article for Nature (vol. 241, p. 384) by Fred Hoyle and Willy Fowler, who state that the problem of the origin of deuterium is now urgent. A way out of the difficulty is to suppose that deuterium may have been produced in significant quantities after the Big Bang.

Normal nuclear reactions inside the stars build up helium from hydrogen and destroy deuterium in the process. Hoyle and Fowler consider the means by which these helium nuclei could subsequently be shaken apart again to produce deuterium. This will happen in circumstances where a low-density gas containing helium is bombarded by energetic protons. If protons moving at about one-third the speed of light slam into helium nuclei substantial deuterium is produced.

....

ON INTERSTELLAR ABUNDANCE ANOMALIES

Wickramasinghe, N. C.; *Observatory*, 88:246-247, 1968.

It has been known for several years that the interstellar medium appears to be anomalously deficient in Ca relative to Na. By comparing the equivalent widths of the interstellar Ca II lines with those of Na I in a large number of high-dispersion stellar spectra, Routly and Spitzer concluded that the average interstellar Ca/Na ratio is at least a factor 10 lower than the solar value. A more recent investigation by Herbig indicates that this deficiency may be as large as a factor $\sim 10^3$ in some cases. Herbig's analysis of ζ Oph also suggests that interstellar Ti is under-abundant by a factor $\sim 10^2$. The possibility exists that other atomic species, for which no absorption lines have been observed, are similarly under-abundant in the interstellar medium. From a failure to detect the ultra-violet Be II lines in the spectra of χ^2 Ori and ζ Persei, Spitzer and Field deduced that the interstellar Be/Na ratio is lower than the solar value by at least a factor of 10.

The surface layers of stars (including the Sun) are expected to be closely similar in composition to the clouds from which they condensed. The explanation of interstellar abundance ratios Ca/Na, Ti/Na, Be/Na which differ markedly from the corresponding solar values is therefore a problem of pressing urgency. Spitzer and Field have suggested that the observational data may be accounted for with normal abundance ratios Ca/Na, Be/Na in the interstellar medium, if it is hypothesized that Ca and Be ions become preferentially attached to the grains. The reason for such selective attachment, however, remained obscure.

A property common to Ca, Be and Ti is that they all form strongly bound oxides which also possess highly refractory crystal structures. The relevant binding energies are set out in Table I.

Table I

	Dissociation Energy (ev)	Sublimation Energy (kcal/mol)
CaO	3.93	140
BeO	4.80	172
TiO	6.80	138

A possibility not hitherto considered is that the enrichment of the interstellar medium in Ca, Be and Ti takes place largely through the ejection of solid particles of CaO, BeO and TiO from cool giant stars. Such particles, once formed, would be virtually indestructible, so that the absence of these metallic elements from the gas phase is readily explained.

The question then arises whether it is reasonable to expect the condensation of CaO, BeO, TiO in a giant star atmosphere, or in a cloud of material escaping from such a star. We carried out molecular equilibrium calculations to determine the temperatures at which these gaseous species saturated with respect to the bulk phase. About 100 different molecular species were included in the calculations, the relevant thermochemical data being taken from the JANAF Tables. The calculations were done for various assumed values of the total density, and of the relative atomic ratios. With a total density of $\sim 10^{15}$ atoms cm^{-3} and with solar relative abundances it turns out that CaO and BeO could condense as solids at $T \simeq 1400^{o}K$, and TiO at $T \simeq 1600^{o}K$. These condensation temperatures were found to be approximately valid for total gas densities in the range 10^{13} - 10^{16} cm^{-3}. The results were also found insensitive to small departures from solar relative abundances, provided C/O remains less than unity.

Boundary temperatures as low as 1400 ^{o}K may be reached during minimum phase in some of the cooler giants, though probably not in the case of stars such as a Ori with effective temperatures $\sim 3000^{o}K$. In all cases, however, a cloud of material ejected from the stellar photosphere would be expected to cool off to a temperature $\sim 1400^{o}K$ before the density drops by a large factor. We may thus expect solid particles of CaO, BeO, TiO to condense in any cloud of material leaving a giant star.

Throughout the preceding discussion it was implied that all heavy elements are synthesized in stars and are injected into the interstellar medium through mass-loss processes from cool stars. As a result of an accident of binding energies some of these elements, such as Ca, Be and Ti, may remain locked away in the form of solid particles, and not be observed by the usual atomic transitions.

• Superclusters (?)

SUPERCLUSTERS: FACT OR FANCY?

Darius, Jon; *New Scientist,* 74:383-385, 1977.

An apocryphal Hindu lengend imagines the Earth to be supported by four elephants, in turn borne on the back of a huge tortoise swimming in the universal ocean. According to ancient Jain cosmology, the mountain Meru occupies the centre of the world, surrounded by concentric rings alternately land and sea, while overhead orbit celestial bodies in ever wider circles, the whole being enclosed in the Cosmic Egg. In the Ptolemaic system, the Universe resembles a transparent onion whose various celestial layers are centred on the Earth.

These world views are not entertained by any modern cosmologist, yet three millennia later we are ill placed to evaluate the very principle on which they were based: the hierarchical structure of the cosmos. Intermediate links in the chain are easy enough to discern: satellites orbit planets revolving about stars, while stars (which may themselves be clustered) comprise galaxies which in turn congregate in clusters. But at the subatomic and extragalactic extremes we have not firmly established whether the inventory peters out---whether elementary particles subdivide into quarks or galaxy clusters group into superclusters. Does our Universe impose structural limits on sub and super?

The concept of a hierarchical universe boasts a few adherents but it is anathema to many cosmologists. It runs counter to the cosmological principle, which posits the spatial (but not the temporal) homogeneity of the universe. Evidence for "small-scale" inhomogeneities in the cosmic density abounds, galaxies being the obvious example; but astronomers cherish the belief that *on a sufficiently large scale* the universe looks the same from any vantage point. "Nature is not economical of structures---only of principles," observed Abdus Salam; it is all the more galling that we cannot empirically deduce the principle but instead can only enshrine it as an *a priori* postulate.

....

The Supercluster Debate. Clustering is an inescapable fact of galactic life; what about superclustering? No evidence can be mustered for, say, a cluster such as the one in Virgo but a hundredfold larger, nor for a vast cloud of 1000 clusters, each containing 1000 galaxies. It was with such entities in mind that Zwicky categorically stated a decade ago, "There are no indications whatsoever for the existence of any systematic clustering of clusters of galaxies." K. Rudnicki has diagnosed the problem in retrospect as a failure to distinguish between discrete clusters, objects readily culled from charts or photographic plates, and the more general notion of clustering, which calls for a statistical approach.

The technique for ferreting out clustering effects sounds straightforward enough: take a sample of clusters (or just galaxies) and determine whether their distribution (to second order for galaxies) shows any departure from a random distribution. Sadly, this conceptual simplicity is hedged with observational difficulties: the sample will be incomplete through obscuration by interstellar dust, which becomes progressively denser toward the galactic plane; and through failure to record or to recognise all galaxies in the field of view, especially sparse specimens of low surface brightness.

The basic approach is as follows: sift the clusters into different distance classes (to compensate for the projection of galaxies in 3-D space onto a 2-D sky); apply a series of grids, progressively increasing the size of the mesh, and in each case determine the probability that the observed distribution is compatible with a random one; then plot this probability as a function of mesh size. If there is a preferred clustering scale, it will show up as a *minimum* at the critical mesh size.

For the sake of analogy, suppose it were necessary to segregate a mixture of walnuts and hazelnuts. A fine sieve and a wide-meshed net would be equally futile; only when the holes are larger than hazelnuts and smaller than walnuts can the process work. Should the probability continue to decrease with larger cell size---the equivalent of discovering a clump of coconuts in the mixture---then one could argue, indeed, for supersuperclustering.

By this technique University of California astronomer George Abell found evidence from the Palomar Sky Survey for superclustering on a scale of up to 40 Mpc. Repeating his study with a "local" rather than a "global" test, I. D. Karachantsev recently claimed that the mean diameter is closer to 20 Mpc, with two clusters per typical supercluster. However, from an earlier 3-D "correlation" analysis of the same material T. Kiang concluded that "clustering of galaxies occurs on all scales" albeit with little physical individuality.

Correlation analysis calls for calculation of an index of clumpiness from galaxy counts. Astronomers at Lick Observatory applied this technique to their Astrographic Survey plates, and found clustering up to 30 Mpc. One of their preliminary papers is often cited with amusement, wherein a synthetic plate was concocted on the premise that all galaxies are clustered. The real and synthetic plates proved very similar---save that the clumpiness was even more severe on the real plate.

For the past eight years, Jim Peebles and his coworkers at Princeton have been a powerhouse of correlation analysis, and its alter ego power-spectrum analysis. In 1969 J. T. Yu and Peebles attacked the Abell catalogue anew and showed that any superclustering tendency must be rather feeble. But a reanalysis by M. G. Hauser and Peebles in 1973 nevertheless revealed "clear and direct evidence of superclusters with small angular scale" containing two or three clusters apiece; while their reanalysis of the Lick data went on to disavow any higher-order clustering up to 200 Mpc. Latest results point to a sharp break in the correlation function at 9 Mpc although larger clumps are not ruled out.

Peebles also inferred that inhomogeneities from 50 kpc to 5 Mpc shade into each other---a partial endorsement of Kiang's "hypothesis of indefinite clustering". The nomenclature of rich and poor clusters becomes arbitrary; even from cluster to supercluster there is no clear gap. This smooth hierarchy is disputed by Paul Wesson, who finds clusters clearly delineated at 2 Mpc and superclusters at 15 Mpc.

....

• The Missing Mass

GALAXY CLUSTERS AND THE MASS ANOMALY

Bouw, Gerhardus, D.; *Creation Research Society Quarterly,*, 14:108-112, 1977.

<u>Introduction</u>. It seems that no matter how hard secular man at-

tempts to describe and affect his environment without reference to the Creator, there is always a vague, uneasy "something's missing" aspect to that endeavour.

This "missing" factor has manifestations in every humanistic search for truth and can be seen in many forms; from the three existential questions of philosophy---Who am I? What do I? and How be I?---to the missing links of evolutionary biology; from the missing volatiles of Mars to that missing factor in each human life which many scientists deem necessary to complete that life. The Christian reader will recognize this vacuous state as the quest for Christ who is the Way, the Truth and the Life.

In this study one particular "missing" property is singled out for a more detailed look. This property is variously called the "missing mass"---by those who hold that the mass is actually present in some undetected form---or the "mass anomaly"---by those who are not sure whether there really is an undetected source of mass or whether modern physics is really applicable in its present form on such a grand scale.

The mass anomaly is made evident by estimating the total mass of a galaxy or galaxy cluster by two different methods. The first of these involves a straightforward count of the members coupled with a mass estimate for each member. The second method is based on the dynamics of the system. In practice the two mass estimates differ by factors of two or more for galaxies, but 10 to 400 for galaxy clusters.

The Problem on the Galactic Scale. The smallest systems for which this phenomenon is currently recognized is for galaxies; in particular, the Milky Way. This is not because of anything special about galaxies or their sizes so much as that the next smallest system for which comparable dynamic data exist is the solar system. For the Galaxy the two methods of estimating the mass are as follows:

(1) The number of stars per unit volume in the neighborhood of the sun is counted. These are grouped into brightness classes via the mass-luminosity function. Hence, given the luminosity (intrinsic brightness), the star's mass can be estimated and summing over all the types of stars per unit volume gives a stellar mass-density estimate. Add to this the gas density from radio and optical interstellar lines and a local density of about 6.0×10^{-24} gm cm^{-3} is arrived at.

(2) The second mass-estimating method involves using the space motions of the local stars relative to the galactic center as based on the analysis of 21-cm radio line profiles in the Galaxy and estimates of the distance from the sun to the Galactic center. This yields a force law or rotation-curve from which the amount of matter necessary dynamically to bind the system can be calculated. Doing so yields a local density of 10.0×10^{-24} gm cm^{-3} which, considering the range of individual density determinations, is about a factor of two greater than the count-mass.

The Problem on the Galaxy Cluster Scale. The mass anomaly does not end there. Two orders of magnitude greater than the scale-length of a galaxy and 10^{10} times that of the solar system lies the scale-length of the galaxy cluster. Here, too, the mass anomaly is evident but to a far greater degree. There are now three methods for determining the mass of a cluster of galaxies.

(1) The first of these is analagous to the luminosity-determined mass-density for the solar neighborhood mentioned above. It involves

a straight-forward count of the number of galaxies of each type in the cluster.

For each of these types one can derive an estimate of the mass-to-light ratio based on observing the rotation-curves of individual galaxies also by statistical velocity considerations of multiple galaxy systems. Given an estimate of the distance to the galaxy a mass-to-light ratio can be calculated. These, in turn, also appear to be on the high side.

It has been noted that the mass-to-light ratio for the Galaxy is about 2. More detailed analysis indicates a value of 3. The Galaxy is of morphological type Sb to Sc and the mass-to-light ratio usually assumed for an Sb galaxy is 20 while that for an Sc is 10. From the count of the galaxies in the cluster a total mass for the cluster can be derived. Intracluster mass is assumed negligible.

(2) The second method of determining the cluster mass is to invoke the virial theorem. The theorem was originally derived for application to the kinetic theory of gases but has found wide application in kinetics and dynamics. Basically, it amounts to balancing the gravitational attraction force by the centripetal force; but instead of using the orbital velocity directly, it is inferred from the radial velocity distribution and dispersion about the mean of the members of the cluster. Given the standard deviation about the mean velocity of the radial velocity of the radial velocity distribution, ρ, the mass, M_V, is given by:

$$M_V = 3\ R\rho^2/KG \qquad (1)$$

where $\underline{R}$ is usually taken to be the cluster radius, $\underline{G}$ is the gravitational constant, and $\underline{K}$ is a near-constant which is a weak function of galaxy size and estimates of which range from 0.29 to 3. Depending only in part upon the value of $\underline{K}$ assumed, the virial theorem mass-estimates can be from 10 to 400 times that of the "number-mass" estimates.

(3) The third method, presented for the first time in the preceding issue of the Quarterly and hereafter referred to as paper 1, involves observation of rotation of the cluster and computation of the amount of matter required to hold the system together. It yields mass estimates which are comparable to those obtained via the virial theorem and, for the Virgo cluster, it yields an empirical value for $\underline{K}$ of 0.96 ± 0.26 in Equation 1.

....

Implications for a Young Earth. In the past, the mass anomaly has been used in support of a young-cosmos hypothesis. It was argued that given the conditions inside a cluster of galaxies revealed by the application of the virial theorem, the mass anomaly indicated that the clusters are not bound and that in a billions-of-years-old universe they would have dissipated long ago. With the discovery reported in paper 1 that clusters of galaxies exhibit a rotation-curve, this approach has suffered a set-back. In paper 1 it was noted, though, that two other problems arose for evolutionists.

At present the mass anomaly appears to mean that astronomers do not understand the cosmos well enough to draw any definite conclusions. The fact that gravity is so poorly understood that it cannot be related to other fields (forces) underscores this point. The cause of a variable G, if confirmed, is totally unknown. To Creationists it would not seem possible that variable G could account for the mass anomaly, since the implication would be that galaxies have been around long enough that such a slow rate of change could occur.

The problem could simply be one of perspective. As Shakespeare put it: "The fault, dear Brutus, is not in our stars, but in ourselves ...." Man tends to view time as a unidirectional flow of events. If, however, it is part of a four-dimensional continuum, there might be longer and shorter time-paths between the same events. It would be somewhat as in a baseball diamond; from home to third the right way around is 270 feet, but only 90 the reverse way.

In any case, no satisfactory suggestion has been made which can account for the mass anomaly. Some seem most attractive, such as the distance-scale and variable G proposals, but the last word is still far from in. It is an extremely fundamental consideration, and there is some evidence that the mass anomaly may even occur in star clusters although the ratio there only appears to be about 1.7 ± 0.2. After all who today is really any closer to understanding what mass and inertia are than Newton was?

PHYSICIST DISCOVERS GALAXY CIRCLED BY STARS

Anonymous; *Astronomy*, 6:70, February 1978.

Using a newly designed photometer, Dr. Dennis Hegyi has reported detecting matter in the form of a spherical halo surrounding a galaxy 80 million light-years from Earth. He says the matter is composed of billions of previously unseen stars.

According to Hegyi, the halo appears to be massive, possibly containing more matter than what had been thought to be in the spiral galaxy itself. If this is characteristic of other galaxies, he said, the halos might contain enough matter to stop the present expansion of the universe.

Hegyi, an assistant professor of physics at the University of Michigan, said that radio observations of the galaxy had indicated the presence of some unseen matter, but neither the nature nor distribution of the matter could be determined. He designed and built a new light measuring instrument---an annular scanning photometer---which enabled him to detect the presence of the halo or cloud surrounding the galaxy.

The finding, if verified, might help settle a debate among scientists about whether the universe will expand forever, or if the pull of gravity among the galaxies eventually will stop the expansion. This would cause all matter to fall together again into one gigantic primeval nucleus which, according to a widely held theory, existed before a tremendous explosion started the universe on its expansion course.

"A fundamental question," Hegyi said, "is whether there is enough matter in the universe to stop the expansion. It is generally believed that the visible portions of galaxies studied from photographs do not contain enough matter. But it is also generally believed there could be a considerable amount of matter in forms which do not radiate much visible light, and would be difficult to detect."

Hegyi said the halo around the galaxy he observed, NGC-4565 in the constellation Coma Berenices, is so faint that it eluded detection on photographs and other sensitive observing systems. These techniques only reveal the brighter, central dislike spiral of stars well known to astronomers for decades and for which these galaxies have been named.

• Supposed Holes of Various Sorts

A JET BLACK HOLE IN A RADIO GALAXY

Anonymous; *Science News*, 113:180, 1978.

Astronomers generally believe that physical processes involving enormous amounts of energy are going on in the centers of galaxies. Evidence to support such a view is plentiful. The nuclei of galaxies look brighter than the outer regions in all ranges of electromagnetic radiation. Radio observation also reveals that many galaxies are associated with large lobes of radio-emitting matter that go far beyond the confines of the visible galaxy. Simply, the geometry of these lobes lends plausibility to the suggestion that they consist of material pumped out of the nucleus of the galaxy by whatever is going on there.

A number of theorists have suggested that the center of activity in the nucleus of a galaxy is a giant black hole, one so massive that its gravitational field dominates the center of the galaxy, compels stars to orbit around it, tears stars apart and swallows them whole (SN: 2/19/77, p. 121). Now that suggestion is supported by three observers, A. C. S. Redhead, M. H. Cohen and R. D. Blandford of the California Institute of Technology. They conclude, as a result of radio observations of the galaxy NGC6251, that the phenomena found in the nucleus of that galaxy can best be explained by the presence there of a black hole with a mass approximately 100 million times the sun's, or more. Their report is in the March 9 Nature.

The observations reported in the March 9 paper concern the small-scale structure of the radio-emitting matter in the nucleus of NGC6251. The most striking feature is a jet of matter that seems to be emerging from the nucleus. The axis of this jet lies along the same line as the axis of two large radio-emitting lobes that lie outside the visible part of NGC6251 on either side of the galaxy. It seems likely that this jet in the nucleus and the two large lobes that lie in the same line are produced by one and the same phenomenon, something in the center of the galaxy that pumps out matter along that line.

The character of the radio waves emitted by the jet and its geometry lead the three observers to make some assumptions about the dynamics of the jet, and then to calculate the probable physical conditions in the region where the pumping takes place. They conclude that these are likely to be "conditions that may exist around an accreting black hole of mass [approximately or greater than 10^8 solar masses]."

BLACK HOLE IN NUCLEUS OF GIANT GALAXY

Anonymous; *New Scientist,* 78:84, 1978.

The giant galaxy M87 almost certainly has a huge black hole at its centre. Observations recently made in the US show that this exotic object, long considered a prime candidate in the search for black holes, has a great concentration of mass in its nucleus. The finding of a nuclear black hole in galaxy M87 may unlock the secrets of the energetic outpouring of quasars.

In the past few years astronomers have come to terms with black holes. It's two centuries since Laplace realised that an object with a sufficiently large gravitational field would be invisible, if the escape velocity exceeded the speed of light. It seemed unlikely that such objects really existed until massive invisible stars were found in certain X-ray binaries. Some of these unseen companion stars may be black holes voraciously gobbling up the gas from visible stars and emitting X-rays in the process.

The galaxy M87 has several features which point to a large invisible mass lurking in the nucleus. First there's the famous jet, a streamer of electric blue light, that has been shot out of the nucleus. Secondly, the galaxy is anomalously energetic, pumping out 10^{35} watts more than it should be. Finally, radio astronomers have located minute centres of activity, some only a dozen light weeks across. It all adds up to a scenario of excessive energy generation in a very compact region.

At the Mount Palomar Observatory, Jerome Kristian and colleagues have searched for evidence of the putative black hole by means of accurate photometry. They have confirmed that M87 has a brilliant pin point of light. For the first time a measure of its brightness is given: it is equal to one hundred million suns.

The Kitt Peak National Observatory, Arizona, granted time on the 4-metre telescope for an investigation of the dynamical properties of the nucleus of M87. The interesting factor in this research is the discovery that the velocity of matter circling the nucleus suddenly increases near the centre. A jump of about 100 kilometres per second is indicated.

The change in the velocity of circulation is the strongest evidence yet uncovered for a super-massive black hole in M87---in a sense the higher speed provides the centrifugal force needed to balance the gravitational pull of the massive centre. This furnishes a method of estimating the mass of the central core, which seems to be of order 10^{10} (10 billion) solar masses.

• • • • • • •

WHITE HOLES—A COMING FASHION?

Gribbin, John; *New Scientist*, 68:199, 1975.

For several years the kind of astrophysicist who enjoyed speculating about black holes has also mused on the possibility that the opposite effect, "white holes" bursting outwards from singularities, might exist in space. While black holes have gained considerable fame through their tentative association with binary X-ray systems, no one seems to have taken the white-hole idea seriously. But now that situation may be changed through the work of Professor J. V. Narlikar and his colleagues at the Tata Institute in Bombay.

Narlikar is well known as a proponent of extreme cosmological views, originally through his association with Fred Hoyle on steady-state models of the universe. The creation of matter in the universe, as required by steady-state theories, is only slightly different from the idea that matter in superdense singularities may be bursting from its gravitational bonds and becoming available for interactions with the rest of the material around, so it is entirely appropriate that Hoyle's former protege should now be the patron of white holes in the universe. Now, working with Professor K. M. V. Apparao, he has come up with a much larger and broad ranging contribution, setting the white-hole phenomenon in the context of the present problems of high-energy astrophysics, and pointing the way for further work (Astrophysics and Space Science, vol 35, p 321).

Cosmic fountains. In fact, the name commonly given to this phenomenon is a little misleading. "White" certainly conveys the opposite of "black", but the whole point about a white hole is that it is not a hole, but a source of matter and energy, as far as the outside universe is concerned---a white fountain. The philosophy behind invoking such cosmic gushers in astrophysics is simply the difficulty of explaining observed very energetic objects in any other way. X- and gamma-ray sources within our own Galaxy are difficult enough to explain; farther away, there is still no really good theory to account for the 10^{62} ergs or more produced in violent outbursts of the nuclei of some galaxies and quasars.

The exact nature of such a phenomenon has been widely argued. Perhaps some parts of the universe were left behind during the expansion following the big bang, and now these retarded cores are catching up in little bangs of their own; perhaps collapsing objects can, instead of becoming black holes, "bounce" at some critical density; or, most entertaining of all, perhaps the cosmic gusher of a white hole is simply the "other end" of a black hole, recycling material which has been swallowed up somewhere else in the universe (or in a different universe altogether).

A Blue-Shifted Spectrum. Narlikar and Apparao have now started to look at the kind of spectrum which should be produced from an exploding white hole, and calculated how the blue-shift effect of the expansion will modify the spectrum to an outside observer. The early stages of such an expansion follow the same equations as those of the Einstein-de Sitter cosmology for the expanding universe. The general effect of the blue shift on radiation from the white hole is to produce a power law spectrum with the exponent -3. Because the more strongly blue-shifted components will arrive "outside" first, this also produces an initially hard spectrum which softens with time---and that is a feature of several of the recently observed high-energy sources.

Starting with the biggest phenomena, Narlikar and Apparao suggest that exploding Seyfert galaxies could be a white-hole effect, and that the sum of such processes could explain the observed cosmic background of X-radiation. That radiation has a power law spectrum with exponent -1, but absorption in the gas around a Seyfert nucleus could explain the softening and, according to the Indian astrophysicists, "a reasonable case can be made for white holes being responsible for the infrared and soft X-ray emission from Seyfert nuclei". It's debatable just how good that case is, and perhaps they weaken their argument by trying to spread it to encompass Seyferts (see New Scientist, vol 66, p 63). Certainly, however, the model works well when applied to smaller events within our own Galaxy.

The transient X-ray sources, for example, show a sudden rise to peak energy with subsequent softening of the spectrum, and a power law spectrum, with exponent in the range -1 to -3. Different detailed properties of their spectra can be explained in terms of black body, free-free or synchrotron radiation modified by the white-hole expansion and the blue-shift effect. And gamma-ray bursts look even better candidates, with exponent -3 power law spectra, softening with time, and even sharper rise times.

Even particle production can be accommodated by the theory. Although the spectrum of high-energy cosmic rays does not quite fit the raw white-hole model, in this case it seems entirely reasonable to argue, as Narlikar and Apparao do, that it has been modified by the interaction of the charged particles with the galactic magnetic field and local magnetism.

What would a white-hole explosion leave behind? For a transient X-ray source the energy requirement suggests a mass a few times that of the Sun, well within the bounds of plausibility; while for gamma-ray bursts the end product could well be a one solar-mass white dwarf. The theory could be developed, as black-hole theory has been, by calculation of effects of charge and non-spherical distribution, so that the success of the, as yet very simple, version in accounting for observed phenomena is far from bad.

Of course, there have been arguments that white holes cannot exist. In particular, Dr. D. M. Eardley suggested last year that any embryonic white hole would soon accrete enough matter to become a black hole (Physical Review Letters, vol. 33, p. 442). But people used to say that black holes cannot exist (some still do!) and Narlikar and Apparao present a cogent case for further investigation of the ideas they outline.

EXTRATERRESTRIAL RADIATION

Solar and extrasolar cosmic rays are spoken of with such familiarity that one would almost believe that science understood them. In truth, the origin(s) and accelerating mechanism(s) of ordinary cosmic rays are as mysterious as they were when extraterrestrial radiation was discovered a scant few decades ago. Our objective, here, however, is the collection of phenomena that are not part and parcel of science. Only exceptional cosmic-ray observations qualify, such as the recently detected anomalous bursts of X-rays and gamma rays. These bursts are so energetic that they transend the emission capacities of current stellar models. Like quasars, cosmic-ray bursts hint at incredibly energetic processes transpiring somewhere beyond the solar system and, probably, beyond our own galaxy.

There is room here to touch briefly upon the subject of gravitational radiation. Strangely enough, gravitational radiation itself is hardly anomalous because General Relativity predicts its existence. The anomaly arises when scientific instruments that do not seem up to the task detect gravitational waves. If gravitational waves are not being detected, what is being recorded? To this category of gravitational glitches, we also assign some peculiar phenomena associated with solar eclipses, although the events are so strange that our categorization is mostly guesswork.

• X-rays, Gamma Rays, and Neutrinos

ANOTHER GIANT BURST FROM AN X-RAY STAR

Anonymous; *New Scientist*, 62:110, 1974.

Cygnus X-3 is an X-ray star with the jitters. On six occasions in the past two years it has produced giant bursts of radio waves, a property unshared by other X-ray sources. An intensity-time graph of the latest outburst, in mid-1973, is reproduced in Nature (vol. 243, p. 319) by K. A. Marsh, Chris Putton and Paul Feldman of York University, Toronto. They used the 46-metre telescope at Algonquin Park, which pioneered [sic] the Cygnus X-3 bursts to follow the sixth event.

The interesting feature of the latest curve is the exponential decay. Such behaviour is incompatible with simple models of the burst, which envision the free expansion of a cloud of high-energy electrons. The

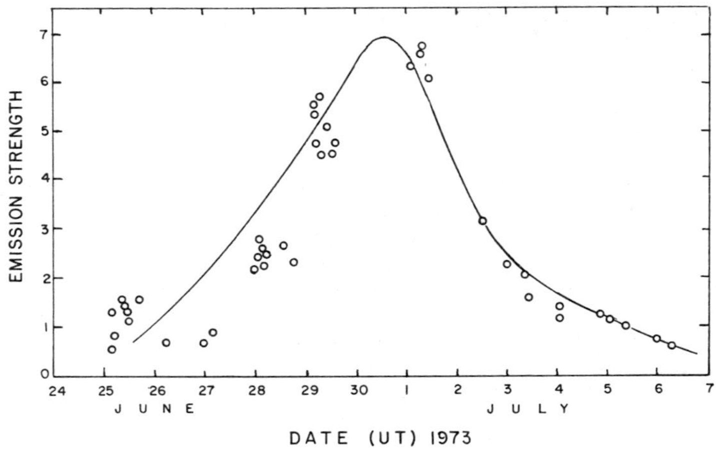

X-ray burst from Cygnus X-3

intensity of the radio emission from electrons in an expanding cloud follows a power-law decay. The three Canadian researchers propose, instead, a model in which cold gas infiltrates the electron cloud, so that the energy loss mechanism is modified to give an exponential death rate. On the rise side the shape of the burst is governed mainly by the energy distribution of electrons emerging from the unseen machine that triggers the bursts. So far the radio and X-ray observations have not drawn back the veils across this mysterious source of energy.

....

A WEIRDLY JITTERY X-RAY SOURCE

Anonymous; *Science News*, 109:101, 1976.

X-ray astronomy piles astrophysical mystery upon astrophysical mystery. The latest is reported by scientists from the Massachusetts Institute of Technology working with data from the Astronomical Netherlands Satellite. It involves a strange pattern of X-ray bursts or pulses coming apparently from a globular cluster of stars in the constellation Sagittarius.

This is another to add to the menagerie of pulsed signals, but it is an extremely weird one. The bursts rise to maximum in about half a second and take ten seconds to die down. They occur on the average of every 15,718 seconds, but the repetition is not exactly precise. There is a "phase jitter" of about 500 seconds one way or the other, the longest recorded discrepancy being about 1,000 seconds. The data were reported at the meeting of the High Energy Astrophysics Division of the American Astronomical Society held recently at MIT by graduate student Jesse G. Jernigan, Jr. His collaborators were George W. Clark, Claude R. Canizares, Satio Hayakawa, a visiting professor from the University of Nagoya, and Fuk Kwok Li.

Such a difference between pulse length and repetition time is unique in pulsed X-ray phenomena. Normally, pulsed signals are attributed to pulsing or rotating bodies, but the difference in the numbers and the jitter make it hard to imagine what kind of body could produce these. If the source is indeed in the globular cluster in Sagittarius from the direction of which the bursts come, the intensity of a burst is a million times the intensity of all radiation from the sun.

COSMIC X-RAY BURSTS

Brecher, Kenneth; *Nature*, 261:542, 1976.

Cosmic X-ray bursts, first observed by the Russian Cosmos 428 satellite in 1971 and reported in a widely overlooked paper by O. P. Babushkina et al. (Sov. Astron. Lett., 1, 32; 1975) have now been detected by many other X-ray satellites. The ANS, SAS 3, Vela 5, and Ariel V groups have reported in rapid fire succession a range of exciting and perplexing observations of these fast transient cosmic events. In this issue of Nature, (page 562) J. Heise et al. report on their observations of "rapidly repetitive X-ray burster" MXB 1730-335 made with the Astronomical Netherlands Satellite. This object, discovered by the Massachusetts Institute of Technology SAS 3 group and reported by Professor Walter Lewin (IAU Circ., 2922, March 5, 1976) is perhaps the most enigmatic "X-ray burster" found to date. Sometimes bursting at several-second intervals, at other times waiting minutes between events, its X-ray burst pattern reminds one very much of a relaxation oscillator. Indeed, as pointed out by Professor George Clark (MIT) the tight correlation found in the SAS 3 data between the X-ray burst size and the time to the following burst is reminiscent of the behaviour of a neon light bulb (rather than, for example, a flush toilet). It is as if a reservoir of energy (or material to provide it) must build up to a fixed level and discharge; the larger the discharge, the longer it takes to build up for the next firing. (In the flush toilet analogue, with filling occurring at a constant rate, the size of a burst would be proportional to the time preceding it.)

. . . .

GAMMA-RAY BURSTS FROM DEEP SPACE

Anonymous; *Sky and Telescope*, 46:146, 1973.

Short, powerful bursts of gamma radiation originating outside the solar system are reported by three scientists at the Los Alamos Scientific Laboratory in New Mexico. Approximately 20 of these events have been recorded since July, 1969. As a wave of gamma rays passes the earth, it deposits a pulse of energy in the upper atmosphere at rates up to 20 megawatts.

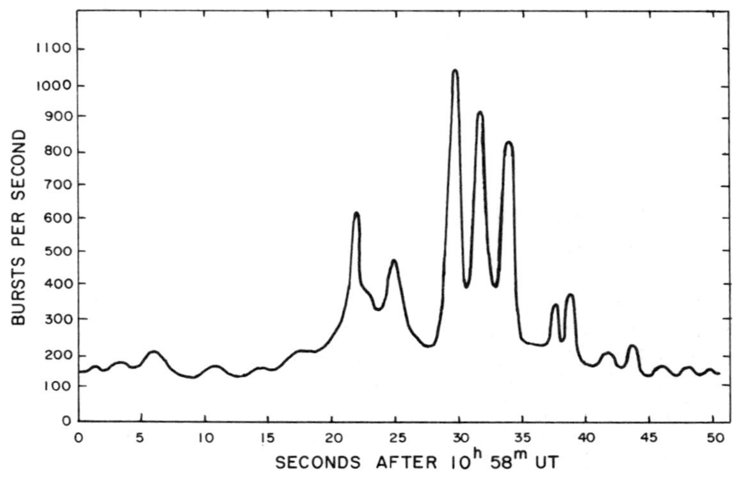

Gamma-ray burst of April 27, 1972, recorded by the Apollo-16 gamma-ray spectrometer

Ray Klebesadel, Ian Strong, and Roy Olson made this discovery while analyzing data from the four most recent Vela satellites, which move around the earth in circular orbits 140,000 miles in diameter. The prime function of the Vela satellites is to act as "watchdogs" over the limited nuclear test-ban treaty signed in 1963. Each craft carries a nondirectional array of cesium iodide sensors which, linked to sophisticated computers, can recognize sudden short bursts of gamma radiation (wavelength 0.01 angstrom and shorter).

Unless the satellites are engaged in some special activity, or are, as occasionally happens, too strongly affected by particles from the sun, all four of them "see" each burst of gamma rays. As the blast, traveling at the speed of light, passes each satellite, the time is recorded to an accuracy of about 0.05 second. From the differences between the arrival times as monitored by the several satellites, it is possible to deduce the direction from which a given burst has come. In this way it has been found that the radiation does not come from the earth, moon, sun, or other planets.

The origin of the gamma-ray bursts is a mystery. In 1968, Stirling Colgate, New Mexico Institute of Mining and Technology, said the explosion in a supernova should cause a brief flash of gamma radiation and X-rays. However, the times and directions of the observed events show little relation to recorded supernovae.

Jonathan Grindlay of Harvard Observatory has suggested that flare stars may emit some bursts of short-wavelength radiation. However, the predicted energies are lower and the durations are longer than the 0.1 to 30 seconds observed by Dr. Klebesadel and his associates. It is not even clear that the bursts are all of a kind.

HEAO RECORDS GAMMA-RAY BURST

Metz, William D.; *Science,* 199:870, 1978.

On 29 October HEAO discovered one of the mysterious gamma-ray bursts that were first seen in 1973. These rapid bursts, a series of emissions lasting less than 1 minute, have characteristics of nuclear explosions and remain unexplained since their discovery. [Excerpt]

WAS A NEUTRON STAR BORN ON 4 JANUARY?

Anonymous; *New Scientist*, 63:500, 1974.

A strange event, recorded by an array of Cerenkov counters deep underground in South Dakota on 4 January, may have represented the birth pangs of a new neutron star collapsing under a strong gravitational field from a normal star. Such, at any rate, seems to be one of the only plausible ways to account for a series of antineutrino pulses lasting for about a microsecond each, and separated by roughly a millisecond. The experiment was described at a conference on neutrino physics held in Downington, Pennsylvania, and is described in Physics Today (vol. 27, no. 7, p. 17).

Kenneth Lande of Pennsylvania University, who reported the result, and his team mates both there and from the Universities of Texas and Torino, arranged a battery of seven water Cerenkov counters 4850 feet deep in the Homestake Gold Mine at Lead. It is here that Dr. Raymond Davis has been looking for solar neutrinos with a huge tank of cleaning fluid---and, embarrassingly, finding far too few. However, whereas Davis seeks neutrinos, Lande and his team were searching for antineutrinos by their interaction with protons to give a neutron and an electron. The counters were simply 500-gallon fuel oil tanks filled with water and fitted out with the requisite electronics.

Theorists have proposed that a collapsing star evolving into a super-dense neutron star will, especially if the collapse is rapid, bounce several times as it condenses. The period of this reverberation, they say, should be around a millisecond. And it would emit antineutrinos with higher energies than the concomitant neutrinos formed in pair-production. That would answer satisfactorily why Dr. Davis failed to see any corresponding unusual event on 4 January among the neutrinos in his cleaning-fluid tank.

• Gravitational Radiation

HOW REAL ARE GRAVITATIONAL WAVES?

Anonymous; *Nature*, 224:411, 1969.

In June this year, there was great excitement at the report in Physical Review Letters (22, 1320; 1969) that Professor J. Weber had finally succeeded in the detection of gravitational waves (see Nature, 222, 1117; 1969). For one thing, people were delighted that Professor Weber's long-standing devotion to the cause of gravitational waves had finally been rewarded. It was also pleasurable to be able to admire the elegance of his experimental design---two groups of detectors separated by the distance of 1,000 km between the Argonne National Laboratory and the University of Maryland in the hope of eliminating spurious coincidences caused by local disturbances. Finally, it was understandable that outsiders should have been enormously excited by what seemed to be the demonstration of the reality of gravitational waves, well on the way to being the Holy Grail of General Relativity. With the feasibility of detection apparently assured, it was possible to think of the experiments which might be carried out to turn this new development to good account. The cosmologists were understandably poised on the edges of their armchairs, devising schemes for the detection of gravitational waves from pulsars and other such phenomena.

This is why it is a pity that enthusiasm is now abating. A part of the trouble is that there has not yet been confirmation of the original result. It is true that equipment like that installed at the Argonne Laboratory and the University of Maryland does not lie about like scrap iron, but there have presumably been further measurements since the last week in March this year, when the original series of observations was carried out. There is no doubt that the technique which must be used for the detection of gravitational waves is exceedingly exacting. The detecting instruments consist of massive cylinders of metal tuned so as to have a natural frequency of vibration at about 1,660 Hz. The immediate task is to tell when the cylinders, carefully isolated from adventitious sources of disturbance, are set in vibration, which can be accomplished by piezoelectric instruments of various kinds. The next step is to send a continuous record of the condition of the Argonne cylinders by radio for comparison with the records at the University of Maryland. In June, it seemed as if there had been seventeen occasions during a period of 81 days on which the cylinders at the two sites were oscillating in unison, and the estimates that were made of the length of time for which it would be necessary to wait for these coincidences to arise by chance were almost staggering by their size. On March 20, 1969, three sets of detectors at the two sites were simultaneously set in oscillation on two occasions separated from each other by two and a half minutes. At the time, the frequency of such a happening was estimated to be once in 70 million years, convincing enough, it may be thought.

But is the interpretation as straightforward as it seems? These multiple correlations are not by themselves compelling tests of the reality of the coincidences. Because local disturbances may set groups of cylinders in motion, it is only coincidence at a distance that really matters. Unhappily, there is good reason to wonder whether these coincidences are as remarkable as they seemed to be at first. It is possible to work out from the figures which have been provided an estimate of the frequency of occurrence of coincidences like those recorded in the 81 days for which the observations were continued. The chance that purely random processes can account for what seems to be the most remarkable coincidence, that of January 1 this year, seems to work out at about one in 250. The other coincidences are most easily accounted for by chance---what seems to have been the weakest of the coincidences could have been expected to occur roughly thirty times in the course of 81 days, and may have stood out from the ruck only because it followed half a minute after a more pronounced event.

The difficulty, of course, is that it is exceedingly difficult to know just what weight to give to these coincidences without knowing more about the sustained pattern of noise in the experimental apparatus. In the long run, it will be interesting to see what can be winnowed away from the seventeen coincidences which seemed only a little while ago to be harbingers of a new experimental science. There is no doubt of course, that the experiment has been entirely worthwhile, and that it has won unalloyed admiration for elegance and ingenuity. People must necessarily look to the University of Maryland for such confirmation as may be forthcoming, at least for the time being. If the experimenters feel people breathing down their necks in the weeks ahead, that is a measure of the importance which is attached to the new ground they are breaking.

MAGNETIC CORRELATION?

Anonymous; *Nature*, 243:439, 1973.

The chief evidence for associating the signals recorded by Weber's gravitational wave detectors with an astronomical source outside the solar system has been the 12-hour periodicity displayed in published plots of event rate as a function of sidereal time. Now a wider analysis of the times of some of Weber's events has been made, and this indicates that there may also be correlations with disturbances of the Earth's magnetic field and with sunspot activity. Such correlations would suggest that Weber's detectors may be influenced by phenomena unconnected with gravitational radiation and may help to resolve some of the problems raised by Weber's observations.

The new analysis has been made by Tyson, Maclennan and Lanzerotti of Bell Laboratories (Phys. Rev. Lett., 30, 1006; 1973), and follows earlier work by Adamyants, Alekseev and Kososnitsyn in the Soviet Union. The Soviet physicists based their analysis on published times of seventeen coincident events recorded by Weber's detectors at Maryland and Chicago in late 1968 and early 1969. They found a correlation between changes of terrestrial magnetic activity and the days when Weber events were recorded and they also found a rather stronger correlation, involving a time lag of 2 days, between sunspot activity and Weber pulses. These results have generally been regarded as interesting, but of limited significance because of the small number of events analysed.

The group at Bell Laboratories was provided by Weber with data on 262 events recorded with his detector in a period of 121 days beginning in August 1969. They have searched for correlations between the daily rates of Weber's events and many other phenomena including geomagnetic activity, sunspot number, averaged earthquake amplitudes at a point between Maryland and Chicago, tidal strain in the Earth and changes in barometric pressure and in temperature. Correlations at levels above 2 standard deviations are found between the Weber data and geomagnetic, sunspot and earthquake activity; with a time lag of 8 days in the sunspot correlation. The strongest geophysical correlation found was with magnetospheric ring-current intensity and was at the level of 2.7 standard deviations. The rate of occurrence of the gravitational wave events throughout the sidereal day was also investigated and a distribution similar to those reported by Weber was obtained, with peaks occurring twice a day at the times when the detectors would be expected to have maximum sensitivity for gravitational radiation coming from the direction of the galactic centre. This correlation was at the level of 3 standard deviations.

....

• Curious Eclipse Phenomena

1970 SOLAR ECLIPSE AS "SEEN" BY A TORSION PENDULUM

Saxl, Erwin J., and Allen, Mildred; *Physical Review D*, 3:3:823-825, 1971.

Introduction. In this paper a study is made of the variations in behavior of a torsion pendulum during the solar eclipse of 7 March 1970. The torsion pendulum is essentially the same as that used in studying the effect of added weights on its period, but with modifications discussed later in this paper. With this setup, when the pendulum is started reproducibly from rest in a precisely defined release position, the time of traversing a constant portion of its path is timed accurately and recorded automatically. To do this, light from a fixed source is reflected from a mirror attached to the torus to fall on a photocell. Over a preamplifier the latter starts a crystal-controlled counter as the light beam travels clockwise across the photocell and stops it on its return counterclockwise trip, to record the times between these two passages of the light beam across the photocell. This gives recordings of the times used in traversing a constant fixed part of the total vibration path of the oscillating torus on the first swing from rest.

Some improvements had been made in the apparatus since the previous work, which increased its precision still further. Notably, a stronger light source was used which made it possible to narrow the vertical slit limiting the width of the light beam falling on the photocell. Preamplification of the signal received by the photocell and other minor changes (such as a constant voltage transformer for a uniform power supply and a nonferrous and nonmetallic manual release mechanism) were made to assure safe and reliable action of the electronic timing and printout mechanism. Earlier observations during other eclipses, taken before these needed improvements, agree qualitatively with the present results, but are not good enough for quantitative comparison.

Furthermore, it was possible to keep the temperature around the isoelastic Ni-Span "C" suspension wire at 21.7°C with a fluctuation of only ± 0.6°C. This suspension has been kept under the given load for some 17 years so that possible creep should have reached equilibrium. Moreover, the operation of the pendulum was started at 4 a.m. EST while the critical readings did not begin until 10:15 a.m. This long running-in period should eliminate for the suspension wire any lack of stability due to repeated twisting or malfunction arising from mechanical hysteresis or sudden slippage in its metallic structure or lack of temperature stability inside the tube surrounding it. In another prolonged set of observations taken to study the effect of electric charge on the period of the pendulum, there was no comparable change in the grounded period. Moreover, the torsional elasticity of the suspension wire studied statically did not change with the position of the moon. Furthermore, the wire was checked statically to see that it followed Hooke's law exactly as well as calculations made to show that the margin of safety for operation within the elastic limit was substantial (1:7). To avoid slipping, not only was the wire tightly clamped at both ends but three pointed set screws were driven, one above the other, solidly into the wire.

Observations in the present case were recorded alternately with the pendulum grounded and with it charged to + 4900 V, but only the grounded results are presented in this paper. Somewhat different, and at times unexpected, effects were noted when the pendulum was charged electrically.

Procedure. Under these carefully guarded conditions, automatic recordings of the times during which the torus rotated through a constant angle were made from 10:15 a.m. until 3:40 p.m. EST on the day of the eclipse in the town of Harvard, Mass. The eclipse in Boston, some 20 m distant, was about 96.5% total.

Significant variations in the recorded times were observed during the course of the eclipse, as is shown by the full line in Fig. 1. Each point in this figure is the average of five consecutive grounded readings. The limited vertical lines indicate the average deviations of the five readings from the average circled values. The beginning of the eclipse at 12:31 p.m., its midpoint at 1:40 p.m. and its end at 2:58 p.m. are also indicated on the graph. It is to be noted that these observed three intervals level off to about 29.581 sec after the end of the eclipse, whereas in the morning they had started at about 29.570 sec, an appreciable difference inasmuch as the times can be read to 0.00001 sec and are significant to about 0.001 sec. The precision of the quartz crystal-controlled oscillator in the Beckman EPUT (events per unit time) counter is one part in 10^8.

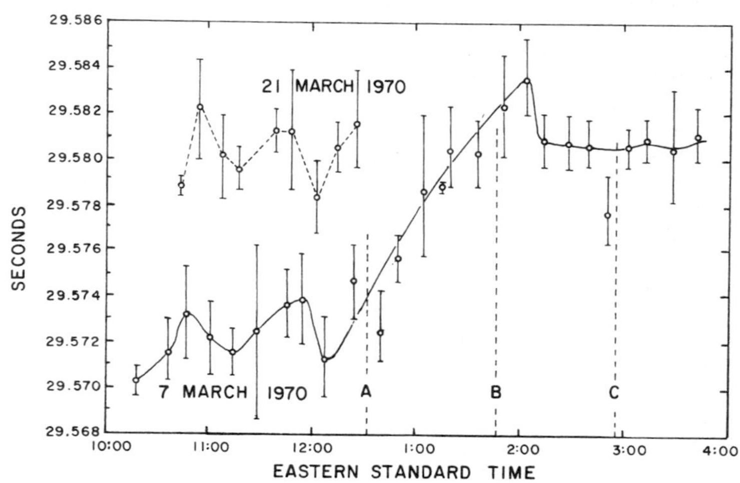

Variations in the period of a torsion pendulum during the 1970 solar eclipse

The irregularities occurring before the start of the eclipse might be considered accidental, except that data taken two weeks later at the same hour of the day (dashed curve) show corresponding humps---an indication, by the way, that the observations are reproducible. These maxima and minima may indicate a kind of gravitational fine structure which is reproducible even when the positions of the sun and moon relative to the earth are quite different. This apparent wavelike structure has been observed over the course of many years at our Harvard laboratory. It cannot be predicted on the basis of classical gravitational theory nor has it been observed in the quasistationary experiments underlying this theory (e.g., spring-operated gravimeters, seismographs, and interferometer devices).

Furthermore, the actual values of these observed times are greater at the later date. On that occasion the sun and moon were on the opposite sides of the earth, whereas during the eclipse they were in conjunction on the same side. This difference in relative position might well explain an increase in the observed times. These times are known to increase with increase in tension of the wire and therefore with gravitational attraction. Thus the moon pulling in the same direction as the earth could be expected to increase the observed times.

The difficulty is that this <u>relative</u> increase of about 2.7×10^{-4} recorded here would require an increase of tension of 1.2 kg. as calculated from the results of our paper on the period of a torsion pendulum. This is 5% of the total weight of the pendulum bob, 23.4 kg (51.5 lb), and is far greater than classical theories of gravitation can explain. Results of this order of magnitude have been consistently observed in Harvard over a period of 17 years. The greatest possible variation in <u>g</u> computed according to the older theories for a given site on the earth's surface is 0.00016 cm sec^2 or 1.6×10^{-5}%, so that <u>our results are about 10^5 times as great</u>. As shown in Fig. 1, the maximum average deviation of our results (which is a measure of our uncertainty) is about 2.5×10^{-2}%.

It is further to be noted that the greatest change occurs between the onset of the eclipse and its midpoint. This agrees qualitatively with the work of Allais with a paraconical pendulum, where the change of azimuth increased substantially in the first half of the eclipse of 30 June 1954. Both these effects would seem to have a gravitational basis which cannot be explained by accepted classical theory.

Both our experimental findings and those of Allais cause one to question whether the classical laws of gravitation hold without modification.

<u>Conclusion</u>. Quantitative observations made with a precise torsion pendulum show, in agreement with many earlier, less precise recordings made at Harvard since 1953, that the times required to traverse a fixed fraction of its total angular path vary markedly during the hours before the eclipse and during its first half, i.e., up to its midpoint. Also the significant changes in these times do not coincide exactly with the astronomically determined onset, midpoint, and endpoint of the eclipse.

These variations are too great to be explained, on the basis of classical gravitational theory, by the relative change in position of the moon with respect to the earth and sun. This leads to the same conclusion arrived at by Allais---that classical gravitational theory needs to be modified to interpret his (and our) experimental results. Moreover, the findings with the torsion pendulum, the significant mass of which moves perpendicularly to the geogravitic vector, seem to indicate the possibility of a fine structure in these observations neither predicted nor recorded using the orthodox methods of quasistationary gravitational investigations.

AN APPARENT DECREASE IN GALACTIC RADIO NOISE DURING A TOTAL SOLAR ECLIPSE

Kingan, S. G., et al; *Nature,* 211:950, 1966.

During the total eclipse of 1965, May 30, observations were made on the island of Manuae (lat., 19° 15' 43" S.; long., 158° 57' 43" W.) of changes in the D layer absorption of radio signals, and also of fluctuations in the galactic noise level at 18.2 Mc/s. The latter observations in particular produced an unexpected effect which is reported here.

The apparatus used in measuring the noise consisted of two half-wave dipoles in phase, with reflectors, the main lobe being directed at R. A. 1h 20m, Dec. 16° 21'S., that is 8° from the zenith in a direction W. 20° S. This aerial fed a low noise receiver of 10 kc/s bandwidth, and the audio-output was amplified, rectified and used to drive a pen recorder. The system was calibrated with a 5722 noise diode.

Observations were made between 18h 45m and 21h 00m U. T. on every day between May 27 and June 2. Time markers were received from station WWVH and placed manually on the chart.

The unexpected feature consisted of two large bursts of noise, each of 13 sec duration, the leading edges of which coincided to 3 sec with the times of optical second and third contacts. The mean power level in each burst was about 2.5 times the mean noise power during the eclipse. No such similar bursts occurred at any other time while observations were being made. Because all expeditions on the Island had collaborated in the detection, identification and elimination of any sources of local radio interference, we are satisfied that the noise originated from a source outside the island.

Reduction of the data on this and other observations is not yet complete.

TESTS OF RELATIVITY

Any phenomenon or measurement that challenges the well-entrenched Special or General Theories of Relativity must be considered an anomaly. Four major types of evidence supporting Relativity are: (1) the advance of Mercury's perihelion (already covered in Chapter 2); (2) Michelson-Morley-type experiments that show negligible ether drift; (3) the bending of light rays in the vicinity of the sun (the Einstein displacement); and (4) red-shift experiments in which signal wavelengths are increased by the action of gravity. The data of this section prove that the last three "proofs" do not provide complete, absolute support of Relativity. Generally speaking, however, Relativity is so secure that scientists question the experiments rather than Relativity itself.

• Ether-Drift Experiments

THE MICHELSON-MORLEY EFFECT

Hicks, W. M.; *Report of the British Association, 1901,* 562, 1902.

In the theory of this experiment, as usually presented, no account is taken of the alteration in wave length produced by reflection from a moving surface, nor of the alteration in the direction of incidence as the drift alters, when the source of light is fixed to the apparatus. When this is done it follows that the phenomena to be expected are not precisely the same as those usually supposed, and in certain cases the displacement of the fringes is subject to a quite different law. The two sets of interfering waves, when there is drift, have not the same wave length in space, although their apparent frequencies at any point moving with the apparatus are equal. Consequently interference fringes are produced on a screen which is fixed to the apparatus, and these fringes are displaced a certain number of bands when the apparatus drifts. Usually, however, the fringe is observed by an optical apparatus which produces an image on the retina. But the two interfering pencils from any point of the actual fringe, when they arrive at the retina, have a different phase-difference from that at the original point. Consequently the image of the central bright line will not itself be a bright line. The central bright band on the retina will be the optical image of another point on the original, and the fringe-image shows the original one displaced by a certain amount which alters with the drift. The observed displacement is therefore the resultant of two others, one of which may in certain circumstances quite mask the other. Supposing the drift of the apparatus to be comparable with that of the earth's orbital motion---say 10^{-4} times that of light---it was shown to be possible that in Michelson's actual experiment the arrangements were such that the effect he expected was quite masked by the other.

MOTION OF THE EARTH THROUGH THE ETHER

Anonymous; *Science,* 61:sup x, May 15, 1925.

Evidence against the validity of the relativity theory was unfolded before the meeting of the National Academy of Sciences by Professor Dayton C. Miller, of the Case School of Applied Science, who, by a much refined and improved repetition of the so-called Michelson-Morley experiment, has shown that there is a definite and measurable motion of the earth through the ether.

Professor Miller has obtained on four occasions a small positive effect at Cleveland, namely, the equivalent of a velocity of about 2 kilometers per second at the altitude of the Case School of Applied Science, and about 3 kilometers per second on the level of the neighboring hills. Whereas at the altitude of the Mount Wilson Observatory in four consecutive experiments spread out over four years he obtained with increasing precision a positive result of 10 kilometers per second, his last result this April justifying him in asserting that the result is correct to within one half kilometer per second.

The technical details of these experiments themselves will be described shortly in special papers by Professor Miller. The purpose of the present note is to say a few words about the implications of these results as viewed from the standpoint of the relativity and the ether theories.

In the first glance then this definite result is entirely antagonistic to the Einstein relativity theory, which in fact could not be adapted to these results of Professor Miller by any conceivable modifications, unless the very fundamental principles of Einstein's theory were given up. This, however, is as much as to say that Professor Miller's results knock out the relativity theory radically.

In the second place, from the point of view of an ether theory, this set of results, as well as all others previously discovered, are easily explicable by means of the Stokes ether concept, as modified by Planck and Lorentz.

Without entering into the mathematical details associated with this statement we may say only that Professor Miller's results as obtained in Cleveland and Mount Wilson are given immediately by the main property of such an ether, namely, to adhere almost completely to the surface of the earth, and therefore share almost entirely its translational motion over its surface and to have a gradually increasing velocity relative to it when we go higher and higher up.

In the third place, the result of the recent rotational terrestrial experiment at Clearing, Ill., near Chicago, which gave a full effect associated with the spinning motion of the earth, can be accounted for by making the natural assumption that our globe, being almost perfectly spherical and having a purely gravitational grip upon the ether, does not appreciably drag it in its rotatory motion. Also the deflection of the light rays around the sun to the amount claimed by the Einstein formula can be easily accounted for by means of a compressible ether provided its dielectric constant is related to its density and pressure by a very simple formula published a few years ago in the Philosophical Magazine.

The amount of additional evidence for the reality of Professor Miller's results afforded by his tables showing the relations of the observed azimuths of drift to the sidereal time is very remarkable. These tables indicate a motion of the solar system in a direction and with a velocity in good accordance with the independent results obtained by Dr. Stromberg and others.

ETHER DRIFT EXPERIMENTS

Miller, Dayton, C.; *Nature*, 117:890, 1926.

The editor has kindly shown me the proof of a letter from Sir Oliver Lodge, relating to my ether drift experiments, and in particular to the account of this work recently given at the Royal Institution. Sir Oliver has in a very courteous manner described the principal argument put forth, and, as he says, the matter is now before the scientific world, sub judice. Notwithstanding this, it may not be out of order to make further explanation of some of the points raised by Sir Oliver.

It seems necessary to direct attention again to the fact that this experiment has never given a true zero or 'null' effect in general; it has only given a 'negative' result, that is, it has answered 'no' to a specific question. Following the report by Morley and Miller, in 1905, Prof. Einstein made the hypothesis that the motion of the observer produces no effect upon the apparent velocity of light. This hypothesis has been given in hundreds of books as the correct interpretation of the experiments, and has been accepted, perhaps without fully examining the original papers. Michelson and Morley, in 1887, said: "Considering the motion of the earth in its orbit only ... the relative velocity of the earth and the ether is ... certainly less than one-fourth of the orbital velocity....In what precedes only the orbital motion of the earth is considered. If this is combined with the motion of the solar system, concerning which but little is known with certainty, the result would have to be modified." In 1905 Morley and Miller, using a much larger interferometer, included in the calculations with the orbital motion of the earth a presumed motion of the solar system of twenty kilometres per second towards the constellation Hercules (still leaving the orbital motion predominant), and came to practically the same conclusion. But, in each of these reports the effect to be expected was calculated in advance, both as to direction and magnitude, and then the observations at two different times of day (as determined from the expected result) were combined so as to add together the expected components, and so as to cancel effects other than those being looked for.

In all of these observations there was a definite positive result, which is observed as a periodic displacement of the interference fringes; but, for the two calculated times of day, the phases of the positive periods differed in such a way that when the two sets of readings were added, they neutralised each other and left only the very small result which was correctly reported as that obtained for the orbital component of any existing ether drift. It is this positive effect, then eliminated, that is now being examined. While it is not large, it is by no means insignificant; the present amount of observed relative motion is 10 kilometres per second with a probable error of ± 0.5 kilometres per second. As this effect was not 'expected' it was not easy to interpret. The history of the experiment has been given in Science for June 19, 1925, and in a more concise report in Nature for July 11, 1925, p. 49.
The re-examination of all of the observations from 1887 to 1926, is the subject of a paper presented to the National Academy of Sciences in Washington on April 26, 1926 and not yet published. It is hoped to give a summary of this paper shortly in Nature.

Sir Oliver Lodge asks what would appear if the results were plotted on the hypothesis that the south side of the housing is warmer than the north side, or with regard to other conditions. It is exactly for answering these questions, and others, that the experiments have been continued over a period of six years, in which time the thousands of readings have been made. Every disturbing cause that could be thought of has been exhaustively studied; among these are: daily and annual variations in temperature, meteorological conditions, radiant heat, magnetism, magneto-striction, differential gravitation, gyrostatic action, influence of method of illumination, transparent and opaque coverings of the light path, speed and direction of rotation, lack of balance in the rotating parts, position of the observer, and other conditions. One after another, these disturbances have been shown not to produce the observed effects. Finally, it has been possible to combine into one logical solution all the readings made to test all these varying conditions, without any omissions (excepting a few readings made under abnormal conditions, such as artificial heating), without any corrections, and without assigning any weights. This solution is entirely consistent with the observations of Michelson and Morley of 1887, and with those of Morley and Miller of 1902-1906.

The observations with the interferometer are made to detect a periodic shift of the fringes, periodic in each half-turn of the instrument; it is not the absolute position of the fringes that is important, but rather the periodic, vibratory change in the position of the fringes, as the interferometer turns through 180°. This can be detected best by having the instrument in slow and uniform movement, while the observer watches the fringes continuously. The instrument turns once in about a minute, and floats so freely that when once started it will continue in rotation for more than two hours without being touched. A series of readings is made in less than fifteen minutes.

The important part of the argument is that the reported effect has always been present, and is evident in every single observation and not merely in the mean; it is not related to the instrument, or to its surroundings; neither is it dependent upon day and night, or summer and winter; it is clearly shown to be directly related to sidereal time, that is, to a cosmical cause. This is explained, with numerous curves showing the results of the 1925 observations, in Science for May 1, 1926. The final arguments are based upon four very complete series of observations made at Mount Wilson for the epochs April 1, August 1, and September 15, 1925, and for February 8, 1926, thus covering various seasonal conditions, and various orbital positions of the earth.

In making the observations, two independent quantities are noted, the direction in which the interferometer points when the effect is a maximum, and the amount of the periodic displacement of the interference fringes. Each of these two sets of readings leads to an independent determination of the right ascension and declination of the apex of the supposed motion of the earth in space. It is very significant that these two determinations are wholly concordant.

THE ETHER DRIFT

Anonymous; *Science*, 70:sup x-x11, November 1, 1929.

Science still must answer the great and fundamental question: "Is there an ether?" And despite the many feats of the Einstein theories of relativity in explaining and predicting observed facts of physics, such as the way the planet Mercury moves in its orbit, they are seriously menaced by having one of their foundations pulled out from under them.

For Professor Dayton C. Miller has reported to the Optical Society of America meeting at Ithaca that he has during the past year laboriously repeated the ether drift experiments that he has been making during the last nine years in a Cleveland laboratory and on high Mount Wilson in California.

Again he finds an observed effect in the light path of his apparatus such as would be produced by a relative motion of the earth and the ether of about ten kilometers (six miles) per second. This is the same result that Dr. Miller has obtained during the past few years. In 1925 his paper on this work won the annual prize of the American Association for the Advancement of Science. This continued ability to obtain the same results over a period of years, whether the apparatus is at normal level in Cleveland or on a California mountain, makes Dr. Miller's results all the more important.

Nor does Dr. Miller feel that his experiments repudiate the famous Michelson-Morley experiments on ether drift performed in 1887. Prevalent opinion holds that this historic test showed that there is no ether drift, that there is not something filling all space, and it was upon this interpretation that Professor Albert Einstein based his special theory of relativity when he enunciated it in 1905. But Dr. Miller, studying the results of his latest experiments performed this year on the campus of the Case School of Applied Science, only about 300 feet from the location of the original Michelson-Morley interferometer of 1887, finds that his results showing the solar system moving through space "fully agree with and confirm the original Michelson-Morley observations, although the present interpretation is different."

In the 1887 Michelson-Morley experiment there was discovered a slight difference in the time that it took light to travel over two paths, one at right angles to the other. But this was attributed to experimental errors, to those slight deviations that enter into all observations. Dr. Miller, by performing hundreds of experiments and by improving the details of the ether-drift interferometer, has by his results demonstrated that the observational differences of the original experiments and his many later tests are real and not due to error in the apparatus. Such refinements as shock-absorbing pads on the supporting piers and extreme precautions to eliminate temperature differences were taken in this year's experiments. The interferometer uses the interference of light waves to measure far more accurately than any mechanical means. Dr. Miller's instrument gives numerical results reliable to the hundredth part of a wave-length of light, although the length of the light path is 130,000,000 wave-lengths. He can detect a relative motion of earth and ether a twentieth that which he actually observed.

The discovered motion of six miles a second is not a mere earthly phenomenon, but a cosmic one. It is fixed with relation to sidereal time, that is, it is toward a fixed place in space. The earth and its millions and the whole solar system is rushing, Dr. Miller declares, "toward the point having a right ascension of 17 hours."

How are the scientists to reconcile with their theories this well-tested motion that the ether-drift experiments demonstrate? Dr. Miller says: "It seems impossible at the present time to account for a cosmic effect of this small magnitude and it will be necessary to continue these experiments and to coordinate them with others before an acceptable theory can be propounded."

REPETITION OF THE MICHELSON-MORLEY EXPERIMENT

Anonymous; *Nature*, 128:729, 1931.

Six years ago, Prof. D. C. Miller announced the results of a series of experiments that he made on Mt. Wilson; these seemed to indicate a variation of some 9 km./sec. in the course of the year, which he ascribed to a drift of the solar system in a direction nearly normal to the ecliptic. In the "Encyclopaedia Britannica." (14th edition, vol. 15, p. 418) and described three series of experiments, by Kennedy and Illingworth, by Piccard and Stahel, and by Michelson, Pease, and Pearson, all made between 1927 and 1929, and all giving a zero effect like the original experiment.

The Scientific American for October gives details of a later series of experiments carried out by Prof. G. Joos, of the University of Jena, using apparatus constructed by the Zeiss works. The path of the light in the apparatus is brought up to 70 feet by repeated reflections. The results are recorded photographically and they are stated to preclude any ether drift exceeding one mile per second, so that it may be assumed to be zero. The large cross of the apparatus is constructed of quartz, for the double reason of its low coefficient of expansion and its freedom from magnetic effects. The cross is suspended by 700 springs of piano-wire, so as to support every part equally and prevent torsion. There is thus a great majority of experiments that indicate a zero effect of ether drift.

THE ETHER-DRIFT EXPERIMENT AND THE DETERMINATION OF THE ABSOLUTE MOTION OF THE EARTH

Miller, Dayton C.; *Nature,* 133:162-164, 1934.

The ether-drift experiment first suggested by Maxwell in 1878 and made possible by Michelson's invention of the interferometer in 1881, though suitable for the detection of the general absolute motion of the earth, was actually applied for detecting only the known orbital component of the earth's motion. For the first time, in 1925 and 1926, I made observations at Mount Wilson of such extent and completeness that they were sufficient for the determination of the absolute motion of the earth. These observations involved the making of about 200,000 single readings of the position of the interference fringes.

The ether-drift observable in the interferometer, as is well known, is a second order effect; and the observations correctly define the line in which the absolute motion takes place, but they do not determine whether the motion in this line is positive or negative in direction.

At the Kansas City meeting of the American Association for the Advancement of Science, in December, 1925, before the completion of the Mount Wilson observations, a report was made showing that the experiment gives evidence of a cosmic motion of the solar system, directed towards a northern apex; but the effects of the orbital motion were not found, though it seemed that the observations should have been quite sufficient for this purpose.

The studies of the proper motions and of the motions in the line of sight of the stars in our galaxy have shown that the solar system is moving, with respect to our own cluster, in the general direction of a northern apex in the constellation Hercules. This apex is near that indicated by the ether-drift observations as just reported, and seemed to be confirmatory evidence of its correctness. Probably it was this that caused the continuation of the analysis of the problem, on the supposition that the absolute motion was to the northward in the indicated line. All possible combinations and adjustments failed to reconcile the computed effects of combined orbital and cosmic motions with the observed facts.

In the autumn of 1932, a re-analysis of the problem was made, based upon the alternative possibility that the motion of the solar system is in the cosmic line previously determined, but is in the opposite direction, being directed southward. This gives wholly consistent results, leading for the first time to a definite quantitative determination of the absolute motion of the solar system, and also to a positive detection of the effect of the motion of the earth in its orbit.

The absolute motion of the earth may be presumed to be the resultant of two independent component motions. One of these is the orbital motion around the sun, which is known both as to magnitude and direction. For the purposes of this study, the velocity of the orbital motion is taken as 30 kilometres per second, and the direction changes continuously through the year, at all times being tangential to the orbit. The second component is the cosmical motion of the sun and the solar system. Presumably this is constant in both direction and magnitude, but neither the direction nor magnitude is known; the determination of these quantities is the particular object of this experiment. The rotation of the earth on its axis produces a velocity of less than four tenths of a kilometre per second in the latitude of observation and is negligible so far as the velocity of absolute motion is concerned but this rotation has an important effect upon the apparent direction of the motion and is an essential factor in the solution of the problem. Since the orbital component is continually changing in direction, the general solution is difficult; but by observing the resultant motion when the earth is in different parts of its orbit, a solution by trial is practicable. For this purpose it is necessary to determine the variations in the magnitude and in the direction of the ether-drift effect throughout a period of twenty-four hours and at three or more epochs of the year. The observations made at Mount Wilson correspond to the epochs April 1, August 1 and September 15, 1925, and February 8, 1926.

The point on the celestial sphere towards which the earth is moving because of its absolute motion is called the apex of its motion. This point is defined by its right ascension and declination, as is a star, and the formulae of practical astronomy are directly applicable to its determination from the interferometer observations. The theoretical consideration of the determination of the apex of the motion of the earth has been given in a paper by Prof. J. J. Nassau and Prof. P. M. Morse.

Table 1 gives the right ascensions and declinations of the apexes of the earth's cosmical motion as obtained from the interferometer observations for the four epochs on the presumption of a southward motion, together with the right ascensions and declinations calculated upon the theory of an ether-drift.

Table 1. Location of resultant apexes

Epoch	a (Obs.)		a (Calc.)		(Obs.)		(Calc.)	
Feb. 8	6^h	0^m	5^h	40^m	-77°	$27'$	-78°	$25'$
April 1	3	42	4	0	76	48	77	50
Aug. 1	3	57	4	10	64	47	63	30
Sept. 15	5	5	5	0	62	4	62	15

Apex of cosmic component $a = 4^h\ 56^m$, $\delta = -70^\circ\ 33'$

From these resultant apexes are determined four values for the apex of the cosmic component, which is the apex of the motion of the solar system as a whole. This apex has the right ascension $4^h\ 56^m$ and the declination $70^\circ\ 33'$ south.

Continuing the astronomical description, having found the elements of the 'aberration orbit', these are used to compute the apparent places of the resultant apexes for the four epochs of observation. On the accompanying chart of the south circum-polar region of the celestial sphere (Fig. 1), the large star indicates the apex of the cosmic motion, and the four circles show the locations of the calculated apexes. These apexes necessarily lie on the closed curve representing the calculated aberration orbit, the centre of which is the apex of the cosmic component of the earth's motion. This aberration orbit is the projection of the earth's orbit on the celestial sphere, which in this case is approximately a circle. The observed apexes for the four epochs are represented by the small stars. The locations of the pole of the ecliptic and of the star Canopus are also shown. The close agreement between the calculated and observed apparent apexes would seem to be conclusive evidence of the validity of the solution of the ether-drift observations for the absolute motion of the earth and also for the effect of the orbital motion of the earth, which hitherto has not been demonstrated.

It may seem surprising that such close agreement between observed and calculated places can be obtained from observations of such minute effects, and effects which are reputed to be of such difficulty and uncertainty. Perhaps an explanation is the fact that the star representing the final result for the February epoch is, in effect, the average of 8,080 single determinations of its location; the star for the August epoch represents 7,680 single determinations, that for September, 6,640, and that for April, 3,208 determinations.

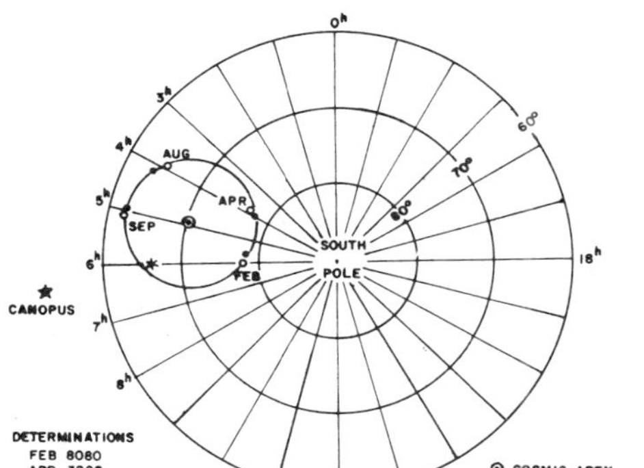

Calculated and observed apexes of solar-system absolute motion. (Fig. 1)

The location of the apex of the solar motion is in the southern constellation Dorado, the Sword-Fish, and is about 20° south of the star Canopus, the second brightest star in the heavens. It is in the midst of the famous Great Magellanic Cloud of stars. The apex is about 7° from the pole of the ecliptic and only 6° from the pole of the invariable plane of the solar system; thus the indicated motion of the solar system is almost perpendicular to the invariable plane. This suggests that the solar system might be thought of as a dynamic disc which is being pulled through a resisting medium and therefore sets itself perpendicular to the line of motion.

It is presumed that the earth's motion in space is projected on to the plane of the interferometer, and the direction of this motion is determined by observing the variations produced in the projected component by the rotation of the earth on its axis and by the revolution around the sun. Both the magnitude and the direction of the observed effect vary in the manner and in the proportion required by an ether-drift, on the assumption of a stagnant ether which is undisturbed by the motion of the earth through it. But the observed magnitude of the effect has always been less than was to be expected, indicating a reduced velocity of relative motion, as though the ether through which the interferometer is being carried by the earth's motion were not absolutely at rest. The orbital velocity of the earth being known, 30 kilometres per second, the cosmical velocity of the solar system, determined from the proportional variations in the observed effects, is found to be 208 kilometres per second.

Table II gives the observed periodic displacement of the fringe system as the interferometer rotates on its axis, and the corresponding velocity of relative motion of the earth and ether.

Table 2. Displacements and velocities

Epoch	Fringe Shift	Velocity (Obs.)	Velocity (Calc.)
Feb. 8	0.104 1	9.3 km./sec.	195.2 km./sec.
April 1	0.123	10.1	198.2
Aug. 1	0.152	11.2	211.5
Sept. 15	0.110	9.6	207.5

The last column gives the velocity to be expected in the stagnant ether theory on the presumption that the cosmic component and the orbital component are both reduced in the same proportion in the interferometer. The mean factor of reduction is $k = 0.0514$. The azimuth of the observed effect is subject to a diurnal variation, produced by the rotation of the earth on its axis. The observed oscillations of the azimuth are in accordance with theory as to magnitude and time of occurrence, but for some unexplained reason, the axis of the oscillations is displaced from the meridian. In order to account for the results here presented, it seems necessary to accept the reality of a modified Lorentz-FitzGerald contraction, or to postulate a viscous or dragged ether as proposed by Stokes.

The results here reported are, notwithstanding a common belief to the contrary, fully in accordance with the original observations of Michelson and Morley of 1887, and with those of Morley and Miller of 1904-5. The history of the ether-drift experiment and a description of the method of using the interferometer, together with a full account of the observations and their reduction has been published elsewhere.

GARDNER'S HYPOTHESIS AND THE MICHELSON-MORLEY EXPERIMENT

Synge, J. L.; *Nature,* 170:243-244, 1952.

When Gardner's hypothesis (see preceding communication) is applied to the Michelson-Morley experiment, performed with an interferometer with horizontal rigid arms, carried on the rotating earth, it yields the following formula for fringe-shifts if we take the drag-point to be the centre of the earth:

$$\delta/w = (D/\lambda)\,(E/c)\,\phi\,\sin 2\theta.$$

Here δ is the fringe-shift, $\underline{w}$ the fringe-width, $\underline{D}$ the arm-length of the interferometer, λ the wave-length used, $\underline{E}$ the velocity of the earth's surface due to the earth's rotation alone, $\underline{c}$ the velocity of light, ϕ the deviation of the plumb line, and θ the azimuth of the telescope arm, from the north eastward. If, on the other hand, the drag-point is near the interferometer, the predicted fringe-shift is below the limits of observation.

In a set of experiments involving a very great number of observations, D. C. Miller observed fringe-shifts resembling those predicted by the above formula.

. . . .

The theoretical amplitude is about four times that observed by Miller; but the variation as sin 2θ is like that which Miller found, for the harmonic analysis of his results gave a strongly predominant term with period 180°, and the azimuths giving maximum shift showed a fairly strong preference for the northeasterly quadrant; indeed, his observations of June 1, 1925, give an average azimuth of 42°, to be compared with the value of 45° given by the above formula. A small shift with period 360°, on the existence of which Miller insisted, can be explained by a slight inclination of the arms to the horizontal, which yields terms in sin θ and cos θ, omitted from the above formula for the sake of simplicity.

The positive results obtained by Miller have long remained an anomaly because experiments by Joos, Kennedy, Illingworth and Michelson (with collaborators) gave essentially null results. I suggest as a possibility that the discrepancy between these results and those of Miller may be found in the mountings of the interferometers. Joos supported his instrument on springs and so insulated it mechanically from the earth; the others used flotation on mercury, as did Miller, but Miller used a substantial pin attached to the interferometer and passing loosely through a bearing. This pin is shown by Miller and I am much indebted to Prof. R. S. Shankland, of the Case Institute of Technology, for further information obtained from the mechanic who worked with Miller.

Although the pin was far from attaching the interferometer rigidly to the earth, it may have provided sufficient constraint to make the interferometer behave, at least partially, as though it were part of the earth itself, and so give us an excuse for taking the earth's centre as drag-point (the most natural point, surely, if we were dealing with the earth itself as a rigid body).

However that may be, the concept of rigid-body motions is important in physics and an adequate theory of these motions is most desirable. It is true that we have the Born-Herglotz theory; but I have been unable to see how that theory leads to any definite prediction regarding the Michelson-Morley experiment performed on the rotating earth. Now that the special theory of relativity is firmly established, we should look on experiments like the Michelson-Morley experiment as experiments on rigid bodies rather than on light propagation. But as a test of Gardner's hypothesis, an interferometer with horizontal arms is very inefficient. When the arms are tilted up, the theory predicts quite a large effect with period 360° in azimuth. Indeed, with one arm inclined to the vertical at 45° and the other horizontal, Gardner's hypothesis predicts a fringe-shift of a full fringe-width with an arm-length of only 20 cm., provided, of course, that the interferometer is firmly mounted on a vertical axis to ensure that it may be regarded as part of the rotating earth and so have the earth's centre as drag-point.

....

AETHER DRIFT DETECTED AT LAST

Rowan-Robinson, Michael; *Nature,* 270:9–10, 1977.

A group of experimenters from Berkeley (Smoot, Gorenstein & Muller, Phys. Rev. Lett. 39, 898), using a U-2 aircraft in the unfamiliar mode of looking upwards, seem to have decisively detected the motion of the Earth with respect to the cosmic microwave background. This radiation, with the spectrum of a 2.7 K blackbody, is believed to be the relic of the 'fireball' (radiation-dominated, optically thick) phase of the Universe. The isotropy of the microwave background to one part in 10^3 has been the strongest argument in favour of the homogeneous, isotropic models of the Universe dreamed up by Einstein, de Sitter, Friedmann, Eddington and Lemaitre in the 1920s. But it has long been expected that a simple dipole anisotropy of order 10^{-4} to 10^{-3} should be found, due to the random motions that galaxies have with respect to each other and to the cosmological frame of reference (the frame in which the expansion of the Universe looks exactly isotropic, according to the models). The radiation should look slightly hotter in the direction we are travelling towards, and slightly colder in the direction we are travelling from, by an amount $\Delta T/T \simeq v/c$, due to the Doppler shift. Failure to detect this effect would put us in the uncomfortable position of happening to be exactly at rest with respect to the cosmological frame. It was Michelson and Morley's failure a century ago to detect a contribution from the motion of the Earth to the local velocity of light that led to the downfall of aether theories and ultimately ushered in the relativistic age. A completely isotropic microwave background might have been equally revolutionary, but the Berkeley workers seem to have saved us from this fate. However the magnitude of the velocity deduced for the Milky Way, 600 km s^{-1}, is so large as to throw existing ideas about our cosmic environment into disarray.

....

What are we to make of this? The authors note that the velocity they have found conflicts with various attempts to measure our velocity with respect to nearby galaxies, but offer no explanation of this. With respect to the Local Group of galaxies, the motion of the Solar System hardly differs from that expected due to our circular motion round the Galaxy. This suggests that the whole Local Group has to be moving along together at this velocity of 600 km s^{-1} with respect to the microwave background. And this velocity is more than ten times the residual random motion of galaxies within 20 Mpc about the Hubble flow, so that most nearby galaxies, including the Virgo cluster of galaxies, would seem to have to move along together at this velocity. The Universe may be much more inhomogeneous than we have realised till now, and we may have to be careful about interpreting the expansion time-scale we measure locally as the age of the Universe.

The Berkeley measurement also conflicts with the velocity of the Earth with respect to more distant galaxies (at several hundred Mpc) determined by Rubin et al. (Astr. J. 81, 687; 1976). They also found a velocity of about 600 km s^{-1}, but in a direction almost at right angles to the velocity with respect to the background. This 'Rubin-Ford effect' may not seem so surprising now. The matter in the Universe may be divided up into quite large 'metaclusters' or 'superclusters' in rapid motion relative to each other and to the cosmological frame.

It is unlikely that this new result will dent the hot big bang picture of the Universe very much but our ideas about our cosmological locality are in for some revision. The Berkeley group have made an important experiment, which is clearly worth repeating. A good test would be to do a similar series of flights in the southern hemisphere but perhaps U-2's are still not all that welcome outside US airspace.

• Einstein-Displacement Measurements

OBJECTIONS TO THE THEORY OF RELATIVITY

Anonymous; *British Astronomical Association, Journal,* 33:81, 1922.

Prof. Pickering considers the three tests offered as confirmation of Einstein's theory and concludes that the results are very unsatisfactory. The rotation of the Sun on its axis in 26 days should cause an excess of the equatorial over the polar diameter of 0''.04, an amount which should produce an advance of the perihelion of the orbit of Mercury of about 3''.05 a century. This reduces the difference between the observed advance and that due to gravitation to 38''.05, a discrepancy of 4''.05 between this figure and Einstein's result. He also quotes Dr. Poor's view, stated in the Scientific American Monthly, June 1921, that the whole of the advance may be accounted for by the assumption of a suitable and not improbable density of matter within the orbit of Mercury. The conclusions drawn from the 1919 eclipse are, Prof. Pickering thinks, very unsatisfactory, because of the rejection of the 19 images taken with the 13-inch telescope at Sobral. If these were accepted the result would show a deviation of 0''.06 from the theoretical Newtonian result, and is, therefore, much better than that obtained by either of the other telescopes. The third test---the displacement of the lines in the solar spectrum---is distinctly unfavourable for the theory, and on the whole the evidence in its favour is by no means convincing.

EINSTEIN DISPLACEMENT ON THE PLATES TAKEN BY THE CANADIAN PARTY AT THE AUSTRALIAN ECLIPSE

Chant, C. A.; *Science*, 57:469, 1923.

At the eclipse of the sun of September last two plates were taken to test for the Einstein effect, by the Canadian party at Wallal, western Australia. The focal length of the camera was eleven feet and the aperture of the lens was six inches. Dr. R. K. Young, of the Dominion Astrophysical Observatory, Victoria, B. C., who was a member of the party, has completed the measurement and computation of the plates, and the final results are given below. Over thirty stars were recognized on the plates and twenty-three were submitted to measurement, but eight of these were discarded in the course of the work, thus leaving fifteen. The measured displacement outward from the sun, as well as the expected amount according to Einstein's theory, are given in seconds of arc:

Measured 0.30, 0.44, 0.28, 0.25, 0.66, 0.22, -0.31, 0.12, -0.11, 0.23, 0.08, 0.06, 0.53, 0.77, -0.05.

Expected 0.48, 0.41, 0.40, 0.30, 0.28, 0.27, 0.24, 0.24, 0.24, 0.22, 0.22, 0.21, 0.21, 0.21, 0.18.

It will be seen that the displacement is undoubted and its amount is approximately that predicted by Einstein, but the results can hardly be considered decisive.

THE DEFLECTION OF LIGHT IN THE SUN'S GRAVITATIONAL FIELD

Trumpler, Robert J.; *Astronomical Society of the Pacific, Publications,* 44:167-173, 1932.

Among the various expeditions sent out to observe the total solar eclipse of May 9, 1929, that of the Potsdam Observatory (Einstein Stiftung) seems to be the only one which obtained photographs suitable for determining the light deflection in the Sun's gravitational field. Two instruments were used, but so far only the results of the larger one, a

28-foot horizontal camera combined with a coelostat, have been published. The three observers, Freundlich, von Kluber, and von Brunn, claim that these observations (four plates containing from seventeen to eighteen star images each) lead to a value of 2''.24 for the deflection of a light ray grazing the Sun's edge; a figure that deviates considerably from the results of the 1922 eclipse, and which is in contradiction to Einstein's generalized theory of relativity. In view of the importance and general interest attached to this problem, a few critical remarks on the Potsdam observations and their reduction should not be out of place here:

1. The accuracy claimed for the Potsdam result is somewhat illusory. A more liberal discussion of the residuals would lead to a probable error of at least ± 0''.10 (mean error ± 0''.15), 50 per cent larger than that given by the observers; but even this does not fully take into account the uncertainty in the adopted scale difference of the plates.

2. The star field of the 1929 eclipse was unfavorable, for of the eighteen stars bright enough to be photographed on the Potsdam plates, seventeen were located on one side of the Sun and only one star on the other side. That is, we may draw a straight line through the center of the Sun such that a single star is on one side of the line and the other seventeen stars are on the other side. This extremely unsymmetrical star distribution has the consequence that the light deflections to be determined depend to a high degree on the plate constants used in the reduction; and five such constants for each plate had to be derived from the star measures.

3. The Potsdam reduction is based on the assumption (in accord with the Einstein theory) that the light deflection is inversely proportional to the star's angular distance from the Sun's center. However, the deflections determined by the Potsdam observers are not in accord with this Einsteinian requirement. When the stars are arranged according to their angular distances from the Sun's center, a systematic run in the radial components of the residuals is apparent. Among the six stars nearest to the Sun, five have negative residuals, while of the six most distant ones five have positive residuals. This leads to the conclusion, either that the Potsdam observations are affected by a systematic error, or that the assumption on which the reduction is based is incorrect. Ludendorff has investigated the question of a systematic measuring error; the evidence presented in the following indicates that the fault lies in the scale determination.

4. In our problem the most important correction to be applied to the differential measures of an eclipse photograph and a photograph of the same (eclipse) group of stars obtained at night for comparison purposes several months before or after the eclipse date is the scale difference between the two plates. The Potsdam observers tried to determine this scale difference independently of the star observations. By means of a collimating telescope, a reseau of fine lines was copied on each of the eclipse plates as well as on the comparison plates, and the observers assumed that the angle corresponding to a reseau interval remained unchanged during the period of from five to seven months which elapsed between the eclipse and the comparison observations. This assumption, however, has no sound foundation. The temperature at the time of the comparison observations (secured at night) was on the average 5°C. lower than during the eclipse, and the tube of the collimating telescope,

and in consequence the intervals between the reseau lines recorded on the eclipse and comparison photographs, must have suffered changes due to temperature changes. In fact, the probable temperature effect neglected by the Potsdam observers would be of the right sign and order of magnitude to account for the excess of the observed light deflection over the requirements of the Einstein theory.

5. Originally the Potsdam expedition in its plans had foreseen the possibility of a change in the collimator-reseau combination. A duplicate horizontal camera was utilized to photograph, at the time of the eclipse, another star field far to one side of the eclipsed Sun, and therefore not subject to appreciable Einsteinian deflections; the photographs thus secured, when compared with photographs of this star field obtained at night, were to be used in checking or measuring the effect of such a change. Unfortunately the optical qualities of this second telescope were somewhat defective (perhaps because of strains in the objective at the time of the comparison observations), so that photographs of the reseau's straight lines show optical distortion. The observations of the second telescope can therefore not serve their purpose, and without this check the Potsdam scale determination deserves no confidence and should be rejected.

....

THE EINSTEIN SHIFT—AN UNSETTLED PROBLEM

Schmeidler, F.; *Sky and Telescope*, 27:217-219, 1964.

One crucial experimental test of the theory of relativity is the deflection of light passing near the limb of the sun. Beginning in 1919, attempts have been made at several total eclipses to measure this deflection, usually called the Einstein shift because its existence was predicted by Albert Einstein shortly after 1900. Many astronomers have undertaken eclipse expeditions to measure the Einstein shift, including F. Dyson, A. S. Eddington, W. W. Campbell, R. J. Trumpler, E. Finlay-Freundlich, H. von Kluber, A. Mikhailov, G. Van Biesbroeck, and the present author. In spite of the great labor spent on the problem, to this day no definitive confirmation of relativity theory has been obtained in this way.

Curiously, the idea that light rays may be deflected by solar gravitation had been examined a century before Einstein. The 1804 volume of the Astronomisches Jahrbuch, edited by J. E. Bode in Berlin, contains an article by Johann Soldner (1777-1833), who later became director of Munich Observatory. He pointed out that light should be influenced by gravitation if light consists of material particles emitted by a source. When these corpuscles come near to a center of gravitation, they must move around that center in an orbit which, owing to the high velocity of light, will be hyperbolic. Consequently, the direction of the light ray will be altered. According to Soldner, a light ray that just grazes the sun should be deflected 0.875 second of arc.

Soldner's paper was forgotten, for about 20 years later A. Fresnel's investigations convinced astronomers and physicists that light consisted of waves rather than particles, and there seemed to be no reason why waves should be influenced by gravitation. Long afterward, when Einstein reexamined the matter in connection with relativity theory, he obviously did not know about Soldner's paper.

According to the theory of relativity, light rays move along geodesic lines in the four-dimensional continuum. These lines are curved in the neighborhood of masses, with consequences similar to those phenomena that are called gravitational in classical theory. Therefore, light rays must move in a curved line when passing near a gravitating body. The deflection of a light ray just grazing the sun was predicted by Einstein's theory to be 1.75 seconds of arc---twice Soldner's value.

At present, total solar eclipses are the only occasions when stars near the sun become visible to us, permitting measurement of the Einstein shift. Of course we cannot photograph stars exactly at the edge of the solar disk, but the deflection for a star at distance r from the sun's center will be 1.75/r seconds, if we express r in units of angular radius of the solar disk. It is easy to see that the amount of this deflection becomes observationally quite inappreciable at distances of more than, say, six radii from the sun's center.

In principle, the deflection is measured by comparing two photographs of the same star field, one taken during the eclipse, the other some months later (or possibly earlier), when the star field is visible at night. On the eclipse plate the stars are shifted, on the night plate they are not. Differences in the measured coordinates of stars on the two photographs should then reveal the existence and numerical amount of the Einstein shift.

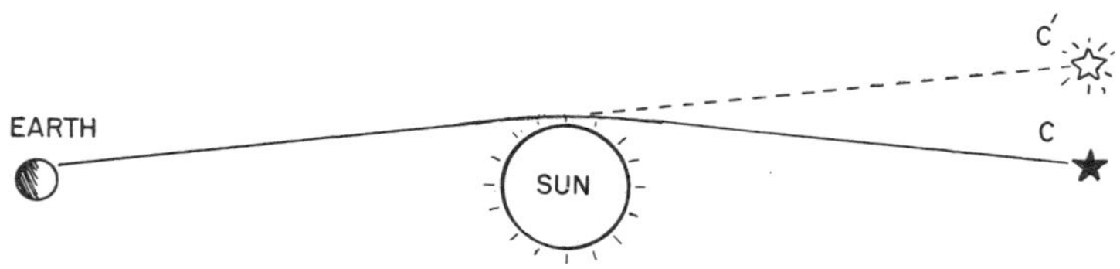

How a star's image is shifted by the sun's gravitational field

In practice, these measurements are very difficult and uncertain, since eclipse observations are made under unusual circumstances. The telescope for photographing the star field must be mounted somewhere within the path of totality, under field conditions rather than at an observatory. Air temperature and humidity change very quickly during an eclipse, and scattered skylight usually produces a general darkening of the plate, blacking out faint stars. For these reasons, the pictures taken during totality never have the same quality as the night exposures.

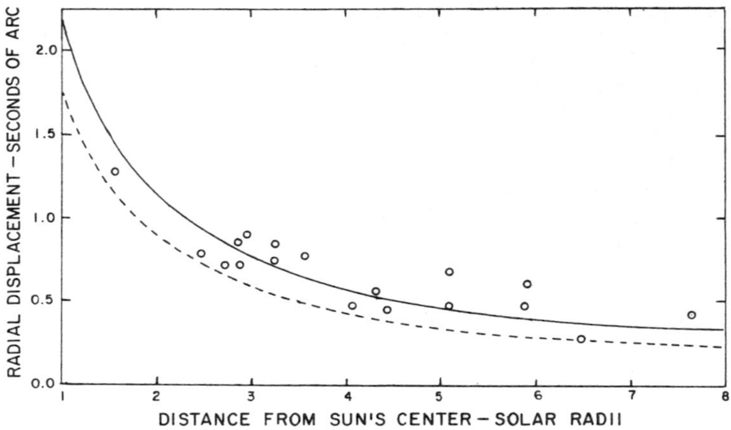

Einstein Displacement of stars as a function of distance from the sun. Circles represent actual measurements. Dashed line is the theoretical displacement.

After the eclipse, the telescope is normally left untouched at the expedition site until a few months later, when the night plates are taken. Theoretically, the night plates can be exposed at some other site, but it has been found advantageous to make them with the telescope in the same situation as during the eclipse.

Working along these lines, several astronomers have tried to evaluate the Einstein shift. All of them found that such a shift exists, but its exact amount is still uncertain. The first successful expeditions were to Brazil in 1919 from Greenwich Observatory (Dyson, Eddington, and M. Davidson) and to Australia in 1922 from Lick Observatory (Campbell and Trumpler). Both expeditions' results seemed to provide good confirmation of the theoretical value, 1.75 seconds, for the deflection at the sun's limb.

However, an expedition sent to Sumatra in 1929 by Potsdam Observatory (Finlay-Freundlich, von Kluber, and A. von Brunn) obtained a much larger result, 2.24 seconds. The small mean error (± 0.10 second) of this determination strongly suggested a real discrepancy between theory and observation.

Further examination of the 1919 and 1922 data---originally interpreted as confirming the theory of relativity---tended to favor the larger shift found by Finlay-Freundlich. The results depended very strongly on the manner of reducing the measurements; for example, small uncertainties in focal length and other instrumental quantities, or optical disturbances, could make large changes in the final answer. Omitting individual stars with particularly poor plate images sometimes altered the result greatly. Several astronomers painstakingly rediscussed the measurements published by earlier expeditions, and in almost every case plausible hypotheses about the uncertain effects made the value of the light deflection larger than originally reported.

At later eclipses, some astronomers found shifts tending to agree with relativity theory, others with Finlay-Freundlich's value. A Russian expedition under Mikhailov at the 1936 eclipse in Siberia even found 2.73 seconds---more than half again as much as predicted! On the other hand, Van Biesbroeck made two expeditions that yielded 2.01 in 1947 (South America) and 1.70 in 1952 (Africa). A new reduction by Mikhailov of Van Biesbroeck's observations, using another method, gave 2.20 and 1.43 respectively.

The present author obtained plates of star fields surrounding the sun at two eclipses, on October 2, 1959, at Kidal in the southern Sahara, and on February 15, 1961, at Monte Conero, near Ancona, Italy. The 1959 plate yielded 2.17 seconds for the deflection of light at the sun's limb. While the 1961 photograph is still unreduced, no very good result can be expected, since the sun was only 15 degrees above the horizon during the eclipse.

....

• Red-Shift Experiments

ANOMALOUS REDSHIFTS EXPLAINED?

Anonymous; *Nature*, 233:371, 1971.

The article by Ferencz and Tarcsai on page 404 of this issue of Nature will be a relief to orthodox workers in general relativity because it provides an explanation, simple in principle, for certain puzzling phenomena for which very speculative explanations have hitherto had to be constructed.

Three years ago, Sadeh, Knowles and Yaplee reported Science, 159, 307; 1968) an anomalous redshift in the 21 cm radiation from Taurus A as it passed near the Sun. The extent of the shift was 150 Hz---about one part in 10^7---but it was too great, they claimed, to be explained either by the interaction of the signal with the Sun's atmosphere, or by any effect attributable to the curvature of space time around the Sun according to the equations of general relativity and similar theories of gravitation. For two reasons this anomaly was either worrying or exciting. First, relativistic theories of gravitation are so elegant and illuminating that theorists would be more loath to part with them than with other esoteric theories. Second, the two sources of redshift in well-understood situations, recessional velocity of the source and gravitational fields, provide the best means of measuring masses and velocities of very distant objects. The first mentioned is the best reason for believing that the universe expands. It is suspected that other causes of redshift exist because it is difficult to explain the huge redshifts of quasi-stellar objects; but to find inexplicable redshifts near to the Sun was most perplexing.

Later in 1968, Sadeh, Knowles and Au reported (Science, 161, 567; 1968) a repetition of this experiment and also another phenomenon of similar type. In this experiment a caesium clock was carried about on the Earth's surface and exhibited a decrease in frequency proportional to the distance moved. Again, a timing system, this time a clock instead of a photon, was showing an irreversible decrease in frequency on being moved through a gravitational field. If such a phenomenon were general it would cause the frequency of photons to diminish as they moved through space and would account for 10 per cent of the observed redshift.

Doubts were cast on the validity of this experiment by Markowitz (Science, 162, 1387; 1968). Sadeh and Knowles, in collaboration with Hollinger and Youmans, had also studied the most precise clock in the sky, the pulsar CP 0950, as it was eclipsed by the Sun (Science, 162, 897; 1968). By now a diminution in the frequency of its flashes was expected as its light grazed the solar limb, but none was observed.

At least three radical theories were invoked to explain the measurements. Sadeh, Knowles and Yaplee in their first article had put forward the hypothesis that photons might resemble electrons in losing the energy that they gain by falling into gravitational fields but requiring energy to climb out of them. Such a picture exaggerates the effect of the first experiment by a factor which they calculate to be about 15, explains the small result of the third but provides no clue to the disputed second.

Szekeres explained (Nature, 220, 1116; 1968) the results by a unified field theory. Theories of this type ascribe the electromagnetic field as well as the gravitational field to curvature in the space-time continuum. Szekeres pointed out that if the electromagnetic field was simply proportional to the difference between the "connexions" (geometrical quantities describing displacements between parallel lines)---one for world structure, the other for the actual displacements of light waves---then frequency changes of the type observed in the first and the second experiments could be ascribed to travel through the magnetic fields of the Sun and of the Earth. Laboratory experiments reported later by Shamir (Nature, 222, 362; 1969) did not support this hypothesis, however,

A third theory was that of Woodward and Yourgrau, who pointed out last year (Nature, 226, 619; 1970) that the crucial point is whether or not the relative frequency shifts due to the gravitational doppler effect is frequency dependent. A comparison between the numerical magnitudes of the results of the first two experiments suggests that it is. If this is so, however, it contradicts general relativity, according to which acceleration cannot normally be distinguished from gravity. Accordingly, although Woodward and Yourgrau's model of a frequency dependent interaction fitted all the experimental data, it gave rise to much controversy.

All would be explained if the first experiment was understandable. The second experiment has not been successfully repeated and the third gave a negative result. Even the first is partially contradicted by the fact that experiments with radar pulses grazing the Sun as they echo from the Earth to the planet Mercury give results in accordance with relativity.

Exactly such an understanding is claimed in the article by Ferencz and Tarcsai in this week's Nature. The authors cite detailed calculations which for the first time take into account the fact that the Sun's plasma is not at rest nor homogeneous. They account for shifts in frequency well within the limits of experimental error of the first experiment. General relativity is thus vindicated.

But Ferencz and Tarcsai have not simply laid a ghost. Their technique of calculating redshifts in inhomogeneous plasmas may explain in the future the large redshifts of quasi-stellar objects. The original statement about the plasma is beginning to look like a fruitful error.

PHYSICAL LAWS AND CONSTANTS

Science is fundamentally a search for order in the cosmos, preferably order expressed in terms of a few immutable physical laws. If laws change or if the constants in the mathematical manifestations of these laws vary, science is built upon quicksand.

Relativity has already been questioned in the preceding section, but it is not as sacrosanct as the Law of Gravitation, which affects us all in everyday life. Yet, even Newton's synthesis has been questioned by experimenters. Does it hold over very long and very short distances? Is it exactly an inverse-square law? Does the constant of gravitation vary with time?

The velocity of light is supposed to be constant. Indeed, its constancy was assumed by Einstein when he formulated his Special Theory of Relativity. Historical records, however, seem to show that light has in reality been slowing down with time.

Whether these quicksands exist is controversial. If they are real, what is at the bottom, if anything? The whole problem is made more disturbing by the apparent relationships among some of the physical constants themselves. If real, these supercosmic connections would seem to tie the universe together with some sort of mysterious glue.

• Validity of the Law of Gravitation

SOME NEW EXPERIMENTS IN GRAVITATION

Brush, Charles F.; *American Philosophical Society, Proceedings,* 63: 57-61, 1924.

The writer outlined "A Kinetic Theory of Gravitation" before the American Association For The Advancement of Science in December, 1910. Since that time he has presented several papers on the subject of gravitation, the last one at the general meeting of this Society in April, 1923.

That paper was chiefly descriptive of apparatus and method for comparing the velocities of freely falling bodies in two aluminum containers, alike in size, shape, and smoothness of surface, and dropped simultaneously, side by side, through exactly the same distance (about 122 cm.). The description was freely illustrated with plates and diagrams.

Each container, at the end of its journey, breaks an electric circuit. But the breaks of both containers are in series in the same circuit, so that the break which occurs first produces a bright spark, while the belated break gives no spark because its circuit is already open.

When the containers are equally loaded with the same metal, there is no visible spark at either break or a very feeble spark at one or the other indifferently. But when they are equally loaded with certain different metals, one container persistently produces a bright spark, though the containers are always reversed in position for each trial. From this it seems clear that the container giving the spark falls a little faster than the other. This sparking condition is clearly manifested when the faster container reaches the end of its free path as little as .0125 mm. (.0005 inch) in advance of its neighbor. This indicates a time difference about 1/400,000 second. (During their half second of falling the containers acquire a velocity about 4900 cm. per second).

Some of the metals show very many times greater difference in falling velocities than the near minimum indicated above (about 1/100,000) and the paper referred to describes a method of approximate measurement of the lag of the slower container.

Following the presentation of the last paper referred to, many comparisons of metals and other substances were made with every precaution that could be thought of to avoid error, and in the light of much experience with the apparatus. The general results are embodied in the "Table of Relative Falling Velocities" herewith.

Table of Relative Falling Velocities

Fast.	Platinum.
	Gold coin, Tungsten forged, Lead.
	Bismuth.
	Zinc, Tin.

Aluminum, Magnesium, Carbon, Silicon, Boron, Water.

<u>Slow</u>. Sulphur, Selenium.

The spacing of the lines indicates approximately the relative velocity differences found. Differences, if any, between substances arranged in the same line were too small to be appreciable.

The observed difference in the falling velocities of platinum and sulphur (lag of the sulphur) is approximately .02 cm. in the 122 cm. of fall; or about 1 part in 6000.

The quantitative value of zinc in the table is somewhat uncertain, for reasons indicated in former papers, and later in the present paper.

It seems significant that the fast-falling metals, platinum, gold, tungsten and lead all have high atomic weight, and density. Lead, though less dense than the others, has higher atomic weight. Yet bismuth, with virtually the same atomic weight as lead, and little less density, is considerably slower.

Much lower down in the scale of velocities we find the elements of comparatively very low atomic weights and small densities, aluminum, magnesium, carbon, silicon, boron (and water) with no appreciable difference in behavior. This also seems significant.

Sulphur and selenium are in a class by themselves, being much slower than the light elements last referred to. Much care was taken to make sure of this. They were always melted in their containers, and each was checked repeatedly, and very satisfactorily, with aluminum and carbon.

It was rather expected that tellurium (also melted in its container) would, on account of its chemical similarity, behave much like sulphur and selenium. But it did nothing of the sort. On the contrary it behaved very much like zinc and tin. Probably this was due to its comparatively high atomic weight and density. This again seems significant.

In general then, it appears that the metals of very high atomic weight and great density have the highest falling velocities. And elements of very low atomic weight and small density have the lowest falling velocities; while elements of intermediate atomic weights and densities (zinc, tin and tellurium) have intermediate falling velocities or mass-weight ratios.

Allotropic condition seems not to affect mass-weight ratio. Thus, Acheson graphite and electric light carbon showed no difference in falling velocities. And the same was true of ordinary yellow sulphur and its gutta percha-like modification obtained by melting at a much higher temperature; though much heat energy must have been absorbed and rendered latent during the change.

In a former paper it was shown that "fusible alloy" behaved the same as its constituent metals mixed but not alloyed.

Only one experiment has been made involving chemical combination. Lead sulphide (galena) was found a little slower than metallic lead. As nearly as could be estimated, it behaved like a mixture of lead and sulphur in the proper proportions to form the sulphide. Apparently, in this case, chemical combination does not affect mass-weight ratio. The writer thinks this is probably true as a general proposition, though of course a single isolated experiment is but a small step toward establishing it.

In the hope that some close relationship might be found between the observed falling-velocities of the elements tested, and their physical or chemical properties or combinations of properties, a table of all the common elements was prepared, arranged in the order of their densities. In parallel columns were placed the corresponding atomic weights, atomic numbers, specific heats and thermal conductivities both in terms of weight and volume. But a careful study of this Table failed to shed any new light on the subject.

It was thought that something of value might be learned by comparing the transparency (or opacity) to very hard X-rays of equal weight-thicknesses of some elements. To this end discs or cylinders of 24 elements, including all those shown in the Table except boron, were prepared of thicknesses inversely proportional to their densities; so that the X-rays must pass through equal weights of the several elements before reaching the photographic plate below them. Thus the platinum disc was only .117 cm. thick, while the magnesium cylinder was 1.439 cm. high. All were mounted in a circle of holes in an aluminum plate or carrier, placed directly over the photographic plate with only the usual orange and black papers intervening. The photographic plate was backed by a thick lead plate, and the whole outfit was revolved in a horizontal plane by clockwork, around the center of the circle of specimens; thus securing equal exposure to all. Several revolutions were made during an exposure.

The hard X-rays were furnished by a Coolidge tube with tungsten target, and excited by 200 Kv. The target was placed 1 meter above the photographic plate, and the rays were filtered through a 3/4 mm. copper plate and an aluminum plate. 15 or 20 seconds exposure was found suitable.

Prints from the negatives were made through a wide range of exposures under a tungsten lamp, while the negatives were revolved as before, to secure uniform exposure.

The prints are beautiful and extremely interesting in several ways; but while they show strong contrast between the platinum and aluminum groups, they do not disclose anything more specific.

Toward the close of last year's paper the writer cited reasons for believing that the mass-weight ratios of some metals, particularly bismuth and zinc, are not constant, but vary slightly with physical condition. In other words, that a constant mass, or quantity of metal may be appreciably changed in weight by changing its physical condition.

The writer has been earnestly at work on this problem for six months or more, and occasionally long prior to that, principally endeavoring to find a practicable method of attack; and has so far succeeded as to leave very little room for doubt that actual weight changes are being effected. The observed effects, though small, are several times larger than the probable experimental error.

The best method of procedure thus far found consists in certain heat-treatments of the metal, to bring about a definite molecular condition, before the first weighing. This is essential. After weighing, the metal is subjected to more or less drastic mechanical treatment, usually in several stages, with occasional periods of rest to permit the spontaneous changes which sometimes occur, and re-weighed after each stage of treatment. The treatments are so conducted as to preclude any loss or gain of substance.

The refinements of weighing and precautions against error, developed through many years of experience in refined weighing, will not be detailed here but reserved for a future paper, after this fascinating subject of weight-change has been more fully exploited.

SHOULD THE LAWS OF GRAVITATION BE RECONSIDERED?

Allais, Maurice F. C.; *Aerospace Engineering*, 18:46-52, September, and 18:51-55, October 1959.

Abstract. The motions of a pendulum, suspended on a ball and resting on an anisotropic support, have statistically significant amplitude and periodic components of periods approaching 24 and 25 hours.

The installation and the experimental technique are briefly described. The observed motions result from four conjugate effects: the Foucault effect, an effect of the suspension release, the aleatory influence of balls, and, finally, a periodic influence.

The observed periodic structure cannot be considered as due to the disturbances of an aleatory order. Neither can it be considered as produced by an indirect influence of known factors (temperature, pressure, magnetism, etc.). Finally, it cannot be identified with periodic lunisolar effects resulting from the actual theory of gravitation.

A remarkable disturbance has also been observed at the time of the total solar eclipse---June 30, 1954. [See also pp. 655-658.]

At this stage of the discussion, the observed effects must be considered as produced by the direct action of a new field.

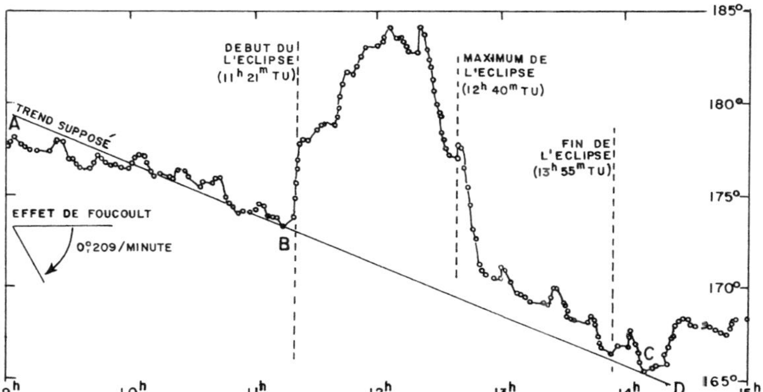

Effect of a solar eclipse on a pendulum

ON THE VALIDITY OF NEWTON'S LAW AT A LONG DISTANCE

Finzi, Arrigo; *Royal Astronomical Society, Monthly Notices,* 127:21-30, 1963.

1. Introduction.---It has been known for a long time that in the great majority of clusters of galaxies the relative velocities of the member galaxies are very large and do not seem, at first, to be compatible with the stability of the clusters. As the reality of this phenomenon seems to have been established beyond doubt, we are confronted with a problem to which no really satisfactory solution has been found so far within the framework of the accepted dynamical laws.

....

It should, perhaps, he permissible to consider the possibility that Newton's law may not be valid for distances of this order, as the main experimental confirmation of the law comes from the study of the solar system, where distances smaller by some eight orders of magnitude are involved. Admittedly, we should then be forced to conclude that general relativity also does not apply when distances of the order of 1 kpc are involved. In fact, for the relatively weak fields and low velocities encountered in the study of clusters, the predictions of Newton's law and those of general relativity are basically the same. The application of general relativity to the cosmological problem would then not be justifiable.

It may perhaps be regretted that we do not give an expression for the gravitational force which is valid for every r and which fits Newton's law when r is small. This is due to the fact that we do not know of any astronomical observation from which the behaviour of the gravitational force at an intermediate distance can be deduced. On the other hand, we could not have deduced this behaviour from theoretical considerations; in fact, equations (1) themselves represent merely an attempt to explain some observational facts, and do not claim to have a theoretical foundation.

If the attractive forces acting between the galaxies are stronger than Newton's law would predict, the large internal velocities in the clusters could be explained. We mention now some other problems which can, perhaps, be solved by introducing the same hypothesis.

1. The estimate of the mass of the Galaxy, deduced from the motion of very distant globular clusters, is about three times larger than the value deduced from the solar motion.

2. If one compares the rotational curve of M_{31} (the great nebula in Andromeda) obtained from radio observations with the curve derived theoretically by Schwarzschild from the distribution of masses, one sees that at a long distance from the centre of the nebula the radio curve runs clearly higher than the theoretical curve.

....

EXPERIMENTAL EXAMINATION OF THE GRAVITATIONAL INVERSE SQUARE LAW

Long, Daniel R.; *Nature,* 260:417-418, 1976.

In a previous paper we pointed out that earlier data on the measurement of the gravitational constant suggest a systematic shift in the value with the separation of the masses. This laboratory has been examining this question for about a year, and the early data gave a discrepancy of 0.4%. New, intensive experiments have given 92 data points, with a resulting discrepancy of 0.37%.

Our experiment compares the gravitational constant at a mass separation of 29.9 cm with the value at a separation of 4.48 cm. To expedite this comparison the gravitational torque signals are kept about the same by using a much larger mass at the greater distance. Following the ingenious suggestion of Faller and Koldweyn, we have used rings for the attracting mass because the field is very flat at the 'Helmholtz' point and thus the positional accuracy need not be very great. The far ring is made of centrifugally cast brass and has a mass of 57,580.83 g. The near ring is made of pure tantalum with a mass of

Table 1. Error contributions to Δ_{thy}

Source	Error
Torsion pendulum assembly balance	0.00007
Big ring floor tilt	0.00005
Big ring-pendulum assembly interaction	0.00036
Small ring-pendulum assembly interaction	0.00004
Big ring mass, dimensions and position	0.00031
Small ring mass, dimensions and position	0.00012
Attracted ball sphericity	0.00001
Ring density inhomogeneity	0.0001
Total error	0.00051

1,225.271 g. The far ring has an outside radius of 27.112 cm, an inside radius of 21.589 cm, a thickness of 7.633 cm, and the field maximum is located on the ring axis 17.415 cm from the face of the ring. The near ring has an outside radius of 4.5536 cm, an inside radius of 2.7513 cm, a thickness of 1.7765 cm, and the field maximum is located on the ring axis 2.6088 cm from the face of the ring.

The gravitational force was detected with a 48.6-cm Cavendish torsion balance suspended by a 0.002-inch tungsten fibre. The attracted ball, with a mass of 50 g. was made of tantalum and hung by two fibres 87 cm below the balance rod. This reduced the gravitational torques caused by the pendulum rod and counter balance ball. The pendulum hung in a vacuum of 1 x 10^{-7} mmHg and the vacuum changer wall temperature was controlled to $\pm$ 0.01°C. Substantial effort went into mounting the apparatus to reduce vibrations (diffusion pumps vibrate badly) and the pendulum hung from a vibration damper suggested by Liebes. The position of the pendulum was detected by a beam of light which was split into a two-photodiode null detector.

Another important aspect of the experiment was the use of an electrostatic puller to balance the torque of the rings. The experiment was

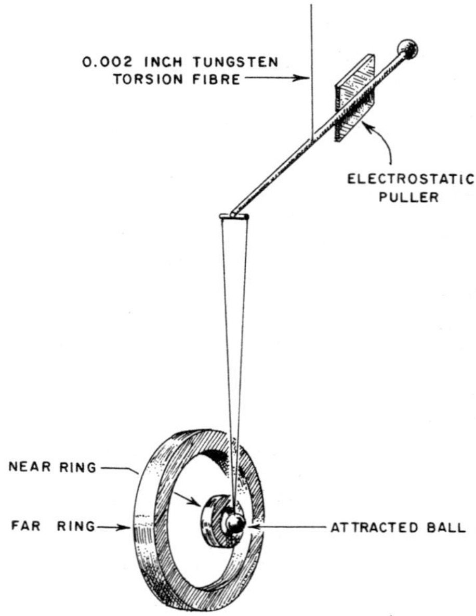

operated in such a way that the attracted ball did not change its position appreciably. The balance voltage (squared) was the measured gravitational torque signal in the experiment; several voltages would be tried which gave an output just slightly off null and the correct null voltage would be interpolated from them.

The quantity measured in the experiment was

$$\Delta = \frac{\tau_{far} - \tau_{near}}{\tau_{near}} \tag{1}$$

where τ_{far} is the total torque produced by the far ring and τ_{near} is the total torque produced by the near ring. The Newtonian value of Δ (Δ_{thy}), was calculated by finding the torque produced by each ring on the attracted ball and each element of the pendulum assembly. Numerical integration over the rings was used to find

$$\Delta_{thy} = 0.03807 \pm 0.0005$$

where the error analysis is given in Table 1.

Table 2. Error contributions to Δ_{exp}

Standard deviation of the mean of the data	0.00015
Signal against voltage proportionality constant	0.00015
Contact potential	0.00028
Small ring torque	0.00027
Total error	0.00044

Δ_{exp} was found by directly measuring the torque difference $\tau_{far} - \tau_{near}$ by first putting one ring and then the other in position. This gave

$$\triangle_{exp} = 0.04174 \pm 0.0004$$

The discrepancy between the two results is

$$\delta = \triangle_{exp} - \triangle_{thy} = 0.0037 \pm 0.0007$$

The experimental values of δ are given in a histogram in Fig. 2.

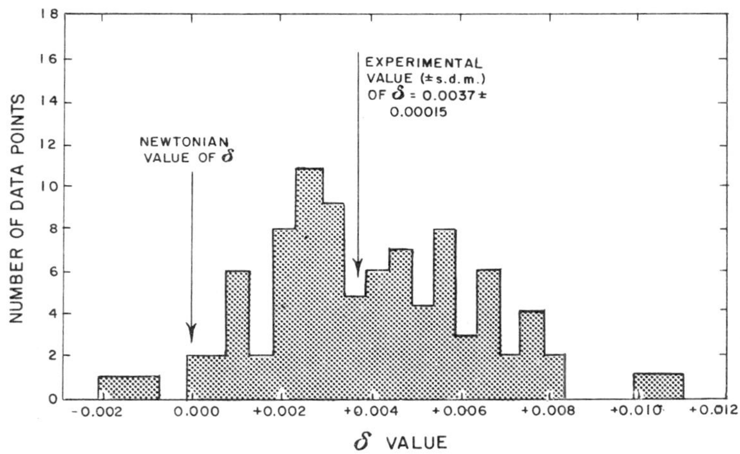

Experimental values of delta. (Fig. 2)

A careful re-examination of the experiment indicates that there is no error present large enough to account for this value of δ. In particular, reversal of the rings showed, as theoretically expected, that there is no density inhomogeneity of the rings sufficient to explain this result.

The slope of G(R) from this experiment is shown with G data from ref. 1 in Fig. 3. It will be seen that the two sets of data agree quite well and suggest that the gravitational force law should be modified at laboratory dimensions by

$$G(R) = G_o [1 + (0.002) \ln R]$$

It should be remarked that while these results indicate an inverse square law failure at small distance, it is also clear that the experiment was carried out in the Earth's gravitational field and that we may, in fact, have a failure of the superposition principle.

. . . .

1 Long, D. R., Phys. Rev., D, 9, 850 (1974).
2 Long, D. R., Phys. Rev., D, 10, 1677 (1974).
3 Boys, C. V., Phil. Trans. R. Soc., A1, (1895).

4 Braun, K., S. J. Deukschr, Akad. Wiss. Wien. Mathnaturw. Cl. 64, 187 (1896).
5 Rose, R. D., Parker, H. M., Lowry, R. A., Kuhlthau, A. R., and Beams, J. W., Phys. Rev. Lett., 23, 655 (1969).
6 Heyl, P., J. Res. natn. Bur. Stand., 5, 1234 (1930).
7 Heyl, P., and Chrzanowski, P., J. Res. natn. Bur. Stand., 29, 1 (1942).
8 Poynting, J. M., Phil. Trans. R. Soc., A182, 565 (1891).
9 Richarz, F., and Krigar-Menzel, O., Nature, 55, 296 (1898).

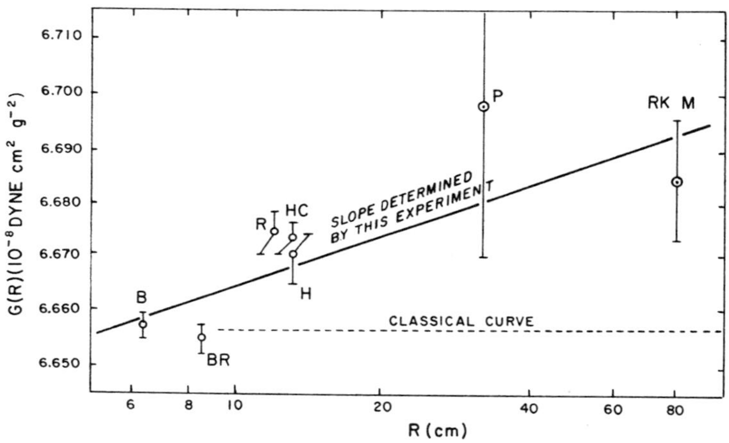

Slope of G(R) from experiment compared with the classical prediction. (Fig. 3)

DIMINISHING GRAVITY IS NO JOKE

Anonymous; *New Scientist*, 63:711, 1974.

When is a constant not a constant? Twentieth-century physicists have slaughtered herds of sacred cows; since the fall of parity their very laws are no longer sacrosanct. Are the constants going to prove variable as well?

So far, satisfactorily low upper limits have been placed on the possible time variation of most of the fundamental physical constants; besides, no compelling theoretical arguments for their variation have been advanced. The universal constant of gravitation, G, is an exception: several cosmologies call for a G which decreases with time (albeit at different rates) and the observational evidence not only has failed to demolish this disturbing possibility but lately has increasingly favoured it. Recent results from T. C. Van Flandern of the US Naval

Observatory may very well put a spoke in the wheel of diehard supporters of constant G, among whose number must be counted not merely old-school Newtonians but general relativists as well.

The Earth rotates much faster than the Moon revolves, and thus produces a tidal torque which very gradually reduces the Earth's spin and transfers angular momentum to the Moon. In consequence, the Earth-Moon distance slowly increases and the motion of the Moon against the stellar background correspondingly decreases. Now several determinations have been made of the deceleration of the Moon's mean longitude, and although they do not all agree, the deduced rate of deceleration is always greater than that which would be expected from tidal interaction alone. The difference has always been attributed with stunning acumen, to non-tidal forces.

Van Flandern set out to obtain the best determination yet of the "non-tidal" contribution and to show that, among the various interpretations, a dwindling G is the strongest contender. What he has done is to assess the rate of change in the mean orbital motion of the Moon in two time systems: Ephemeris Time and Atomic Time. Since Ephemeris Time is an astronomical scale derived ultimately from the apparent motion of the Sun across the sky, any quirk of gravity affecting the Sun and Moon in the same fashion would simply not show up. Suppose for a moment that all matter were radially expanding at the rate of one inch per year. A tailor could measure off a yard of material with his yardstick and a year later remeasure it: both values would read 36 inches, not 36 and 37, because his unit of measure would be augmenting as fast as his bolt of cloth.

Atomic Time, on the other hand, ostensibly provides an absolute scale and could no longer conceal such a gravitational change. (Of course, Atomic Time might not be uniform itself, but there are no grounds for that supposition and it must fall under Occam's Razor.) Van Flandern analysed lunar occultation data from 1955 to 1973 on the Atomic Time scale with the help of an improved lunar ephemeris and obtained, as expected, a larger value for the deceleration than the one measured in Ephemeris Time. Taking the difference between the two values, he concludes that the rate of change in G is -1.2 ± 0.3 parts in 10^{10} per year.

....

NEWTON'S APPLE FELL FASTER

Anonymous; *New Scientist*, 66:364, 1975.

Gravity is decreasing. Preliminary results to that effect (see New Scientist, vol. 63, p. 711) have now been confirmed. The change is very slow, so it will not assist slimmers: a person weighing about 10 stones would lose one millionth of the weight of a paper clip each year!

Dr. Thomas C. Van Flandern of the US Naval Observatory, Washington, has compiled records dating back 20 years of the occultations of stars by the passing Moon as timed by atomic clocks. He finds a drift in the timings which can be explained by assuming that the constant of gravitation, G, is falling by (7.5 ± 2.7) x 10^{-11} parts per year.

This figure is comparable with the Hubble constant, (5.6 ± 0.7) x 10^{-11} per year, which measures the rate at which the universe is expanding, so it is perhaps not surprising that the result fits neatly with certain cosmological theories that relate the strength of gravity to the existence of the rest of the universe. In particular, the result supports Dirac's "large number" theory of the rate of change of G and a theory due to Hoyle and Narlikar, but conflicts with predictions of the Brans and Dicke "scalar tensor" theory which requires a slower rate of change.

Van Flandern finds support also in the ancients, whose measurements of solar eclipses 20 centuries ago are slightly out of step with the present data---unless it is assumed that G has decreased by (12 ± 7) x 10^{-11} parts per year in the interim, a figure which agrees with Van Flandern's measurement.

Van Flandern, by using lunar occultation rather than laser ranging, has dealt with a problem facing ranging that any other effect than a decreasing G which causes the Moon to recede has to be subtracted from the raw figures. One of these is tidal friction, which is a messy thing to tot up accurately and has an effect of the same order as Van Flandern's result. His measurement leaves Einstein's general relativity precariously at variance with experiment (Bulletin of the American Physical Society, vol. 20, p. 543).

• Constancy of the Velocity of Light Questioned

THE VELOCITY OF LIGHT

Anonymous; *Science,* 66:sup x, September 30, 1927.

M. E. J. Gheury de Bray writing in L'Astronomie, the official journal of the Astronomical Society of France, ventures the daring speculation that the velocity of light is decreasing at such a rate that each year it darts through space about four kilometers a second slower than it did a twelvemonth earlier. He cites in support of his claim the results of determinations of the velocity of light during a period of over three quarters of a century, of which only one, made in 1855 with apparatus which may have been faulty, is really notably out of step.

The velocity of light is usually stated as 186,000 miles, or 300,000 kilometers, per second, which is fast enough to take it seven times around the earth while the clock ticks once. But for exact work in astronomy, physics and other sciences, determinations to fill out the three blank ciphers usually ignored in ordinary statement are desired, and these have been made a number of times. The most recent research was that of Dr. A. A. Michelson, of the University of Chicago, in 1926, which set the figure at 299,796 kilometers a second. This, according to M. de Bray, is the lowest velocity ever observed, but the new determination, on which Dr. Michelson is working now, should turn out even lower.

The series of determinations, in order to their dates, are given by M. de Bray as follows, the figures indicating velocity in kilometers per second

1849	313,300
1855	298,000
1855	305,650
1871	300,400
1885	299,940
1902	299,895
1906	299,880
1924	299,802
1926	299,796

The differences between these determinations are insignificant from the practical point of view, but if the present claims receive support from subsequent determinations, the accepted ideas in theoretical physics, especially those on which relativity is based, are in for a revolutionary upsetting.

THE VELOCITY OF LIGHT

de Bray, M. E. J. Gheury; *Nature*, 133:464, 1934.

In 1927 there was published in these columns a table of all the determinations of the velocity of light which I compiled from the original memoirs, together with a discussion, and I pointed out that except a pair of practically simultaneous values obtained in 1882 the final values (printed in heavy type) indicate a secular decrease of velocity. The last (and lowest) value given is 299,796 ± 4 km./sec. for 1926.

Since then, two determinations have been made: the first by Karolus and Mittelstaedt (1928) using a Kerr cell, to the terminals of which an alternating potential was applied, for interrupting periodically the luminous beam, instead of a toothed wheel. A frequency can be obtained in this way, of the order of a million per second, which can be accurately calculated, thus permitting a very short base to be used (41.386 metre) without any loss of accuracy. The value found (mean of 755 measurements) was 299,778 ± 20 km./sec. The second recent determination is mentioned in Nature of February 3, p. 169 it gives for the velocity of light in 1933 the value 299,774 ± 1 or 2 km./sec.

The determinations of this so-called constant made during the last ten years (the most accurate of the whole series) are therefore:

1924	299,802 ± 30 km./sec.
1926	299,796 ± 4 km./sec.

1928	299,778 ± 20 km./sec.
1933	299,774 ± 1 or 2 km./sec.

No physicist, looking at the above table, can but admit that the alleged constancy of the velocity of light is absolutely unsupported by observations. As a matter of fact, the above data, treated by Cauchy's method, give the linear law:

V km./sec. = 299,900 -4T (1900) years.

When I first pointed out this fact (in 1924) it was objected that the data available were inconclusive because the probable errors of the observations were greater than the alleged rate of change. Sir Arthur Eddington has dealt the death blow to the theory of errors and "this theory is the last surviving stronghold of those who would reject plain fact and common sense in favour of remote deductions from unverifiable guesses, having no merit other than mathematical tractability". Even "die-hards", however, may fruitfully meditate over the 2nd and the 4th values of the above table.

THE VELOCITY OF LIGHT

Birge, Raymond T.; *Nature*, 134:771-772, 1934.

F. K. Edmondson has recently stated that the observed values of the velocity of light are well represented by the equation

$$c = 299,885 + 115 \sin (2\pi\ 40)\ (t - 1901),$$

and de Bray has given a plot of this equation, together with certain experimental points. By comparing these points with the complete table of values published earlier by de Bray, I find that they include each one of the seven final declared values of c, from 1875 to the present time, as well as the preliminary value of Pease and Pearson. The agreement with Edmondson's equation is so remarkable that it seems desirable to tabulate the actual figures. The first eight items in the accompanying table comprise this information.

Date	Investigator	Obs. Velocity km./sec.	Obs.-Calc. km./sec.
1874.8	Cornu-Helmert	299,990 ± 200	+10
1879.5	Michelson	299,910 ± 50	-2
1882.7	Newcomb	299,860 ± 30	+5
1882.8	Michelson	299,853 ± 60	±0
1902.4	Perrotin	299,901 ± 84	-9
1926.5	Michelson	299,796 ± 4	-1.6
1928.0	Mittelstaedt	299,778 ± 10	-4.5
1932.5	Pease and Pearson	299,774	+0.8
1923.0	Mercier	299,782 ± 30	-67
1906.0	Rosa and Dorsey	299,781 ± 10	-185

In addition to Edmondson's 40-year period, Pease and Pearson have found evidences of two shorter periods, one of 14-3/4 days, and the other of one year. Each had an amplitude of about 20 km./sec., although the shorter period fluctuation nearly vanished during December 1932 and January 1933, reappearing again in February 1933. The origin of these short period fluctuations is still obscure, but if we admit their reality, it is not improbable that a large fluctuation of the type postulated by Edmondson *may* also exist. On the other hand, it is important to notice that these apparent variations occurred only in the direct measurements of *c*, for which the apparatus extended over one to twenty-five miles. When we turn to the indirect methods, for which a very compact apparatus is used, there is no evidence of a variation.

One of these indirect methods is the measurement of the velocity of electric waves guided by wires (standing waves). J. Mercier has carried out the latest and by far the most accurate work by this method. His result is 299,700 ± 30 km./sec., and I quoted this value in my 1929 discussion of the general physical constants. N. E. Dorsey has, however, recently noted that this value is for air, and when reduced to vacuum becomes 299,782, as given in the table. This revised figure agrees surprisingly well with Mittelstaedt's result, which was obtained with a Kerr cell and with a base line of only 41.4 metres, and with Pease and Pearson's final *average* result. It disagrees, however, by 67 km./sec. with Edmondson's predicted value for the epoch 1923.

The second indirect method for determining *c* is by means of the ratio of the electrostatic to the electromagnetic unit of electricity. The best experimental value of this ratio, by E. B. Rosa and N. E. Dorsey, was obtained at the mean epoch 1906.0. Their direct result is 299,710 ± 10, but this is in terms of international electric units. Using *p* 1.00051 (one int. ohm. = *p* abs. ohm), I obtained 299,790 ± 10 km./sec. The best value of *p* is, however, now 1.00046, giving 299,781 ± 10 km./sec. This last value, which appears in the table, is in complete agreement with the results obtained with other relatively compact apparatus (Mercier and Mittelstaedt). On the other hand, it disagrees violently with Edmondson's calculated value for 1906.

To this last conclusion the objection may properly be raised that it is the value of *p* that should vary with time, rather than the experimental result in terms of international units, and that the 299,781 value properly applies to 1932, the epoch at which *p* = 1.00046 was observed, and for which the calculated value of *c* is 299,773. But the experimental values of *p* do not show the predicted variation with time. Thus F. E. Smith, in 1914, obtained 1.00052, and E. Gruneiser and E. Giebe, in 1920, obtained 1.00051. In order to satisfy Edmondson's equation, the value of *p*, from 1914 to 1932, should vary by 0.00144, an amount twenty-four times the observed variation.

It is thus probable that the fluctuations observed in the directly measured value of *c* are related in some way to the long base-line employed in the apparatus. That such fluctuations are instrumental, rather than real, is indicated also by the evidence listed by O. C. Wilson and by R. J. Kennedy, although the general situation outlined by them is not so clean-cut as their letters might indicate. In brief, this situation is as follows.

The apparent observed change in *c* is in terms of the standard metre and the mean solar day. Now if we assume the wave theory of light, and if we assume further that there is no dispersion in empty space, so that the group (that is, the measured) velocity is identical with the wave velocity, then we can write the fundamental equation $\underline{c} = \lambda v$, where v is the frequency of the atomic oscillator producing the light of wavelength λ. It has, however, been shown experimentally, as Wilson points out, that the length of the standard metre, in terms of λ, did not change by a measurable amount from 1892 to 1906. Furthermore, the short-period fluctuations observed by Pease and Pearson should have produced as much as one entire fringe shift per day in Kennedy's apparatus, whereas his observed shifts were always less than 10^{-4} fringe per day.

Hence if the value of c, in terms of the standard metre and mean solar day, is actually changing with time, but the value of λ in terms of the standard metre shows no corresponding change, then it necessarily follows that the value of every atomic frequency, in terms of the mean solar day, must be changing. Such a variation is obviously most improbable, but unfortunately it is not possible to make a direct test, since one cannot compare directly an atomic frequency with any macroscopic standard of time. A real variation in the measured group velocity c, if such variation exists, might then be interpreted as due to a real variation in v, or as due to a failure of the measured c to equal the wave velocity λv. Parenthetically, it may be noted that any such general variation in v would not affect the value of the non-dimensional quantity, index of refraction, contrary to Wilson's statement.

The last four values of c given in the table represent respectively the best result by each of four quite different methods, and agree remarkably well. As a final weighted average of these results, I suggest 299,776 ± 4 km./sec.

LIGHT VELOCITY DEPENDENT ON SPEED OF SOURCE?

Anonymous; *New Scientist*, 16:276, 1962.

According to the second postulate of Einstein's Special Theory of Relativity the speed of light is independent of the uniform motion of its source. Direct experimental evidence by W. Kantor of the US Navy Electronics Laboratory, San Diego, leads him to the surprising conclusion that it may be untenable (Journal of the Optical Society of America, Vol. 52, No. 8, p. 978).

In the experiment light from a collimated xenon source was split with a beam splitter by reflection and transmission. The two beams traversed an interferometer in opposite directions recombining at the beam splitter to produce vertical interference fringes which were observed through a telescope. The lower half of each beam passed through a thin glass window (0.005 in. thick), the two windows being mounted on opposite sides of a rapidly rotating disc. The upper part of each beam passed over the windows through the interferometer to produce a set of reference fringes in the telescope. By use of a mirror system the transmitted light was directed through the windows in opposition to their motion while the reflected beam was concurrent with it.

The xenon source was triggered through a pick-up coil by a magnet rotating with the disc so that, when the windows were normal to the light beams, a 15 microsecond flash was produced. It can be shown that, in fact, the glass absorbs and re-emits the incident light and thus the windows acted effectively as moving light sources.

If Einstein's postulate is correct there should be no displacement between the two sets of interference fringes on spinning the disc, because the light from the approaching and receding windows, respectively, should all have the same velocity. In fact, an unambiguous, easily noted shift of the fringes was apparent when the mirrors were in motion (maximum linear velocity: 4,690 cm per sec) and Kantor deduces that Einstein's second postulate is incorrect. The fringe shift, moreover, appeared to depend on the speed of the disc. If the present work turns out, on more rigorous research, to be flawless and free from experimental artefacts, and if there is no obvious alternative explanation for the observed effects, there may be a need to reconsider some basic ideas in physics.

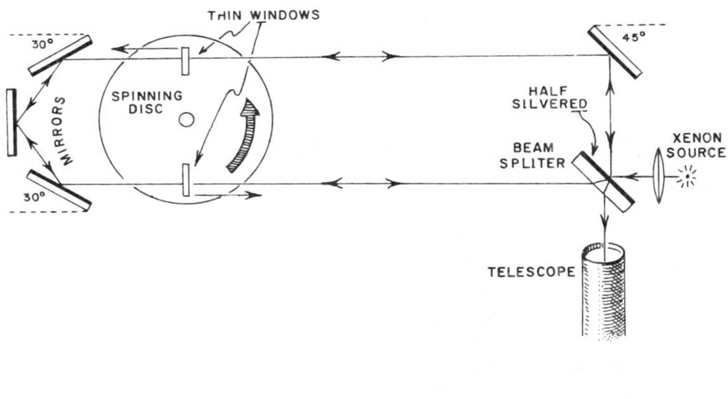

• Curious Relations Among Physical Constants

RELATIONS BETWEEN FUNDAMENTAL CONSTANTS

Mills, J. E.; *Science*, 75:243, 1932.

A numerical relation has been found between the fundamental physical constants shown below and the velocity of light. This relationship is of such a nature that the constants can be calculated from a single equation $1/2\pi c = C^{12}$ and the power of the velocity of light shown, provided the decimal point is ignored. A complete solution of the relation given, enabling the decimal point to be properly placed, has not yet been found. It is, however, not possible that any merely accidental agreement could produce the numerical agreement shown below. The facts concerned will be presented in a more technical paper.

Calculations of Fundamental Physical Constants

		Observed	Calculated
C	C	$(2.99796 \pm 0.00004) \times 10^{10}$	2.99960
C^2	m_o	$(8.994 \pm 0.014) \times 10^{-28}$	8.99761
C^6	$1/k$	$(7.294 \pm 0.0074) \times 10^{15}$	7.28415
C^8	h	$(6.547 \pm 0.008) \times 10^{-27}$	6.55402
C^{14}	ε	$(4.770 \pm 0.005) \times 10^{-10}$	4.77401
C^{34}	m_P	$(1.6610 \pm 0.0017) \times 10^{-24}$	1.66014
C^{36}	$\frac{1}{G}$	$(1.5006 \pm 0.0005) \times 10^{7}$	1.49373

π = Geometrical Constant; C = Velocity of light; m_o = Mass of electron; m_P = Mass of proton; h = Planck's Constant; ε = Electronic charge; G = Gravitation Constant; k = Boltzmann Constant.

RELATIONS BETWEEN FUNDAMENTAL PHYSICAL CONSTANTS

Birge, Raymond, T.; *Science,* 75:383, 1932.

In an article with the above title, J. E. Mills gives a number of numberical coincidences, such as $(2\pi c)\, m_o{}^6 = 10^{-151}$, $(2\pi c)\, h^{3/2} = 10^{-28}$, $(2\pi c)\, e^{6/7} = 10^{23/7}$, where c = velocity of light, m_o = mass of electron, h = Planck's constant, e = electronic charge. Since the numerical values of the constants that he quotes are those suggested by me in 1929[2] I may perhaps be permitted a few words on the subject.

It seems necessary, from time to time, to call attention to the fact that the numerical value of any constant, or combination of constants, is entirely arbitrary if the constant, or combination, possesses dimensions. The value in such cases depends directly upon the unit adopted for each dimension. Mills uses values in the C. G. S. system, and these values accordingly depend upon the arbitrarily chosen units of length, mass and time---the cm. gram and second. It is almost inconceivable that there should be an accidental theoretically significant relation between these three units, whose origin is too well known to restate. All the combinations of constants given by Mills have dimensions, and all are equated to 10^n, where n is an integer or fraction.

It is only in the case of dimensionless combinations that the numerical value can have theoretical significance. The two most famous combinations of this character are the fine structure constant a $(= 2\pi e^2/hc)$ and the ratio of the mass of the proton to that of the electron. There are in the literature several articles discussing the possible theoretical significance of these two pure numbers (approximately 1/137 and 1840, respectively). As a much more striking example of the type of numerical coincidence found by Mills, one may quote the known values, mass of electron = 9.035 x 10^{-28} gram, angular momentum (spin) of electron = 9.02 x 10^{-28} erg. sec (see, for instance, Pauling and Goudsmit, "Structure of Line Spectra" page 54). But one can not equate grams to erg. sec any more than one can equate horses to oranges---to use a homely but correct analogy.

DIRAC COMPLETES HIS THEORY OF LARGE NUMBERS

Davies, P. C. W.; *Nature*, 250:460, 1974.

Dirac's theory begins with the question: what is the age of the Universe expressed in some naturally occurring units? A fundamental unit of time is the interval required for light to cross a characteristic atomic dimension, say 10^{-13} cm. Then the age of the Universe comes out at about 10^{39} in these units. The significance of this number is that it turns out to be simply related to other large dimensionless numbers formed from astronomical and atomic data; for example, the total number of particles in the observable Universe is around 10^{78}, that is $(10^{39})^2$, and the ratio of the electric to gravitational attraction between an electron and proton, e^2/Gm_em_p is also about 10^{39}. In view of the fact that the first number obviously increases with time, the coincidences involved by these simple relationships seems great. Dirac's explanation of this curiosity is that all such large quantities are connected, so that as the first increases with time so do the others. This is already a radical departure from conventional physics, which in addition imposes severe restrictions on cosmology. Dirac calls this connection the large numbers hypothesis, in which he expresses "great confidence".

In his 1937 article, Dirac was led by this hypothesis to a model universe which expands with time t like $\underline{t^{1/3}}$, making it about half as old as the more conventional big-bang models. In his recent work, however, the bald hypothesis is accompanied by a detailed gravitational theory which bears some similarity to the work of Hermann Weyl and E. A. Milne. A central feature of this work is the proposal that there are really two space time metrics, one, which is unmeasurable directly, enters into the Einstein equations (which remain valid) and the other is what is actually measured in laboratory experiments involving atomic apparatus. A connection between these two metrics is provided by a consideration of the motion of the Earth around the Sun, which in Newtonian approximation is essentially determined through the relation $\underline{GM = v^2r}$ connecting the Newtonian constant of gravitation $\underline{G}$ with the solar mass $\underline{M}$ velocity of the Earth $\underline{v}$ and radius of the Earth's orbit $\underline{r}$.

There then follows an argument which is typical of that used throughout Dirac's work. In terms of Einstein units the quantities in this equation are constant. But in atomic units one must satisfy the large numbers hypothesis in that $\underline{G}$ must decrease as $\underline{t^{-1}}$ in order that $\underline{e^2/Gm_em_p}$ increases proportionally to $\underline{t}$. Theories involving a changing constant of gravitation are not new: for example, Brans-Dicke theory predicts a similar effect. The change is, however, very small and barely detectable with current technology.

Once again, the number of particles in the Universe $(10^{39})^2$ at present, is required by the large numbers hypothesis to increase as t^2, which is interpreted by Dirac as a form of continual creation along the lines of the steady state theory.

Two possible creation mechanisms are proposed, with the new matter either appearing concentrated in existing masses (multiplicative creation) or spread out in the intergalactic spaces (additive creation). In the latter case the mass M in the above equation is constant, so that r in atomic units varies like t^{-1}. In the former case $M \propto t^2$ and $r \propto t$. It follows that the ratio of the two metrics can be t^{-1} or t^2, and that the Solar System must be either expanding or contracting with time.

In terms of Einstein units, if multiplicative creation is adding continuously to the mass M of the Sun, conservation of energy requires that the nucleons in the Sun decrease in mass as t^{-2} to compensate. Through the relation $e^2/Gm^2 \propto t$ this further requires $e \propto t^{-3/2}$ and h (Planck's constant) $\propto t^{-3}$. A number of interesting cosmological consequences follow. First Dirac argues that any universe consistent with the large numbers hypotheses must be static, to avoid the introduction of a characteristic cosmological epoch (that is a constant large number). The red shift of distant galaxies normally interpreted as a recessional effect, is instead accounted for by the decrease in the unit of atomic time interval. The Universe is prevented from gravitational collapse by the introduction in Einstein's equations of the cosmological constant which may be non-zero for multiplicative creation because of the correct time dependence of the metric ratio in this case. Dirac thus arrives at a version of Einstein's original static model universe.

In the case of additive creation Dirac proposes to conserve energy by the introduction of an unobservable negative energy field spread throughout the Universe, along the lines of the C-field of Hoyle and Narlikar. The total energy density of the Universe thus vanishes, so that the global geometry is just the Minkowski space of special relativity. In comparing the two models, Dirac inclines to the latter on the basis that the creation of new atoms in existing material would lead to insuperable difficulties concerning the crystaline structure of very old rocks.

Dirac's article is written in his usual lucid and direct style. The ideas it contains, a mixture of old and new are probably unpalatable for most modern cosmologists. Yet they are the product of a lively imagination, challenging the fundamental principles on which modern theories of astronomy, cosmology and physics itself are founded. Coming from a physicist of Dirac's stature, that is at the very least thought provoking.

STATISTICAL PHYSICS, PARTICLE MASSES AND THE COSMOLOGICAL COINCIDENCES

Lawrence, J. K., and Szamosi, G.; *Nature*, 252:538, 1974.

For many years, cosmologists have been intrigued by possibly coincidental relationships between physical constants of the microscopic and cosmic domains[1]. Examples of these are the approximate equality of such large dimensionless numbers as the ratio of the electromagnetic to the gravitational interaction strengths of particles, the ratio of the size of the Universe to the size of an elementary particle and the square root of the number of particles in the Universe. There has been controversy, not only over the possible meaning of these 'cosmological coincidences', but also over the existence of any meaning at all. Here we would like to show how, by means of simple assumptions about the nature of physical parameters and of the Universe, the cosmological coincidences may be derived in a logically coherent scheme. Equally important, this procedure yields a very reasonable limit for the mass of an elementary particle expressed in terms of fundamental constants. An extension of the procedure leads also to an expression for the fundamental electric charge.

The usual view of elementary particles is that similar particles are absolutely identical---that they have absolutely identical masses, electrical or nuclear interaction strengths, magnetic moments, and so on. A less restrictive view, and one perhaps more in keeping with our views of the physics of the early Universe, might be that the properties of elementary particles are determined stochastically. Thus, some physical parameters of a set of elementary particles, particularly their masses, might not be absolutely identical, but rather form statistical distributions about average values, which depend upon the overall thermal properties of the Universe. According to the usual principles of statistics, the standard deviation of the equilibrium mass would be

$$\delta m/m \approx 1/\sqrt{n} \qquad (1)$$

where $\underline{n}$ is the number of particles in the Universe. Whether $\underline{n}$ is considered to be the total number of particles in the Universe or the number of a given species, such as electrons or nucleons, is not important within the accuracy considered here. In this same spirit, we regard the (average) masses of all elementary particles to be essentially equal.

Variations of mass. We can now further restrict the mass fluctuations, using an anthropocentric argument similar to ones used in the past by Dicke[2] and Carter in unpublished work. The existence of human life and awareness requires that the Pauli exclusion principle act over cosmological times. The nuclei of the various elements from which we are constructed must have existed for at least the age of the Solar System, and the structure of these nuclei depends on the Pauli principle, which depends on the indistinguishability of protons and of neutrons. If the masses of these nucleons differed by a detectable amount, the particles would no longer be indistinguishable and the Pauli principle would break down. According to quantum mechanics, in the age $\underline{T}$ of the Universe (or of the Solar System) it is possible to detect a mass difference H/c^2T. Thus we conclude that the mass spread of elementary particles must be less than this critical value. Then,

$$\delta m \approx h/K_1c^2T \tag{2}$$

where $K_1 > 1$ is a dimensionless constant.

We next make the assumption that the Universe, at least approximately, is just gravitationally bound. This allows us to relate the gravitational constant G, the mass M, the radius R and the age of the Universe by

$$R \approx cT \approx GM/c^2 \approx Gmn/c^2 \tag{3}$$

This relationship, sometimes connected with Mach's principle, is roughly compatible with observations and is widely assumed by cosmologists.

It is now possible to combine equations (1), (2) and (3) to obtain an expression for the mass of an elementary particle in terms of fundamental constants:

$$m \approx (h^2/K_1GeT)^{1/3} \tag{4}$$

Taking $T \approx 10^{10}$ yr gives the numerical relation

$$m \approx K_1^{-1/3} \times 10^{-25} \text{ g} \tag{5}$$

This gives an upper limit for elementary particle masses which is consistent with the existence of human life in the Universe:

$$m \lesssim 10^{-25} \text{ g} \tag{6}$$

A wide range of values of K_1, covering many orders of magnitude, will give values of m lying within the range of elementary particle rest masses ($\sim 10^{-27}$ g to $\sim 10^{-24}$ g). A recent study by Reines and Sobel[3] indicates that the lifetime of the Pauli principle for inner shell electrons in iodine may be $\sim T \times 10^{10}$ yr. Since the breakdown rate would be proportional to the square of the perturbation δm, this result implies a value $K_1 \approx 10^5$ for these particles. This, when substituted into equation (5) implies $m \simeq 10^{-27}$ g, the electron mass. It would be of great interest to measure the analogous Pauli breakdown rate for heavier particles, such as nuclear protons or neutrons. A value $K_1 \approx 1$ also gives a mass well within the range of known particle masses, and for convenience we will now adopt this value. It should be kept in mind that both the theory presented here and the corresponding observational results are very rough. Thus large changes in the value of K_1 will have no fundamental effect on any of the following equations.

Cosmological Coincidence. Apparently, Weinberg[4] first noted the existence of equation (4) as a numerical coincidence and discussed its possible importance as a clue for a connection between microphysics and the Universe as a whole. The above considerations appear to give a possible rationale for its existence.

If we use $K_1 \approx 1$ in equation (5), we find $n \approx 10^{80}$, which is compatible with present data, and $\delta m \approx 10^{-65}$ g. By eliminating G from equations (3) and (4), we obtain

$$R/(h/mc) \approx \sqrt{n} \tag{7}$$

Elimination of R yields

$$m \approx m_*/n^{1/4} \tag{8}$$

with $m_* = (hc/G)^{1/2}$, the Planck mass. Also one finds the gravitational fine structure constant

$$hc/Gm^2 = (e^2/Gm^2)\ (1/a) \approx \sqrt{n} \qquad (9)$$

Equations (7) to (9) have often been noted in the literature as the 'cosmological coincidences.'

It is of further interest to consider the effect induced on particle self-energies by the variation δ m. By requiring the induced variations in self-energy to remain too small for detection in the age of the Universe, as in the argument leading to equation (2), we obtain constraints on the values of the interaction coupling constants. We consider the formula for the lowest order electromagnetic self-energy of an electron given by Salam *et al.*[5] where the ultraviolet divergence has been removed by inclusion of gravitational self-energy:

$$E \approx (me^2c/h)\ (\ln n) \qquad (10)$$

Requiring the variation induced in this by δ m to be undetectable gives for the fine structure constant

$$e^2/hc \approx (K_1/K_2)/(\ln n) \qquad (11)$$

Where $K_2 > 1$ is another dimensionless constant. This result is roughly correct if $K_2 \approx K_1$.

1 Harrison, E. R., *Physics Today*, 30-34 (December, 1972).
2 Dicke, R. H., *Nature*, 192, 440-441 (1961).
3 Reines, F., and Sobel, H. W., *Phys. Rev. Lett.*, 32, 954 (1974).
4 Weinberg, S., *Gravitation and Cosmology*, 619-620 (Wiley, New York, 1972).
5 Isham, C., Salam, A., and Strathdee, J., *Phys. Rev. D*, 3, 1805-1817 (1971).

INDEX